Periodic Table of the Elements

GROUP

Group	1 IA	2 IIA	3 IIIB	4 IVB	5 VB	6 VIB	7 VIIB	8 VIII	9 VIII	10 VIII	11 IB	12 IIB	13 IIIA	14 IVA	15 VA	16 VIA	17 VIIA	18 VIIIA
Period 1	2.20 (1) 1 **H** 1.01																	- 2 **He** 4.00
Period 2	0.98 (1) 3 **Li** 6.94	1.57 (2) 4 **Be** 9.01											2.04 (3) 5 **B** 10.81	2.55 ($\pm$4,2) 6 **C** 12.01	3.04 ($\pm$3,4,5) 7 **N** 14.01	3.44 (-2) 8 **O** 16.00	3.98 (-1) 9 **F** 19.00	- 10 **Ne** 20.18
Period 3	0.93 (1) 11 **Na** 22.99	1.31 (2) 12 **Mg** 24.31											1.61 (3) 13 **Al** 26.98	1.90 (4) 14 **Si** 28.09	2.19 ($\pm$3,4,5) 15 **P** 30.97	2.58 ($\pm$2,4,6) 16 **S** 32.07	3.16 ($\pm$1,3,5,7) 17 **Cl** 35.45	- 18 **Ar** 39.95
Period 4	0.82 (1) 19 **K** 39.10	1.00 (2) 20 **Ca** 40.08	1.36 (3) 21 **Sc** 44.96	1.54 (3,4) 22 **Ti** 47.88	1.63 (2,3,4,5) 23 **V** 50.94	1.66 (2,3,6) 24 **Cr** 52.00	1.55 (2,3,4,5,6) 25 **Mn** 54.94	1.83 (2,3) 26 **Fe** 55.85	1.88 (2,3) 27 **Co** 58.93	1.91 (2,3) 28 **Ni** 58.69	1.90 (1,2) 29 **Cu** 63.55	1.65 (2) 30 **Zn** 65.39	1.81 (3) 31 **Ga** 69.72	2.01 (4) 32 **Ge** 72.61	2.18 ($\pm$3,5) 33 **As** 74.92	2.55 (-2,4,6) 34 **Se** 78.96	2.96 ($\pm$1,5) 35 **Br** 79.90	- 36 **Kr** 83.80
Period 5	0.82 (1) 37 **Rb** 85.47	0.95 (2) 38 **Sr** 87.62	1.22 (3) 39 **Y** 88.91	1.33 (4) 40 **Zr** 91.22	1.6 (3,5) 41 **Nb** 92.91	2.16 (2,3,4,5,6) 42 **Mo** 95.94	1.9 (7) 43 **Tc** (99)	2.2 (2,3,4,6,8) 44 **Ru** 101.07	2.28 (2,3,4) 45 **Rh** 102.91	2.20 (2,3,4) 46 **Pd** 106.42	1.93 (1) 47 **Ag** 107.87	1.69 (2) 48 **Cd** 112.411	1.78 (3) 49 **In** 114.82	1.96 (2,4) 50 **Sn** 118.71	2.05 ($\pm$3,5) 51 **Sb** 121.75	2.1 (-2,4,6) 52 **Te** 127.60	2.66 ($\pm$1,5,7) 53 **I** 126.90	- 54 **Xe** 131.29
Period 6	0.79 (1) 55 **Cs** 132.91	0.89 (2) 56 **Ba** 137.33	1.10 (3) 57 **La** 138.91	1.3 (4) 72 **Hf** 178.49	1.5 (5) 73 **Ta** 180.95	2.36 (2,3,4,5,6) 74 **W** 183.85	1.9 (-1,2,4,6,7) 75 **Re** 186.21	2.2 (2,3,4,6,8) 76 **Os** 190.2	2.20 (2,3,4,6) 77 **Ir** 192.22	2.54 (2,4) 78 **Pt** 195.08	2.54 (1,3) 79 **Au** 196.97	2.00 (1,2) 80 **Hg** 200.59	2.04 (1,3) 81 **Tl** 204.38	2.33 (2,4) 82 **Pb** 207.2	2.02 (3,5) 83 **Bi** 208.98	2.0 (-2,4) 84 **Po** (209)	2.2 ($\pm$1,3,5,7) 85 **At** (210)	- 86 **Rn** (222)
Period 7	0.7 (1) 87 **Fr** (223)	0.9 (2) 88 **Ra** (226)	1.1 (3) 89 **Ac** (227)	(4) 104 **Rf** (261)	(5) 105 **Db** (262)	(6) 106 **Sg** (263)	(7) 107 **Bh** (262)	108 **Hs** (265)	109 **Mt** (266)	110 (269)	111 (272)	112 (277)						

Lanthanide series:

1.12 (3,4) 58 **Ce** 140.12	1.13 (3,4) 59 **Pr** 140.91	1.14 (3,4) 60 **Nd** 144.24	1.13 (3) 61 **Pm** (147)	1.17 (2,3) 62 **Sm** 150.36	1.2 (2,3) 63 **Eu** 151.97	1.20 (3) 64 **Gd** 157.25	1.2 (3,4) 65 **Tb** 158.93	1.22 (3) 66 **Dy** 162.50	1.23 (3) 67 **Ho** 164.93	1.24 (3) 68 **Er** 167.26	1.25 (2,3) 69 **Tm** 168.93	1.1 (2,3) 70 **Yb** 173.04	1.27 (3) 71 **Lu** 174.97

Actinide series:

1.3 (4) 90 **Th** (232)	1.5 (4,5) 91 **Pa** (231)	1.38 (3,4,5,6) 92 **U** (238)	1.36 (3,4,5,6) 93 **Np** (237)	1.28 (3,4,5,6) 94 **Pu** (244)	1.3 (3,4,5,6) 95 **Am** (243)	1.3 (3) 96 **Cm** (247)	1.3 (3,4) 97 **Bk** (247)	1.3 (3) 98 **Cf** (251)	1.3 (3) 99 **Es** (252)	1.3 (3) 100 **Fm** (257)	1.3 (2,3) 101 **Md** (258)	1.3 (2,3) 102 **No** (259)	(3) 103 **Lr** (260)

Legend (key):
- Electronegativity — 2.20
- Oxidation States — 1
- Atomic number — 1
- Element symbol — H
- Atomic mass* — 1.01

- Metals
- Semimetals
- Non Metals

*The mass number of an important radioactive isotope—not the atomic mass—is shown in parenthesis for an element with no stable isotopes.

Coordination Numbers and Effective Ionic Radii (Å)*

(H)

Li⁺ — Li^+: 4 0.68, 6 0.82
Be²⁺ — Be^{2+}: 3 0.25, 4 0.35

B³⁺ — B^{3+}: 3 0.10, 4 0.20

(C)
(N)
O²⁻ — O^{2-}: 2 1.27, 3 1.28, 4 1.30, 6 1.32, 8 1.34
F⁻ — F^-: 2 1.285, 3 1.30, 4 1.31, 6 1.33

(He)
(Ne)

Na⁺ — Na^+: 4 1.07, 6 1.08, 9 1.40
Mg²⁺ — Mg^{2+}: 4 0.66, 6 0.80, 8 0.97

Al³⁺ — Al^{3+}: 4 0.47, 5 0.56, 6 0.61
Si⁴⁺ — Si^{4+}: 4 0.34, 6 0.48

P⁵⁺ — P^{5+}: 4 0.25
S²⁻ — S^{2-}: 6 1.72
Cl⁻ — Cl^-: 6 1.72

(Ar)

K⁺ — K^+: 4 1.46, 8 1.59, 9 1.63, 10 1.67, 12 1.68
Ca²⁺ — Ca^{2+}: 6 1.08, 8 1.26, 9 1.36, 10 1.36, 12 1.43

Sc³⁺ — Sc^{3+}: 6 0.83, 8 0.95
Ti⁴⁺ — Ti^{4+}: 6 0.69
V⁵⁺ — V^{5+}: 4 0.44, 6 0.62
Cr³⁺ — Cr^{3+}: 6 0.07; **Cr⁴⁺** — Cr^{4+}: 4 0.52, 6 0.63; **Cr⁶⁺** — Cr^{6+}: 4 0.38
Mn²⁺ — Mn^{2+}: 6 0.83, 8 1.01; **Mn³⁺** — Mn^{3+}; **Mn⁴⁺** — Mn^{4+}: 6 0.62
Fe²⁺ — Fe^{2+}: 4 0.71, 6 0.77; **Fe³⁺** — Fe^{3+}: 4 0.57, 6 0.68
Co²⁺ — Co^{2+}: 4 0.65
Ni²⁺ — Ni^{2+}: 6 0.77
Cu⁺ — Cu^+: 2 0.54; **Cu²⁺** — Cu^{2+}: 4 0.70, 6 0.81
Zn²⁺ — Zn^{2+}: 4 0.68, 6 0.83

Ga³⁺ — Ga^{3+}: 4 0.55, 6 0.70
Ge⁴⁺ — Ge^{4+}: 4 0.48, 6 0.62
As⁵⁺ — As^{5+}: 4 0.42, 6 0.58
Se²⁻ — Se^{2-}: 6 1.88
Br⁻ — Br^-: 6 1.88

(Kr)

Rb⁺ — Rb^+: 6 1.57, 8 1.68, 12 1.81
Sr²⁺ — Sr^{2+}: 6 1.21, 8 1.33, 10 1.40, 12 1.48

Y³⁺ — Y^{3+}: 6 0.98, 8 1.10, 9 1.18
Zr⁴⁺ — Zr^{4+}: 6 0.80, 8 0.92
Nb⁵⁺ — Nb^{5+}: 4 0.40, 6 0.72
Mo⁴⁺ — Mo^{4+}: 6 0.75; **Mo⁶⁺** — Mo^{6+}: 4 0.35
Tc⁴⁺ — Tc^{4+}: 6 0.72
Ru³⁺ — Ru^{3+}: 6 0.75; **Ru⁴⁺** — Ru^{4+}: 6 0.71
Rh³⁺ — Rh^{3+}: 6 0.76; **Rh⁴⁺** — Rh^{4+}: 6 0.70
Pd²⁺ — Pd^{2+}: 4 0.72, 6 0.94
Ag⁺ — Ag^+: 4 1.10, 6 1.23, 8 1.38
Cd²⁺ — Cd^{2+}: 4 0.88, 6 1.03, 8 1.15, 12 1.39

In³⁺ — In^{3+}: 6 0.88, 8 1.00
Sn⁴⁺ — Sn^{4+}: 6 0.77
Sb³⁺ — Sb^{3+}: 4 0.85; **Sb⁵⁺** — Sb^{5+}: 6 0.69
(Te)
(I)

(Xe)

Cs⁺ — Cs^+: 6 1.78, 8 1.89, 10 1.89, 12 1.96
Ba²⁺ — Ba^{2+}: 6 1.44, 8 1.50, 10 1.60, 12 1.68

La³⁺ — La^{3+}: 6 1.13, 8 1.18, 10 1.36, 12 1.40
Hf⁴⁺ — Hf^{4+}: 6 0.79, 8 0.91
Ta⁵⁺ — Ta^{5+}: 6 0.72, 8 0.77
W⁴⁺ — W^{4+}: 6 0.73; **W⁶⁺** — W^{6+}: 4 0.50, 6 0.68
Re⁴⁺ — Re^{4+}: 6 0.71; **Re⁶⁺** — Re^{6+}: 6 0.60; **Re⁷⁺** — Re^{7+}: 4 0.48
Os⁴⁺ — Os^{4+}: 6 0.71
Ir³⁺ — Ir^{3+}: 6 0.81; **Ir⁴⁺** — Ir^{4+}: 6 0.71
Pt²⁺ — Pt^{2+}: 6 0.68
Au³⁺ — Au^{3+}: 4 0.78
Hg²⁺ — Hg^{2+}: 4 1.04, 6 1.10, 8 1.22

Tl³⁺ — Tl^{3+}: 6 0.75
Pb²⁺ — Pb^{2+}: 6 1.26, 8 1.37, 9 1.41, 12 1.57
Bi³⁺ — Bi^{3+}: 6 1.10, 8 1.19
Po⁴⁺ — Po^{4+}: 8 1.16
(At)

(Rn)

(Fr)
Ra²⁺ — Ra^{2+}: 8 1.48, 12 1.64

(Ac)

*The first number in each entry is the coordination number; the second is the ionic radius in coordination with oxygen having a radius of 1.32Å. Values for elements in parentheses were not determined. These values are modified from those in Zoltai and Stout (1984) *Mineralogy: Concepts and Principles*, Burgess Publishing Co., New York.

Mineralogy

Mineralogy

Dexter Perkins

University of North Dakota

PRENTICE HALL

Upper Saddle River, New Jersey 07458

Library of Congress Cataloging in Publication Data

Perkins, Dexter.
  Mineralogy/Dexter Perkins.—2nd ed.
    p. cm
  Includes bibliographical references and index.
  ISBN 0-13-062099-8
  1. Mineralogy I. Title

QE636.2 .P42 2002
549—dc21

2001055488

Senior Editor: Patrick Lynch
Assistant Editor: Amanda Griffith
Marketing Manager: Christine Henry
Assistant Managing Editor: Beth Sturla
Production/Composition: WestWords, Inc.
Manufacturing Manager: Trudy Pisciotti
Assistant Manufacturing Manager: Michael Bell
Art Director: Jayne Conte
Cover Designer: Bruce Kenselaar
Managing Editor, Audio/Visual Assets: Grace Hazeldine
Art Editor: Adam Velthaus
Editorial Assistant: Sean Hale

© 2002, 1998 by Prentice-Hall, Inc.
Upper Saddle River, New Jersey 07458

Printed in the United States of America
10 9 8 7 6 5 4 3 2 1

ISBN 0-13-062099-8

Pearson Education Ltd., *London*
Pearson Education Australia Pty., Limited, *Sydney*
Pearson Education Singapore, Pte. Ltd.
Pearson Education North Asia Ltd., *Hong Kong*
Pearson Education Canada, Ltd., *Toronto*
Pearson Educación de Mexico, S.A. de C.V.
Pearson Education—Japan, *Tokyo*
Pearson Education Malaysia, Pte. Ltd.

Contents

3

Mineral Properties: Hand Specimen Mineralogy

4

Optical Mineralogy

5

Igneous Rocks and Silicate Minerals

6

Sedimentary Minerals and Sedimentary Rocks   **117**

7

Metamorphic Minerals and Metamorphic Rocks   **133**

8

Ore Deposits and Economic Minerals   **155**

12

X-ray Diffraction 251

13

Atomic Structure 271

PART III: MINERAL DESCRIPTIONS 295

14

Descriptions of Minerals 297

Preface

Several excellent mineralogy texts are available today. They are well written, contain good figures and tables, and are complete. In short, they make excellent reference books, and I am glad I have them on my shelf. However, in my experience they are not appropriate for undergraduate mineralogy courses because they do not stimulate students or present information in ways that help students learn. Of course, the most enthusiastic and self-motivated students always do well and enjoy learning, and they may enjoy any well-written book, but many of my students are not of this ilk. They are good students, but many of them have, over the years, expressed frustration and dissatisfaction with mineralogy texts and, consequently, the way that mineralogy is taught.

As I see it, the major problem is one of thinking. In particular, it is a problem stemming from scientific minds that picture the world as a bunch of facts that, when combined, add up to big pictures. I find that most mineralogy students are bored by facts and often have not developed the imagination or perseverance needed to see their implications. As a scientist, I don't have a problem starting with atoms and atomic theory and building to molecules, crystals, rocks, regions, continents, and the Earth. I have no problem spending time discussing symmetry before I discuss minerals. It doesn't bother me if a class or an article never gets beyond interesting details and abstractions, or if a particular topic is never fully related to any other. However, as a teacher, I find that the scientific way of thinking is not the students' way of learning. Most students, in fact, seem to learn best by starting with the big things they know and understand—a rock or a pretty crystal, for instance—and then focusing on details and, finally, abstractions. They are interested and stimulated only when they understand the context and implications of the material they are learning. This means that the order and presentation of subjects in available mineralogy books are in many ways opposite from what can best promote learning.

Most of today's students won't be mineralogists and few will be petrologists. They don't need to know all the details of crystallography, crystal chemistry, and many other things we have taught in the past. Instead, they need to know how to think, they need to appreciate science and how it works, and, if they are to go on to careers in the Earth sciences, they need to know how minerals fit into a bigger picture.

This is the second edition of *Mineralogy* but, like the first, it approaches the subject of mineralogy from a student's perspective. My goal is to provide a book that students will enjoy reading and that will help them learn and become excited about the science I have made my career. I have tried to emphasize ideas and thinking and to relate mineralogy to other sciences. Consequently, I have deemphasized facts and sacrificed some completeness. Most, but not all, of the same material found in other mineralogy books is included, but the order, presentation, and depth of coverage are different. Mineralogical purists may say that I have strayed into different disciplines or that I have omitted some important details. Of these crimes I am guilty; but I have not done this without thought, and I hope that my thinking has been consistent with my goals.

When I wrote the first edition of this book, I was asked what would make it different and successful. I am not sure what makes a book successful, but the most important things that make this book different from others are

- With the exception of the first chapter, topics are covered beginning with the big, easy-to-see picture and ending with the details and theory.
- Topics are not completely divided into separate chapters as in many books; there is overlap and some redundancy.
- In an attempt to put mineralogy in context, I have placed more emphasis on petrology, chemistry, and other sciences not normally considered mineralogy.

- The history and human side of mineralogy—individuals and their contributions—have received more emphasis and are placed in better context than in other books.
- Boxed material relates mineralogy to things that are relevant to our daily lives.
- Jargon, classification schemes, and other vocabulary are only mentioned when important, and they are never emphasized.
- This book includes a glossary of over 1000 mineralogical terms.
- This book is not a complete mineralogy reference book; some things have been omitted or covered only briefly.
- I have tried to write in a style that is easy to read and less rigid than many science texts.
- Every chapter includes some "Questions for Thought." Most of the questions do not have absolutely correct answers. Instead, they require thinking about the material in the chapter and combining it with material learned elsewhere. They are intended to stimulate student thought and discussion and to inspire students to look in other books or journals for information.
- With some additions, and with emphasis on Chapters 1–8, this text could form the basis for a combined mineralogy/petrology course.

In this, the second edition, I have made some significant changes. Many photos and line drawings have been replaced, and new topical boxes have been added. I have omitted some of the more tedious parts of the first edition, replacing them with more information relating mineralogy to the world around us. Additionally, this edition is accompanied by a CD-Rom that contains over 400 high-quality mineral photographs.

This book would not have been possible without an incredible amount of help from my friends and others. So many people have contributed that listing them all would add another chapter. Al Falster and others provided some photos for this book, but I am especially indebted to Alan Kantrud who took many photographs for me and taught me how to take my own. I am also indebted to Eric Dowty and Shape Software for providing graphics software and data files to create crystal structure drawings.

Thanks go the following reviewers: Penelope Morton, University of Minnesota-Duluth; William P. Leeman, Rice University; David L. Smith, La Salle University; Jeffery Ryan, University of South Florida; Jennifer A. Thomson, Eastern Washington University; Julia Nord, George Mason University; Philip Goodell, University of Texas at El Paso.

In the final analysis, the success or failure of any textbook depends on how it is received by students and teachers who use it. I hope you like *Mineralogy* and will tell me of things I can do to make it more useful and enjoyable.

Dexter Perkins

Mineralogy

Mineral Properties and Occurrences

Dolomite from Penfield, New York (lower left), and Joplin, Missouri (lower right), and cerrusite from Whim Creek, Australia.

Elements and Minerals

The Earth is composed of minerals. If we are to understand the properties of the Earth and its minerals, we must understand basic chemistry and the elements. In this chapter we define minerals and elements and discuss why they are important. We review the development of modern chemistry and how it relates to mineralogy and discuss the behavior of elements in minerals and the significance of mineral formulas.

MINERALS

The word **mineral** means different things to different people. In ancient times, people divided all things on Earth into the animal, vegetable, or mineral kingdoms, so a mineral was any natural inorganic substance. Today, dieticians use the term to refer to nutritional elements such as calcium, iron, or sodium, while miners often use it for anything they can take out of the ground—including coal, sand, or gravel. Mineralogists and geologists of the twentieth century developed a more specific definition: minerals must be crystalline solids formed as a result of geological processes. They are generally **inorganic,** although some people consider biogenic substances such as calcium phosphate (apatite) in bones or teeth, or calcium carbonate (calcite) in shells, to be minerals. They must also have a well-defined chemical composition. Later chapters will discuss crystals and mineral compositions in more detail. For now it is sufficient to know that **crystalline** means "having an orderly and repetitive atomic structure," and well defined means "varying within limits." Mineralogists have named and described more than 3,000 minerals; they discover about 50 more each year.

Many mineral-related substances exist but do not fit the definition for one reason or another (Figure 1.1). Synthetic diamonds and rubies, for exam-ple, are not considered true minerals. Ice is considered a mineral, but its liquid counterpart, water, is not. Elemental mercury, in its natural state, is not considered a mineral because it is liquid. Refined sugar is crystalline but is not considered a mineral because it is human-made and organic. Window glass, made mostly from quartz, is not a mineral because it is not crystalline. The rust that forms on our cars is not considered a mineral, although the mineral goethite has nearly the same composition and properties.

Two major factors set most minerals apart from other crystalline materials: time and temperature. Many minerals form at high temperatures, and many form over long periods of time. Igneous and metamorphic minerals grow at temperatures hundreds or even a 1,000 °C hotter than normal Earth surface conditions. Their crystallization and metamorphism may take millions of years. Sedimentary minerals, while not forming at high temperatures, often crystallize or recrystallize over long periods of time. High temperatures and long periods of time lead to highly ordered crystal structures that synthetic and organic processes cannot normally imitate. Scientists can make synthetic gems in the laboratory; many of these are beautiful and valuable. However, if we examine the fine details of their structures, we find they are generally not as well ordered as their natural equivalents. In addition, there are many subtle differences

▶**FIGURE 1.1**
Natural materials that are not minerals: coal (upper left), opal (top center), oil (right), and mercury (bottom). These are not minerals because they are not crystalline. Oil and mercury also flunk the definition of a mineral because they are liquids. The watch glass and beaker flunk because they are not crystalline and they are not natural.

between minerals and their synthetic or organic counterparts.

THE IMPORTANCE OF MINERALS

An understanding of minerals is important for many reasons. A geologist considers minerals as the basic building blocks comprising nearly the entire crust of the Earth. If geologists and others are to understand the Earth, its formation, and its dynamics, they must understand minerals. Knowledge of minerals is es-

sential in many engineering fields. Constructing an office building would later prove foolish when the building collapsed because the underlying rock did not provide a sound foundation, or if construction materials were not adequate (Box 1.1). Miners and gem dealers view minerals as commodities to process and sell, while industrial manufacturers see minerals as the raw materials for marketable products (Figure 1.2). In the electronics industries, minerals or their synthetic equivalents are used to make computer chips, diodes, capacitors, superconductors, and other crucial components. A basic and won-

Box 1.1 The Amoco Building, Chicago

The Amoco Building, formerly known as the Standard Oil Building, is the third tallest building in Chicago. At the time of its original construction, and for a short time after, it was the largest marble-clad building in the world. It was also a hazard for people walking beneath it.

The thin slabs of marble, made nearly entirely of the mineral calcite, were unable to withstand Chicago's harsh climate, and they began to warp and fall off as soon as the building was completed. In a two-year project ending in 1992, the marble was replaced with granite. The K-feldspar and quartz of the granite are better able to handle the weather, and pedestrians can once again walk beneath the building safely.

▶**FIGURE 1.2**
(a) Two diamonds, the largest about 3 carats (600 mg). Diamonds have traditionally been the most popular gemstones because of their beauty and durability.
(b) Morganite, a variety of beryl named after J. Pierpont Morgan, normally has a pink color. Here it has been faceted to produce an almond shape.

drous asset of our natural world, historically minerals have been the foundation of much, if not all, technology.

ELEMENTS: THE BASIC BUILDING BLOCKS

Historical Views of Elements and Matter

All matter is composed of fundamental building blocks called **elements** that exist as discrete atoms, and may combine to form molecules, crystals, and minerals (Figure 1.3). While we accept this statement as true today, its significance and meaning have been debated for more than two thousand years. Greek philosophers recognized that there must be some order and structure to Earth materials. Empedocles (490–430 B.C.) summarized current thought when he spoke of "four fundamental elements": earth, air, fire, and water. Aristotle (384–322 B.C.) described matter, as opposed to something he called *essence,* and he supported Leucippos's (500 B.C.) theory of atoms as the smallest, indivisible building blocks. Democritus adopted and advanced the theory, describing atoms as small, invisible hard particles with no color, taste, or smell. A student of Aristotle's, Theophrastus (372–287 B.C.), applied the atomic theory to rocks and minerals. His short book *Concerning Stones* is considered by many to be the first mineralogy book

▶**FIGURE 1.3**
This diagram shows the relationships between atoms; molecules, which are compounds of several atoms; unit cells made up of several molecules; a collection of unit cells; and mineral crystals.

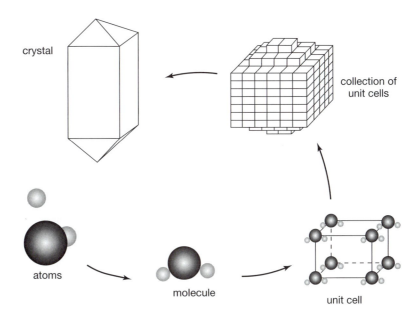

ever written. After the Greeks, there seems to have been a period of more than fifteen hundred years when little progress was made in the science of mineralogy. While people collected minerals and tabulated their properties, no one understood how or why they possessed those properties.

The processing and isolation of individual minerals and chemical elements dates as far back as 2900 B.C., when furnaces for melting gold were used in Egypt, and the Bronze Age began in Greece. The refining of gold and silver for coins and art was widespread. Other elements, such as lead and antimony, were known at the time of the Sumerians and in ancient China. Gemstones such as jade and turquoise were collected and hoarded as early as 3000 B.C. Practitioners of alchemy and related "arts" appeared in Greece, China, and India as early as 1500 B.C. They developed laboratory techniques for isolation and purification of elements from minerals, and began to study their properties. The Semitic Chetites processed iron to make tools in 1500 B.C., and Asian Indians forged iron before 1000 B.C. Mercury has been found in Egyptian tombs of about the same age.

Pliny (A.D. 23–79) expanded on the work of Theophrastus and others in his book *Natural History,* but major advances in mineralogy did not occur until the middle ages. The apparent slow development of all sciences during this time was due to a dichotomy. Philosophers and theoreticians, who tried to understand science, were isolated from the practical world. At the same time, understanding eluded practitioners who improved upon techniques and applications. The utility of combining theory and practice was not realized until the beginnings of the industrial revolution. Rapid advances in science soon followed. Atomic theory was developed and accepted, and acclaimed as "modern," leading rapidly to new developments in the sciences of chemistry and mineralogy.

Modern Views of Elements and Matter: A Review of Basic Chemistry

In 1660 the term *element* was first used in the same sense as it is today. Robert Boyle, in his book *The Sceptical Chemist (sic),* said elements "are certain primitive and simple, or perfectly unmingled bodies, which not being made of any other bodies or of one another, are the ingredients of which all. . .bodies are immediately compounded, and into which they are ultimately resolved." This definition launched the search for a complete list of elements. In the early and mid-eighteenth century, Joseph Proust and others observed that elements combine in definite proportions to form **compounds.** By 1800 chemists such as William Higgins and John Dalton extended this **law of definite proportions** and began to discuss molecules as combinations of discrete numbers of atoms. However, some fundamental errors in their ideas made experimental confirmation of their theories impossible.

In the early 1800s Sweden's Jöns Jakob Berzelius invented new techniques in analytical chemistry. Berzelius, who had nearly flunked out of high school, snubbed theoreticians and devoted himself to analyzing and isolating chemical elements in the laboratory. He started mostly with mineralogical materials, many from rich ore deposits. Berzelius determined atomic weights for many elements, including lead, chlorine, potassium, sulfur, silver, and nitrogen, basing his values on weights determined for oxides. Working before understanding of the periodic table and ionic charge, Berzelius had to speculate about the formulas for some oxides. Because he used incorrect oxide formulas, some of his values were off by factors of 2 or 4. After corrections, his results (Table 1.1) are remarkably close to those accepted today. Berzelius also is responsible for the one- and two-letter chemical symbols, based on elemental names, used today.

Grouping of Elements and the Periodic Chart

The successes of Berzelius spawned further study of the elements. Nineteenth-century scientists isolated and described tellurium, zirconium, titanium, cerium, chromium, beryllium, selenium, and vanadium; 55 elements were known by 1830. Chemists began to see similarities between various groups of elements and sought theories to explain the relationships, if for no other reason than because they wanted to know just how many more there were. In 1829 Doebereiner, for example, described four "triads" of elements with similar behavior:

Triad 1: calcium (Ca), strontium (Sr), barium (Ba)
Triad 2: lithium (Li), sodium (Na), potassium (K)
Triad 3: chlorine (Cl), bromine (Br), iodine (I)
Triad 4: sulfur (S), selenium (Se), tellurium (Te)

In each triad the properties of the elements were similar, and the difference in atomic weights between the first and second elements was the same as that between the second and third. In 1862 Alexandre Beguyer de Chancourtois devised a "telluric helix," in which he arranged all known elements on a spiral. A few years later, in separate studies, Julius Lothar Meyer and John A. R. Newlands presented the ele-

►**TABLE 1.1**
Atomic Weights of Berzelius

Name of Element Today	Name of Element in Berzelius's Time	Symbol	$100 \times wt_{element}/wt_{oxygen}$ (Berzelius, 1813)	$100 \times wt_{element}/wt_{oxygen}$ (Today)
oxygen	oxygen	O	100.00	100.0
hydrogen	hydrogen	H	6.64	6.87
carbon	carbon	C	75.1	75.1
sulfur	sulfur	S	201.0	200.4
molybdenum	molybdenum	Mo	601.6	599.6
chromium	chromium	Cr	354.0	325.0
platinum	platinum	Pt	1206.7	1219.3
silver	argentum	Ag	672.1	674.2
mercury	hydrargyrum	Hg	1265.8	1253.7
copper	cuprum	Cu	403.2	397.2
lead	plumbum	Pb	1298.7	1295.0
tin	stannum	Sn	735.3	741.8
iron	ferrum	Fe	346.8	349.1
calcium	calcium	Ca	255.1	250.5
sodium	natrium	Na	144.8	143.7
potassium	kalium	K	244.5	244.4

Modified from values in Farber, 1952.

ments in rows and columns based on their atomic weights and similar properties.

The great breakthrough came in 1868 when Dmitri Ivanovich Mendeleyev wrote his handbook of chemistry. Mendeleyev used atomic weights and charges to devise a **periodic chart** remarkably like those in use today. In 1870 he published the chart in his *Natural System of the Elements*, which had some blanks and predicted the future discoveries of scandium (Sc), gallium (Ga), and germanium (Ge) 10 to 15 years later (Table 1.2). Mendeleyev's chart and papers explained the periodicity of elemental properties in such detail that few contested it. Less than a year later, J. L. Meyer published a graph of atomic volumes versus atomic weights that confirmed most of Mendeleyev's findings (Figure 1.4). The periodic

►**TABLE 1.2**
Arrangement of Elements and Atomic Weights in Mendeleyev's First Periodic Table

						Ti	50	Zr	90			
						V	51	Nb	94	Ta	182	
						Cr	52	Mo	96	W	186	
						Mn	55	Rh	104.4	Pt	197.4	
						Fe	56	Ru	105.4	Ir	198	
						Co	59	Pd	106.6	Os	199	
H	1					Cu	63.4	Ag	108	Hg	200	
		Be	9.4	Mg	24	Zn	65.2	Cd	112			
		B	11	Al	27.4	(Ga)		Ur	116	Au	197	
		C	12	Si	28	(Ge)		Sn	118			
		N	14	P	31	As	75	Sb	122	Bi	210	
		O	16	S	32	Se	79.4	Te	128			
		F	18	Cl	35.5	Br	80	I	127			
Li	7	Na	23	K	39	Rb	85.4	Cs	133	Tl	204	
				Ca	40	Sr	87.6	Ba	137	Pb	207	
				(Sc)		Ce	92					
				Er	56	La	94					
				Yt	60	Di	95					
				In	57.6	Th	118					

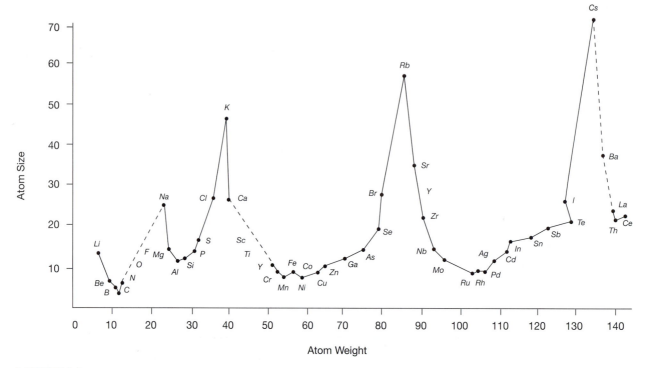

▶**FIGURE 1.4**

A modern version of Meyer's (1870) graph of elemental weights. Meyer's study, which showed the periodic variation of elemental weights, confirmed the basic validity of Mendeleyev's periodic chart.

table allowed mineralogists and chemists to explain why the properties of some minerals were similar and to predict the existence of minerals not yet found.

ATOMS AND ELEMENTS

Mendeleyev developed his chart using empirical methods. He relied on what he could measure and left the theories for others to develop. In spite of many uncertainties, throughout this period most physicists and chemists, including Sir Isaac Newton, continued to believe that atoms were hard, spherical particles. A major change in thinking occurred around 1897, when J. J. Thomson discovered **electrons,** and hypothesized both negatively and positively charged particles in atoms. Soon, Ernest Rutherford completed a series of experiments that led him to predict the existence of small **nuclei** of high mass in the center of atoms. He subsequently predicted the existence of **protons** and **neutrons,** which led to the first modern model of the atom.

The Bohr Model

Niels Bohr made the most significant breakthrough in the development of modern atomic theory. Following in the footsteps of Max Planck and others, Bohr introduced the idea of **energy levels.** He envis-

aged the atom as a small planetary system and used equations analogous to Kepler's astronomical laws to describe the orbits (Figure 1.5). His first major publications appeared in 1913 and, like all radical ideas, found little acceptance in the scientific community. A prominent physicist, Otto Stern, expressed his disdain: "If this nonsense is correct, I will give up being a physicist" (Holton, 1986). However, by 1922, when most physicists agreed with the Bohr's basic ideas, he was awarded the Nobel Prize in Physics.

In the **Bohr model,** electrons are like small, spherical particles orbiting around the nucleus. Positively charged protons in the nucleus attract and hold onto the negatively charged electrons. The number of protons is called the **atomic number** (Z). Each element has a different atomic number, beginning with hydrogen, H ($N = 1$), and continuing to the named element with the highest number, lawrencium, Lw ($N = 103$). The protons in the nucleus are accompanied by neutrons; together, they add up to nearly all the mass of the atom, because electrons, the only other substantial particles in an atom, have little mass. (Electrons, protons, and neutrons have atomic weights of 0.0005486, 1.007276, and 1.008665 mass units, respectively.) Bohr showed that electrons can only occupy discrete energy levels. He designated the levels by a number n, called the first quantum number. For the known elements, n ranges from 1 to 7, from lowest to highest energy

►**FIGURE 1.5**
Bohr's model of the atom showing protons and neutrons in the nucleus, with electrons circling them to form an enclosing cloud.

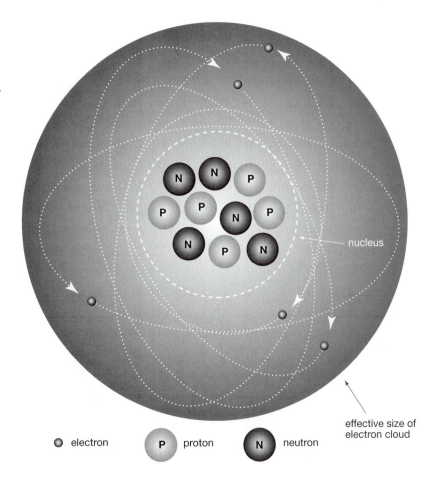

nucleus

effective size of
electron cloud

○ electron   ⬤ P proton   ⬤ N neutron

levels. The seven levels are also designated, in order, K, L, M, N, O, P, and Q. When an electron moves from a higher energy level to a lower one, the energy difference is given off as electromagnetic radiation, often as X rays.

Within each energy level there are sublevels characterized by electron orbitals of different shapes (Figure 1.6). From the lowest to the highest energy, the sublevels are designated by letters: *s*, *p*, *d*, and *f*. *s*-orbitals are spherical, *p* are dumbbell-shaped, *d* are shaped like four-leaf clovers, and *f* are more complicated. A shorthand notation is used to indicate **orbitals** (specific sublevels within a level). For example, 1*s* stands for the lowest energy level and the *s* sublevel; 2*p* indicates the second lowest energy level and the *p* sublevel, and so on. Different orbitals contain different numbers of electrons. *s*, *p*, *d*, and *f* orbitals may hold up to 2, 6, 10, and 14 electrons, respectively.

The electron configuration of atoms is, in general, predictable and quite regular. Available electrons fill orbitals from lowest to highest energy. This process is known as the **Aufbau principle.** Elemental Sr, for example, has 38 electrons occupying the 1*s* through 5*s* orbitals. The outermost electrons in an atom are called **valence electrons.** The valence electrons strongly affect chemical properties; elements with valence electrons in similarly shaped orbitals are chemically similar.

Modifications to the Bohr Model

Bohr's model worked well for the hydrogen atom, but not for heavier elements. A fundamental problem with Bohr's mathematical model was that it predicted that electrons should not remain in orbitals, but instead should be attracted to the center of the nucleus. The solution to this problem became possible when, in 1923, Louis de Broglie showed that electrons are not simple particles but instead have wavelike properties. Werner Heisenberg soon pointed out that, due to wave motion, electrons could not follow simple orbits around a nucleus, and their exact location must be uncertain. The **Heisenberg uncertainty principle** was incorporated into the **Schrödinger wave model** published in 1926. The Schrödinger model, which defines the probable energy distribution in an atom, improved on Bohr's work and is the basis for all modern treatments of atomic properties and structure.

▶**FIGURE 1.6**
Shapes of electron orbitals:
(a) *s*-orbitals are spherical;
(b) *p*-orbitals are dumbbell-
shaped; (c) *d*-orbitals are generally shaped like a four-leaf clover.

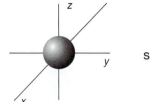

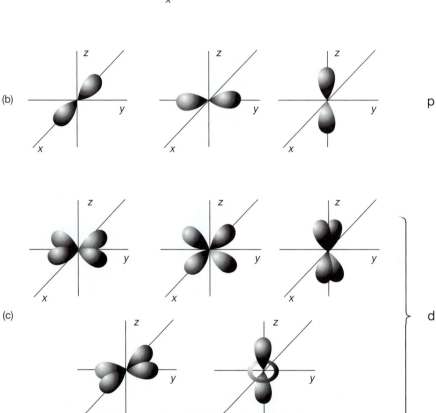

IONS

An important corollary to the Aufbau principle is that atoms are more stable if electrons completely fill energy levels and sublevels. Atoms may give up or borrow electrons, becoming **ions,** to obtain this stability. As a result, they may become **cations,** which have a net positive charge because there are more protons than electrons, or **anions,** with a net negative charge because there are more electrons than protons. The **ionic charge,** also called the **valence,** is the number of protons less the number of electrons. We usually indicate valence by a superscript number after the elemental symbol (for example, Fe^{2+} indicates iron with two more protons than electrons). We call all ions with a charge of +1 or −1 **monovalent.** Those with a charge of +2 or −2 are **divalent. Trivalent** and **tetravalent** refer to charges of ±3 and ±4. Since hydrogen, with atomic number 1, has only one electron in the first energy level (1*s*), it may either borrow an electron from a

donor to fill the first energy level and become a monovalent anion H^-, or (more commonly) donate an electron to another atom and become a monovalent cation H^+. Helium, with atomic number 2, has two protons and two electrons that completely fill the first energy level. Because the level is filled, helium has no tendency to accept or donate an electron, and so maintains a valence of 0. Elements such as helium, in which electrons completely fill the *s*-orbitals and *p*-orbitals, are called **noble** or **inert** elements because they are extremely stable and unreactive. Other noble gases include neon (Ne), argon (Ar), krypton (Kr), xenon (Xe), and radon (Ra). They have atomic numbers 10, 18, 36, 54, and 88, respectively. Although filling or emptying the 1*s* orbital in hydrogen to achieve stability is a simple concept, predicting the valence of other elements is not always straightforward. The formation of ions is often complicated because elements may achieve at least partial electronic stability in different ways. This is especially true when *d*-electrons and *f*-electrons are involved. For most ions, however, the number of electrons is close to the number of protons, and the ionic charge is small.

The process of losing an electron is called **oxidation.** Gaining an electron is called **reduction.** Oxidation leaves metals with a positive charge. They may combine with oxygen anions (O^{2-}) to form oxides. Fe-metal (Fe^0), for example, may oxidize to become Fe^{2+} or Fe^{3+}. If Fe^{2+} combines with oxygen, it may form wüstite (FeO); if Fe^{2+} combines with oxygen, it may form hematite (Fe_2O_3). Wüstite and hematite are both minerals, but hematite is more common because under normal Earth surface conditions, iron easily oxidizes. Many elements besides hydrogen and iron may exist in more than one valence state. Carbon may reduce to become C^{4-} or oxidize to become C^{4+}, although the cationic form is more common. Adding further complication, carbon may at times have a charge of +2. The complexities of predicting electronic configurations and valence are many. Simplifying the matter, however, is that for ions with more than one potential ionic state, a single state is usually much more common. Figure 1.7 shows the most common valences of elements in minerals. Those with no indication of valence may ionize but have no common ionic forms in minerals.

THE MODERN PERIODIC TABLE

Although the Periodic Table of the Elements has appeared in many forms, the basic relationships are the same today as they were in 1870. Development of the Schrödinger wave model led to a slight rearrangement of the table so that it mirrors the order in which electrons occupy orbitals. The present version, depicted in Figure 1.8, separates the elements into **periods** (rows) and **groups** (columns).

Periods

Elements are ordered by increasing atomic weight. The number of the period (row) indicates the orbitals occupied by the electrons. For example, elements in the first period contain electrons in the $1s$ orbital; elements in the second period have electrons in the $2s$ and $2p$ orbitals; elements in the third period have electrons in the $3s$ and $3p$ orbitals; and so on. The first period has only 2 elements, while the second and third have 8 each. The fourth through seventh periods contain 10 extra elements having electrons in *d*-orbitals. The sixth through seventh periods contain an additional 14 elements with electrons in *f*-orbitals (which are listed separately at the

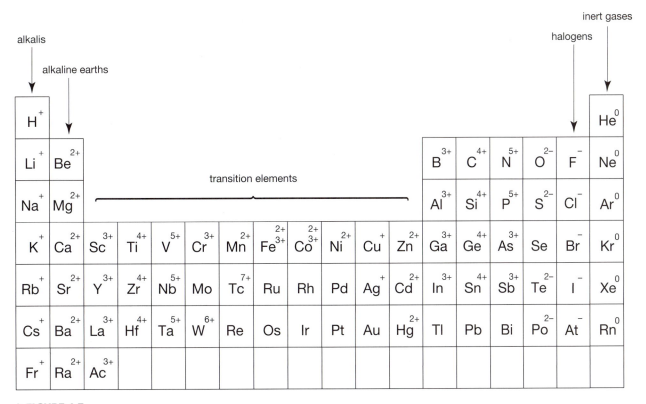

▶**FIGURE 1.7**
Typical ions in minerals. Elements with no indication of valence have no common ionic state in minerals.

Periodic Table of the Elements

GROUP

PERIOD

Legend (key):
- Electronegativity
- Oxidation States
- Atomic number
- Element symbol
- Atomic mass*

```
2.20          1
        1
        H
       1.01
```

Legend blocks: Metals | Semimetals | Non Metals

Group 1 IA	2 IIA	3 IIIB	4 IVB	5 VB	6 VIB	7 VIIB	8 VIII	9 VIII	10 VIII	11 IB	12 IIB	13 IIIA	14 IVA	15 VA	16 VIA	17 VIIA	18 VIIIA
2.20 1 H 1.01																	2 He 4.00
0.98 1 3 Li 6.94	1.57 2 4 Be 9.01											2.04 3 5 B 10.81	2.55 ±2,±4 6 C 12.01	3.04 ±3,4,5 7 N 14.01	3.44 -2 8 O 16.00	3.98 -1 9 F 19.00	10 Ne 20.18
0.93 1 11 Na 22.99	1.31 2 12 Mg 24.31											1.61 3 13 Al 26.98	1.90 4 14 Si 28.09	2.19 ±3,4,5 15 P 30.97	2.58 ±2,4,6 16 S 32.07	3.16 ±1,3,5,7 17 Cl 35.45	18 Ar 39.95
0.82 1 19 K 39.10	1.00 2 20 Ca 40.08	1.36 3 21 Sc 44.96	1.54 3,4 22 Ti 47.88	1.63 2,3,4,5 23 V 50.94	1.66 2,3,6 24 Cr 52.00	1.55 2,3,4,5,6 25 Mn 54.94	1.83 2,3 26 Fe 55.85	1.88 2,3 27 Co 58.93	1.91 2,3 28 Ni 58.69	1.90 1,2 29 Cu 63.55	1.65 2 30 Zn 65.39	1.81 3 31 Ga 69.72	2.01 2,4 32 Ge 72.61	2.18 ±3,5 33 As 74.92	2.55 -2,4,6 34 Se 78.96	2.96 ±1,5 35 Br 79.90	36 Kr 83.80
0.82 1 37 Rb 85.47	0.95 2 38 Sr 87.62	1.22 3 39 Y 88.91	1.33 4 40 Zr 91.22	1.6 3,5 41 Nb 92.91	2.16 2,3,4,5,6 42 Mo 95.94	1.9 7 43 Tc (99)	2.2 2,3,4,6,8 44 Ru 101.07	2.28 2,3,4 45 Rh 102.91	2.20 2,4 46 Pd 106.42	1.93 1 47 Ag 107.87	1.69 2 48 Cd 112.411	1.78 3 49 In 114.82	1.96 2,4 50 Sn 118.71	2.05 3,5 51 Sb 121.75	2.1 -2,4,6 52 Te 127.60	2.66 ±1,3,5,7 53 I 126.90	54 Xe 131.29
0.79 1 55 Cs 132.91	0.89 2 56 Ba 137.33	1.10 3 57 La 138.91	1.3 4 72 Hf 178.49	1.5 5 73 Ta 180.95	2.36 2,3,4,5,6 74 W 183.85	1.9 -1,2,4,6,7 75 Re 186.21	2.2 2,3,4,6,8 76 Os 190.2	2.20 2,3,4,6 77 Ir 192.22	2.28 2,4 78 Pt 195.08	2.54 1,3 79 Au 196.97	2.00 1,2 80 Hg 200.59	2.04 1,3 81 Tl 204.38	2.33 2,4 82 Pb 207.2	2.02 3,5 83 Bi 208.98	2.0 2,4 84 Po (209)	2.2 ±1,3,5,7 85 At (210)	86 Rn (222)
0.7 87 Fr (223)	0.9 88 Ra (226)	1.1 89 Ac (227)	4 104 Rf (261)	105 Db (262)	106 Sg (263)	107 Bh (262)	108 Hs (265)	109 Mt (266)	110 (269)	111 (272)	112 (277)						

Lanthanide series:

1.12 3,4 58 Ce 140.12	1.13 3,4 59 Pr 140.91	1.14 3 60 Nd 144.24	1.13 3 61 Pm (147)	1.17 2,3 62 Sm 150.36	1.2 2,3 63 Eu 151.97	1.20 3 64 Gd 157.25	1.2 3,4 65 Tb 158.93	1.22 3 66 Dy 162.50	1.23 3 67 Ho 164.93	1.24 3 68 Er 167.26	1.25 3 69 Tm 168.93	1.1 2,3 70 Yb 173.04	1.27 3 71 Lu 174.97

Actinide series:

1.3 4 90 Th (232)	1.5 4 91 Pa (231)	1.38 3,4,5,6 92 U (238)	1.36 3,4,5,6 93 Np (237)	1.28 3,4,5,6 94 Pu (244)	1.3 3,4,5,6 95 Am (243)	1.3 3 96 Cm (247)	1.3 3,4 97 Bk (247)	1.3 3 98 Cf (251)	1.3 3 99 Es (252)	1.3 3 100 Fm (257)	1.3 2,3 101 Md (258)	1.3 2,3 102 No (259)	103 Lr (260)

*The mass number of an important radioactive isotope—not the atomic mass—is shown in parenthesis for an element with no stable isotopes.

▶ **FIGURE 1.8**
The modern Periodic Table of the Elements.

12

bottom of the table as the **lanthanides** [also called **rare earth elements**] and **actinides** to keep the chart from becoming too wide).

Groups

Elements in the same groups (columns) have valence electrons in similar orbitals. Hence they have similar chemical properties. Elements in groups on the left side of the chart have a few "extra" electrons in outer orbitals, and these readily give up electrons to become cations. Elements in group 1 **(alkali elements)** generally have a valence of +1. Those in group 2 **(alkaline earth elements)** usually have a valence of +2. Both alkalis and alkaline earths have valence electrons in *s*-orbitals. Elements on the right side of the Periodic Table have nearly full outer orbitals and easily accept extra electrons to become anions. Elements in group 17 **(halogens)** typically have a valence of −1 because they acquire an extra electron to fill their outermost *p*-orbital. The properties of elements in the central portion of the chart, called **transition elements,** with partially filled *d*-orbitals and *f*-orbitals, are less predictable. Elements in groups 3, 4, and 5 usually have valences of +3, +4, and +5, respectively. The rest of the transition elements exist in a number of valence states, typically +2 or +3. The three most abundant elements in the Earth's crust (O, Si, and Al) are on the right-hand side of the Periodic Chart, in groups 13, 14, and 16. Oxygen typically has a charge of −2, Si of +4, and Al of +3.

Atomic Number and Mass

An element's atomic number is the number of protons in its nucleus (Z). It is equal to the number of electrons orbiting the nucleus in neutral (non-ionized)

atoms, and is close to the number of electrons in most ions. Fe^0, for example, has 26 protons ($Z = 26$) in its nucleus and 26 electrons, creating an **electron cloud** around the nucleus. Fe^{2+} has 26 protons and 24 electrons. Because the size of its electron cloud controls the diameter of an atom, elements with many protons, having many electrons, tend to be large; those with few electrons are small. According to the Aufbau principle, electrons fill orbitals in a predictable manner. Thus atomic number determines the number and nature of valence electrons and atomic or ionic size. Atomic number is the most significant factor controlling elemental properties.

The number of neutrons (N) in the nucleus of a given element may vary (Figure 1.9). This leads to isotopes of different **mass numbers** *(A):*

$$A = Z + N \qquad (1)$$

Most chemical elements have several different naturally occurring **isotopes.** Thus, for example, oxygen may be ^{16}O, ^{17}O, or ^{18}O, where the preceding superscript denotes A. Examination of Equation 1 indicates that the three isotopes of oxygen must have 8, 9, and 10 neutrons, respectively, because all must have 8 protons.

The **atomic weight of an atom** is calculated by summing the weights of its atomic particles, and subtracting a small amount to account for mass lost to energy holding the nucleus together. The **atomic weight of an element** is the sum of the masses of its naturally occurring isotopes weighted in accordance with their abundances. Although isotope mass numbers are always integers, atomic weights are not. For example, many tables and charts give the atomic weight of oxygen as 15.9994 and that of Fe as 55.847. The scale used to measure atomic masses has changed slightly over time. Today, it is standardized so that the mass of $^{12}C \equiv 12.0000$. All

▶**FIGURE 1.9**

Comparing the atomic structures of ^{12}C and ^{13}C. ^{13}C has one "extra" neutron in its nucleus.

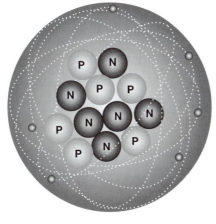

^{12}C

^{13}C

atomic masses are given in **atomic mass units** (amu), defined as $\frac{1}{12}$ of the mass of ^{12}C. Atomic weights of elements are dimensionless numbers because they are all calculated relative to the atomic weight of ^{12}C.

A **mole** of an element is defined as the amount of that element that has its weight in grams equal to its atomic weight. A mole of an element always contains 6.022×10^{23} atoms. The number 6.022×10^{23} is known as **Avogadro's number.** One mole of ^{12}C, equal to 6.022×10^{23} carbon atoms, by definition weighs exactly 12 grams. Using atomic weights and Avogadro's number, it is possible to convert from weight to number of atoms and vice versa. Atomic weights of compounds (combinations of elements) are just the sum of the weights of the elements (Box 1.2).

Most elements have very small isotopic variation, no matter where they are found. Thus, all quartz (SiO_2) contains about the same relative amounts of ^{16}O, ^{17}O, and ^{18}O. Furthermore, isotopic variations have extremely small effects on the properties of minerals. Most mineralogists, therefore, do not worry very much about isotopes. The small isotopic variations, however, may be significant to a geochemist trying to determine the genesis of a particular mineral or rock.

Most common isotopes are **stable isotopes.** In 1896 Henri Becquerel discovered **unstable isotopes** and radioactivity when he unintentionally conducted an experiment. He put some radioactive samples in a drawer along with a photographic plate and subsequently found that the plate had recorded the image of a key that had been sitting on top of it. Although Becquerel didn't know it, X rays given off by uranium-rich minerals had caused the image. During the following decade, researchers including Marie and Pierre Curie and Ernest Rutherford determined that some elements, termed **radioisotopes,** emit α and β particles and γ radiation as they decay, forming **daughter isotopes.** A few **remnant heavy isotopes** are decaying so slowly that they remain from before the creation of the solar system. Others decay so rapidly that they only exist for brief periods in nuclear reactors or explosions. Many radioisotopes decay in one step to a stable daughter isotope. Others decay in many steps from an original radioactive **parent isotope** to a final stable (not decaying) daughter element. Figure 1.10 shows such a decay chain for ^{238}U. Geochemists use radioisotopes to estimate the ages of minerals and rocks.

BONDING IN MINERALS

Negatively and positively charged particles attract each other. Protons (positive charge) attract electrons (negative charge) in atoms. Similarly, positively charged cations attract negatively charged anions, producing **ionic bonds** in minerals. In fact, the forces

---

 # Box 1.2 What Is a Mole of Quartz?

Quartz, SiO_2, is one of the most common and well-known minerals. We might wonder: How much quartz comprises a mole? To answer this question, we use the atomic weights of silicon and oxygen as well as some crystallographic data.

Silicon and oxygen have atomic weights of 28.0855 and 15.9994, respectively. The atomic weight of quartz, SiO_2, is therefore 60.0843 (28.0855 + 15.9994 + 15.9994). A mole of quartz, 6.022×10^{23} SiO_2 molecules, weighs 60.0843 grams.

Crystallographers have determined that quartz crystals are made of fundamental unit cells shaped like hexagonal prisms (discussed in detail in later chapters) containing three SiO_2 molecules. Each unit cell has a volume of 112.985 Å^3. So we may calculate the volume of a mole of quartz as:

$$V = N \times v/Z$$
$$= 6.022 \times 10^{23}\ SiO_2/mole$$
$$\times\ 112.986\ \text{Å}^3/\text{unit cell} \div 3\ SiO_2/\text{unit cell}$$
$$= 2.268 \times 10^{25}\ \text{Å}^3$$
$$= 22.68\ \text{cm}^3\ \text{(slightly smaller than}$$
a golf ball)

where V, N, v, and Z are the molar volume, Avogadro's number, the unit cell volume, and the number of molecules per unit cell, respectively.

We can, if we wish, then calculate the density (ρ) of quartz from the molar data:

$$\rho = \text{molar weight/molar volume}$$
$$= 60.0843\ \text{gm/mole} \div 22.68\ \text{cm}^3/\text{mole}$$
$$= 2.649\ \text{gm/cm}^3$$

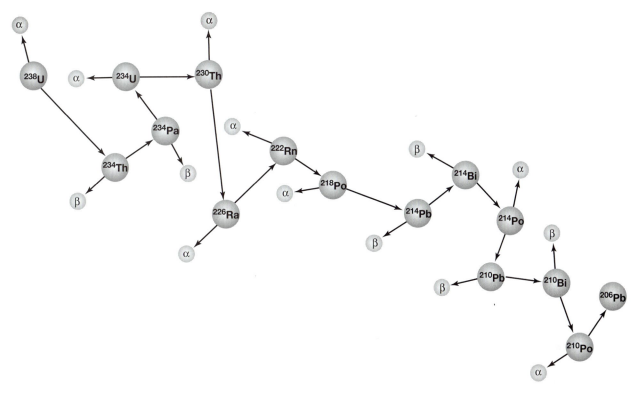

▶**FIGURE 1.10**
Decay of ^{238}U to ^{206}Pb (after Richardson and McSween, 1989). α and β particles are emitted during decay.

Box 1.3 Ionic, Covalent, and Metallic Crystals

Most minerals are neither 100% ionic, 100% covalent, nor 100% metallic, but some come close. In halite (NaCl), Na^+ and Cl^- exhibit such a minor amount of electron sharing that for all practical purposes halite can be considered perfectly ionic. Fluorite (CaF_2), too, is nearly completely ionic. In contrast, diamond (C) is nearly completely covalent, while copper (Cu) represents end-member metallic bonding.

Chemical bonding in minerals controls many properties. Covalent bonds are stronger than ionic bonds, which in turn are stronger than metallic bonds. As a consequence, covalent minerals are hard and tenacious, while metallic ones are usually not, and ionic minerals fall in between. Melting temperatures follow a similar pattern (covalent compounds generally melt at the highest temperatures due to their stronger bonds), while solubility in water is greatest for ionic crystals (because of their weak bonds and easy ionization).

Crystal symmetry is also affected by bond type. Both ionic and metallic bonds are nondirectional so bonding can occur equally in all directions. In contrast, covalent bonds involve pairs of atoms and are linear. Consequently, metallic and ionic minerals generally have high symmetry compared to covalent ones.

that bond elements together to form compounds are all electrical. Chemists have identified different types of bonds, and three are most significant for minerals: **ionic, covalent, and metallic** (see Box 1.3). The general properties of minerals with ionic, covalent, and metallic bonds are predictable (Table 1.3).

▶**TABLE 1.3**
General Characteristics of Ionic, Covalent, and Metallic Bonds

	Ionic Bonds	**Covalent Bonds**	**Metallic Bonds**
Common Elements Involved	**Elements from Opposite Sides of the Periodic Chart, such as Alkalis and Halides, or Alkaline Earths and Oxygen**	**Elements Close Together in Periodic Chart**	**Transition Metals**
electrical conductivity	low	low	high
thermal conductivity	low	low	high
solubility in water	high	low	very low
melting temperature	moderate to high	high	variable
hardness	medium to hard	very hard	variable, often minerals are malleable
breaking	brittle, good cleavage	brittle, often conchoidal fractures	variable
crystal form	high symmetry	low symmetry	very high symmetry
transparency	fully to partially transparent	partially transparent or opaque	opaque
how common?	most minerals	some minerals	most metals
examples	halite, NaCl; calcite, $CaCO_3$	diamond, C; sphalerite, ZnS	copper, Cu; silver, Ag

Because atoms are unstable if electrons do not completely fill energy levels and sublevels, most elements ionize. Some ionize more easily than others, and we call those that easily ionize to become cations **metallic elements.** The degree to which elements are metallic generally decreases from left to right in the periodic table. We call elements on the right-hand side of the table, which ionize to become anions, **nonmetallic.** The formation of cations and anions go hand in hand. Nonmetals becoming anions gain electrons given up by metals becoming cations. Thus, Na and Cl react to produce the ions Na^+ and Cl^-. Because cations and anions have opposite charges, they attract each other, forming ionic bonds. For Na and Cl, this results in the mineral halite, NaCl.

Rather than giving up or gaining electrons to become ions, atoms may become stabilized by sharing electrons. We call the sharing of electrons between atoms a **covalent bond.** If the sharing is complete, the bond is 100% covalent. This is the case for diatomic gases such as H_2, N_2, or O_2. Covalent bonds are stronger than ionic bonds. Many elements form both covalent and ionic bonds. As mentioned, Cl^- forms ionic bonds with Na^+ to make the mineral halite. Two Cl atoms also bond covalently to form a stable gas, Cl_2. Bonds may also be partially ionic and partially covalent if electron sharing is limited. Thus, a continuum exists between substances such as halite, in which the bonding is nearly 100% ionic, and diatomic gases or diamond, C, in which it is 100% covalent. The degree to which a particular bond is ionic depends on both elements involved. Generally, elements from the outer columns (groups) of the periodic chart form more ionic bonds compared with elements from the center. Because alkali elements (Group 1) have a very strong tendency to become cations, and halogens (Group 17) have an equal tendency to become anions, halite (NaCl) and other alkali halides form crystals in which bonds are nearly 100% ionic (Figure 1.11). Alkaline earth oxides such as MgO (periclase) or CaO (lime), involving cations from Group 2 and O^{2-} from Group 16, are about 75% ionic. The Si-O bonds in **silicates,** minerals with structures based on SiO_4 groups, is generally about 50% ionic and 50% covalent. Silicates are the most common kind of mineral in the Earth's crust.

In covalent bonding, pairs of atoms share electrons. Sometimes, many atoms share the same electrons. This produces a third type of bonding, **metallic bonding,** which is especially common for minerals involving transition metals such as Cu, Au, or Zn. Good examples of metallic crystals are the minerals gold, silver, and copper. Since valence electrons move easily throughout the structure, metallically bonded compounds are good conductors of heat and electricity. They may be malleable and have only low to moderate hardness, reflecting the loose nature of their bonds.

Many ore minerals have little ionic character, but most other minerals are partially to dominantly ionically bonded (Figure 1.11). Some minerals are combinations of ionic and covalent bonding. Oth-

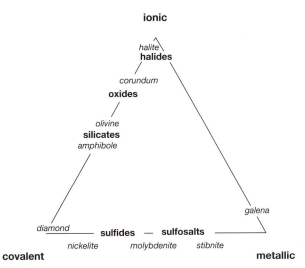

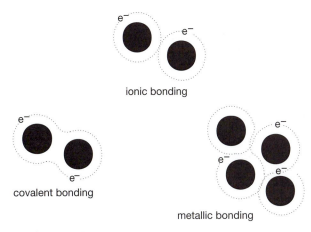

▶**FIGURE 1.11**
Different minerals contain different kinds of atomic bonds; most minerals have combinations of ionic and covalent bonds, or combinations of covalent and metallic bonds.

▶**FIGURE 1.12**
In ionic bonds, electrons (e^-) are localized around individual ions (black circles); in covalent bonds they are shared between two atoms (black circles); in metallic bonds, they are shared within a structure.

ers, especially sufides and sulfosalts, are combinations of covalent and metallic bonding. Metallic and ionic bonds do not often combine, although galena, PbS, may be one example in which they do. Figure 1.12 illustrates the three types of bonds, emphasizing the locations of the valence electrons that set them apart.

Besides the three mentioned, other types of bonds, such as **van der Waals** and **hydrogen** exist in minerals. These types of bonds do not involve valence electrons but instead result from relatively weak electrostatic forces due to uneven charge distribution in a crystal structure. Very weak van der Waals bonds are important in graphite and some clay minerals, for example. This explains why graphite is much softer than diamond, which has the same composition but covalent bonds. In the mineral brucite, combinations of hydrogen and van der Waals bonds hold sheets of Mg^{2+} together. Clays have excellent cleavage because, while covalent and ionic bonds create strongly bonded layers, weak van der Waals and hydrogen bonds hold the layers together. Fortunately for mineralogists and mineralogy students, we may ignore van der Waals, hydrogen, and dative bonds for most purposes. In fact, we will concentrate only on ionic bonds in our further discussions. We do this because:

- Consideration of ionic bonds explains most mineral properties even if the minerals are not completely ionic.
- Most minerals have predominantly ionic bonds.
- Ionic bonds are simpler to understand.

ORIGIN OF THE ELEMENTS AND THE EARTH

Hubble's Contributions

How did elements form in the first place? How did they get where we find them today? Perhaps the first real insight into these questions was made in 1929 by Edwin Hubble. Hubble determined the speed at which some galaxies (in the Virgo cluster) were moving away from Earth. Hubble and others then noticed that the universe seems to be expanding; all matter is moving away from a central location. This led scientists to hypothesize that the expansion started at one place and time. Based on his observations, Hubble later estimated that one nearly instantaneous event created the universe about two billion years ago. Hubble's calculations were not widely accepted, particularly because they conflicted with radioactive age dates, which suggested that the Earth was older than two billion years. In spite of the contradictions, Arno Penzias and Robert Wilson confirmed the basic validity of Hubble's ideas in 1964, for which they subsequently won the Nobel Prize in Physics. Errors have since been found in some constants used in Hubble's calculations. Current best estimates are that the universe originated in a fraction of a second, during the big bang, about 15 billion years ago (Figure 1.13). All of its mass and energy were created nearly instantaneously.

In the Beginning

At its creation, the universe and all matter in it were at extremely high temperatures. As the universe expanded, it cooled to temperatures near a

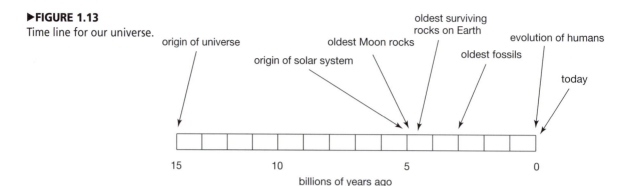

▶**FIGURE 1.13**
Time line for our universe.

origin of universe

origin of solar system

oldest Moon rocks

oldest surviving
rocks on Earth

oldest fossils

evolution of humans

today

15   10   5   0

billions of years ago

billion degrees and subatomic particles combined to form nuclei of hydrogen and helium, the lightest elements. This process was incredibly brief; most present-day chemists and cosmologists estimate it continued for only a half hour or so. No elements heavier than helium ($Z = 2$) formed. After another 750,000 years, scientists estimate that temperatures had cooled to 3,000 °F, and electrons began to attach themselves to nuclei.

The cooler temperatures overall permitted some clumps of matter to begin to come together, even as the universe as a whole continued to expand. Stars, nebulae, and galaxies began to form, increasing their mass and gravitational attraction and the temperatures in their cores. Ultimately, temperatures became hot enough to sustain a "nuclear furnace" powered by hydrogen fusion. The nuclear reactions created heavier elements through a series of complicated reactions occurring in steps at different temperatures. Many stars acquired planets and other orbiting bodies. Today, the space between stars is filled with hydrogen and helium, dating from the original creation of the universe, and remnant-heavier elements formed in the interior of stars that have exploded.

The Formation of Our Solar System

Our own solar system formed when a nebula condensed about six billion years ago. The process was very fast at first; hydrogen fusion in our Sun began during the first 100,000 years. The solar nebula contained primordial H and He, and for the most part, these elements dominate our solar system today. As the nebula condensed, refractory elements (unreactive elements) remained in hotter regions. Pressure and temperature gradients led to differentiation of the solid/gas nebula—elements that easily vaporized only remained in the cooler outer parts. Various clumps of matter condensed to form planetesimals (with radii ranging from a few meters to 1,000 km) with compositions varying predictably from the center of protosolar system to the outside (Figure 1.14). Minerals formed, with oxides and Fe-Ni alloy minerals collecting in the center of the protosolar system and Mg-Fe silicates concentrating farther out. Water, methane, and other volatiles concentrated in the outermost sections.

Today, remnant-heavy elements from previously existing stars are concentrated in the terrestrial planets (Mercury, Venus, Earth, and Mars) and parent bodies of meteorites (now asteroids). The Earth and other terrestrial planets seem to have condensed in stages. The core formed first from Fe-Ni–rich planetesimals. More planetesimals, richer in Si, were added to the outside, thus giving the sharp compositional boundary between the core and mantle. Finally, planetesimals rich in volatile elements were added, ultimately leading to the Earth's early atmosphere. It seems that, in its early history, the Earth was entirely molten, but it soon cooled and developed a crust.

ABUNDANT ELEMENTS AND MINERALS

Goldschmidt's Classification

V. M. Goldschmidt and other geochemists in the early twentieth century believed that the Earth formed by processes involving interactions between various molten material, solids and gases. They hypothesized that the Earth was originally completely homogeneous and molten, subsequently cooling and separating into the layered structure we know today. Although many of their ideas are now known to be incorrect, their basic observations were correct.

As a model for the internal processes of the Earth, Goldschmidt made observations of Cu-smelting furnaces at Mansfeld, Germany, and found that molten materials often divided into several different sorts of liquid melts: one rich in Fe alloys, another rich in sulfide compounds, and one containing silicates. In most smelting processes, the latter two were considered waste, forming **matte** and **slag,** respectively. Goldschmidt also studied meteorites and igneous rocks, where he found minerals falling into the same three general chemical groups.

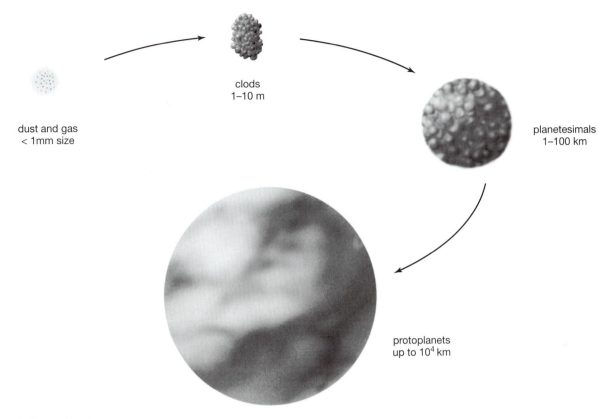

dust and gas
< 1mm size

clods
1–10 m

planetesimals
1–100 km

protoplanets
up to 10^4 km

▶**FIGURE 1.14**
Planets and planetesimals.

Goldschmidt and his colleagues devised a classification scheme for elements based on their behavior in a hypothetical Fe alloy-silicate-sulfide-gas system (Figure 1.15). They divided elements into four groups: **siderophile** (elements that concentrate in an iron-rich liquid), **chalcophile** (elements that concentrate in a sulfur-rich liquid), **lithophile** (elements that concentrate in a silica-rich liquid), and **atmophile** (elements that form a gas). Although the divisions shown in Figure 1.15 are not unambiguous, this classification scheme is significant because it implies that properties and behavior of elements are not random. Through normal Earth processes, we can predict that certain groups of elements will accumulate in certain places and behave in certain ways.

Abundance of Elements

Most minerals seen in daily life formed in the crust of the Earth. The major elements in the crust are the same as those in the mantle, but the mantle is richer in Fe, Mg, and O, and the crust is enriched in Si, Al, Ca, Na, and K. Silicate minerals dominate the crust of the Earth, reflecting the abundance of oxygen and silicon (Table 1.4).

Some elements are common in many different minerals. Oxygen (O) and silicon (Si) are perhaps the best examples. Many sedimentary rocks and nearly all igneous and metamorphic rocks are composed of multiple minerals containing O and Si. In contrast, because of their properties, some other elements tend to be found mainly in only a few distinct minerals. For example, titanium (Ti) may occur as a minor component in biotite, amphibole, or other minerals. In many rocks, however, Ti is concentrated in Ti-rich minerals such as rutile (TiO_2), titanite ($CaTiSiO_5$), and ilmenite ($FeTiO_3$). Rocks rich in Ti always contain one of these latter three minerals. Similarly, rocks containing significant amounts of phosphorous usually contain apatite, $Ca_5(PO_4)_3(OH, F, Cl)$, or monazite, $(Ce, La, Th, Y)PO_4$.

Anionic Complexes

While thinking of individual elements coming together to form minerals is convenient, in reality, atoms are seldom unbonded to others. Single atoms are very reactive. They tend, when possible, to bond to other atoms to form compounds. Sometimes, they bind to other atoms of the same element. For example, N_2, composed of molecules containing

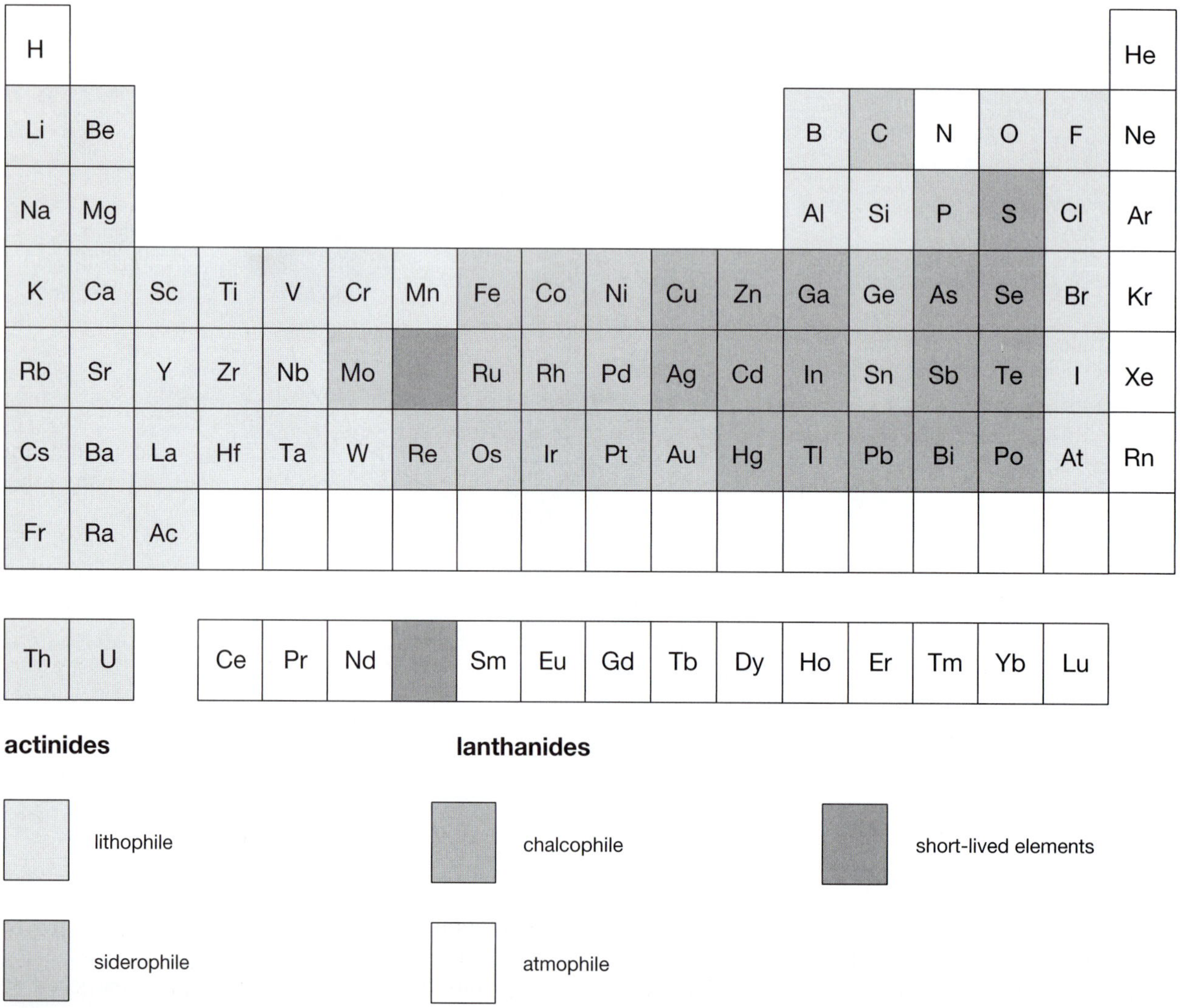

▶**FIGURE 1.15**
Goldschmidt's classification scheme for the elements.

two nitrogen atoms, dominates the Earth's atmosphere. Small atoms with several valence electrons, such as silicon or carbon, are especially reactive. They seldom exist by themselves, readily combining with oxygen, and sometimes other elements, to form **anionic complexes** called **radicals.** Thus, silicon, carbon, phosphorous, nitrogen, and sulfur are usually found in silicate, carbonate, phosphate, nitrate, or sulfate minerals (Table 1.5). Radicals are strongly bonded units and behave just like individual anions in many minerals. Most mineralogical and geological texts (this one included) classify minerals based on their anions or anionic complexes because the properties of minerals with the same anions are generally very similar (see Chapter 2). Mineral formulas are usually written to emphasize any anionic groups that exist (Box 1.4).

Mineralogists can acquire mineral analyses in many ways. In the past, most chemical analyses were

▶**TABLE 1.4**
The Most Abundant Elements in the Earth's Crust by Weight

Element	Weight %
O	46.60
Si	27.72
Al	8.13
Fe	5.00
Ca	3.63
Na	2.83
K	2.59
Mg	2.09
Ti	0.44
H	0.14
P	0.10
Mn	0.09
Ba	0.04

determined by titration and other "wet chemical" techniques. Today we use sophisticated analytical machines, including **atomic absorption spectrophotometers** and **electron microprobes.** We normally report analytical results by listing oxide weight percents. These values must be normalized if we wish to have **mineral formulas** (Box 1.5).

▶**TABLE 1.5**
Common Radicals in Minerals

Element	Radical
silicon (Si)	$(SiO_4)^{4-}$
carbon (C)	$(CO_3)^{2-}$
nitrogen (N)	$(NO_3)^{-}$
sulfur (S)	$(SO_4)^{2-}$
phosphorous (P)	$(PO_4)^{3-}$

MAJOR ELEMENTS, MINOR ELEMENTS, AND TRACE ELEMENTS

Most natural minerals are formed from combinations of many elements. Unlike mechanical mixtures (such as two things being ground up together) the elements are ordered and intimately bonded together. The resulting crystalline solutions are **solid solutions.** Common olivine, for example, is a solution of Fe_2SiO_4 and Mg_2SiO_4 (Box 1.4). Some minerals, such as hornblende, contain many elements in solid solution and have long and complicated formulas, while others such as fluorite (CaF_2) or quartz (SiO_2) vary little from their ideal compositions. We must use sophisticated analytical instruments to determine the composition of natural hornblendes and many other solid solution minerals.

The elements that comprise a mineral may be broadly classified in one of three categories: **major**

Box 1.4 Chemical Formulas of Minerals

Throughout this book, we follow standard chemical conventions. We list elements with subscripts to indicate the relative numbers of atoms present. We list cations before anions and anionic complexes, with the largest cations coming first. Following are some examples of formulas:

marialite	$Na_4(AlSi_3O_8)_3Cl$
skutterudite	$(Co,Ni)As_3$
clinohumite	$Mg_9(SiO_4)_4(OH,F)_2$
olivine	$(Mg,Fe)_2SiO_4$
natrolite	$Na_2Al_2Si_3O_{10} \cdot 2H_2O$
montmorillonite	$(Na, Ca)(Al,Mg)_2(Si_4O_{10})$ $(OH)_2 \cdot nH_2O$

Subscripts outside parentheses apply to everything within if no commas are present. The formula unit of marialite indicates that 4 Na, 3 Al, 9 Si, 24 O, and 1 Cl are in one formula of marialite. Commas show an either-or situation. In one formula of clinohumite, for example, there are two atoms of OH or of F, or of the two combined. If we had omitted the comma, it would indicate that there were both 2 OH and 2 F per formula. Elements separated by commas, then, can be thought of as substituting for each other. For example, montmorillonite may contain either Al or Mg, or both; olivine may contain either Fe or Mg, or both.

Parentheses surround complexes such as SiO_4 or CO_3 when it helps with clarity. In clinohumite, parentheses around SiO_4 emphasize clinohumite's chemical similarity to forsterite and other olivines, all of which have SiO_4 in their formulas. In montmorillonite, the Si_4O_{10} is in parentheses to emphasize that the structure is that of a sheet silicate, many of which have (Si_4O_{10}) in their formula.

Loosely bonded interstitial components (such as Cl in marialite, or OH and F in clinohumite) are on the right. We indicate loosely bonded H_2O, often called nonstructural water, by a dot preceding nH_2O at the far right in a formula. The n in the formula for "clay" indicates that an unknown amount of nonstructural water is present. Natrolite, a zeolite, has H_2O in holes in its structure. When completely hydrated, there are two moles of H_2O for each $Na_2Al_2Si_3O_{10}$ formula.

When useful, we use superscripts to indicate ionic charge. $(OH)^-$ indicates the hydroxyl radical, which has a charge of −1. Similarly $(SiO_4)^{4-}$ indicates an Si atom bonded to 4 O, with a net charge of −4. Sometimes showing coordination (the number of bonds) of an atom is useful. We do this with superscript roman numerals; they are discussed later.

Box 1.5 How To Normalize a Mineral Analysis

We normally report mineral analyses in oxide weight %, but mineral formulas are written in terms of numbers of atoms. **Normalization** is the process of converting an analysis to a formula. Normalization is a tedious but straightforward arithmetical operation when done by hand.

The table below gives the chemical analysis of a feldspar from Grorud, Norway. Column A lists the oxides, columns B through D give chemical data for the oxides, and column E displays the actual results of the mineral analysis. The Grorud feldspar contains 65.90 wt. % SiO_2, 19.45 wt. % Al_2O_3, 1.03 wt. % Fe_2O_3, 0.61 wt. % CaO, 7.12 wt. % Na_2O, and 6.20 wt. % K_2O. We can think of the values in column E as being the number of grams of each oxide in 100 grams of the feldspar: 65.90 grams of SiO_2, 19.45 grams of Al_2O_3, 1.03 grams of Fe_2O_3, 0.61 grams of CaO, 7.12 grams of Na_2O, and 6.20 grams of K_2O.

To convert from grams of oxides to moles of oxides, we divide the oxide wt. % values (column E) by oxide atomic weight (column B); column F shows the results. The values in column F are relative values only; they total to 1.48569, which has no scientific meaning. In column G, we have multiplied the values in column F by a fudge factor so the total is 100%. We can see then, that the Grorud feldspar is 73.829 mole % SiO_2, 12.840 mole % Al_2O_3, 0.434 mole % Fe_2O_3, 0.732 mole % CaO, 7.735 mole % Na_2O, and 4.430 mole % K_2O. These values vary from the weight % values in column E because the different oxides have different atomic weights.

To convert the mole % values to numbers of cations, we multiply the values in column G (moles of oxides) by the number of cations in each oxide (column C). To convert the mole % values to numbers of oxygen, we multiply the values in column G by the number of O^{2-} in each oxide (column D). Columns I and J give the results of these calculations. Using these numbers, we could write the formula of the feldspar as $Ca_{0.732}Na_{15.469}K_{8.860}Fe_{0.868}Al_{25.680}Si_{73.829}O_{200.377}$, but normal feldspar analyses are written with eight atoms of O^{2-}. So, in column K we have multiplied all the atom numbers in columns I and J by $(8 \div 200.377)$ so that the number of atoms of O^{2-} is exactly eight. The values in column K give us the normalized formula for the feldspar: $Ca_{0.03}Na_{0.62}K_{0.35}Fe_{0.03}Al_{1.03}Si_{2.95}O_8$, equivalent to $(Ca_{0.03}Na_{0.62}K_{0.35})(Fe_{0.03}Al_{1.03}Si_{2.95})O_8$. Ideal feldspar formulas have stoichiometry $(Ca,Na,K)(Fe,Al,Si)_4O_8$; the Grorud feldspar comes close. The small discrepancy is due to analytical error.

A	B	C	D	E	F		G	H	I	J	K
											$K =$ Normalized values from columns G and J
					G = Normalized values from column F						
				F = E/B					I = G × C	J = G × D	
Oxide	Atomic wt. of oxide (gm/ mole)	# of cations in oxide	# of O^{2-} in oxide	Oxide wt. % in the mineral (determined by analysis)	Number of moles of oxide in the mineral			Cation	Relative number of moles of cation in the mineral	Relative number of moles of O^{2-} contributed by each cation	Number of moles of ion in the mineral
					Relative value	Normalized to 100%					
SiO_2	60.08	1	2	65.90	1.09687	73.829		Si^{4+}	73.829	147.658	2.95
Al_2O_3	101.96	2	3	19.45	0.19076	12.840		Al^{3+}	25.680	38.520	1.03
Fe_2O_3	159.68	2	3	1.03	0.00645	0.434		Fe^{3+}	0.868	1.302	0.03
CaO	56.08	1	1	0.61	0.01088	0.732		Ca^{2+}	0.732	0.732	0.03
Na_2O	61.96	2	1	7.12	0.11491	7.735		Na^+	15.469	7.735	0.62
K_2O	94.20	2	1	6.20	0.06582	4.430		K^+	8.860	4.430	0.35
Total				100.31	1.48569	100.000		Σ cations = 125.438		$\Sigma O^{2-} =$ 200.377	$\Sigma O^{2-} =$ 8.00

elements, minor elements, and **trace elements.** Major elements are those fundamental to the mineral, and control its basic atomic structure and gross properties. Minor elements are those present in small amounts, usually as replacements for some of the major elements. Such elements, present in amounts up to a few weight %, may affect color and some other properties, but the basic atomic structure is fixed by its major element chemistry. Minerals also contain extremely small amounts of other elements called *trace elements.* Trace elements are present in all minerals and provide valuable information for geologists attempting to determine how, when, and where specific minerals formed. They have little effect on most mineral properties. A notable exception to this is sometimes color; even trace amounts of some elements can have major effects on a mineral's color.

As an example of chemical variability in minerals, let's consider olivine. Major and minor element analyses of some olivines from different geologic environments are given in Tables 1.6 (element weight %), 1.7 (oxide weight %), and 1.8 (number of atoms). The three tables are redundant; values from one can be converted to another by an arithmetic process called **normalization** (Box 1.5). Table 1.6 lists element weight %, but most mineralogists and petrologists prefer to consider oxide weight % (for reasons explained in Chapter 5) or numbers of atoms because numbers of atoms directly translate into mineral formulas. As can be seen from the tables, Mg, Fe, and Si are major elements in olivines. Olivine may also contain Ca, Mn, Ti, alkalis, and alkaline earths, but except in extremely rare circumstances, only as minor or trace elements. Many other elements may be present at trace levels in olivine as well; Ni and Cr are good examples.

The weight % values in Tables 1.6 and 1.7 show a wide range of chemical composition for olivine. The major elements Si, Mg, and Fe are especially variable. How can we make sense of it all? The answer is to look at numbers of atoms (Table 1.8) rather than weight %. The values in Table 1.8 yield mineral formulas. Thus, the olivine from New Zealand has a formula $Ca_{0.004}Mg_{1.859}Mn_{0.003}Fe_{0.158}Al_{0.006}Si_{0.985}O_4$. The ideal **stoichiometry** of olivine is R_2SiO_4, where R is usually Fe, Mg, Mn, or Ca; all analyses in Table 1.8 have close to ideal stoichiometry. Apparent deviations from ideal stoichiometry can mostly be attributed to inclusions of other minerals within the olivine, or to analytical error.

While chemical substitutions are common in most minerals, in some minerals the substitutions are complex. In olivine they are quite simple. We can describe the major and minor elemental compositions of most natural olivines as combinations of the **end members** (represented by ideal formulas) forsterite (Mg_2SiO_4), fayalite (Fe_2SiO_4), tephroite (Mn_2SiO_4), and calcio-olivine (Ca_2SiO_4). The sample from Burma is nearly 100% forsterite (Mg_2SiO_4), while that from Germany is nearly 100% fayalite (Fe_2SiO_4). The other olivines fall between. In contrast with olivines, amphiboles, micas, and some other minerals have complex formulas and may have many elements substituting in their structures. Consequently, choosing appropriate end members is difficult and arbitrary.

The degree to which elements may substitute for each other depends on the elements and on the mineral. In olivine, Fe and Mg mix freely, so any composition between fayalite and forsterite is possible. Olivines can incorporate only minor amounts of Ca, however, so no compositions midway between calcio-olivine and forsterite are found in nature. In contrast, Ca, Mg, and Fe mix freely in garnets; natural garnets can have any composition between end members grossular ($Ca_3Al_2Si_3O_{12}$), almandine ($Fe_3Al_2Si_3O_{12}$), and pyrope ($Mg_3Al_2Si_3O_{12}$).

▶**TABLE 1.6**

Analyses of Major and Minor Elements in Some Olivines (Element Weight %)

Rock:	Marble	Marble	Peridotite	Mafic Sill	Fe-gabbro	Fe-formation
Location:	Burma	Finland	New Zealand	Minnesota	East Greenland	Germany
Si	19.50	19.21	19.15	14.22	14.09	13.94
Ti	0.00	0.03	0.01	0.72	0.12	0.00
Al	0.00	0.29	0.11	0.26	0.04	0.00
Fe	0.86	3.39	6.11	44.79	50.84	54.01
Mn	0.00	0.18	0.10	0.00	0.78	0.22
Mg	34.88	32.60	31.27	4.93	0.63	0.00
Ca	0.00	0.00	0.11	0.94	1.56	0.00
Na	0.00	0.00	0.01	0.00	0.00	0.00
O	45.42	44.73	44.30	33.37	32.05	31.42
Total	100.66	100.43	101.17	99.23	100.11	99.59

▶**TABLE 1.7**
Analyses of Major and Minor Elements in Some Olivines (Oxide Weight %)

Rock:	Marble	Marble	Peridotite	Mafic Sill	Fe-gabbro	Fe-formation
Location:	Burma	Finland	New Zealand	Minnesota	East Greenland	Germany
SiO_2	41.72	41.07	40.96	30.42	30.15	29.83
TiO_2	0.00	0.05	0.01	1.20	0.20	0.00
Al_2O_3	0.00	0.56	0.21	0.50	0.07	0.00
Fe_2O_3	0.00	0.65	0.00	0.00	0.43	0.00
FeO	1.11	3.78	7.86	57.62	65.02	69.48
MnO	0.00	0.23	0.13	0.00	1.01	0.28
MgO	57.83	54.06	51.84	8.17	1.05	0.00
CaO	0.00	0.00	0.15	1.32	2.18	0.00
Na_2O	0.00	0.00	0.01	0.00	0.00	0.00
Total	100.66	100.43	101.17	99.23	100.11	99.59

▶**TABLE 1.8**
Analyses of Major and Minor Elements in Some Olivines (Ions per 4 Oxygen)

Rock:	Marble	Marble	Peridotite	Mafic Sill	Fe-gabbro	Fe-formation
Location:	Burma	Finland	New Zealand	Minnesota	East Greenland	Germany
Si^{4+}	0.978	0.979	0.985	0.971	1.002	1.011
Ti^{4+}	0.000	0.001	0.000	0.029	0.005	0.000
Al^{3+}	0.000	0.016	0.006	0.019	0.003	0.000
Fe^{3+}	0.000	0.012	0.000	0.000	0.011	0.000
Fe^{2+}	0.022	0.075	0.158	1.538	1.807	1.970
Mn^{2+}	0.000	0.005	0.003	0.000	0.028	0.008
Mg^{2+}	2.021	1.919	1.859	0.389	0.052	0.000
Ca^{2+}	0.000	0.000	0.004	0.045	0.078	0.000
Total	3.021	3.007	3.016	2.991	2.986	2.989

▶QUESTIONS FOR THOUGHT

Some of the following questions have no specific correct answers; they are intended to promote thought and discussion.

1. Why are minerals important in our daily lives?
2. Understanding chemistry is a key to understanding minerals. Yet the field of mineralogy predates chemistry by well over a thousand years. Explain this apparent contradiction. Why were there mineralogists before there were chemists?
3. The Bohr model of the atom, the Aufbau principle, and related discoveries were a key to understanding bonding in minerals. Why?
4. Why is the Periodic Table of the Elements useful when considering mineral chemistry?
5. Different ions may have different numbers of protons, neutrons, and electrons. How do these variations relate to, or control, mineral chemistry and crystal structure?
6. There are relatively few common minerals. Why?
7. What is the difference between major elements, minor elements, and trace elements? Can there be a major element in one mineral, a minor element in another, and a trace element in a third?
8. All olivines are not the same. Why do we use one name to describe minerals, such as olivine, that do not have a fixed chemical composition?

▶RESOURCES

Burbidge, E. M., E. R. Burbidge, W. A. Fowler, and F. Hoyle. Synthesis of the elements in stars. *Review of Modern Physics* 29 (1957): 547–650.

Farber, E. *The Evolution of Chemistry: A History of Its Ideas, Methods, and Materials.* New York: Ronald Press, 1952.

Faure, G. *Inorganic Geochemistry.* New York: Macmillan, 1991.

Holton, G. Niels Bohr and the integrity of science. *American Scientist* 74 (1986): 237–243.

Moore, F. J. A *History of Chemistry.* New York and London: McGraw Hill, 1939.

Richardson, S. M., and H. Y. McSween, Jr. *Geochemistry: Pathways and Processes.* Englewood Cliffs, NJ: Prentice Hall, 1989.

▶GENERAL REFERENCES

Blackburn, W. H., and W. H. Dennen. *Principles of Mineralogy,* 2nd ed. Dubuque, IA: William C. Brown, 1994.

Caley, E. R., and J. F. C. Richards. *Theophrastus on Stones.* Columbus, OH: Ohio State University, 1956.

Deer, W. A., R. A. Howie, and J. Zussman. *Rock Forming Minerals.* 5 vols. New York: John Wiley & Sons, 1962.

Gaines, R. V., H. C. W. Skinner, E. E. Foord, B. Mason, and A. Rosenzweig. *Dana's New Mineralogy.* New York: John Wiley & Sons, 1997.

Klein, C., and C. S. Hurlbut, Jr. *Manual of Mineralogy,* 21st ed. New York: John Wiley & Sons, 1993.

Putnis, A. *Introduction to Mineral Sciences.* Cambridge and New York: Cambridge University Press, 1992.

Ribbe, P. H., ed. *Reviews in Mineralogy.* Washington, DC: Mineralogical Society of America, 1974–1996.

Zoltai, T., and J. H. Stout. *Mineralogy: Concepts and Principles.* Minneapolis: Burgess, 1984.

2

Crystallization and Classification of Minerals

In the previous chapter we discussed elements and the definition of a mineral. We said that all minerals are crystalline. In this chapter we look in more detail at what *crystalline* means and how crystals form. We discuss crystal growth and examine why some minerals are more abundant than others. We also introduce and discuss a mineral classification scheme that will be used in later chapters.

CRYSTALS AND CRYSTALLIZATION

The last chapter used the term *crystal* without really defining it. To most people, a crystal is a sparkling gemlike solid with well-formed faces and a geometric shape. For many scientists, including mineralogists, *crystal* and *crystalline* also refer to any solid compound having an ordered, repetitive, atomic structure. The atomic structure may or may not result in flat crystal faces and a gemmy appearance. We use the term *crystal* in both ways. When a mineralogist refers to a quartz crystal, the reference is usually to a six-sided prismatic shape with pyramidal ends. On the other hand, petrologists and mineralogists may refer to crystals of quartz in a rock, such as a granite. Such crystals rarely have perfectly developed smooth faces. All minerals are crystalline by definition, but perfectly formed crystals are rare. When faces on a mineral are flat and fully developed, giving the mineral a geometric shape, we say the crystal is **euhedral.** When no crystal faces are visible, we say it is **anhedral.** Those minerals that fall between are **subhedral** (Figure 2.1).

Some mineral-like substances are **amorphous,** which means they have a random atomic structure. Natural volcanic glass, obsidian, is an example (Figure 2.2). Window glass, too, is amorphous. Glass mak-

►**FIGURE 2.1**
In this group of garnets, the sample on the left is anhedral; the others are euhedral. The one on the right is the most perfectly formed; it has 12 nearly identically shaped faces.

▶FIGURE 2.2
Obsidian (with patches of "snowflake" cristobalite), a glass plate used to determine mineral hardness, and two quartz crystals. The obsidian exhibits conchoidal fracture, the quartz has flat crystal faces that formed as the crystals grew, and the glass plate has manufactured flat surfaces. All materials shown in this photograph are made predominantly of SiO_2, but only the cristobalite is a mineral because everything else is amorphous (not crystalline).

ers melt mixtures, containing mostly quartz sand, and allow the melt to solidify quickly so atoms can't arrange themselves in a regular, repetitive atomic structure. The process produces a glass which, unlike a mineral, is **noncrystalline.** Noncrystalline materials, and a few minerals, are **isotropic,** meaning they have the same properties in all directions. Most minerals are **anisotropic,** meaning they have different properties in different directions. For this reason, glass makes better windows than most minerals because light passes through it equally well in all directions.

Artisans make geometrical crystal shapes and imitation gems from glass, but that does not make them minerals. Because minerals and glass do not have identical properties, a gemologist can often identify imitation gems by the way light passes through them.

The formation of crystals involves the bringing together and ordering of constituent elements. For example, potassium and chlorine combine in an ordered way to form the mineral sylvite (Figure 2.3). Crystals grow from a small single molecule to their final visible form. If the conditions are right, crystals

▶FIGURE 2.3
Atoms come together to produce crystals in igneous, metamorphic, or sedimentary systems. Here we see K^+ and Cl^- combined to form KCl, a salt mineral called *sylvite* found in some sedimentary rocks. The ions are arranged in an overall cubic pattern, so sylvite sometimes forms cubic crystals. For simplicity, ions in the interior of the cube have been omitted.

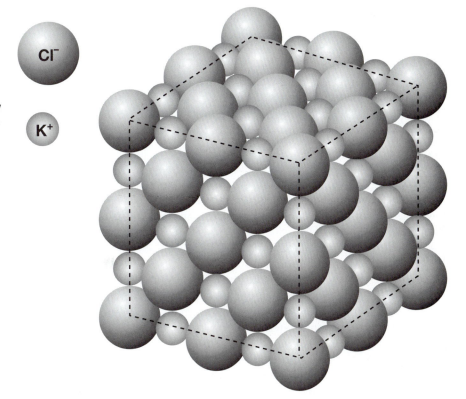

Box 2.1 Pegmatites Contain the Largest Crystals

Pegmatites are extremely coarse grained igneous rocks, usually of granitic or granodioritic composition (Figure 2.4). Granitic and granodioritic pegmatites consist largely of quartz, alkali feldspars, biotite, muscovite, and perhaps almandine garnet. They probably form during the end stages of crystallization, when remaining magma is enriched in volatiles and elements that do not easily enter into common rock-forming minerals. Such elements include boron, cesium, beryllium, zirconium, niobium, uranium, thorium, tantalum, tin, rare earth elements, chlorine, fluorine, lithium, and phosphorus. The presence of abundant fluids and uncommon elements often results in the formation of large, unusual, and sometimes valuable minerals. Gemstones from pegmatites include tourmaline, bright green feldspar (amazonite), beryl, chrysoberyl, topaz, cryolite, and spodumene.

The largest crystals in the world have been found in pegmatites. A single crystal of phlogopite from Ontario, Canada, was described as being 14 feet wide and 33 feet long. A quartz crystal from a Russian pegmatite weighed more than 2,000 lbs. The largest quartz crystal on record, however, was from Brazil and weighed more than five tons. The Etta Mine in the Black Hills of South Dakota is famous for its "logs" of white spodumene, which are up to 50 feet long. Elsewhere in the Black Hills, mineralogists have found tourmaline crystals that are 9 feet long, topaz crystals that weigh hundreds of pounds, and amblygonite crystals that are several feet across.

Gem-grade tourmalines in pegmatites come in a variety of colors. The most common form of tourmaline is the black or iron-rich variety, called *schorl.* Brown tourmaline (dravite) contains magnesium. Lithium is primarily responsible for the green (verdelite), yellow, pink (rubellite), and blue (indicolite) varieties. Colorless tourmaline, called *achroite,* is rare. Individual tourmaline crystals may also contain zones of different colors, which resulted from chemical changes during the growth of the crystal (Plate 1.1). The most famous multicolor tourmalines are the "watermelon" variety, which consist of a red core with green "rinds."

Most spodumene in pegmatites is white and opaque. However, the rare gem variety, kunzite, is pink and transparent. Emerald-green spodumene is called *hiddenite* and was named after W. E. Hidden, who first discovered the mineral when he overturned a tree on a farm near Stony Point, N. C. Beryl from pegmatites is usually yellow-green to blue-green. One variety of beryl is aquamarine, which varies from transparent blue to sea-green. Heliodor refers to the golden-yellow variety of beryl that is found in Namibia. Chrysoberyl, whose color ranges from white to yellowish green or green, is found in a number of different lithologies, but some of the best specimens are found in pegmatites. One variety of chrysoberyl, alexandrite, has the usual property of being emerald green in daylight and red in artificial light.

---

may grow to be very large. Some crystals in **pegmatites** (very coarse grained igneous rocks; see Box 2.1) are tens of meters in the longest direction. Many crystals, however, are so small that it takes a microscope to see them.

The most important factors controlling crystal size and perfection are temperature, time, abundance of necessary elements, and the presence or absence of a flux. All work together, but we can make some generalizations. *Temperature* is important because at high temperatures atoms are very mobile. Crystals can grow quickly; large and well-formed crystals may be the result. Principles of thermodynamics tell us that crystals that form at high temperatures have simpler atomic structures than those

that are stable at low temperatures, which may relate to their ability to be large and well ordered. *Time* is important because if a crystal has a long time to grow, it will naturally be larger and better ordered than one that grows quickly. More atoms have time to migrate to the growing crystal and to order themselves in a regular way. This explains why intrusive igneous rocks, which cool slowly underground, are coarser grained than extrusive igneous rocks of the same compositions. Some extrusive igneous rocks, such as obsidian, cool so quickly that they contain glass. Whatever the time and temperature, crystals cannot grow large if the necessary elements are not available. In most rocks, a dozen elements or less account for 99% of the composition. Minerals composed of those elements will usually be larger than those composed of rarer elements. Even if time, temperature, and atoms are right, crystals may not grow large. Diffusion of atoms through solids is slow, and atoms may not be able to migrate to spots where crystals are growing. However, if a fluid such as interstitial water or a magma is present it acts as a **flux,** transporting atoms to growing crystals. If a flux is present, elements may be carried long distances to sites of mineral growth, and even minerals composed of rare elements may grow to be large. This explains why some minerals of unusual composition are large in pegmatites, coarse-grained rocks that crystallize from water-rich melts left over after most of a magma has crystallized (Figure 2.4).

▶**FIGURE 2.4**
Large crystals of quartz, K-feldspar, and biotite in a pegmatite from the Black Hills, South Dakota. This sample is about 25 cm across.

▶**FIGURE 2.5**
An image of the atomic structure of crocidolite, an asbestiform amphibole, obtained with a transmission electron microscope. The black and white colors indicate atomic units composed of a small number of atoms. The entire view shows an imperfect grain composed of multiple subgrains with slightly different atomic orientations, as shown by the letters and vectors labeling crystallographic axes (see Chapter 10). *Zipper* faults show lines along which the atomic structure is defective. In some places, especially along subgrain boundaries, a coarsening of texture indicates small areas that have atomic structure dissimilar from that of normal crocidolite. Photo courtesy of the Mineralogical Society of America and *American Mineralogist* from "Microstructures and fiber-formation mechanisms of crocidolite asbestos" by Jung Ho Ahn and Peter R. Buseck, *American Mineralogist*, 76 (1991), 1467–1478.

CRYSTAL IMPERFECTIONS: DEFECTS

A hypothetical perfect crystal has a perfectly ordered atomic structure with all atoms in the correct places, and it contains no elements other than those described by its chemical formula. As pointed out by C. G. Darwin in 1914, such crystals cannot exist. While a crystal may look perfect on the outside, atomic structures always contain some flaws, called **defects.** In 1921 A. R. Griffiths showed that the low strength of many crystals was due to structural defects. Subsequently, mineral physicists found that plastic deformation of crystals always depends on the presence of defects. During the last few decades, advanced X-ray, transmission electron microscope (TEM), and, most recently, high-resolution transmission electron microscope (HRTM) studies have allowed mineralogists to image such defects (Figure 2.5).

No mineral is perfectly pure; minerals always contain minor or trace amounts of elements not described by their formula, often at levels that we cannot detect using standard analytical techniques. Perhaps the simplest type of defect is an **impurity defect,** occurring when a foreign atom is present in a mineral's atomic structure. The atom may replace one normally in the structure, or it may occupy an interstitial site (Figure 2.6a). We call both **point defects** because they occur at one, or a few, points in the structure. Other types of point defects are **Schottky** and **Frenkel defects** (Figures 2.6b, c). Frenkel defects occur when an atom is displaced from the position it normally occupies to an interstitial site. Frenkel defects may affect both cations and anions, but cation defects are more common because anions are larger and usually more securely bonded in place. Schottky defects occur when an atom is displaced from a structure altogether. Schottky defects involve both cations

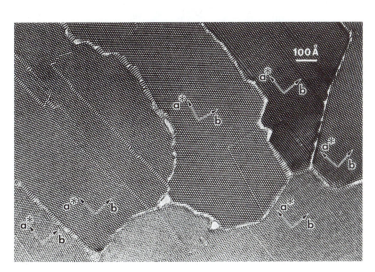

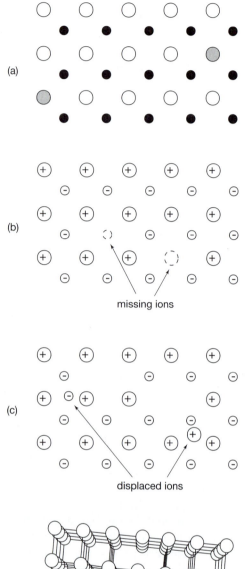

missing ions

displaced ions

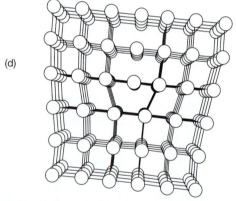

▶FIGURE 2.6
Diagram illustrating different types of crystal defects. (a) Impurity defects occur when a foreign atom (gray) replaces one normally in a crystal structure. (b) Schottky defects occur when atoms are missing from a structure; (c) Frenkel defects occur when atoms are displaced to interstitial sites not normally occupied. (d) Edge dislocations occur when a plane of atoms in a structure terminates at a line in the crystal's interior.

and anions. To maintain charge neutrality, missing anions must be compensated by missing cations.

Other types of defects include **line defects** and **plane defects** (Figure 2.6d). Line defects, such as **edge dislocations** and **screw dislocations** are defects that end at lines in a structure. Plane defects, as their name implies, are planes along which a crystal's structure is displaced or distorted. On a macroscopic scale, grain boundaries can be considered plane defects. At the atomic level, irregular **lineage structures** may separate slightly misoriented portions of a crystal structure or, in layered atomic structures, **stacking faults** may separate layers that are out of order. Because of defects, many crystals consist of **domains** of slightly different atomic orientation (Figure 2.5).

CRYSTAL IMPERFECTIONS: ZONING

Crystallizing magmas may produce uneven mineral distribution (Plate 1.4). On a smaller scale, individual minerals develop **compositional zoning** if different parts of a mineral have different compositions. Zoning is present in many minerals but often on such a small scale that we have difficulty detecting it. Occasionally, zoning results in distinct color changes, such as those shown in Plates 1.1 and 1.2. In many cases zoning can be seen with a petrographic microscope because zones of different composition have different optical properties. In still other cases detailed chemical analyses are needed to detect its presence.

Most zoning is an artifact of crystal growth. It may result from changes in pressure or temperature during crystallization. It may also result from changes in magma or fluid composition as crystals grow. The principles of thermodynamics dictate that zoned minerals are unstable, but they are common in nature because diffusion of elements is often not fast enough for growing minerals to remain homogeneous. Most zoning is concentric, representing growth rings about an original crystal seed. In some minerals it is more complex and results in compositional zones that are difficult to explain and interpret.

CRYSTAL IMPERFECTIONS: TWINNING

In ideal crystals, atoms are in repetitive arrangements that are oriented the same way in all parts of the crystal. **Twins** result when different domains of a crystal have different atomic orientations. The domains share atoms along a common surface, typically a plane called the **composition plane.** Twins differ

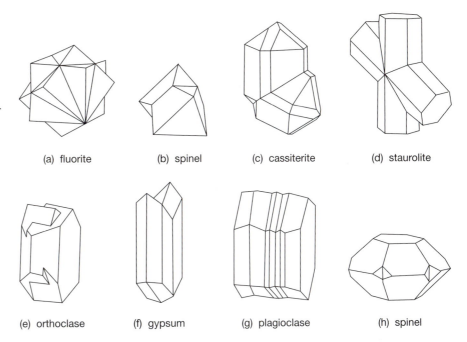

►**FIGURE 2.7**
Examples of different types of twinning (and a mineral in which it occurs): (a) penetration twin (fluorite); (b) contact twin (spinel); (c) contact twin (cassiterite); (d) penetration twin (staurolite); (e) penetration twin (orthoclase); (f) contact swallowtail twin (gypsum); (g) polysynthetic twin (plagioclase); and (h) cyclic twin (spinel).

(a) fluorite   (b) spinel   (c) cassiterite   (d) staurolite

(e) orthoclase   (f) gypsum   (g) plagioclase   (h) spinel

from crystal **intergrowths** composed of crystals that grow next to each other. In a twinned crystal the structure and bonds continue across the composition plane; in intergrowths they are discontinuous. **Simple twins** comprise two domains that share common planes of atoms that separate them. **Complex twins** comprise more than two individual domains.

Figure 2.7a–f show some simple twins composed of two individual domains. In some the twins appear to be crystals in contact; we call these **contact twins.** In others (Figures 2.7a, d, and e), the crystals appear to have grown through each other. In such twins, called **penetration twins,** a volume of atoms is shared by both domains, not just a plane. Plate 1.5 shows orthoclase crystals with penetration twins; compare this plate with Figure 2.7e. Figure 2.7g shows another type of twinning, **polysynthetic twinning** composed of alternating domains. Such twins are common in plagioclase, calcite, and some other minerals but usually cannot be seen without a microscope. In Na-rich plagioclase they may result in fine lamellae on crystal faces that can be seen with the naked eye. The presence of twin lamellae is one way to tell plagioclase from K-feldspar in hand specimen. Figure 2.7h shows **cyclic twinning,** which is rare but sometimes spectacular. Some minerals twin to form domains with irregular shapes and boundaries. Quartz is a good example, but the twins in most quartz crystals are difficult to see.

Different types of twins in different minerals are given names, referred to as **twin laws.** The laws refer to the way **twin domains** are related.

Feldspars exhibit many twin laws. K-feldspars often contain simple twins related by the **Carlsbad, Baveno, albite,** or **pericline** laws. Plate 1.5 shows orthoclase twinned according to the Carlsbad Law. Albite and pericline twinning combine to give the characteristic scotch plaid pattern that can be seen when microcline is viewed under crossed polars with a petrographic microscope. Plagioclase may twin according to the same laws as orthoclase, and according to the albite law. Albite and pericline twins are often polysynthetic and give plagioclase a characteristic zebra stripe appearance when viewed under crossed polars with a petrographic microscope.

Twinning comes at all scales and may be difficult to detect. Sometimes we can see it with the naked eye, sometimes we can only see it with a microscope, and sometimes we can't detect it without more sophisticated devices. Identifying twins in hand specimen can be difficult, especially in poorly formed crystals. One diagnostic feature is the presence of **re-entrant angles,** visible in Figure 2.7a–h. Two crystal faces intersect to form a re-entrant angle when they produce an angular concavity that points toward the interior of a crystal instead of the (normal) exterior.

IGNEOUS MINERALS

Several different processes lead to crystal growth. **Igneous minerals** form from molten rock **(magma).** When magma is at high temperature, it is completely

Box 2.2 Making Glass from Minerals

Glass manufacturers make glasses by melting ingredients and cooling the melt so quickly that crystals cannot form. We call the most common kind of glass, which is used to make bottles or window panes, *soda-lime-silica glass*. People have manufactured similar glasses since Roman times. Soda-lime-silica glasses typically contain about 75 wt. % SiO_2, 18 wt. % Na_2O, and 7 wt. % CaO. Raw ingredients are usually the minerals quartz (SiO_2), trona [$Na_3H(CO_3)_2 \cdot 2H_2O$], and calcite ($CaCO_3$). In the United States, we obtain high-quality quartz and calcite from many places. Most of the trona comes from the Green River Formation in Wyoming, where trona beds are up to 30 feet thick. Manufacturers heat the mineral ingredients to between 1550 and 1600°C, producing a homogeneous melt; CO_2 and H_2O escape into the atmosphere. Quenching, to produce a final glass, is done in various ways depending on the product wanted.

Manufacturers change glass properties by adding small amounts of other ingredients. Addition of boron produces heat-resistant glass such as Pyrex® or Vycor®. Lead gives glass the optical properties needed for making imitation crystals. Aluminum can make glass resistant to weathering. Fluorine makes glass opaque. Lithium reduces the melting point of glass. Trace amounts of metals change glass color: iron makes glass green, nickel makes it brown or orange, and cobalt makes it blue.

liquid because high kinetic energy ensures that no solid is stable. As magma cools, atoms slow down and bond together, resulting in the formation of crystals. Because of high temperatures and the molten state of magma, atoms are quite mobile and easily move to surfaces of growing crystals. The result may be a rock with crystals forming a mosaic pattern if crystal growth is uniform (Figure 2.8). The crystals are often large, euhedral, and homogeneous. Olivine, pyroxenes, feldspars, and many other minerals form in this way. Table 2.1 lists some common minerals in igneous rocks.

Igneous processes are quite variable. Some magmas cool slowly underground, so mineral crystals grow to be large. Other magmas are extruded as lavas and cool quickly to form an extrusive rock. Mineral crystals in extrusive rocks may be so small that they cannot be seen with the naked eye or even with a microscope. Igneous rocks are generally dominated by silicate minerals because most magmas are rich in silicon and oxygen. However, magma compositions vary, so igneous rocks have variable composition and consequently the minerals are not the same in all igneous rocks.

▶**FIGURE 2.8**
Thin section showing a mosaic texture in an igneous rock.

▶**TABLE 2.1**
Common Minerals in Igneous Rocks

Mineral Class or Group	Examples of Important Minerals or Mineral Series	Chemical Formula
olivine	olivine	$(Mg,Fe)_2SiO_4$
pyroxene	diopside	$CaMgSi_2O_6$
	augite	$(Ca,Mg,Fe,Na)(Mg,Fe,Al)(Si,Al)_2O_6$
	orthopyroxene	$(Mg,Fe)_2SiO_6$
amphibole	hornblende	$(K,Na)_{0-1}(Ca,Na,Fe,Mg)_2(Mg,Fe,Al)_5(Si,Al)_8O_{22}(OH)_2$
micas	biotite	$K(Mg,Fe)_3(AlSi_3O_{10})(OH)_2$
	muscovite	$KAl_2(AlSi_3O_{10})(OH)_2$
feldspar	orthoclase	$KAlSi_3O_8$
	microcline	$KAlSi_3O_8$
	sanidine	$(K,Na)AlSi_3O_8$
	plagioclase	$(Ca,Na)(Al,Si)_4O_8$
feldspathoid	leucite	$KAlSi_2O_6$
	nepheline	$(Na,K)AlSiO_4$
	sodalite	$Na_3Al_3Si_3O_{12} \cdot NaCl$
silica	quartz	SiO_2
oxides	magnetite	Fe_3O_4
	ilmenite	$FeTiO_3$
	rutile	TiO_2
sulfides	pyrite	FeS_2
	pyrrhotite	$Fe_{1-x}S$
other	titanite	$CaTiSiO_5$
	zircon	$ZrSiO_4$
	apatite	$Ca_5(PO_4)_3(OH,F,Cl)$

MINERALS THAT PRECIPITATE FROM AQUEOUS SOLUTIONS

Many minerals grow from **aqueous solutions** (water containing dissolved elements). If the solutions are not saturated, kinetic energy ensures that dissolved atoms do not bond together. If temperature decreases, or if water evaporates to increase the concentration of dissolved material, the solution may become saturated. Atoms will bond together and solid **chemical precipitates,** usually crystalline, will form. Calcium carbonate precipitates to form calcite ($CaCO_3$) if concentrations of Ca^{2+} and CO_3^{2-} in water are high enough. Inland lakes or seas commonly precipitate calcite, halite, gypsum, and other minerals. In some places evaporating waters have deposited salt beds thicker than 300 meters. On a much smaller scale, minerals precipitating from slowly moving groundwater can fill fractures and cracks in rocks (Figure 2.9).

Table 2.2 gives examples of minerals that precipitate from water. The most common ones are minerals with high solubilities in water, such as calcite, halite, and other salts. Other minerals, having lower solubility but composed of elements in great

▶**FIGURE 2.9**
Smithsonite ($ZnCO_3$) crystals that grew in a vug in a carbonate rock. Smithsonite is often formed by precipitation from aqueous solution. This photo is about 5 cm across.

abundance, also form from aqueous solutions. Quartz is an example. At low temperature it may precipitate as Herkimer diamonds or as quartz crys-

tals in **geodes** (Figure 2.10b). Low-temperature precipitation typically results in very fine-grained or massive minerals with few well-developed crystal faces. Fine-grained limestones (made of calcite) form at low temperatures. Occasionally, however, coarsely crystallized minerals form at low temperature. Some spectacular calcite and quartz crystals are examples.

If chemical precipitation occurs at elevated temperatures, we call the process **hydrothermal.** Hydrothermal waters may be of meteoric, magmatic, metamorphic, or oceanic origin. Circulation is driven by high heat flow, as hot water flows toward regions of lower temperature. In some instances, waters flow long distances before precipitating minerals.

Travertine and tufa deposited by hot springs are examples of hydrothermal deposits. Both are primarily composed of calcite. Hydrothermal minerals are also deposited underground and make up many **ore deposits** (Figure 2.11). Ore deposits include oxides, sulfides, and other ore minerals in sufficient concentration to make mining profitable. Table 2.3 gives examples of some common hydrothermal ore minerals.

▶TABLE 2.2

Examples of Minerals Formed by Low-Temperature Precipitation from Surface Water or from Groundwater in Sediments and Rocks

Mineral Class or Group	Examples of Important Minerals or Mineral Series	Chemical Formula
silica	quartz	SiO_2
carbonate	calcite	$CaCO_3$
	magnesite	$MgCO_3$
halide	halite	$NaCl$
	sylvite	KCl
sulfate	gypsum	$CaSO_4 \cdot 2H_2O$
	anhydrite	$CaSO_4$
native element	sulfur	S

▶FIGURE 2.10a

Herkimer diamonds from New York. These crystals, about 1 cm in longest dimension, are actually made of quartz, not diamond.

▶FIGURE 2.10b

Quartz crystals in a geode about 15 cm across.

METAMORPHIC AND DIAGENETIC MINERALS

A third way minerals may crystallize is through **metamorphism.** Metamorphism sometimes involves recrystallization and coarsening of a rock with no change in mineralogy. Often, however, it involves replacement of preexisting minerals by new ones (Figure 2.12). Bonds are broken and atoms migrate by **solid state diffusion** or are transported short distances by **intergranular fluids** to sites where new minerals crystallize and grow. In sedimentary rocks a

▶TABLE 2.3

Examples of Minerals Common in Hydrothermal Ore Deposits

Mineral Class	Examples of Important Minerals or Mineral Series	Chemical Formula
sulfide	pyrite	FeS_2
	pyrrhotite	$Fe_{1-x}S$
	chalcopyrite	$CuFeS_2$
	galena	PbS
	sphalerite	ZnS
	molybdenite	MoS_2
tungstate	wolframite	$(Fe,Mn)WO_4$
oxide	cassiterite	SnO_2
	pyrolusite	MnO_2

▶FIGURE 2.11
Examples of hydrothermal ore minerals. The dark mineral (top center) is sphalerite (ZnS), the shiny mineral with high reflectivity (bottom) is pyrite (FeS_2). Also present are prismatic quartz crystals (SiO_2).

▶FIGURE 2.12
Some common metamorphic minerals: (a) twinned staurolite; (b) garnet; and (c) kyanite.

(a) Staurolite

(b) Garnet

(c) Kyanite

low-temperature form of metamorphism, called **diagenesis,** takes place.

Metamorphism may involve replacement of one mineral by another. For example, calcite may become aragonite or vice versa. Both minerals are $CaCO_3$, but their atomic structures differ. Mineralogical changes due to metamorphism, however, usually involve several different minerals *reacting* together. Calcite ($CaCO_3$) and quartz (SiO_2) may react to form wollastonite ($CaSiO_3$) if a limestone containing quartz is metamorphosed at high temperature. Adding further complications, during metamorphism, a rock's composition may change as fluids carry in or remove soluble materials. Such a process, called **metasomatism,** often leads to large, well-developed metamorphic minerals.

The mineralogy of metamorphic rocks is much more diverse than sedimentary or igneous rocks. Nearly all the minerals found in igneous rocks can be present in metamorphic rocks. Many found in sedimentary rocks may be present as well. In addition, other minerals, uncommon or nonexistent in igneous

and sedimentary rocks, form through metamorphism. Table 2.4 lists only a few of the more common metamorphic minerals.

COMMON ROCKS AND MINERALS

We classify rock-forming minerals in many ways. Often, petrographers contrast **primary** and **secondary minerals.** Primary minerals are those that are present from the time a rock first forms. Secondary minerals form later by chemical or physical reaction within the rock. Often, such secondary reactions involve H_2O or CO_2 and occur during low-temperature alteration of a preexisting rock.

We further divide primary minerals into **essential** and **accessory minerals.** Essential minerals are those whose presence is implied by the name of the rock. All limestones, for example, contain calcite or dolomite, and all granites contain quartz and K-feldspar. Accessory minerals are present in minor amounts and do not affect most rock properties. They often involve

▶TABLE 2.4
Examples of Metamorphic Minerals

Silicate Mineral Subclass	Examples of Important Minerals or Mineral Series	Chemical Formula
framework silicate	cordierite	$(Mg,Fe)_2Al_4Si_5O_{18}$
chain silicate	tremolite wollastonite	$Ca_2Mg_5Si_8O_{22}(OH)_2$ $CaSiO_3$
isolated tetrahedral silicate	andalusite kyanite sillimanite staurolite chloritoid garnet	Al_2SiO_5 Al_2SiO_5 Al_2SiO_5 $Fe_2Al_9Si_4O_{23}(OH)$ $(Fe,Mg)Al_2SiO_5(OH)_2$ $(Ca,Fe,Mg)_3(Al,Fe)_2Si_3O_{12}$
paired tetrahedral silicate	zoisite	$Ca_2Al_3Si_3O_{12}(OH)$

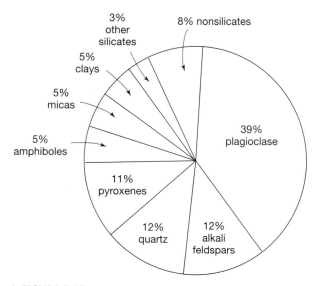

▶FIGURE 2.13
Estimated volume percentages for the common minerals in the Earth's crust. Modified from Ronov and Yaroshevsky, 1969.

incompatible elements, elements that are not easily incorporated into essential minerals. Zr, for example, often concentrates in zircon, $ZrSiO_4$, a common accessory mineral in many rocks. Phosphorus (P) may lead to formation of phosphate minerals such as apatite, $Ca_5(PO_4)_3(OH,F,Cl)$, equally common. The distinction between primary and secondary minerals is not exact. A primary mineral in one rock may be an alteration product in another, or vice versa. In Table 2.1, the minerals in the top of the chart are often essential minerals in igneous rocks, while the oxides, sulfides, and others at the bottom are rarely essential in igneous rocks.

More than 90 elements are found in nature, and they could combine to form minerals in many ways. Nevertheless, they don't. Mineralogists have identified only 3,000 to 4,000. Of these, less than 200 are common, and less than 50 are common enough to be considered essential minerals. Plagioclase is the most abundant mineral in the Earth's crust; other silicates are abundant, too (Figure 2.13). In contrast, nonsilicates make up less than 10% of the crust. Why is this?

First, although more than 90 elements exist, many are present in such small amounts that they only exist as minor diluents for major elements. Occasionally they may be concentrated in accessory minerals. Table 1.4 lists the most abundant elements in the Earth's crust; 13 elements (O, Si, Al, Fe, Ca, Na, K, Mg, Ti, H, P, Mn, Ba) make up nearly 99.5 wt. %. We can expect the common minerals to be made from these elements. Oxygen and silicon, being the most abundant elements in the Earth's crust, naturally lead to an abundance of silicate minerals. It is no wonder that

other types of minerals, or minerals that include rarer elements as key components, are rarer.

Second, as described in Chapter 1, elements have specific properties that dictate the kinds of bonds they may form and the kinds of mineral structures they may enter. A third reason the number of common minerals is small is that there are only a few basic mineral-forming processes. Although all igneous events are not identical, they all involve the same processes: melting to form a magma and later recrystallization. Exceptions exist, but most igneous events involve magmas of similar major-element composition. Most chemical precipitation involves water and elements with high solubilities, and there aren't many of them. Most metamorphism occurs over a relatively small range of pressure and temperature and affects rocks belonging to a small number of distinct chemical groups. It is not surprising then, that these processes yield a limited number of common minerals.

LIFE SPANS OF MINERALS

Many minerals commonly occurring in modern sediments and rocks are too unstable to survive in great abundance in older terrestrial rocks. For example, olivine was once abundant in many terrestrial Precambrian mafic rocks, but since the Pre- cambrian, most of the olivines have been altered by oxygen, carbon dioxide, and water to serpentine, iron oxides, and magnesite. Because olivines crystallize in hot and dry magmas, they are thermodynamically unstable under much cooler and wetter surface and near surface conditions. Because of their tendency to weather rapidly, detrital olivines and pyroxenes are largely restricted to Cenozoic sediments and sedimentary rocks. However, Precambrian olivines and even glasses occur in Moon rocks and meteorites that have been isolated from oxygen and water.

Other examples of minerals generally absent from older terrestrial rocks include tridymite, a high-temperature polymorph of quartz, and aragonite, a high-pressure polymorph of calcite. Tridymite is common in Cenozoic siliceous volcanics, including rhyolites, obsidian, and andesites. However, except in stony meteorites and lunar basalts, the mineral changes to quartz over time and is rarely found in rocks that are older than Tertiary.

Many marine organisms excrete shells that consist of aragonite rather than calcite. Unless aragonite fossils are deeply buried, they will alter to calcite over time. The oldest known aragonite fossil is from an organic-rich shale of Mississippian age. Geologists have only found Paleozoic aragonite fossils in three localities. One of the rocks is tuffaceous, while the others are black shale and asphaltic limestone. The presence of abundant organic matter in three of the four known rocks with Paleozoic aragonite is probably responsible for the preservation of the aragonite. The organic matter coated the fossils and probably prevented water from reaching them and promoting their conversion to calcite.

Some nonmineral materials are unstable and invert to minerals over time. Opal and volcanic glass are amorphous materials and not minerals. Over time, both weather or alter into more stable crystalline compounds, such as quartz. Obsidian is rarely found in rocks older than the Miocene. The oldest known volcanic glass is in a 70-million-year-old welded tuff. Opal is slightly more stable than obsidian. Reaction rate calculations indicate that opal will entirely convert to quartz in about 180 million years at 20°C, approximately 4.3 million years at 50°C, and in only about 47 years at diagenetic temperatures of 200°C. Not surprisingly, the oldest known opal is Lower Cretaceous, or about 125 million years old.

THE LAWS OF THERMODYNAMICS

A fourth and perhaps most important reason the number of common minerals is limited relates to the **laws of thermodynamics.** There are several different laws of thermodynamics. J. Willard Gibbs pointed out their most important consequences in 1878. Gibbs defined a form of energy that defines compound stability. We now call it the **Gibbs free energy** and indicate it by the variable G. Natural chemical systems are most

Box 2.3 Scientific Laws, Theories, and Hypotheses

Scientific **laws** are general observations that have never been found to be violated. Examples are Snell's Law (dealing with the refraction of light) and Fick's Law (dealing with diffusion). Laws are empirical, dealing only with observations and not explanations. In contrast, scientific **theories** are explanations for observed events. Theories are often quite general, explaining a large number of related phenomena. They are developed from **hypotheses** through the scientific method. A hypothesis, which is really a tentative guess, only becomes a theory after exhaustive scientific tests support its validity.

stable when energy is minimized, so minerals and mineral assemblages with low Gibbs energy are more stable than those with high energy. Unstable minerals break down to form different minerals, with lower Gibbs free energy, over time. Minerals with relatively low Gibbs energies are more common than others.

An important corollary to the laws of thermodynamics, the **phase rule,** says that the number of stable phases that can coexist in any chemical system must be small. Thus, not only are stable minerals predictable, they are limited to a small number in any given rock. For a given rock, the stable minerals may not be the same under all conditions. If a rock is metamorphosed due to pressure or temperature changes, minerals may react to produce new minerals with lower Gibbs free energy. When they stop reacting, they have reached **equilibrium.** If the Gibbs free energy is minimized, the system is at **stable equilibrium.**

Consider a chemical system containing Fe and O. Possible solid phases that could exist include iron (Fe), wüstite (FeO), magnetite (Fe_3O_4), and hematite (Fe_2O_3). Under normal Earth surface conditions, however, hematite has the lowest Gibbs free energy, and thus Fe and other Fe oxides react to hematite if they can (Figure 2.14).

Although the laws of thermodynamics tell us what the most stable mineral(s) will be, they do not tell us how long it will take to reach stable equilibrium. We all know from experience with cars, for example, that it may take a while for the iron in steel to rust; the same is true of wüstite and magnetite. Some low-temperature systems may never reach stable equilibrium. Such systems may obtain an intermediate stage

▶**FIGURE 2.14**
Relative Gibbs free energies of Fe and Fe-oxides. Mixtures of Fe-metal and magnetite have Gibbs free energies indicated by line segment A, mixtures of magnetite and hematite have Gibbs free energies indicated by line segment B, and mixtures of hematite and oxygen have Gibbs free energies indicated by line segment C. Because the free energy of wüstite lies above the three line segments, wüstite is generally not stable. Similarly, because the Gibbs free energy of hematite is relatively low, it tends to form from other Fe-oxides if sufficient oxygen is present.

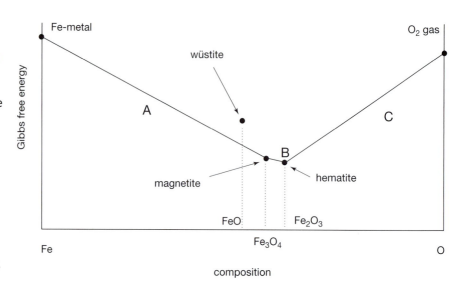

called **metastable equilibrium** if reactions cease, but most natural systems approach stable equilibrium given enough time.

CLASSIFICATION OF MINERALS

In order to talk about minerals in a convenient manner, it is useful to classify or group them in some way. In 1768 Carolus Linnaeus suggested that minerals may be classified in much the same way that biologists were classifying plants and animals. The external shape of mineral crystals was the basis for his classification scheme. Linnaeus made detailed measurements and drawings of many different minerals, and is often credited with being the founder of the science of crystallography.

Linnaeus's classification scheme, and others based on crystal morphology, is often ambiguous. Instead, some petrologists, and other geologists, find it convenient to group minerals according to the types of rocks that contain them. Igneous petrologists, for example, know that granite contains quartz and K-feldspar. To them, quartz and K-feldspar are granitic minerals. Unfortunately, ambiguity remains (Figure 2.15). A sedimentary petrologist associates the same two minerals with arkosic sandstones, while a metamorphic petrologist knows they occur in a wide range of metamorphic rocks.

In contrast with petrologists, mineralogists are less concerned about the occurrences of minerals. They are more interested in understanding how and why certain elements bond together to form certain minerals. They look for patterns in mineral structures and chemistry that will help them understand why minerals form as they do. Rather than consider all minerals individually, they study them in groups. Mineralogists know, for example, that mica and other sheet silicates all have similar atomic structures (Figure 2.16).

Many mineral classification schemes could be used by mineralogists. The standard ones are based on principles developed by Berzelius. The system used in this text is modified slightly from that presented by Zoltai and Stout (1984). It is similar to Dana's System of Mineralogy described by Palache et al. (1944) and to the classification of silicates presented by Strunz (1957). We assign minerals to **classes** based on their anions or anionic complexes; Table 2.5 lists the most important classes. We group some classes together because minerals of those classes have similar properties. The native metal, sulfide, and sulfosalt classes include minerals in which the dominant type of bonding is not ionic. For that reason we give them their own classes, and no ionic charges are given for them in Table 2.5.

Dividing minerals into classes based on anion or anionic complex is convenient because we can determine class from chemical formula. Besides being con-

▶**FIGURE 2.15**
Feldspathic sandstone (left) and granite (right). Both samples contain predominantly quartz and K-feldspar. However, the sandstone is a sedimentary rock and the granite is an igneous rock. Both samples are about 8 cm wide.

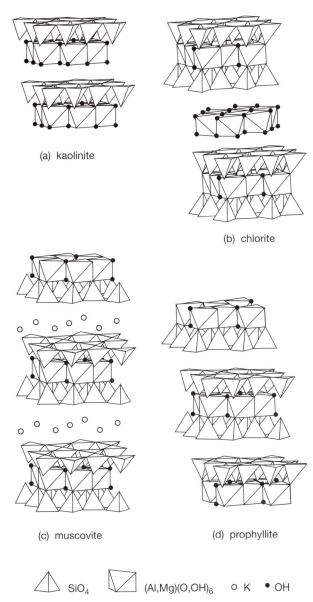

○ K   ● OH

▶FIGURE 2.16
The structures of several micaceous minerals. Polyhedra are used to represent SiO_4 and $Al(O,OH)_6$ groups with O and OH at the corners, and open circles indicate K^+ and solid dots OH^-. (a) The two-layer structure of kaolinite; (b) the three-layer structure of chlorite; (c) the three-layer structure of muscovite with K^+ between the layers; and (d) the three-layer structure of pyrophyllite.

venient, however, this classification scheme makes sense in other ways. Within each class, the type of structure and bonding are similar. This means that minerals within a class often have similar physical properties, making the classes useful in mineral identification. Such would not be the case if we divided minerals into groups based on cations. For example pyrite, FeS_2, and fayalite, Fe_2SiO_4, both contain Fe, but have few properties in common. Minerals within a single class are often found together. Silicates make up more than 99% of the minerals found in igneous rocks and ac-

count for more than 90% of the Earth's crust and mantle. Carbonates are similarly dominant in limestones.

Most common minerals belong to the silicate, oxide, hydroxide, or sulfide and sulfosalt classes. Oxides and hydroxides together account for about 500 species. Sulfides and sulfosalts also account for about 500 species. The silicate class contains the largest number of minerals, with more than 800 known. In contrast, the native element class contains few members.

Because there are so many important minerals in the silicate class, we divide it into subclasses (Table 2.6). In all silicates, except for some very rare high-pressure minerals, four O^{2-} anions surround every Si^{4+} cation (Figure 2.17), forming a **tetrahedron,** a pyramid shape with four identical faces. We name the subclasses according to how the tetrahedra are linked (polymerized) in the atomic structure (Table 2.6). In the **isolated tetrahedral silicates** (also called *island silicates*), the $(SiO_4)^{4-}$ tetrahedra are not polymerized; they are all separated (like a bunch of islands) with cations between them. The Si:O ratio for the structure is 1:4. In contrast, in **framework silicates,** each oxygen is shared between two tetrahedra; the collection of tetrahedra make up a three-dimensional framework with an Si:O ratio of 1:2. Other subclasses have Si:O ratios between 1:4 and 1:2. In Table 2.6, we give the Greek names (in parentheses) of the subclasses because some mineralogists use them. Unless you know Greek, however, they are not particularly meaningful. Some silicate minerals with complex structures do not fit conveniently into one of the six first subclasses listed in Table 2.6, but most common silicates do.

We further divide classes and subclasses of minerals into **groups** based on structural or chemical similarity. For example, we group the feldspar minerals together because, although their compositions vary, they all have similar atomic structures. We group the serpentine minerals (antigorite, chrysotile, lizardite) together because they all have the same composition. When two or more minerals have the same formula but different crystal structures, as in the case of serpentine, we call them **polymorphs.**

Groups may also be further divided into **subgroups** or **series.** Subgroups are minerals that naturally group together for chemical or other reasons. The K-feldspars, for example, are a subgroup because they all have composition $KAlSi_3O_8$. Series involve minerals with compositions we can describe in terms of two end members. The plagioclase feldspar series consists of feldspars that are solutions of the two end members albite ($NaAlSi_3O_8$) and anorthite ($CaAl_2Si_2O_8$). Albite and anorthite form a **solid solution series,** and plagioclase can have any composition between pure albite and pure

▶TABLE 2.5
Mineral Classes

Class	Anion, Anionic Complex, or Key Elements	Example Mineral	Chemical Formula
silicates	$(SiO_n)^{4-2n}$	quartz	SiO_2
halides	Cl^-, F^-, Br^-, I^-	halite	$NaCl$
oxides	O^{2-}	corundum	Al_2O_3
hydroxides	$(OH)^-$	gibbsite	$Al(OH)_3$
carbonates	$(CO_3)^{2-}$	calcite	$CaCO_3$
nitrates	$(NO_3)^-$	nitratite	$NaNO_3$
borates	$(BO_3)^{3-}$ or $(BO_4)^{5-}$	sinhalite	$MgAlBO_4$
sulfates	$(SO_4)^{2-}$	gypsum	$CaSO_4 \cdot 2H_2O$
chromates	$(CrO_4)^{2-}$	crocoite	$PbCrO_4$
tungstates	$(WO_4)^{2-}$	scheelite	$CaWO_4$
molybdates	$(MoO_4)^{2-}$	wulfenite	$PbMoO_4$
phosphates	$(PO_4)^{3-}$	apatite	$Ca_5(PO_4)_3(OH,F,Cl)$
arsenate	$(AsO_4)^{3-}$	scorodite	$FeAsO_4 \cdot 4H_2O$
vanadate	$(VO_4)^{3-}$	vanadinite	$Pb_5(VO_4)_3Cl$
native elements	single elements	copper	Cu
sulfides	S	pyrite	FeS_2
sulfosalts	S, As, Sb	niccolite	$NiAs$

▶FIGURE 2.17
Silicon and four oxygen atoms. (a) "Marbles" forming a teterahedron; (b) and (c) are ball and stick drawings showing a top view and a side view.

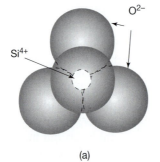
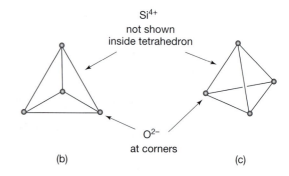

(a)   (b)   (c)

anorthite (Box 2.4). Usually, atomic structure or composition are identical for all minerals in subgroups or series. They are always very similar. Individual mineral species, which make up series and subgroups, are occasionally further divided into **varieties** if the varieties have some special properties. For example, rose quartz, smoky quartz, amethyst, and citrine are all varieties of quartz.

A CLASSIFIED LIST OF MINERALS

Appendix A lists all the common minerals and some less common ones that are discussed in this book. It is a good place to find mineral formulas or to determine how a particular mineral is classified. About 200 minerals are included. Many are uncommon and some are not particularly significant, so

trying to memorize them all is not worthwhile. However, noting the similarities of the formulas for minerals within groups is worthwhile. It should not be surprising that the properties of minerals in a particular group, subgroup, or series are similar. By learning the properties of 10 to 20 key minerals, it is possible to make inferences about others based on their composition.

Table 2.7 shows the portion of Appendix A that includes the most abundant framework silicates. The list includes quartz and the feldspars, the most common minerals in the Earth's crust. It shows most of the features of the classification scheme. Twenty minerals belong to five main groups; two others are classified separately. Five SiO_2 polymorphs are included in the silica group; three $KAlSi_3O_8$ polymorphs are included in the potassium feldspar group. Solid solution series are represented by plagioclase feldspar

Box 2.4 Plotting Mineral Compositions

One way to depict mineral compositions is to plot them on composition diagrams. In simple chemical systems, where the minerals being discussed are only made of two components, they can be plotted on a line as shown (Figure 2.18). Pure Fe plots on one end of the line and pure oxygen on the other. Neither exist as minerals, but the three compositions in the middle (wüstite, magnetite, and hematite) do. The ratio Fe:O determines where on the line the minerals plot. Hematite, for example, has the composition Fe_2O_3, so it is 2/5 iron, and it plots 2/5 of the way from pure oxygen toward iron.

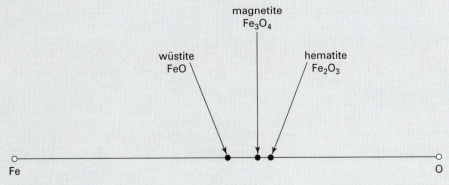

▶**FIGURE 2.18**
The system Fe-O.

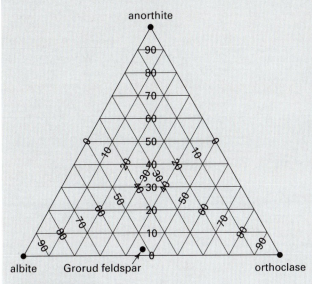

▶**FIGURE 2.19**
The system albite-anorthite-orthoclase.

Most minerals have more than two components. Figure 2.19 shows a three-component system. We can use any chemical components we want when plotting mineral compositions. In this figure we have used the components anorthite ($CaAl_2Si_2O_8$), albite ($NaAlSi_3O_8$,) and orthoclase ($KAlSi_3O_8$) because we want to show feldspar compositions. The Grorud feldspar, discussed in Box 1.5, is 3% anorthite, 62% albite, and 35% orthoclase (ignoring the minor Fe impurity). It plots as shown on the triangular diagram. The vertical row of numbers within the triangle shows the amount of anorthite. The two diagonal rows show the amounts of albite and orthoclase. If a feldspar contained no anorthite, it would plot on the bottom side of the triangle (between albite and orthoclase). But, the Grorud feldspar contains 3% anorthite so it plots 3% of the way from the bottom toward the anorthite apex at the top. Similarly, it plots 62% of the way from the anorthite-orthoclase side toward albite (in the lower left corner), and 35% of the way from the albite-anorthite side toward orthoclase (in the lower right corner).

▶TABLE 2.6
Silicate Mineral Subclasses

Subclass	Si:O Ratio	Si Radical	Example	Mineral Formula	Arrangement of SiO_4 Tetrahedra
framework silicates (tectosilicates)	1:2	SiO_2 or $(Al_xSi_{1-x})O_2$	quartz albite	SiO_2 $Na(AlSi_3)O_8$	
sheet silicates (phyllosilicates)	2:5 = 4:10	$(Si_4O_{10})^{4-}$	pyrophyllite	$Al_2Si_4O_{10}(OH)_2$	
single-chain silicates (inosilicates)	1:3 = 2:6	$(SiO_3)^{2-}$ or $(Si_2O_6)^{4-}$	enstatite	$Mg_2(SiO_3)_2$ or $Mg_2Si_2O_6$	
double-chain silicates (inosilicates)	4:11 = 8:22	$(Si_8O_{22})^{12-}$	tremolite	$Ca_2Mg_5Si_8O_{22}(OH)_2$	
ring silicates (cyclosilicates)	1:3 = 6:18	$(Si_6O_{18})^{12-}$	tourmaline	$(Na,Ca)(Fe,Mg,Al,Li)_3Al_6(BO_3)_3Si_6O_{18}(OH)_4$	
isolated tetrahedral (island) silicates (nesosilicates or orthosilicates)	1:4	$(SiO_4)^{4-}$	forsterite	Mg_2SiO_4	
paired tetrahedral silicates (sorosilicates)	2:7	$(Si_2O_7)^{6-}$	akermanite	$Ca_2MgSi_2O_7$	
more complex silicates	1:4 and 2:7	$(SiO_4)^{4-}$ and $(Si_2O_7)^{6-}$	vesuvianite	$Ca_{10}(Mg,Fe)_2Al_4(SiO_4)_5(Si_2O_7)_2(OH)_4$ $= Ca_{10}(Mg,Fe)_2Al_4Si_9O_{34}(OH)_4$	

(albite-anorthite) and scapolite (marialite-meionite). Table 2.7 is not a complete list of framework silicates. We could, for example, list many other zeolite minerals. Some minerals in Table 2.7, such as the feldspars, are common. Others, such as the scapolite minerals, are rare.

▶**TABLE 2.7**
Framework Silicates

Class (Subclass)	Group	Series or Subgroup	Species	Chemical Formula
silicate (framework silicates)	silica group		quartz	SiO_2
			cristobalite	SiO_2
			tridymite	SiO_2
			coesite	SiO_2
			stishovite	SiO_2
	feldspar group	potassium feldspars	orthoclase	$KAlSi_3O_8$
			sanidine	$(K, Na)AlSi_3O_8$
			microcline	$KAlSi_3O_8$
		plagioclase feldspar series	albite	$NaAlSi_3O_8$
			anorthite	$CaAl_2Si_2O_8$
	feldspathoid group		analcime	$NaAlSi_2O_6 \cdot H_2O$
			leucite	$KAlSi_2O_6$
			nepheline	$(Na, K)AlSiO_4$
	scapolite group	scapolite series	marialite	$Na_4(AlSi_3O_8)_3Cl$
			meionite	$Ca_4(Al_2Si_2O_8)_3(CO_3,SO_4)$
	zeolite group		natrolite	$Na_2Al_2Si_3O_{10} \cdot 2H_2O$
			chabazite	$CaAl_2Si_4O_{12} \cdot 6H_2O$
			heulandite	$CaAl_2Si_7O_{18} \cdot 6H_2O$
			stilbite	$CaAl_2Si_7O_{18} \cdot 7H_2O$
			sodalite	$Na_3Al_3Si_3O_{12} \cdot NaCl$
	other		beryl	$Be_3Al_2Si_6O_{18}$
			cordierite	$(Mg,Fe)_2Al_4Si_5O_{18}$

▶**QUESTIONS FOR THOUGHT**

Some of these questions have no specific correct answers; they are intended to promote thought and discussion.

1. What is the definition of a mineral, and how do minerals differ from other solid substances? Why do mineralogists use such a restricted definition of a mineral?
2. Why do rocks only contain a few major elements?
3. All minerals have defects. What are the major kinds of defects? We can't normally see defects when we look at a mineral sample, so how do we know they are present?
4. Although there is a great deal of variation, why are sedimentary minerals generally fine-grained compared with igneous minerals?
5. What factors lead to the precipitation of minerals from aqueous solutions? Which minerals are the most common aqueous minerals? Why?
6. Why did we make the debatable assertion that metamorphic minerals are more diverse than igneous or sedimentary minerals?
7. What is the difference between primary and secondary minerals? How do primary minerals form in igneous, sedimentary, and metamorphic rocks? How do secondary minerals form in these same rock types?
8. Why are there so few common minerals?
9. What is the basis for the mineral classification scheme used by mineralogists? What are mineral classes, subclasses, groups, series, subgroups, and species?

▶**RESOURCES**

(See also the general references listed at the end of Chapter 1.)

Blatt, H., G. Middleton, and R. Murray. *Origin of Sedimentary Rocks.* Englewood Cliffs, NJ: Prentice Hall, 1980.

Bloss, F. D. *Crystallography and Crystal Chemistry.* Washington, DC: Mineralogical Society of America, 1994.

Bunn, C. W. *Crystals: Their Role in Nature and Science.* New York: Academic Press, 1964.

Hartman, P., ed. *Crystal Growth, An Introduction.* Amsterdam: North-Holland, 1973.

Hurlbut, C., Jr. *Minerals and Man.* New York: Random House, 1968.

Hyndman, D. W. *Petrology of Igneous and Metamorphic Rocks.* New York: McGraw-Hill, 1985.

Palache, C., H. Berman, and C. Frondel. *Dana's System of Mineralogy,* 7th ed. New York: John Wiley & Sons, 1944.

Phillips, W. R., and D. T. Griffen. *Optical Mineralogy: The Nonopaque Minerals.* San Francisco: W. H. Freeman, 1981.

Ronov, A. B., and A. A. Yaroshevsky. Chemical composition of the Earth's crust. *American Geophysical Union Monograph no. 132,* 1969.

Strunz, H. *Mineralogische Tabellen,* 5th ed. Leipzig: Akademische Verlagsgesellschaft, Geest und Portig K. G., 1970.

Zoltai, T., and J. H. Stout. *Mineralogy: Concepts and Principles.* Minneapolis: Burgess, 1984.

Mineral Properties: Hand Specimen Mineralogy

The properties of minerals derive from their composition and crystal structure, so we can identify minerals, both in the laboratory and in the field, on the basis of their physical properties. In this chapter we look at the most important physical properties, including appearance, crystal shape, hardness, the way in which a mineral breaks or cleaves, and density. We discuss some of the things that control the properties, and we discuss how the properties may be used for mineral identification. We also consider other properties that can be keys to mineral identification in some special cases.

Mineral names are based on mineral appearance, mineral chemistry, where the mineral is found, a famous scientist, or anything else deemed important by a mineral's discoverer. The Commission on New Minerals and New Mineral Names of the International Mineralogical Association reviews proposed new names and descriptions and judges their appropriateness. The Commission also occasionally discredits old names. Absolute identification and classification of minerals requires knowledge of their composition and atomic structure. Such information must be included when mineralogists submit names to the Commission for approval.

Determining mineral composition and structure may require time and equipment unavailable to most mineralogists or to mineralogy students, but fortunately we can use other methods to tell minerals apart. Differences in composition and structure lead to differences in appearance and in many other properties of minerals we use for identification. Most of the properties discussed in this book fall into two general groups: **hand specimen** properties, which are easily determined using large samples, and **optical properties,** which can only be seen with specially prepared slides and a **polarizing light microscope,** also called a **petrographic microscope.** This chapter reviews hand specimen properties and discusses their use in identification. Optical properties are covered in the next chapter.

IDENTIFICATION OF MINERALS

Give a mineral specimen to a nongeologist and ask them to describe it. Generally, the appearance, especially color, is mentioned first. With a little prodding, they may go on to describe the shape and nature of visible crystals. For example, pyrite might be described as being metallic, gold in color, and forming cubic crystals. A piece of rose quartz might be described as being pink, glassy, transparent, and having a hexagonal shape. *Metallic* and *glassy* are terms describing

luster. *Gold* and *pink* describe **color.** *Transparent* describes **diaphaneity.** *Cubic* and *hexagonal* describe **symmetry,** a property relating to **shape.** These four properties (luster, color, diaphaneity, and shape) are basic for mineral identification. Other properties, including **streak** (the color of a mineral when powdered), the way a mineral breaks (**cleavage, parting, fracture**), and **hardness,** can be even more important.

Given a single property, perhaps luster, it is possible to divide minerals into groups. In the case of luster, we usually divide minerals into those that are metallic and those that are nonmetallic. There are, however, many metallic and many nonmetallic minerals; other properties must be considered if minerals are to be identified. Nonmetallic minerals can, for instance, be divided further based on more subtle luster differences (Table 3.1). Other properties, such as color, diaphaneity, crystal shape, cleavage, parting, fracture, and hardness are also diagnostic. Ultimately, minerals can be identified by name or at least placed in small groups. It is tempting, then, to come up with a standard list of properties that should be evaluated when identifying minerals. However, most mineralogists know that, depending on the sample and circumstances, some properties are more important than others. Rather than going through a long list or filling out a standard table, experienced mineralogists focus on the properties that are most diagnostic. In some cases, a single property, such as strong effervescence by hydrochloric acid (diagnostic of calcite), may serve for mineral identification. At first, mineral identification may seem tedious, but with a little experience, it is possible to find shortcuts to make the process more efficient. (Appendix B includes tables that may be used for mineral identification.)

MINERAL APPEARANCE

Luster

Luster refers to the general appearance or sheen of a mineral. It refers to the way in which a mineral reflects light. Minerals that have the shiny appearance of polished metal are said to have a **metallic luster.** Some could be used as mirrors. Well-crystallized pyrite is a good example (Figure 3.1). Other examples of metallic minerals are shown in Plates 1.3, 2.7, 7.2, 7.3, and 7.5. Minerals that do not appear metallic are said to have a **nonmetallic luster.** Those that appear only partially metallic are called **submetallic.**

Mineralogists use many terms to describe nonmetallic minerals. **Vitreous** minerals are those with a glassy appearance. Quartz, many examples of which are shown in Plates 2.1 through 2.8, is an excellent example. The term **adamantine** is used to describe crystals that sparkle or appear brilliant; the Herkimer diamonds in Plate 2.8 are adamantine. In contrast, the pectolite in Plate 3.1 is **dull** and the sphalerite in Plate 7.4 is **resinous.** Some of the more common terms describing luster of nonmetallic minerals are given in Table 3.1.

Diaphaneity

Diaphaneity refers to a mineral's ability to transmit light. Some minerals are **transparent.** When they are thick, some distortion may occur, but light passes relatively freely through them. Uncolored quartz is often transparent (Plates 2.1, 2.4, 2.6, 2.7, and 2.8). Minerals that do not transmit light as well as clear quartz may be **translucent.** Although it is not possible to see through them as with transparent minerals, translucent minerals, if thin enough, transmit light. Still a third type of mineral, called an **opaque mineral,**

▶TABLE 3.1

Terms Used To Describe Luster of Nonmetallic Minerals

Luster	Meaning	Minerals That Sometimes Exhibit the Luster
vitreous	having a glassy appearance	quartz, tourmaline
resinous	having the appearance of resin	sphalerite, sulfur
greasy	reflecting light to give a play of colors; similar to oil on water	chlorite, nepheline
silky	having surfaces appearing to be composed of fine fibers	chrysotile (asbestos), gypsum
adamantine	a bright, shiny, brilliant appearance similar to that of diamonds	diamond, cerussite
pearly	appearing iridescent, similar to pearls or some seashells	muscovite, talc
dull	not reflecting significant amounts of light or showing any play of colors	kaolinite (clay), niter

▶FIGURE 3.1
Pyrite typically has a metallic luster and so reflects light well.

does not transmit light unless the mineral is exceptionally thin. Most opaque minerals have metallic lusters. Pyrite and many other sulfide and oxide minerals are good examples. Their opacity sets them apart from most other minerals.

Color

Color is often used for quick identification of minerals. In some cases, it can be diagnostic, but in many it is ambiguous or even misleading. The deep red color of rubies may seem distinctive. Ruby is, however, just one variety of the mineral corundum. Sapphires are different colored varieties of the same mineral. To add to the confusion, other minerals, such as spinel or garnet, may have the same deep red color as does ruby. Color is ambiguous because many things can give a mineral its color.

Color is one of the most misunderstood mineral properties. It's easy to look at a ruby illuminated by white light and say it has a red color. If the ruby is illuminated by light of a different color, it may not appear red. Color, then, is not a property of a mineral. It is instead the result we observe when light and a mineral interact. When we see that something has color, what we are really observing is the color of the light that is being reflected or transmitted to our eye. Normal light, called **white light,** includes many different colors. When white light strikes a mineral surface, if all of the colors are reflected back to our eyes, the mineral will appear white. If none of the colors is reflected back to our eye, the mineral will appear black. Most minerals, like ruby, appear to have color because only one or a few

wavelengths make it back to our eye. The other wavelengths of light are scattered in other directions or are absorbed or transmitted by the mineral in some way.

Metallic minerals, especially sulfides, tend to be constant in their coloration (Plates 7.1 through 8.2), so mineralogists use color as a key tool for sulfide identification. However, metallic minerals easily tarnish, so a fresh surface is needed to see the true color. Plate 7.3 shows pyrite altering and tarnishing to produce purple colors characteristic of covellite. Color is a poor property to use for identifying many nonmetallic minerals because there are so many factors that affect coloration. Quartz may be colorless, rosy (rose quartz), yellow (citrine), purple (amethyst), milky, smoky, or black. Some examples are shown in Plates 2.1 through 2.8. Corundum may be just about any color, including colorless, black, brown, pink, yellow, blue, purple, or red. Gem varieties of corundum are ruby (red), oriental topaz (straw yellow), and sapphire (other colors); Plate 8.5. shows several corundum varieties.

The most significant control on color is a mineral's chemical composition. Elements that give a mineral its color are called **chromophores.** It does not take large amounts of chromophores to color a mineral. Minor amounts, less than 0.1 wt. % of transition metals such as Fe and Cu, may control a mineral's color because electrons in the *d*-orbitals of transition metals are extremely efficient at absorbing certain visible wavelengths of light. The remaining wavelengths are reflected and give minerals their color.

If the elements controlling the selective reflection of certain wavelengths are major components in a mineral, the mineral is called **idiochromatic,** or "self-coloring." Sphalerite, for example, is an idiochromatic mineral. It changes from white to yellow to brown to black as its composition changes from pure ZnS to a mixture of ZnS and FeS. Many copper minerals are green or blue, while many manganese minerals are pinkish. These colors derive from selective absorption of certain colors by copper and manganese. Idiochromatic elements may have different effects in different minerals. Malachite is green and azurite is blue, but in both minerals the color is due to copper.

Ruby and sapphire are examples of **allochromatic** varieties of corundum. In allochromatic minerals, minor or trace elements determine the color. Very small amounts of Fe and Ti give sapphire a deep blue color. Small amounts of Cr give ruby and some other gemstones deep red colors. The effects of allochromatic elements may be different in different minerals. Allochromatic Cr is also responsible for the striking green color of emerald (a variety of the mineral beryl), chrome diopside, and some tourmalines.

Structural defects in minerals may also give them color. Radiation damage gives quartz, for example, a purple (Plate 2.3), smoky, or black color. The purple color of many fluorites is caused by Frenkel defects. Other causes of coloration include the oxidation or reduction of certain elements (especially Fe), and the presence of minute inclusions of other minerals.

Streak

Although it would never occur to many people to check a mineral's streak, streak is in some cases a key diagnostic property. It is not a useful property for identifying most silicates but is especially useful for distinguishing oxide and sulfide minerals. The streak of a mineral is the color it has when finely powdered. For mineral identification, it is much more reliable than mineral color, and it is easy to determine. The usual method of determining streak is to rub the mineral against a ceramic streak plate (Figure 3.2). Because the mineral is finely powdered, structural and other nonchemical effects are minimized. Streak is routinely used by mineralogists both in the laboratory and in the field. It can be extremely useful for telling dark-colored minerals, especially metallic ones, apart. For example, hematite may be red, gray, or black in hand specimen and may or may not have a metallic luster. It always, however, has a diagnostic red streak.

Luminescence

Some minerals will emit light when they are activated by an energy form other than visible light. Such an effect is called **luminescence.** Examples of luminescence include **fluorescence, phosphorescence,** and **thermoluminescence.** Fluorescent minerals give off visible light when they are struck by energy of short-er wavelength. The invisible radiation from ultraviolet lamps, for example, may cause scheelite, willemite, or fluorite to appear to glow in the dark. If the visible emission continues after the energy source is turned off, the mineral is phosphorescent. Pectolite is an example of a phosphorescent mineral. Thermoluminescent minerals, such as some tourmalines, give off visible light in response to heating. Some varieties of fluorite, calcite, and apatite also have this property.

Play of Colors

Opalescence (exhibited by opals and a variety of K-feldspar called *moonstone*) and **pearly luster** (exhibited by labradorite feldspar and by some talc) are two examples of **play of colors.** White light is separated into individual wavelengths of varying intensities emitted in different directions. The play of colors is a form of light scattering due to very fine particles in the minerals. **Chatoyancy** and **asterism** are two other scattering effects; they are most easily seen in gemmy minerals. Chatoyant minerals show a bright band of scattered light, usually perpendicular to the long direction of a crystal. Such minerals are sometimes said to have a cat's-eye (chrysoberyl) or tiger's-eye (crocidolite) appearance. The satinspar variety of gypsum is chatoyant. Chatoyancy is caused by closely packed parallel fibers or inclusions of other minerals within a mineral. Asterism, a property sometimes visible in rubies, sapphires, garnets, and some other gems, refers to scattered light appearing as a "star." As in chatoyancy, asterism results from the scattering of light by small inclusions of a different mineral.

Some minerals exhibit **iridescence,** similar to the play of colors that is sometimes seen on an oily surface. Metallic minerals such as bornite may acquire iridescence as they tarnish. Labradorite and some other feldspars exhibit an internal iridescence (also called **labradorescence** or **schiller**) due to fine compositional layering.

CRYSTAL SHAPE

For well-developed crystals, crystal **form** and **habit** are excellent diagnostic properties. Habit refers to the overall shape of a crystal or aggregate of crystals. To mineralogists, the term *form* refers specifically to a group of crystal faces, related by the crystal's symmetry, that have identical chemical and physical properties. Although museum specimens and pictures of minerals in textbooks often show distinctive habits and forms, most mineral samples do not. Small irregular crystals without flat faces, or massive aggregates, are typical, often rendering habit and form of little use for hand specimen iden-

►**FIGURE 3.2**
Graphite and a streak plate. Graphite is a very soft mineral that has a black-gray streak.

tification. Because form and habit reflect the internal arrangement of atoms in a crystal, when visible they are important diagnostic properties (Box 3.1).

Faces of a single crystal form have identical properties because they contain identical atoms in identical arrangements. Some minerals, such as chabazite, halite, and garnet, normally contain only one form; others contain more (Figure 3.3). Chabazite crystals grow as rhombohedrons, cubes "squashed" along one main diagonal. Halite crystals are typically cubic, having six square faces. Garnet crystals commonly have 12 diamond-shaped faces. Other minerals, such as ilmenite, corundum, gehlenite, vesuvianite, and datolite (Figure 3.3), may contain multiple forms and have more complicated shapes. Some minerals—for example, calcite—have many common forms. But, as we shall see later, they all have a common property called *symmetry*.

▶**FIGURE 3.3**
Forms and combinations of forms of six minerals. Different samples of the same mineral may crystallize with different forms, but those shown here are typical. The lines on the crystal faces show orientations of prominent cleavages.

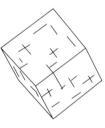

chabazite

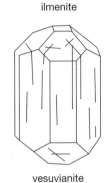

ilmenite

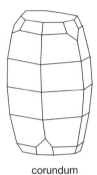

corundum

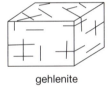

gehlenite

vesuvianite

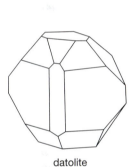

datolite

Box 3.1 What's Wrong With This Picture?

In the 1995 movie *Congo,* an exploration team goes to Africa to seek large, flawless diamonds. When the diamonds are shown, the movie immediately loses credibility with mineralogists because the crystals are hexagonal prisms (long crystals with a hexagonal cross section). Mineralogists know that diamond habit does not include hexagonal prisms (Figure 3.4).

▶**FIGURE 3.4**
These quartz crystals with hexagonal prismatic habit are not diamonds because diamonds cannot form hexagonal crystals of this sort.

A single crystal's habit is controlled by the forms that are present, the way the forms combine, the relative sizes of crystal faces, and other features relating to crystal growth. The most useful terms describing habit are self-explanatory (Table 3.2). Common ones used to describe habit of single crystals include **equant** (equidimensional), **acicular** (needlelike), **tabular,** and **bladed** (Figure 3.5). For a group of crystals, habit includes the shape of the crystals and the way they are intergrown. The terms **massive, granular, radiating,** and **fibrous** are typical of the terms used to describe crystal aggregates (Figure 3.5).

The color plates in this book show a wide range of crystal habits. The orthoclase crystals in Plate 1.5 are **blocky;** the celestite in Plate 1.6 is **tabular;** the okenite needles in Plate 1.7 and the selenite needles in Plate 3.5 are **acicular;** the smithsonite in Plate 1.8 and pectolite in Plate 3.1 are **botryoidal.** Most of the quartz crystals in Plate 2 are **prismatic,** as are the beryl and tourmaline in Plate 3.3. The actinolite in Plate 3.2 and the kyanite in Plate 3.6 are **bladed,** while the anthophyllite (Plate 3.4) and chrysotile (Plate 3.7) are **fibrous** aggregates. In Plate 3.8, **drusy** pyrite has grown on top of calcite.

STRENGTH AND BREAKING OF MINERALS

The color and shape of minerals are obvious to anyone, but there are other, more subtle, properties that a mineralogist will notice. Several relate to the strengths of bonds that hold crystals together. These properties are especially reliable for mineral identification because they are not strongly affected by chemical impurities or defects in crystal structure.

Tenacity

The term *tenacity* refers to a mineral's toughness and its resistance to breaking or deformation. Those that break, bend, or deform easily have little tenacity. In contrast, strong unbreakable minerals have great tenacity. Jade, composed of either the pyroxene jadeite or the amphibole actinolite, is one of the most tenacious natural materials known. It does not easily break or deform, even when under extreme stress. Table 3.3 contains some of the terms typically used to describe tenacity.

The tenacity of a mineral is controlled by the nature of its chemical bonds. Ionic bonding often leads to rigid, brittle minerals. Halite is an excellent

▶**TABLE 3.2**
Terms Used To Describe Crystal Habit

Terms Generally Used To Describe
Individual Crystals (With Example Minerals)

equant	having approximately the same dimensions in all directions (garnet, spinel)
blocky	equant crystals with approximately square cross sections (halite, galena)
acicular	needlelike (actinolite, sillimanite)
tabular or platy	appearing to be plates or thick sheets stacked together (gypsum, graphite)
capillary or filiform	hairlike or threadlike (serpentine, millerite)
bladed	elongated crystals that are flattened in one direction (kyanite, wollastonite)
prismatic or columnar	elongated crystals with identical faces parallel to a common direction (apatite, beryl)
foliated or micaceous	easily split into sheets (muscovite, biotite)

Terms Generally Used To Describe
Crystal Aggregates

massive	solid mass with no distinguishing features
granular	composed of many individual grains
radiating or divergent	crystals emanating from a common point
fibrous	appearing to be composed of fibers
stalactitic	stalactite shaped
lamellar or tabular	flat plates or slabs growing together
stellated	aggregate of crystals giving a starlike appearance
plumose	having a feathery appearance
arborescent or dendritic	having a branching treelike or plantlike appearance
reticulated or latticelike	slender crystals forming a lattice pattern
colloform or globular	spherical or hemispherical shapes made of radiating crystals
botryoidal	having an appearance similar to a bunch of grapes
reniform	having a kidney-shaped appearance
mammillary	breastlike
drusy	having surfaces covered with fine crystals
elliptic or pisolitic	very small or small spheres

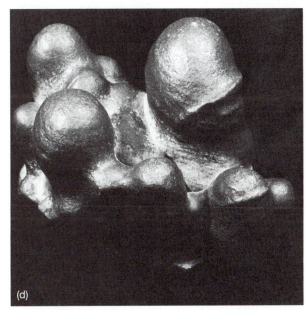

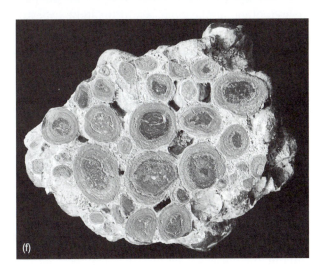

▶FIGURE 3.5
Examples of crystal habits: (a) galena showing a blocky habit;
(b) gypsum showing a bladed habit (and scratch marks caused
by tests for hardness); (c) serpentine showing a fibrous habit;
(d) goethite showing a stalactitic habit; (e) hematite (on
quartz) showing a radiating habit; and (f) limonite showing a
pisolitic habit.

Box 3.2 Asbestiform Minerals and Health Risks

We use the term **asbestiform** to describe a mineral habit characterized by small, strong, and flexible fibers, equivalent to hairs or whiskers (for example, serpentine, as shown in Figure 3.5c and Plate 3.7; and anthophyllite, as shown in Plate 3.4). **Asbestos** is a commercial name for any marketable asbestiform mineral. Mineralogists have described many asbestiform mineral varieties, but most are rare and only a few are produced for sale. "White asbestos," composed of the mineral chrysotile, accounts for about 95% of the commercial market. Chrysotile, a member of the serpentine group, is a widespread but minor mineral in many altered ultramafic rocks. In North America, production occurs in a few large deposits in Quebec and California. Other commercial asbestos is composed of crocidolite ("blue asbestos") and amosite ("brown asbestos"), varieties of the amphiboles riebeckite and grunerite, respectively. Other minerals with asbestiform varieties include other amphiboles (anthophyllite, tremolite, actinolite), clays (sepiolite, palygorskite), and members of the zeolite group, including roggianite, mazzite, erionite, mordenite, and okenite.

Historically, asbestos has had many uses. Since around 1878, it has been mined in large quantities because it is tough, flexible, and fire and chemical resistant. Between 1900 and 1986, builders sprayed asbestos on walls, ceilings, and pipes in many buildings in the United States. Industries have used asbestos in brake linings, roof shingles, and other applications. Unfortunately, asbestos easily crumbles to make a fine dust that people can inhale. Fibers can become embedded in lung tissue and cause asbestosis (a chronic breathing disorder that may be fatal), lung cancer, or mesothelioma (another form of cancer). For the most part, epidemiologists have documented these diseases in workers exposed to high levels of asbestos over long times.

In 1907 health workers reported the first asbestos-related diseases, but it was not until around 1960 that the threat posed by asbestos was accepted as serious. In 1974 the Environmental Protection Agency (EPA) banned asbestos for most commercial use in the United States, and soon afterward launched a vigorous program to remove asbestos from commercial structures. However, American companies still ship many products containing asbestos to developing countries.

Despite the ban and efforts to eliminate asbestos from our environment, it is still common in many buildings and as a component in urban dust. Fortunately, current studies suggest that exposure to low levels of chrysotile, the most common form of asbestos, may not pose as serious a health threat as once thought. Furthermore, the EPA has found that removing asbestos that is not crumbling or releasing fibers can increase asbestos concentrations in the air and cost a great deal of money without significantly decreasing health threats. For these reasons, Congress modified asbestos laws in 1986.

example of a **brittle** mineral. It shatters into many small pieces when struck. Quartz, too, is brittle, although the bonding in quartz is only about 50% ionic. Many metallically bonded minerals, such as native copper, are **malleable.** Other minerals, such as gypsum, are **sectile.** Some minerals, including talc and chlorite, are **flexible** due to weak van der Waals and hydrogen bonds holding well-bonded layers of atoms together. When force is applied, slippage between layers allows bending. When pressure is released, they do not return to their original shape. Still other minerals, notably the micas, are **elastic.** They may be bent but resume their original shape after pressure is released if they were not too badly deformed. In micas and other elastic minerals, the bonds holding layers together are stronger than those in chlorite or clays.

▶**TABLE 3.3**
Terms Used to Describe Tenacity

flexible	bendable
elastic	a bendable mineral that returns to its original shape after release
malleable	capable of being hammered into different shapes
ductile	capable of being drawn into a wirelike shape
brittle	easily broken or powdered
sectile	capable of being cut into shavings with a knife

Fracture, Cleavage, and Parting

Fracture is a general term used to describe the way a mineral breaks or cracks. Terms used to describe fracture include *even, conchoidal,* and *splintery* (Table 3.4). Because atomic structure is not the same in all directions and chemical bonds are not all the same strength, most crystals break along preferred directions. The orientation and manner of breaking are important clues to crystal structure. If the fractures are planar and smooth, the mineral is said to have good **cleavage** (Figure 3.6). Cleavage involves minerals breaking parallel to planes of atoms. There are a few exceptions, such as quartz, that break only along curved surfaces to form **conchoidal fractures** (see Plate 2.1), but the majority of minerals exhibit cleavage.

If a mineral cleaves along one particular plane, a nearly infinite number of parallel planes are equally prone to cleavage. This is due to the repetitive arrangement of atoms in atomic structures. The spacing between planes is the repeat distance of the atomic structure, on the order of Ångstroms ($1 Å = 10^{-10}$ m). The whole set of planes, collectively referred to as a cleavage, represents planes of weak bonding in the crystal structure. Biotite (Plate 6.6) is an excellent example of a mineral with one excellent cleavage. Minerals that have more than one direction of weakness will have more than one cleavage direction (Figure 3.7). The direction and angular relationships between cleavages, therefore, give valuable hints about atomic structure.

Minerals that are equally strong in all directions, such as quartz, fracture to form irregular surfaces (Plate 2.1). Minerals with only one direction of weakness, such as gypsum and micas, have one direction of cleavage and usually break to form thick slabs or sheets. We say they have *basal cleavage.* Kyanite (Plate 3.6) and anthophyllite, which have two good cleavages, easily break into splintery shapes. Other minerals may have three (halite), four (fluorite), or even six cleavages (Figure 3.7). We use geometric terms such as cubic, octahedral, or prismatic to describe cleavage when appropriate. The ease with which a mineral cleaves is not the same for all minerals or for all the cleavages in a particular mineral. Mineralogists describe the quality of a particular cleavage with qualitative terms: *perfect, good, distinct, indistinct,* and *poor.* Quartz has poor cleavage in all directions, while micas have one perfect cleavage.

Crystal faces and cleavage surfaces may be difficult to tell apart. A set of parallel fractures indicates a cleavage, but if only one flat surface is visible, there can be ambiguity. However, crystal faces often display subtle effects of crystal growth. **Twinning** (oriented intergrowths of multiple crystals) and other **striations** (parallel lines on a face), growth rings or layers, pitting, and other imperfections make a face less smooth than a cleavage plane and give it lower reflectivity and a drabber luster. In some minerals, principal cleavage directions are parallel to crystal faces, but in most they are not. Plates 1.3 and 2.7 show pyrite with well-developed striations on its crystal faces.

Cleavage is an excellent property for mineral identification. Often the quality and number of cleavages may be seen in hand specimen. Sometimes a hand lens is used to identify the set of fine parallel cracks, more irregular than twinning and striations, which indicate a cleavage that is too poorly developed to be seen with the naked eye. Angles between cleavages may be estimated or, if accurate angular measurements are needed, techniques involving a **petrographic microscope** or a device called a **goniometer** may be used to measure them.

Some mineral specimens exhibit **parting,** a phenomenon that looks like cleavage. Parting is not due to atomic structure weaknesses, but to crystallographic imperfections such as **twin planes** (planes that separate

▶**TABLE 3.4**
Terms Used to Describe Fracture and Cleavage (and Examples)

Fracture Terms

even	breaking to produce smooth planar surfaces (halite)
uneven or irregular	breaking to produce rough and irregular surfaces (rhodonite)
hackly	jagged fractures with sharp edges (copper)
splintery	forming sharp splinters (kyanite, pectolite)
fibrous	forming fibrous material (chrysotile, crocidolite)
conchoidal	breaking with curved surfaces as in the manner of glass (quartz)

Cleavage Terms

basal	also sometimes called *platy*; refers to cleavage in minerals such as micas that have one well-developed planar cleavage
cubic	geometric term used to describe three cleavages at 90° to each other (galena)
octahedral	geometric term used to describe four cleavages that produce octahedral cleavage fragments (fluorite)
prismatic	multiple directions of good cleavage all parallel to one direction in the crystal

►**FIGURE 3.6**

Fracture and cleavage examples: (a) Microline (a K-feldspar variety) typically exhibits two cleavages that intersect at about 90°. It fractures in other directions. The cleavage surfaces are smoother than the fracture surfaces; (b) Obsidian, in contrast with microcline and most minerals, does not cleave because it is not crystalline. Instead it breaks to form conchoidal fractures; (c) Micas have only one good cleavage, so they break into sheets; (d) Tremolite has two cleavages which allow it to break to form long splintery pieces; (e) Quartz (top) has no cleavages and fractures conchoidally. Halite (left) has two cleavages that intersect at exactly 90°. Calcite (right) has two cleavages, but they do not intersect at 90°; (f) Fluorite has four excellent cleavages, allowing it to be broken into octahedra.

A. Cleavage in One Direction. Example: muscovite.

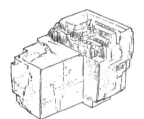

B. Cleavage in Two Directions at Right Angles. Example: feldspar.

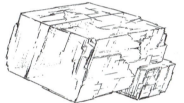

C. Cleavage in Two Directions Not at Right Angles. Example: amphibole.

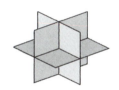

D. Cleavage in Three Directions at Right Angles. Example: halite.

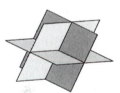

E. Cleavage in Three Directions Not at Right Angles. Example: calcite.

F. Cleavage in Four Directions. Example: fluorite.

G. Cleavage in Six Directions. Example: sphalerite.

▶**FIGURE 3.7**
Cleavage in minerals. Compare these drawings with the photos in Figure 3.6.

domains with different atomic-structure orientations), stress, or chemical alteration. In contrast with cleavage, parting is restricted to one or a few distinct planes rather than an infinite set. Unlike cleavage, parting will not be present in all specimens of a particular mineral, and parting surfaces are usually less smooth than cleavage planes.

Hardness

Hardness is a mineral's resistance to abrasion or scratching. **Relative hardness** is determined by trying to scratch a surface of one mineral with an edge or corner of another. If a scratch or abrasion results, the first mineral is the softer. Although rarely done by

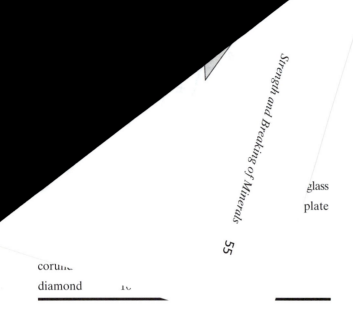

glass
plate

corun...

diamond   10

Box 3.3 Hard and Soft Minerals

Appendix E contains a list of the most common minerals ordered by hardness. Most have a hardness greater than 2 and less than 7. Following is a list of some of the more common minerals with hardnesses outside this range.

Name	Formula	Hardness
talc	$Mg_3Si_4O_{10}(OH)_2$	1
molybdenite	MoS_2	1 to $1^1/_2$
graphite	C	1 to 2
pyrophyllite	$Al_2Si_4O_{10}(OH)_2$	$1^1/_2$
covellite	CuS	$1^1/_2$ to 2
orpiment	As_2S_3	$1^1/_2$ to 2
realgar	AsS	$1^1/_2$ to 2
gypsum	$CaSO_4 \cdot 2H_2O$	2
stibnite	Sb_2S_3	2
sylvite	KCl	2

(Most other minerals fall between these values)

cordierite	$(Mg, Fe)_2Al_4Si_5O_{18}$	7
quartz	SiO_2	7
andalusite	Al_2SiO_5	$7^1/_2$
zircon	$ZrSiO_4$	$7^1/_2$
beryl	$Be_3Al_2Si_6O_{18}$	$7^1/_2$ to 8
spinel	$MgAl_2O_4$	$7^1/_2$ to 8
topaz	$Al_2SiO_4(F,OH)_2$	8
chrysoberyl	$BeAl_2O_4$	$8^1/_2$
corundum	Al_2O_3	9
diamond	C	10

mineralogists, absolute values of hardness may be determined in several ways; the easiest is to use an indenting tool similar to ones used to determine the hardness of steel. The indenting tool measures the force necessary to produce a permanent indentation in a flat surface. The results will be almost the same as those determined using scratch tests.

The symbol H is used for hardness. Table 3.5 gives the relative hardness scale used by mineralogists. Based on 10 well-known minerals, it is called the Mohs hardness scale, named after Austrian mineralogist Friedrich Mohs who developed it in 1812. In terms of absolute hardness, the Mohs hardness scale is not linear but is close to being exponential (Figure 3.8a). The hardnesses of the softest minerals are more similar than the hardness of the four hardest ones (quartz, topaz, corundum, diamond). Gypsum $(H = 2)$ is only slightly harder than talc $(H = 1)$, but diamond $(H = 10)$ has a hardness five times greater than corundum $(H = 9)$.

Relative hardness can be determined by conducting scratch tests to compare the hardness of an unknown mineral to the minerals in the Mohs hardness scale. Alternatively, good approximations of hardness may be made by comparing mineral hardness to a fingernail, penny, pocketknife, glass, or several other common objects (Table 3.5). Figure 2.2 shows a glass plate that has been used for hardness tests. However, there can be complications. Mineral specimens may be too small or too valuable to scratch. Large samples may be made of many grains loosely cemented together so that scratch tests are not possible. Others may cleave or fracture when tests are performed. In still other cases, the results of scratch tests may be ambiguous.

The hardness of a mineral relates to its weakest bond strength. Graphite has a hardness of $1^1/_2$, and diamond has a hardness of 10. Both are made of carbon, but in diamond the carbon atoms are uniformly spaced and tightly bonded together, while in graphite bonding is very weak in one direction. Because bonds are usually not the same in all directions in minerals, hardness may vary depending on the direction a mineral is scratched. In kyanite, for example, hardness varies from $4^1/_2$ to $6^1/_2$ depending on the direction of the scratch test. In most minerals, however, hardness is about the same in all directions. While the general relationship between hardness and bond strength is known, mineralogists have difficulty predicting hardness for complex atomic structures. For some simple ionic compounds, however, theoretical calculations match measurements well. Minerals with high density, highly charged ions, small ions, or covalent bonding tend to be hardest.

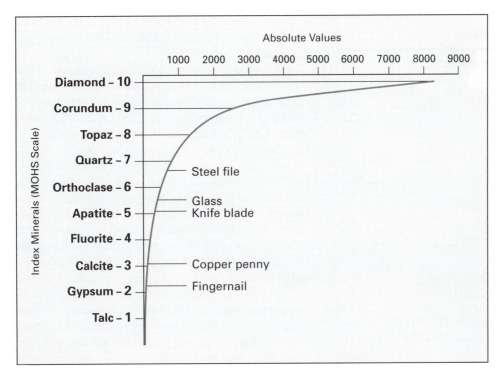

▶**FIGURE 3.8a**

Mohs hardness scale compared with an absolute hardness scale and some common everyday objects. Geologists rarely take a complete hardness kit in the field and so often use their fingernail, a penny, or pocket knife to estimate hardness.

▶**FIGURE 3.8b**

Gypsum has a hardness of 2 on the Mohs hardness scale and so can be scratched with a fingernail to estimate its hardness.

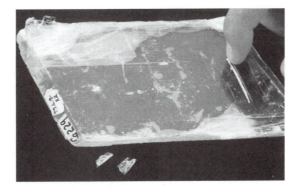

DENSITY AND SPECIFIC GRAVITY

The Greek letter ρ (rho) symbolizes **density.** The density of a mineral is usually given in units of grams/centimeter^3 (gm/cm^3). Density varies slightly depending on pressure or temperature, but most minerals have values between 2 and 8 gm/cm^3. Accurate determination of density can be difficult or impossible because it requires knowing the volume of a crystal, which can be difficult to measure with accuracy. A related property, **specific gravity (G),** is often used instead. Specific gravity is the ratio of the mass of a mineral to the mass of an equal volume of water at 1 atm and 4°C. Because mass and weight are proportional, we normally determine specific gravity by comparing weights. If a mineral is at 4°C, density

(gm/cm^3) and specific gravity (unitless) have identical values because the density of water is 1 gm/cm^3:

$$G_{4°C} = \frac{we}{}$$
$$= \frac{\rho_{\text{mineral}}}{\rho_{\text{water}}} = \rho_{mineral} \qquad (3.1)$$

At room temperature, the density of water is still approximately 1, and most scientists ignore the minor differences between density and specific gravity. In general, borates, halides, sulfates, and some other mineral classes have lower densities than silicates or carbonates, which are of moderate density. Native metals, oxides, and sulfides, in contrast, are denser than most other minerals. There are many exceptions

nerals

pecific

	$a_2B_4O_5(OH)_4 \cdot 8H_2O$	1.8
	KCl	2.0
	S	2.1
graphite	C	2.1
halite	NaCl	2.2
gypsum	$CaSO_4 \cdot 2H_2O$	2.3
orthoclase	$KAlSi_3O_8$	2.6
kaolinite	$Al_2Si_2O_5(OH)_4$	2.6
albite	$NaAlSi_3O_8$	2.6
quartz	SiO_2	2.6
calcite	$CaCO_3$	2.7
anorthite	$CaAl_2Si_2O_8$	2.8
muscovite	$KAl_2(AlSi_3O_{10})(OH)_2$	2.8
talc	$Mg_3Si_4O_{10}(OH)_2$	2.8
dolomite	$CaMg(CO_3)_2$	2.8
biotite	$K(Mg,Fe)_3AlSi_3O_{10}(OH)_2$	3.0
chlorite	talc+brucite combinations	3.0
andalusite	Al_2SiO_5	3.2
fluorite	CaF_2	3.2
apatite	$Ca_5(PO_4)_3(OH,F,Cl)$	3.2
olivine	$(Mg,Fe)_2SiO_4$	3.3
diopside	$CaMgSi_2O_6$	3.4
sillimanite	Al_2SiO_5	3.2
diamond	C	3.5
kyanite	Al_2SiO_5	3.6
staurolite	$Fe_2Al_9Si_4O_{23}(OH)$	3.7
corundum	Al_2O_3	4.0
sphalerite	ZnS	4.0
chalcopyrite	$CuFeS_2$	4.2
rutile	TiO_2	4.2
goethite	FeO(OH)	4.3
smithsonite	$ZnCO_3$	4.4
barite	$BaSO_4$	4.5
ilmenite	$FeTiO_3$	4.5
pyrolusite	MnO_2	4.5
covellite	CuS	4.6
pyrrhotite	$Fe_{1-x}S$	4.6
zircon	$ZrSiO_4$	4.7
hematite	Fe_2O_3	5.0
pyrite	FeS_2	5.1
magnetite	Fe_3O_4	5.2
chalcocite	Cu_2S	5.8
cuprite	Cu_2O	6.0
bornite	Cu_5FeS_4	6.0
scheelite	$CaWO_4$	6.1
cobaltite	(Co,Fe)AsS	6.3
anglesite	$PbSO_4$	6.4
cerussite	$PbCO_3$	6.5
cassiterite	SnO_2	7.0
galena	PbS	7.6
cinnabar	HgS	8.1
copper	Cu	8.8
silver	Ag	10.3
platinum	Pt	16.5
gold	Au	17.2

to these generalizations, some of which are apparent from the list of values in Table 3.6; a more complete list appears in Appendix F.

The specific gravity of a mineral depends on the atoms within it and how closely packed they are. The polymorphs diamond and graphite are both made of carbon (C). Diamond has specific gravity of 3.5, while graphite's is 2.2, due to differences in atomic structure. Graphite forms under Earth surface conditions, but diamond, with its high specific gravity, only forms deep in the Earth where pressures are great. The laws of thermodynamics tell us that high pressures favor dense minerals, which makes sense because at high pressure things are squeezed together. Table 3.7 compares the specific gravities of SiO_2 and of Al_2SiO_5 polymorphs. For both the SiO_2 and Al_2SiO_5 minerals, the densest polymorphs are stable at highest pressures.

The effect of composition on G can be seen by comparing values for **isostructural minerals,** minerals with identical atomic structures but different compositions. Minerals in the garnet group are isostructural. They are often divided into two subgroups, the **pyralspites** (*py*rope-*al*mandine-*spe*ssartine) with composition $(Mg,Fe,Mn)_3Al_2Si_3O_{12}$, and the **ugrandites** (*u*varovite-*gr*ossular-*and*radite) with composition $Ca_3(Cr,Al,Fe)_2Si_3O_{12}$. The names of the two groups are based on the names of garnet species within the groups.

Table 3.8 compares the specific gravities of garnets within the two groups. (For comparison, grossular has been included with the pyralspite garnets, although it is not normally considered a member of that group.) Within the pyralspite series, the increase in G from pyrope to grossular to spessartine to almandine is consistent with the different atomic weights of Mg, Ca, Mn, and Fe. Similarly, the increase in G from grossular to uvarovite to andradite of the ugrandite series mirrors the increase in atomic weight from Ca to Cr to Fe.

Some minerals have specific gravity less than 2.0. Others have values greater than 10 (Table 3.6). Because of great variability, it is possible to distinguish between minerals with high, moderate, or low specific gravity simply by holding a specimen in your hand. We use the term **heft** for estimations of G made by holding hand specimens; heft can be very useful in mineral identification. For example, the mineral barite ($BaSO_4$) sometimes exists as massive white material that is easy to confuse with feldspars. Its high density, easily discerned by picking it up, helps identify it. Similarly, cerussite ($PbCO_3$) can be distinguished from other carbonate minerals by its heft.

Density differences can also help in the separation of minerals. In the laboratory, crushed rock is often separated into mineral components by "floating" samples in liquids of different densities. In

▶**TABLE 3.7**
Specific Gravity (G) and Occurrence for SiO_2 and Al_2SiO_5 Polymorphs

Mineral	Composition	Specific Gravity	Occurrence
tridymite	SiO_2	2.26	volcanic rocks
cristobalite	SiO_2	2.32	volcanic rocks
quartz	SiO_2	2.65	many Earth surface rocks
coesite	SiO_2	3.01	meteor impact craters and rare high-pressure metamorphic rocks
stishovite	SiO_2	4.35	meteor impact craters
andalusite	Al_2SiO_5	3.18	low-pressure metamorphic rocks
sillimanite	Al_2SiO_5	3.23	low- to high-pressure metamorphic rocks
kyanite	Al_2SiO_5	3.60	high-pressure metamorphic rocks

▶**TABLE 3.8**
Specific Gravity (G) Values for Pyralspite and Ugrandite Garnets

Mineral	Composition	Key Element	Atomic Weight of Key Element	G of Mineral
Pyralspites				
pyrope	$Mg_3Al_2Si_3O_{12}$	Mg	24.304	3.54
grossular	$Ca_3Al_2Si_3O_{12}$	Ca	40.08	3.56
spessartine	$Mn_3Al_2Si_3O_{12}$	Mn	54.938	4.19
almandine	$Fe_3Al_2Si_3O_{12}$	Fe	55.847	4.33
Ugrandites				
grossular	$Ca_3Al_2Si_3O_{12}$	Al	26.982	3.56
uvarovite	$Ca_3Cr_2Si_3O_{12}$	Cr	51.996	3.80
andradite	$Ca_3Fe_2Si_3O_{12}$	Fe	55.847	3.86

these **heavy liquids,** which are much denser than water, minerals separate as some float and others sink according to their specific gravities. In mining operations, ore minerals are often separated from uneconomical minerals by using gravity separation techniques that depend on density differences. This occurs in natural systems, too. Placer gold deposits are formed when gold from weathered rock, because of its high specific gravity, concentrates in stream beds.

A number of techniques can be used to determine a mineral's specific gravity. A **Jolly balance** or a **Berman balance** determines a mineral's weight when suspended in air and when suspended in water (Figure 3.9). The specific gravity is then calculated:

$$G = \frac{weight_{air}}{weight_{air} - weight_{water}} \quad (3.2)$$

Balance techniques require relatively large samples, at least 0.5 cm^3 or more, to be accurate. The samples must also be homogeneous. When such material is not available, an alternative method, which uses a powdered sample and a small bottle called a **pycnometer,** is employed. (Box 3.4; Figure 3.10).

MAGNETISM OF MINERALS

Magnetism derives from a property of electrons called the *magnetic moment* that results from their spinning and orbiting motions. The sum of all the magnetic moments of all the atoms in a mineral gives it magnetism. Minerals are classified as **ferromagnetic, diamagnetic,** or **paramagnetic.** If the moments of a mineral's atoms interact in a constructive way, the mineral will have properties similar to those of a magnet. Such is the case for a few minerals, including magnetite and pyrrhotite. Magnetite, pyrrhotite, and other magnetic minerals are called *ferromagnetic* because they have the same magnetic properties as metallic iron. Most minerals exhibit little magnetic character but may be weakly repelled by a strong magnetic field; they are *diamagnetic*. Pure feldspars, halite, and quartz all exhibit

▶**FIGURE 3.9**
A Berman balance, a modern precise scale used to determine specific gravity of minerals.

Box 3.4 What Is a Pycnometer?

A pycnometer (Figure 3.10) is a small bottle fit with a ground glass stopper containing a small hole. The bottle is filled with water and the top placed on it. It is carefully wiped dry after excess water squeezes out of the hole, and then weighed. A small amount of mineral, of known weight, is put in the bottle with the water. The top is replaced and excess water is wiped off. It is weighed again and the mineral's specific gravity is calculated:

▶**FIGURE 3.10**
Pycnometer.

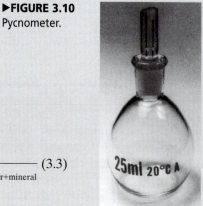

$$G = \frac{weight_{mineral}}{weight_{bottle+water} + weight_{mineral} - weight_{bottle+water+mineral}} \quad (3.3)$$

weak diamagnetism. An impure feldspar, however, may contain iron (Fe), which results in *paramagnetism,* attraction to a strong magnet. Other paramagnetic minerals include garnet, hornblende, and many pyroxenes.

For a few minerals, magnetic properties aid in identification. In the field, ferromagnetic minerals, such as magnetite, may be identified because a magnet will be attracted to them. In the laboratory, subtle differences in the magnetic properties of minerals are used to separate different minerals in crushed rock samples (Figure 3.11).

ELECTRICAL PROPERTIES

Some minerals may conduct electricity. Electrical conduction occurs when a mineral's electrons can move throughout its structure. Such will be the case in structures containing metallic bonds. The native metals, such as copper, are the best examples. Small amounts of electrical conduction may also occur in minerals with defects or other imperfections in their structures. Other minerals, while being unable to conduct electricity, may hold static charges for brief periods of time. They may be charged by exposure

►**FIGURE 3.11**
Magnetic separator.

to a strong electric field, a change in temperature, or an application of pressure. A mineral charged by temperature change is **pyroelectric;** a mineral charged by pressure change is **piezoelectric.** Because they are difficult to measure, electrical properties are not often used for mineral identification.

REACTION TO DILUTE HYDROCHLORIC ACID

One chemical property, the reaction of minerals to dilute (5%) hydrochloric acid (HCl), is included here because it is diagnostic for calcite, one of the most common minerals of the Earth's crust. Drops of acid placed on coarse samples of calcite cause obvious bubbling or fizzing, called *effervescence* (Figure 3.12). Dolomite, a closely related carbonate mineral, effervesces when finely powdered but not when coarse. Other carbonate minerals (smithsonite, aragonite, strontianite) effervesce to different degrees. They are distinguished by crystal form, color, and other properties. Although acid tests have limited use, most mineralogy labs are equipped with small bottles of HCl and eyedroppers to aid in carbonate identification. Many geologists carry a small bottle of dilute hydrochloric acid when they go in

the field so they may distinguish between rocks that contain calcite and rocks that do not.

OTHER PROPERTIES

Minerals possess many other properties (for example, radioactivity or thermal conduction). Because they are of little use for mineral identification in most cases, they will not be discussed individually here.

►**FIGURE 3.12**
A drop of dilute hydrochloric acid on calcite causes it, and a few other minerals, to effervesce.

►QUESTIONS FOR THOUGHT

Some of these questions have no specific correct answers; they are intended to promote thought and discussion.

1. Why do all samples of a given mineral (for example, quartz) have similar physical properties?
2. One systematic approach to mineral identification would be to fill out a table that has columns for luster, color, streak, cleavage, hardness, and so on. Mineralogists rarely do this. Why?
3. When identifying an unknown mineral, color is often a poor property to use. Form and habit, however, are often good properties. What makes color unreliable? What controls a crystal's form? Are there additional things that control its habit?
4. We know that asbestiform minerals lead to health problems if they get in our lungs. Why don't all minerals cause the same sort of health problems?

5. Why do different minerals break in different ways? Quartz has a conchoidal fracture; biotite has one excellent cleavage. What does this tell you about the difference between quartz and biotite?

6. If you look in the table in Appendix E you will find that only a few minerals have hardness greater than 8.0. Why?

7. Appendix F lists minerals by specific gravity. Many of the least dense minerals are evaporite minerals (they form by precipitation from water). Why? Many of the most dense minerals are metals. Why?

▶RESOURCES

(See also the general references listed at the end of Chapter 1.)

Battey, M. H. *Mineralogy of Students,* 2nd ed. New York: Longman, 1981.

Fleischer, M. *A Glossary of Mineral Species 1980.* Tuscon: Mineralogical Record, 1980.

Guthrie, G. D., Jr., and B. T. Mossman, eds. Health effects of mineral dusts. *Reviews in Mineralogy,* vol. 28. Washington, DC: Mineralogical Society of America, 1993.

Nassau, K. The origin of color in minerals. *American Mineralogist* 63 (1978): 219–229.

Proctor, P. D., P. R. Peterson, and U. Kackstaetter. *Mineral-Rock Handbook: Rapid-Easy Mineral-Rock Determination.* New York: Macmillan, 1991.

Strunz, H., and C. Tennyson. *Mineralogische Tabellen,* 5th ed. Leipzig: Akademische Verlag, 1970.

Zussman, J. *Physical Methods in Determinative Mineralogy.* New York: Academic Press, 1977.

Optical Mineralogy

This chapter discusses the interaction of minerals and light, and the properties of minerals in thin section. It discusses the most practical aspects of *optical mineralogy,* which is the branch of mineralogy that deals with the optical properties of minerals.

A fundamental principle of optical mineralogy is that most minerals, even dark-colored minerals and others that appear opaque in hand specimens, transmit light if we slice them thinly enough. We use a *polarizing microscope* to examine them by *transmitted light microscopy* (Figure 4.1). We look at small mineral grains (powdered samples) or specially prepared thin sections (0.03-mm-thick specimens of minerals or rocks mounted on glass slides) to determine properties that are otherwise not discernible. Minerals with metallic luster, and a few others, are termed *opaque minerals.* They don't transmit light even if they are thin-section thickness. For these minerals, transmitted light microscopy is of no use. *Reflected light microscopy,* a related technique, can reveal some of the same properties. It is an important technique for economic geologists who deal with metallic ores but is not used by most mineralogists or petrologists, so we discuss it only briefly in this book.

Most minerals can be identified when examined with a polarizing microscope, even if unidentifiable in hand specimen. Optical properties also allow a mineralogist to estimate the composition of some minerals. For example, the Mg:Fe ratio of olivine, $(Mg,Fe)_2SiO_4$, can be distinguished based on optical properties. Plagioclase feldspar, $CaAl_2Si_2O_8$–$NaAlSi_3O_8$, composition can be similarly distinguished. Box 4.1 summarizes the optical properties used for mineral identification and gives the properties of some common minerals. We divide minerals into those that will not transmit light unless the sections are much thinner than normal thin sections (*opaque minerals*) and those that will (*nonopaque minerals*). Nonopaque minerals are further divided into those that are *isotropic* (having the same properties in all directions) and those that are *anisotropic* (having different optical properties in different directions). Finally, the anisotropic minerals are divided according to whether they are *uniaxial* or

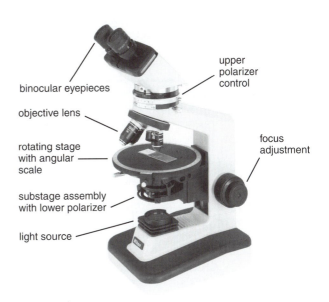

upper polarizer control

binocular eyepieces

objective lens

rotating stage with angular scale

substage assembly with lower polarizer

light source

focus adjustment

▶FIGURE 4.1

A polarizing microscope with main features labeled. From Nikon, Inc., Melville, New York. Photo used with permission.

biaxial, and according to whether they have a positive or negative *optic sign.* The details of these, and other diagnostic properties, will be discussed later.

Besides mineral identification, the polarizing microscope reveals important information about rock-forming processes (*petrogenesis*). When we use thin sections, distinguishing igneous, sedimentary, and metamorphic rocks is often easier than when we use hand specimens. More significantly, it is possible to identify minerals and distinguish among different types of igneous, sedimentary, and metamorphic rocks. The microscope allows us to see textural relationships in a specimen that give clues about when and how different minerals in the rock formed. Microscopic relationships between mineral grains allow us to determine the order in which minerals crystallized from magma, and we can identify minerals produced by alteration or weathering long after magma cooling. Similar observations are possible for sedimentary or metamorphic rocks. Only the microscope can give us

such information, information that is essential if rocks are to be used to interpret geological processes and environments.

WHAT IS LIGHT?

The Properties of Light

Before starting a discussion of optical mineralogy, it is helpful to take a closer look at light and its properties. Light is one form of *electromagnetic radiation* (Figure 4.2). Radio waves, ultraviolet light, and X rays are other forms of electromagnetic radiation. All consist of propagating (moving through space) electric and magnetic waves. The interactions between electric waves and crystals are normally much stronger than the interactions between magnetic waves and crystals (unless the crystals are metallic). Consequently, this book only discusses the electrical waves of light. In principle, however, much of the dis-

---

Box 4.1 *Optical Classification of Minerals*

Mineralogists often classify minerals according to the mineral's optical properties. The table below shows the basic classification scheme and gives examples of minerals belonging to each of six categories. At the highest level, we divide minerals into two groups: opaque minerals and nonopaque

minerals. We further divide the nonopaque minerals into those that are isotropic and anisotropic, and then we divide the anisotropic minerals by other properties discussed later in this book—all of which can be determined using a polarizing microscope (see Box 4.2).

Opacity	Isotropy	Optic Class	Optic Sign	Examples of Common Minerals
opaque				gold, copper, pyrite, pyrrhotite, magnetite, ilmenite
nonopaque	isotropic			garnet, diamond, halite, fluorite, periclase, spinel
	anisotropic	uniaxial	(+)	quartz, zircon, ice, brucite, rutile, leucite
			(−)	apatite, calcite, dolomite, beryl, tourmaline, corundum, nepheline
		biaxial	(+)	enstatite, diopside, sillimanite, gypsum, plagioclase, barite
			(−)	K-feldspar, muscovite, hornblende, plagioclase, fayalite, epidote

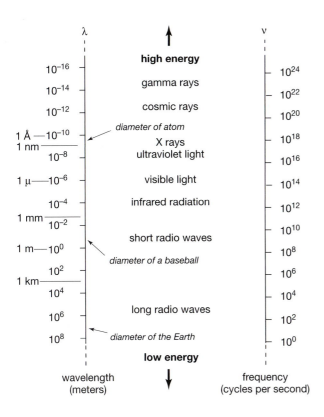

►FIGURE 4.2

The electromagnetic spectrum. Visible light is a form of electromagnetic radiation with wavelengths and energies that fall in the middle of the spectrum.

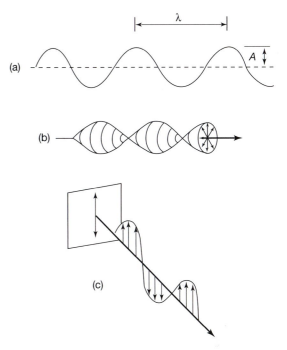

►FIGURE 4.3

Properties of light rays; (a) Different colors of light are characterized by different wavelengths (λ); the intensity of a wave is proportional to its amplitude (*A*); (b) the electric vectors of unpolarized light (arrows) vibrate in all directions perpendicular to the direction of travel; (c) the electric vectors of plane polarized light are constrained to vibrate in a plane.

cussion applies to the magnetic waves as well. Light waves, like all electromagnetic radiation, are characterized by a particular wavelength, λ, a frequency, ν, and a polarization state (Figure 4.3). The velocity, v, of the wave is the product of λ and ν:

$$\text{v} = \lambda \, \nu \qquad (4.1)$$

In a vacuum, the velocity of light is 3×10^8 meters per second. Light velocity is slightly less when passing through air, and can be considerably less when passing through crystals. When the velocity of light is altered as it passes from one medium (for example, air) to another (perhaps a mineral), the wavelength changes, but the frequency remains the same.

Visible light has wavelengths of 390 to 770 nanometers, which is equivalent to 3,900 to 7,700 Ångstroms, or $10^{-6.1}$ to $10^{-6.4}$ meters. Different wavelengths correspond to different colors of light (Figure 4.4). The shortest wavelengths, corresponding to violet light, grade into invisible ultraviolet radiation. The longest wavelengths, corresponding to red light, grade into invisible infrared radiation. Light composed of multiple wavelengths appears as one color to the human eye. If wavelengths corresponding to all the primary colors are present with nearly equal intensities, the

light appears white. White light is *polychromatic* (many colored), containing a range, or spectrum, of wavelengths. Polychromatic light can be separated into different wavelengths in many ways. When one wavelength is isolated, the light is *monochromatic* (single colored).

Interference

Besides λ and ν, an amplitude and a phase characterize all waves. *Amplitude* (*A*) refers to the height of a wave. *Phase* refers to whether a wave is moving up or down at a particular time. If two waves move up and down at the same time, they are *in phase;* if not, they are *out of phase.* When two waves interact, traveling in the same direction simultaneously, they interfere with each other. The nature of the interference depends on the relationships between their wavelengths, amplitudes, and phases. Light waves passing through crystals can have a variety of wavelengths, amplitudes, and phases that are affected by atomic structure in different ways. They yield interference phenomena, giving minerals distinctive optical properties.

In Figure 4.5a, two in-phase waves of the same wavelength are going in the same direction. If we could measure the intensity of the two waves together,

▶**FIGURE 4.4**
The wavelengths of visible light. The wavelength of violet light is about half that of red light. The boundaries between visible light and invisible radiation are not precisely defined, but visible light grades into ultraviolet radiation at short wavelengths and into infrared radiation at long wavelengths.

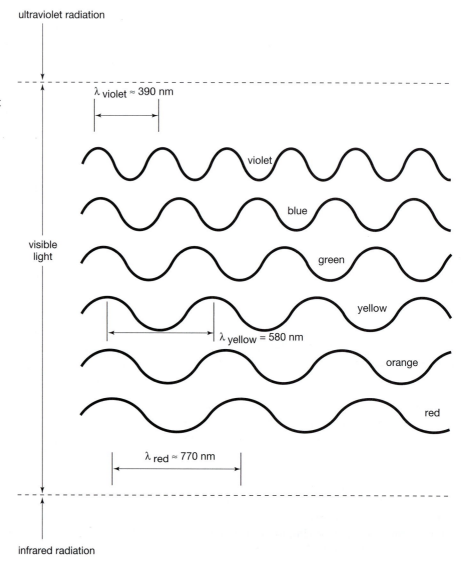

▶**FIGURE 4.5**
Waves in phase and out of phase, and the results when they combined: (a) Waves are in phase if their peaks and wavelengths correspond, so they interfere constructively to produce one wave with greater amplitude; (b) when waves are partially in phase, their peaks do not exactly correspond; so combination results in some loss of energy; (c) when waves are completely out of phase, they interfere destructively their motions cancel and addition leads to complete energy loss.

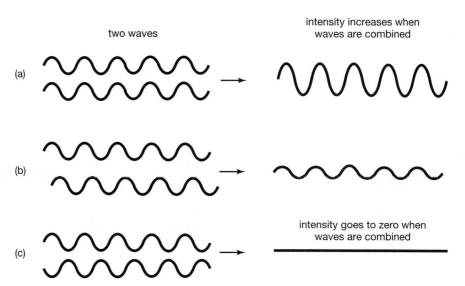

we would find that it is about twice the intensity of each individual wave. When waves are in phase, no energy is lost; this is *constructive interference.* In contrast, Figure 4.5b shows two waves that are partially out of phase, and Figure 4.5c shows two waves that are completely out of phase. When waves are out of phase, wave peaks and valleys do not correspond. If they are completely out of phase, the peaks of one wave correspond to the valleys of the other. Consequently, addition of out-of-phase waves can result in *destructive interference,* a condition in which the waves "consume" some or all of each other's energy. For perfect constructive or destructive interference to occur, waves must be of the same wavelength. Interaction of waves with different wavelengths is more complicated.

POLARIZATION OF LIGHT AND THE POLARIZING MICROSCOPE

Polarized Light

The vibration motion of a light wave is perpendicular, or nearly perpendicular, to the direction it is propagating. In normal unpolarized beams of light, waves vibrate in many different directions, shown by arrows in Figure 4.3b. However, we can filter or alter a light beam to make all the waves vibrate in one direction parallel to a particular plane (shown by arrows in Figure 4.3c). The light is then *plane-polarized,* sometimes called just *polarized.* Light becomes polarized in different ways. Reflection from a shiny surface can partially or completely polarize light because light

vibrating in planes parallel to the reflecting surface is especially well reflected, while light vibrating in other directions is absorbed. This is why sunglasses with polarizing lenses help eliminate glares.

Suppose light passes through a polarizing filter that constrains it to vibrate in a north-south (up-down) direction. The polarized beam, although perhaps decreased in intensity, appears the same to our eyes because human eyes cannot determine whether light is polarized. If, however, another polarizing filter is in the path of the beam, we can easily determine that the beam is polarized (Figure 4.6). If the second filter allows only light vibrating in a north-south direction to pass, the polarized beam will pass through it (Figure 4.6a). If we slowly rotate the second filter to an east-west direction, it will gradually transmit less light, and eventually no light (Figure 4.6b).

Polarizing Microscopes

Polarizing microscopes, also called *petrographic microscopes,* are in many respects the same as other microscopes (Figures 4.1 and 4.7). They magnify small objects so we can see them in greater detail. A bulb provides a white light source. The light passes through several filters and diaphragms before it reaches the stage and interacts with the material being observed. One of the most important filters is the lower polarizer, which ensures that all light striking samples on the stage is plane polarized (vibrating, or having wave motion, in only one plane). The presence of a lower polarizer sets polarizing microscopes apart from others. In most modern polarizing microscopes, the lower polarizer only allows light

▶**FIGURE 4.6**
Several small polarizing filters on top of a large polarizing sheet. The amount of light transmitted depends on the relative orientations of the polarization of the sheet and the small filter. (a) When polarization directions of the two are parallel, the maximum amount of light possible is transmitted; (b) when polarization directions are perpendicular, no light is transmitted. At other orientations, the two filters transmit intermediate amounts of light.

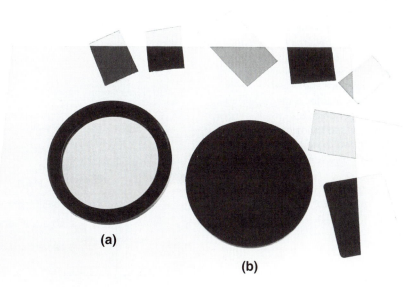

(a)

(b)

vibrating in an east-west direction to reach the stage. Older microscopes, however, have the lower polarizer oriented in a north-south direction. A fixed condensing lens and a diaphragm in the substage help concentrate light on the sample. For most purposes, we use *orthoscopic illumination,* in which an unfocused beam travels from the substage through the sample and straight up the microscope tube. The light rays travel orthogonal to the stage and to a sample or thin section on the stage. However, we can insert a special lens, a *conoscopic lens,* between the lower polarizer and stage to produce *conoscopic illumination* when needed (Figure 4.7). The conoscopic lens, also called a *condenser lens,* causes the light beam to converge (focus) on a small spot on the sample and illuminates the sample with a cone of nonparallel rays.

We can rotate the microscope stage to change the orientation of the sample relative to the polarized light. Because most minerals are anisotropic, the interaction of the light with a mineral varies with stage rotation. A calibrated angular scale allows us to make precise measurements of crystal orientation. The scale is also useful for measuring angles between cleavages, crystal faces, and twin orientations, and for measuring other optical properties.

Above the stage, a rotating turret holds several *objective lenses.* They usually range in magnification from about 2× to 50×. Different objective lenses can have different *numerical apertures* (N.A.), a value that describes the angles at which light can enter a lens, which is an important consideration when making certain measurements. *In the discussion of inter-ference figures later in this book, we have assumed that the objective lens being used has an N.A. of 0.85, since this is by far the most common today. If you use a lens with a different N.A., some of the given angular values may be in error.* The ocular, an additional lens usually providing 8× or 10× magnification, is in the eyepiece. Binocular microscopes, such as the one in Figure 4.1, have two eyepieces and two oculars. Oculars have cross hairs that aid in making angular measurements when we rotate the stage. The total magnification, the product of the objective lens magnification and the ocular magnification, varies from about 16× to 500×, depending on the lenses used.

We can insert several other filters and lenses between the objective lens and the ocular when needed (Figure 4.7). The *upper polarizer,* sometimes called the *analyzer,* is a polarizing filter oriented at 90° to the lower polarizer, which we can insert or remove from the path of the light beam. If no sample is on the stage, light that passes through the lower polarizer cannot pass through the upper polarizer. If a sample is on the stage, it usually changes the polarization of the light so that some can pass through the upper polarizer. We can also insert an *accessory plate* above the upper polarizer. The most common kind of accessory plate used today is called a "full wave" plate. In the past, all full wave plates were made of gypsum and are still often referred to as "gypsum plates," but today they are made of quartz. Above the accessory plate, most polarizing microscopes have a *Bertrand lens* and diaphragm. We use them with the *substage* conoscopic lens to view minerals in conoscopic illu-

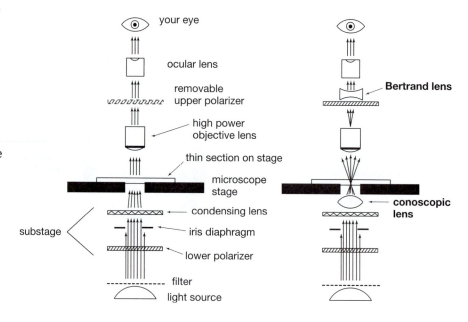

▶**FIGURE 4.7**
The most important components of a polarizing microscope. For normal orthoscopic illuminations, light from a bulb passes through a filter, the lower polarizer, a diaphragm, and a condensing lens in the substage before it hits the sample on the stage. Above the stage the objective and ocular lenses magnify and focus the light. The upper polarizer, Bertrand lens, and conoscopic lens are inserted to view a sample using conoscopic illumination.

Box 4.2 Identifying Minerals and Mineral Properties With a Polarizing Microscope

You will often find diagnostic properties or shortcuts that help speed mineral identification, but the process below covers all the key properties. You may wish to look at a number of different grains, and you should get used to changing back and forth between plane polarized (PP) and crossed polarized (XP) light. For routine mineral identification, it is often unnecessary to obtain an interference figure and identification can be quite rapid.

Look at the whole thin section (and at a hand specimen if available):
If you are looking at a thin section of a rock: What kind of rock is it? How many different major minerals does it contain? What are the associated minerals?

Examine several grains of the same mineral under PP light:
Is the mineral opaque or nonopaque?
What color is the mineral? Rotate the stage. Is it pleochroic? What is the color variation?
What is the crystal shape and habit?
What cleavage does it display, if any?
What is its apparent relief?

Now cross the polars to examine several grains of the mineral under XP light:
Rotate the stage. Is it isotropic or anisotropic?

If anisotropic:
Rotate the stage. What is the range of interference colors? Estimate the maximum birefringence by looking at grains with the highest order of interference colors.
For minerals with a long dimension of principal cleavage: What is the maximum extinction angle? Is it length-fast or length-slow?
Is the mineral twinned? If so, what kinds of twins?

If necessary, obtain an interference figure. You may have to try several different grains to get one that is useful:
Is the mineral uniaxial or biaxial?
What is its optic sign?

If biaxial:
Estimate *2V.*

mination, allowing us to make some special kinds of measurements.

Petrologists and mineralogists use polarizing microscopes with or without the upper polarizer inserted (Box 4.2). Without the upper polarizer, we see the sample in *plane polarized light* (*PP* light); with the upper polarizer, we see it in *crossed polars* (*XP* light). Grain size, shape, color, cleavage, and other physical properties are best revealed in PP light. The optical properties refractive index and pleochroism are also determined using PP light. We use XP light, sometimes focused with conoscopic and Bertrand lenses, to determine properties including retardation, optic sign, and *2V*. These properties are discussed in detail later in this book.

We examine minerals or rocks in *grain mounts* or in *thin sections* (Figure 4.8). For determining some mineral properties, a small amount of a powdered mineral sample is placed on a glass slide to produce a grain mount. The grains must be thin enough so that light can pass through them without a significant loss

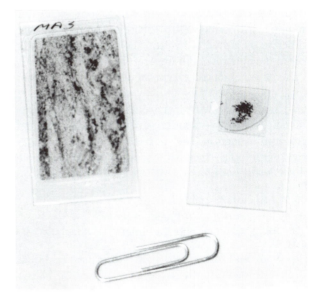

▶**FIGURE 4.8**
Thin section and grain mount.

of intensity, usually 0.10 to 0.15 mm in longest dimension. A small amount of liquid (often referred to as a *refractive index oil*) surrounds them, and a thin piece of glass, called a cover slip, is placed over the grains and liquid. Grain mounts and refractive index oils are absolutely necessary for making some types of measurements. Petrologists use thin sections, however, for routine mineral identification and other petrographic work. For more information about studying minerals in grain mounts, see the optical mineralogy texts listed in the references at the end of this chapter.

Colors in Plane Polarized (PP) Light and Crossed Polarized (XP) Light

In hand specimens, many minerals appear strongly colored, but minerals viewed with a microscope using PP light generally display a weak color or appear colorless. Many minerals in thin section or in grain mount are not thick enough to absorb significantly or enhance specific wavelengths of light. If minerals do appear colored, the color may change when we rotate the microscope stage because rotating the stage changes the orientation of the mineral's crystal structure with respect to the polarized light. Some minerals absorb different wavelengths of light depending on light vibration direction. We call this property *pleochroism*. Biotite is an example of a mineral that normally displays marked pleochroism (Plates 5.3 and 5.4).

Pleochroism is an especially useful diagnostic property when identifying some minerals, but it can be overlooked. In thin sections, orthopyroxenes are commonly colorless, but some show a faint pleochroism from pink to green. Pleochroism of pyroxenes is an important property because it distinguishes the two major pyroxene subgroups: orthopyroxene and clinopyroxene. For minerals with noticeable pleochroism, reference tables describe the property by listing colors seen when looking at the mineral in different directions. For pleochroic uniaxial minerals, color varies between two hues. For biaxial minerals, color varies between three hues.

In contrast with pyroxenes, many amphiboles display strong colors and a very noticeable pleochroism in thin section. The biaxial mineral glaucophane (an amphibole) has pleochroism described by its *pleochroic formula:*

X = colorless or pale blue
Y = lavender-blue or bluish green
Z = blue, greenish blue, or violet

X, Y, and Z refer to light vibrating parallel to each of three mutually perpendicular vibration directions in the crystal. In thin sections, glaucophane's colors vary within the limits described for X, Y, and Z, depending on the crystal orientation, as we rotate

the microscope stage. The biotite in Plates 5.3 and 5.4 is pleochroic in green and brown, but the standard pleochroic formula for biotite might be:

X = colorless, light tan, pale greenish brown, or pale green
Y ≅ Z = brown, olive brown, dark green, or dark red-brown

When we insert the upper polarizer, we see minerals in *crossed polarized* (*XP*) light, and we may see colors that are brighter and more pronounced than when we view the same grain in PP light. These are *interference colors*. They do not result from absorption of different wavelengths by the mineral (which is how minerals get their normal color). Instead, they result from the interference of light rays passing through the upper polarizer. They rarely resemble the true color of the mineral. Interference colors depend on grain orientation, so different grains of the same mineral in one thin section normally display a range of interference colors. Because different minerals can display different ranges of interference colors, interference colors are useful for mineral identification. Interference colors also vary with the thickness of the grains, so it is important that thin sections be of uniform thickness. Additionally, the edges of some grains, grains near the edge of a thin section, or grains adjacent to holes in a thin section (places where the sample is thin), may display abnormal interference colors.

THE VELOCITY OF LIGHT IN CRYSTALS AND THE REFRACTIVE INDEX

When electromagnetic radiation passes near an atom, the electric wave causes electrons to oscillate. The oscillations absorb energy from the light, and the wave slows down. A wave's velocity through a crystal is described by the crystal's refractive index (n), which depends on chemical composition, crystal structure, and bond type in the crystal. The refractive index (n) is the ratio of the velocity (v) of light in a vacuum to the velocity in the crystal:

$$n = v_{vacuum}/v_{crystal} \qquad (4.2)$$

Because light passes through a vacuum faster than through any other medium, n always has a value greater than 1. High values of n correspond to materials that transmit light slowly. Under normal conditions, the refractive index of air is 1.00029. Because it is much easier to work with air than with a vacuum, this is a common reference value.

As light passes from air into most nonopaque minerals, its velocity decreases by a third or a half.

Because the frequency of the light remains unchanged, we know that the wavelength must decrease by a similar fraction (Equation 4.1). Most minerals have refractive indices between 1.5 and 2.0. Fluorite, borax, and sodalite are examples of minerals that have a very low (<1.5) index of refraction. At the other extreme, zincite, diamond, and rutile have very high indices (>2.0). The refractive index is one of the most useful properties for identifying minerals in grains mounts but is less valuable when we examine minerals in thin sections.

The refractive index of most materials varies with the wavelength of light. In other words, the velocity of light in a crystal varies with the light's color. This property, called *dispersion,* is a property of minerals that can sometimes be seen in thin sections but is not discussed in detail in this book. An excellent but nonmineralogical example of dispersion is the separation of white light into colored "rainbows" when refracted by a glass prism. When a beam of white light enters a prism, different wavelengths (colors) are refracted at different angles, resulting in the production of the "rainbow." For a mineralogical example, we may consider diamond. Diamond's extreme dispersion accounts, in part, for the play of colors ("fire") that diamonds display. Minerals with low dispersion, such as fluorite, appear dull no matter how well cut or faceted. They may, however, be useful as lenses when dispersion causes unwanted effects.

A mineral's refractive index and dispersion profoundly affect its *luster.* Minerals with a very high refractive index and dispersion, such as diamond or cuprite, appear to sparkle and are termed *adamantine.* Minerals with a moderate refractive index, such as spinel and garnet ($n = 1.5 - 1.8$), may appear vitreous (glassy) or shiny, while those with a low refractive index, such as borax, will appear drab because they do not reflect or refract as much incident light. Refractive index depends on many things, but a high *n* value suggests minerals composed of atoms with high atomic numbers, or of atoms packed closely together.

Most minerals are *anisotropic,* so their refractive index varies with direction. In contrast, a glass, such as window glass or obsidian, is *isotropic* because it has a random atomic structure. Randomness means that, on the average, the structure and refractive index are the same in all directions. Isotropic minerals are relatively easy to spot in thin sections. When viewed with a polarizing microscope and XP light, they remain *extinct,* appearing black as the stage rotates, no matter what their orientation is on the microscope stage. There are few common isotropic minerals, but the most common are garnet, sphalerite, and fluorite (Appendix C). Sometimes thin sections contain holes, places with no mineral and

only epoxy. They appear isotropic and can occasionally be mistaken for isotropic minerals. Usually we can tell isotropic minerals apart by looking at color, relief, habit, and cleavage.

Anisotropic minerals normally do not appear extinct under XP light, but, as the microscope stage rotates, they go extinct briefly every 90°. However, if we orient an anisotropic crystal so that light passes through it parallel to a special direction called an *optic axis,* it will appear isotropic. It remains extinct when we rotate the stage. Fortunately, anisotropic minerals can only have one (*uniaxial minerals*) or two (*biaxial minerals*) optic axes, so the odds of the optic axis being exactly parallel to the light beam are small, and confusing isotropic and anisotropic minerals is rarely a problem. When in doubt, we can distinguish them using conoscopic illumination because anisotropic minerals will transmit some conoscopic light and display interference figures (discussed later), while isotropic minerals do not.

Snell's Law and Light Refraction

We have all seen objects that appear to bend as they pass from air into water. A straw in a glass of soda, or an oar in the water, seem bent when we know they are not. We call this phenomenon *refraction.* Refraction occurs when a beam of light passes from one medium to another with a different refractive index (Figure 4.9). If the light strikes the interface at an angle other than 90°, it changes direction. Consider a beam traveling from air into water (Figure 4.9a, b). The side of the beam that reaches the interface first will be slowed as it enters the water. The beam bends toward the water, the medium with a higher refractive index, because one side of the beam moves faster than the other. Figure 4.9c and 4.9d show the opposite case: a beam traveling from a medium with a high refractive index to another with a lower refractive index. The beam refracts toward the medium with a higher index, as in Figure 4.9a. A beam traveling at 90° to an interface, whether from a medium with a high refractive index to a low or vice versa, is not refracted at all.

The angle between the incoming beam and a perpendicular to the interface is the *angle of incidence* (θ_i). The angle between the outgoing beam and a perpendicular to the interface is the *angle of refraction* (θ_r). The relationship between the angle of incidence (θ_i) and the angle of refraction (θ_r) is

$$\sin \theta_i / \sin \theta_r = v_i/v_r = n_r/n_i \qquad (4.3)$$

where v_i and v_r are the velocities of light through two media, and n_r and n_i are the indices of refraction of the two media. This relationship, *Snell's Law,* is named after Willebrod Snell, the Dutch scientist who first derived it in 1621.

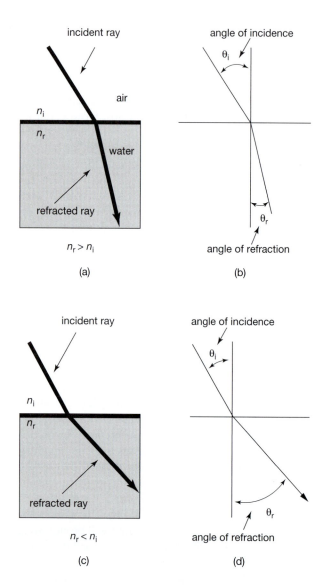

▶FIGURE 4.9
Refraction of a beam of light: (a) A light ray is bent as it crosses the boundary from air into water (or from any medium to another with higher refractive index); (b) the geometry of refraction shown in (a); (c) a light ray is bent as it crosses the boundary from one medium to another with lower refractive index: (d) the geometry of refraction shown in (c).

Rearranging Snell's Law tells us that we can calculate the angle of refraction:

$$\theta_r = \sin^{-1}[n_i/n_r \times \sin\theta_i] \qquad (4.4)$$

By definition, sin values can never be greater than 1.0. Suppose a light beam is traveling from a crystal into air. In this case, $n_i > n_r$ and, because the term in square brackets on the right-hand side of Equation 4.4 must be less than or equal to 1.0, for some large values of θ_i there is no solution. The limiting value of θ_i is the *critical angle* of refraction. If the angle of incidence is

greater, none of the light will escape; the entire beam will be reflected inside the crystal (Figure 4.10). This is the reason crystals with a high refractive index, such as diamond, exhibit internal reflection that gives them a sparkling appearance. Measuring the critical angle of refraction is a common method for determining refractive index of a mineral. Instruments called *refractometers* simplify such measurements.

Relief and Becke Lines

If we immerse an isotropic mineral grain in a liquid with the same refractive index, we will have difficulty seeing it unless it is one of the few minerals with very strong coloration. The edges of the mineral grain will not stand out. However, if a grain has an index of refraction that is significantly different from the liquid, light refracts and reflects at the edges of the grains. As the difference between the index of the liquid and the mineral increases, the boundary between the two becomes more pronounced (Plate 4.1). The term *relief* describes the contrast between the mineral and its surroundings (in this case, liquid). Grains with low relief are barely visible, while those with high relief stand out clearly (Figure 4.11; Plates 4.1 and 4.2).

Minerals in thin sections also show relief. The relief depends on the difference in the indices of refraction of the mineral and the material (today usually a special type of epoxy) in which it is mounted. As the difference in indices increases, relief becomes more noticeable. Minerals with high refractive indices show high ("positive") relief because their index of refraction is *greater* than that of the epoxy. They also tend to show structural flaws, such as scratches, cracks, or pits, more than those with low refractive indices. Some minerals (fluorite, for example) with very low refractive indices also show high relief (termed "negative" relief) because their index of refraction is *lower* than that of the epoxy. We do not differentiate between positive and negative relief in this book; for most purposes, we need only to know whether a mineral displays high, medium, or low relief. A few minerals (such as calcite) display variable relief with stage rotation; variable relief is a useful diagnostic property. We can see relief with either a monocular or a binocular microscope, but more easily with the latter.

As pointed out before, when we immerse a grain in liquid, some light rays bend toward the medium with the refractive higher index. Other light rays are completely reflected because they hit the mineral-liquid interface at an angle greater than the critical angle of refraction. The light interacts with the grain as if it were a small lens (Figure 4.12). If $n_{mineral} > n_{liquid}$, light rays are refracted and converge after passing through the grain. If $n_{mineral} < n_{liquid}$, light rays are refracted and diverge after passing through the grain. If we slowly lower the microscope stage, shifting the focus to a point

▶**FIGURE 4.10**
Refraction and reflection of a beam of light passing from a crystal into air for various angles of incidence. The solid rays show the incident and refracted beams; the dashed rays are the reflected beams. In (a), (b), and (c), the angle of incidence is less than the critical angle, so a refracted beam escapes the crystal. In (e) and (f), the angle of incidence is greater than the critical angle, so all the light is reflected back into the crystal.

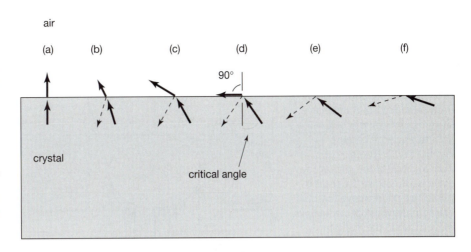

▶**FIGURE 4.11**
Minerals appear to have high relief when in a liquid if their index of refraction differs greatly from that of the liquid. Minerals have low relief, and may nearly disappear, if the mineral and the liquid have similar indices of refraction. This figure shows grossular (low relief, $n = 1.750$) and fluorite (high relief, $n = 1.430$) in a liquid with $n = 1.720$.

above the mineral grain, a bright narrow band of light called a *Becke line* appears at the interface and moves toward the material with higher refractive index (Figure 4.12; Plate 4.3). A complementary, but more difficult to see, dark band moves toward the material with lower refractive index. Although not as straightforward, we can also use Becke lines to compare the relief of minerals in thin sections by purposely focusing and defocusing the microscope while we examine a grain boundary. We also compare relief by noting how well a mineral appears to stand out above another.

INTERACTION OF LIGHT AND CRYSTALS

Double Refraction

In most modern polarizing microscopes, polarized light leaves the lower polarizer vibrating in the east-west direction. If it encounters an isotropic mineral on the stage, it slows as it passes through the mineral, but is still east-west polarized when it emerges. Upon entering an anisotropic crystal, however, light is normally split into two polarized rays, each traveling through the crystal along a slightly different path with a slightly different velocity and refractive index (Figure 4.13a). For uniaxial minerals, we call the two rays the *ordinary ray* (O ray), symbolized by ω, and the *extraordinary ray* (E ray), symbolized by ϵ'. The O ray travels a path predicted by Snell's Law, while the E ray does not. The O ray and E ray vibration directions depend on the direction the light is traveling through the crystal structure, but the vibration directions of the two rays are always perpendicular to each other (Figure 4.13b, c).

We call the splitting of a light beam into two perpendicularly polarized rays *double refraction*. All randomly oriented anisotropic minerals cause double refraction. We can easily observe it by placing clear calcite over a piece of paper on which a line, dot, or

▶**FIGURE 4.12**
Mineral grains behave like lenses when immersed in a liquid of different index of refraction, causing light to converge or diverge. In this figure, the fluorite grains (left) cause light to diverge and the grossular grains (right) cause light to converge because
$n_{fluorite} < n_{liquid} < n_{grossular}$.
Consequently, if the microscope stage is lowered so the focus is raised to a plane above the grains, narrow bright Becke lines move out into the liquid from the fluorite-liquid boundary, and in toward the center of the grain from the grossular-liquid boundary. When the stage is lowered, Becke lines always move toward the crystal or liquid with the greater index of refraction.

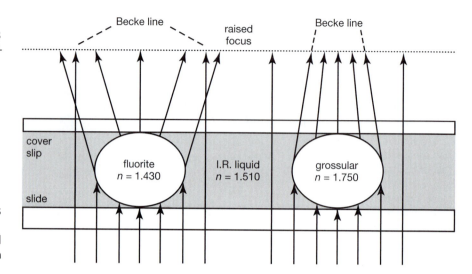

other image has been drawn (Figure 4.14). Two images appear, one corresponding to each of the two rays. A thin piece of polarizing film placed over the calcite crystal would verify that the two rays are polarized and vibrating perpendicular to each other. If we rotate either the film or the crystal, every 90° one ray becomes extinct, and we will see only one image. Calcite is one of the few common minerals that exhibits double refraction that is easily seen without a microscope, but even minerals that exhibit more subtle double refraction can be tested using polarizing filters. Gemologists use this technique to tell gems from imitations made of glass. Glass, like all isotropic substances, does not exhibit double refraction.

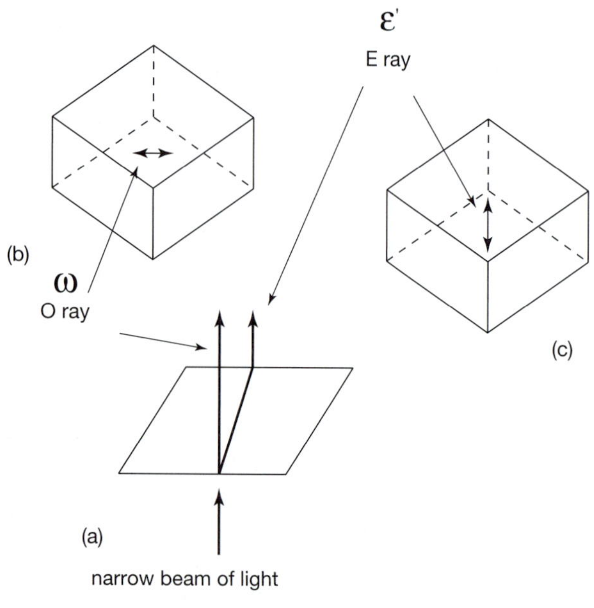

As the two rays pass through an anisotropic crystal, they travel at different velocities unless they are traveling parallel to an *optic axis*. We call the two rays the *slow ray* and the *fast ray*. Because the rays travel at different velocities, their refractive indices must be different. The difference in the indices of the fast ray and the slow ray $(n_{slow} - n_{fast})$ is the *apparent birefringence* (δ'). It varies depending on the direction light is traveling through the crystal and ranges from zero to some maximum value (δ) determined by the crystal structure. The maximum birefringence (δ) is a diagnostic property of minerals.

When the slow ray emerges from an anisotropic crystal, the fast ray has already emerged and traveled some distance. This distance is the *retardation*, Δ. Re-

▶**FIGURE 4.13**
Splitting of polarized light into an ordinary (O) and extraordinary (E) ray by calcite: (a) A side view showing that the paths of the two rays are different; (b) the O ray vibrates parallel to the long axis of the calcite rhomb. (c) the E ray vibrates perpendicular to the long axis of the calcite rhomb.

▶**FIGURE 4.14**
Calcite crystal showing double refraction.

tardation is proportional to both the thickness (t) of the crystal and to the birefringence in the direction the light is traveling (δ'):

$$\Delta = t \times \delta' = t \times (n_s - n_f) \qquad (4.5)$$

The birefringence and retardation of isotropic crystals are always zero. No double refraction occurs, and all light passes through isotropic crystals with the same velocity because the refractive index is equal in all directions. Most anisotropic minerals have birefringence between 0.01 and 0.20. Appendix D contains a list of minerals ordered by interference colors, which are a function of birefringence.

Crystals Between Crossed Polars

When viewed with the upper polarizer in place [under crossed polar (XP) light], we can differentiate isotropic and anisotropic crystals. Suppose we are viewing an isotropic crystal using XP light. It will remain dark through 360° of stage rotation. This is because the light emerging from the mineral retains the polarization it had on entering and will always be east-west polarized. It cannot pass through the upper polarizer, oriented at 90° to the lower polarizer. The effect is the same as if no mineral were on the stage.

When we view an anisotropic crystal with XP light, light is split into two rays unless we are looking down an optic axis. The two rays, after emerging from the crystal, travel on to the upper polarizer where they are resolved into one ray with north-south polarization. Because the vibration directions of both the rays are normally not perpendicular to the upper polarizer, components of both pass through the upper polarizer and combine to produce the light reaching our eye. As we rotate the microscope stage, however, the relative intensities of the two rays emerging from the crystal vary. Every 90°, the intensity of one is zero, and the other is vibrating perpendicular to the upper polarizer. Consequently, no light passes through the upper polarizer and the crystal appears extinct every 90°.

If we used a monochromatic light source in our microscope and looked at an anisotropic crystal under XP light, it would go from light to complete darkness as we rotate the stage. Extinction would occur every 90°, and maximum brightness would be at 45° to the extinction positions. However, most polarizing microscopes use polychromatic light. Because of dispersion, double refraction is slightly different for different wavelengths. Minerals with high dispersion may never appear completely dark, but most come close.

Interference Colors

When white light passes through an anisotropic mineral, all wavelengths are split into two polarized rays vibrating at 90° to each other. Different colors have different wavelengths, so when the rays leave the crystal, some colors may be retarded an even number of wavelengths, but most will not. Consequently, when the north-south components of the two rays are combined at the upper polarizer, constructive interference occurs for some colors, and destructive interference for others. If we look at a mineral of uniform thickness under XP light, we see one color, the *interference color*. Interference colors depend on the retardation of different wavelengths, which in turn depends on the orientation, birefringence, and thickness of a crystal. Interference colors change intensity and hue as we rotate the stage; they disappear every 90°, when the mineral goes extinct.

Normal interference colors are shown in a *Michel-Lévy Color Chart* (see Plate 4.11). Very low-order interference colors, corresponding to a retardation of less than 200 nm, are gray and white. The interference color of a mineral with very low birefringence, then, changes from white (or gray) to black every 90° as we rotate the microscope stage. For minerals with slightly greater birefringence, yellow, orange, or red interference colors will appear when we rotate the stage. These colors, corresponding to retardation of 200 nm to 550 nm, are called *first-order* colors. As retardation increases further, colors repeat every 550 nm. They go from violet to red (*second order*) and then from violet to red again (*third order*). They become more pastel (washed out) in appearance as order increases. Fourth-order colors are often so weak that they appear *"pearl" white* and may occasionally be confused with first order white. When describing an interference color, it is important to state both the color and the order; for mineral identification, the order is often more important than the color. The difference in retardation between "orders" is 550 nm, the average wavelength of visible light.

Plates 5.1 and 5.2 show calcite ($\delta = 0.172$) viewed under crossed polars; the interference colors are pastels of high order. In contrast, Plate 5.8 shows plagioclase ($\delta = 0.011$) with first-order gray and white interference colors. Other minerals in Plates 5.6 and 5.8 show interference colors between those of plagioclase and calcite. The interference colors in Plates 5.6 and 5.8 can be compared with the actual colors of the mineral grains viewed under PP light (Plates 5.5 and 5.7).

Appendix D lists minerals in order of their interference colors, often a key property for identifying minerals in thin section. Minerals with very low birefringence that display first-order white, gray, or yellow interference colors in thin section include leucite, nepheline, apatite, beryl, quartz, and feldspar. At the other extreme, minerals such as titanite (sphene), calcite, dolomite, and rutile display extreme retardation; interference colors are light pastels of high order. They may have such weak colors that it is hard to determine the retardation and birefringence with certainty.

Anisotropic minerals have different refractive indices depending on the path light travels when passing through them. Their optical properties, including birefringence, and thus interference colors, depend on their orientation. For identification purposes, the maximum birefringence (δ), corresponding to the highest order interference colors, is diagnostic. This may be hard to estimate in grain mounts because mineral thickness varies, making it difficult or impossible to estimate birefringence from interference colors. In thin sections, the task is easier because thickness is known (0.03 mm), so we can use the Michel-Lévy Chart to determine birefringence from interference color. Randomly oriented mineral grains may not show maximum interference colors; often we must rotate the stage and look at many grains of the same mineral. Because it is difficult to be exact, we normally use qualitative terms such as "low," "moderate," "high," or "extreme" to describe retardation and birefringence.

Some minerals have *anomalous interference colors,* colors that are not represented on the Michel-Lévy Color Chart. Anomalous interference colors may result if minerals have highly abnormal dispersion, if they are deeply colored, or for a number of other reasons. Minerals that commonly display anomalous inference colors include chlorite, epidote, zoisite, jadeite, tourmaline, and sodic amphiboles.

Uniaxial and Biaxial Minerals

Isotropic minerals have the same light velocity and therefore the same refractive index (n) in all directions. This is not true for anisotropic minerals, whether they are uniaxial or biaxial. For uniaxial minerals, we need two indices of refraction (ϵ and ω) to describe the mineral's refractive index. For biaxial minerals, we need three (α, β, and γ).

Optic axes are directions that light can travel through a crystal without being split into two rays (Figure 4.15). In some *uniaxial minerals,* the optic axis is parallel or perpendicular to crystal faces; in *biaxial minerals,* the two optic axes rarely are. Light traveling

▶**FIGURE 4.15**
Sketches of crystals showing orientation of optic axes:
(a) Anatase, TiO_2, uniaxial ($-$); (b) xenotime, YPO_4, uniaxial ($+$); (c) aragonite, $CaCO_3$, biaxial ($-$); (d) diaspore, $AlO(OH)$, biaxial ($+$)

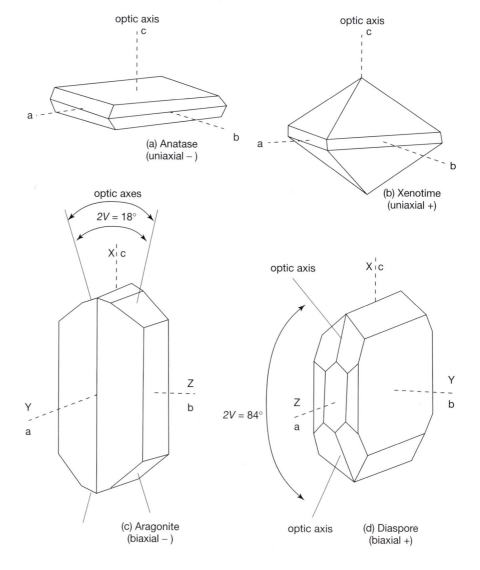

(a) Anatase (uniaxial $-$)

(b) Xenotime (uniaxial $+$)

$2V = 18°$

(c) Aragonite (biaxial $-$)

$2V = 84°$

(d) Diaspore (biaxial $+$)

▶FIGURE 4.16
Geometric relationships between X, Y, Z, optic plane, optic axes (OA) and *2V* in biaxial positive and negative crystals. (a) In positive crystals, the acute angle between the optic axes is bisected by Z. (b) In negative crystals, the acute angle between the optic axes is bisected by X.

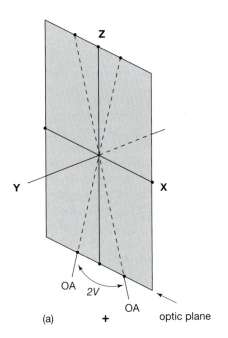

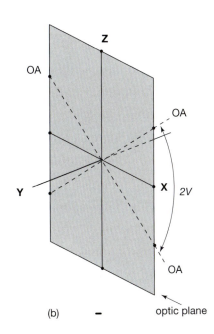

parallel to the single optic axis of a uniaxial mineral travels as an ordinary ray and has a unique refractive index, designated ω. Light traveling in other directions is doubly refracted, splitting into two rays with one having refractive index ω. The other has refractive index ϵ', which varies depending on the direction of travel. ϵ' may have any value between ω and ϵ, a limiting value corresponding to light traveling perpendicular to the optic axis. If $\omega < \epsilon$, the mineral is uniaxial positive ($+$). If $\omega > \epsilon$, the mineral is uniaxial negative ($-$). We sometimes use the mnemonics POLE (positive = omega less than epsilon) and NOME (negative = omega more than epsilon) to remember these relationships. The maximum possible value of birefringence in uniaxial crystals, δ, is $|\omega - \epsilon|$ (Table 4.1). We can only see maximum birefringence if the optic axis is parallel to the microscope stage.

Most minerals are biaxial, having two optic axes (Figures 4.15c and 4.15d). Light passing through a *biaxial crystal* experiences double refraction unless it travels parallel to an optic axis. We describe the optical properties of biaxial minerals in terms of three mutually perpendicular directions: X, Y, and Z (Figure 4.16). The vibration direction of the fastest possi-

ble ray is designated X, and that of the slowest is designated Z. The indices of refraction for light vibrating parallel to X, Y, and Z are α, β, and γ. α is therefore the lowest refractive index, and γ is the highest. β, having an intermediate value, is the refractive index of light vibrating perpendicular to an optic axis.

Normally, light passing through a randomly oriented biaxial crystal is split into two rays, neither of which is constrained to vibrate parallel to X, Y, or Z, so their refractive indices will be some values between α and γ. However, if the light is traveling parallel to Y, the two rays have refractive indices equal to α and γ, vibrate parallel to X and Z, and the crystal will display maximum retardation. If the light travels parallel to an optic axis, no double refraction occurs, and it has a single refractive index, β. There is no birefringence or retardation, and the mineral appears extinct.

In biaxial minerals, we call the plane that contains X, Z, and the two optic axes the *optic plane* (Figure 4.16). The acute angle between the optic axes is *2V*. A line bisecting the acute angle must parallel either Z (in *biaxial positive* crystals) or X (in *biaxial negative* crystals). In biaxial positive minerals, the intermediate refractive index β is closer in value to α

▶TABLE 4.1
Indices of Refraction and Birefringence for Light Passing Through Isotropic and Anisotropic Minerals

	Principal Indices of Refraction	Index of Refraction for Light Traveling Parallel to an Optic Axis	Indices of Refraction in a Random Direction	Birefringence in a Random Direction	Maximum Possible Birefringence
Isotropic crystals	n	n	n	0	0
Uniaxial crystals	ω, ϵ	ω	ω, ϵ'	$\delta' = |\omega - \epsilon'|$	$\delta = |\omega - \epsilon|$
Biaxial crystals	α, β, γ	β	α', γ'	$\delta' = \gamma' - \alpha'$	$\delta = \gamma - \alpha$

Box 4.3 Determining the Extinction Angle and the Sign of Elongation

Viewed with crossed polars, anisotropic grains go extinct every 90° as we rotate the microscope stage. We can measure the *extinction angle,* the angle between a principal cleavage or direction of elongation and extinction (Figure 4.17). Minerals with cleavages that exhibit *parallel extinction* go extinct when their cleavages or directions of elongation are parallel to the upper or lower polarizer (Figure 4.17a). Many monoclinic and all triclinic crystals exhibit *inclined extinction* and go extinct when their cleavages or directions of elongation are at angles to the upper and lower polarizer (Figure 4.17b). Some minerals exhibit *symmetrical extinction;* they go extinct at angles symmetrical with respect to cleavages or crystal faces (Figure 4.17c).

Because an extinction angle depends on grain orientation, determining an extinction angle for minerals in thin sections requires measurements on a number of different grains, or on one grain in the correct orientation (determined by looking at interference figures, discussed later). If the grains are randomly oriented in the thin section and if the sample size is large enough, the maximum value of the measurements should approximate the actual extinction angle of the mineral.

Some anisotropic crystals have a prismatic habit or a well-developed cleavage that causes them to break into elongated fragments. Polarized light passing through an anisotropic prismatic crystal with polarization parallel to the long dimension

▶**FIGURE 4.17**
Extinction of minerals viewed with XP light in this section: (a) Parallel extinction of minerals such as orthopyroxene occurs every 90° when the long axis of grains or the cleavage are perpendicular to either polarizer; (b) inclined extinction in many minerals, including most micas, occurs when the cleavage is at an angle to both polarizers; (c) symmetrical extinction of minerals, such as calcite, occurs when cleavages are oriented symmetrically to cross hairs and polarizers.

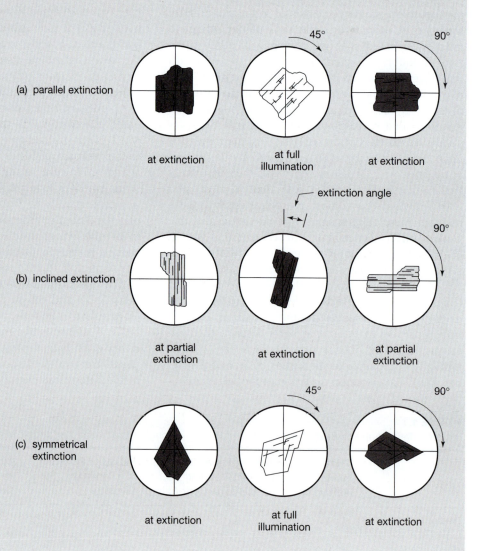

will not travel at the same velocity as light polarized in other directions. This distinction allows the prismatic minerals to be divided into two *signs of elongation:* "*length fast*" (faster light vibrates parallel to the long dimension) and "*length slow*" (slower light vibrates parallel to the long dimension). The sign of elongation cannot be determined on anisotropic crystals that cleave to produce equidimensional fragments.

Determining the sign of elongation (length fast or length slow) is usually straightforward for tetragonal and hexagonal prismatic crystals (all of which are uniaxial). We orient the long dimension of a grain in a southwest-northeast direction (45° to the lower polarizer) and note the interference colors. Many minerals, especially if grains are small, exhibit low first order interference colors (grays). We can insert a full wave accessory plate (having a retardation of 550 nm, equivalent to first-order red interference colors). After insertion, the slow direction of most accessory plates will be oriented southwest-northeast. If gray interference colors are added to first-order red, first-order blue results. If gray interference colors are subtracted from first-order red, first-order yellow results. For a grain oriented in the southwest-northeast direction, addition or subtraction is often just a matter of looking for blue or yellow. If we see blue or other higher order interference colors when the plate is inserted, the mineral is length slow. If we see yellow or other lower order colors, it is length fast (Figure 4.18).

If the interference colors for a mineral grain are not mostly gray, determining addition or subtraction may not be quite so simple. Sometimes it will be necessary to rotate the stage 90° to see the colors that appear when the mineral is oriented northwest-southeast. The effects in that orientation will be opposite to those seen when the mineral is oriented southwest-northeast. When oriented northwest-southeast, higher order colors correspond to length fast, lower order to length slow. A quartz wedge can be useful in determining the sign of elongation if a grain contains several color bands rather than just gray. As the wedge is inserted into the accessory slot, the color bands on a southwest-northeast-oriented grain will move toward the thicker portions of the grain (usually the center of the grain) if the retardation is being subtracted (length fast). If the bands move away from the thicker portions of the grain, the retardation is being added (length slow).

Determining the sign of elongation for an orthorhombic, monoclinic, or triclinic mineral can be problematic or impossible. We can sometimes determine it for orthorhombic or monoclinic minerals with parallel extinctions, but it may vary with the orientation of the mineral. Mineral identification tables sometimes list whether a given mineral is likely to provide a sign of elongation and whether the sign may vary with the mineral's orientation. If the extinction angle of the monoclinic or triclinic grain is only a few degrees, we can often determine a sign of elongation. If they have inclined extinction, we often cannot because the sign may vary in a complicated way with the orientation of the grain, and because determining the orientation of the grain on the microscope stage is difficult.

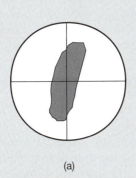   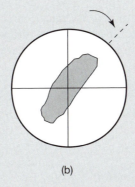   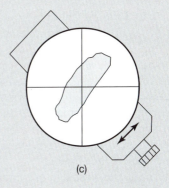

(a)   (b)   (c)

▶**FIGURE 4.18**
Determining the sign of elongation: (a) An elongated grain at extinction position (XP light); (b) the grain is rotated until it is at 45° to the cross hairs and oriented southwest-northeast: (c) the accessory plate (slow direction marked with a double arrow) is inserted to determine whether the interference colors add or subtract. If they add, producing blue or other higher order interference colors, the grain is "length slow": if they subtract, producing yellow or other lower order interference colors, the grain is "length fast."

than to γ. In biaxial negative minerals, it is closer in value to γ. Retardation and *apparent birefringence* vary with the direction light travels through a crystal, but the maximum possible value of birefringence (δ) in biaxial crystals is always $\gamma - \alpha$.

Accessory Plates and the Sign of Elongation

Polarizing microscopes have *accessory plates* we can insert above the objective lens. When inserted, the slow and fast vibration directions of the plate are at 45° to the lower and upper polarizers. A double-headed arrow on accessory plates usually marks the slow direction. A standard full wave plate has a retardation of 550 nm (equal to the average wavelength or visible light), equivalent to first-order red interference colors. A quartz wedge is sometimes a useful alternative to a full wave plate. The wedge has a variable thickness, with retardation ranging from 0 nm to 3,500 nm.

By inserting a plate when we are viewing a crystal on the microscope stage, we can add or subtract retardation. Accessory plates make it possible to learn which vibration direction in the crystal permits polarized light to travel the fastest. If crystals have a long dimension, we can learn whether the mineral is "*length fast*" (also sometimes called "negative elongation") or "*length slow*" ("positive elongation"). Determining the sign of elongation is often straightforward and can be helpful when identifying a mineral (Box 4.3).

Uniaxial Interference Figures

Optic sign (positive or negative) is another useful characteristic for identifying anisotropic minerals. The easiest way to learn whether a uniaxial mineral is positive or negative is to examine an interference figure. Examples are shown in Plates 4.5 through 4.10. We obtain interference figures by passing conoscopic light through a mineral (Box 4.4). The *conoscopic lens* focuses light into the crystal from many different converging directions. After the light leaves the crystal and passes through the upper polarizer, we can insert a *Bertrand lens* to refocus the rays and magnify the interference figure.

To determine the optic sign of a uniaxial mineral, it is best to look down, or nearly down, the optic axis (Box 4.5). The figure obtained is an *optic axis figure* (OA figure). Finding grains that give an OA figure is normally not difficult. Grains oriented with the optic axis vertical appear isotropic because they have no retardation when being viewed down an optic axis. Grains oriented with the optic axis close to vertical have low birefringence and, therefore, low-order in-

terference colors. An ideal uniaxial OA figure has a black cross that does not move much when the stage rotates (Figure 4.19a–c; Plates 4.8 and 4.9). Even if the cross is somewhat off center, we can use it to determine the optic sign. The center of the cross, called the *melatope,* corresponds to the direction of emergence of the optic axis. We call the dark bands forming the cross *isogyres.* The surrounding colored rings, if present, are *isochromes* (Plate 4.8). They are bands of equal retardation caused by the light entering the crystal at slightly different angles. Minerals with low birefringence, like quartz, may not show isochromes (Plate 4.9). When viewing an OA figure, we use an accessory plate to learn whether a uniaxial mineral is positive or negative (Box 4.5).

Most uniaxial mineral grains do not exhibit a perfectly centered OA figure when viewed in conoscopic light. The optic axis is only one direction in the crystal and grains are unlikely to be oriented with the optic axis vertical. If we choose a random grain, we typically get an off-center figure (Figure 4.19d–g). If the optic axis of a grain lies parallel to the stage of the microscope, we get an *optic normal figure,* also called a *flash figure.* The flash figure appears as a vague cross or blob that nearly fills the field of view when the grain is at extinction (when the optic axis is perpendicular to one of the polarizers). Upon stage rotation, it splits into two curved isogyres flashing in and out of the field of view with a few degrees of stage rotation. Plate 6d–g shows a diffuse biaxial obtuse bisectric figure (Bxo). Uniaxial flash figures have the same general appearance.

Biaxial Interference Figures

We obtain *biaxial interference figures* in the same way as uniaxial interference figures. However, complications arise with biaxial minerals because it is more difficult to find and identify grains oriented in a useful way. We can get interference figures from all grains, but interpreting them can be difficult or impossible. Four types of interference figures are commonly identified (Box 4.6, page 84). When we observe an acute bisectric figure (Bxa) for a grain in an extinction orientation (Y perpendicular to one of the polarizers), it appears as a black cross, similar in some respects to a uniaxial interference figure. When we rotate the stage, the cross splits into two isogyres that move apart and may leave the field of view (Figure 4.20; Plates 4.5, 4.6, and 4.7). After a rotation of 45°, the isogyres are at maximum separation; they come back together to reform the cross with further stage rotation. The maximum amount of isogyre separation depends on *2V* (Box 4.7, page 85). If *2V* is less than about 60°, the isogyres stay in the field of view as we rotate the stage. If *2V* is

Box 4.4 Obtaining an Interference Figure

Any uniaxial or biaxial mineral (whether in a grain mount or a thin section) will, in principle, produce a visible interference figure; isotropic minerals will not. Care must be taken to choose grains without cracks or other flaws so light can pass through without disruption. In addition, for some purposes it is necessary to find grains with a specific orientation.

Having chosen an appropriate grain, obtaining an interference figure is relatively straightforward. Carefully focus the microscope using PP light and high magnification. (If perfect focus is ambiguous, it may help to focus first at low magnification.) Insert the upper polarizer to get XP light and, if the microscope is properly aligned, the grain will still be in focus. Fully open the substage diaphragm, insert the substage conoscopic lens, and then insert

the Bertrand lens above the upper polarizer. You will now see an interference figure, perhaps similar to one of those shown in Plates 4.5 to 4.10.

The conoscopic lens focuses light so it enters the crystal from many different angles simultaneously. The Bertrand lens focuses the light so it is parallel again when it reaches the eyepiece. In effect, the two lenses together permit the examination of light traveling through a crystal in many different directions. Without the lenses, we would have to look at many different crystals of known optical orientation to obtain the same information. Some older microscopes do not have Bertrand lenses, but interference figures can still be obtained by removing an ocular and inserting a peep sight, or by just peering down the tube. The figures, however, will be quite small.

▶**FIGURE 4.19**
Uniaxial interference figures: In (a), (b), and (c) the figure is centered; in (d), (e), (f), and (g) it is off center but the cross is visible. In (a), the small crosses show the vibration directions of ϵ' and ω; ϵ' always vibrates radially and ω vibrates perpendicular to ϵ'. In (b), one set of isochromes is shown, while in (c) there are two sets due to higher birefringence. (d) through (g) show how the melatope (the center of the cross) of an off-center figure precesses as the microscope stage is rotated. See also Plates 4.8 and 4.9.

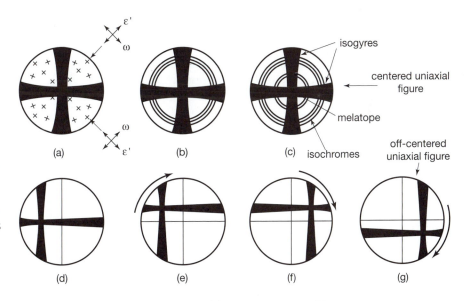

▶**FIGURE 4.20**
Biaxial acute bisectrix (Bxa) interference figures for a biaxial ($+$) crystal: In (a) the crystal is oriented with X and Y parallel to the polarizers; in (b), (c), and (d) the stage has been rotated 45° so that X (the trace of the optic plane) is oriented northwest-southeast and the isogyres are at maximum separation. The amount of separation is proportional to *2V*. Similar relationships apply to biaxial ($-$) crystals except that X and Z are interchanged.

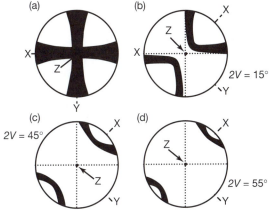

Box 4.5 Determining the Optic Sign of a Uniaxial Mineral

To determine the optic sign of a uniaxial mineral, it is necessary to know whether $\omega > \epsilon$ or $\epsilon > \omega$. We do this by examining an optic axis figure and using an accessory plate with known orientation of the fast and slow rays. Standard accessory plates have their slow direction oriented southwest-northeast at 45° to both polarizers. The plate is inserted, and we observe interference color changes or isochrome (color ring) movements in the southwest and northeast quadrants of the interference figure (Figure 4.21).

Uniaxial optic axis figures appear as black crosses (Figures 4.19, 4.21; Plates 4.8, 4.9). In all parts of a uniaxial optic axis interference figure ϵ' vibrates radially (along the radius of the interference figure) and ω vibrates tangent to isochromes (see Figure 4.19a). One way to remember these relationships is with the mnemonic WITTI (ω is tangent to isochrome). If interference colors in the southwest and northeast quadrants shift to higher orders when the plate (or wedge) is inserted, addition of retardation has occurred. ϵ' is the slow ray and the crystal is uniaxial positive. Subtraction (lower-order interference colors) in the southwest and northeast quadrants indicates that ϵ' is the fast ray, and the crystal is optically negative. Because ϵ' vibrates parallel to the prism axis (*c*-axis) in prismatic uniaxial minerals, the optic sign is the same as the sign of elongation.

Although we could use a quartz wedge, we normally use a full wave accessory plate when the optic axis figure shows low-order interference colors. In positive crystals, addition produces blue in the southwest and northeast quadrants of the interference figure, while subtraction produces yellow in the northwest and southeast quadrants. In

negative crystals, the effect is the opposite. If several different color rings (isochromes) are visible, the blue and yellow colors will only appear on the innermost rings near the melatope (the center of the black cross).

We can sometimes use a full wave plate, but normally use a quartz wedge, to determine an optic sign for minerals that exhibit high-order interference colors. The retardation of the quartz will add to the retardation of the unknown mineral in two quadrants as the wedge is inserted. It will also subtract from the retardation in the other two. The result will be color rings (isochromes) moving inward in quadrants where addition occurs, and outward in quadrants where subtraction occurs (Figure 4.21). For positive minerals, this means that colors move inward in the southwest and northeast quadrants and outward in the northwest and southeast quadrants. For negative minerals, the motion is opposite.

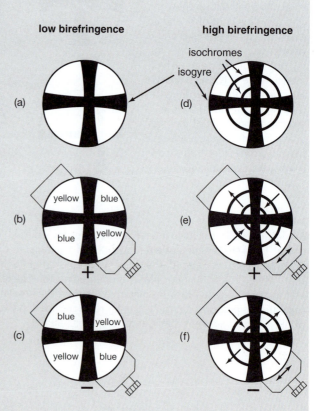

▶**FIGURE 4.21**
Determining the sign of a uniaxial mineral: (a) A slightly off-center uniaxial interference figure showing no isochromes (low birefringence); (b and c) when we insert the accessory plate yellow and blue colors seen in different quadrants allow us to determine whether the mineral is positive or negative; (d) a slightly off-center uniaxial interference figure with isochromes (due to high birefringence); (e and f) when we insert a quartz wedge the isochromes move in and out in different quadrants depending on optic sign.

greater than 60°, the isogyres completely leave the field of view. Plate 4.10 shows an interference figure for biotite; *2V* is 20–25°.

The points on the isogyres closest to the center of a Bxa or Bxo, the *melatopes,* are points corresponding to the orientations of the optic axes (Figure 4.24) of different colors. If the retardation of the crystal is great, *isochromes* circle the melatopes; none are apparent in Plate 4.10. Interference colors increase in order moving away from the melatopes because retardation is greater as the angle to the optic axes increases.

The isogyres in a Bxo figure always leave the field of view because, by definition, an obtuse angle separates the optic axes in the Bxo direction. Thus, if the isogyres remain in view, the figure is a Bxa figure. For standard lenses, if they leave the field of view, the figure may be a Bxo or a Bxa for a mineral with high *2V* (greater than about 60°). Getting a perfectly centered Bxo or Bxa figure is difficult. It may only be possible to see one isogyre clearly. Experienced mineralogists can tell a Bxa from a Bxo figure by the speed with which the isogyre leaves the field of view on rotation. For the rest of us, it is probably best to search for another grain with a better orientation.

Determining optic sign from a Bxa figure is equivalent to asking whether the Bxa corresponds to the fast direction (biaxial negative crystals) or to the slow direction (biaxial positive crystals). We can make the determination in much the same manner as for a uniaxial interference figure (Box 4.7, page 87). However, for figures where the isogyres leave the field of view, this is not the recommended way to learn optic sign because Bxa and Bxo figures are so easily confused when *2V* is greater than 70°–80°.

An alternative way to determine optic sign and *2V* is to find a grain that yields a centered optic axis figure (Box 4.8). Finding such grains is often not difficult because they show very low-order interference colors, or may appear isotropic under crossed polars.

OTHER MINERAL CHARACTERISTICS IN THIN SECTIONS

Besides the optical properties already discussed, we can use several other mineral characteristics to aid mineral identification. These characteristics give certain minerals a distinctive property in thin section. An experienced microscopist can, for example, often identify amphibole because of its cleavage and plagioclase feldspar because of its twinning. Besides cleavage and twinning, other important characteristics include alteration, compositional zonation, exso-

lution, anomalous extinction, or the presence of inclusions. They are, however, beyond the scope of this book.

Cleavage

Many minerals exhibit *cleavage*; when it can be seen with a microscope, it can be an important diagnostic tool. We use qualitative terms such as *perfect, good, fair,* and *poor* to describe the ease with which a mineral cleaves in different directions. Cleavage appears as fine parallel cracks in mineral grains when viewed with a microscope. Minerals with one or more good or perfect cleavages can be expected to show cleavage most of the time, while those with only poor cleavage may not. Additionally, minerals with low relief do not show cleavage as readily as those with high relief. This problem can be overcome somewhat by closing down the substage diaphragm, which narrows the cone of light hitting the thin section and increases contrast.

Minerals may have zero, one, two, three, four, or even more cleavages, but, because thin sections provide a view of only one plane through a mineral grain, we rarely see more than three at a time. Minerals that have elongate habits generally exhibit different cleavage patterns when viewed in a cross section than they do when viewed in a longitudinal section. Amphibole, for example, shows two good cleavages intersecting at about 60° and 120° in a cross section, but only one good cleavage in a long section. The number of different cleavages and the angles between them aid mineral identification. Hornblende and other amphibole cross sections often show a typical diamond cleavage pattern (Plate 5.5), which serves to distinguish amphiboles from other similar minerals. However, it is important to remember that cleavage angles depend on grain orientation. If a mineral has two cleavages at 60° to each other, the cleavages will appear to intersect at any angle from 0° to 60° depending on grain orientation. So we must often examine many grains (or one with a known orientation determined by examining an interference figure) to determine the maximum, and true, cleavage angle.

Twinning

Many minerals *twin,* and sometimes we can see the twins with a microscope (see the feldspars in Plate 5.8, for example). They manifest themselves as different regions of a grain that have different crystallographic orientations, so they do not go extinct at the same time when the microscope stage is rotated. *Contact twin* domains are separated by a sharp line, the trace of the twin plane. *Penetration twins* generally have

Box 4.6 The Four Kinds of Oriented Biaxial Interference Figures

Biaxial interference figures are more difficult to interpret than uniaxial interference figures. With care, however, it is possible to identify three different kinds of useful figures: optic axes, acute bisectrix, and optic normal. Each corresponds to a different light path through the crystal, relative to the orientation of the optic axes (Figure 4.22). A fourth type of figure (Bxo) is described below for completeness, but Bxo figures are generally not useful.

Optic axis figure (OA): We obtain an OA figure by looking down an optic axis (Figures 4.22 and 4.23). We can locate grains oriented to give an OA figure because they have zero or extremely low retardation. Under orthoscopic XP light, they remain dark even if the microscope stage is rotated. An interference figure will show only one isogyre unless *2V* is quite small (less than 30°). We can use the curvature of the isogyre to estimate *2V* (see Box 4.8; Figure 4.23).

Acute bisectrix figure (Bxa): We obtain Bxa figures, such as those in Figures 4.21 and 4.24 and Plates 4.10, by looking down the *acute bisectrix*, the line bisecting the acute angle between the two optic axes (Figure 4.22). In biaxial positive crystals, Bxa corresponds to a view along Z; in biaxial negative crystals, it corresponds to a view along X. For small values of *2V*, the isogyres form a black cross similar to a uniaxial interference figure when Y is parallel

to the lower polarizer or upper polarizer. On rotation of the stage, the isogyres will split into hyperbolas, reforming the cross every 90° (Figure 4.24). For *2V* values greater than about 60°, the isogyres will completely leave the field of view before coming back together to re-form the cross.

Optic normal figure: An ON is the figure obtained when looking down Y, normal to the plane of the two optic axes (Figure 4.22). Grains that yield an optic normal figure are those that have maximum retardation. The interference figure resembles a poorly resolved Bxa, but the isogyres leave the field of view with only a slight rotation of the stage. Biaxial optic normal figures appear similar to uniaxial flash figures.

Obtuse bisectrix figure (Bxo): Bxo figures are generally not useful for mineral identification, but they can be confused with other, more useful figures. A Bxo is the figure obtained when looking down the *obtuse bisectrix*, a line bisecting the obtuse angle between the optic axes (Figure 4.22). A Bxo looks superficially like a Bxa, but the isogyres will always leave the field of view on a stage rotation (because the angle between the optic axes is greater than 90° along the Bxo; see Plate 6d–g). Distinguishing a Bxo from a Bxa is difficult or impossible for large values of *2V* (greater than 70°). Bxo figures are also easily confused with ON figures in minerals with small *2V*.

▶**FIGURE 4.22**
Relative orientations of optic axis (OA), Bxa, Bxo, and optic normal (ON) directions in a biaxial (+) crystal: (a) schematic showing optic axes and optic plane; (b) drawing of a crystal in a similar orientation. In biaxial (−) crystals the relationships are similar but the X and Z axes are interchanged.

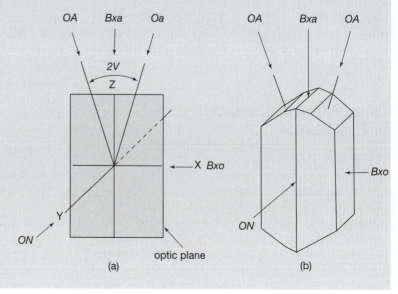

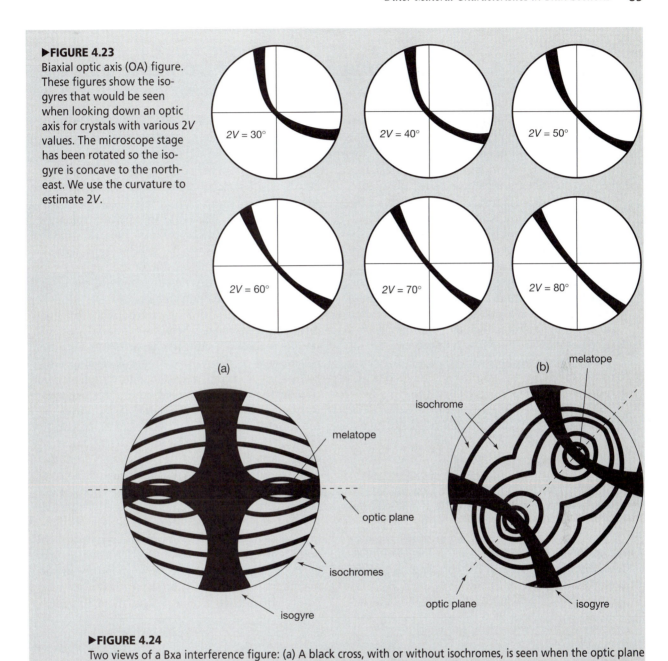

▶**FIGURE 4.23**
Biaxial optic axis (OA) figure. These figures show the isogyres that would be seen when looking down an optic axis for crystals with various 2*V* values. The microscope stage has been rotated so the isogyre is concave to the northeast. We use the curvature to estimate 2*V*.

$2V = 30°$ $2V = 40°$ $2V = 50°$

$2V = 60°$ $2V = 70°$ $2V = 80°$

(a) (b)

melatope

isochrome

melatope

optic plane

isochromes

optic plane isogyre

isogyre

▶**FIGURE 4.24**
Two views of a Bxa interference figure: (a) A black cross, with or without isochromes, is seen when the optic plane is parallel to a polarizer; (b) the cross separates into two curved isogyres when the optic plane is parallel to neither polarizer. The optic plane is oriented northeast-southwest. See also Plate 4.10.

irregular domain boundaries. Simple twins consist of two individual *domains*, but *lamellar twins* are characterized by multiple parallel bands called *twin lamellae*. The plagioclase in Plate 5.8 shows well-developed twin lamellae. Some minerals, such as plagioclase, exhibit *polysynthetic twinning* caused by many parallel twin lamellae, often quite narrow, related by parallel twin planes. Still others, such as andalusite may exhibit *cyclic twinning*, although it is rarely seen in thin sections. The feldspars are excellent examples of minerals that twin. Plagioclase is characterized by

polysynthetic twinning, orthoclase often by simple penetration or contact twins, and microcline by two types of lamellar twins with different orientations (that combine to produce "scotch plaid" twinning). Calcite is characterized by polysynthetic twins parallel to the long diagonal of its rhombohedral shape (Plate 5.2). Other carbonates have no twins or have twins parallel to the short diagonal. Thus, for the feldspars, the carbonates, and for other minerals, twinning can be a key to identification.

Box 4.7 Determining Sign and 2V from a Bxa Figure

Finding a grain that yields a Bxa is easiest for minerals with small *2V*. You can begin by searching for a grain that has minimal retardation. Such a grain will be oriented with one of the optical axes near vertical. Then obtain an interference figure, rotate the stage, and note how the isogyres behave. After checking several grains, you should find a Bxa. For minerals with low to moderate *2V* (0 to 60°), both isogyres will stay in the field of view, but the interference figure may not be perfectly centered (Plate 4.10). If one of the isogyres leaves the field of view, check other grains until you are sure you are looking at a nearly centered Bxa. For minerals with high *2V*, the search for a Bxa sometimes becomes frustrating. It may be difficult or impossible to distinguish a Bxa from a Bxo. For this reason, the optic sign often is best determined using an OA figure (Box 4.8).

Having found a Bxa, rotate the stage so the isogyres are in the southwest and northeast quadrants (Figure 4.24b; 4.25a–b). The Y direction in the crystal is now oriented northwest-southeast. The points corresponding to the optic axes (the melatopes) are the points on the isogyres closest to each other, and either X or Z is vertical, depending on optic sign.

To determine optic sign, we must know which direction (X or Z) is vertical, corresponding to Bxa. If the slow direction (Z) corresponds to Bxa, the crystal is positive. If the fast direction (X) is Bxa, the crystal is negative. To make the determi-

nation, insert the full wave accessory plate (with slow oriented direction southwest-northeast) and note any changes in interference colors on the concave sides of the isogyres. If the interference colors add on the concave sides of the isogyres (and subtract on the convex sides), the mineral is positive (Figure 4.25a–b). In positive minerals with low to moderate retardation, the colors in the center of the figure will be yellow (subtraction), and those on the concave side of the isogyres will be blue (addition). In a negative mineral, the color changes will be the opposite.

For minerals with high retardation, it may be difficult to determine whether a full wave accessory plate adds or subtracts retardation because isochromes of many repeating colors circle the melatopes. A quartz wedge facilitates determination. As the wedge is inserted, color rings move toward the melatopes if there is addition of retardation, or the rings move away from the melatopes if there is subtraction. If the interference colors move toward the optic axes from the concave side of the isogyres, and away on the convex side, the mineral is positive. The opposite effect is seen for a negative mineral.

You can estimate *2V* from a Bxa figure by noting the degree of separation of the isogyres when you rotate the stage. For standard lenses, if the isogyres just leave the field of view when they are at maximum separation, *2V* is 60–65°. If the isogyres barely separate, *2V* is greater than about 10°.

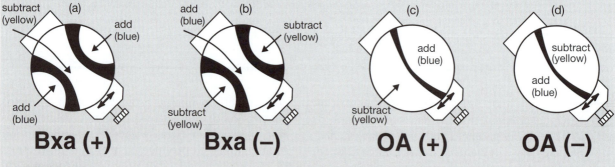

▶**FIGURE 4.25**
Determining optic sign from Bxa and OA figures: The accessory plate (with slow direction oriented southwest-northeast) is inserted and interference color changes are noted. For a biaxial positive mineral, interference colors add (increase to higher order) on the concave side of the isogyres; for a biaxial negative mineral, they subtract. For grains with low retardation, addition produces blue interference colors and subtraction produces yellow. (a) Bxa figure for a biaxial positive mineral; (b) Bxa figure for a biaxial negative mineral; (c) OA figure for a biaxial positive mineral; (d) OA figure for a biaxial negative mineral.

Box 4.8 Determining Sign and 2V from an Optic Axis Figure

Determining the optic sign from an optic axis figure can often be simpler than looking for a Bxa. To find an OA figure, look for a grain displaying zero or very low retardation. Obtain an interference figure and rotate the stage. It should be an OA figure, showing one centered or nearly centered isogyre (Figures 4.23 and 4.25c–d). If you see two isogyres that stay in the field of view when you rotate the stage, you are looking at a Bxa for a mineral with low *2V*. Rotate the stage so the isogyre (or the most nearly centered if there are two) is concave to the northeast (Figure 4.25). Note that the isogyre rotates in the opposite sense from the stage. Insert the full wave ac-

cessory plate. If the retardation increases on the concave side of the isogyre (and decreases on the convex side), the mineral is positive (Figure 4.25c–d). In minerals with low to moderate retardation, it is often necessary only to look for yellow and blue. Blue indicates an increase and yellow a decrease in retardation. The increase or decrease of retardation will be opposite if the mineral is negative.

You can estimate *2V* by noting the curvature of the isogyre. If *2V* is less than 10° to 15°, the isogyre will seem to make a 90° bend. If *2V* is 90°, it will be straight. For other values, it will have curvature between 90° and 0° (Figure 4.23).

▶QUESTIONS FOR THOUGHT

Some of these questions have no specific correct answers; they are intended to promote thought and discussion.

1. Why do we use a polarizing light microscope to examine minerals? Why not use a standard microscope?
2. Minerals that grow cubic crystals are isotropic when viewed under a polarizing microscope. What does isotropic mean? What is the connection between crystal shape and optical properties?
3. When calcite is viewed with polarized light and a petrographic microscope, its relief may vary as the stage is rotated. Why?

4. In general, high-density minerals have high refractive index compared to low-density minerals. Why do density and refractive index correlate?
5. Minerals that do not cause double refraction are not birefringent. What do we call such minerals? What is the connection between double refraction and birefringence?
6. Some anisotropic minerals are uniaxial and some are biaxial. What is the difference between them? Do you suppose it is possible to distinguish uniaxial and biaxial minerals simply by looking at their crystal forms? What would you look for? Must we always use a microscope?

▶RESOURCES

Bambauer, H.U., F. Taborszky, and H.D. Trochim. *Optical Determination of Rock-Forming Minerals.* Stuttgart: E. Schweizerbart'sche Verlagsbuchhandlung, 1979.

Bloss, F.D. *An Introduction to the Methods of Optical Crystallography.* New York: Holt, Rinehart and Winston, 1961.

Deer, W.A., R.A. Howie, and J. Zussman. *Rock Forming Minerals.* Five vols. New York: John Wiley & Sons, 1962 et seq.

Gaines, R.V., H.C.W. Skinner, E.E. Foord, B. Mason, and A. Rosenzweig. *Dana's New Mineralogy,* 8th ed. New York: John Wiley & Sons, 1997.

Kerr, P.F. *Optical Mineralogy,* 4th ed. New York: McGraw Hill, 1977.

Klein, C., and C.S. Hurlbut, Jr. *Manual of Mineralogy,* 21st ed. New York: John Wiley & Sons, 1993.

Nesse, W.D. *Introduction to Optical Mineralogy,* 2nd ed. New York: Oxford University Press, 1991.

Perkins, D. *Mineralogy.* Upper Saddle River: Prentice Hall, 1998.

Phillips, W.R. *Mineral Optics: Principles and Techniques.* San Francisco: W.H. Freeman & Co., 1971.

Phillips, W.R., and D.T. Griffen. *Optical Mineralogy: The Nonopaque Minerals.* San Francisco: W.H. Freeman & Co., 1981.

Pichler, H., and C. Schmitt-Riegraf, translated by L. Hoke. *Rock-Forming Minerals in Thin Section.* Chapman and Hall, 1997.

Shelley, D. *Optical Mineralogy,* 2nd ed. New York: Elsevier, 1985.

Stoiber, R.E., and Morse, S.A. *Microscopic Identification of Crystals.* Krieger, 1972.

Zoltai, T., and J.H. Stout. *Mineralogy: Concepts and Principles.* Minneapolis: Burgess, 1984.

Igneous Rocks and Silicate Minerals

Igneous minerals crystallize from a magma to form igneous rocks. In this chapter we discuss crystallization processes and the resulting minerals. We point out the chemical differences between igneous rocks of various types and how their mineralogies vary. Particular emphasis is placed on silicate minerals because of their dominance in the Earth's crust, and in igneous rocks in general. A simple scheme for naming igneous rocks is introduced, and some common types of igneous rocks are discussed.

MAGMAS AND IGNEOUS ROCKS

Magmas are complex liquids that vary greatly in composition and properties. They have temperatures as great as 1,400 °C (2,500 °F), and often originate in regions 50 km to 200 km deep in the Earth. They may be partially crystalline, containing crystals of high-temperature minerals such as leucite, olivine, or clinopyroxene. Because magma has a lower density than the solid upper mantle and crust of the Earth, buoyancy moves it upward. The race between upward movement and cooling ultimately determines whether magma becomes an **intrusive** or **extrusive** igneous rock (Figure 5.1). Magma solidifies as an intrusive rock if it crystallizes before it reaches the surface. Intrusive rocks often form **plutons** (a general term given to any intrusive igneous rock body), so geologists sometimes use the terms *intrusive* and *plutonic* interchangeably. If magma reaches the surface while molten, or partially molten, we term it **lava.** The cooling lava forms an **extrusive rock,** also called a **volcanic rock** (although it does not necessarily form a volcano).

Cooling rate directly affects grain size of an igneous rock. The common plutonic rock granite contains crystals of quartz and potassium feldspar that are easily seen with the naked eye. A common volcanic rock, rhyolite, may contain the same minerals, but we need a microscope to see the crystals (Figure 5.1). The difference is due to cooling rate. Intrusive rocks cool and crystallize slowly; other very hot rocks surround and insulate them as crystallization occurs. Crystals in granites have a long time to form and grow. In contrast, volcanic rocks, such as rhyolite, crystallize rapidly because extrusion exposes the lava to the cool atmosphere at the surface of the Earth. Sometimes extrusive igneous rocks cool so quickly that no crystals form. This is especially likely to occur if lava meets surface water. The result is a rock composed of glass called **obsidian.** In other cases, different minerals may grow to distinctly different sizes. The result is a **porphyry** in which coarse crystals called **phenocrysts** are floating in a sea of fine-grained crystals called **ground mass.**

Quickly cooled plutonic rocks may be very fine grained and difficult to tell from volcanic rocks. Some petrologists, therefore, prefer to classify and name igneous rocks based on their grain size rather than their genesis (origin). They divide rocks into those containing very fine grains **(aphanitic),** rocks containing very coarse grains **(phaneritic),** and rocks containing combinations of large and small crystals **(porphyritic).**

▶**FIGURE 5.1**

▶**FIGURE 5.1**
Three igneous rocks: coarse-grained granite (left), medium-grained granite (center), and rhyolite (right). Granite and other intrusive rocks are generally coarser grained than extrusive rocks such as rhyolite.

COMPOSITIONS OF IGNEOUS ROCKS

Mafic and Silicic Magmas

Melting of rocks occurs at many places in the Earth, and magma compositions reflect their sources. Although compositions cover a wide spectrum, most contain 40 to 75 wt. % SiO_2 (Box 5.1). Magmas richest in SiO_2 also tend to be rich in Al_2O_3. They may contain appreciable amounts of FeO and Fe_2O_3, but are usually deficient in MgO. We term such magmas **silicic** (Si rich), **sialic** (Si and Al rich), or **felsic.** Light-colored minerals dominate felsic rocks, so many geologists use the term *felsic* to refer to any light-colored igneous rock, even if the chemical composition is unknown. At the other end of the spectrum, magmas with <50 wt. % SiO_2 are usually rich in MgO and contain more FeO and Fe_2O_3 than silicic magmas. Thus, we call them **mafic** (Mg and Fe rich) and, in some extreme cases, **ultramafic.** They are usually dark in color. The term **intermediate** describes rocks with compositions between mafic and silicic. Besides distinctions between mafic, intermediate, and silicic rocks, petrologists often classify rocks based on the alkali ($K_2O + Na_2O$) and alkaline earth (CaO) contents; alkalic rocks are those with high ($K_2O + Na_2O$):CaO ratios. Some rare and unusual magma types produce igneous rocks rich in nonsilicate minerals such as carbonates or phosphates, but we will not consider them here.

Rocks of different compositions have different melting temperatures because some elements combine to promote melting. Silicon and oxygen, in particular, promote melting because they form very stable molten **polymers** (long chains of Si and O). Silicic minerals, and SiO_2-rich rocks, therefore, melt at lower temperatures than mafic minerals and SiO_2-poor rocks. Conversely, magmas of different compositions crystallize at different temperatures. Temperatures measured in flowing lavas generally range from 900 to 1,100 °C, with higher temperatures corresponding to basaltic (mafic) lavas and lower temperatures to andesitic or rhyolitic (intermediate to silicic) lavas. Eventually, as a magma cools, it crystallizes; the first crystals form at the **liquidus** temperature. With further cooling of up to 200 °C, the magma completely solidifies. The last drop of melt crystallizes at the **solidus** temperature. Some magmas crystallize at temperatures well above 1,000 °C, but granites and other silicic magmas may crystallize at temperatures as low as 700 °C.

Volatiles

Magmas also contain **volatiles** (gas, liquid, or vapor). H_2O and CO_2 are the most common, but compounds of sulfur, chlorine, and several other elements may also be present. Consequently, although igneous minerals crystallize at high temperatures, they may contain H_2O, CO_2, or other gaseous components. Volatiles may separate from a melt to form bubbles, most commonly in cooling lava, creating **vesicles** as the magma solidifies (Figure 5.2). Water is especially important in the crystallization process. A small amount will appreciably lower melting and crystallization temperatures, change magma viscosity, and produce large amounts of vapor or steam. Steam is often responsible for explosions such as those that took place at Mount St. Helens on May 18, 1980. The vesicle shown in Plate 1.7 may have been filled with liquid or steam after the surrounding basalt initially crystallized.

Box 5.1 Compositions of Magmas Producing Igneous Rocks (oxide wt. %)

The following table gives compositions of some common magma types, identified by the name of the plutonic rock they produce. Rock and mineral compositions are normally given in terms of oxides rather than individual element because (1) oxygen is the only significant anion in most minerals and igneous rocks; (2) oxygen is not normally directly analyzed, so its content must be inferred; and (3) rocks must be charge-balanced (cation charge balanced by anion charge), so it is convenient to list components that are charge-balanced. It should be emphasized that the chemical oxides listed in this table do not refer to specific oxide minerals; oxides are listed to describe the chemical composition only.

| wt. % Oxide | SiO₂-Rich Magmas | | | SiO₂-Poor Magmas | | | | |
| | Alkali Rich | | Alkali Poor | Alkali Rich | | | | Alkali Poor |
	Alkali Granite	Granite	Tonalite	Alkali Syenite	Syenite	Diorite	Gabbro	Peridotite
SiO_2	73.86	72.08	66.15	55.48	59.41	51.86	50.78	43.54
TiO_2	0.20	0.37	0.62	0.66	0.83	1.50	1.13	0.81
Al_2O_3	13.75	13.86	15.56	21.34	17.12	16.40	15.68	3.99
Fe_2O_3	0.78	0.86	1.36	2.42	2.19	2.73	2.26	2.51
FeO	1.13	1.67	3.42	2.00	2.83	6.97	7.41	9.84
MnO	0.05	0.06	0.08	0.19	0.08	0.18	0.18	0.21
MgO	0.26	0.52	1.94	0.57	2.02	6.12	8.35	34.02
CaO	0.72	1.33	4.65	1.98	4.06	8.40	10.85	3.46
Na_2O	3.51	3.08	3.90	8.86	3.92	3.36	2.14	0.56
K_2O	5.13	5.46	1.42	5.35	6.53	1.33	0.56	0.25
H_2O	0.47	0.53	0.69	0.96	0.63	0.80	0.48	0.76
P_2O_5	0.14	0.18	0.21	0.19	0.38	0.35	0.18	0.05

*Analyses from Nockolds (1954).

▶**FIGURE 5.2**
Vesicles (0.5–1.0 cm across) in basalt that formed as the basalt solidified. Percolating pore waters have subsequently filled many of the vesicles with white calcite.

Slowly cooling magmas do not crystallize all at once. After partial crystallization, the remnant melt may contain water and dissolved **incompatible elements** that did not enter any of the minerals already formed. When the remnant melt finally crystallizes, **pegmatites** containing minerals rich in incompatible elements such as K, Rb, Li, Be, B, or rare earth elements (REEs) may result. Pegmatites often contain large euhedral crystals because the water acts as a flux and promotes crystal growth (Figure 5.3; Plate 1.4). Many spectacular and valuable mineral specimens come from pegmatites.

CRYSTALLIZATION OF MAGMAS

Equilibrium Between Crystals and Melt

Every mineral has a characteristic melting temperature. This leads to an orderly and predictable process

▶**FIGURE 5.3**
Euhedral tourmaline from a pegmatite at Paraibo, Brazil (5 cm across). Compositional zoning has given different parts of the crystal different colors. Note especially the dark-colored rim. (See also Plate 1.1.)

as magma solidifies. For many mafic magmas, olivine and Ca-rich plagioclase, which have high melting temperatures, will be the first minerals to crystallize. As temperature falls, olivine and Ca-plagioclase may become unstable. Some or all of the olivine may disappear as it reacts with the remaining melt to produce pyroxene by reactions such as:

$$(Fe,Mg)_2SiO_4 + SiO_2 = (Fe,Mg)_2Si_2O_6 \quad (5.1)$$
$$\text{(olivine)} \quad \text{(in melt)} \quad \text{(pyroxene)}$$

At the same time, the plagioclase, which has the general formula $(Ca,Na)(Al,Si)_4O_8$, may exchange

some of its Ca and Al for Na and Si in the remaining melt:

$$CaAl_2Si_2O_8 + Na + Si = NaAlSi_3O_8 + Ca + Al \ (5.2)$$
$$\text{(in plagioclase) (in melt) (in plagioclase) (in melt)}$$

Thus, plagioclase crystals often become richer in albite ($NaAlSi_3O_8$) and poorer in anorthite ($CaAl_2Si_2O_8$) as temperature falls. With continued falling temperature, the pyroxene may become unstable and react with the melt to produce amphibole and then mica, while the plagioclase continues to become more Na-rich. Thus we see that the minerals present during crystallization may change as crystallization continues. However, when crystallization is complete, the minerals present will reflect the overall magma composition.

The first minerals to crystallize from silicic magmas are not normally olivine or pyroxene, but instead are more likely to be amphibole, mica, or feldspar. Reactions between minerals and melt take place as cooling continues. In the final stages of crystallization, many silicic magmas crystallize K-feldspar (K-rich alkali feldspar), muscovite, and quartz. When the rock is completely solidified, the relative abundances of the various minerals depend on the overall magma composition.

Bowen's Reaction Series

N. L. Bowen pioneered in the study of magma crystallization, for which he received the Roebling Medal from the Mineralogical Society of America in 1950. Through study of naturally occurring igneous rocks and laboratory experimentation, he derived an idealized model for equilibrium crystallization in a magmatic system. We call the model **Bowen's reaction series** (Figure 5.4). Although we cannot discuss some magma types using Bowen's series, it is an excellent model to describe the process of magmatic crystallization.

▶**FIGURE 5.4**
Bowen's reaction series. The names in the continuous series refer to specific compositions of plagioclase. The minerals of the discontinuous series are all solid solutions and so, too, may change composition as crystallization occurs. The residual phases are those often crystallizing from water-rich remnant melts. We frequently find them in pegmatites.

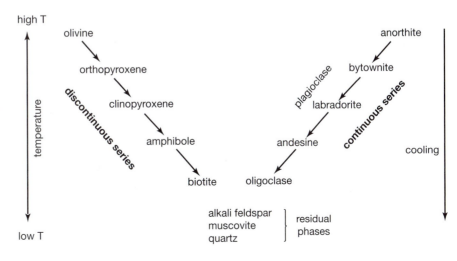

We could develop similar series for magmas of any composition.

Bowen's reaction series shows the order in which minerals crystallize from cooling magma. We call the left-hand side of the series the **discontinuous side** because abrupt changes occur when one mineral reacts with the melt to produce another (such as olivine reacting to form pyroxene). We call the right-hand side the **continuous side** because plagioclase is continually present during crystallization; as cooling progresses, plagioclase reacts with the melt to maintain an equilibrium composition. The minerals at the bottom of Bowen's series are sometimes called **residual phases** because they crystallize at the end of the process from whatever is left in the melt. Just like plagioclase, minerals on the discontinuous side of the series change composition as cooling proceeds. At higher temperatures, for example, olivine usually has a greater Mg:Fe ratio than at lower temperatures.

Mineral crystallization temperatures depend on mineral composition. Although Bowen created his reaction series by observing the temperatures at which minerals crystallize, it reflects general changes in mineral chemistry during solidification of a magma. Minerals at the top of the series, such as olivine, pyroxene, and Ca-rich plagioclase, are silica-poor. Those at the bottom are silica-rich. Silica content is the most significant factor controlling melting temperature. The mafic minerals at the top of the discontinuous series also are deficient in Al and alkalis and rich in Fe and Mg, compared with minerals at the bottom.

▶FIGURE 5.5
Zoned crystals. (a) Plagioclase crystals sometimes show zoning as light and dark bands when viewed in thin section (photo is about 1 mm across). (b) Zoning in a columbite crystal (photo is about 0.35 mm across) viewed with a scanning electron microscope.

It is important to remember that Bowen's reaction series is an idealized model. If the process always went to completion, all igneous rocks would be combinations of K-feldspar, quartz, and possibly muscovite. Such might be the case for a magma of felsic composition but is not true for most igneous rocks. Consider, for example, a hypothetical melt that is 100% SiO_2. It cannot crystallize any minerals in Bowen's reaction series except quartz because it does not contain the necessary elements. It will skip all the other minerals. On the other hand, if a melt were 100% Mg_2SiO_4, it would crystallize forsterite (olivine of composition Mg_2SiO_4) and be completely solidified. These two extreme examples do not exist in nature, but compositions of natural magmas do control the extent to which crystallization follows Bowen's reaction series. Although all magmas crystallize different minerals at different temperatures, few, if any, follow the complete series.

Mafic magmas, which crystallize at high temperature, may completely solidify after olivine, pyroxene, and some plagioclase have formed. Some rare ultramafic rocks crystallize to nearly 100% olivine. If no melt is left to react with the existing crystals, reactions will stop and the rock will continue to cool with no change in mineralogy. Silicic magmas, which crystallize at lower temperatures, yield rocks rich in K-feldspar and quartz. The melt completely resorbs any mafic minerals that may have initially crystallized.

Disequilibrium

Minerals do not always remain in equilibrium with magmas as they crystallize. Several things may cause disequilibrium. The crystals may become separated from most of the magma as a whole if they settle or float to form a layer at the bottom or top of the magma chamber. The layer of crystals, termed a **cumulate** if it is sufficiently well developed, may not take part in additional reactions with the melt. Many minerals, including feldspars, pyroxenes, and oxides, can be found in thick, nearly **monomineralic** cumulates. Even if a cumulate layer does not develop, the composition of the upper part of the magma chamber may not be the same as the lower part. When a melt gets separated from early formed crystals, we call the process **partial crystallization** or **fractional crystallization.** It is a significant process in the Earth; it caused the chemical differentiation making the Earth's crust more silicic than the mantle. Goldschmidt and others hypothesized that the process of differentiation due to melting and crystallization explained all compositional layering in the Earth. Today, most geologists believe that differentiation has been important throughout Earth's history but

that the Earth may not have been completely homogeneous at its start.

Besides separation of crystals and melt, disequilibrium occurs for other reasons. Sometimes large mineral grains do not remain in equilibrium with a surrounding magma. Because diffusion of elements through solid crystals is slow, the central parts of large crystals may not have time to maintain equilibrium compositions. In compositionally **zoned** crystals, only the outermost zones remain in equilibrium with the melt as crystallization takes place. Marked chemical zonation often occurs if a magma begins to cool at one depth and then rapidly moves upward to cooler terrane. The result will be a porphyritic rock with large zoned phenocrysts. The zones may be visible with the naked eye if color or textural variations mirror the compositional variations (Figure 5.3; Plates 1.1 and 1.2). If the zoning is not visible to the naked eye, it may be visible with a petrographic microscope or with a scanning electron microscope (Figure 5.5a,b). Sometimes, however, we can only discern zoning with sophisticated analytical instruments.

SILICATE MINERALS

Silicate minerals dominate igneous rocks because Si and O are the most common elements in the source regions for magmas. The following discussions systematically consider the important silicate minerals and groups. We begin by considering the SiO_2 polymorphs, the feldspars, and the feldspathoids, mineral groups within the **framework silicate** subclass. Silicates have structures containing polymerized SiO_4, or combinations of SiO_4 and AlO_4, tetrahedra (Figure 5.6). In framework silicates, all oxygen atoms are shared between two tetrahedra, creating a three-dimensional network. In **feldspars** and **feldspathoids,** alkalis and alkaline earths occupy large sites between tetrahedra. Figure 5.6 shows the structure of tridymite, one of the SiO_2 minerals. Other framework silicates have tetrahedra arranged differently.

SiO$_2$ Polymorphs

Quartz, like many other minerals, is polymorphic (meaning "having many forms"); mineralogists and chemists have identified more than 10 different SiO_2 polymorphs, some of which do not occur as minerals. Common quartz, more properly called *low quartz* (because it has lower symmetry than high quartz), is the only common polymorph stable under normal Earth surface conditions, but it has many different appearances (Plates 2.1 through 2.8). Other polymorphs, which exist metastably at

▶**FIGURE 5.6**
The structure of tridymite shows one way tetrahedra join to create a framework silicate. In tridymite, all tetrahedra contain Si^{4+} at their centers, but in feldspars, and some other framework silicates, some tetrahedra contain Al^{3+}.

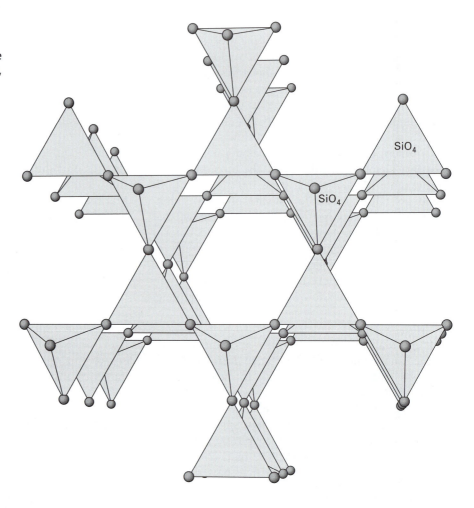

the Earth's surface, often change into low quartz given enough time. Still others only exist at high temperatures or high pressures. If all rocks maintained and stayed at equilibrium, we would have no samples of any silica polymorphs other than low quartz to study.

Quartz may crystallize from magmas, from silica-saturated groundwaters, or from hydrothermal solutions. It is an essential mineral in silicic and intermediate igneous rocks, many sediments, and many metamorphic rocks. In some sedimentary rocks, such as chert or sandstone, it may be the only mineral present. In silicic and intermediate igneous rocks, quartz is commonly found with K-feldspar. Both are Si-rich minerals that form at relatively low temperature. Quartz is not normally found in mafic igneous rocks; SiO_2 is too scarce to remain after crystallization of mafic minerals such as olivine or pyroxene. Quartz cannot exist in rocks containing corundum, Al_2O_3, because the two would react to form an Al_2SiO_5 mineral. It cannot exist in rocks containing feldspathoids (leucite, nepheline, or analcime) because quartz and feldspathoids react to give feldspars. For similar reasons, quartz is absent or minor in many

alkali-rich igneous rocks and in rocks containing the oxide mineral spinel, $MgAl_2O_4$.

For several decades, petrologists have understood that different silica polymorphs occur in different geological settings because they are stable under different pressure-temperature conditions. Figure 5.7, a phase diagram, shows the stability relationships of some SiO_2 polymorphs. The vertical scale on the left gives pressures in kilobars (thousands of atmospheres), and the scale on the right shows the depths in the Earth corresponding to those pressures.

Pressure-temperature (P-T) phase diagrams such as Figure 5.7 show which mineral or mineral assemblage is stable for any combination of P-T (Box 5.2). Low quartz is the stable phase over a wide range of P-T, including normal Earth surface conditions; it is therefore the most common polymorph. Stishovite and coesite are dense minerals, only stable at very high pressures—pressures not normally encountered in the Earth's crust. They are usually associated with meteorite impact craters. Tridymite and cristobalite only exist in certain high-temperature silicic volcanic rocks. They require temperatures greater than 900°C to form. Although

Box 5.2 Experimental Petrology and Phase Diagrams

Mineralogists and petrologists use many types of phase diagrams. All are designed to show variation in mineral, or mineral assemblage, stability as physical or chemical conditions change. The most common kind of mineralogical phase diagram, referred to as a P-T diagram, has pressure and temperature as its two axes. Normally, pressure corresponds to the vertical axis, and temperature to the horizontal axis. P-T diagrams show the stability fields for specific minerals, or mineral assemblages, in pressure-temperature space. Figure 5.7 is a P-T diagram for the chemical system SiO_2, a system that includes quartz and its polymorphs. Pressure increases upward and temperature increases to the right. A few petrologists flip the vertical axis and draw P-T diagrams with pressure increasing downward, because the pressure within the Earth increases with depth.

Geologists generally derive phase diagrams by conducting experiments in the laboratory. They combine minerals or chemicals, allow them to react under different conditions, and then examine the results. We call this area of research *experimental petrology*. After many careful experiments, experimental petrologists can construct phase diagrams. Like most scientific results, phase diagrams are refined or modified when experimental petrologists

gain more information or conduct further experiments. Experimental petrology is expensive and painstaking work. Consequently, only a few laboratories are producing most of the best results today.

Figure 5.7 shows which of the SiO_2 polymorphs are stable under different pressure-temperature conditions. To determine the stable polymorph at any P-T, draw a horizontal line at the pressure of interest and a vertical line at the temperature of interest. They intersect within a specific *stability field*, labeled with the stable SiO_2 mineral name. Although the temperature scale does not go below 500 °C, at room temperature and pressure (0.001 Kilobars and 25 °C,) low quartz is stable. At the highest pressures, stishovite is stable because, as we shall see, it is the most dense of the SiO_2 minerals. At the highest temperatures, SiO_2 melts.

Although P-T diagrams, such as in Figure 5.7, are often useful and informative, they only apply to specific chemical systems. Sometimes elemental substitutions in minerals cause significant changes in size and location of stability fields. Furthermore, phase diagrams only apply to mineral systems that stay in equilibrium, and many geological systems do not. For example, cristobalite, a mineral normally stable only at high temperature, sometimes crystallizes as chalcedony in low-temperature sedimentary rocks.

▶**FIGURE 5.7**
Phase diagram showing the stability fields of some of the SiO_2 polymorphs, and the melting field at high temperature.

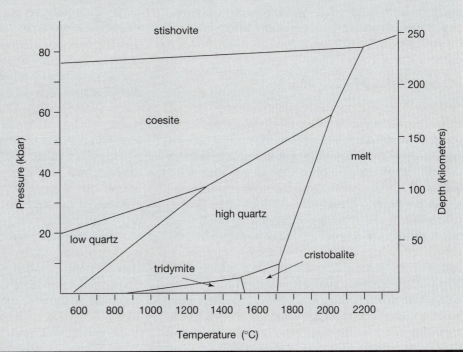

▶**FIGURE 5.8**
Plagioclase and alkali feldspar in hand specimen (photos a,b are about 5 cm across) and thin section (0.7 mm field of view): (a) plagioclase feldspar; (b) orthoclase crystals, each about 1.5 cm in longest dimension; (c) large plagioclase crystal in thin section showing characteristic gray and white stripes indicative of twinning; (d) microcline (an alkali feldspar) in thin section showing characteristic scotch plaid twinning.

not shown in Figure 5.7, as with quartz, tridymite and cristobalite have both high- and low-symmetry polymorphs.

Feldspar Group

Feldspars are the most abundant minerals in the Earth's crust, in part because they contain six of the seven most abundant elements in the crust. They are usually easy to distinguish in thin section but may be difficult to tell apart in hand specimen (Figure 5.8). Feldspars are essential minerals in many igneous, metamorphic, and sedimentary rocks and are stable over a wide range of pressure-temperature conditions, although they will eventually break down when subjected to weathering. They are solid solution minerals and have the general formula $(Ca,Na,K)(Al,Si)_4O_8$. Rarely, they contain significant amounts of other elements such as Ba^{2+}, Sr^{2+}, B^{3+}, or Fe^{3+}. For most purposes, we consider them as ternary solutions, which means we can describe their composition in terms of three end members (Figure 5.9). Box 5.3 discusses in

more detail how we use triangular diagrams to plot feldspar compositions.

The important feldspar end members have compositions $NaAlSi_3O_8$ (albite), $CaAl_2Si_2O_8$ (anorthite), and $KAlSi_3O_8$ (orthoclase). We call any feldspar near $NaAlSi_3O_8$ in composition *albite*, and one near to $CaAl_2Si_2O_8$ *anorthite*. Plate 6.8 shows albite from a classic locality at Amelia Courthouse, Virginia. We term feldspars having compositions close to $KAlSi_3O_8$ *K-feldspar* or *alkali feldspar*. If we know the crystal structure, we may be more specific and distinguish among the different K-feldspar polymorphs: orthoclase, microcline, and sanidine. Plate 1.5 shows twinned orthoclase crystals.

Feldspars form two distinct series, shown on the triangular composition diagram in Figure 5.9. Plagioclase, sometimes called *plagioclase feldspar*, is generally a solid solution of $CaAl_2Si_2O_8$ and $NaAlSi_3O_8$, anorthite and albite, although it may contain up to 10 wt. % $KAlSi_3O_8$. Alkali feldspars, mostly solutions of $KAlSi_3O_8$ and $NaAlSi_3O_8$, sometimes contain up to 15 wt.% $CaAl_2Si_2O_8$. All feldspars are

►**FIGURE 5.9**

The feldspar ternary. Natural feldspars rarely have compositions that fall within the shaded region (the miscibility gap). The gap gets smaller at high temperature. We normally describe feldspar compositions using the three end members at the corners of the diagram: anorthite, albite, and orthoclase. Less commonly we use bytownite, labradorite, andesine, and oligoclase to refer to intermediate plagioclase compositions.

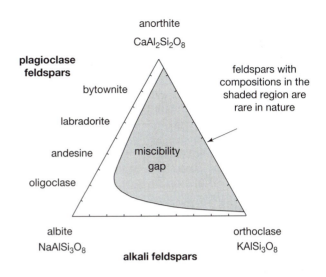

Box 5.3 How To Use Triangular Diagrams

Triangular diagrams (**ternary diagrams**) may be used to plot the mineralogy or chemistry of rocks or magma. The diagrams provide a concise picture of composition. We may use them to differentiate different rocks and mineral assemblages, or to identify possible reactions that may take place between minerals or between minerals and magma.

Figure 5.10 is a CaO-Al_2O_3-SiO_2 ternary diagram; such diagrams are often used to plot the compositions of minerals in Ca-Al-rich metamorphic rocks. Each apex represents a chemical component. Mole percentages are generally used to plot mineral compositions. Quartz, 100 mole % SiO_2, plots at the SiO_2 apex; compositions containing no SiO_2 plot on the side opposite the SiO_2 apex, on the line joining the CaO and Al_2O_3 apices. Compositions containing some SiO_2 plot between the SiO_2 apex and the CaO-Al_2O_3 join, depending on the amount of SiO_2 they contain. Similarly, compositions containing all three components plot in the triangle's interior; the distance from the component apex is proportional to the amount of the component they contain.

Wollastonite, $CaSiO_3$, plots halfway between CaO and SiO_2 because it contains an equal number of moles of CaO and SiO_2 in its formula. Grossular, $Ca_3Al_2Si_3O_{12}$, has $CaO:Al_2O_3:SiO_2$ ratio $3:1:3$, equivalent to 43% CaO, 14% Al_2O_3, and 43% SiO_2. Any two of the three percentages

can be used to plot grossular on the diagram. Because it contains an equal number of moles of CaO and SiO_2, it plots at equal distances from the CaO and SiO_2 apices.

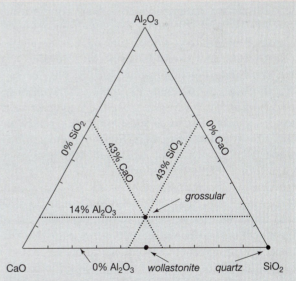

►**FIGURE 5.10**

CaO-Al_2O_3-SiO_2 ternary diagram. We use ternary (triangular) diagrams to plot the compositions of minerals or rocks containing three main chemical components. Such diagrams provide a visual picture of compositional relationships. This figure is useful for showing the compositions of minerals in some marbles and marls.

▶**FIGURE 5.11**
Schematic diagram showing alkali feldspar polymorphs stable at different temperatures at 1 atm. pressure. The vertical axis shows temperature and the horizontal axis shows feldspar composition. Mineralogists give names to feldspars of different compositions formed at different temperatures. Each feldspar species has its own stability field, and each has an atomic arrangement slightly different from the others. For compositions and temperatures that fall within the shaded area (the miscibility gap), no single feldspar is stable.

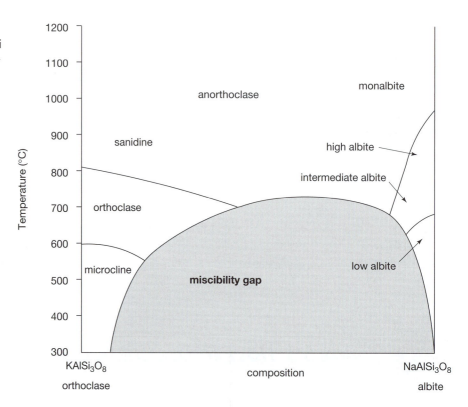

framework silicates containing both SiO_4 and AlO_4 tetrahedra.

Usually, we describe the compositions of feldspars by using abbreviations with subscripts. Ignoring Fe^{3+}, the feldspar we discussed in Box 1.5 has composition about $(Ca_{0.03}Na_{0.62}K_{0.35})(Al_{1.03}Si_{2.97})O_8$, so it contains 3 mole % anorthite $(CaAl_2Si_2O_8)$, 62 mole % albite, $(NaAlSi_3O_8)$, and 35 mole % orthoclase $(KAlSi_3O_8)$. We abbreviate its composition $An_3Ab_{62}Or_{35}$ or say $X_{An} = 3$, $X_{Ab} = 62$, and $X_{Or} = 35$. The symbol X stands for mole fraction when we use it to describe a solid solution. Because feldspars vary greatly in composition, mineralogists have divided the composition triangle into small ranges, each with its own name (Figure 5.9). Labradorite, for example, is plagioclase with composition between $An_{70}Ab_{30}$ and $An_{50}Ab_{50}$. Small amounts of $KAlSi_3O_8$ may be present in labradorite as well. Confusion arises because the names of some composition ranges are the same as the feldspar end members (albite, anorthite, orthoclase). Albite, for example, is the name given to end member $NaAlSi_3O_8$. It is also the name we give to any feldspar that is $>90\% NaAlSi_3O_8$.

As with the SiO_2 minerals, some feldspar compositions have more than one polymorph. Figure 5.11 is a temperature-composition phase diagram showing how alkali feldspars vary with temperature. Sodic (Na-rich) feldspars have different atomic structures at high, medium, and low temperatures. At high temperatures, just below their melting points, we call them *monalbite*. At lower temperature, three other polymorphs exist, all generally called albite. At the other end of the alkali feldspar series, the stable K-feldspars are sanidine, orthoclase, and microcline at high, medium, and low temperatures, respectively. Unlike the SiO_2 polymorphs, the differences in atomic structure between the K-feldspar polymorphs are not great and the boundaries on phase diagrams are not well known. When high-temperature feldspars change into low-temperature ones because of cooling or reequilibration at the Earth's surface, twinning often results.

In principle, alkali feldspars can have any composition between albite and K-feldspar ($Ab_{100}Or_0$ to Ab_0Or_{100}), but intermediate compositions, called *anorthoclase*, are stable only at high temperatures, where the two end members are sanidine ($KAlSi_3O_8$) and monalbite ($NaAlSi_3O_8$). At low temperatures, two separate feldspars form instead because a **miscibility gap** exists between albite and orthoclase (Box 5.4). A miscibility gap is a composition range within which no single mineral is stable under a particular set of pressure-temperature conditions. Many mineral series show complete miscibility at high temperatures, meaning that all compositions are stable. At low temperatures, partial or complete **immiscibility** restricts possible compositions. We might make an analogy to a pot of homemade chicken soup that separates into two compositions (fat and chicken stock) after cooling.

We call the process of a single feldspar separating into two compositions **exsolution,** equivalent to **unmixing.** If an intermediate composition alkali feldspar

Box 5.4 Miscibility Gaps

Figure 5.12 shows the same temperature-composition *(T-X)* phase diagram as Figure 5.11. For a given composition, and a given temperature, the diagram shows what mineral or minerals will be stable. The miscibility gap in the center of the diagram means that feldspars of intermediate composition are not stable except at very high temperatures ($> 700\ °C$). Instead, low-temperature rocks contain two separate feldspars (for example, X' and X''), usually a K-feldspar and plagioclase. The curves, outlining the miscibility gap and showing the compositions of the two separate feldspars, are called the **solvus**.

Consider a feldspar of composition $Ab_{60}Or_{40}$ (marked with an X on the diagram). At high temperature, it will exist as one alkali feldspar (anorthoclase). At low temperature it separates into two feldspars, as shown, one called *albite* (plagioclase, shown by X'') and one called *microcline* (a K-feldspar shown by X'). The phase diagram indicates that the composition of the albite will be very close to end member $NaAlSi_3O_8$, while that of the microcline will be about Ab_7Or_{93}.

▶**FIGURE 5.12**
1 atm. *T-X* diagram for alkali feldspars. The compositions of stable feldspar species below about 700 °C can be found by drawing a horizontal line at the temperature of interest and noting the two points (and corresponding compositions) where it intersects the solvus.

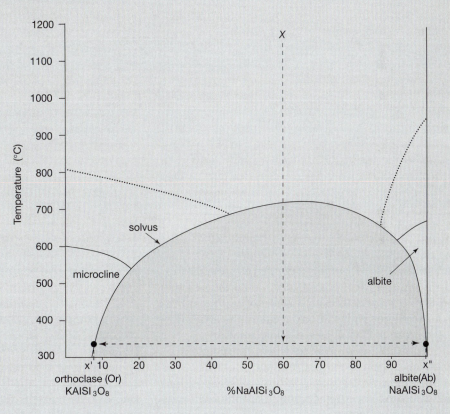

cools rapidly, it may not have time to exsolve. Thus, we have examples of anorthoclase to study. On the other hand, if cooling is slow, the feldspar will unmix. This may result in separate grains of K-feldspar (orthoclase or microcline) and Na-rich feldspar. More often, it results in alternating layers or irregular zones of albite and orthoclase within the original grain. If the layers or zones are planar or nearly so (appearing long and thin in thin section), we call them **exsolution lamellae** (see Figure 3.6). The layering or zones are sometimes visi-

ble with the naked eye, but frequently require a microscope to detect (see Figure 5.21a). If the original feldspar was K-rich, most of the layers will be K-feldspar. After it exsolves, we call it *perthite*. If the original feldspar was Na-rich, most of the layers will be albite. After exsolution, we call it *antiperthite*. Plate 6.7 shows salmon-colored perthite; the exsolution lamellae are thin, and it takes an active imagination to see them with the naked eye. If the lamellae in perthite or antiperthite are microscopic, they may give mineral grains a

Box 5.5 The Plagioclase Phase Diagram and Fractional Crystallization

Figure 5.13 is a *T-X* phase diagram describing plagioclase melting and crystallization at 1 atm. Composition varies across the diagram because plagioclase is normally a mixture of two end members, albite and anorthite. Plagioclase that is 100% albite

plots on the left-hand side, and albite content decreases across the diagram until 100% anorthite is reached on the right-hand side. Temperature increases from the bottom of the diagram to the top. At the highest temperatures, above the liquidus,

▶**FIGURE 5.13**
1 atm plagioclase *T-X* phase diagram. (See text for explanation.)

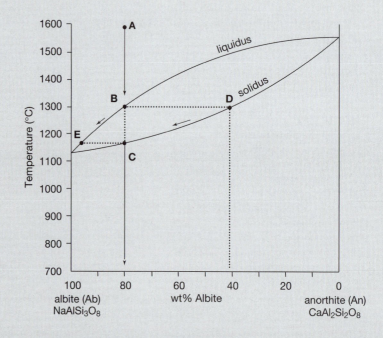

schiller effect that can help with mineral identification. Schiller is a play of colors caused by light interference due to very fine exsolution lamellae.

In contrast with alkali feldspars, plagioclase compositions spanning the entire albite-anorthite series are common. At high temperatures all compositions are stable. At lower temperatures several small solvi exist which, if cooling is slow enough, may result in microscopic exsolution. The exsolution may give the feldspars an iridescence or a play of colors that helps to identify them. Labradorite, for example, often shows a bluish schiller in hand specimen. Because exsolution, if present, in plagioclase is on a very fine scale, mineral properties are relatively homogeneous. For most purposes, we can ignore the presence or absence of exsolution in plagioclase. In thin section, we can usually identify plagioclase by its characteristic twinning (see Figure 5.8), usually visible in thin section and sometimes visible in hand specimen.

The melting temperatures of the two plagioclase end members are different, as depicted by Bowen's reaction series; albite melts at lower temperature than anorthite (Box 5.5). For intermediate compositions, melting relationships are complicated because when a feldspar melts, the first melt produced is more sodic than the feldspar itself. In effect, the albite part of the feldspar melts faster than the anorthite. If an intermediate plagioclase is partially melted, the melt will be more albite-rich than the original feldspar. The remaining unmelted feldspar will therefore have to be more anorthite-rich than the original feldspar. Similarly, if a magma begins to crystallize plagioclase, the first crystals to form will be more calcic than the melt. Box 5.5 discusses the melting and crystallization relationships of plagioclase in more detail.

Both K-feldspar and plagioclase exist in a wide range of igneous rocks. Feldspars with compositions

plagioclase of any composition will melt. At the lowest, below the solidus, any plagioclase will be solidified. In between, some compositions will melt and some will be solid.

Point A represents an albite-rich feldspar melt at 1,600 °C. The melt has the same composition as plagioclase that is 80% albite and 20% anorthite ($Ab_{80}An_{20}$) but is completely molten. As the melt cools, at about 1,300 °C, the first crystals begin to form. A horizontal line at 1,300 °C indicates the composition of the melt and the first crystals. It intersects the liquidus at the melt composition (Point B) and the solidus at the crystal composition (Point D). When this first crystallization occurs, the liquid still has its original composition, but the crystals that start to form are considerably more anorthite-rich ($Ab_{42}An_{58}$).

As the melt continues cooling, crystals continue to form, but they change composition from D to C as they do so. Previously crystallized plagioclase reacts with the melt so all crystals have the same composition. A horizontal line can be drawn at any temperature. The intersection of the line with the solidus indicates plagioclase composition and the intersection of the line with the liquidus indicates melt composition, provided the melt and crystal stay in equilibrium. Because the crystals

are more anorthite-rich than the melt, the melt becomes more albite-rich as crystallization continues. As crystallization goes to completion, plagioclase crystals change in composition from $Ab_{42}An_{58}$ (Point D) to $Ab_{80}An_{20}$ (the composition of the original melt). Melt composition changes from $Ab_{80}An_{20}$ to an extremely albite-rich composition ($Ab_{96}An_{04}$), indicated by Point E on the diagram. At about 1,165 °C, the last drop of melt will crystallize. All plagioclase crystals have the same composition, which must equal that of the original melt.

If the crystallization process always went to completion, plagioclase crystals would always end up the same composition as the original melt. However, several things can disrupt this equilibrium. Often crystallization occurs so fast that crystals do not have time to react with the melt as they form. In such cases, the first plagioclase to crystallize is preserved in the centers of large crystals. The crystals are compositionally zoned, the outer zones being more albite-rich. Another complication arises if crystals and melt get separated during the crystallization process. This may occur due to crystal settling, or because remaining melt "squirts off" after the melt is partially solidified. In such cases, crystal-melt equilibrium is not maintained.

in the middle of the feldspar ternary are almost never found because they unmix to form coexisting alkali feldspar and plagioclase (Figure 5.14). In most silicic rocks, such as granite, plagioclase is subordinate to K-feldspar. Similarly, in most mafic rocks, K-feldspar is not normally present. Intermediate compositions nearly always contain both feldspars. Plagioclase composition varies predictably with rock composition. Silicic rocks, which are usually relatively rich in Na, contain albitic plagioclase. Most mafic

▶**FIGURE 5.14**
Ternary feldspar miscibility gap. The gap varies with temperature as shown by the curves for 650 °C, 750 °C, and 900 °C. Feldspars that plot in the center of the diagram are unstable at all temperatures, but note that even some alkali feldspars (plotting on the bottom of the diagram) fall in the gap and are unstable at 650 °C.

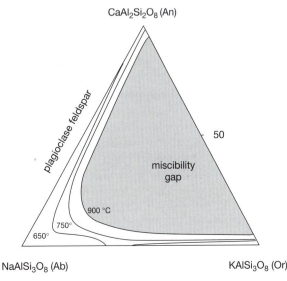

rocks, being relatively rich in Ca, contain anorthitic plagioclase.

Feldspathoid Group

The feldspathoid minerals are similar to feldspars in many of their properties because they have similar atomic structures and chemistries. They are, however, much less common. The most important feldspathoids are analcime ($NaAlSi_2O_6 \cdot H_2O$), leucite ($KAlSi_2O_6$), and nepheline ($NaAlSiO_4$). They are restricted to rocks in which quartz is not present because they will react with quartz to produce feldspars:

$$NaAlSi_2O_6 \cdot H_2O + SiO_2 = NaAlSi_3O_8 + H_2O \quad (5.3)$$

analcime + quartz = albite + vapor

$$KAlSi_2O_6 + SiO_2 = KAlSi_3O_8 \quad (5.4)$$

leucite + quartz = orthoclase

$$NaAlSiO_4 + 2\,SiO_2 = NaAlSi_3O_8 \quad (5.5)$$

nepheline + 2 quartz = albite

Leucite is a rare mineral found in K-rich volcanic rocks. It is unknown in plutonic, metamorphic, or sedimentary rocks. Nepheline is a common mineral in syenite and other silica-poor volcanic or plutonic igneous rocks. Leucite and nepheline are usually associated with K-feldspar. Analcime also crystallizes from a magma and it may form as a secondary mineral in vugs, cracks, or veins. Occasionally it is found as a secondary mineral in sandstones or tuffs. When it crystallizes from a magma, it is commonly associated with olivine, leucite, and perhaps sodalite. When it is secondary, other low-temperature minerals such as zeolites or prehnite often accompany it.

Mica Group

The two most important groups of minerals within the sheet silicate subclass are micas and clays. We will consider only the micas in this chapter and defer discussion of clay minerals to the chapter on sedimentary rocks and minerals. All sheet silicates have atomic structures composed of sheets (layers) of $(Si,Al)O_4$ tetrahedra with alkalis and other metals between (Figure 5.15a). Bonding within a sheet is much stronger than bonding between sheets. Consequently, as shown in Figure 5.15b and Plate 6.6, well-developed basal cleavage characterizes mica and most other sheet silicates. Common micas contain K^+ as an interlayer cation between the tetrahedral layers. Fe^{2+}, Mg^{2+}, or Al^{3+} occupy positions between the apices of the tetrahedra. Within the tetrahedral layers themselves, Al^{3+} commonly substitutes for some Si^{4+}. A general formula for micas is $K(Al,Mg,Fe)_{2-3}(AlSi_3O_{10})(OH)_2$. So, we can de-

scribe most mica compositions in terms of three end members:

muscovite	$KAl_2(AlSi_3O_{10})(OH)_2$
annite	$KFe_3(AlSi_3O_{10})(OH)_2$
phlogopite	$KMg_3(AlSi_3O_{10})(OH)_2$

The two most important micas are biotite and muscovite (Figure 5.16). Other mica minerals exist, but they are not common. Muscovite is usually close to the end member composition given above. Biotite may have any composition between end members annite and phlogopite, and may incorporate a small amount of muscovite as well. The name *annite* refers only to the ideal end member, but the name *phlogopite* is often used in a more general sense to describe any Mg-rich biotite. Other elemental substitutions (such as F^- substituting for OH^-) occur in micas, but they are generally minor. Lepidolite, an especially Li-rich mica similar to muscovite, is a common large and euhedral mineral in pegmatites. It is often associated with the lithium-aluminum pyroxene, spodumene.

Muscovite is more common than biotite, but both occur in a wide variety of igneous and metamorphic rocks. Micas react to form clays and other minerals when exposed to prolonged weathering, and are therefore absent from most sedimentary rocks. Biotite, and especially muscovite, are relatively Al- and Si-rich compared with many other igneous minerals. Muscovite, therefore, is found in silicic igneous rocks such as granite, but rarely in rocks of intermediate or mafic composition. Biotite is found in rocks ranging from granitic to mafic composition. Phlogopite is occasionally found in ultramafic rocks.

Chain Silicates

Pyroxenes and **amphiboles** are the two most important groups of minerals within the chain silicate subclass. All chain silicates are characterized by single or double chains of SiO_4 tetrahedra linked by monovalent or divalent cations. In most pyroxenes, the linking cations occupy two distinct kinds of sites between the chains; in amphiboles, they occupy four or five kinds of sites between the chains. In both pyroxenes and amphiboles, one interchain site is significantly larger than the other(s). Figure 5.17 compares the structures of pyroxenes and amphiboles. Figure 13.17 (Chapter 13) shows a more detailed comparison. Pyroxenes are anhydrous minerals having simple formulas compared with amphiboles, all of which are hydrous. The similarity between the structures of pyroxenes and amphiboles results in some similarity in physical properties.

▶**FIGURE 5.15**
Mica structure and cleavage:
(a) The atoms in muscovite are arranged in layers. K^+ ions link layers of SiO_4 tetrahedra sandwiched around Al^{3+} ions; (b) due to its layered structure, biotite has one excellent cleavage and easily breaks into sheets and flakes.

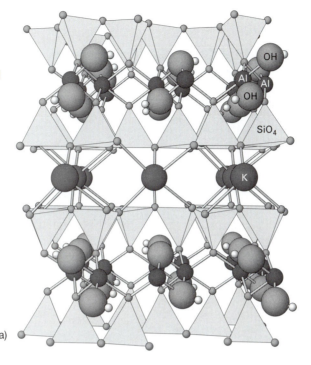

(a)

(b)

Pyroxene Group

Plates 6.1–6.4 show some pyroxenes. Pyroxenes may contain many different elements, but we describe the compositions of most pyroxenes in terms of five end members:

wollastonite	$Ca_2Si_2O_6 = CaSiO_3$
ferrosilite	$Fe_2Si_2O_6 = FeSiO_3$
enstatite	$Mg_2Si_2O_6 = MgSiO_3$
diopside	$CaMgSi_2O_6$
hedenbergite	$CaFeSi_2O_6$

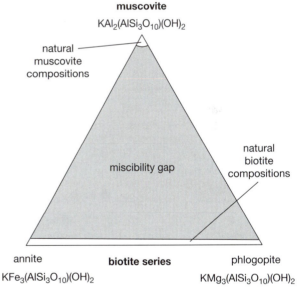

▶**FIGURE 5.16**
Ternary diagram showing the principal mica end members: muscovite, phlogopite, and annite. Micas termed muscovite are generally close to end member $KAl_2(AlSi_3O_{10})(OH)_2$ in composition. In contrast, biotite compositions range between annite and phlogopite. Compositions between muscovite and biotite do not exist due to the large miscibility gap.

A general formula, then, would be $(Ca^{2+}, Fe^{2+}, Mg^{2+})_2Si_2O_6$. Figure 5.18 shows a wollastonite-ferrosilite-enstatite ternary diagram, with some important end members indicated. The mineral wollastonite, although used as an end member to

▶**FIGURE 5.17**
Comparison of the structures of pyroxenes and amphiboles. (a) Schematic drawing showing how silica tetrahedra are polymerized in both to form chains. (b) An oblique view of the chains. (c) Schematic view looking down the chains to show the locations of interchain cations (solid black dots). The small open circles in the amphibole drawing show sites occupied by K or Na in hornblende and a few other amphiboles. Note that in these drawings, as in most drawings of silicates, the Si^{4+} cations at the centers of the tetrahedra are omitted for clarity. For another view of these structures, see Figure 13.17.

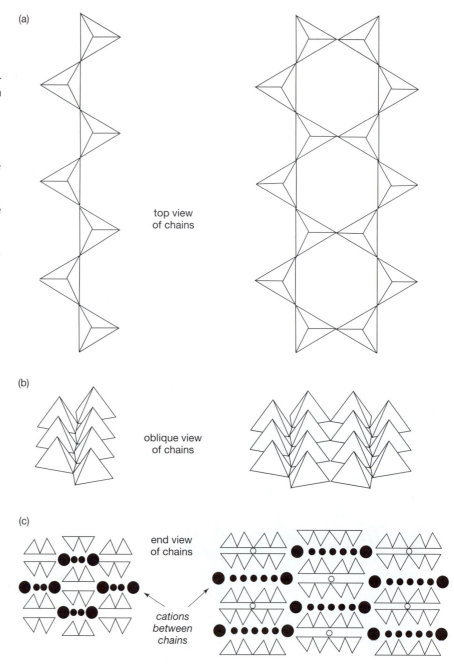

(a)

top view of chains

(b)

oblique view of chains

(c)

end view of chains

cations between chains

describe pyroxene compositions, is not a pyroxene. It has a different structure and belongs to the pyroxenoid group. As with the feldspars, we use abbreviations to give composition. A pyroxene of composition $Wo_{10}Fs_{83}En_{07}$, for example, has the formula $(Ca_{0.10}Fe_{0.83}Mg_{0.07})SiO_3$, equivalent to $(Ca_{0.10}Fe_{0.83}Mg_{0.07})_2Si_2O_6$.

Natural pyroxenes fall into two main series distinguished by different atomic arrangements and different crystal shapes. **Orthopyroxene,** predominantly solid solutions of end members ferrosilite and enstatite, has general formula $(Mg,Fe)_2Si_2O_6$. **Clinopyroxene,** predominantly solid solutions of diopside and hedenbergite, has the general formula $Ca(Mg,Fe)Si_2O_6$. Figure 5.19 shows two examples of clinopyroxene; all the pyroxenes in Plates 6.1–6.4 are clinopyroxene. No pyroxenes have compositions more calcic than clinopyroxene because Ca is limited to only the larger two of the four sites between the chains (Figure 5.17c). We call lines between end members on diagrams such as Figure 5.18 **joins.** The four-sided polyhedron bounded above by the diopside-hedenbergite (clinopyroxene) join, and below by the enstatite-ferrosilite (orthopyroxene) join, is the **pyroxene quadrilateral** (Figure 5.20). It encompasses the compositions of all natural Ca-Mg-Fe pyroxenes.

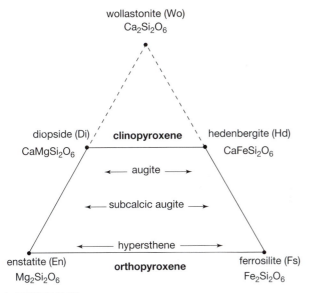

►FIGURE 5.18

Ternary diagram showing the principal pyroxene end members. We term the bottom half of the diagram, cornered by diopside, hedenbergite, ferrosilite, and enstatite, the *pyroxene quadrilateral*. End member clinopyroxenes have compositions that fall on the diopside-hedenbergite join, and end member orthopyroxenes fall on the enstatite-ferrosilite join. We call intermediate pyroxenes augite, subcalcic augite, or hypersthene. Natural pyroxenes are never more calcic than ideal clinopyroxene. Wollastonite, which is often used as an end member to describe pyroxene composition, is not a true pyroxene.

►FIGURE 5.19

Two examples of clinopyroxenes: (a) Augite, principally a Ca-Mg-Fe clinopyroxene, forms prismatic crystals and displays two prominent cleavages that intersect at 87° and 93°. The cleavage angles help distinguish clinopyroxene from hornblende: (b) a large euhedral crystal of hedenbergite (Ca-Fe clinopyroxene) in a pyroxene-feldspar matrix.

Clinopyroxenes often contain appreciable amounts of Na^+ and Al^{3+} besides the elements already discussed. *Augite* is the name given to a "dirty" clinopyroxene with the general formula $(Ca,Mg,Fe,Na)(Mg,Fe,Al)(Si,Al)_2O_6$. An end member pyroxene, jadeite, has formula $NaAlSi_2O_6$. It exists in high-pressure metamorphic rocks. Other Na-bearing pyroxenes such as acmite $(NaFeSi_2O_6)$ are found in some igneous rocks. Li, Cr, and Ti also are found in clinopyroxene, but are normally minor elements. However, spodumene, $LiAlSi_2O_6$, is an important pyroxene in some pegmatites. Al^{3+} and some other elements can substitute into orthopyroxene, but the extent of the substitutions is usually minor.

Natural clinopyroxene can have any composition between diopside and hedenbergite, and often is somewhat deficient in $CaSiO_3$, giving it a composition that plots within the pyroxene quadrilateral. Similarly, orthopyroxene may have any composition between enstatite and ferrosilite, and often contains a small amount of $CaSiO_3$. However, orthopyroxene and clinopyroxene have different crystal structures and there is a complex solvus between them, in the middle of the quadrilateral. Consequently, pyroxenes with compositions that plot in the middle of the quadrilateral are only found in some high-temperature rocks; augite, subcalcic augite, and pigeonite are names given to them when they occur (Figures 5.18 and 5.20). Augite and subcalcic augites are clinopyroxenes with Ca:(Mg + Fe) values significantly less than 1.0. Pigeonite, first found at Pigeon Point, Minnesota, are very high temperature pyroxenes that have an atomic structure dissimilar from both clinopyroxene and orthopyroxene.

Subcalcic augite and pigeonite are, in general, only stable at high temperature. Analogous to the feldspars, a homogeneous high-temperature pyroxene of intermediate composition may become unstable and unmix to form two pyroxenes at low temperature. This sometimes leads to exsolution similar to that of feldspars; Figure 5.21 compares the two. The compositions of exsolution lamellae depend on the temperature at which exsolution occurred. So, both the pyroxene and feldspar solvi are sometimes used as

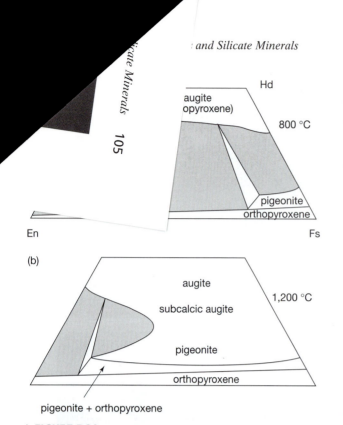

(b)

▶**FIGURE 5.20**
Two views of the pyroxene quadrilateral showing the miscibility gap between orthopyroxene and clinopyroxene: (a) 800 °C and (b) 1,200 °C. Note that the gap is much larger at 800 °C than at 1,200 °C. Both augite and pigeonite are clinopyroxenes.

geothermometers to learn the temperature at which a rock formed (Box 5.6).

Amphibole Group

The atomic structure of amphibole is more complex than pyroxene, although they are both chain silicates (Figure 5.17). Amphibole chemistry is highly variable and yields many different end member formulas (Table 5.1). The amphibole ternary in

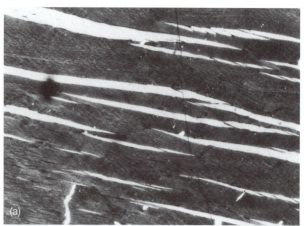

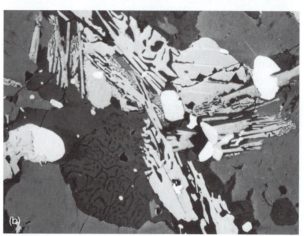

▶**FIGURE 5.21**
Two examples of exsolution: (a) photograph (1 mm wide) of a thin section showing exsolution in alkali feldspar and (b) scanning electron microscope image (0.4 mm wide) of pyroxene showing bleb and lamellar exsolution.

Figure 5.23 shows some of the simpler end members. Along the base of the diagram, the figure is complicated because Ca-free amphiboles form two

▶**TABLE 5.1**
Comparison of Amphibole and Pyroxene End Members and Series

Pyroxene End Member or Series		Amphibole End Member or Series	
enstatite	$Mg_2Si_2O_6$	magnesio-anthophyllite	$Mg_7Si_8O_{22}(OH)_2$
orthopyroxene	$(Mg,Fe)_2Si_2O_6$	anthophyllite	$(Mg,Fe)_7Si_8O_{22}(OH)_2$
ferrosilite	$Fe_2Si_2O_6$	ferroanthophyllite	$Fe_7Si_8O_{22}(OH)_2$
diopside	$CaMgSi_2O_6$	tremolite	$Ca_2Mg_5Si_8O_{22}(OH)_2$
Mg-Fe clinopyroxene	$Ca(Mg,Fe)Si_2O_6$	actinolite	$Ca_2(Fe,Mg)_5Si_8O_{22}(OH)_2$
hedenbergite	$CaFeSi_2O_6$	ferroactinolite	$Ca_2Fe_5Si_8O_{22}(OH)_2$
jadeite	$NaAlSi_2O_6$	glaucophane	$Na_2Mg_3Al_2Si_8O_{22}(OH)_2$
acmite	$NaFeSi_2O_6$	riebeckite	$Na_2Fe_5Si_8O_{22}(OH)_2$
augite	$(Ca,Mg,Fe,Na)(Mg,Fe,Al)$ $(Si,Al)_2O_6$	hornblende	$(K,Na)_{0-1}(Ca,Na,Fe,Mg)_2$ $(Mg,Fe,Al)_5(Si,Al)_8O_{22}(OH)_2$

Box 5.6 Diopside-Enstatite Solvus and Geothermometry

If a mineral system contains a miscibility gap, we can sometimes use mineral compositions to determine the temperature at which a rock equilibrated. Petrologists call such mineral systems *geothermometers*. Geothermometers are based on a fundamental consequence of thermodynamics: at high temperatures, solid-solution minerals may have intermediate compositions, but at low temperatures they tend to unmix so that compositions are relatively close to end members.

The miscibility gap between orthopyroxene (OPX) and clinopyroxene (CPX) is often used to calculate the temperature at which igneous or metamorphic rocks formed. Figure 5.22 is a schematic showing the OPX-CPX gap between enstatite and diopside. The solvus, the line that shows compositions of coexisting OPX and CPX at different temperatures, is narrower at high temperature than at low temperature. Consequently, low-temperature pyroxenes are closer to end member enstatite and diopside than high-temperature pyroxenes.

A pyroxene of composition X will be stable at high temperature (above T_0), but if it crystallizes or equilibrates at lower temperature, it will unmix to form two pyroxenes. Unmixing may produce separate grains of OPX and CPX in the same rock, or it may produce single grains of pyroxene containing blebs or exsolution lamellae of different compositions (Figure 5.21). By analyzing the compositions of coexisting OPX and CPX, petrologists can estimate the temperature of equilibration.

Pyroxenes are not the only minerals that can be used as geothermometers. Feldspars, carbonates, and others can serve the same purpose. Geothermometry is not, however, always as simple as Figure 5.22 might imply. Many things besides temperature affect the compositions of coexisting minerals. Among others, petrologists must be concerned with the effects of pressure, of minor elements in minerals, and of disequilibrium.

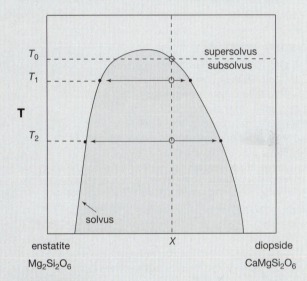

▶FIGURE 5.22

Schematic diagram showing the diopside-enstatite solvus: A pyroxene of composition X (indicated by the vertical dashed line) will be stable above T_0 (at super-solvus temperatures). Below T_0 (at subsolvus temperatures) it will unmix to produce two pyroxenes (shown by solid dots). The degree to which it unmixes depends on the temperature: at T_2 pyroxene compositions will be closer to end members than at T_1.

distinct series, having different atomic structures. We call the most Mg-rich amphiboles *anthophyllite* or, if they contain appreciable amounts of Fe, *ferroanthophyllite*. Plate 3.4 shows a fibrous variety of anthophyllite. We give aluminous anthophyllite (not shown on Figure 5.23) the name *gedrite*. The cummingtonite-grunerite series covers most intermediate and Fe-rich compositions. Calcic amphiboles (such as the actinolite in Plate 3.2) can have any composition between an Mg end member (tremolite) and an Fe end member (ferroactinolite). A miscibility gap exists between the calcic amphiboles

and the Ca-poor amphiboles, analogous to the one between clinopyroxene and orthopyroxene.

Amphibole chemistry also varies in ways not shown by Figure 5.23. The diagram depicts variations in Ca, Mg, and Fe content well, but many amphiboles contain K, Na, Al, Ti, and other elements in significant amounts. Additionally, some amphiboles do not contain the same number of atoms in their formulas as quadrilateral amphiboles. We generally call such amphiboles *hornblende*. Hornblende, the most common amphibole, has an idealized formula, $(K,Na)_{0-1}(Ca,Na,Fe,Mg)_2(Mg,Fe,Al)_5(Si,Al)_8$

$O_{22}(OH)_2$. Even this complicated formula does not do justice to all the possible compositional variations. Without chemical analyses, telling different amphiboles apart is difficult, and the name hornblende is often used to refer to any black amphibole (Figure 5.24). Many amphibole end members have analogs in the pyroxene group; we compare some of the more important end members in Table 5.1.

Amphiboles are important and essential minerals in many kinds of igneous rocks. Hornblende is absent from many granitic rocks but is common in rocks of intermediate to mafic composition where it coexists with plagioclase. Besides hornblende, other amphiboles are also found in igneous rocks. Amphiboles also exist in many metamorphic rocks, including marbles and metamorphosed mafic igneous rocks. They are especially common in high-temperature metamorphic rocks called *amphibolites*, which contain predominantly hornblende and plagioclase. *Glaucophane*, a blue sodic amphibole, has the general formula $Na_2 Al_2 Si_8 O_{22}(OH)_2$. Its presence is often associated with rocks formed in subduction zones under high pressures and moderate temperatures. Other Ca-free amphiboles (anthophyllite, gedrite, cummingtonite, and grunerite) are found in metamorphic rocks and occasionally in extrusive igneous rocks. Tremolite is common in high-temperature marbles. As with some other silicates that exist in marbles,

▶**FIGURE 5.23**
The principal amphibole end members. Calcic amphiboles generally have compositions falling near the tremolite-ferroctinolite join and noncalcic amphiboles have compositions near the anthophyllite-grunerite join. Natural amphiboles are never more calcic than tremolite-ferroactinolite.

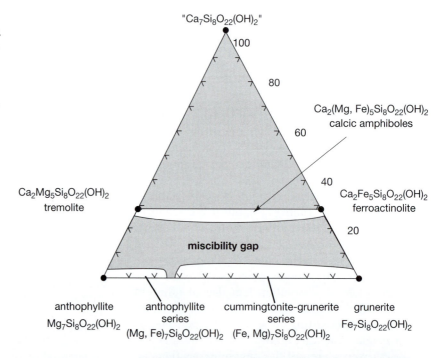

▶**FIGURE 5.24**
Hornblende (a) can be confused with black pyroxene, but the 60° and 120° angles between cleavages (shown in the enlarged drawing) help distinguish it. Seeing cleavage in hand specimen may require a hand lens. (b) clevage is more easily seen in thing section views such as this one.

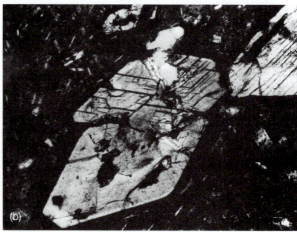

identifying it may be difficult because of its inconspicuous white color. Amphiboles are absent in the highest temperature marbles and other metamorphic rocks.

Olivine Group

Olivines, belonging to the isolated tetrahedra silicate subclass, have structures comprising individual SiO_4 tetrahedra linked by divalent cations (Figure 5.25). In contrast with some of the other silicates previously discussed, olivine chemistry is quite simple. Its general formula is $(Mg,Fe,Mn,Ca)_2SiO_4$ but often we omit Mn and Ca because they are generally minor components. End members are:

forsterite	Mg_2SiO_4
fayalite	Fe_2SiO_4
tephroite	Mn_2SiO_4
larnite	Ca_2SiO_4

Natural olivines are generally solid solutions of fayalite and forsterite with only minor tephroite and larnite. Other end members and elemental substitutions in olivine are very minor. We usually use abbreviations and subscripts to indicate olivine compositions. $Fo_{88}Fa_{09}Te_{02}La_{01}$, for example, refers to an olivine of composition $(Mg_{0.88}Fe_{0.09}Mn_{0.02}Ca_{0.01})_2SiO_4$. This

might be the composition of the forsterite-ri[ch] shown in Plate 6.5.

Olivine is primarily an igneous mineral, cryst[alliz]ing from high-temperature magmas. Its two most i[m]portant end members, forsterite and fayalite, melt at different temperatures, so olivine's melting behavior is similar to that of plagioclase (Figure 5.26). Olivine occurs predominantly in mafic and ultramafic igneous rocks. Its very mafic composition restricts its occurrence, and other mafic minerals, containing the same elements but more Si, are more common. In some basalts Mg-rich olivine forms large green phenocrysts in a fine-grained plagioclase-pyroxene groundmass (Plate 6.5). Mg-rich olivine is never found in silicic rocks, but Fe-rich olivine is occasionally found in some granites. Olivine occurs in some metamorphic rocks, including marble and metamorphosed igneous rocks.

Other Minerals in Igneous Rocks

Besides the important silicates, igneous rocks contain many other minerals (Table 5.2). Most are only common as accessory minerals because they are composed of rare elements or of elements that easily fit into other minerals. Magnetite and ilmenite, for example, are generally minor because they comprise elements that also easily fit into other, more stable, essential minerals. Rocks that contain F, Zr, or P may contain fluorite, zircon, or apatite, but F, Zr, and P are very minor elements in all but the most unusual rocks. Although not abundant, zircon often contains uranium and lead, which we

▶**FIGURE 5.25**
Schematic view of the atomic arrangement of olivine showing isolated silica tetrahedra, some pointing up and some pointing down, with metal cations (black dots) between. In most olivines, Mg^{2+} and Fe^{2+} are the principal cations linking silica tetrahedra.

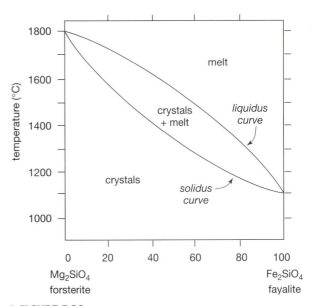

▶**FIGURE 5.26**
Phase diagram showing the melting and crystallization temperatures of forsterite-fayalite olivines at 1 atm. Compare this diagram with Figure 5.13.

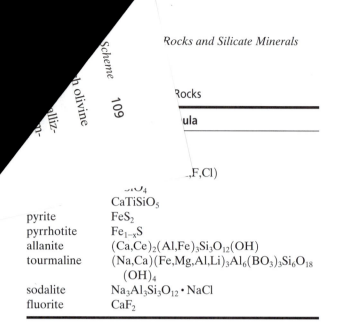

Scheme

109

	$CaTiSiO_5$
pyrite	FeS_2
pyrrhotite	$Fe_{1-x}S$
allanite	$(Ca,Ce)_2(Al,Fe)_3Si_3O_{12}(OH)$
tourmaline	$(Na,Ca)(Fe,Mg,Al,Li)_3Al_6(BO_3)_3Si_6O_{18}$ $(OH)_4$
sodalite	$Na_3Al_3Si_3O_{12} \cdot NaCl$
fluorite	CaF_2

may analyze to learn radiometric ages of igneous rocks.

THE NAMING OF IGNEOUS ROCKS

Dividing igneous rocks into groups that give them names is an arbitrary process. Igneous rocks vary greatly in mineralogy and texture, and rock types grade from one to another. If we invoke too many names, terminology becomes too complicated to be useful. If we invoke too few names, they are too broad to have much meaning. The task is somewhat simplified because, with rare exceptions, the major elements in all igneous rocks are the same (see Box 5.1, page 90). In addition, certain minerals have affinities for each other because of their chemical compositions. Mafic igneous rocks, for example, generally contain pyroxene and Ca-plagioclase. Silicic igneous rocks contain predominantly quartz, alkali feldspar, and micas. Mafic rocks, therefore, have a characteristic dark color, while silicic rocks are white or pink. Despite these generalizations, however, igneous petrology is complex and highly variable.

SIMPLE CLASSIFICATION SCHEME

Most igneous rocks contain feldspar and quartz. Figure 5.27 depicts a simple classification scheme for igneous rocks based on the relative amounts of quartz, plagioclase, and alkali feldspar that are present. Figure 5.27 works well for most igneous rocks, and we can often identify quartz, plagioclase, and alkali feldspar in the field. The classification scheme does not work well for rocks that do not contain significant amounts of quartz or feldspar,

and it ignores compositional variations in feldspars. Magmas poor in SiO_2 but rich in alkalis may crystallize nepheline or leucite instead of feldspars. Similarly, ultramafic rocks (those without any quartz or alkali feldspar) are not well represented on this diagram. Box 5.1 gives the chemical compositions of some of the rocks in Figure 5.27. We can verify that the relative proportions of quartz, alkali feldspar, and plagioclase reflect variations in silica (SiO_2), alkali (K_2O and Na_2O), and alkaline earth (CaO) content.

MINERAL MODES

Figure 5.28 is an idealized diagram showing the relative amounts (by volume), also called the *modes,* of key minerals present in some common igneous rocks. This figure emphasizes the affinities that some minerals have for others. Mafic rocks often contain Mg-rich olivine or pyroxene along with Ca-rich plagioclase. They contain little quartz or K-feldspar. Silicic rocks are more likely to contain hydrous minerals such as muscovite or biotite and are often rich in quartz, K-feldspar, or Na-rich plagioclase. Olivine, nepheline, and other feldspathoid minerals can never coexist with quartz. Thus, certain minerals can be associated with each other in nature, while others cannot. Different igneous rock types, formed in different environments, have characteristic mineral assemblages. Such relationships are primarily controlled by the composition of the magma, which in turn reflects the process and source that generated it.

COMMON TYPES OF IGNEOUS ROCK

Silicic Igneous Rocks ($>$20% Quartz)

Silicic igneous rocks include the plutonic rocks granite, tonalite, and granodiorite, and their volcanic equivalents (rhyolite, andesite, and dacite). Table 5.3 lists some varieties. All contain $>$20% quartz; we name them based on their feldspar content. Biotite, hornblende, and muscovite also may be present as minor minerals. Some granitic rocks contain no plagioclase, and some tonalitic rocks contain no K-feldspar, but feldspars of some sort are always present. Plagioclase in granitic rocks may be nearly pure albite (the Na end member), but in other silicic igneous rocks plagioclase is more Ca-rich. In extrusive rocks, K-feldspar may be sanidine instead of orthoclase or microcline, and cristobalite or tridymite may replace quartz. Accessory minerals include magnetite, ilmenite, rutile, pyrite, pyrrhotite, zircon, sphene, and apatite. Because they contain large amounts of quartz and feldspars, granitic

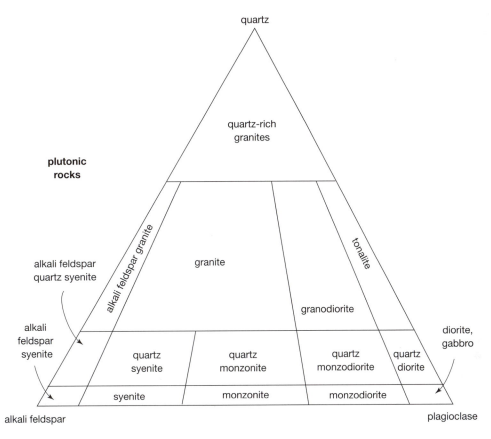

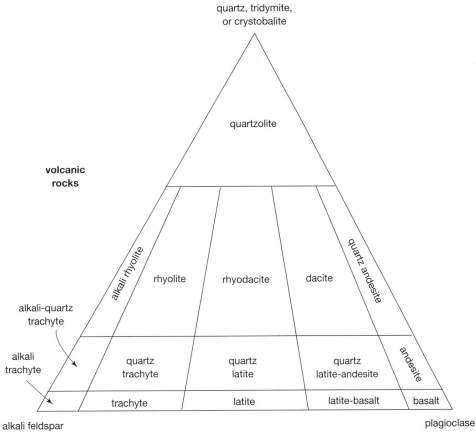

▶**FIGURE 5.27**
An igneous rock classification scheme. Adapted from Hyndman, 1972, and Klein and Hurlbut, 1993.

▶TABLE 5.3
Examples of Silicic Igneous Rocks and Minerals

Feldspars Present	Plutonic Rock Name	Volcanic Rock Name	Major Minerals	Minor Minerals
more K-feldspar than plagioclase	granite	rhyolite	K-feldspar quartz plagioclase	biotite hornblende
more plagioclase than K-feldspar	granodiorite	dacite	plagioclase K-feldspar quartz hornblende	biotite
much more plagioclase than K-feldspar	tonalite	quartz andesite	plagioclase quartz	K-feldspar biotite hornblende

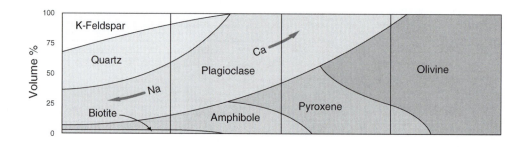

▶FIGURE 5.28
The major minerals and their proportions (modes) in igneous rocks. Granite and basalt are the two most common igneous rocks, andesite and rhyolite slightly less common, and the others relatively rare.

and granodioritic rocks are light in color (Figure 5.29). Granite and rhyolite may have a pinkish color due to oxidized Fe in K-feldspar. Tonalite may be darker in color due to large amounts of biotite, hornblende, and, occasionally, pyroxene. Often, however, telling silicic rocks apart without using a microscope is difficult.

Granitic melts form at temperatures as low as 650 °C. While some granitic magmas come from deep within the Earth, partial melting **(anatexis)** of shallow crustal rocks produces others. Silicic melts may intrude to form large plutons or dikes. If they reach the surface, violent volcanic activity may re-

sult. More commonly, they crystallize to form large plutons. Huge **batholiths,** plutons of extreme size, are common in many mountain belts around the world.

Intermediate and Mafic Igneous Rocks (0–20% Quartz)

Intermediate igneous rocks are those in which quartz accounts for 5% to 20% of the rock. They include quartz syenite, quartz monzonite, and quartz diorite, and their volcanic equivalents quartz

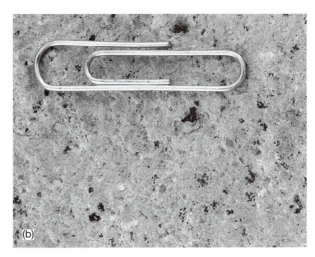

▶FIGURE 5.29
Silicic igneous rocks in hand specimen: 6 cm wide samples of (a) granite and (b) rhyolite. The granite contains black crystals of biotite, white K-feldspar and gray quartz. The rhyolite contains the same minerals but they are too fine grained to be seen. Both photos are at the same scale.

trachyte, quartz latite, and quartz andesite. K-feldspar dominates syenitic rocks (Table 5.4). Plagioclase is the only feldspar in most diorites. The plagioclase varies from an Na-rich composition in syenite to intermediate Na-Ca compositions in diorite. Hornblende, biotite, and pyroxene are common in these rocks, but usually in small amounts. As with silicic rocks, intermediate rocks with more plagioclase (diorite and some monzonite) tend to have more mafic minerals and thus a darker color. Magnetite, ilmenite, and apatite are common as accessory minerals.

A dark color, due to an abundance of hornblende, clinopyroxene, or olivine, characterizes mafic rocks (Figure 5.30). Examples in Table 5.4 include basalt (volcanic), and diorite and gabbro (plutonic).

K-feldspar and quartz are minor or absent in all mafic rocks. Plagioclase varies from an intermediate Ca-Na composition (in diorite) to one that is very Ca-rich (in gabbro). While all intermediate and mafic igneous rocks may contain hornblende and clinopyroxene, hornblende is more common in diorite and andesite, and clinopyroxene is more common in gabbros and basalts. Gabbros and basalts may also contain olivine. Some spectacular basalts contain large green olivine crystals in a fine matrix of plagioclase and clinopyroxene (Plate 6.5). Magnetite, ilmenite, apatite, sphene, and zircon are common accessory minerals in all mafic rocks.

Mafic magmas form at very high temperatures ($>1,100\,°C$). Melting of crustal rocks cannot produce them; most come from great depth. If they reach the

▶TABLE 5.4
Examples of Intermediate and Mafic Igneous Rocks

Feldspars Present	Plutonic Rock Name	Volcanic Rock Name	Major Minerals	Minor Minerals
more K-feldspar than plagioclase	syenite	trachyte	K-feldspar plagioclase ±quartz	biotite hornblende
equal amounts of K-feldspar and plagioclase	monzonite	latite	plagioclase K-feldspar ±quartz	biotite hornblende clinopyroxene
plagioclase only	diorite	andesite	plagioclase hornblende	hornblende biotite K-feldspar
plagioclase only	gabbro	basalt	plagioclase clinopyroxene	orthopyroxene olivine

(a)   (b)

▶**FIGURE 5.30**

Mafic rocks in hand specimen: (a) gabbro and (b) vesicular basalt. The plutonic rock (gabbro) is coarser grained than the volcanic rock (basalt). Both rocks are dominated by plagioclase. They contain lesser amounts of clinopyroxene and very minor amounts of olivine. The vesicles (bubbles) in the basalt formed from gas that was present when the rock crystalized.

surface, the result may be lava flows, called **flood basalts,** that cover large areas. Excellent examples are found on the Snake River Plain of eastern Idaho and eastern-central Washington. They also produce quiescent volcanoes, such as those found in Hawaii. Scientists studying the Hawaiian volcanics sometimes closely approach vents and lava fountains, something that is never done on Mount St. Helens or other potentially violent mountains characterized by silicic volcanics.

ULTRAMAFIC IGNEOUS ROCKS

Ultramafic rocks are especially poor in Si and have high Mg:Fe ratios; Table 5.5 gives some examples. Peridotite is a general term used for all of them. Pyroxene, either orthopyroxene or clinopyroxene, olivine, and plagioclase are the dominant minerals in peridotites. We call rocks composed nearly entirely of pyroxene *pyroxenite,* and those composed nearly

entirely of olivine *dunite* (Figure 5.31). Many dunites look like light-green equivalents of sandstone. Because of their very high melting temperatures and mafic compositions, ultramafic magmas must come from deep within the Earth. They rarely reach the surface to produce volcanic rocks, but some Precambrian terranes contain spectacular examples of ultramafic lava flows called *komatiites.*

Ultramafic minerals are most stable at high temperature and tend to alter by reaction with water or carbon dioxide when exposed to normal Earth surface conditions. Consequently, finding fresh, unaltered ultramafic rock is difficult. Variable amounts of secondary serpentine, chlorite, talc, brucite, or calcite are nearly always present. Rocks called *serpentinites,* in which serpentine has replaced all mafic minerals, often result. Many ultramafic rocks, including *kimberlites,* ultramafic rocks associated with diamonds, have been altered so much that we cannot determine the original mineralogy.

▶**TABLE 5.5**

Ultramafic Igneous Rocks

Plutonic Rock Name	Volcanic Rock Name	Major Minerals	Minor Minerals
peridotite	komatiite	olivine clinopyroxene orthopyroxene	hornblende
pyroxenite	does not exist	clinopyroxene orthopyroxene	olivine hornblende
dunite	does not exist	olivine	spinel

▶**FIGURE 5.31**
Dunite in hand specimen and thin section. (a) A 0.5 cm wide view of dunite in hand specimen, showing only olivine. (b) Dunite in thin section showing high relief olivine grains. (1 mm wide view).

Box 5.7 *Minerals from the Moon*

Moon rocks collected by the *Apollo* astronauts and unmanned Soviet missions include anorthosites, gabbros, and basalts. Most are basalts. Although these rocks indicate an active volcanic history, significant eruptions ceased on the moon about 3.2 billion years ago. Petrologists classify the ancient basalts exposed at the Moon's surface into several groups: olivine basalts, silica-rich basalts, aluminum-rich basalts, titanium-rich basalts, and so on. All have equivalents on the Earth, but lunar basalts are generally richer in TiO_2, FeO, MgO, and Cr_2O_3, and poorer in volatiles (H_2O, CO_2, O_2) than terrestrial basalts.

Lunar basalts and other moon rocks contain the same minerals found in terrestrial rocks, including pyroxenes, plagioclase, olivine, ilmenite, and small amounts of K-feldspar. However, the chemical differences between Earth and Moon rocks result in slight differences in mineral chemistries. For example, lunar anorthosites typically consist of very calcium-rich plagioclase ($An_{97-98}Ab_{2-3}$) with lesser amounts of orthopyroxene, olivine, clinopyroxene, pigeonite, and Mg-rich spinel. In contrast, terrestrial anorthosites tend to have more sodium in plagioclase ($An_{34-77}Ab_{23-66}$), and spinels are normally rich in iron. Study of lunar rocks also disclosed some previously undiscovered minerals resulting from excess titanium and other chemical differences. Two new silicates include pyroxferroite, $CaFe_6(SiO_3)_7$, and tranquillityite, $Fe_8(Zr,Y)_2Ti_3Si_3O_{24}$, which is named after the Sea of Tranquility, the site of the first Moon landing. Armalcolite, $(Fe,Mg)Ti_2O_5$, is named after the three *Apollo 11* astronauts, Neil Armstrong, Ed Aldrin, Jr., and Michael Collins.

▶QUESTIONS FOR THOUGHT

(Some of these questions have no specific correct answers; they are intended to promote thought and discussion.)

1. Why are igneous rocks dominated by silicate minerals?
2. Most igneous rocks contain either oxide or sulfide minerals (or both) as accessory minerals. Why?
3. Why do some elements become concentrated in pegmatites?
4. All granites contain quartz and K-feldspar. Why?

5. Suppose some olivine or pyroxene crystallize from a mafic magma. The remaining magma then moves upward, leaving the crystals behind. If this process repeats, how will the magma's composition change?
6. The Earth's upper mantle is roughly gabbroic (mafic) in composition and acts as a source region for magmas that move up into the crust. Partial crystallization, alluded to in question 5, commonly occurs. If crystals are left behind in the mantle and the magma then

crystallizes in the crust, what effect does this have on the relative compositions of the crust and mantle?

7. Quartz has conchoidal fracture. Biotite has perfect planar cleavage. Why?

8. Suppose olivine begins to crystallize from a magma at high temperature. As cooling progresses, the olivine crystal grows larger as its composition changes. If the olivine stays in equilibrium, it will always be homogeneous. What compositional changes would you expect to occur as cooling takes place? Suppose the cooling is too rapid for the olivine crystal to stay in equilibrium. It will then develop compositional growth rings. How will the composition of the center of the grain compare to the composition of the margins?

9. Plagioclase is the most abundant mineral in the Earth's crust. However, if you wander outside in most places, even if you could see the bedrock you probably wouldn't see a lot of rocks containing plagioclase. What kind of rocks contain lots of plagioclase? Where is all the crustal plagioclase?

▶RESOURCES

(See also the general references listed at the end of Chapter 1.)

Hatch, F. H., A. K. Wells, and M. K. Wells. *Petrology of the Igneous Rocks.* London: Thomas Murby and Co., 1980.

Hyndman, D. W. *Petrology of Igneous and Metamorphic Rocks.* New York: McGraw-Hill, 1972.

Nockolds, S. R. Average chemical compositions of some igneous rocks. *Geological Society of America Bulletin* 65 (1954): 1007–1032.

Thorpe, R., and G. Brown. *The Field Description of Igneous Rocks.* New York: Halsted/Wiley, 1985.

6

Sedimentary Minerals and Sedimentary Rocks

Sedimentary rocks and minerals cover most of the surface of the Earth. The processes that produce them are more variable than those that produce igneous rocks and minerals, but the number of common sedimentary minerals is small. In this chapter we talk about sedimentary processes and common sedimentary minerals and introduce a classification scheme for sedimentary rocks.

Sediments and **sedimentary rocks** cover about 80% of the Earth's surface but are less than 1% of the volume of the Earth's crust. They are, in effect, a thin blanket on top of igneous and metamorphic **basement rocks** (Figure 6.1). Sediments, and thus sedimentary rocks, are mostly recycled materials derived from preexisting igneous, metamorphic, or sedimentary rocks.

Petrologists usually divide sedimentary rocks into two main groups: **detrital rocks** and **chemical rocks.** Detrital sedimentary rocks are formed by **compaction** and **cementation** of **clastic** sediments composed of individual mineral grains or **lithic fragments** (pieces of rock). The mineral grains and lithic fragments vary greatly; we call them collectively **clasts** (from *klastos,* the Greek word meaning "broken"), so we often call detrital sedimentary rocks **clastic rocks.** Because their mineralogy varies so much, we generally classify detrital sedimentary rocks based on grain size rather than composition. **Conglomerate** contains large rounded clasts (>2 mm in longest dimension) separated by a fine-grained material called **matrix. Sandstone** contains sand-sized (0.062 to 2 mm in longest dimension) quartz or

feldspar grains, and sometimes lithic fragments. **Mudstone** and **shale** primarily contain microscopic (<0.062 mm in longest dimension) clay and quartz grains. Sedimentary Petrologists use the term **clay** to refer to clastic grains smaller than 0.004 mm in longest dimension; in this text, however, we use it to refer to minerals of the clay mineral group, regardless of grain size. In the coarser-grained sedimentary rocks, the compositions of lithic fragments give clues to the origin of the sediment. In the finer-grained rocks, mineralogical composition is often difficult to determine and interpret.

Chemical sedimentary rocks are formed by precipitation of minerals from water, or by alteration of already precipitated material. Many **limestones, dolostones, evaporites,** and **cherts** form this way. In contrast with most detrital sedimentary rocks, petrologists name chemical sedimentary rocks based on chemical composition. Chemical sedimentary rocks usually include only one or a few minerals because the chemical processes that form them tend to isolate certain elements. The most common precipitated minerals consist of elements of high solubility (for example, Na or K) or elements of great abundance (for example, Si).

▶**FIGURE 6.1**
Sedimentary rocks over basement. In some parts of North America, the layers of sedimentary rocks are thousands of feet thick. In other places, notably Precambrian shields and mountain ranges, the layers of sedimentary rocks are completely missing.

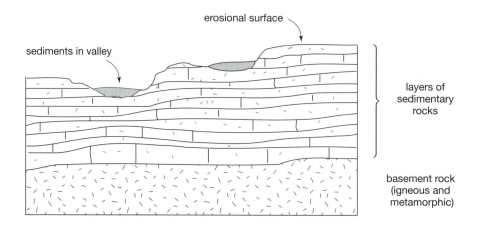

Some limestones, cherts, and other rocks are formed largely from **biogenic** (organic) debris. Petrologists often classify them separately from chemical and detrital sedimentary rocks. We will not, however, consider them separately here. Much overlap exists between chemical, detrital, and organic sedimentary rocks. Many chemical sedimentary rocks contain clastic material, and many detrital sedimentary rocks are held together by chemical cements precipitated from water. Both chemical and detrital rocks may contain biogenic components.

Weathering

Figure 6.2 shows the two parallel processes that result in sedimentary rocks. An original source rock (igneous, metamorphic, or sedimentary) is exposed to weathering. The weathering forces may be **mechanical** (water, wind, gravity, glaciers, waves, and frost) or **chemical** (dissolution by water, perhaps containing acids). Mechanical weathering, which produces clastic material called **detritus,** is of much less significance than chemical weathering. Even apparently dry climates have enough water to promote chemical weathering on exposed surfaces, although the weathering rate may be slow.

Chemical weathering produces dissolved material, called the **hydrolysate,** and leftover rock and mineral detritus that did not dissolve. We sometimes call the undissolved material the **resistate** because it resists dissolution. More easily dissolved elements, especially the alkalis and alkaline earths, go into solution, while resistate remains to become sediment (Table 6.1). Typical resistates include quartz, clay, K-feldspar, garnet, zircon, rutile, or magnetite.

Chemical weathering often produces clay minerals, the most important being montmorillonite, illite, and kaolinite. Reactions that produce clays are complex, involving the reaction of water with previously existing minerals, such as feldspars, to produce clays and dissolved elements. We call such reactions **hydrolysis reactions.** Figure 6.3 depicts a K-feldspar grain undergoing hydrolysis. Mechanical and chemical decomposition produces kaolinite and dissolved ions, including K^+ and Si^{4+}. The dissolved material is carried away and will eventually precipitate somewhere else. The **residual** kaolinite may remain where

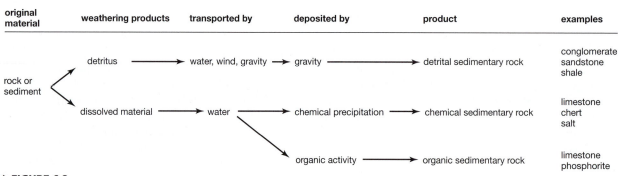

original material	weathering products	transported by	deposited by	product	examples
rock or sediment	detritus	water, wind, gravity	gravity	detrital sedimentary rock	conglomerate sandstone shale
	dissolved material	water	chemical precipitation	chemical sedimentary rock	limestone chert salt
			organic activity	organic sedimentary rock	limestone phosphorite

▶**FIGURE 6.2**
Processes that form sedimentary rocks. The weathering products—rock debris and dissolved material—are transported and then deposited by gravity, chemical precipitation, or organic activity to produce sediments and eventually sedimentary rocks.

▶TABLE 6.1
Products of Weathering of Some Common Igneous Minerals

Mineral	Formula	Resistate Minerals	Formula	Hydrolysate Cations
quartz	SiO_2	quartz	SiO_2	
feldspar	$(Ca,Na,K)(Al,Si)_4O_8$	quartz	SiO_2	K^+, Na^+, Ca^{2+}
		muscovite	$KAl_2(AlSi_3)O_{10}(OH)_2$	
		clays	complex chemistry	
olivine	$(Mg,Fe)_2SiO_4$	hematite	Fe_2O_3	Mg^{2+}
		goethite	$FeO(OH)$	
		magnetite	Fe_3O_4	
		quartz	SiO_2	
		muscovite	$KAl_2(AlSi_3)O_{10}(OH)_2$	
		clays	complex composition	
pyroxene	$(Ca,Mg,Fe)_2Si_2O_6$	Same as for olivine		Ca^{2+}, Mg^{2+}
amphibole	$(K,Na)_{0-1}(Ca,Na,Fe,Mg)_2(Mg,$ $Fe,Al)_5(Si,Al)_8O_{22}(OH)_2$	Same as for olivine		K^+, Na^+, Ca^{2+}, Mg^{2+}

▶FIGURE 6.3

Hydrolysis reaction. Feldspar reacts with water as it undergoes both mechanical and chemical weathering to produce kaolinite and dissolved material.

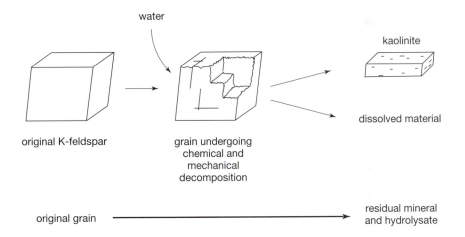

it forms, but water, gravity, or wind can transport clays produced by hydrolysis, just like any other detrital material.

If we examine fresh outcrop in a road cut, rock often appears hard and shiny. Examination with a hand lens reveals that minerals have well-defined boundaries and jagged outlines and may show good cleavage or crystal faces. Minerals have their normal diagnostic colors: quartz is clear, feldspars are white or pink, muscovite is silvery and sparkly, magnetite appears metallic, and biotite and other mafic minerals appear black. The picture is not the same in outcrops exposed to weathering for a long time. After weathering, rock and most minerals have a dull or drab appearance. Grain boundaries and cleavages are obscured. Red, brown, and gray material coats all surfaces, obscuring diagnostic minerals.

Goldich (1938) made such observations, publishing a well-known weathering series showing the ease with which some common igneous minerals break down (Figure 6.4). He derived his series from studying the formation of clays on outcrops of granite, diabase, and amphibolite in the Minnesota River Valley. The keen reader will notice that Goldich's series is nearly identical to Bowen's reaction series. Minerals that crystallize from a magma at high temperature—minerals poor in Si and O—are generally less resistant to weathering than those that crystallize at low temperature. Fe-Mg silicates, such as olivine or pyroxene, calcic feldspars, and many minerals with high solubilities in water, break down relatively easily. Quartz, some feldspars, and some nonsilicate minerals are relatively resistant to weathering because they contain more Si-O bonds, which do not break easily. It should not be surprising that minerals stable in high-temperature igneous rocks, or those most often precipitated from water, are the first to decompose under Earth surface conditions.

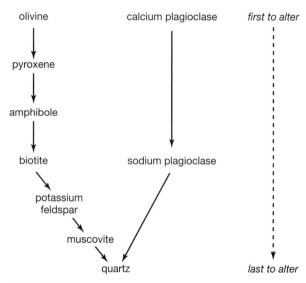

▶FIGURE 6.4
A modified version of Goldich's weathering series describing the relative order in which minerals decompose due to weathering. Olivine and Ca-plagioclase weather most rapidly, while quartz is the most resistant.

TRANSPORTATION, DEPOSITION, AND LITHIFICATION

Flowing water transports detrital sediments to a place where they collect, perhaps a river or lake bottom. Wind, gravity, and other agents can move sediments as well. Eventually, they will be deposited when forces of gravity overcome those trying to move them. After deposition, the loose sedimentary material will, over time, change into a detrital sedimentary rock by **lithification** (from *lithos*, the Greek word meaning stone), a process that involves compaction and cementation. Common cementing agents include the minerals quartz, calcite, and hematite.

Water transports dissolved material produced by chemical weathering until precipitation of chemical sediment occurs. Several things may cause precipitation; most common are changes in temperature, acidity (pH), and biological activity. Hot springs deposit a form of calcite called **travertine,** for example, when cooling water becomes oversaturated with $CaCO_3$. In freshwater streams or lakes, a pH change due to biological activity may cause dissolution or precipitation of another form of calcite called **marl.** Many reef-building organisms have shells or skeletons made of organic calcite. Calcite and other chemical sedimentary minerals, then, precipitate in many ways. In contrast with detrital sediments, chemical sediments are usually lithified as they are deposited.

After deposition and lithification, sedimentary rocks undergo textural or chemical changes due to heating, compaction, or reaction with groundwaters.

We call the changes **diagenesis.** Diagenesis is a low-temperature/pressure form of metamorphism. New minerals formed by this process are termed **authigenic minerals.** Zeolites (Box 6.1), clays, feldspar, pyrite, and quartz can all be authigenic minerals. The Herkimer diamonds (quartz) in Plate 2.8 are authigenic minerals.

SEDIMENTARY MINERALS

Silicates

The last chapter discussed silicate minerals common in igneous rocks. In principle, they could all exist in sediments and sedimentary rocks. In practice, most break down so quickly that they cannot be weathered or transported very much before completely decomposing. Quartz is the most resistant to weathering. It is also a common component of many igneous and metamorphic rocks found at the Earth's surface. It is no surprise, therefore, that quartz is the main component of many clastic sediments. Feldspars are also found in clastic sediment rocks. They are usually subordinate to quartz, and are rare in sediments transported long distances or weathered for long times. Mafic silicate minerals occasionally exist in detrital rocks but are never major components. Besides quartz, the other common silicates in sedimentary rocks are clays and zeolites (Box 6.1). Important nonsilicate minerals found in sedimentary rocks include carbonates, sulfates, and halide minerals.

Clay Minerals

Clays account for nearly half the volume of sedimentary rocks. They are usually very fine grained, often less than 1 μm (10^{-6} m) in size, which makes identification of individual clay species difficult (Figure 6.5). In contrast with quartz and feldspar, clays do not form in igneous and metamorphic environments. Clays crystallize in a sedimentary environment and their compositions depend on the sources of the sediment. The clay mineral group includes many different minerals, all sheet silicates. They have layered atomic structures, similar in many respects to micas. Clays are hydrous, some containing as much as 15 to 20 wt.% H_2O. Their layered structure and the weak bonding between layers give them a characteristic slippery feel when wet. Clays are common in shales and other sedimentary rocks. Although usually fine grained, they form thick beds or layers. They also form as coatings on other minerals undergoing weathering. These generalizations are true of all clay minerals, but many different kinds exist. In part, the many varieties are due to the low temperatures at which clay minerals form. At

Box 6.1 Zeolites

The zeolite group includes more than 40 minerals. All are framework silicates containing open cavities capable of holding loosely bonded large cations and water. Different zeolites have different-sized openings, and in some zeolites the openings connect to form channels. Because of the cavities, zeolites can be used as molecular sieves or as ion exchange media. Manufacturing industries also use zeolites as drying agents, catalysts, and washing materials. Synthetic zeolites have, for some purposes, replaced natural ones. The most common zeolites include:

natrolite	$Na_2Al_2Si_3O_{10} \cdot 2H_2O$
laumontite	$CaAl_2Si_4O_{12} \cdot 4H_2O$
chabazite	$CaAl_2Si_4O_{12} \cdot 6H_2O$
clinoptilolite	$(Na,K)Al_2Si_7O_{18} \cdot 6H_2O$
heulandite	$CaAl_2Si_7O_{18} \cdot 6H_2O$
stilbite	$CaAl_2Si_7O_{18} \cdot 7H_2O$
sodalite	$Na_3Al_3Si_3O_{12} \cdot NaCl$

The compositions of zeolites seem highly variable, but the ratio $(Ca,Sr,Ba,Na_2,K_2)O : Al_2O_3$ is always $1:1$, and the ratio $(Al + Si):O$ is always $1:2$. Zeolites and feldspathoids are closely related, but zeolites have more open structures and contain loosely bonded H_2O. Sodalite is sometimes grouped with the feldspathoids rather than the zeolites, but it has a zeolite-type structure and contains loosely bonded $NaCl$. Analcime, $NaAlSi_2O_6 \cdot H_2O$, is also sometimes considered a zeolite, but is closer to leucite and other feldspathoids in structure.

Zeolites are best known for their occurrences in vugs and other open cavities in volcanic rocks. Large volumes are found, however, in volcanic ashes and saline lake deposits. Zeolites also may be present as products of diagenesis or low-grade metamorphism in sediments and rocks.

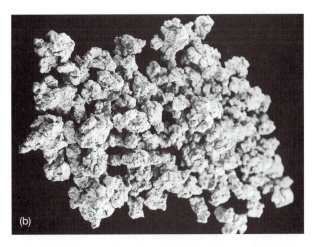

▶**FIGURE 6.5**
Three views of clay minerals: (a) 2–3 cm wide pieces of attapulgite, also called Fuller's Earth, from Ramsden County, Florida; (b) montmorillonite (0.5-cm fragments) formed as a surface deposit in western North Dakota; (c) kaolinite from the Palermo #1 Mine, North Groton, New Hampshire, seen with a scanning electron microscope. The kaolinite cluster is about 0.1 mm across.

▶FIGURE 6.6
Atomic arrangement in calcite. The large spheres represent Ca^{2+} ions and the triangles represent CO_3 groups. Many other carbonates have similar structures.

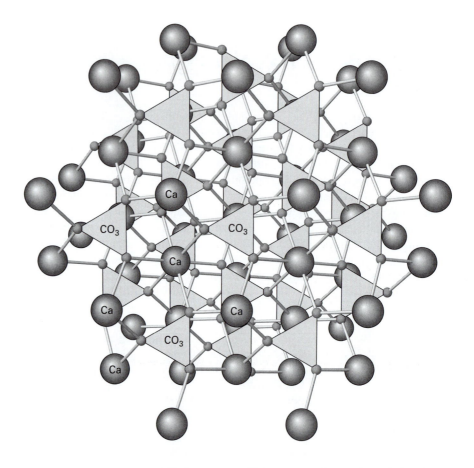

▶TABLE 6.2
Important Clay Minerals

Clay	Chemical Formula
kaolinite	$Al_2Si_2O_5(OH)_4$
illite	$K_{1-1.5}Al_4(Si,Al)_8O_{20}(OH)_4$
smectite group:	
montmorillonite	
	$(Ca,Na)_{0.2-0.4}(Al,Mg,Fe)_2(Si,Al)_4O_{10}(OH)_2 \cdot nH_2O$
vermiculite	
	$(Mg,Ca)_{0.3-0.4}(Mg,Fe,Al)_3(Al,Si)_4O_{10}(OH)_2 \cdot nH_2O$

high temperatures, minerals and mineral structures tend to be simple, and a relatively small number of stable minerals exist. At low temperatures, structures are often more complex or disordered, and many different minerals may exist. Montmorillonite dominates modern clay-rich sediments and sedimentary rocks; illite dominates most sedimentary rocks that are older than about 100 million years. Geologists ascribe this change to changes in tectonic activity resulting in changes in sediment sources, changes in biological activity, and diagenesis.

Fine grain size, complex chemistry, and structural variations can make identification of clay mineral species difficult. X-ray analysis is often necessary to tell them apart. Talc, a secondary mineral that forms when Mg-silicates such as olivine or pyroxene are altered, and pyrophyllite, an uncommon metamorphic mineral, are often grouped with the clays. They are, however, less variable in their structure and composition and contain less H_2O. They are transitional between clays and micas in structure and, when seen in hand specimen or thin section, are typically easier to identify than clays.

The three most important kinds of clays are illite, montmorillonite, and the clays of the kaolinite group (Table 6.2). Kaolinites, also called *kandites,* vary less in composition and structure than other clays, although several kaolinite polymorphs are known. Kaolinite is the principal clay used to make ceramic ware (Box 6.2). Illite is similar to muscovite in some ways, but contains more Si and less K. Montmorillonites, which belong to the smectite group, can take up extra water or other fluids between the layers of their atomic structure. In the process they expand; thus, we sometimes call them *expandable* or *swelling clays.* Because they absorb liquids so well, gas station operators use them to clean up spilled oil, and homeowners use them as kitty litter. They are the major components of earthy material called *bentonite,* sometimes prized for its water-absorbing and cation exchange properties. Another clay of the smectite group, vermiculite, is often used to lighten up potting soil.

Box 6.2 Clays Used in Industry, Arts, and Ceramics

Clays are widely used to make bricks, tile, paper, rubber, water pipes, and china. They are even used today by some restaurants to thicken milkshakes. Early peoples made bowls and other artifacts by shaping clay and allowing it to dry in the sun. Many different clays have industrial uses. Porcelain and china makers commonly use kaolinite. If baked to temperatures above 500 °C, kaolinite will dehydrate to metakaolinite ($Al_2Si_2O_7$), a nonmineral that is relatively hard. Such temperatures can be obtained over open fires, and many early pottery consists of metakaolinite. Although metakaolinite is porous, it will not soften when wetted, in contrast to sun-baked clays.

Porcelain refers to a special type of high-grade white ceramic. The white color is only obtained by using extremely pure kaolinite. Porcelain is baked, or fired, at very high temperatures. At temperatures of 925 °C and above, metakaolinite converts to a mixture of cristobalite, mullite, and other phases, but it does not melt unless the temperature exceeds 1,600 °C. Prior to firing, feldspar or talc is mixed with the kaolinite in small amounts. When fired at 1,200° to 1,450 °C, the feldspar or talc melts to form a glass, which gives porcelain its glassy luster without melting the kaolinite and destroying the object being made.

Box 6.3 Lime Kilns, Mortar, and Cement

Calcite is an important component in the production of cement and mortar. Both have been used since at least 2100 to 2700 B.C., when ancient Egyptians built one of the Seven Wonders of the World: the great pyramids. Subsequently, Greeks and Romans used cement produced from volcanic ash and lime in much of their building construction.

To make cement or mortar, calcite is **calcined** at temperatures above 900 °C in a **lime kiln** to get lime (CaO) and carbon dioxide. Mortar is then produced by mixing the lime with sand and water. When lime reacts with water, portlandite, $Ca(OH)_2$, forms, consuming much of the water. The formation of portlandite (also called *slaked*

lime) releases heat, which aids in hardening or "setting" as the remaining water evaporates.

Portland cement, the most widely used cement, consists of a mixture of lime, silica, alumina, and iron oxides. Silica, alumina, and iron oxides may be obtained from clays, blast furnace slag, or even coal fly ash. Gypsum is typically added to Portland cement to control setting time. The mixture is combined with water and, depending on its composition, hardens rapidly. After setting, Portland cement consists of a number of Ca-Al-Fe-Si-O phases, some of which resemble minerals. The strength of cement derives from the presence of the nonmineral phases $Ca_3Al_2O_6$ and Ca_2SiO_4.

Carbonate Minerals

Mineralogists have identified more than 50 different carbonate species; all contain $(CO_3)^{2-}$ but some contain other anions or anionic radicals. Plate 1.2 shows smithsonite, $ZnCO_3$, one of the rarer carbonates. Few carbonate minerals are common, and many have complex compositions and structures. The atomic structures of calcite and dolomite, the most common

carbonates by far, contain $(CO_3)^{2-}$ groups alternating with divalent cations (Figure 6.6). Both calcite and dolomite may exist as spectacular individual crystals, as essential minerals in limestones, or as clastic fragments in other sedimentary rocks (Figure 6.7). Calcite also exists in metamorphic rocks such as marbles, in some hydrothermal deposits, and in rare igneous rocks called carbonatites. Dolomite is common in limestones and dolostones (limestones containing dolomite instead of

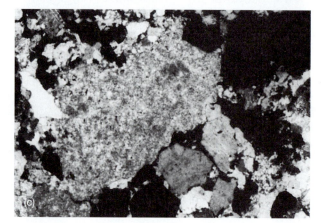

▶**FIGURE 6.7**
Dolomite and calcite: (a) 0.5-cm dolomite crystals on a carbonate rock; (b) 3-mm-wide thin section view (XP) of calcite in a limestone (note concentric growth lines); (c) 2.5-mm-wide thin section view (XP) of a large grain of calcite (in center of photograph) in a sandstone. Note the twinned plagioclase, quartz, and other detrital grains.

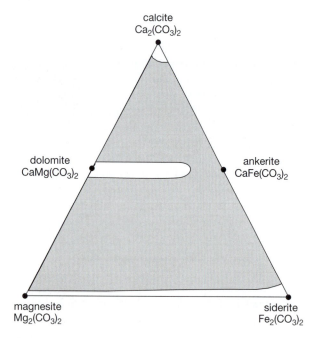

▶**FIGURE 6.8**
The principal end members of carbonate minerals. Natural compositions falling in the shaded region are rare due to immiscibility. Here and in Figure 6.9, the formulas of calcite, magnesite, and siderite have been doubled to be consistent with the formulas of dolomite and ankerite.

calcite), and occurs in metamorphic and hydrothermal deposits. Most dolomite is secondary and forms by reaction of calcite with Mg-rich solutions during diagenesis. Magnesite ($MgCO_3$), a related carbonate, forms as an alteration product of mafic and ultramafic rocks.

We can plot the compositions of Ca-Mg-Fe carbonates on a triangular diagram similar to that used for pyroxenes (Figure 6.8). Miscibility gaps exist between the Ca-bearing carbonates and the Ca-free ones, similar to the gaps between wollastonite, clinopyroxene, and orthopyroxene. This can best be seen by looking at a temperature composition diagram (Figure 6.9). Compositions intermediate between calcite and dolomite are not stable, except at high temperature where calcite may contain extra Mg. Compositions intermediate between dolomite and magnesite are never stable. At temperatures above about 900 °C, calcite breaks down to produce quicklime (CaO) and CO_2 gas. The other carbonates also break down at high temperature.

Sulfate Minerals

More than 100 sulfate minerals are known. They fall into two main groups: those that contain no water (anhydrous sulfates) and those that do (hydrous sul-

▶**FIGURE 6.9**

Temperature composition diagram for calcite-dolomite-magnesite carbonates. Two large miscibility gaps restrict carbonate compositions. The boundary limiting the dolomite field at high temperature is poorly known. At high temperature, calcite may contain excess magnesium and approach dolomite in composition, but carbonates intermediate between dolomite and magnesite are never stable.

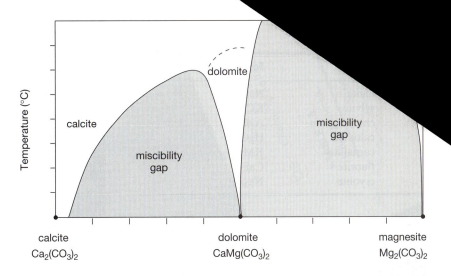

calcite
$Ca_2(CO_3)_2$

dolomite
$CaMg(CO_3)_2$

magnesite
$Mg_2(CO_3)_2$

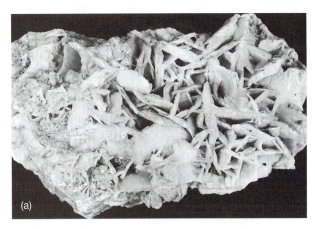

(a)

▶**FIGURE 6.10**

Sulfate minerals. All photographs are about 10 cm across: (a) barite, $BaSO_4$; (b) celestite, $SrSO_4$; (c) gypsum, $CaSO_4 \cdot 2H_2O$.

(b)

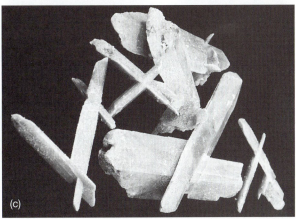

(c)

fates). Examples in the first group include anhydrite ($CaSO_4$), barite ($BaSO_4$), celestite ($SrSO_4$) and anglesite ($PbSO_4$) (Figure 6.10a, b). Plate 1.6 shows a spectacular example of celestite from Ohio. In the second are gypsum ($CaSO_4 \cdot 2H_2O$; Figure 6.10c) and some rare species. Rare green gypsum from Australia is shown in Plate 3.5. Gypsum and anhydrite can both be major rock-forming minerals. Thick deposits of gypsum or anhydrite, or both, are associated with limestone, dolostone, or halite beds. They may also associate with native sulfur (S) deposits. All the sulfate minerals can be found in vugs or fractures in a variety of rock types, and many occur in soils. Some sulfates are common as minor, and sometimes major, minerals in ore deposits—typically as replacements for primary sulfides. Anglesite ($PbSO_4$), for example, forms during weathering or alteration as a replacement for galena (PbS).

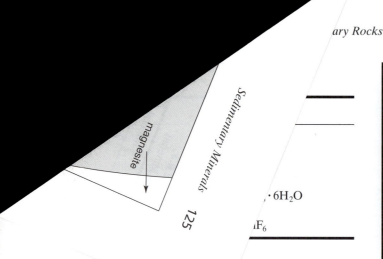

▶**FIGURE 6.11**
Fluorite (CaF_2) and halite (NaCl). Fluorite crystals are typically cubic, but cleavage fragments form octahedra as shown here. Halite has excellent cubic cleavage (three cleavages at 90°).

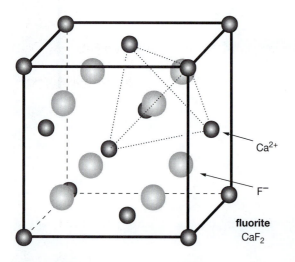

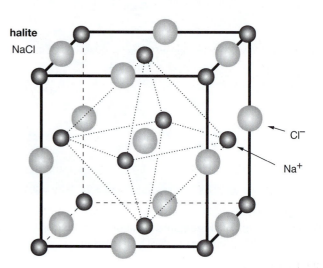

halite
NaCl

Cl^-

Na^+

Ca^{2+}

F^-

fluorite
CaF_2

▶**FIGURE 6.12**
Atomic structures of fluorite and halite. Both have cations in cubic arrangement. In fluorite, the anions (F^-) are within the cubic cell and each is surrounded by four Ca^{2+} cations. In halite the anions (Cl^-) are at the center of the cell and the centers of the edges; each anion is surrounded by six Na^+ cations.

Halides

The halide group consists of minerals containing a halogen element, most commonly chlorine or fluorine, as an essential anion (Table 6.3; Figure 6.11). Halide atomic structures are highly ionic. In halite, Na^+ and Cl^- alternate in a cubic three-dimensional arrangement. In fluorite, the arrangement of atoms is also cubic, but two F^- are present for every Ca^{2+} (Figure 6.12). Although many halides exist, only halite and sylvite are common in sedimentary rocks. Except fluorite (CaF_2), the other halide minerals are rare in rocks of all types. Halite is typically found as rock salt in massive salt beds, often occurring with other evaporite minerals such as gypsum or anhydrite, and sometimes with sulfur. Sylvite is much less

common than halite. When found, however, it is usually associated with halite.

Chert

Chert is a fine-grained (microcrystalline) form of quartz. It is also the name given to rock composed primarily of fine-grained quartz. It may be massive and well layered, and often forms as nodules or concretions in limestone. Sometimes chert seems to have formed by recrystallization of amorphous silica. Many varieties exist. Jasper is chert with a characteristic red color due to hematite inclusions. Flint, a darker form of chert, contains organic matter. Opal and chalcedony, two other types of SiO_2, are often associated with chert deposits. From the Stone Age until the In-

▶**FIGURE 6.13**
Several detrital sedimentary rocks having different grain sizes: (a) sandstone; (b) an enlarged view of quartz grains from sandstone. The grains have been rounded by abrasion
during transportation; (c) mudstone showing ripple marks on its surface; (d) a conglomerate with pebbles up to 2 cm across.

dustrial Revolution, chert and flint were highly valued
as weapons, tools, and fire starters.

COMMON SEDIMENTARY ROCKS

Detrital Sedimentary Rocks

Grain size in clastic sediments and detrital rocks
varies from huge clasts and boulders in gravels and
conglomerates, to fine "clay size" (<0.004 mm) particles in muds and shales (Figure 6.13). Table 6.4
gives a simple classification scheme for clastic rocks
based on clast/grain size.

In the coarsest clastic rocks, a fine-grained **matrix** separates the clasts, which may be of many different types. If the clasts are angular, the rock is a
breccia; if they are rounded, it is a conglomerate.
Lithic fragments often dominate the clasts in conglomerate and breccia. Sandstones of intermediate
clast size may contain both lithic fragments and individual detrital mineral grains. We call sandstones
containing significant amounts of matrix material
and lithic fragments *wackes* or *greywackes,* and those
lacking in matrix and lithic fragments *arenites.* In the

finer-grained clastic rocks, siltstone and mudstone,
lithic fragments are rare or lacking. The finest grained
rocks are called *shale* if they exhibit **fissility.** Fissility,
the ability to cleave into very thin layers, results
from parallel alignment of clay grains. Conglomerates and sandstones account for 20% to 25% of all
sedimentary rocks; the finer-grained rocks are even
more common.

Since detrital sediments can be derived from any
preexisting rocks, they may contain a variety of minerals and rock fragments. However, only a few are
common. Quartz, feldspar, and lithic fragments containing quartz and feldspar dominate all but the finest
grained rocks because they are resistant to weathering. Rocks dominated by quartz are called *quartzose,*
while those containing large amounts of feldspar are
termed *feldspathic,* or (if they are rich in K-feldspar)
arkosic. In mudstones and shales, clay minerals dominate. Other minerals such as micas, magnetite, rutile,
ilmenite, sphene, zircon, apatite, or garnet may occasionally be significant components of clastic sediments. Carbonate grains and organic material may be
present as well. Most minerals break down to clay and
quartz if exposed at the Earth surface for sufficient
time. It should be no surprise that mudstone and

▶**TABLE 6.4**
Classification Scheme for Detrital Sedimentary Rocks

Rock Type	Sediment	Grain Size	Typical Clast Material	Varieties
conglomerate or breccia	gravel	>256 mm	rock fragments, quartz, feldspar	boulder conglomerate
		64 to 256 mm	rock fragments, quartz, feldspar	cobble conglomerate
		4 to 64 mm	quartz, feldspar, rock fragments	pebble conglomerate
		2 to 4 mm	quartz, feldspar	granule conglomerate
sandstone	sand	0.062 to 2 mm	quartz	quartzose sandstone
			quartz, feldspar	feldspathic sandstone
			rock fragments, quartz, much matrix	greywacke sandstone
			quartz, little matrix	arenite sandstone
siltstone	silt	0.004 to 0.062 mm	quartz	siltstone
mudstone	"clay"	<0.004 mm	clays	shale
			clays	mudstone

Box 6.4 Gypsum: Ingredient of Plaster and Sheetrock

Early Romans used a lime-based plaster, but **plaster of Paris,** gypsum-based plaster, was introduced around 1254. At the time, the best and most productive gypsum quarries were in Montmartre, a section of Paris. Modern plasters and sheetrock both contain plaster of Paris. Manufacturers produce it by calcining (grinding and heating) gypsum ($CaSO_4 \cdot 2H_2O$) to reduce its water content. Most plasters contain about 25% of the gypsum's original water. Complete dehydration of gypsum would produce anhydrate ($CaSO_4$), which is not useful as a plaster because it does not recombine easily with water. However, when plaster of Paris is mixed with water, reaction occurs quickly, giving off heat and promoting drying and hardening.

shale, which are composed primarily of clay and quartz, are the most abundant sedimentary rocks.

Identification of minerals in clastic rocks can be difficult if the rocks are not coarse grained. Suppose we look at a sandstone with a hand lens. Quartz, and perhaps K-feldspar, may be easy to find. However, lithic fragments can be confusing and small grains of chert or magnetite may be impossible to identify. For fine sandstones and finer-grained rocks, mineral identification can be problematic, even with a petrographic microscope; distinguishing quartz from feldspar and telling clay minerals apart can be impossible without using an X-ray diffractometer or other sophisticated equipment.

Chemical Sedimentary Rocks

All minerals are soluble in water to some extent. Sulfates, halides, and other salts have very high solubilities. Carbonate minerals such as calcite and dolomite have moderate solubilities (Figure 6.14).

Silicate minerals have relatively low solubilities. If solutions become oversaturated, they precipitate chemical sediments, decreasing concentrations of dissolved material, until the solution and sediments achieve equilibrium. This process produces chemical sediments and chemical sedimentary rocks (Table 6.5). Because of their high solubility, large amounts of evaporation may be necessary before salts such as halite precipitate. In contrast, carbonate minerals (calcite and dolomite) and silica (chert) often precipitate early in an evaporation event. Silica (SiO_2), as chert, is the only silicate mineral that commonly forms a chemical sedimentary rock. Chert beds several hundred meters thick occur in much of Arkansas, Oklahoma, west Texas, and California. Thick chert beds are much more common in Precambrian rocks than in Phanerozoic rocks.

Salts such as gypsum ($CaSO_4 \cdot 2H_2O$), anhydrite ($CaSO_4$), halite (NaCl), and sylvite (KCl) have high

Box 6.5 Disposal of Radioactive Waste in Salt Domes

The disposal of highly radioactive wastes generated by nuclear power plants is an unsolved problem. After four decades of study, scientists still can't agree on the best disposal method. One plan is to bury the waste underground after reprocessing removes the longest lived isotopes. To this end, the federal government investigated the possibility of disposing of nuclear waste in thickly bedded salt deposits. Salt is an attractive host because it has a high melting temperature and low porosity and permeability, and it flows under pressure rather than fracturing. Thus, the mineral can be expected to help seal the waste repository. Pilot studies were conducted in New Mexico, but there are no current plans to begin disposing of radioactive waste in salt domes.

▶**FIGURE 6.14**
Smithsonite, $ZnCO_3$, a carbonate mineral with relatively high solubility in water. This sample from Larium, Greece, is about 7 cm across; it precipitated with a "stalactitic" texture. See also Plates 1.2 and 1.8. The selenite in Plate 3.5 has even higher solubility than smithsonite.

▶**TABLE 6.5**
Chemical Sedimentary Rocks

Rock	Major Mineral Components
limestones	calcite
dolostones	dolomite
phosphorites	phosphates such as apatite
chert	quartz
evaporites	chlorides, sulfates, carbonates, borates
iron formation	iron oxides, carbonates, silicates
travertine	calcite

Coast areas of the United States. Massive salt beds exist in many parts of the world (Figure 6.15). In all, petrologists have reported nearly 100 minerals from them. Less than a dozen are common (Table 6.6).

solubilities in water, so their chemical components are common as dissolved species in natural waters. As the water evaporates, often in a closed inland basin, these minerals may precipitate to form thick beds of evaporite minerals (Box 6.6). Geologists estimate that 35% of the United States is underlain, at some depth, by thick evaporite beds. Most evaporite minerals are rare at the Earth's surface because they are so soluble that they dissolve away in all but the most arid climates. The most common evaporite mineral in outcrop is gypsum ($CaSO_4 \cdot 2H_2O$) because it is less soluble than others, and because it forms from any anhydrite ($CaSO_4$) exposed at the surface. In the subsurface, massive gypsum and halite beds are common, as are the **salt domes** found in Texas and other Gulf

▶**FIGURE 6.15**
Salt crystals and Morton salt. Commercial salt is typically mined by dissolving it out of thick underground beds that formed when oceans or salt lakes evaporated.

Box 6.6 Did the Mediterranean Sea Repeatedly Dry up in the Past?

Many scientists have evaporated seawater to study the precipitates that form. If seawater is evaporated, after 65% of it is gone, gypsum begins to crystallize. When 90% is gone, the first halite crystals form. When 95% is gone, sylvite and magnesium salts precipitate.

Recent deep-sea drilling and geophysical studies in the Mediterranean Basin indicate the widespread presence of thick beds of gypsum and halite. In some places the beds are up to 2,000 meters thick. The deposits resulted from tectonic events that periodically closed and reopened the connection between the Mediterranean Sea and the Atlantic Ocean (the present-day Straits of Gibraltar) over the past six to seven million years. When the Mediterranean Basin was isolated from the Atlantic, evaporation over several thousand years led to the deposition of thick layers of gypsum followed by halite. If enough evaporation occurred, sylvite and magnesium salts also precipitated.

When connection to the Atlantic Ocean was reestablished, the briny waters of the Mediterranean were replaced by normal seawater. Recent studies suggest that refilling was much faster than evaporation, taking only hundreds of years to complete. Some of the salt and gypsum redissolved, but much was preserved. Studies indicate that the evaporation-drying cycle took place at least 30 times.

While *limestone* is a general term given to all carbonate rocks, we use the name *dolomite* or *dolostone* for rocks in which dolomite is the dominant mineral. Limestone and dolostone account for 10% to 15% of all sedimentary rocks. They are usually quite pure, containing mostly calcite or dolomite. Surprisingly, rocks with significant amounts of both calcite and dolomite are rare. Telling the two carbonates apart may be difficult on a fresh surface, but weathering sometimes alters dolomite to a yellow-brown color. Field geologists carry dilute hydrochloric acid, which reacts only with calcite or powdered dolomite, or chemical stains to help identification of carbonate minerals. Aragonite, a polymorph of calcite, exists in some very young carbonate deposits, but never in old rocks because it changes to calcite over time. In contrast, for reasons that are not completely clear, dolomite is rare in modern carbonates but is common in Paleozoic and Precambrian rocks.

When $CaCO_3$ precipitates to form limestones, it may be as fine-grained lime muds, resulting in microcrystalline calcite, called *micrite*. It also precipitates as coarser sparry calcite, which may be clear and easily visible with the naked eye. Many carbonate rocks also contain carbonate minerals that originated as clastic material or carbonate fossils (see Plate 5.1). Fossils can be of many different sorts. Pellets, mostly formed from small animal feces, are small agglomerations of microcrystalline calcite, usually well under a millimeter in longest dimension. In addition, carbonate rocks often contain detrital quartz or clay

▶ TABLE 6.6
Some Minerals Common in Evaporite Deposits

Mineral	Chemical Formula
halite	NaCl
sylvite	KCl
anhydrite	$CaSO_4$
gypsum	$CaSO_4 \cdot 2H_2O$
calcite	$CaCO_3$
dolomite	$CaMg(CO_3)_2$
sulfur	S

minerals and may contain authigenic minerals of many kinds.

Other Chemical Sedimentary Rocks

Many other types of chemical sedimentary rocks exist. They are not abundant, but they may be important as sources of ore. Collophane, a cryptocrystalline form of apatite, $Ca_5(PO_4)_3(OH,F,Cl)$, found in small amounts in many kinds of sedimentary rocks, has an organic origin. In rocks called *phosphorites*, apatite may comprise nearly the entire rock. Major phosphorite deposits are found in Wyoming and Idaho, where mines produce phosphate, an important component of fertilizers.

Iron formations, most Precambrian in age, are mined for iron in the Mesabi Range of Minnesota and elsewhere (Box 6.7). Fe oxides and hydroxides (hematite, Fe_2O_3; goethite, $FeO(OH)$; and mag-

Box 6.7 Iron Formation

Iron formation is a general term given to iron-rich chemical sedimentary rocks. Sedimentary iron deposits, our most important source of iron, are mostly of Precambrian age. They consist of chert with variable amounts of iron oxides (hematite and magnetite), iron sulfides (pyrite), iron carbonates (siderite, ankerite), and iron silicates (greenalite, minnesotaite, stilpnomelane, chamosite). Typical iron formations are banded, having iron-rich layers and iron-poor layers. North American mining companies have produced iron from iron formations in Ontario, Minnesota, Michigan, and a few other places. Iron formations are also mined at Thabazimbi, South Africa, and in the Hamersley Range of western Australia.

The mineralogy of iron formations seems to reflect the chemistry of the oceans and atmosphere at the time of deposition. Most geologists believe deposition started because of an increase in O_2 in the Earth's atmosphere during the late Archean and early Proterozoic periods. Carbonate minerals were deposited from waters relatively rich in $(CO_3)^{2-}$, sulfide minerals from water relatively rich in S, and oxides and silicates elsewhere.

Box 6.8 Laterites and Bauxites

Consider a tropical area with warm weather and abundant rainfall. Weathering and leaching will be extreme, and even clay minerals may decompose. Normally soluble elements, and even relatively insoluble silica, will be dissolved and carried away. The remaining material, called a *residual deposit*, is often composed primarily of aluminum oxides and hydroxides, the least soluble of all common minerals. We term such deposits **laterites** (if not lithified) or **bauxites.** They are our most important source of aluminum. The mineralogy of a laterite depends on the composition of rocks weathered to produce it; laterites can also be important sources of iron, manganese, cobalt, and nickel, all of which have low solubilities in water.

Most laterites are aluminous. The most important aluminum ore in laterites is bauxite, a mixture of several minerals, including the polymorphs boehmite and diaspore, $AlO(OH)$, and gibbsite, $Al(OH)_3$. Bauxite is mined in large amounts in Australia and Indonesia, and in smaller quantities in the Americas and in Europe. In some places, relatively young laterites produce ore, but in Australia they are more than 65 million years old.

netite, Fe_3O_4) or Fe-carbonate (siderite, $FeCO_3$) are the most common Fe-minerals in iron formations. Occasionally, Fe-silicates or Fe-sulfides may be present. Iron formations form from sediments originally deposited in shallow marine conditions. Young manganese deposits, similar to iron formations, have been dredged from ocean floors.

DIAGENESIS

Diagenesis refers to chemical, mineralogical, or textural changes occurring in sediments or sedimentary rocks after deposition, but before metamorphism.

Diagenesis creates many minerals, but most are so fine grained that they cannot be identified without X-ray analysis. Diagenesis occurs before, during, and after lithification, taking place at, or very near, the surface of the Earth. Changes taking place at greater depth, where pressure and temperature are both significantly elevated, are usually considered metamorphism. The most important agents causing diagenesis are pressure, heat, and water. Biological agents, such as small animals or bacteria, also can be important, as can chemical agents brought in by flowing water. Textural changes, including compaction and loss of pore space, are part of diagenesis. Nearly all sediments and sedimentary rocks undergo this kind of diagenesis. **Recrystallization,** changing fine-grained rocks into coarser ones, is another form of diagenetic textural change.

Chemical changes during diagenesis usually involve water because the mobility of elements at Earth surface temperatures is very low if they are not dissolved in water. Dissolution of minerals **(leaching)** and the formation of clay minerals are both common during diagenesis. When leaching is extreme, laterites and bauxites may form (Box 6.8). Precipitation of authigenic minerals is also common.

The processes of sedimentation, lithification, diagenesis, and low-grade metamorphism form a continuum. Lithification changes unconsolidated sediment into a rock. Cementation by quartz, calcite, or hematite may be part of the lithification process. It also may be considered a diagenetic process. Similarly, the formation of many low-temperature minerals such as zeolites, a normal part of diagenesis, overlaps with the beginnings of metamorphism. Metamorphic petrologists often define the onset of metamorphism by the first occurrence of metamorphic minerals. This definition can be hard to apply because many diagenetic minerals are also metamorphic minerals. Furthermore, laumontite, often considered the lowest temperature of all metamorphic minerals, is a zeolite hard to distinguish from those that form diagenetically.

▶QUESTIONS FOR THOUGHT

(Some of the following questions have no specific correct answers; they are intended to promote thought and discussion.)

1. What are the most common minerals in clastic sediments? Why?
2. What are the most common minerals in chemical sediments? Why?
3. Why do ceramicists use clays, instead of other minerals such as quartz and feldspar, to make porcelain and other ceramics?
4. The many different carbonate minerals share some common properties. What are those properties? Why?
5. Sulfates and halides are both common evaporite minerals. Why? What physical properties do they have in common? Why?
6. Why are clastic sedimentary rocks generally classified according to grain size rather than chemistry or mineralogy?
7. When groundwater percolates down through soil, it eventually reaches the water table. As it percolates, it leaches (dissolves) soluble minerals from upper soil horizons and deposits them in lower soil horizons. Consequently, how do the mineralogical and chemical compositions of upper and lower soil horizons differ?

▶RESOURCES

(See also the general references listed at the end of Chapter 1.)

Blatt, H. *Sedimentary Petrology,* 2nd ed. San Francisco: W. H. Freeman and Co., 1992.

Ehlers, E. G., and H. Blatt. *Petrology: Igneous, Sedimentary and Metamorphic.* San Francisco: W. H. Freeman and Co., 1980.

Goldich, S. M. A study in rock weathering. *Journal of Geology* 46 (1938): 17–58.

Manning, D. A. C. *Industrial Minerals.* London: Chapman & Hall, 1995.

Stewart, F. H. Marine evaporites. *United States Geological Survey Professional Paper 440-Y.* Washington, DC: United States Department of Interior, 1963.

7

Metamorphic Minerals and Metamorphic Rocks

Metamorphic minerals and metamorphic rocks form when preexisting rocks undergo changes in chemistry, texture, or composition without melting. The changes are caused primarily by pressure and temperature. In this chapter we discuss metamorphic minerals and rocks and how they develop in response to such changes. We discuss mineral reactions and look at some of the basic laws of thermodynamics we use to interpret and predict metamorphic minerals and metamorphic reactions.

The term **metamorphism** describes a change in a rock's mineralogy, texture, or composition without melting. It occurs when minerals undergo metamorphic reactions, when texture changes due to recrystallization or realignment of minerals, or when a rock's chemistry is changed by flowing fluids. The principal agents of metamorphism are heat, pressure, and fluids composed primarily of H_2O and CO_2. All types of rocks can be metamorphosed. Some, such as clay-rich sediments, change greatly when metamorphosed; others, such as granite, change to a lesser extent. The nature of the changes depends on rock composition and the conditions at which metamorphism occurs.

THE CAUSES OF METAMORPHISM

A variety of tectonic processes bury and carry rocks deep into the Earth, including subduction and mountain building. As they move downward or have more material piled on top of them, they experience increased temperature and increased pressure (Figure 7.1). This produces **regional metamorphism,** so called because it covers large regions. Today and in the past, regional metamorphism has occurred in orogenic belts, but the most significant examples of regional metamorphic terranes are found in **Precambrian shields,** sometimes flat-lying areas that may be thousands of kilometers across.

Heat from an igneous intrusion may also cause metamorphism. We call this **contact metamorphism** because it is usually localized at the contact between the intrusion and surrounding rock called **country rock** (Figure 7.1). Contact metamorphism leads to the development of metamorphic zones called **contact aureoles** or **skarns,** which wrap around an intrusion and may be anywhere from a few centimeters to many kilometers thick. The formation of skarns may involve **metasomatism,** a change in rock composition due to flowing metamorphic fluids. The width of skarns mainly depends on the size of the intrusion and how much fluid (mostly H_2O and CO_2) it gives off. Aureoles often develop in concentric zones or layers, each containing a distinct mineral assemblage that reflects the maximum temperature attained and the degree of metasomatism.

▶**FIGURE 7.1**
Regional and contact metamorphism. (a) Regional metamorphism occurs when rocks are buried and subjected to increases in temperature and pressure. (b) Crustal rocks experience regional metamorphism in tectonic belts where the crust is highly deformed. (c) High temperatures and pressures of metamorphism also result from large-scale faulting, which carries crustal rocks to great depth. (d) Contact metamorphism occurs adjacent to plutons when heat and fluids from the pluton alter surrounding country rock.

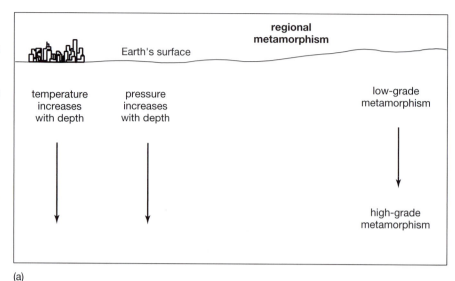

(a)

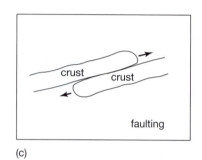

(b)

(c)

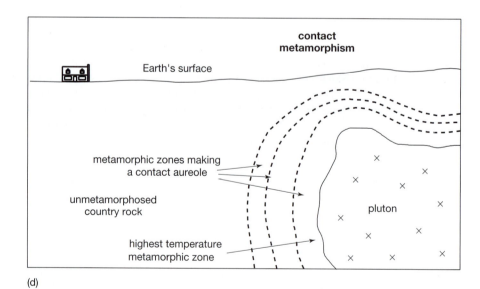

(d)

Regional and contact metamorphism account for most metamorphic rocks. Some geologists have described another kind of metamorphism, called **burial metamorphism,** but it is really just high-temperature diagenesis. **Dynamic metamorphism,** also called **shock metamorphism** or **cataclastic metamorphism,** is an additional unique and uncommon form of metamorphism. It occurs because of sudden pressure exerted by faults or meteorite impacts. The results are often the fracturing and granulation of rocks and sometimes the creation of high-pressure minerals such as coesite or stishovite, polymorphs of quartz.

Metamorphism, which may affect any kind of rock, occurs over a wide range of pressure and temperature conditions. This leads to tremendous variation in metamorphic rocks and the minerals they contain. Most of the metamorphic minerals we see form at temperatures of 100 to 800 °C, and at pressures of 1 bar (10^5 Pa) to 10 Kbar (10,000 atm, equivalent to 10^9 Pa). Exceptions, however, do exist (Box 7.1). The term

Box 7.1 Extremes of Metamorphism

Most metamorphic rocks we see were formed at temperatures of less than 800 to 850 °C and pressures less than 10 Kbar. The temperature limit is because most rocks melt above about 850 °C; at higher temperatures they become at least partially molten. The pressure limit is because it is necessary to go deep in the Earth to reach pressures greater than 10 Kbar. Although metamorphic rocks do form at higher pressure, there are few mechanisms to bring them to the surface of the Earth for examination.

Petrologists do occasionally find rocks that were metamorphosed at pressures greater than 10 Kbar or temperatures greater than 850°C. The talc-kyanite schists from the Dora Maira Massif, Italy, and the metapelites (metamorphosed clay-rich sediments) of the Napier Complex, Enderby Land, Antarctica, are examples of extreme metamorphism. The Dora Maira rocks have been called **whiteschists** because of their very light color. Besides talc and kyanite, they contain pyrope (white to pink Mg-garnet), phengite (a white mica related to muscovite), and quartz as major minerals. Coesite (a high-pressure polymorph of SiO_2) and ellenbergite (another high-pressure mineral) are found as small inclusions in some of the pyrope crystals. A number of papers discuss the formation of these unusual rocks; most significant is the seminal paper by Chopin (1984). Chopin and others concluded that the Dora Maira rocks may have originated as shallow crustal rocks but were subsequently metamorphosed at about 35

Kbar pressure and temperatures of 700 to 750 °C. While the temperature is not extreme, the pressure is; 35 Kbar is equivalent to a depth in the Earth of more than 100 km. There are few places where rocks metamorphosed at that depth are found at the Earth's surface—especially if they originally came from the shallow crust. The Dora Maira rocks were apparently uplifted during the Alpine Orogeny in southern Europe, but exactly how and why are unknown.

Rocks metamorphosed at extreme temperatures have been reported from a number of locations, including some in Uganda, South Africa, and Labrador, Canada. The best studied, however, are Mg-rich pelites from the Napier Complex of Antarctica. Carrington and Harley (1995) summarize the previous studies of the Napier rocks. The rocks contain distinctive high-temperature mineral assemblages, which include the relatively rare minerals pigeonite (a high-temperature pyroxene), ossumilite, and sapphirine. Several different studies have concluded that the Napier rocks were metamorphosed at pressures of 7 to 8 Kbar and temperatures of more than 1,000 °C, perhaps as high as 1,075 °C. How did these rocks get so hot, and how did they escape melting? The answer to the first question is not known. Some investigators believe that a magma must have been involved. The answer to the second is that the rocks are of a composition that does not melt easily when water is absent. The Napier rocks were apparently metamorphosed in the absence of water.

---

metamorphic grade describes the temperature at which a particular metamorphism occurs. Low-grade metamorphic rocks form at low temperatures, generally between 150 and 450 °C. At the lower end of this range, diagenesis overlaps metamorphism. Low-grade metamorphic rocks are often fine grained. Because they are hard to study and frequently do not represent chemical equilibrium, many metamorphic petrologists prefer to study higher grade rocks. Medium-grade metamorphism, occurring at temperatures of between 400 and 650 °C, often results in rocks containing conspicuous metamorphic minerals we can easily study with a microscope. High-grade metamorphic rocks, which form at temperatures greater than about 600 °C, are usually quite coarse grained, containing minerals

easily identified in hand specimen. Depending on its composition, a high-grade metamorphic rock may undergo **partial melting,** also called **anatexis,** so both metamorphic and igneous processes contribute to its evolution. We call the resulting partially melted rocks **migmatites,** which means "mixed rocks."

Prograde metamorphism occurs when low-grade or unmetamorphosed rocks change mineralogy or texture in response to a temperature increase. If the metamorphism is gradual and predictable, we call it **progressive metamorphism.** During progressive metamorphism, a series of reactions occur as the degree of metamorphism increases; rock mineralogy changes multiple times before equilibrating at the highest temperature conditions. While this idea makes a convenient conceptual model, it cannot be correct for all metamorphic rocks. For example, many metamorphic rocks from deep in the Earth were initially formed at high temperature. Other rocks go from low temperature to high temperature, perhaps because of rapid intrusion of a pluton, so rapidly that they skip intermediate stages. Finally, some metamorphic rocks form by **retrograde reactions** (metamorphism in response to temperature decrease). This is especially true of basalts and other high-temperature volcanic rocks involved in low-grade regional metamorphism.

One of the most intriguing questions about metamorphic rocks is: Why do we find high-grade minerals at the surface of the Earth where theoretically they are unstable? The laws of thermodynamics say that rocks will change mineralogy in response to increasing temperature (prograde metamorphism), so why don't they undergo opposite changes when temperature decreases (retrograde metamorphism)? Several facts help answer these questions:

- Prograde metamorphic events are usually of much longer duration than retrograde, giving minerals more time to achieve equilibrium.

- Prograde metamorphism liberates fluids not present when retrogression occurs. The fluids act as fluxes to promote prograde metamorphism; their absence may hinder retrogression.
- At low temperature, reactions are very sluggish; they may not have time to reach equilibrium.
- More complex, low-grade minerals often have difficulty nucleating and growing.

METAMORPHIC TEXTURES

Textural changes take place as rocks undergo prograde metamorphism. Deformation fabrics may develop (Figure 7.2a), and a general coarsening of grain size is typical as small mineral grains recrystallize to form larger ones. Metamorphic minerals may grow and modify rock texture. If they form large crystals and are surrounded by a sea of smaller crystals, we call them **porphyroblasts.** The garnets in Figure 7.2b, d are good examples of porphyroblasts. If the smaller crystals are small and uniform in size, we call them the *ground mass* (Figure 7.2d). Porphyroblasts may form by recrystallization of minerals already in a rock. **Augen**

▶**FIGURE 7.2**
Examples of metamorphic textures in rocks: (a) a deformed biotite schist; (b) a garnet gneiss showing foliation due to alternating light and dark bands; (c) close-up view of 1–2-cm wide feldspars surrounded by biotite in an augen gneiss; (d) thin section photograph (XP) showing an 8 mm garnet poikiloblast (black hexagon) in a ground mass of primarily quartz and feldspar.

gneisses, for example, contain large recrystallized **augen** (German for "eyes") of feldspar (Figure 7.2c).

In some deformed rocks, mineral grains assume a distinctive arrangement that gives metamorphic rocks a **lineation,** long mineral grains all pointing in the same direction, or a **foliation,** minerals lining up to give a planar fabric (Figure 7.2a, b). Lineation occurs when kyanite, sillimanite, and other minerals that form long, thin crystals lie parallel in a rock. Alignment of clays, micas, graphite, or other platy minerals, or the separation of a rock into light and dark layers, leads to foliation. Rocks lacking lineation or foliation are termed **hornfels.** Hornfels typically form at low pressure from contact metamorphism. We call foliated rocks **slates, phyllites, schists,** or **gneisses.** Slates, which normally result from low-grade metamorphism of shales, comprise primarily microscopic clay grains. Phyllites, which form at higher metamorphic grades, sparkle because clay minerals have metamorphosed to produce small grains of micas. Schists, which form under medium-grade metamorphic conditions, contain medium to coarse crystals of mica we can easily see with our naked eye (Figure 7.2a). At higher grades, metamorphic rocks may develop compositional layering. We call such

rocks *gneisses.* A gneiss is distinctive because of its alternating light and dark layers or bands (Figure 7.2b). The light bands contain quartz and feldspar, while the dark bands contain mafic minerals. At high temperatures ($>700\ °C$), some minerals in pelitic rocks (metamorphosed clay-rich sediments) may melt. Partial melting often produces a migmatite, a rock containing veins and patches composed of quartz and feldspar that crystallized from the partial melt.

METAMORPHIC MINERALS

Metamorphic rocks contain a great variety of minerals. Nearly all the minerals common in sedimentary and igneous rocks may be present, as well as many minerals exclusive to metamorphic rocks. The two most important factors controlling mineralogy are the composition of the rock and the grade of metamorphism. We generally describe rock composition by listing a rock's chemical composition or, more simply, by dividing the most common rock types into general compositional classes. Table 7.1 lists some common rock classes used by metamorphic petrologists.

▶**TABLE 7.1**
Compositional Classes of Metamorphic Rocks

Metamorphic Rock Class	Common Kinds of Metamorphic Rocks	Essential Minerals	Unmetamorphosed Equivalents
pelitic	slate phyllite schist gneiss	quartz micas	clay-rich sediments
psammitic	quartzite quartzofeldspathic gneiss	quartz	sandstones feldspathic sandstones
mafic (metabasite)	greenstone amphibolite mafic gneiss mafic granulite eclogite	amphiboles plagioclase	basaltic rocks
carbonate	marble	carbonates	limestone dolomite
marl	slate phyllite schist gneiss	micas carbonates	calcareous shale shaly limestone
iron formation	banded iron formation	quartz jasper magnetite hematite	iron-rich chemical sediments
ultramafic	serpentinite	serpentine talc Mg-rich amphiboles	ultramafic igneous rocks
granitic	granitic gneiss granulite	K-feldspar quartz micas	granitic rocks

Table 7.2 lists common metamorphic minerals for each rock class. Minerals at the top of the table dominate low-grade rocks; those at the bottom dominate high-grade rocks. Metapelites exhibit the most mineralogical variation. Before metamorphism, pelitic rocks contain a variety of low-temperature minerals, many of which become unstable on heating. In contrast, granitic rocks show few mineralogical changes when metamorphosed because they comprise predominantly quartz and K-feldspar, which are stable under most metamorphic conditions.

METAMORPHIC FACIES

Pentti Eskola, a geology professor at the University of Helsinki, introduced the idea of **metamorphic facies** in 1920. He observed that the equilibrium mineral assemblage in metamorphosed mafic rocks **(metabasites)** varies with pressure and temperature. Therefore, the mineral assemblage records the pressure and temperature at which the rock was metamorphosed. Eskola defined **facies** as general ranges of pressure and temperature characterized by a distinct mineral assemblage in mafic rocks. Facies diagrams, such as Figure 7.3, are similar to phase diagrams because they divide P-T space into small areas associated with specific minerals or mineral assemblages. The main differences between facies diagrams and phase diagrams are that facies diagrams involve many chemical components, the locations of different facies in P-T space are not precise, and we often do not know the exact reactions that relate one facies to another.

Eskola originally identified eight facies. Other petrologists have divided some to more precisely represent pressure and temperature ranges (Figure 7.3). Each facies name comes from its most characteristic metabasite minerals or rock types. Table 7.3 summarizes key mineral assemblages for each facies. The **zeolite facies** represents the lowest grade of metamorphism, at times hard to distinguish from diagenesis. The formation of zeolite minerals and clays characterize the zeolite facies. As temperature rises, the zeolite facies gives way to the **prehnite-pumpellyite facies,** the **greenschist facies,** the **amphibolite facies,** and the **granulite facies.** Contact metamorphism produces two low-pressure, high-temperature facies, the **pyroxene-hornfels facies** and the **sanidinite facies.** The **lawsonite-albite facies,** the **blueschist facies,** and the **eclogite facies** occur at high pressure.

Although Eskola based his facies names on mineral assemblages in mafic rocks, petrologists use the same names when talking about rocks of other compositions. This leads to some confusion. Table 7.3 lists key mineral assemblages for different metamorphic

▶**FIGURE 7.3**
Metamorphic facies. The labeled fields indicate relative pressure-temperature ranges for each of the 10 facies, but their locations vary somewhat with rock composition. There are no facies at very low temperatures and high pressures because those conditions are not reached in nature. The upper temperature limit corresponds to conditions at which most rocks melt. The upper pressure limit has been arbitrarily chosen; some eclogites and other rocks form at pressures in excess of 14 Kbar.

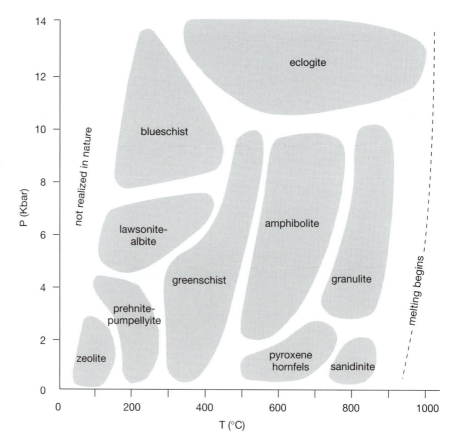

Some Important Minerals in Metamorphic Rocks of Different Compositions

Grade direction: Low Grade → High Grade

Pelitic	Mafic	Carbonate	Iron Formation	Ultramafic	Granitic
quartz SiO_2	zeolites (variable Ca-Al silicates)	calcite $CaCO_3$	quartz SiO_2	talc $Mg_3Si_4O_{10}(OH)_2$	quartz SiO_2
kaolinite $Al_2Si_2O_5(OH)_4$	prehnite $Ca_2Al(AlSi_3O_{10})(OH)_2$	dolomite $CaMg(CO_3)_2$	hematiter Fe_2O_3	brucite $Mg(OH)_2$	K-feldspar $KAlSi_3O_8$
pyrophyllite $Al_2Si_4O_{10}(OH)_2$	pumpellyite (similar to epidote but contains Mg and Fe^{2+})	quartz SiO_2	magnetite Fe_3O_4	serpentine $Mg_6Si_4O_{10}(OH)_8$	biotite $KMg_3(AlSi_3O_{10})(OH)_2$
chlorite (variable combinations of talc + brucite)	plagioclase (Ca-rich) $(Ca,Na)(Al,Si)_4O_8$	biotite (phlogopite) $KMg_3(AlSi_3O_{10})(OH)_2$	minnesotaite $Fe_3Si_4O_{10}(OH)_2$	olivine (forsterite) Mg_2SiO_4	hornblende (complex amphibole)
muscovite $KAl_2(AlSi_3O_{10})(OH)_2$	epidote $Ca_2(Al,Fe)_3Si_3O_{12}(OH)$	tremolite $Ca_2Mg_5Si_8O_{22}(OH)_2$	actinolite $Ca_2(Fe,Mg)_5Si_8O_{22}(OH)_2$	anthophyllite $(Mg,Fe)_7Si_8O_{22}(OH)_2$	garnet (almandine) $Fe_3Al_2Si_3O_{12}$
kyanite Al_2SiO_5	chlorite (variable combinations of talc + brucite)	garnet (grossular) $Ca_3Al_2Si_3O_{12}$	grunerite $Fe_7Si_8O_{22}(OH)_2$	clinopyroxene (diopside) $CaMgSi_2O_6$	orthopyroxene $(Mg,Fe)_2Si_2O_6$
andalusite Al_2SiO_5	actinolite $Ca_2(Fe,Mg)_5Si_8O_{22}(OH)_2$	olivine (forsterite) Mg_2SiO_4	hedenbergite $CaFeSi_2O_6$	garnet (pyrope-almandine) $(Mg,Fe)_3Al_2Si_3O_{12}$	
biotite $K(Mg,Fe)_3(AlSi_3O_{10})(OH)_2$	hornblende (complex amphibole)	clinopyroxene (diopside) $CaMgSi_2O_6$	fayalite Fe_2SiO_4	orthopyroxene (enstatite) $Mg_2Si_2O_6$	
chloritoid $(Fe,Mg)Al_2SiO_5(OH)_2$	garnet (almandine-pyrope) $(Fe,Mg)_3Al_2Si_3O_{12}$	periclase MgO			
garnet (almandine) $Fe_3Al_2Si_3O_{12}$	biotite $K(Mg,Fe)_3(AlSi_3O_{10})(OH)$	wollastonite $CaSiO_3$			
staurolite $Fe_2Al_9Si_4O_{23}(OH)$	clinopyroxene (diopside) $CaMgSi_2O_6$				
cordierite $(Mg,Fe)_2Al_4Si_5O_{18}$	orthopyroxene (enstatite) $Mg_2Si_2O_6$				
K-feldspar $KAlSi_3O_8$					
sillimanite Al_2SiO_5					
orthopyroxene $(Mg,Fe)_2Si_2O_6$					

▶**TABLE 7.3**
Key Mineral Assemblages in Metamorphic Facies

Kind of Metamorphism	Metamorphic Facies	Key Mineral Assemblages in Metabasites
contact metamorphism (very low pressure)	pyroxene hornfels facies	orthopyroxene + clinopyroxene + plagioclase + quartz (without garnet)
	sanidinite facies	sanidine or tridymite or pigeonite or glass
low-pressure metamorphism	zeolite facies	zeolite + quartz
	prehnite-pumpellyite facies	prehnite or pumpellyite + quartz (without zeolite)
	greenschist facies	chlorite + epidote + albite + quartz (without prehnite or pumpellyite)
	amphibolite facies	hornblende + plagioclase + quartz
	granulite facies	orthopyroxene + clinopyroxene + garnet + plagioclase + quartz
high-pressure metamorphism	lawsonite-albite facies	lawsonite + albite + quartz (without glaucophane or jadeite)
	blueschist facies	glaucophane
	eclogite facies	omphacite + garnet + quartz

facies, but the assemblages will never be present in some rock types. For example, pelitic or calcareous rocks do not form greenschists (green mafic schists) or amphibolites (mafic rocks dominated by amphibole and plagioclase) even when metamorphosed at conditions within the greenschist or amphibolite facies. In addition, for some rock classes, several different mineral assemblages may be stable within a single facies. Further confusion arises because petrologists use some facies names in a more restricted sense, referring to particular rock types with important tectonic significance. The greatest confusion derives from the rock names *granulite*, an orthopyroxene-bearing, coarse-grained, high-temperature rock, and *eclogite*, a rock dominated by garnet and Na-rich clinopyroxene called *omphacite*. Despite these problems, the facies concept provides a convenient way to discuss general ranges of pressure and temperature, and it receives wide use. Sometimes, however, the number and name of facies are different from those in Figure 7.3, depending on who is using them.

SOME COMMON TYPES OF METAMORPHIC ROCK

Metamorphosed Pelitic Rocks (Metapelites)

slate
phylite
schist
gneiss

Metapelites are rich in Al, Si, and K and may contain substantial amounts of Fe and Mg, so minerals containing these elements dominate metapelitic rocks. Many metapelitic rocks contain an Al_2SiO_5 polymorph: andalusite, kyanite, or sillimanite. Plate 3.6 shows a spectacular example of kyanite from Minas Gerais, Brazil. Besides aluminosilicates, many other minerals are common in metapelites (Table 7.2). Often they form large porphyroblasts with micas and other minerals packed around them; the garnet in Plate 8.4 is an example.

Between 1890 and 1910, George Barrow studied metapelites in Scotland and showed that the presence of certain **index minerals** correlated with the temperature of metamorphism. From low temperature to high, these minerals are chlorite, brown biotite, almandine, staurolite, kyanite, and sillimanite. Barrow showed that the occurrences of these minerals allow a metamorphic terrane to be divided into **metamorphic zones,** separated by lines called **isograds.** Each zone is a different metamorphic grade.

Metapelites derive from the metamorphism of shale and other clay-rich sediments, and less commonly from aluminous igneous rocks. Quartz is an essential mineral, and either muscovite or biotite is always present (Figure 7.4). When metamorphosed, clay minerals dehydrate to produce new minerals containing less H_2O. At low grade this leads to the formation of pyrophyllite and muscovite, and at medium grades to biotite. As micas form from the breakdown of clays, foliated textures develop. Figure 7.5 illustrates how shale may be metamorphosed and eventually melted to produce an igneous rock.

▶FIGURE 7.4

Metapelitic rocks. (a) Kyanite forms bladed crystals. The crystals seen here are about 3 cm long. (b) Andalusite, a polymorph of kyanite, sometimes displays chiastolite crosses. (c) A close up of a metapelite reveals rounded equant garnets and long rectangular staurolite in a sea of fine grained micas. (d) Thin section view (PP light) of a metapelite. The blocky mineral at the top of the photo is garnet; the small grains with square cross sections are end views of sillimanite needles; the dark grains are biotite; the white minerals are plagioclase and quartz.

▶FIGURE 7.5

A modified portion of the rock cycle showing how progressive metamorphism may change shales to slate, phyllite, schist, and gneiss before melting begins and, eventually, the cycle starts over again.

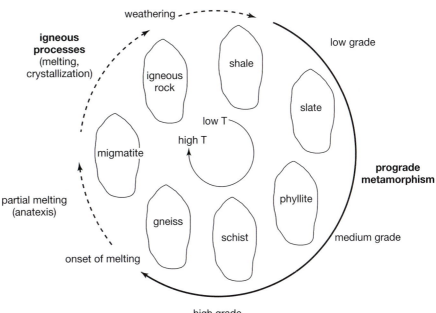

►**FIGURE 7.6**
Metapsammite cut by a granitic dike: the fine-grained rock contains primarily quartz, plagioclase, and biotite with minor garnet. The granitic dike contains K-feldspar and quartz. Note the dark selvage, the dark layer where the two rock types meet, caused by contact metamorphism.

Metamorphosed Psammitic Rocks (Metapsammites)

> quartzite
> quartzofeldspathic rocks

Metapsammites are rocks rich in quartz and alkali feldspar, containing more alkalis and less Al than metapelites (Figure 7.6). They form by metamorphism of sandstones and feldspathic sandstones. Because quartz and feldspar are stable at all metamorphic grades, metapsammites may undergo few obvious metamorphic reactions. Sometimes small amounts of clay minerals in sandstones produce aluminous minerals through dehydration reactions, as in pelites. Foliation, typical of metapelitic rocks, is usually lacking because of a lack of micas. At low grades, metapsammites typically appear massive and homogeneous, containing light-colored quartz and feldspar with small micas and other dark minerals scattered evenly throughout. At higher grades they may recrystallize and become coarser, sometimes becoming gneisses.

Metamorphosed Mafic Rocks (Metabasites)

> greenstone
> amphibolite
> mafic gneiss
> mafic granulite
> eclogite

Compared to metapsammites and metapelites, metabasites are relatively poor in Al and Si and rich in Ca, Mg, and Fe. They are equivalent to mafic igneous rocks such as basalt. Many different minerals may form, and metamorphic reactions are complex (Figure 7.7), but the most important metamorphic minerals are Ca and Mg silicates (Table 7.2). Metabasites tend to be more massive and less foliated than pelitic rocks, but at higher grades they do form schist and gneiss.

We call fine-grained, very low-grade metabasites **greenstones** because of their conspicuous light to dark green color. The characteristic green color comes from fine-grained chlorite and epidote. They also may contain Na-rich plagioclase (albite), quartz, carbonates, and zeolites. At grades equivalent to the greenschist facies, they become greenschists, obtaining schistosity from parallel arrangements of the green amphibole actinolite and chlorite.

In the amphibolite facies, chlorite, epidote, and actinolite break down, producing a specific kind of rock called an **amphibolite.** Amphibolites contain large grains of black hornblende and whitish plagioclase in subequal proportions. Garnet, biotite, and light-colored amphiboles such as anthophyllite or cummingtonite may also be present. At the highest grades, all amphiboles become unstable. Assemblages including garnet and clinopyroxene, or orthopyroxene, are diagnostic of the granulite facies. Minor minerals at all grades include many that are present in mafic igneous rocks.

Metamorphosed Limestones and Dolostones (Marbles)

> marble
> skarn

To building contractors, marble refers to any kind of rock that can be slabbed and polished. Commercial marble is made from many different rock types, including intrusive igneous rocks such as gabbro. To geologists, however, marble refers only to metamorphosed limestone or dolostone (Figure 7.8; Plate 6.3). Marbles are usually massive, lacking in conspicuous lineation, foliation, or compositional layering.

The metamorphism of a limestone or dolostone composed only of carbonate minerals produces few mineralogical changes at any but the highest grades. A general increase in grain size may take place, but no diagnostic minerals can form because of the limited chemical composition and the high stabilities of both calcite and dolomite. However, most limestones contain some quartz and other minerals besides carbonates. In such rocks, a series of interesting Ca-silicates, Ca-Mg-silicates, and Ca-Al-silicates form as metamorphism progresses (Table 7.2). Talc, tremolite, forsterite, diopside (Figure 7.8c; Plate 6.3), and grossular (garnet) are common in marbles. At very high grades, periclase,

▶**FIGURE 7.7**

(a) This finely layered mafic gneiss is composed almost entirely of hornblende (dark mineral) and plagioclase (light mineral). The sample is 8 cm across. (b) Thin section of a metabasite (PP light) showing white plagioclase, dark gray hornblende (with two cleavages at 60°), and biotite (with one cleavage parallel to the edges of the flakes).

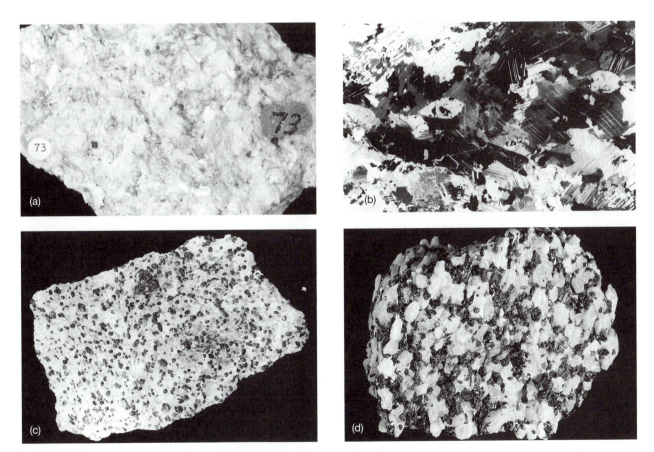

▶**FIGURE 7.8**

Marble in hand specimen and thin section: (a) Dolomite marble from Franklin, New Jersey (10 cm across); (b) 3 mm crossed polars view of a marble in thin section showing conspicuously twinned calcite (striped pattern); (c) marble from the Adirondack Mountains, New York, containing abundant dark colored diopside; (d) marble from the Adirondack Mountains, New York, containing wollastonite (white tabular mineral) and garnet (darker mineral).

▶**FIGURE 7.9**
(a) Metamorphosed ultramafic rock (serpentinite) about 30 cm across. This sample is made of serpentine, chlorite, and talc; it is impossible to tell what it was before metamorphism. (b) Closeup of serpentinite (field of view is 1 cm).

wollastonite, and several other diagnostic minerals may appear (Figure 7.8d). Some sedimentary rocks are rich in carbonates but contain appreciable amounts of pelitic or mafic minerals. We call such rocks *marls.* Many of the same reactions and minerals that characterize metapelites, metabasites, and marbles also occur in marls.

The metamorphic reactions in marbles frequently involve the breakdown of carbonates to release CO_2. If a pluton intrudes a limestone or dolostone, H_2O often flows out of the pluton and into the surrounding carbonate, forming a skarn. The presence of both H_2O and CO_2 and their relative proportions control the formation of many key minerals. Flowing CO_2-H_2O fluids may result in significant **metasomatism** (a change in rock chemistry) in contact aureoles, so different zones in the aureole may have significantly different compositions. Metasomatism is responsible for the formation of many spectacular minerals and some important ore deposits.

Metamorphosed Ultramafic Rocks and Iron Formations

Because of their chemistry, Mg-silicates such as talc, serpentine, anthophyllite, forsterite, diopside, and enstatite dominate ultramafic rocks (Table 7.2, Figure 7.9). Through metamorphism, magnesium oxides, hydroxides, and carbonates may form. Because ultramafic compositions usually start as high-temperature igneous rocks, they do not normally undergo progressive metamorphism. Unless metamorphic temperatures are very

high, metamorphism produces retrograde, rather than prograde, minerals in ultramafic rocks.

Low-grade metamorphism or alteration of olivine-bearing rocks often produces one of the serpentine polymorphs (Plate 3.7). We call rocks rich in serpentine **serpentinites.** Talc, brucite, and chlorite are common low-temperature minerals as well. At higher grades, olivine, anthophyllite, enstatite, periclase, and spinel may be found.

Iron formations are rich in iron and silicon. When metamorphosed, they commonly contain the iron equivalents of the magnesium minerals found in metamorphosed ultramafics (Table 7.2). These include Fe-amphibole, Fe-pyroxene, and Fe-garnet.

Metamorphosed Granitic Rocks

granitic hornfels
granitic gneiss

Metamorphism of granites may be boring because the quartz, K-feldspar, and plagioclase that make up most granites and intermediate igneous rocks are stable at all grades of metamorphism (Figure 7.10). However, mafic minerals such as biotite and hornblende are not; they may react to produce metamorphic minerals at medium and high grade. These minerals include garnet and orthopyroxene. Accessory minerals found in unmetamorphosed granites and intermediate igneous rocks may also be present after metamorphism. At high grade, granitic rocks sometimes develop gneissic banding, even if mineralogy has not significantly changed.

►FIGURE 7.10
Close view (5 cm across) of a granitic gneiss. The black mineral is biotite, the white mineral is K-feldspar, and the gray mineral is quartz. Note the foliation (parallel alignment of grains).

High-Pressure Rocks

blueschist
eclogite

Because of their special significance, high-pressure metamorphic rocks are often classed by themselves. They include mainly blueschists and eclogites and are rare (Figure 7.11). Special tectonic conditions are required to create them and bring them to the surface of the Earth.

Blueschist is a name given to one type of rock that forms at conditions within the blueschist facies, a facies characterized by high pressure and relatively low temperature. Compositionally, blueschists are variable, lying somewhere between pelitic sediments and metabasites. They have conspicuous mineralogy. A blue amphibole, called *glaucophane*, is responsi-

►FIGURE 7.11
Glaucophane schist from Sonoma County, California. This rock is sufficiently fine grained so mineral identification is difficult in hand specimen. Many blueschists are even more fine grained, so thin sections are needed when studying them.

ble for the name of the facies. Other diagnostic minerals include a colorless to green Na-pyroxene called *jadeite* ($NaAlSi_2O_6$), green or white lawsonite ($CaAl_2Si_2O_7(OH)_2 \cdot H_2O$), and pale aragonite (the high-pressure polymorph of calcite). Epidote, garnet, zoisite, quartz, and other accessory minerals may also be present. Because they form at low temperature, blueschists are often fine grained, poorly crystallized, and difficult to study.

Eclogites are mafic rocks metamorphosed at high pressure and moderate to high temperature. They contain the essential minerals pyrope (Mg-rich garnet) and the green Na-rich clinopyroxene called *omphacite.* Orthopyroxene may also be present in significant quantities. Accessory minerals include kyanite, quartz, spinel, titanite, and many others. Eclogites originate in the deep crust or in the mantle. Many mantle **xenoliths,** carried up as nodules within magma, are eclogites. Eclogites are also found as layers or bands associated with some peridotites.

METAMORPHIC REACTIONS

Under any pressure and temperature, the most stable mineral assemblage is the one with the lowest **Gibbs free energy.** Similarly, the most stable arrangement of crystals is the one with the lowest **strain energy.** If temperature, pressure, or composition of a rock change, mineralogy or texture, or both, may change so that energy remains minimized. When metamorphosed at high pressure, graphite reacts to form diamond. When a mica-bearing rock is stressed, micas line up parallel to one another. In these ways, rocks attain chemical and textural equilibrium. In this chapter, we are mostly concerned with chemical equilibrium and the formation of metamorphic minerals.

Box 7.2 Examples of Metamorphic Reactions

Solid-solid reactions:

andalusite = sillimanite   (Reaction 7.1)

$$Al_2SiO_5 = Al_2SiO_5$$

grossular   + quartz =

$$Ca_3Al_2Si_3O_{12} + SiO_2 =$$

anorthite   + 2 wollastonite   (Reaction 7.2)

$$CaAl_2Si_2O_8 + 2\ CaSiO_3$$

Dehydration reactions:

muscovite   + quartz =

$$KAl_2(AlSi_3)O_{10}(OH)_2 + SiO_2 =$$

K-feldspar + sillimanite + vapor   (Reaction 7.3)

$$KAlSi_3O_8 + Al_2SiO_5 + H_2O$$

kaolinite   + 2 quartz =

$$Al_2Si_2O_5(OH)_4 + 2\ SiO_2 =$$

pyrophyllite   + vapor   (Reaction 7.4)

$$Al_2Si_4O_{10}(OH)_2 + H_2O$$

Decarbonation reactions:

calcite   + quartz = wollastonite + CO_2

$$CaCO_3 + SiO_2 = CaSiO_3 + CO_2 \qquad \text{(Reaction 7.5)}$$

dolomite   + 2 quartz =

$$CaMg(Co_3)_2 + 2\ SiO_2 =$$

diopside   + 2 CO_2 (Reaction 7.6)

$$CaMgSi_2O_6 + 2\ CO_2$$

Carbonation reaction:

forsterite + 2 CO_2 =

$$Mg_2SiO_4 + 2CO_2 =$$

2 magnesite + quartz   (Reaction 7.7)

$$2\ MgCO_3 + SiO_2$$

Hydration reaction:

enstatite   + 2 H_2O =

$$Mg_2Si_2O_6 + 2H_2O =$$

2 brucite   + 2 quartz   (Reaction 7.8)

$$2Mg(OH)_2 + 2SiO_2$$

Box 7.2 gives examples of different types of metamorphic reactions. **Solid-solid reactions** are those that involve no H_2O, CO_2, or other vapor phase. Reaction 7.1 involves only two minerals, both Al_2SiO_5 polymorphs. Reaction 7.2 involves four minerals, but is a solid-solid reaction. **Dehydration** and **decarbonation reactions** give off H_2O and CO_2, respectively. **Hydration** and **carbonation reactions** consume H_2O and CO_2, respectively. Most metamorphic reactions involve more than two minerals, and many involve H_2O or CO_2.

Mineralogical equilibrium is easier to interpret and model than textural equilibrium. The Laws of Thermodynamics (Box 7.3) allow us to predict which minerals are stable under particular conditions. Conversely, we can sometimes estimate the pressure and temperature at which a rock formed from the minerals it contains. In contrast, textural changes often involve recrystallization, causing a change in grain size and shape that may be difficult to interpret. Nevertheless, textural changes can reveal information about deformation and stress that is otherwise unavailable. Consequently, textures are important to

structural geologists and others interested in mechanical processes and tectonism, while mineralogists and many petrologists generally focus on the minerals present rather than how they got there or what they look like.

EQUILIBRIUM

Metamorphic reactions involve changes in mineralogy or in mineral composition. A mineral assemblage is at chemical equilibrium if no such changes are occurring. If the assemblage has the lowest Gibbs free energy possible for the given conditions, it is at **stable equilibrium.** In principle, all rocks tend toward stable equilibrium. Whether they reach it depends on many things, including temperature, grain size, and reaction kinetics. If reactions cease before a rock has reached stable equilibrium, the rock is at **metastable equilibrium.**

We call a stable mineral assemblage representative of a given set of pressure-temperature conditions

Box 7.3 Thermodynamic Laws and Definitions

The First Law of Thermodynamics: The first law of thermodynamics defines the **internal energy (E)** of a chemical system. If the composition of a system does not change (the system is closed), E is constant unless heat flows or work is done. E is usually expressed in units of energy/mole: joules/mole (J/mol) or calories/mole (cal/mol or Kcal/mol). One calorie is equivalent to 4.184 joules.

The Second Law of Thermodynamics: The second law of thermodynamics defines **entropy (S).** Entropy is a thermodynamic value expressing the degree of disorder of a system. Minerals with simple atomic structure and simple chemistry have low entropy. Those with complex structure or chemistry have high entropy. S is usually expressed in units of energy/mole-degree: joules/mole-K (J/mol-K) or calories/mole-K (cal/mol-K or Kcal/mol-K).

The Third Law of Thermodynamics: Entropy varies with temperature. The third law of thermodynamics states that the entropy of crystals tends toward zero as the absolute temperature approaches 0 °K (-273.15 °C).

Molar Volume: The **molar volume** is the volume occupied by one mole of a compound. It is usually expressed in units of cm³/mole (cc/mol). For some calculations, it is expressed in units of joules/bar-mole (J/bar-mol). (1 J/bar = 10 cc.)

Enthalpy: Minerals of high volume are less stable at high pressure than at low pressure. Conversely, minerals of low volume are less stable at low pressure than at high pressure. **Enthalpy (H)** is a thermodynamic quantity that reflects this concept. It includes the internal energy (E) and energy associated with the volume of a substance:

$$H = E + PV \tag{7.9}$$

As with E, H is usually expressed in units of energy/mole: joules/mole (J/mol) or calories/mole (cal/mol or Kcal/mol).

Gibbs Free Energy: Minerals of high entropy *(S)* are more stable at high temperature than at low temperature. Minerals of low entropy are more stable at low temperature than at high temperature. The Gibbs free energy reflects this relationship by adding an entropy term to enthalpy *(H):*

$$G = H - TS = E + PV - TS \tag{7.10}$$

As with H and E, G is usually expressed in units of energy/mole: joules/mole (J/mol) or calories/mole (cal/mol or Kcal/mol).

Clausius-Clapeyron Equation: The Clausius-Clapeyron equation relates the volume and entropy of a reaction to its slope on a pressure-temperature phase diagram:

$$dP/dT = \Delta S/\Delta T \tag{7.11}$$

Gibbs Phase Rule: The Gibbs phase rule says that the number of minerals that may stably coexist is limited by the number of chemical components:

$$p + f = c + 2 \tag{7.12}$$

where p is the number of mineral phases, c is the number of chemical components, and f is the number of degrees of freedom. On a phase diagram, $f = 0$ corresponds to a point, $f = 1$ corresponds to a reaction line, and $f = 2$ corresponds to a general area.

a **paragenesis.** When conditions change, metamorphic reactions may create a new paragenesis as some minerals disappear and others grow. If reactions occur as temperature increases, we call them **prograde reactions.** Reactions also occur due to temperature decrease (retrograde metamorphism), changes in pressure, or changes in chemistry. Reactions 7.7 and 7.8 in Box 7.2 are examples of retrograde reactions that often affect mafic rocks.

Prograde metamorphism involves the breakdown of minerals stable at lower temperature to form minerals stable at higher temperature. Some prograde reactions are solid-solid reactions, but most involve vapors of H_2O or CO_2 that flow along cracks or grain boundaries. As temperature increases, minerals containing H_2O or CO_2 become increasingly unstable, causing dehydration or decarbonation, and the release of H_2O or CO_2 as intergranular fluid. If we ignore H_2O and CO_2, we find that most prograde metamorphism is nearly **isochemical,** meaning that the rock is the same composition before and after metamorphism. In some cases, however, flowing fluids and metasomatism can be the dominant forces controlling metamorphism.

Retrograde metamorphism is in many ways just the opposite of prograde metamorphism. H_2O- and CO_2-free minerals react with fluids to produce hydrous or carbonate minerals. Mg-silicates such as

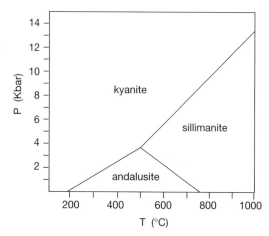

▶FIGURE 7.12
Al_2SiO_5 phase diagram.

forsterite, Mg_2SiO_4, and enstatite, $Mg_2Si_2O_6$, for example, may react to form talc, $Mg_3Si_4O_{10}(OH)_2$; serpentine, $Mg_6Si_4O_{10}(OH)_8$; brucite, $Mg(OH)_2$; or magnesite, $MgCO_3$, at low temperature. In contrast with prograde reactions, retrograde reactions are often quite sluggish. They may not go to completion and frequently do not reach stable equilibrium. Sometimes retrogression only affects parts of a rock or parts of some grains in a rock.

METAMORPHIC PHASE DIAGRAMS AND THE PHASE RULE

Pressure-temperature (P-T) **phase diagrams** show the pressures and temperatures at which metamorphic reactions occur. Phase diagrams for simple chemical systems with few stable minerals may only contain a few reactions. Figure 7.12 shows a phase diagram depicting the phase relationships for a one-component chemical system. The chemical component is Al_2SiO_5; the system includes only polymorphs of Al_2SiO_5 (andalusite, kyanite, sillimanite). Solid lines in Figure 7.12 show the location of three reactions in P-T space:

$$\begin{array}{cc} \text{kyanite} & = \text{andalusite} \\ Al_2SiO_5 & = Al_2SiO_5 \end{array} \qquad \text{(Reaction 7.13)}$$

$$\begin{array}{cc} \text{andalusite} & = \text{sillimanite} \\ Al_2SiO_5 & = Al_2SiO_5 \end{array} \qquad \text{(Reaction 7.14)}$$

$$\begin{array}{cc} \text{kyanite} & = \text{sillimanite} \\ Al_2SiO_5 & = Al_2SiO_5 \end{array} \qquad \text{(Reaction 7.15)}$$

Reactants and products are stable on separate sides of the reaction lines, which separate P-T space into small areas called **stability fields,** each characterized by a specific Al_2SiO_5 mineral. The phase diagram shows us that each of the three polymorphs has its own stability field: Rocks containing kyanite form

at low temperature and high pressure, rocks containing andalusite form at low pressure, and those containing sillimanite form at high temperature. The diagram also allows us to make predictions: If a rock containing andalusite is metamorphosed at high temperature, the andalusite will change into sillimanite (Figure 7.12).

Pressure and temperature are the two most important **intensive variables** (variables controlled by factors outside a rock) that determine the mineral assemblage in a metamorphic rock. Sometimes if pressure or temperature change, the mineral assemblage will change; other times it will not. We can think of a rock as a chemical system. If pressure and temperature can change independently without a change in mineral assemblage, the rock system has 2 **degrees of freedom.** If neither pressure nor temperature can change without a change in mineral assemblage, the rock system has 0 degrees of freedom. If the two intensive variables must change together to maintain a mineral assemblage, the rock system has 1 degree of freedom. The Gibbs phase rule relates the number of possible stable minerals to the number of chemical components and the degrees of freedom of a system:

$$p + f = c + 2 \qquad (7.16)$$

p is the number of phases (number of minerals plus fluid species), c is the number of components (in the case of andalusite-kyanite-sillimanite, $c = 1$), and f is the number of degrees of freedom. The constant 2 in the phase rule refers to the two intensive variables, pressure and temperature, that normally vary and control mineralogy. At an **invariant point,** a point where two or more reactions intersect on a phase diagram, $f = 0$. All three Al_2SiO_5 polymorphs can coexist at about 4 Kbar and 500 °C, the invariant point at which Reactions 7.13, 7.14, and 7.15 intersect (Figure 7.12). Along a **univariant line** ($f = 1$) corresponding to a reaction line on a phase diagram, two of the three polymorphs may be stable. In a general region of P-T space, $f = 2$, and only one Al_2SiO_5 mineral is stable.

Figure 7.13 depicts more complicated reactions in the three-component ($c = 3$) system Al_2O_3-SiO_2-H_2O. The reactions are univariant ($f = 1$), involving four phases ($p = 4$):

$$\begin{array}{cc} \text{kaolinite} & + 2\text{ quartz} = \\ Al_2Si_2O_5(OH)_4 & + 2\ SiO_2 = \end{array}$$

$$\begin{array}{cc} \text{pyrophyllite} & + \text{ vapor} \\ Al_2Si_4O_{10}(OH)_2 & + H_2O \end{array} \qquad \text{(Reaction 7.17)}$$

$$\begin{array}{cc} \text{pyrophyllite} & = \text{ kyanite } + \\ Al_2Si_4O_{10}(OH)_2 & = Al_2SiO_5 + \end{array}$$

$$\begin{array}{cc} 3\text{ quartz } + \text{ vapor} \\ 3\ SiO_2 + H_2O \end{array} \qquad \text{(Reaction 7.18)}$$

Reaction 7.17 is a dehydration reaction that takes place in some metapelitic rocks at low temperature. At low temperature, kaolinite and quartz are stable together, but as temperature increases, kaolinite and quartz react to produce pyrophyllite and vapor. Pyrophyllite itself breaks down at slightly higher temperatures by Reaction 7.18, so pyrophyllite is only stable over a range of about 100 °C (Figure 7.13). It is important to emphasize that Reaction 7.17 does not limit the stability of kaolinite if quartz is not present. When a rock containing kaolinite + quartz reacts according to Reaction 7.17, reaction will cease when either kaolinite or quartz is completely consumed. The resulting assemblage will be kaolinite + pyrophyllite + vapor or quartz + pyrophyllite + vapor unless both kaolinite and quartz are exhausted at the same time.

The reactions in Figure 7.13 are only three of many that can take place in the Al_2O_3-SiO_2-H_2O system. Figure 7.14 is a more complete phase diagram for the system, including all stable minerals and reactions. Eight minerals are involved but they cannot all be stable together. At temperatures over about 450 °C, the stable minerals are corundum (Al_2O_3), quartz (SiO_2), and the three Al_2SiO_5 polymorphs (andalusite, kyanite, sillimanite), but corundum and quartz cannot be found together. The phase diagram permits prediction of the pressures and temperatures at which individual minerals and specific mineral assemblages will form. Conversely, it allows us to estimate the pressure and temperature of formation for some metapelitic rocks containing those minerals and assemblages.

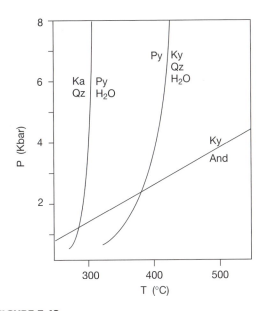

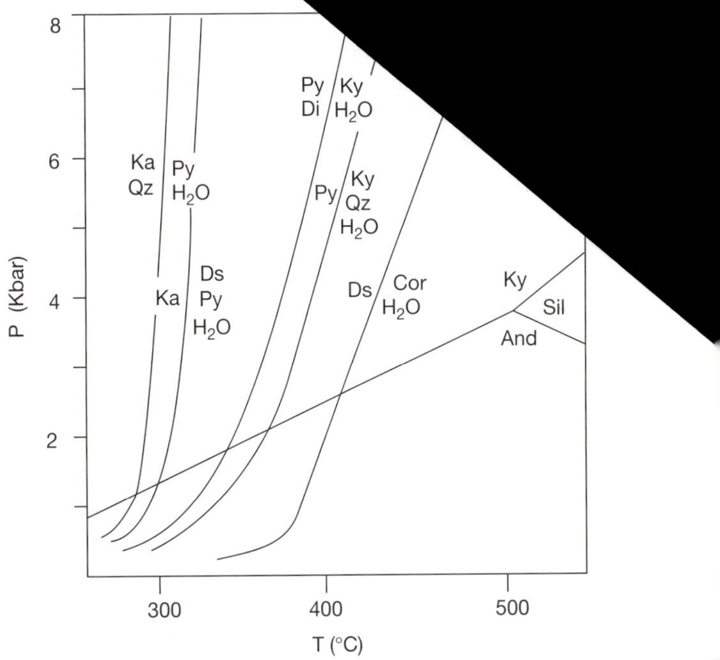

▶ FIGURE 7.14
Phase diagram for the Al_2O_3-SiO_2-H_2O system. Abbreviations: Ka-kaolinite; Qz-quartz; Py-pyrophyllite; Ds-diaspore; Ky-kyanite; Cor-corundum; Sil-sillimanite; And-andalusite.

For example, rocks containing kaolinite and quartz are constrained to have formed at temperatures below about 300 °C. If andalusite accompanies the kaolinite and quartz, pressure is restricted to less than 1 Kbar (Figure 7.14).

Figure 7.15 is a more complex diagram showing other important reactions that take place in metapelitic rocks and involving minerals from several different chemical systems. Some reaction curves only have one side labeled. For instance, the garnet curve at about 450° shows the lowest temperatures at which garnet can be found. The diagram does not give the exact reaction that produces garnet because the reaction varies depending on rock composition. Whatever the exact reaction, it takes place at about the temperatures shown. We can use Figure 7.15 to estimate the P-T conditions of formation of many metapelitic rocks. We could construct similar diagrams for other composition metamorphic rocks. Such diagrams are useful for estimating the pressure and temperature at which metamorphism occurred.

THE THERMODYNAMICS OF REACTIONS

The Gibbs Free Energy

In the preceding discussions, we said that the mineral or mineral assemblage with least energy is more stable than others of the same composition. The Gibbs free energy (often just called *Gibbs energy*) of a mineral is

▶ FIGURE 7.13
Phase diagram showing some reactions involving kaolinite (Ka), quartz (Qz), pyrophyllite (Py), kyanite (Ky), and andalusite (And). The two nearly vertical reactions limit pyrophyllite stability to 300 to 400 °C when water is present.

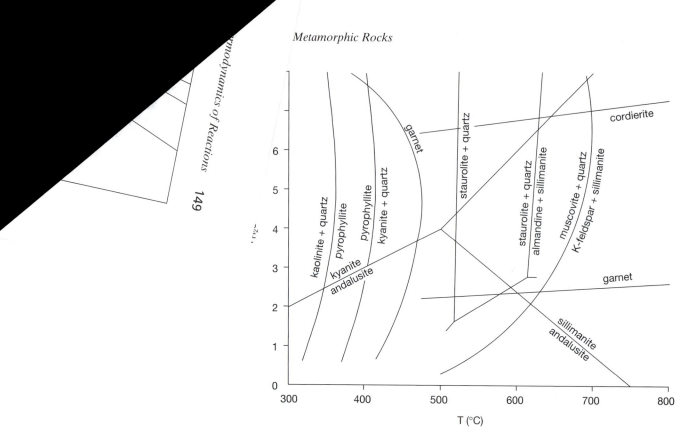

the numerical value describing a mineral's stability (see Box 7.3). Minerals do not have absolute values of Gibbs energy; values are always relative, usually referenced to the elements that comprise a mineral. The Gibbs free energy of formation of calcite from the elements (ΔG_f°), for example, is the Gibbs energy released by the reaction of Ca, C, and O to produce calcite:

$$\begin{array}{ccccc} \text{calcium} & + & \text{carbon} & + & 3\ \text{oxygen} & = & \text{calcite} \\ \text{Ca} & + & \text{C} & + & 3\,\text{O} & = & CaCO_3 \end{array} \quad \text{(Reaction 7.19)}$$

Gibbs energies can be used to calculate the Gibbs free energy of a reaction (ΔG_{rxn}), the difference in Gibbs energy between the products and the reactants, to determine if the reaction will take place. The Gibbs free energy of formation of calcite from the elements is the Gibbs free energy of Reaction 7.19:

$$\Delta G_{f,\,elements}^\circ\,(\text{calcite}) = G_{calcite} - G_{Ca} - G_C - 3\,G_O \quad (7.20)$$

As another example, consider the equilibrium between aragonite and calcite, both having composition $CaCO_3$. The reaction relating the two is:

$$\begin{array}{ccc} \text{calcite} & = & \text{aragonite} \\ CaCO_3 & = & CaCO_3 \end{array} \quad \text{(Reaction 7.21)}$$

The Gibbs energy of this reaction is:

$$\Delta G = G_{aragonite} - G_{calcite} \quad (7.22)$$

Under most P-T conditions, the Gibbs energies of aragonite and calcite are not equal; the one with the least Gibbs energy is stable. If $G_{aragonite} < G_{calcite}$, then ΔG for Reaction 7.21 is negative and aragonite

is stable. If equilibrium is maintained, calcite will react to produce aragonite as the reaction goes to the right. If $G_{calcite} < G_{aragonite}$, then ΔG for Reaction 7.21 is positive and calcite is stable. The reaction goes to the left. Calcite and aragonite may stably coexist only when $G_{calcite} = G_{aragonite}$ and $\Delta G = 0$, conditions represented by the line in Figure 7.16. At Earth surface conditions, $G_{calcite} < G_{aragonite}$, so most $CaCO_3$ is calcite. Some metastable aragonite can be

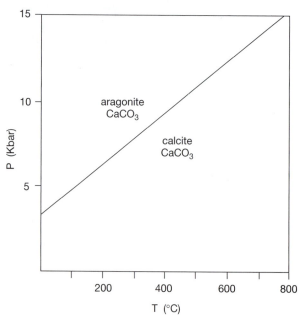

▶**FIGURE 7.16**

Phase diagram showing the reaction calcite = aragonite.

found, most notably in seashells and other biogenic material, but given enough time, it will change into calcite.

Thermodynamic tables in reference books list Gibbs free energies and other thermodynamic values of minerals and related compounds (Box 7.3). We can use the values to calculate phase diagrams and to predict mineral stability. We usually give Gibbs free energies in units of joules per mole (J/mol) or, in older books, kilocalories per mole (Kcal/mol). A reference might say ΔG_f°(calcite) = −1,128,842 J/mol, meaning that 1,128,842 joules of energy are released when pure elements react to produce one mole of calcite. Further, because ΔG_f°(calcite) is negative, calcite is more stable than the elements by themselves. We can combine equations such as 7.20 with 7.22 to verify that we can calculate the Gibbs free energy of any reaction (ΔG_{rxn}) from the ΔG_f° values of the products and reactants:

$$\Delta G_{rxn} = \Sigma\, \Delta G_f \text{ (products)} - \Sigma\, \Delta G_f \text{ (reactants)} \quad (7.23)$$

and therefore for Reaction 7.21:

$$\Delta G_{rxn} = \Delta G_f \text{ (aragonite)} - \Delta G_f \text{ (calcite)} \quad (7.24)$$

Changes in Gibbs Energy With Pressure and Temperature

Gibbs energy values vary with pressure and temperature (Box 7.3). The variations depend on the internal energy *(E)*, molar volume *(V)*, and molar entropy *(S)*, and are directly related to pressure (P) and temperature (T):

$$G = E + PV - TS \quad (7.25)$$

Equation 7.25 describes the Gibbs energy of any mineral. A similar equation describes the Gibbs energy of a reaction:

$$\Delta G_{rxn} = \Delta E_{rxn} + P\Delta V_{rxn} - T\Delta S_{rxn} \quad (7.26)$$

where the internal energy of reaction (ΔE_{rxn}) is:

$$\Delta E_{rxn} = \Sigma\, \Delta E_f \text{ (products)} - \Sigma\, \Delta E_f \text{ (reactants)} \quad (7.27)$$

the volume of reaction (ΔV_{rxn}) is:

$$\Delta V_{rxn} = \Sigma\, \Delta V_f \text{ (products)} - \Sigma\, \Delta V_f \text{ (reactants)} \quad (7.28)$$

and the entropy of reaction (ΔS_{rxn}) is:

$$\Delta S_{rxn} = \Sigma\, \Delta S_f \text{ (products)} - \Sigma\, \Delta S_f \text{ (reactants)} \quad (7.29)$$

Equations 7.25 and 7.26 have some important implications. If the molar volume of a mineral is large and the pressure is high, *G* will be large and the mineral will be relatively unstable. Therefore, at high pressure, minerals of low molar volume (high density) are most stable. Similarly, at high temperature, high-entropy minerals are most stable because high entropy and high temperature lead to low *G*. These observations explain why graphite reacts to form diamond, a denser polymorph, at high pressure, and why sillimanite, a high-entropy polymorph of kyanite, is only stable at high temperature.

If a reaction is at equilibrium, the minerals on both sides of the reaction are stable simultaneously. So, when calcite and aragonite coexist stably, they have equal Gibbs energies and the Gibbs energy of Reaction 7.21 is zero:

$$0 = \Delta E_{rxn} + P\Delta V_{rxn} - T\Delta S_{rxn} \quad (7.30)$$

Equation 7.30 must hold at all points along any reaction curve on a P-T phase diagram. Application of calculus to Equation 7.30 yields a relationship, called the *Clausius-Clapeyron equation,* used for calculating the slope of a reaction on a P-T diagram:

$$\text{slope} = \frac{dP}{dT} = \frac{\Delta S_{rxn}}{\Delta V_{rxn}} \quad (7.31)$$

Volumes and entropies of minerals depend mostly on chemical composition, and, to a lesser extent, on temperature and pressure. Variations with temperature and pressure are predictably small and similar in most minerals. Thus, for solid-solid reactions, the volume and entropy of reaction are nearly constant over a wide range of pressure and temperature, and the Clausius-Clapeyron equation tells us the slopes of reaction lines are nearly constant; they often plot as straight lines. Dehydration and decarbonation reactions, however, usually plot as curves because the volumes and entropies of fluids like H_2O and CO_2 vary greatly with pressure and temperature, leading to great variation in ΔV_{rxn} and ΔS_{rxn}.

The example considered above (Reaction 7.21) involved only two minerals. We can analyze more complex reactions in the same way. For example, consider the following reaction:

grossular (Gr) + quartz (Qz) =
$$\text{Ca}_3\text{Al}_2\text{Si}_3\text{O}_{12} + \text{SiO}_2 \qquad =$$
$$\qquad\qquad\qquad (\text{Reaction } 7.32)$$
anorthite (An) + 2 wollastonite (Wo)
$$\text{CaAl}_2\text{Si}_2\text{O}_8 + 2\,\text{CaSiO}_3$$

For this reaction, we may calculate ΔG_{rxn} as

$$\Delta G_{rxn} = \Delta G_f^\circ(\text{An}) + 2\,\Delta G_f^\circ(\text{Wo})$$
$$- \Delta G_f^\circ(\text{Gr}) - \Delta G_f^\circ(\text{Qz}) \quad (7.33)$$

Box 7.4 Using Phase Diagrams To Interpret Rocks

In this chapter we have talked mostly about meta-morphic minerals and mineral stability in meta-morphic rocks. We have described how minerals change with metamorphism and we have dis-cussed thermodynamics and how we can predict mineral changes.

Petrologists often look at metamorphism and metamorphic minerals in a slightly different way. They find a rock containing a certain mineral or mineral assemblage, and they want to know what the conditions were when that rock formed. Such information allows them to reconstruct the geo-logical history of the rock and, by extension, of the area from which it comes.

Figure 7.17 shows a schematic phase dia-gram depicting the stability fields for some key minerals in pelitic rocks. Reactions that limit the stabilities of pyrophyllite (Py), chloritoid (Cld), chlorite (Chl), garnet (Gt), staurolite (St), kyan-ite (Ky), sillimanite (Sil) and andalusite (And) can give us information about the metamorphic conditions when the rock formed. For simplicity, we have not shown the products for some reac-

tions. They involve multiple minerals and would needlessly complicate this discussion.

Consider a rock that contains garnet. The phase diagram shows that the rock must have formed at temperatures greater than about 500 °C. Garnet does not, however, have an upper tempera-ture limit. On the other hand, if a rock contains staurolite, it must have formed in the temperature window between about 460 and 700 °C (depending on pressure) because staurolite has both a lower and an upper temperature limit. Suppose a rock contains the common assemblage garnet, staurolite, and sillimanite. It must have formed in the shaded window between reactions limiting those minerals.

Figure 7.17 shows only reactions limiting the sta-bility of individual minerals. If we considered reac-tions that limit the stability of mineral assemblages (two or more minerals together) the diagram would become very complex and would include many more reactions. Such diagrams, which petrologists call *pe-trogenetic grids,* divide P-T space into small areas. They allow petrologists to use mineral assemblages to estimate the conditions at which a rock formed.

▶**FIGURE 7.17**
P-T diagram for pelites.

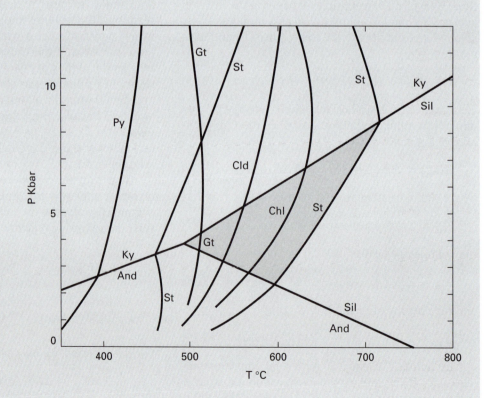

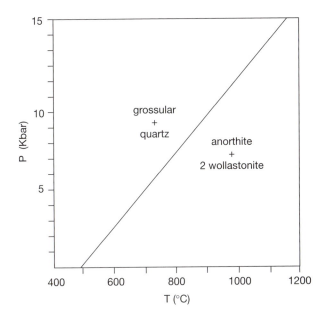

▶FIGURE 7.18
The reaction grossular + quartz = anorthite + 2 wollastonite and a solid-solid reaction.

Note that we multiply the Gibbs energy of wollastonite by 2 because the reaction involves 2 moles of wollastonite. We can calculate ΔV_{rxn}, ΔS_{rxn}, and other thermodynamic quantities for Reaction 7.32 the same way we did for ΔG_{rxn}. As with most solid-solid reactions, Reaction 7.32 plots as a straight line in P-T space (Figure 7.18).

If we knew the thermodynamic properties of all minerals, we could program computers to calculate phase diagrams for any chemical system. Many programs exist for this purpose. For simple systems involving simple minerals, the calculations are relatively straightforward, and all programs yield similar results. However, for some complex systems and minerals, we have inadequate thermodynamic data at present.

▶QUESTIONS FOR THOUGHT

(Some of the following questions have no specific correct answers; they are intended to promote thought and discussion.)

1. If a rock contains quartz and feldspar prior to metamorphism, it will often contain quartz and feldspar after metamorphism. Why? How can we tell, then, that a granitic rock has been metamorphosed?

2. Many of the same minerals characterize both regional and contact metamorphic terranes. Why? On the other hand, some minerals such as sanidine or pigeonite are generally thought to characterize only contact metamorphic terranes. Why are such minerals generally absent from regional metamorphic terranes?

3. Why do many ultramafic rocks experience retrograde metamorphism? What are some of the typical retrograde minerals found in ultramafic rocks? What chemical characteristics do they share?

4. What evidence might you look for in a rock to determine that progressive metamorphism has occurred?

5. What evidence might you look for in a rock to determine that retrograde metamorphism has occurred?

6. When granites are metamorphosed, they rarely develop schistosity. Why?

7. Why is it convenient to divide metamorphic rocks into classes based on chemistry?

8. Most metapelites are formed by metamorphism of clay-rich sediments. They are rarely formed by metamorphism of aluminous igneous rocks. How can they form from both? How can you distinguish the two?

9. What is the significance of Eskola's facies? Why is the facies concept widely used today?

10. Metamorphism of a basalt, or of a greywacke, produces highly variable mineralogy, depending on the conditions of metamorphism, so the minerals in such rocks can be used as indicators of metamorphic grade. On the other hand, metamorphism of a sandstone or of a granite produces less varied mineralogy. Why?

11. If all metamorphic rocks represented equilibrium, we could theoretically enter their chemical and mineralogical compositions into a computer program, and the program could tell us the conditions at which the rock formed. However, we don't do this, and we can't do this. Why?

12. Consider the P-T diagram in Figure 7.17. Under what conditions did a rock form if it contains:
    a. staurolite and andalusite
    b. kyanite and chloritoid
    c. garnet and chloritoid
    Use words, or make a sketch of the phase diagram with appropriate shading, to explain your answers. What does the diagram tell you about the likelihood of finding pyrophyllite and sillimanite together?

▶RESOURCES

(See also the general references listed at the end of Chapter 1.)

Barrow, G. On an intrusion of muscovite-giotite gneiss in the southeast Highlands of Scotland. *Geological Society of London Quarterly Journal* 49 (1893): 330–358.

Barrow, G. On the geology of lower Dee-side and the southern Highland border. *Geological Association Proceedings* 23 (1912): 268–284.

Carrington, D. P., and S. L. Harley. The stability of ossumilite in metapelitic granulites. *Journal of Metamorphic Petrology* 13 (1995): 613–626.

Chopin, C. Coesite and pure pyrope in high-grade blueschists of the Western Alps: a first record and some consequences. *Contributions to Mineralogy and Petrology* 86 (1984): 107–118.

Correns, C. W. Zur geochemie der digenese. *Geochimica Cosmochimica Acta* 1 (1950): 49–54.

Eskola, P. The mineral facies of rocks. *Norsk Geologisk Tidsskrift* 6 (1920): 143–194.

Fry, N. *The Field Description of Metamorphic Rocks.* New York: Halsted/Wiley, 1984.

Hyndman, D. W. *Petrology of Igneous and Metamorphic Rocks,* 2nd ed. New York: McGraw-Hill, 1985.

Miyashiro, A. *Metamorphic Petrology.* New York: Oxford University Press, 1994.

Robie, R. A., R. S. Hemingway, and J. R. Fisher. Thermodynamic properties of minerals and related substances at 298.15K and 1 bar (105 Pascals) pressure and at higher temperature. *United States Geological Survey Bulletin* 1452 (1978).

Winkler, H. G. F. *Petrogenesis of Metamorphic Rocks.* New York: Springer-Verlag, 1976.

Yardley, B. W. D. *An Introduction to Metamorphic Petrology.* New York: John Wiley and Sons, 1989.

PLATE 1

PLATE 1.1 Tourmaline from Paraiba, Brazil, with colors showing compositional zonation.

PLATE 1.2 Smithsonite from Rush Creek, Arkansas, with colors showing compositional zonation.

PLATE 1.3 Pyrite crystals with pyritohedron form showing growth striations on faces.

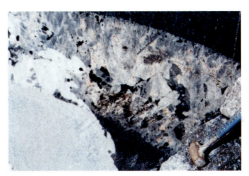

PLATE 1.4 Small pegmatite exposed in Larvikite quarry near Larvik, Norway.

PLATE 1.5 Orthoclase from Petschau, Bohemia, showing penetration twins.

PLATE 1.6 Celestite from Lime City, Ohio.

PLATE 1.7 Acicular okenite in a geode from Bombay, India.

PLATE 1.8 Botryoidal purple smithsonite from Choix, Mexico.

PLATE 2

PLATE 2.1 Rutile needles in quartz from Brazil showing conchoidal fractures.

PLATE 2.2 Rose quartz from Custer, South Dakota.

PLATE 2.3 Amethyst, a purple variety of quartz.

PLATE 2.4 Quartz showing play of colors and internal reflection due to internal fractures.

PLATE 2.5 Brown quartz from California.

PLATE 2.6 Clear quartz crystals from Hot Springs, Arkansas, showing growth striations.

PLATE 2.7 Prismatic quartz crystals on pyrite; sphalerite is the dark mineral behind the quartz crystals (sample from Huaron, Peru).

PLATE 2.8 Herkimer diamonds (in reality, quartz crystals) from Herkimer, New York.

PLATE 3

PLATE 3.1 Pectolite from Patterson, New Jersey, showing radiating habit and scratch marks caused by students checking hardness.

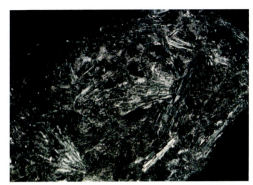

PLATE 3.2 Actinolite from the Adirondack Mountains, New York, showing bladed habit.

PLATE 3.3 Prismatic crystals: beryl from Nigeria (left); rubellite, a pink variety of tourmaline, from San Diego County, California (center); green tourmaline from Minas Gerais, Brazil (right).

PLATE 3.4 Fibrous anthophyllite from Gainesville, Georgia.

PLATE 3.5 Green selenite, a variety of gypsum, from Mt. Gudson, South Australia.

PLATE 3.6 Bladed blue kyanite from Minas Gerais, Brazil.

PLATE 3.7 Chrysotile, the asbestiform variety of serpentine, from Waldheim, Saxony.

PLATE 3.8 Calcite with drusy pyrite from the Campbell Mine, Red Lake, Ontario, Canada.

PLATE 4

PLATE 4.1 Fluorite showing strong relief in liquid with *n*=1.512.

PLATE 4.2 Fluorite showing weak relief in liquid with *n*=1.452.

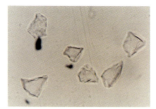

PLATE 4.3 Fluorite in liquid with *n*=1.452; the stage has been slightly lowered to show bright Becke line moving into the liquid.

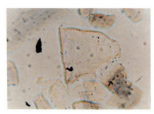

PLATE 4.4 Barite in liquid with *n*=1.636, nearly identical to that of barite.

PLATE 4.5 Olivine interference figure at extinction.

PLATE 4.6 Olivine interference figure 15° from extinction.

PLATE 4.7 Olivine interference figure 45° from extinction.

PLATE 4.8 A nearly centered uniaxial interference figure exhibited by calcite.

PLATE 4.9 A nearly centered uniaxial interference figure exhibited by quartz.

PLATE 4.10 A nearly centered biaxial interference figure exhibited by biotite.

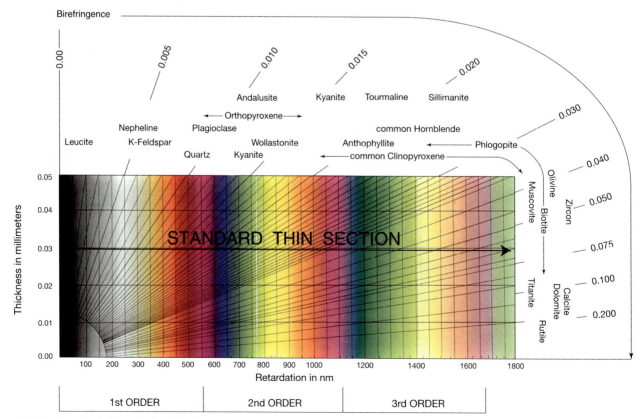

PLATE 4.11 Michel Lévy Color Chart.

PLATE 5

PLATE 5.1 Thin section view of calcite replacing a foraminifera fossil.

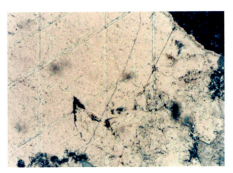

PLATE 5.2 Calcite in thin section.

PLATE 5.3 Thin section with brown biotite, green hornblende (with indistinct 60° cleavage), and clear quartz (PP light).

PLATE 5.4 The same thin section as in Plate 5.3 after stage rotation; note pleochroism.

PLATE 5.5 Sodic hornblende (blue to green with 60° cleavage), quartz (large, clear grains), chlorite (large, light green flake in top center of photo), and garnet (high-relief clear mineral in bottom center and right of photo) in thin section (PP light).

PLATE 5.6 The same thin section as in Plate 5.5; yellow to pink interference colors in hornblende, undulatory extinction in quartz, anomalous gray interference colors in chlorite, opaque garnets (XP light).

PLATE 5.7 Quartz, muscovite, biotite, and polysynthetically twinned plagioclase in thin section (PP light).

PLATE 5.8 The same thin section as Plate 5.7 (XP light).

PLATE 6

PLATE 6.1 Massive green-black augite from Monzoni, Italy.

PLATE 6.2 Chrome diopside (green) and quartz from Minas Gerais, Brazil.

PLATE 6.3 Marble from Lake Placid, New York, containing green diopside and white calcite.

PLATE 6.4 Aggregate of hedenbergite crystals from Nordmarken, Sweden.

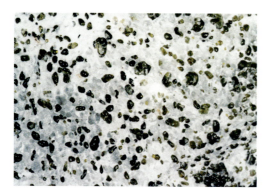

PLATE 6.5 Green olivine crystals in a basalt from San Carlos, Arizona.

PLATE 6.6 A book of biotite from Sioux Lookout, Ontario, Canada.

PLATE 6.7 Perthite, a variety of K-feldspar, from Keystone, South Dakota.

PLATE 6.8 Albite from Amelia Courthouse, Virginia.

PLATE 7

PLATE 7.1 Acicular boulangerite from the Noche Buena Mine, Zacatecas, Mexico.

PLATE 7.2 Arsenopyrite in a biotite schist from the Homestake Mine, Lead, South Dakota.

PLATE 7.3 Covellite (purple tarnish) and pyrite (gold) from Sudbury, Ontario, Canada.

PLATE 7.4 Sphalerite from Tennessee.

PLATE 7.5 Galena in ore from Joplin, Missouri.

PLATE 7.6 Orpiment (yellow) and realgar (orange-pink).

PLATE 7.7 Native copper from the Keweenaw Peninsula, Michigan.

PLATE 7.8 Chalcopyrite from Timmins, Ontario, Canada.

PLATE 8

PLATE 8.1 Pentlandite (silver) with minor pyrrhotite (yellow-gold) from Cobalt, Ontario, Canada.

PLATE 8.2 Molybdenite in quartz.

PLATE 8.3 Amazonite, a gemmy blue variety of microcline (K-feldspar) from Norway.

PLATE 8.4 Almandine in a schist from western Massachusetts.

PLATE 8.5 A collection of gem minerals. Back row, left to right: blue aquamarine (beryl), topaz, pink tourmaline; middle row: yellow beryl, blue-green tourmaline, red ruby (corundum); front row: emerald (beryl), kunzite (a variety of spodumene), yellow sapphire, blue sapphire.

PLATE 8.6 Drusy goethite crystals in cavities in iron formation from the Steep Rock Mine, Atikokan, Ontario, Canada.

PLATE 8.7 Gold ore from Montana. This specimen, 5 cm across, contains visible gold flakes, wires and stringers. It is exceptionally high-grade gold ore.

PLATE 8.8 Ipé Pegmatite Mine in Minas Gerais, Brazil.

Ore Deposits and Economic Minerals

We mine many minerals from the Earth. They include industrial minerals of many sorts, ore minerals that are processed to recover elements they contain, and gems. The best ore deposits are those that contain large amounts of ore minerals. The best ore minerals are those that contain large amounts of elements of interest. Most ore minerals are native elements, sulfides, sulfosalts, oxides, or hydroxides. In this chapter we discuss the many kinds of ore deposits and the minerals they contain. We also consider gems, beautiful and rare varieties of (often common) minerals.

The Earth gives us many mineralogical resources (Figure 8.1). We mine industrial minerals such as halite, gypsum, clays, calcite, asbestos, micas, and zeolites to make salt, plaster, ceramics, construction materials, electronic components, chemical filters, and many other things (Table 8.1). We mine gems for jewelry and a few industrial applications, and we mine many minerals for the elements they contain. We value industrial minerals and gems for their mineralogical properties, but we value most ores because they produce elements that we can use to make other things.

MINERAL USE AND PROFITABILITY

Less than a dozen minerals and eight or nine elements dominate the crust. We use most of them in our daily lives. Other elements and minerals that exist only in small amounts and have uneven distributions are equally vital to our society. If a rock or deposit contains a mineral or element in an amount sufficient to be mined at a profit, we call it an **ore deposit. Ore,** then, is an economic term describing any-

thing that we can profitably take from the ground. Sometimes we mine ores from quarries or from open pits, sometimes we mine them underground, and sometimes we remove them from the Earth by drilling holes and circulating water through beds of soluble minerals.

People have practiced mining and quarrying since ancient times. The first mineral known to be mined was flint, a fine-grained variety of quartz used to make weapons. Early peoples mined other things, such as ochre, for use as pigment in art and religious ceremonies. Egyptians mined native metals, including gold, silver, and copper, from stream beds as early as 3000 to 3700 B.C. Around 2600 B.C. they began to quarry stone to build the great pyramids. By the Middle Ages, mining was common in Europe. Georgius Agricola, a German physician, wrote the first widely read book about mining, *De re Metallica,* in 1556. Agricola's work is said by some to represent the beginning of the science of mineralogy.

We mine some ores because they contain elements that have the metallic properties of conductivity, strength, or shiny appearance, but many ores are valuable for other reasons. We conveniently divide

▶**FIGURE 8.1**

Miner working on a gold bearing quartz vein in the Hishikari mine, Japan, the richest gold mine in the world. The vein, which is slightly less than a meter wide here, consists of thin layers of quartz, some of which contain gold.

Photo from S. E. Kesler (1994), *Mineral Resources, Economics and the Environment,* MacMillan, New York.

▶**TABLE 8.1**
Groups of Ore Commodities

Group	Examples
metallic and semimetallic elements	gold, silver, copper, iron, manganese, nickel, aluminum
nonmetallic elements	potassium, sodium, phosphorous, sulfur
gems	diamond, sapphire, agate
industrial materials: construction and manufacturing	sand, clay, building stone, asbestos, diatomite, talc, pyrophyllite, mica, zeolites
industrial materials: fertilizer and chemicals	limestone, dolomite, phosphate, potash, salt, nitrates, fluorite
energy resources	coal, oil, gas, uranium

ore minerals into several main commodity groups: metallic and semimetallic elements, nonmetallic elements, gems, construction and manufacturing materials, fertilizer and chemical minerals, and energy resources (Table 8.1). We take energy resources and construction materials from the Earth in the greatest quantities. We also mine large amounts of salt and fertilizer components. Of the metals, only iron is removed from the Earth in comparable amounts. In this chapter we focus on ore minerals that contain metallic, semimetallic, and nonmetallic elements, and on gems, because mineralogy is perhaps the most significant factor determining their value.

Many things control the profitability of an ore deposit. We call the amount of known ore in a deposit the **reserves.** The concentration of a commodity in the ore (the "richness" of the ore) determines the **ore grade.** When calculating the profitability, amount of reserves and ore grade are the most significant geological factors, although economic factors such as extraction costs, processing costs, and market price are often more decisive. A high-grade ore deposit may be uneconomical to mine if the reserves are low, because start-up costs could consume all profits. A large high-grade deposit may be uneconomical to mine if it is in a remote area. Even large developed deposits can become uneconomical if the market price falls, perhaps due to the discovery of a better deposit somewhere else.

The elements O, Si, Al, Fe, Ca, Na, K, Mg, and Ti make up 99% of the Earth's crust. It is no wonder, then, that humans have developed processes so we may use these elements in industry, agriculture, and manufacturing. Less abundant elements have also be-

come important to modern society. These include metals, radioactive elements such as uranium or thorium, and fertilizer components such as nitrogen and phosphorous. As shown in Table 8.2, some of these elements make up very small percentages of the Earth's crust; nevertheless, natural processes concentrate them in particular minerals and in particular places. The term **concentration factor** refers to the ratio of minimum economical ore concentration to average crustal concentration (Table 8.2). For example, the average crustal abundance of chromium is about 0.01 wt. %. Chromium ore can sometimes be profitable if it contains 30 wt. % Cr. The necessary concentration factor is therefore 3,000; chromium must be concentrated at least 3,000 times to create profitable ore.

Because profit is related to volume, the minimum concentration factor necessary to make an element economical to mine is generally inversely proportional to crustal abundance. Elements that occur in high abundance do not need a high concentration factor to make mining economical. Iron deposits, for example, are profitable if concentration factors are 5 or 6. In contrast, economical tin, chromium, and lead mining require concentration factors of 2,500 or more. We mine relatively common elements, such as Fe and Al, in many places worldwide, while rarer elements, such as Sn, Cr, or Pb, are mined in far fewer places. Exceptions to the inverse relationship between abundance and economical grade are precious metals such as Au, Ag, and Pt. High market value ensures large profits even when the concentration factor is low. Many gold mines can remain profitable in spite of a concentration factor of less than 750, which amounts to less than 0.1 ounces of gold in a ton of rock. Table 8.2

► **TABLE 8.2**
Mineral Commodities: Minable Grade and Ore Minerals

Element	Average Crustal Abundance (wt. %)	Average Minimum Economical Grade	Average Minimum Economical Concentration Factor	Some Important Ore Minerals	
Al	8	30	3.75	gibbsite	$Al(OH)_3$
				boehmite	$AlO(OH)$
				diaspore	$AlO(OH)$
Fe	5	25	5	magnetite	Fe_3O_4
				hematite	Fe_2O_3
				goethite	$FeO(OH)$
				siderite	$FeCO_3$
				pyrite	FeS_2
Cu	0.005	0.4	80	chalcopyrite	$CuFeS_2$
				bornite	Cu_5FeS_4
				chalcocite	Cu_2S
				covellite	CuS
Ni	0.007	0.5	71	pentlandite	$(Ni,Fe)_9S_8$
				garnierite	$(Ni,Mg)_3Si_2O_5(OH)_4$
Zn	0.007	4	571	sphalerite	ZnS
				wurtzite	ZnS
				zincite	ZnO
				franklinite	$ZnFe_2O_4$
Mn	0.09	35	389	hausmannite	Mn_3O_4
				polianite	MnO_2
				pyrolusite	MnO_2
Sn	0.0002	0.5	2,500	cassiterite	SnO_2
Cr	0.01	30	3,000	chromite	$FeCr_2O_4$
Pb	0.001	4	4,000	galena	PbS
				cerussite	$PbCO_3$
Au	0.0000004	0.0001	250	gold	Au
				calaverite	$AuTe_2$

compares economical concentration factors and lists ore minerals of some important metals.

Ores and ore minerals vary greatly in quality. Ideal ores contain 100% of the commodity of interest. Native copper, for example, is an ideal copper ore. Ideal ores are, however, uncommon. Ores containing lots of a commodity of interest, and in a form that can be processed inexpensively, are considered good ores. Sulfide minerals often comprise good ores because they contain large amounts of the metals of interest, and are easily separated into metal and sulfur. For example, we mine Cu and Cu-Fe sulfides for their copper content. Silicate minerals, in contrast, are poor ore minerals. Although aluminum is found in many common silicates, tight bonding makes producing metallic aluminum from silicates uneconomical. We obtain most aluminum from Al-hydroxides found in bauxite deposits.

After mining, ore goes through a process to separate the valuable minerals from others. This involves crushing the ore, followed by gravity and chemical sep-aration. The unwanted minerals are called **gangue.** Along with other processing and mining waste, gangue is usually discarded in **tailings** piles (Figure 8.2). In some mines, miners return wastes to abandoned portions of a mine to fill voids left by ore removal. Disposal of mine waste sometimes leads to environmental problems, including water, air, and soil contamination. For example, sulfides in mine wastes can react with water and oxygen in air to produce acid mine drainage. The resulting sulfuric acid may kill vegetation and fish in nearby lakes and streams. (See Box 8.5, page 171)

TYPES OF ORE DEPOSITS

The geological processes that concentrate ore are unusual, so good ore deposits are rare. If they were not, market prices would fall, decreasing profits and putting some mines out of business. Because the geology of the Earth varies, the distribution of ore deposits around the globe is uneven. For example, in

▶**FIGURE 8.2**
Abandoned mine and tailings pile.

North America, we could produce the silver, tungsten, sulfur, zinc, and gold to meet our needs. But nickel, chromium, aluminum, and tin must be imported from elsewhere (Box 8.1). Furthermore, we can import many mineral commodities at less than it would cost us to mine them in our own country; tungsten is a good example. Some regions contain most of the supply of certain commodities, which can affect international politics. The United States controls about half the world's molybdenum, Australia about a quarter of the world's aluminum, and Zaire half the cobalt. South Africa, a country rich in mineral commodities, controls 90% of the world's platinum, half the world's gold, and 75% of the chromium. Much of the world's tin comes from Bolivia and Brazil.

Most South African ore deposits are associated with terranes called Precambrian **greenstone belts,** ancient volcanic terranes. Similar greenstone belts in Canada account for many of North America's metallic ore deposits. Various other types of geological terranes are associated with ore deposits. Most economical metal and semimetal deposits are found near margins of continents, or the former margins of continents, where igneous activity has occurred. Many other types of deposits are found in continental interiors.

Ore deposits form in many different ways, and we find them in many different rock types. Some high-temperature ore deposits form in igneous rocks by direct crystallization of a magma, but most metal ores precipitate from **hydrothermal fluids.** These solutions of hot water and dissolved metals deposit minerals in **host rocks,** which may be igneous, metamorphic, or sedimentary.

Magmatic Ore Deposits

In igneous rocks, elements of economic value are often concentrated in **accessory minerals.** In some rare circumstances, these same minerals exist in sufficient abundance to make mining profitable. If the minerals are scattered throughout a host rock, we call the deposit a **disseminated deposit.** Disseminated deposits produce most of the world's diamonds, copper, and molybdenum and also large percentages of the available tin, silver, and mercury. Often, disseminated ores consist of minerals scattered randomly in a host rock. If ore is distributed in many small veins, geologists call the deposit a **lode deposit.**

Magmatic processes can concentrate minerals in different ways. During crystallization of a magma, early forming, dense minerals sometimes sink and accumulate at the bottom of magma chambers (Figure 8.3). **Cumulate** ore deposits form in this way. Chromite, magnetite, and platinum are examples of minerals mined from cumulates. **Pegmatites,** late-stage coarse-grained igneous rocks, also may concentrate economic minerals (Plate 1.4). As discussed in Chapter 5, after much crystallization has occurred, residual magmas often become enriched in **incompatible elements** (those that don't easily fit into crystal structures). These magmas, often water-rich, crystallize as pegmatites, forming dikes or tabular intrusions. Elements of value in pegmatites include Li, Ce, Be, Sn, and U.

Hydrothermal Ore Deposits

As magmas crystallize, hot, water-rich hydrothermal fluids may be released from the melt. The fluids are rich in S, Na, K, Cu, Sn, W, and other elements with relatively high solubilities. Hydrothermal fluids dissolve other elements as they flow through rocks and eventually cool to deposit minerals in **hydrothermal deposits.** Hydrothermal deposits vary in size from huge networks of veins covering many square kilometers to small veinlets only centimeters in width (Figure 8.4). Often, hydrothermal deposits are next to plutons.

Box 8.1 Strategic Minerals and Metals

Although the United States has many ore deposits, some strategic minerals and metals must be imported from other nations. These minerals and metals are not only important to industry and the military, but they also have vast applications in manufacturing. During the Cold War, many of the metals came from the former Soviet Union, its allies, and other nations that were politically unstable, and supplies were often problematic. The current list of important imported metals and their source nations is shown in the table below. The list does not include Cuba, Iran, Iraq, Libya, or North Korea, all of which may export metals but are currently under a trade embargo with the United States. Some of the source nations are importers themselves; Great Britain, for example, imports mineral commodities that it then exports to the United States.

Mineral Commodities and Their Source Nations (1994)

Metal or Mineral	% Imported by the U.S.	Source Nations	Principal Uses
aluminum (Al) /bauxite	100	Australia, Guinea, Jamaica, Brazil	foil, commercial/industrial products
barite	53	China, India, Mexico, Morocco	oil-drilling muds
cadmium (Cd)	49	Canada, Mexico, Australia, Germany	batteries, metal coatings
chromium (Cr)	74	South Africa, Turkey, Zimbabwe	stainless steel
cobalt (Co)	76	Zaire, Zambia, Canada, Norway	alloys, chemicals, magnets
niobium (Nb)	100	Brazil, Canada, Germany	steel manufacturing
diamonds (industrial)	98	Zaire, Great Britain, South Africa, Ireland	grinding and cutting
fluorite	87	Mexico, South Africa, China, Canada	metallurgy and chemicals
graphite	100	Mexico, China, Brazil, Madagascar	metallurgy
manganese (Mn)	100	South Africa, France, Gabon	steel manufacturing
muscovite	100	India, Belgium, Brazil, Japan	electronics
nickel (Ni)	64	Canada, Norway, Australia, Dominican Republic	stainless steel
platinum (Pt)	94	South Africa, Great Britain, Russia	catalytic converters, electronics
potash (KCl)	67	Canada, Israel, Russia, Germany	fertilizer
silver (Ag)	64	Mexico, Canada, Great Britain, Peru	photography, electronics, jewelry, silverware
strontium (Sr)	100	Mexico, Spain, Germany	TV picture tubes, fireworks
tantalum (Ta)	87	Germany, Thailand, Australia	electronics
tin (Sn)	73	Brazil, Bolivia, China, Indonesia	electronics
tungsten (W)	85	China, Bolivia, Germany, Peru	lamp filaments

Large copper and molybdenum deposits, called **porphyries,** are good examples. Some of the world's largest porphyry copper deposits formed when hydrothermal fluids flowed through fractures, dissolving and altering oxide minerals and replacing them with sulfides. **Skarn deposits,** formed by fluids associated with contact metamorphism, may develop when fluids given off by a crystallizing pluton react with adjacent

▶**FIGURE 8.3**
A chromite layer in the Stillwater Complex of Montana. Geologists believe the chromite became concentrated because it crystallized and settled to the bottom of a magma chamber while much of the magma was still molten. Subsequently, the layer was uplifted to the surface and tilted. The Stillwater chromite deposits are not presently being mined, but other chromite cumulates are.

carbonate rocks. When hydrothermal fluids create ore deposits at the Earth's surface, we call the deposits **ex-halitives.**

When a hydrothermal deposit is not directly associated with a pluton, we call it an **epigenetic** deposit. Often, the hydrothermal fluids have traveled so far that their original source is unknown. Many gold and antimony deposits formed in this manner. The lead-zinc deposits of limestones in the central United States, called *Mississippi Valley-type deposits,* are epigenetic (Figure 8.5).

Sedimentary Ore Deposits

Gravity may be an important force concentrating economic minerals. Heavy minerals, weathered from igneous, sedimentary, or metamorphic rocks, often become concentrated in stream or valley bottoms. Rivers may transport them long distances before they become concentrated in **placers** (Figure 8.6). Placer deposits set off the historically important California Gold Rush of 1849. The original sources of minerals in placer deposits are often difficult to determine. The California gold was weathered from extensive vein deposits in the Sierra Nevada Mountains, called the **mother lode.** Many prospectors told stories of finding and mining the mother lode, but in actuality, most of it eroded away long before the "Forty-Niners" arrived. Tin, titanium, zirconium, and diamond are examples of other placer-mined commodities.

Sedimentary ore deposits also form by chemical precipitation. **Banded iron formations** (BIF), found in Precambrian shields, are examples (Figure 8.7). Banded iron formations include oxides, silicates, and carbonates of iron. They were deposited because of changes in the Earth's atmosphere more than two billion years ago and contain very old fossil algae. Iron is not the only commodity deposited by shallow seas. When a body of seawater is trapped, evaporation leads to precipitation of halite and other salts. Thick **evaporite deposits** of halite, sylvite, gypsum, and sulfur have formed in this way. Chemical precipitation of such salts is occurring today around the margins of the Great Salt Lake and other inland seas. In the past, evaporation of seas has led to economical borate deposits in California's Mojave Desert and elsewhere. Nitrates and, more rarely, phosphates may also be deposited from water.

The weathering of preexisting rock may expose and concentrate valuable minerals. Over time, water leaches rocks and soils, dissolving and carrying away soluble material. The remains, called **residuum,** may be rich in Al, Ni, Fe, or other insoluble elements. In tropical climates extreme leaching has produced soils called **laterites,** which are rich in aluminum or, sometimes, nickel. If aluminum-rich laterite lithifies to become rock, we term it *bauxite* (see Box 6.8). We mine laterites and bauxites from open pits to produce nickel and aluminum and, sometimes as a secondary commodity, iron. Leaching of rocks at great depth in the Earth may lead to deposition of economic minerals at shallower depths. We mine such deposits, called **supergene** deposits, for gold, silver, and sometimes copper.

ORE MINERALS

Ore minerals belong to all the mineral classes presented in Chapter 2, but most fall into just four classes: native elements, sulfides and sulfosalts, oxides, and hydroxides. Table 8.2 contains examples of each.

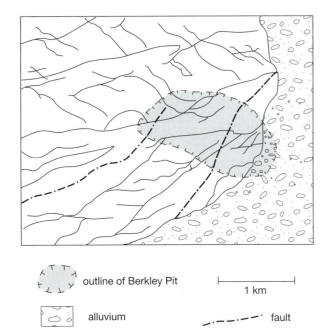

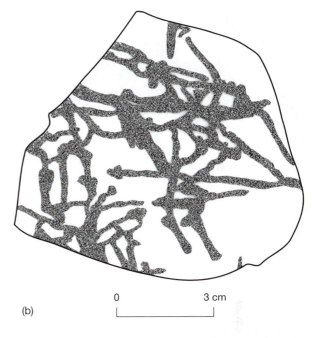

(a)

outline of Berkley Pit

alluvium

granodiorite

1 km

fault

vein

(b)

0　　　3 cm

(c)

▶**FIGURE 8.4**

Hydrothermal deposits and ores come in all sizes:
(a) schematic map of the Butte, Montana, district showing
the many hydrothermal veins; (b) close-up of molybdenite-
bearing veins in a granite hand sample from Climax, Colorado; (c) crystals of wolframite (up to 6 cm in longest
dimension), a tungsten ore mineral, in a quartz vein.

(Illustrations modified from those in Evans, 1980.)

▶**FIGURE 8.5**

Galena, PbS, is one of the principal ore minerals in Mississippi
Valley deposits. These samples,
from near Joplin, Missouri, contain imperfectly formed cubes
and octahedra (8-sided crystals).

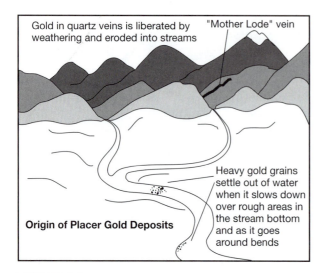

▶FIGURE 8.6

Origin of placer deposits, which are concentrations of heavy clastic grains in stream, lake, or ocean sediment. As shown here, minerals liberated from veins by weathering and erosion are carried downstream and deposited where the water slows. From S. E. Kesler, *Mineral Resources, Economics and the Environment*, New York: Macmillan, 1994. Used by permission.

▶FIGURE 8.7

Banded iron formation from near Hibbing, Minnesota. This sawn slab is about 20 cm wide.

▶FIGURE 8.8

5 cm wide sample with conspicuous arborescent native silver on quartz.

Minerals in these four classes contain relatively large amounts of elements of value. Furthermore, processing and element extraction are usually straightforward and relatively inexpensive.

Native Elements: Metals, Semimetals, and Nonmetals

Native elements are highly valued because they may require no processing before being used in manufacturing, as currency, or for other purposes. The first metals ever used by humans were native minerals. Only later did humans develop refining techniques for the extraction of elements from more complex minerals. Ore geologists divide native elements into metals, semimetals, and nonmetals based on their chemical and physical properties.

Within the metal group, the principal native minerals are gold, silver, copper, and platinum (Plates 7.7 and 8.7, Figure 8.8). These four minerals all contain weak metallic bonds. Gold, silver, and copper have further commonality in their chemical properties and are in the same column of the periodic table. One mineral, an Au-Ag solution called *electrum,* can even vary in composition from pure gold to pure silver. Since copper atoms differ in size from gold and silver atoms, solutions are limited between copper and the precious metals. Native gold, silver, and copper may contain small amounts of other elements. Native copper frequently contains arsenic, antimony, bismuth, iron, or mercury.

Native platinum may contain small amounts of other elements, especially palladium. Native platinum is much rarer than gold, silver, or copper. The native semimetals arsenic, antimony, and bismuth are also rare.

Native copper, gold, silver, and platinum have atomic structures with atoms arranged in a cubic pattern (Figure 8.9). Iron does, too, although native iron is rare except in meteoritcs. Consequently, euhedral crystals may be cubic or, as we will explain in the next chapter, octahedral. More typically, however, these minerals crystallize in less regular shapes (Figure 8.8). Copper is found as branching sheets, plates, and wires, and as massive pieces (Plate 7.7). Silver sometimes occurs in a wirelike or **arborescent** (treelike) form (Figure 8.8). Gold, sometimes mined as nuggets or flakes (Plate 8.7), is also found as wires or scales. These are, however, unusual examples; most precious metals in ores are fine subhedral grains,

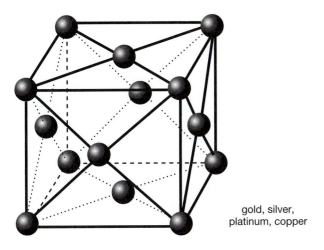

gold, silver,
platinum, copper

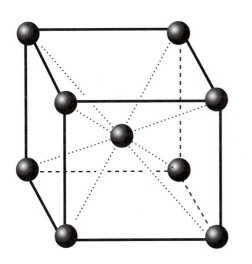

iron

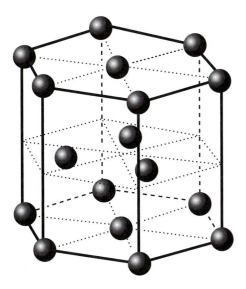

zinc

▶FIGURE 8.9

The atomic structures of some native metals. (a) Gold, silver, platinum, and copper have cubic structures with an extra atom at the center of each face. (b) Iron has a cubic structure with an extra atom at the center of the cube. (c) Zinc has a hexagonal structure.

often microscopic. (all have similar at(although still cub rare mineral, has ure 8.9).

Most hard quartz-rich hy and other su gold and silver in u. hardrock deposits, gold and s... placers (Box 8.2), and native silver is 10u other types of deposits. We mine native platinum primarily from ultramafic igneous rocks, but platinum is also found in placers or is produced as a secondary product of Cu- or Ni-sulfide refining. Native copper occurs in a variety of ore deposits associated with mafic volcanics and in some sandstones.

Graphite, diamond, and sulfur are examples of nonmetallic native elements; both graphite and diamond consist of carbon. Graphite (Figure 8.11a) is common as a minor mineral in many kinds of metamorphic rocks, including marbles, schists, and gneisses. The origin of the carbon is usually organic material in the original sediments. Graphite also occurs in some types of igneous rocks and in meteorites. Native sulfur is associated with volcanoes, often concentrated by fumaroles. It is also found in veins associated with some sulfide deposits and in sedimentary rocks associated with halite, anhydrite, gypsum, and calcite. Native sulfur deposits only account for about half the world's supply. Most of the rest is separated from sulfides during processing to recover metals.

Diamond (Figure 8.11b) only forms at very high pressures associated with the lowermost crust or mantle of the Earth. We mine it from **kimberlite** pipes, where rapidly moving, sometimes explosive,

▶FIGURE 8.10

Pyrite crystals. Although called "fool's gold," pyrite can be distinguished from gold because gold has a duller, more butter-yellow color than pyrite. Spectacular samples of pyrite, such as the one shown here, contain perfect cubic crystals, often twinned (multiple crystal domains sharing atoms) or in aggregates.

Box 8.2 The Witwatersrand Gold Deposits

Gold occurs in many different ore deposits. The world's largest, the Witwatersrand deposits of northeastern South Africa, dominate world production. The Witwatersrand deposits are paleoplacer deposits, meaning that they were placers when originally deposited. They occur in an area about 60 by 25 miles. The size of the Witwatersrand deposits is puzzling because placers form when hard-rock deposits are eroded and sedimentary processes concentrate ore. Yet today there are no hard-rock gold deposits of anywhere near

sufficient size to account for the volume of the Witwatersrand placers.

The Witwatersrand deposits were discovered in 1886. The discovery led to the founding of Johannesburg, a central town in frontier South Africa. When miners reached a zone of pyrite in 1889, the new town began to die because it was not known how to extract gold from sulfides at that time. In the early twentieth century, three Scotsmen discovered the **cyanide method** for gold extraction, and Johannesburg flourished once more.

(a)

(b)

▶FIGURE 8.11

Both graphite (a) and diamond (b) are essentially pure carbon. They are diferent minerals, with different properties, because the carbon atoms are not bonded together in the same way in both. Diamond photo from J. Banfield, used by permission.

mafic magmas have carried it up to the surface. After formation, diamond sometimes concentrates in river and stream beds where we mine it from placer deposits. Although some diamonds are of gem quality, most are not. We call lower-quality diamonds **bort.**

Sulfides and Sulfosalts

True sulfide minerals (such as pyrite, Figures 8.10, 8.12) contain one or several metallic elements and sulfur as the only nonmetallic element. Some examples are shown in Plates 7.1 through 7.6, 7.8, 8.1, and 8.2. Bonding is either covalent, metallic, or a combination of both. Other minerals grouped with the sulfides, because of similar properties, contain selenium (the selenides), tellurium (the tellurides), or bismuth instead of sulfur. A related group of minerals, the sulfosalts, contain the semimetals arsenic and antimony in place

of metal atoms. Many of the sulfides and related minerals have similar atomic structures, so solid solutions are common. Although many sulfides and sulfosalts are found in nature, only a few are abundant.

Sulfide minerals often form distinct **associations.** Pyrite, sphalerite, and pyrrhotite are frequently found together, as are chalcopyrite, pyrite, and bornite or pyrrhotite. In some carbonate-hosted deposits, sphalerite and galena occur together. We can depict sulfide associations using triangular composition diagrams such as that shown in Figure 8.13. The tie lines (straight lines joining two compositions, such as the line from sphalerite to pyrite) in that figure indicate that in Fe-Zn-S ore, the ore-mineral association will be either pyrite-sphalerite-sulfur, pyrite-sphalerite, pyrite-pyrrhotite-sphalerite, depending on the amount of elemental sulfur present. Box 8.3 presents a more detailed discussion of Cu-Fe-S ore minerals

►**FIGURE 8.12**
Pyrite is one of the most common sulfide minerals and often forms spectacular cubic crystals. In this photo, the cubes (about 1 cm on a side) have growth striations on their faces. Prismatic quartz crystals can also be seen.

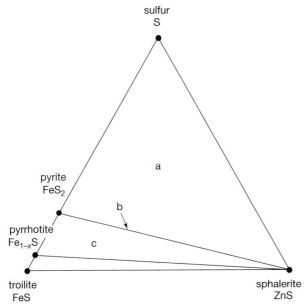

►**FIGURE 8.13**
Triangular diagram for the system FeS-ZnS-S showing the principal end member minerals (solid dots). Triangles and lines indicate possible mineral assemblages. Most Fe-Zn-S ores contain assemblages of sulfur-pyrite-sphalerite (triangle a), pyrite-sphalerite (line b), or pyrite-pyrrhotite-sphalerite (triangle c). Troilite is rare in terrestrial rocks but is common in some meteorites.

and explains how we can use triangular diagrams to show solid solution compositions.

Oxides and Hydroxides

We often group oxides and hydroxides together because they have similar compositions and atomic structures (Table 8.3). They often have similar properties, and most have relatively simple and related formulas. Oxide minerals consist of one or several metal ions bonded to O^{2-}. Hydroxide minerals contain $(OH)^-$ anions in place of all or some of the O^{2-}.

A primary difference between oxides and hydroxides are the temperatures at which they form and are stable. Hydroxides are unstable at high temperature; they exist in low-temperature environments and are commonly products of alteration and weathering. We usually group quartz, the most common oxide, with silicate minerals. Other oxide minerals—magnetite and ilmenite, for example—are high-temperature minerals generally associated with igneous or metamorphic rocks. In fact, most igneous and metamorphic rocks contain oxide minerals. Typically they are present in minor amounts, are easily overlooked, and may be difficult to identify.

Oxides and hydroxides have properties distinct from silicates and sulfides. They are dominated by ionic bonding, and anions (O^{2-} or OH^-) do not control their structure and properties as they do for other mineral groups. Oxides and hydroxides also have distinct properties from carbonates, sulfates, and other ionic minerals that have high solubilities in water.

Examples of oxide and hydroxide minerals are given in Table 8.3. Simple oxides contain one metal element and have formulas R_2O, RO, or R_2O_3, where R is the metal cation. The different formulas reflect different valences of the metals. More complex oxides contain two different metal cations and have formulas XYO_3 or XY_2O_4, where X and Y denote the metals. Oxides with general formula XY_2O_4 (spinel, chromite, and franklinite in Table 8.3) belong to the spinel group; they all have similar atomic structures. Magnetite, with a formula that can be written $FeFe_2O_4$, also has a spinel structure and belongs to this group.

Box 8.3 Another Look at the Phase Rule and Cu-Fe Sulfide Minerals

We use triangular diagrams to plot the compositions of ore minerals and to show ore mineral assemblages, just as we did for metamorphic and igneous minerals (Box 5.3). The phase rule (discussed in Chapter 7) tells us that the number of chemical components (c) in a system is related to the number of degrees of freedom (f) and the number of minerals (p) by:

$$c + 2 = p + f$$

Figure 8.14a is a ternary diagram showing the Cu-Fe-S system ($c = 3$), and the 10 principal minerals within that system. The center of the diagram is enlarged in Figure 8.14b. Minerals such as chalcopyrite, which do not vary much in chemical composition, appear as points on the diagrams. Bornite and pyrrhotite, which have variable compositions, appear as bars. The lengths of the bars indicate the range of compositions possible for each mineral.

Tie lines connect minerals that coexist in ore deposits formed at a variety of pressures and temperatures. Normally, P and T may vary, $f = 2$, and

the phase rule tells us that up to three minerals may be stable together. Possible three-mineral assemblages are shown as triangles (with "3" inside) in Figure 8.14b. Several two-mineral assemblages are also stable. For example, several tie lines connect chalcopyrite with the pyrrhotite bar, showing that chalcopyrite may coexist with different pyrrhotite compositions to make a two-mineral assemblage. Other possible two-mineral assemblages are chalcopyrite-bornite, bornite-chalcocite, and covellite-bornite.

Cu-Fe sulfide mineralogy is complex because many minerals have similar compositions. Figure 8.14b shows there are more than 10 possible assemblages. The assemblage present in a specific deposit depends on the Cu:Fe:S ratio. In Fe-poor ore deposits, for example, sulfide assemblages will include covellite, digenite, or chalcocite, but not pyrite or pyrrhotite. Diagrams such as Figure 8.14 are useful ways to describe complex mineral relationships without words, and we use them to predict and interpret ore deposit mineralogy.

▶**FIGURE 8.14**
(a) A ternary diagram showing the Cu-Fe-S system and the 10 principal minerals within that system. (b) An enlargement of part of the ternary diagram. See text for explanation.

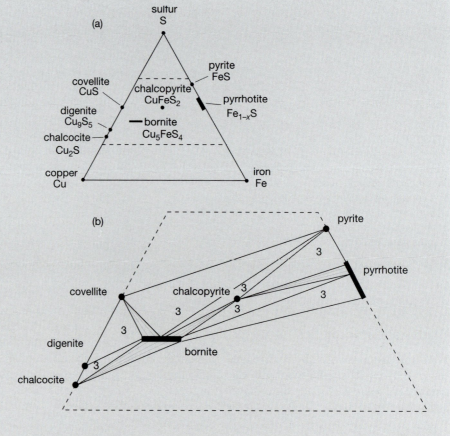

▶TABLE 8.3
Examples of Oxide and Hydroxide Minerals

	Oxide Minerals		Hydroxide Minerals	
		Oxides/Hydroxides Containing One Metal Element		
Al minerals	corundum	Al_2O_3	gibbsite	$Al(OH)_3$
			diaspore	$AlO(OH)$
Mg minerals	periclase	MgO	brucite	$Mg(OH)_2$
Fe minerals	magnetite	Fe_3O_4	goethite	$FeO(OH)$
	hematite	Fe_2O_3	lepidochrosite	$FeO(OH)$
Mn minerals	pyrolusite	MnO_2	manganite	$MnO(OH)$
Zn minerals	zincite	ZnO		
Ti minerals	rutile	TiO_2		
Sn minerals	cassiterite	SnO_2		
		Oxides Containing Two Metal Elements		
Fe-Ti oxide	ilmenite	$FeTiO_3$		
Mg-Al oxide	spinel	$MgAl_2O_4$		
Fe-Cr oxide	chromite	$FeCr_2O_4$		
Zn-Fe oxide	franklinite	$ZnFe_2O_4$		

GEMS AND GEM MINERALS

Gems are precious or semiprecious stones and related substances that can be cut or polished to be used for ornamentation. Gems may be natural or synthetic; the term *gemstone* is sometimes used to refer to gems that are minerals. It is not the composition of gems that makes them valuable, but rather their appearance. Most gems are varieties of common minerals, but they exhibit spectacular color, clarity, brilliance, or fire. Some gems, such as opal, show a **play of color.** Cutting and polishing enhance these desirable features. We value gems such as amber or pearls, which we call gems although they are not minerals, for their unique lusters. The hardest gems are most highly valued because they are most durable. Others are prized because of their rarity.

Many natural materials, including minerals and nonminerals, have been used as gems. Because gemstones have different appearances than common minerals, many have names different from the mineral names. Diamond, emerald (a variety of beryl), and ruby (a variety of corundum) have been, historically, the most valuable gemstones. Sapphire (another variety of corundum) and alexandrite (a variety of chrysoberyl) are nearly as valuable. Table 8.4 presents a list of some common gems and the countries that produce them. Only one or two countries dominate production of many gems, including diamond (Box 8.4). Plate 8.5 shows some examples of minerals that are gems.

Synthetic Gems and Color Alteration

Although by definition minerals must be natural, we can synthesize equivalents of them in the laboratory. However, most "synthetic minerals" are so fine grained they are unusable as gems. The exceptions to this are some synthetic beryl, chrysoberyl, corundum, diamond, garnet, opal, quartz, rutile, spinel, and turquoise. Several synthetic compounds with no natural equivalents are also used as gems. Foremost among them are yttrium aluminum garnet (YAG) and cubic zirconia (CZ), both used as imitation, or "genuine *faux*" (French, meaning "fake"), diamonds. Besides their value as gems, synthetic minerals are often important in such industrial applications as laser optics.

Synthetic minerals are made by several different methods. The Verneuil technique and the Czochralski process both involve crystallizing them from molten material (Figure 8.15). In the more common Verneuil technique, powdered chemicals pass through a hot flame and crystallize as a **boule,** a single elongated crystal. This technique produces large synthetic rubies, sapphires, spinels, and other gems. The synthetic rubies are used in lasers. In the Czochralski process, a seed crystal is placed in contact with a melt and

▶**TABLE 8.4**
The Most Common Gemstones in Approximate Order of Importance

Mineral	Gem	Most Important Producing Countries
diamond	diamond	Australia, South Africa, Namibia, Russia
beryl	emerald	Colombia, Brazil, Russia, Egypt, East Africa
	aquamarine	Brazil, Afghanistan, Pakistan
corundum	ruby	Cambodia, Myanmar, Afghanistan, India
	sapphire	Australia, Thailand, Sri Lanka, Brazil
opal	opal	Australia, Hungary, Mexico
jadeite	jade	Myanmar, China
chrysoberyl	alexandrite	Russia, Brazil
quartz	amethyst	Russia, Sri Lanka, India, Uruguay, Brazil
	citrine	many locations
topaz	topaz	Brazil, Sri Lanka, Russia, India
tourmaline	tourmaline	Namibia, Brazil, United States, Russia
turquoise	turquoise	United States, Egypt, Australia
tremolite-actinolite	nephrite jade	Russia, China, Taiwan, Canada
olivine	peridot	Egypt, Myanmar, Australia
zircon	zircon	Sri Lanka

Box 8.4 The World's Major Diamond Producers

Until recently, the largest diamond-producing countries were in Africa. During the last decade, diamond production has increased in Australia and Russia; both now produce more diamonds than any African nation. Most diamonds are mined from **alluvium** (sediments deposited by flowing water), some from glacial tills, and some from igneous intrusions called *kimberlite pipes*. Diamonds have also been identified in some rare high-pressure metamorphic rocks, but generally in uneconomical quantities. Geologists believe that kimberlites were the primary sources of most of the world's diamonds; they infer that diamonds in alluvium and glacial till originated in kimberlites that were subsequently weathered, transported and deposited.

Besides Australia and Russia, other major diamond producers are Zaire, Botswana, and South Africa. Together, these five countries produce about 111 million carats a year, equivalent to 49,000 pounds. About half the diamonds they produce are gem quality; the other are of a low quality called *bort*, which is used as an abrasive and grinding agent in industrial applications.

The largest producers of gem-quality diamonds are Australia, Russia, Angola, South Africa, and Namibia. In South Africa some of the diamond production comes from volcanic pipes, including the pipe at Kimberley (from which the rock name *kimberlite* is derived). Gem-quality diamonds are mined from pipes in many other places, including Tanzania, which has the largest pipe mine in Africa. In recent years there has been a renewed interest in diamond prospecting in North America; a few mines will soon begin diamond production.

allowed to grow. Rubies up to 40 cm long have been grown using the Czochralski process.

A third approach, the **flux method,** has sometimes yielded large synthetic crystals, notably quartz, ruby, sapphire, alexandrite, and emerald. Flux is a material that promotes reaction and crystal growth but is not incorporated in crystal structures. A mixture of chemicals, including those needed to make the desired mineral, and others to act as the flux, is ground together and heated above its melting point. As the temperature is lowered, crystals begin to grow. After the melt solidifies, water or other reagents remove the

▶**FIGURE 8.15**
(a) Verneuil process. Fine powder passes from the hopper through a sieve with the aid of a tapper and then through the flame fed by oxygen and hydrogen. As the boule crystallizes, it is slowly lowered in the furnace. (b) Czochralski process. A crucible is filled with powder and heated in a furnace so the powder melts. A rod with a seed crystal attached to the bottom end is lowered until it touches the melt and is rotated and slowly withdrawn, "pulling" the crystal from the melt.

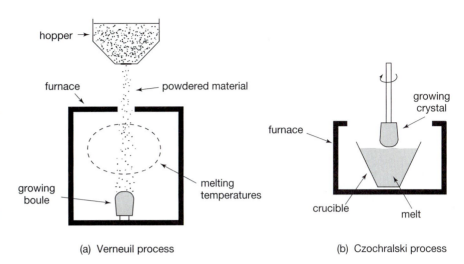

(a) Verneuil process

(b) Czochralski process

flux, leaving the desired crystals. Most synthetic rubies and emeralds are created this way.

Mineralogists use other synthesis techniques, but with a few exceptions they do not produce gems of great value. Synthetic minerals may be grown from hydrothermal solutions in high-pressure reactor vessels called **bombs.** Synthetic quartz, ruby, and emerald have all been made this way. Synthetic diamond and a few other high-pressure crystals are made using a **solid state** (no melt or water) approach. A cylinder of starting material, enclosed in a graphite heater, is squeezed between two pistons. Electricity passing through the graphite heats the material to temperatures at which crystals will grow. Chemists at General Electric have perfected this technique for making gem-quality synthetic diamonds.

Gems and other minerals derive their color from many different things (see Chapter 3). Common minerals may have little value as gems, but if we can alter or enhance colors, even common quartz may become valuable. Gemologists, therefore, often treat gems and minerals, natural or synthetic, to change their color and increase their value. For instance, quartz crystals from the Hot Springs area of Arkansas are irradiated to disrupt their atomic structure and give them a smoky or black coloring (Figure 8.16). Irradiation is also used to induce color changes in diamond and topaz. Gem color can also be changed through dyeing or heat treatment, although dyeing only affects gems that are porous, such as jade and chalcedony. Gemologists have successfully used heat treatment, which

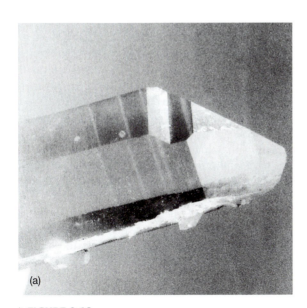

(a)

(b)

▶**FIGURE 8.16**
Quartz from near Hot Springs, Arkansas. (a) The quartz is clear and colorless when it is mined.
(b) After irradiation, it darkens and becomes artificial smoky or black quartz.

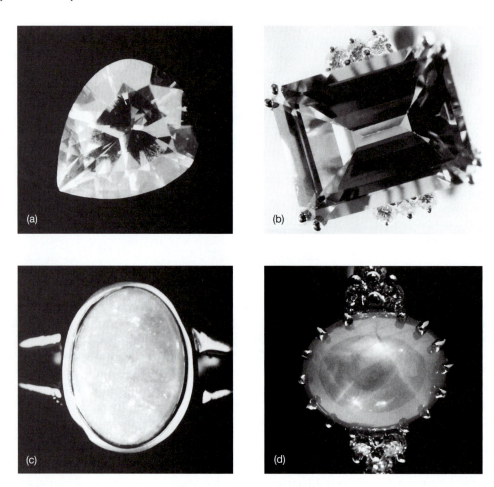

▶FIGURE 8.17
(a) Tourmaline faceted to produce a pear shape; (b) topaz faceted with an "emerald" cut;
(c) an opal cabochon showing play of light; (d) a star sapphire cabochon showing play of light.
These photos are black and white, but the tourmaline is really pink, the topaz is blue, the opal
has rainbow colors, and the sapphire is grey.

changes elemental valences or alters atomic structures, on quartz, beryl, zircon, and topaz, although the results are not always predictable.

Cutting and Polishing Gems

Most gems are sold in shapes that do not resemble natural crystals (Figure 8.17a, b). Ruby, a red variety of corundum, generally grows as hexagonal crystals, but when rubies are incorporated into jewelry, gem cutters shape and polish them to increase their beauty and value. The cutting of a gem affects its value as much as the quality of the raw material. Cutting and polishing takes place in many parts of the world. Israel and Belgium dominate diamond cutting; the United States, India, Hong Kong, Thailand, and Brazil also cut significant amounts of various gems.

Gemologists shape and cut gems in several ways. Some minerals, including agate, opal, chalcedony, and onyx, are **tumbled** in a cylinder with a polishing/abrading agent. The cylinder rotates until the stones have smooth surfaces. They become polished, but often have irregular shapes. Alternatively, gemologists shape gemstones using a **cabochon** cut, with a smooth, domed top and, usually, a flat base (Figure 8.17c, d). Cabochons are shaped using a wet grinding wheel. Softer minerals are ground with a quartz sandstone wheel; harder ones require wheels impregnated with diamonds. As a final step, cabochons are finely polished. Until the Middle Ages, most gems were cut *en cabochon* (Figure 8.17c, d).

While most soft gemstones receive a cabochon cut, many hard gemstones are faceted. **Facets** are small, polished, planar surfaces that give the stones attractive shapes and light properties. With proper cutting, originally dull stones can sparkle. Facets are usually symmetrically arranged in geometric shapes (Figure 8.17a, b). Gemologists create them by mounting the stones on a holder, called a **dop,** and grinding the stone with a diamond-impregnated wheel. As with cabochons, they are polished after being ground.

Box 8.5 Mining, the Environment and Politics

Mining, especially public land mining, is a controversial environmental issue. The 1872 Mining Law gives U. S. citizens and corporations the right to stake claims and mine public lands if they can show that the mining is profitable. The claims last forever, provided miners invest $100 in the property each year. Claim holders have the right to purchase the surface and mineral rights at very low cost through a process called *patenting,* often at $5 an acre or less. Miners pay no royalties, in contrast with those who develop coal, oil, or gas on public lands. In 1999, mining corporations took more than $1 billion worth of minerals from public lands without paying anything back to taxpayers. So, the 1872 law really is a federal subsidy for mining.

Many environmental groups want to see Congress change the 1872 law. They argue that public lands are for the public, not mining corporations, and they point out that mining is incompatible with other uses such as wildlife habitat, hiking, and camping, Furthermore, mining leaves scars on the land and may cause long-term environmental degradation. According to the watchdog economic organization Green Scissors, "Nationally, mines have polluted 12,000 miles of rivers and streams, and more than 550,000 abandoned hardrock mines scar the American landscape. The Mineral Policy Center estimates the cost of cleaning up such sites at $32 billion to $72 billion."

The mining industry argues that we need the 1872 law to ensure a flow of mineral resources to our citizens. They point out the importance of mining to some local western economies, and say they can mine in an environmentally friendly way. They also accuse many mining opponents of being NIMBYs (not-in-my-backyard).

The mining industry is correct when they argue that we need mineral resources and they have to come from somewhere. Despite industry's claims, however, mining always has a cost to the environment. A visit to active or abandoned mines confirms this. Besides scars on the land, less obvious problems include air, water, and soil pollution. All these problems can be limited, but not eliminated, by following the best mining practices.

In today's industrial world we need mineral resources. The real questions are where are they going to come from and how much are we willing to pay? Those who seek reform of the 1872 law argue that some areas should be off limits to mining, that mining companies should pay more in royalties, and that there should be strict antipollution and land reclamation requirements. If Congress enacts these changes, they might affect the price of mineral commodities, but most economists think the effect would be very small.

▶QUESTIONS FOR THOUGHT

(Some of the following questions have no specific correct answers; they are intended to promote thought and discussion.)

1. Why are some minerals valuable as ore minerals, while others are not? In particular, why do many ore minerals belong to the sulfide, oxide, and native element groups?
2. In comparison with many other elements, such as tin, aluminum requires a low concentration factor to be economically mined. Why? What are the main factors that dictate economical concentration factors?
3. Why are actively mined ore deposits unevenly distributed around the world?
4. What kinds of minerals would you expect to find in cumulate deposits? Give examples. Why these minerals and not others?
5. What kinds of minerals would you expect to find in hydrothermal deposits? Give examples. Why these minerals and not others?
6. What kinds of minerals would you expect to find in placer deposits? Give examples. Why these minerals and not others?
7. What kinds of minerals would you expect to find in evaporite deposits? Give examples. Why these minerals and not others?
8. The crystal structures of copper, gold, silver, platinum, and iron are all similar. Why?
9. Most diamonds are not mined from the original rocks in which they formed. Where do diamonds come from and how do they become concentrated where we mine them?
10. Why do ore deposits typically contain few ore minerals?
11. What are the properties that make a gem valuable?
12. Some mineralogists would argue that synthetic gems are inferior to natural ones. Are they just being snobs? What, if any, real differences are there between synthetic and natural gems? What is the origin of such differences?

▶RESOURCES

(See also the general references listed at the end of Chapter 1.)

Bauer, M. *Precious Stones*. New York: Dover Publications Inc., 1928.

Brookins, D. G. *Mineral and Energy Resources*. Columbus, Ohio: Merrill, 1990.

Evans, A. M. *An Introduction to Ore Geology*. New York: Elsevier, 1980.

Guilbert, J. M., and C. F. Park, Jr. *The Geology of Ore Deposits*. New York: Freidman and Co., 1986.

Hurlbut, C. S., Jr., and G. S. Switzer. *Gemology*. New York: Wiley Interscience, 1979.

Jensen, M. L., and A. M. Bateman. *Economical Mineral Deposits,* 3rd. ed. New York: John Wiley and Sons, 1981.

Kesler, S. E. *Mineral Resources, Economics and the Environment*. New York: Macmillan, 1994.

Manning, D. A. C. *Industrial Minerals*. London: Chapman & Hall, 1995.

United States Bureau of Mines. *Minerals Yearbook. Vol. 1, Metals and Minerals*. Washington, DC: United States Bureau of Mines, 1994.

PART

II

Symmetry, Crystallography, and Atomic Structure

Large fluorite cubes in calcite from Rosiclair, Illinois. The field of view is 18 cm.

Crystal Morphology and Symmetry

This chapter is about symmetry. The symmetry of an entity is the relationship between its parts. The symmetry of a crystal is a reflection of internal atomic structure. The way in which atoms are arranged in an atomic structure dictates the way a crystal's properties are distributed. If a crystal has symmetry, the symmetry is common to all of its properties. Crystal faces, magnetic properties, optical properties, and others must all be distributed in the same way. Consequently, by studying physical properties to determine crystal symmetry, crystallographers can make inferences about internal atomic order. Although crystals may have any of an infinite number of shapes, the number of possible symmetries they may have is limited. In this chapter we discuss crystal symmetry and its relationship to crystal shape.

SYMMETRY

The external shape of a crystal reflects its internal atomic structure. Of most importance is a crystal's **symmetry.** As defined by Aristotle, symmetry refers to the relationship between parts of an entity. Zoltai and Stout (1984) give an excellent practical definition of symmetry as it applies to crystals: "Symmetry is the order in arrangement and orientation of atoms in minerals, and the order in the consequent distribution of mineral properties." Figure 9.1 shows the relationship between halite's atomic structure and its typical crystal shape. A cubic mineral, such as halite, can occur as a cubic crystal with six identical faces because its atoms are arranged in a cubic pattern with identical structure in three perpendicular directions. Although the relationship between atomic structure and crystal shape shown in the figure may seem clear, the relationship is not always so simple. In fact, many cubic halite samples are cleav-

age fragments from larger crystals with complex shapes.

Looking for symmetry in natural materials can be complicated. Although crystal structures may be symmetrical, many things control crystal growth, so crystals may not reflect their internal order. Sometimes few or no crystal faces may form. Sometimes several different crystals may become intergrown. Sometimes crystals contain structural imperfections, and still other crystals may be too small to see clearly. An experienced eye and an active imagination are often necessary to see the symmetry of a crystal.

Mirror Planes

Most animals, including humans, appear symmetrical: an imaginary mirror down their center relates the appearance of their right side and left side (Figure 9.2). We call such symmetry **reflection,** and we call the

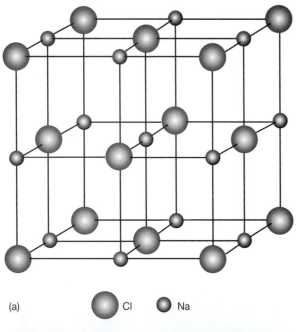

(a)   ● Cl   ● Na

(b)

▶**FIGURE 9.1**
(a) The atomic structure of halite showing the cubic arrangement of its atoms. (b) Cubic halite crystals.

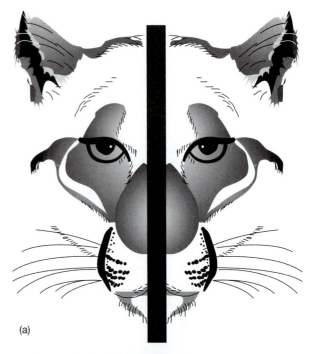

(a)

(b)

▶**FIGURE 9.2**
Animal, including human, faces are often quite symmetrical. (a) Because the right and left sides of the cougar's face are identical, there is a mirror plane running down the center. (b) The visage of country singer Lyle Lovett, on the other hand, appears asymmetrical. His twisted lip, the tilt of his neck and eyes, and other features mean that there is no mirror plane of symmetry down the center of his face.

plane of the imaginary mirror the **mirror plane.** In shorthand notation, we use the letter *m* to designate a mirror of symmetry. Reflection is the symmetry **operation,** and the mirror plane is the symmetry **operator.** Reflection often relates identical faces on a crystal. A face, or any object, on one side of a mirror has an equivalent at an equal distance on the other side of the mirror. The two faces or objects are the same perpendicular distance from the mirror and have opposite "handedness"; if one points to the right, the other points to the left.

Two-dimensional drawings may have many mirror planes or no mirror planes. Circles, for example, contain an infinite number of mirror planes; any line drawn through the center of the circle is a mirror of symmetry (Figure 9.3a). Irregular blobs, on the other hand, have no mirror planes (Figure 9.3b). Rectangles have two mirror planes, while squares have four (Figure 9.3c, d). Three-dimensional ob-

jects, too, may have zero to many mirror planes. A perfect sphere has an infinite number of mirror planes, but, as we will see later, crystals cannot have more mirror planes than a cube. Cubes have nine mirrors (Figure 9.3e-g). Three are parallel to pairs of opposite faces (Figure 9.3f); six intersect opposite

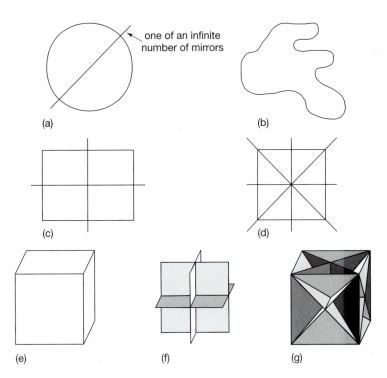

one of an infinite number of mirrors

(a)

(b)

(c)

(d)

(e)

(f)

(g)

▶FIGURE 9.3
Two- and three-dimensional objects may have many mirror planes (shown by lines) or none: (a) a circle has an infinite number of mirror planes; (b) a blob has no mirror planes; (c) a rectangle has two mirror planes; (d) a square has four mirror planes; (e) a cube has the maximum number of mirror planes possible for a crystal, nine; (f) the mirrors of a cube parallel to its faces; (g) the mirrors of a cube parallel to face diagonals.

faces along their diagonals (Figure 9.3g). Thus, a cube has more symmetry than a human or a cougar.

Rotational Symmetry

Reflection by a mirror is but one of several fundamental types of symmetry operations. A second common type of symmetry in crystals, called **rotational symmetry,** is symmetry with respect to a line called a **rotation axis.** In two dimensions, a lens shape appears unchanged when rotated 180° (Figure 9.4). We say it has 2-fold symmetry because two repeats of the rotation operation return it to its original position. Other symmetry operators are shown in Figure 9.4. Equilateral triangles have 3-fold rotational symmetry (corresponding to rotation of 120°), squares have 4-fold rotational symmetry (rotation of 90°), and hexagons have 6-fold rotational symmetry (rotation of 60°). For reasons we will see later, although we can draw shapes that have 5-fold symmetry, or greater than 6-fold symmetry, minerals never possess such symmetry. Some natural materials, including plants and animals such as a starfish, display apparent symmetries not possible for minerals. All objects have 1-fold rotational symmetry because they remain unchanged after rotation of 360°.

We call all the rotation axes we have just discussed **proper rotation axes,** to contrast them with **rotoinversion axes,** which we will discuss later. In shorthand notation, we symbolize proper rotation axes using the numbers 1, 2, 3, 4, or 6, corresponding to rotations of 360°, 180°, 120°, 90°, and 60°, respectively. Figure 9.4

shows examples of both two- and three-dimensional objects with 1-, 2-, 3-, 4-, and 6-fold symmetry.

A cube has 2-fold, 3-fold, and 4-fold rotational symmetry (Figure 9.5). When rotated 180° about an edge diagonal, it appears unchanged. When rotated 120° about a main diagonal, it appears unchanged. When rotated 90° about a line perpendicular to a face, it appears unchanged. Thus, we see that different rotation axes can combine in crystals.

Many minerals grow as **prisms,** crystals having a set of identical faces parallel to one direction. Typically, prismatic crystals are elongated in one direction. Figure 9.6a shows prismatic crystal of scapolite. Although not clear from the photograph, ideal scapolite crystals have square cross sections and flat ends perpendicular to the prism faces. Quartz, in comparison, forms hexagonal prisms, complicated by sloping **terminating faces** at the ends of the crystals (Figure 9.6b). Prism faces are, in principle, identical in shape and size, although accidents of growth often lead to minor differences. Some minerals have multiple, nonidentical faces, all parallel to a common line. We call the collection of faces a **zone** and the line the **zone axis.** Zones, common in crystals, sometimes correspond to axes of symmetry.

Rotational symmetry, even if present, may be hard to see. For example, the 3-fold axes of symmetry in a cube are difficult to see without turning the cube in your hand. Problems may be even more complicated in natural crystals because of growth imperfections in crystal faces, or the presence of many

symmetry operator	2-dimensional example	3-dimensional example

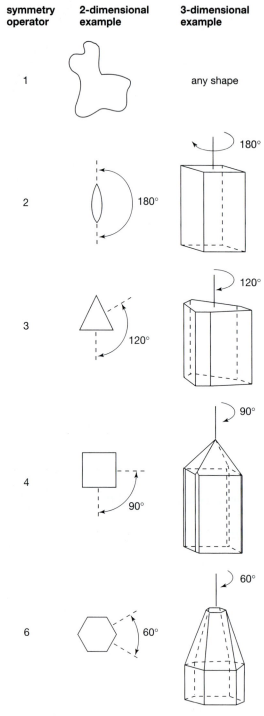

▶**FIGURE 9.4**
Two- and three-dimensional examples of 1-, 2-, 3-, 4-, and 6-fold axes of symmetry.

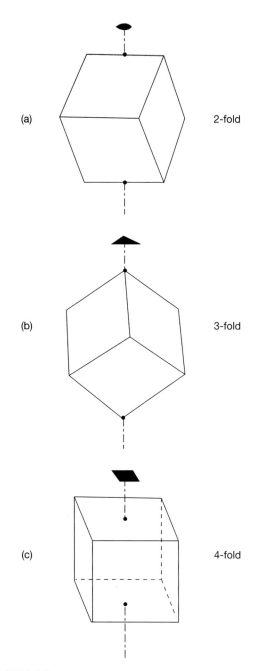

▶**FIGURE 9.5**
Rotation axes of a cube. (a) The edge diagonals of a cube are 2-fold rotational axes of symmetry. (b) The main diagonals of a cube are 3-fold rotational axes of symmetry. (c) The face perpendiculars of a cube are 4-fold rotational axes of symmetry.

differently shaped crystal faces. **Anhedral crystals** exhibit no external symmetry, while **euhedral crystals** may exhibit a lot of it.

Inversion Centers

Inversion, a third type of symmetry, is symmetry with respect to a point. We call the point the **inversion center** and often designate it with the lowercase letter *i*. As with mirror planes, inversion relates two identical faces on a crystal, but while mirror planes "reflect" faces and change their "handedness," inversion centers invert them. Inversion produces faces related in the same way that a lens may yield an upside down and backward image (Figure 9.7). In two dimensions, inversion centers give the same results as 2-fold axes of symmetry (Figure 9.7a, b); in three

▶**FIGURE 9.6**
Prismatic crystals: (a) subhedral scapolite crystals showing tetragonal prisms and flat terminating faces perpendicular to the crystal's long dimension; (b) euhedral quartz crystals displaying typical hexagonal prisms with sloping terminating faces. Note the growth lines on crystal faces.

dimensions, the symmetry is different (Figure 9.7c). Some crystals have combinations of mirror planes, rotation axes, and inversion centers, such as the prismat-

ic crystal with terminating pyramids in Figure 9.7d. The small letters on the crystal faces in Figures 9.7c and 9.7d indicate faces related by symmetry.

▶**FIGURE 9.7**
Examples of inversion symmetry.
(a) The inversion operation relates two individuals through an inversion center, here the letter *B*.
(b) Two fish related by an inversion center (indicated by a small circle). (c) A crystal having an inversion center as its only symmetry (symmetrically related identical faces indicated by letters). (d) A crystal having an inversion center and other symmetry (symmetrically related faces indicated by letters).

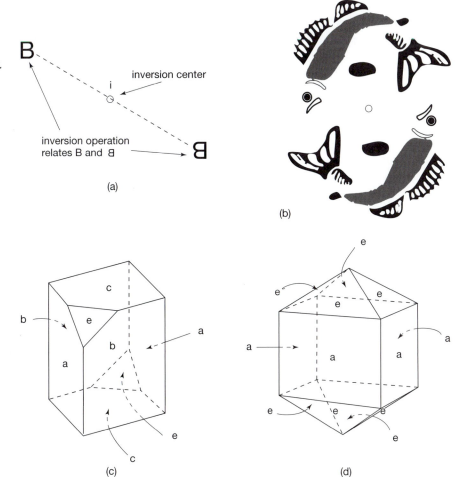

Symmetry Is a Property

Mineralogists often speak of a crystal's **symmetry elements,** the different kinds of symmetry it has, in quantitative terms. We can count mirror planes, rotation axes, and inversion centers. If objects have only a few symmetry elements, we say they have **low symmetry.** Those that display many have **high symmetry.** In addition, we say that objects with 6-fold rotation axes have higher symmetry than those with 4-fold, 3-fold, or 2-fold axes, and so on. References to high and low symmetry are necessarily vague because symmetry manifests itself in many different ways. **Cubes** have the highest symmetry possible for crystals: three 4-fold axes, four 3-fold axes, six 2-fold axes, nine mirror planes, and an inversion center (Figure 9.8). Examination of Figures 9.8a and 9.8b should convince you that a regular **octahedron** has this same symmetry, although appearing dissimilar to a cube. The shape of an object depends on its symmetry, the size of its faces, and on the angles between its faces. Cube faces are at 90° to each other; octahedron faces are at about 55° to each other.

Comparison of the cube and octahedron emphasizes that symmetry is not a physical characteristic. It is a property that objects may possess. With an object in hand, a mineralogist can discuss its symmetry, but a description of an object's symmetry does not unambiguously reveal its appearance. Many objects have mirror planes, yet their overall shapes are quite different. While symmetry does not determine appearance, it does include all aspects of a crystal. For example, if a crystal has six identical faces, it must have six directions of identical atomic structure and, therefore, six directions with the same hardness, reflectivity, and so on. The 6-fold rotation axis present in a hexagonal prism relates six crystal faces, six crystal edges, and six directions with identical atomic structure (Figure 9.9). Hexagonal crystals are optically uniaxial; they have one unique direction (parallel to the 6-fold rotation axis) that corresponds to the optic axis. Symmetry affects everything in a crystal, including faces, edges, corners, physical properties, optical properties, and atomic arrangement.

STEREO DIAGRAMS

Symmetry on Stereo Diagrams

A convenient way to look at the symmetry of a crystal is to use a stereographic projection, also called a **stereo diagram.** Although stereo diagrams depict mirror planes, inversion centers, rotational symmetry, and their relationship to crystal faces, they do not show the exact shape of the faces. Stereo diagrams can be quantitative and very complex. For now, considering their qualitative aspects is sufficient. Figure 9.10a shows stereographic projections of 1-, 2-, 3-, 4-, and 6-fold rotation axes perpendicular to the page. The geometric symbols at the center of the diagrams show the kind of rotation axis. In the five drawings the rotation axis has operated on a single black dot, producing 0, 1, 2, 3, or 5 other dots by rotation of 360°, 180°,

▶**FIGURE 9.8**
Cube and octahedron: (a) The rotation axis in a cube and an octahedron have identical orientations. Lenses, triangles and squares show 2-fold, 3-fold and 4-fold axes of symmetry. (b) The mirror planes in a cube and octahedron are also oriented identically.

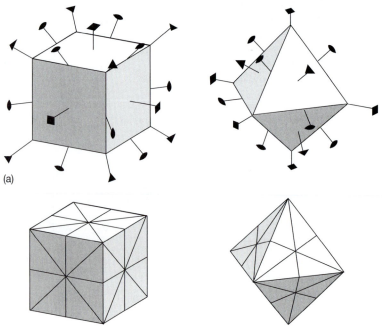

(a)

(b)

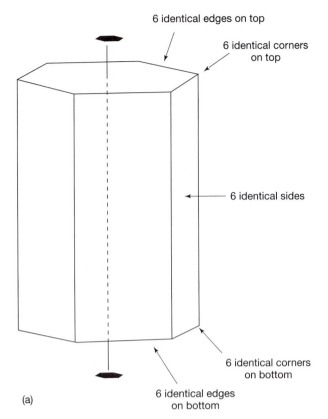

6 identical edges on top

6 identical corners on top

6 identical sides

6 identical corners on bottom

6 identical edges on bottom

(a)

(b)

▶**FIGURE 9.9**
Hexagonal prism. (a) The 6-fold axis relates faces, edges, and corners of the prism.
(b) A 20-cm-across hexagonal crystal of beryl.

120°, 90°, and 60° around the center of the diagram. Box 9.1 (page 188) discusses these operators in detail.

The diagrams in Figure 9.10a show symmetry, and repetition of points, in the plane of the page quite nicely. However, symmetry operations also work in three dimensions. To accommodate the third dimension, we need a way to show points below and above the page. By convention, solid dots represent points above the plane of the page, and circles represent points below the plane of the page. A point above the page that is directly above one below the page is symbolized by a "bull's-eye" formed by a small circle around a dot (⊙). Points within the plane of the page plot as solid dots on the outside circle of the stereo diagram. Figure 9.10b shows a 2-fold and 4-fold axis of symmetry lying in the plane of the page. They operate on points above the page to produce points below. For the 4-fold axis, the points below and above the page project on top of each other and are shown as a bull's-eye. Crystallographers have developed shorthand symbols to describe combinations of mirror planes and rotation axes when the two are perpendicular. We represent them by the symbols $1/m$, $2/m$, $3/m$, $4/m$, or $6/m$. Box 9.2 (page 190) discusses and depicts these combinations of symmetry operators.

Special Points and General Points

In the preceding examples of stereo diagrams, we have applied symmetry elements to points to produce other points. The initial points chosen do not lie at the center, on the edges, or at any other **special points** in the diagram. Figure 9.11 shows three stereo diagrams depicting points related by the symmetry $2/m$, and drawings showing the same symmetry using crystal faces. A single point was chosen to start, and application of the symmetry operators generated all additional points. In Figure 9.11a, a **general point** (a point at some nonspecial location) was chosen as a starting point. The result is four points—two above the mirror plane and two below—equivalent to the four faces shown below the stereo diagram. We call a group of identically shaped faces related by symmetry a **form.** We can easily verify that any general point (a point noncoincident with any symmetry elements) will produce a similar pattern. In the second two examples, special points within the plane of the page and lying on the rotation axes were chosen. After application of the symmetry operators, only two points result. The equivalent two faces are shown below the stereo diagrams. Thus, although all three drawings depict the same symmetry, the distribution of points and the

▶**FIGURE 9.10**

Rotation axes in stereo diagrams showing equivalent dots: (a) 1-, 2-, 3-, 4-, and 6-fold axes of symmetry perpendicular to the page; (b) 2-fold and 4-fold axes in the plane of the page.

(a) rotational axes perpendicular to page:

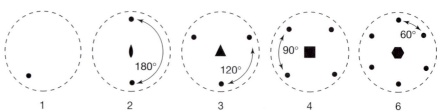

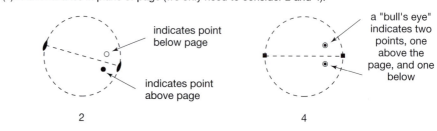
(b) rotational axes in plane of page (we only need to consider 2 and 4):

(a) prism   (b) pinacoid   (c) pinacoid   (d) combined forms

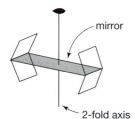

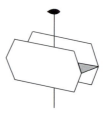

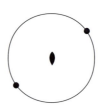

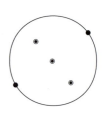

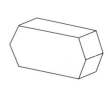

▶**FIGURE 9.11**

Three examples of stereo diagrams and forms related by $2/m$ symmetry; the drawings below the stereo diagrams confirm that the forms are consistent with the rotation axis and mirror plane. We call the form depicted in (a) a *prism;* the other two forms (b and c) are both *pinacoids.* In (b), the 2-fold axis symbol has been omitted for clarity. In (d), the three forms have been combined to create a crystal with $2/m$ symmetry.

equivalent faces are different, depending on whether we start with general points or special points. This observation explains, in part, why crystals of different appearance can have the same symmetry. Different-shaped faces can combine; the combination in Figure 9.11 produces a monoclinic crystal of eight faces (Figure 9.11d). The eight faces belong to three forms.

Rotoinversion

Rotoinversion, a combination of rotation and inversion, is a symmetry operation distinct from the others. The symbols $\bar{1}, \bar{2}, \bar{3}, \bar{4},$ and $\bar{6}$ represent rotoinversion axes. They are articulated as "bar-1,"

"bar-2," etc. In a rotoinversion operation, we apply rotation and inversion simultaneously. For example, in Figure 9.12a, we apply a $\bar{4}$ axis to a solid point (above the page) to produce an open point (below the page). The original point is rotated 90° and inverted to produce the second. A $\bar{4}$ axis relates four points in total (Figure 9.12b). Figure 9.12c shows a crystal with $\bar{4}$ symmetry. The four faces are related by 90° rotation followed by inversion. Box 9.3 (page 191) discusses and depicts all possible rotoinversion operators.

Note that $\bar{4}$ rotoinversion is the only rotoinversion operation distinct from other symmetry operations. A $\bar{2}$ axis is equivalent to *m*, a $\bar{3}$ axis is equivalent to a 3-fold axis and an inversion center applied separately, and a $\bar{6}$ is equivalent to $3/m$. The crystal depicted in Figure 9.12 has a $\bar{4}$ axis, but it does not have an inversion center, nor does it have a normal 4-fold rotation axis of symmetry. Figure 9.13 compares a proper 4-fold axis with a 4-fold rotoinversion axis, with a 4-fold axis and inversion center combined. The operators do not yield the same pattern of points, even when we use the same point as a starting point, nor do they describe the same crystal shapes. The $\bar{4}$ axis produces four points (or faces) in total, but they are not related by a proper 4-fold axis of rotation. Presence of a $\bar{4}$ axis does not imply presence of a 4-fold axis. Box 9.3 (page 191) discusses all possible rotoinversion axes in more detail. $\bar{2}$, $\bar{4}$, and $\bar{6}$ axes are not equivalent to rotation and inversion operating separately, nor are they equivalent to proper 2-fold, 4-fold, or 6-fold rotation axes. However, a $\bar{3}$ operation is equivalent to proper 3-fold rotation and inversion operating separately.

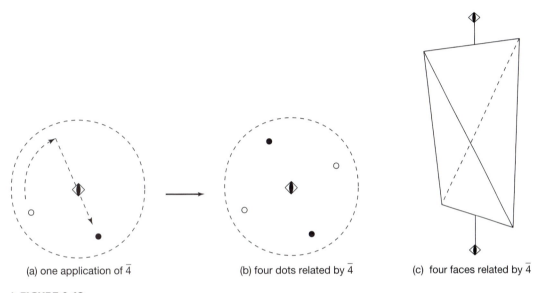

(a) one application of $\bar{4}$ (b) four dots related by $\bar{4}$ (c) four faces related by $\bar{4}$

▶**FIGURE 9.12**
(a) $\bar{4}$ axis operating on a dot; (b) all the dots related by a $\bar{4}$ axis; (c) crystal faces related by a $\bar{4}$ axis. We call this form a *disphenoid*.

▶**FIGURE 9.13**
Comparison of a 4-fold axis (a) with a 4-fold rotoinversion axis (b), and with a 4-fold axis combined with an inversion center (c). Stereo diagrams and form drawings are shown for all three.

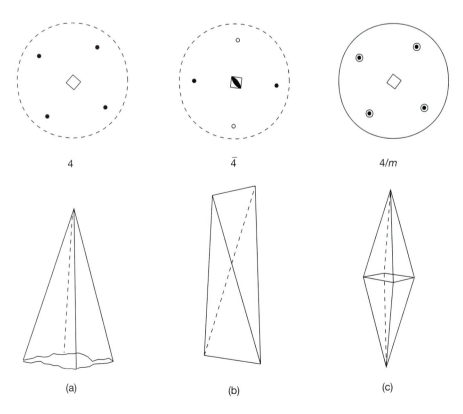

4       $\bar{4}$ 4/m

(a)   (b)   (c)

Because some of the symmetry operators are redundant, we do not need them all to describe the symmetry of crystals. By convention, crystallographers use the 13 operators in the left-hand column of Table 9.1. In mathematical terms, these 13 are *sufficient* to describe symmetry in any crystal.

Combinations of Symmetry Elements: Point Groups

The stereo diagram in Figure 9.14a contains two 2-fold axes. One is perpendicular to the page and one lies within the page. Starting with one point, application of the two symmetry operators yields three additional points. It also results in a pattern that has a third 2-fold axis perpendicular to the first two (Figure 9.14b). Thus, we see that all objects that have two perpendicular 2-fold axes of symmetry must have a third 2-fold axis. It does not matter with which two 2-fold axes we start; the third must be there. Figure 9.14 demonstrates a very important principle of symmetry operators: they operate on each other.

The shorthand notation 222 describes the symmetry depicted in Figure 9.14b, consisting of three mutually perpendicular 2-fold axes. Figure 9.14c shows one possible crystal with 222 symmetry. The three 2-fold axes pass through the centers of edges of the crystal. We call distinct combinations of symmetry, such as 222, **point groups;** they relate the points in

▶**TABLE 9.1**
Symmetry Operations

Standard Symmetry Operations	Equivalent Rotoinversion Axes	Equivalent Rotation Axes with Mirror Planes
i	$\bar{1}$	
m	$\bar{2}$	$1/m$
1		
2		
3		
4		
6		
$\bar{3}$		
$\bar{4}$		
$\bar{6}$		$3/m$
$2/m$		
$4/m$		
$6/m$		

a stereo diagram to each other. The word *group* is used because we may treat the principles of symmetry using mathematical group theory. The terms *operator* and *operation* also derive from group theory. In group theory, the 13 operators in the left-hand column of Table 9.1 form a **basis.**

As another example of symmetry operators acting on each other, consider Figure 9.15a, in which two 2-fold axes intersect at 60°. Starting with one point and

▶**FIGURE 9.14**
(a) Stereo diagram showing two 2-fold axes at 90° (one perpendicular to the page and one in the plane of the page) and points related by them; (b) stereo diagram showing the same points and the third 2-fold axes that relates them; (c) a disphenoid, a crystal belonging to point group 222, having the symmetry shown in (b).

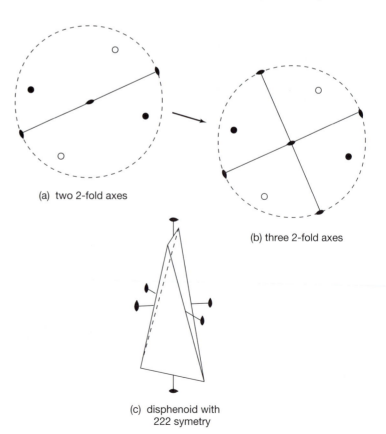

(a) two 2-fold axes

(b) three 2-fold axes

(c) disphenoid with 222 symetry

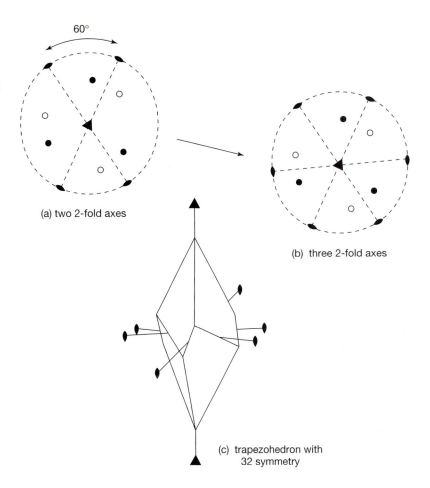

▶**FIGURE 9.15**
(a) Stereo diagram showing two 2-fold axes at 60° and points related by them; (b) stereo diagram showing the same points and axes, and the additional 2-fold and 3-fold axis that must be present; (c) a crystal belonging to point group 32, having the symmetry shown in (b).

60°

(a) two 2-fold axes

(b) three 2-fold axes

(c) trapezohedron with 32 symmetry

applying the symmetry operators, we soon generate five more equivalent points. Examination of the resulting pattern shows that a third 2-fold axis lies at 60° to the first two. A 3-fold axis is perpendicular to the two folds (Figure 9.15b). Another way of looking at the symmetry in Figure 9.15b is to notice that the 3-fold rotation axis acts on the 2-fold axes. It requires that if one 2-fold axis of symmetry is present, two others must be present as well, and that the three are separated by angles of 120°. Consequently, the point group is designated 322 or, more commonly, 32. Figure 9.15c shows an example of a crystal belonging to point group 32.

In Figure 9.14 we showed that two perpendicular 2-fold axes required the presence of a third mutually perpendicular one. In Figure 9.15, we showed that if two 2-fold axes intersect at 60°, another 2-fold and a 3-fold will also be present. Similarly, if we start with a 4-fold axis and one perpendicular 2-fold, we will find other 2-folds perpendicular to the 4-fold and at 45° to each other (Figure 9.16c). If we start with a 6-fold axis and a perpendicular 2-fold, we will find six 2-fold axes in all (Figure 9.16d). Figure 9.16 shows the symmetry of point groups 222, 322, 422, and 622 and drawings of crystals having those symmetries. Note that mirror planes are absent in all four cases. The top and bottom of the crystals do not mirror each other, and the crystal

faces do not have mirror planes down their centers (or else an *m* would be included in the symbol for the point group). The four examples in Figure 9.16 point out that symmetry operators cannot combine in random ways. The presence of two rotation axes requires a third and perhaps more. We can also show that a combination of a rotation axes and a mirror, at angles other than 0° or 90° to each other, requires other rotation axes to be present. Seeing symmetry on the complicated crystal drawings in Figure 9.16 is difficult. (That's why we have stereo diagrams!) A better way to examine symmetry of crystals is to study models in the laboratory.

Special Angles and General Angles

Angles such as 30°, 45°, 60°, 90°, or 120° are called **special angles.** They all divide evenly into 360°. We call nonspecial angles **general angles.** In the examples in Figure 9.16, we started with rotation axes that intersected at special angles. Suppose we start with axes that intersect at general angles. What will be the result?

In Figure 9.17a, we show two 2-fold axes. The angle between them is small and does not divide evenly into 360°. We may apply the 2-fold axes to each other and generate more 2-fold axes. If we continue

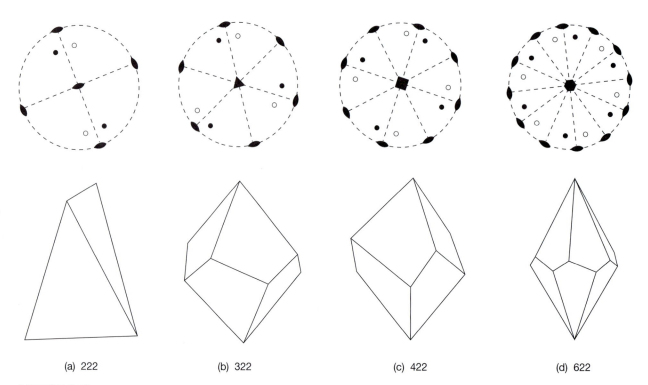

(a) 222   (b) 322   (c) 422   (d) 622

▶**FIGURE 9.16**
Stereo diagrams and crystal drawings for point groups: (a) 222; (b) 32; (c) 422; (d) 622.
Although not shown, the 2-fold axes emerge from the crystals at the center of edges
closest to the crystals' equators.

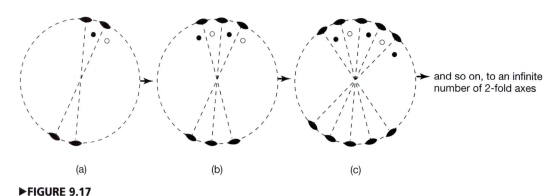

(a)   (b)   (c)   and so on, to an infinite
                                 number of 2-fold axes

▶**FIGURE 9.17**
(a) Two 2-fold axes and two points related by one of them; (b) other points and a 2-fold axis implied
by the presence of the first; (c) moving toward an infinite number of 2-fold axes.

this operation all the way around the circle, we will not end back where we started. The number of 2-fold axes becomes infinite, and an infinite-fold axis of symmetry must be perpendicular to the plane of the page. This is equivalent to the symmetry of a circle. Since crystal structures consist of a discrete number of atoms, we know that infinite symmetry is not possible. We may therefore conclude that if crystals have two 2-fold axes, they must intersect at a special angle so that they are finite in number.

The preceding discussion suggests that rotation axes only combine in a limited number of ways. In fact,

only the six combinations depicted in Figure 9.18 are possible. We have already discussed four of them: 222, 32, 422, 622. (By convention, 322 is abbreviated 32.) The other two (Figure 9.18e, f) are best envisioned by thinking about a cube. The angles between rotation axes are all special angles (to avoid infinite symmetry). In Figures 9.18e and 9.18f, the angles are equal to angles between various diagonals in a cube and perpendiculars to cube faces. If we carried out the exercise, we would find that in crystals with both rotation axes and mirror planes, the angles between the rotation axes and the mirror planes are limited to only a few special

▶**FIGURE 9.18**
The six possible combinations of rotation axes. The angles between axes in (a), (b), (c), and (d) are shown on the drawings. For (e) and (f), based on a cube, the angles are given in the tables below the drawings. Some rotation axes have been omitted in all six drawings; a sufficient number are shown to clarify angular relationships.

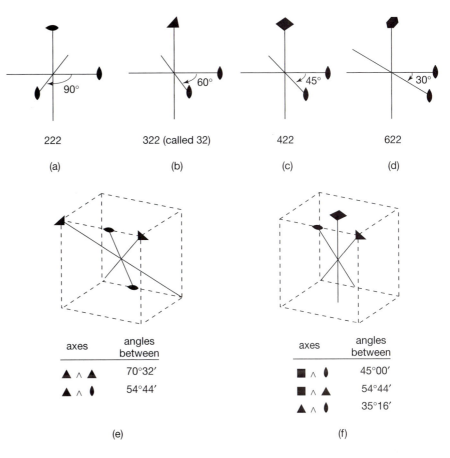

angles as well. Otherwise, we have infinite symmetry. In many crystals the angles are 0° (the rotation axis lies within the plane of the mirror) or 90° (the rotation axis is perpendicular to the mirror).

FORMS AND CRYSTAL MORPHOLOGY

Forms

As pointed out before, we call a set of identical faces related by symmetry a *form*. Use of this term is unfortunate, because most of us think of *form* as meaning "shape." Further confusion derives from the fact that a single crystal may contain multiple forms, and a single form may have as many as 48 faces. Two crystals with identical forms may look different if the forms are of different sizes. Nonetheless, the term is firmly established in crystallography and unlikely to disappear. Mineralogists use the term *habit* to refer to the common or characteristic shape of a crystal. Habit depends on the symmetry, number, and size of forms that are present and how a crystal grows in relation to other crystals around it. For example, we say pyrite has a *cubic habit* because pyrite crystals frequently grow as cubes or near cubes. Crystals may contain multiple forms, but most minerals have only one common habit.

Nature does not always produce perfectly shaped crystal faces. Accidents of growth and other imperfections lead to differences among faces of a form, but forms are often quite easily identified. Because the forms present in particular crystals, and the resulting habits, are dependent on atomic structures, they are consistent for a given mineral. Uncertainty arises, however, because some structures can have several different forms, and it is not always clear why one may be preferred to another. Kinetic factors, associated with the time of crystallization, undoubtedly play a key role.

Some crystals, such as those depicted in Figure 9.20, have only one form, so all the faces are the same shape and size. Other crystals, such as the ones in the right-hand columns of Figures 9.19, 9.21, and 9.22, contain more than one form. Figure 9.23 shows a photograph of a pyrite crystal having more than one form.

We call the forms shown in Figure 9.16 *closed forms* because they enclose space without other forms being present. In contrast, we call the forms in Figure 9.24 *open forms* because they do not completely enclose space. Some forms in Figures 9.19, 9.21, and 9.22 are open forms, too. Because crystals cannot be open sided, additional crystal faces must terminate open forms. For example, in Figure 9.4, pyramids (with four or six faces) and pedions (single

Box 9.1 Proper Rotation Axes and Stereo Diagrams

Crystallographers use conventional symbols to represent symmetry operators on a stereo diagram (Figures 9.19 and 9.20). Figure 9.19 shows stereo diagrams and crystal drawings for each of the five kinds of rotation axes we have considered. The location of the symbols indicates the orientation of the rotation axis. Small lenses, triangles, squares, and hexagons at the center of the diagram represent 2-, 3-, 4-, and 6-fold axes of rotation perpendicular to the plane of the page (Figure 9.19):

Symbol	Axis/Mirror
⬮	2-fold
▲	3-fold
■	4-fold
⬢	6-fold
——	mirror

The third column in Figure 9.19 shows identical crystal faces related by the rotational symmetry. We call such a set of faces a **form.** The faces are depicted by dots in the stereo diagrams. In some cases the faces do not make an entire crystal. The fourth column shows more complicated complete crystals (which do not correspond to the stereo diagrams) with the same symmetry.

It is sometimes not possible or desirable to orient rotation axes perpendicular to the page. We indicate axes that lie parallel to the page by straight lines, passing through the center of the diagram—solid if they are parallel to a mirror, dashed if they are not—with the appropriate symbols at each end (Figure 9.20a, b).

Axes that are neither parallel nor perpendicular to the page plot as symbols between the center

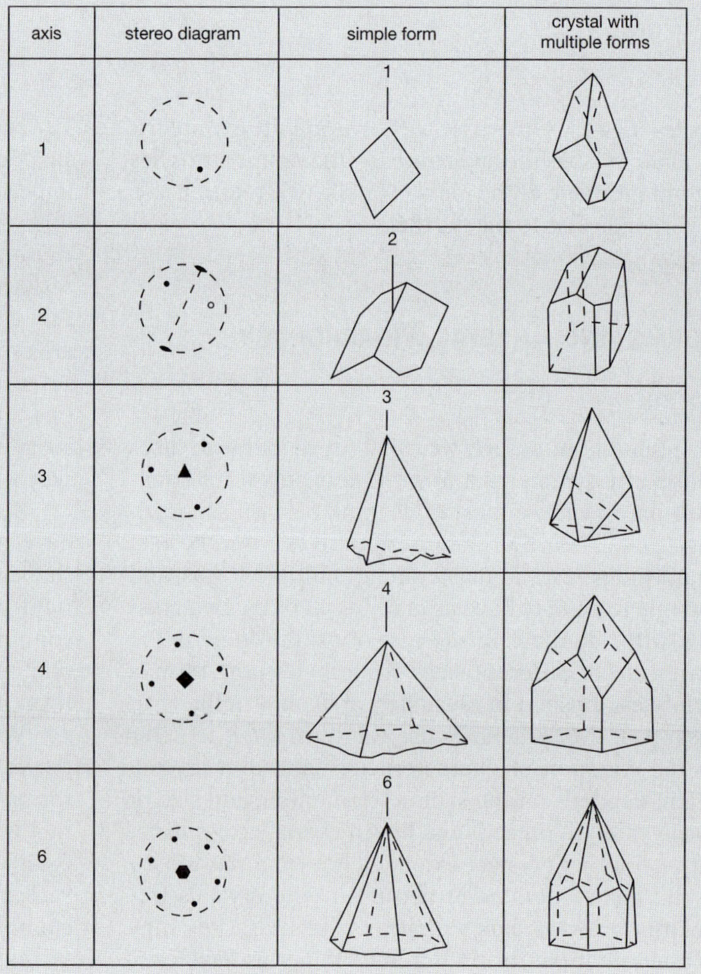

▶**FIGURE 9.19**
Proper rotation axes in stereo diagrams and equivalent crystal faces: The stereo diagrams depicting the axes contain dots corresponding to crystal faces shown in the drawing in the third column. The fourth column shows more complicated crystals with the same symmetry.

axis	stereo diagram	simple form	crystal with multiple forms
1		1	
2		2	
3		3	
4		4	
6		6	

and periphery of the stereo diagram (note the 3-fold axes in Figure 9.20a). The distance from the center to the symbol is proportional to the angle between the axis and a line perpendicular to the page. If a crystal has a single axis of symmetry, by convention we usually orient it vertically and place the appropriate symbol at the center of the diagram. However, for crystals that contain only one 2-fold axis of symmetry, the axis is sometimes oriented horizontally, as shown in the second row of Figure 9.19.

Solid lines show mirror planes. If a crystal has a mirror plane parallel to the page, the circle of the stereo diagram is solid (Figure 9.20a); if not, it is dashed (Figure 9.20b). We indicate mirror planes perpendicular to the page by straight lines passing through the center of the diagram (Figure 9.20a,b). If a mirror is inclined to the page, we plot it as a great circle whose chord shows the intersection of the mirror with the horizontal equatorial plane (Figure 9.20a). Note that Figure 9.20a contains mirror planes in the plane of the page, perpendicular to the page, and inclined to the page.

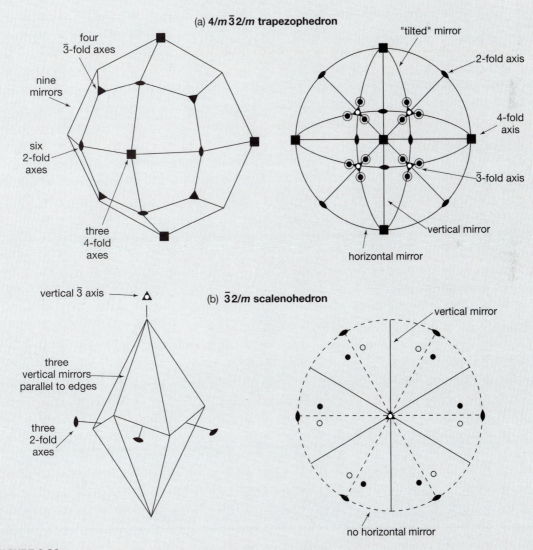

▶FIGURE 9.20

Trapezohedron and scalenohedron, and their stereo diagrams. (a) A trapezohedron containing 4-fold, 3-fold, and 2-fold axes of symmetry. The 3-fold axes are really $\bar{3}$ axes and so are indicated by triangles with dots in their centers (see Box 9.3). (b) A scalenohedron containing a 3-fold axis and three 2-fold axes of symmetry. Both crystals have mirror planes, but in the scalenohedron the mirror planes are all parallel to the long dimension of the crystal (oriented vertically in the stereo diagram).

Box 9.2 Rotation Axes with Perpendicular Mirror Planes

In many crystals, rotation axes are perpendicular to mirror planes. When listing symmetry elements, we use the special symbols $1/m, 2/m, 3/m, 4/m$, and $6/m$ to indicate such combinations. Figure 9.21 shows stereo diagrams for the five combinations. Recall that a solid circle around the outside of the diagram indicates a mirror plane within the page. For $2/m$, we have plotted the 2-fold axis in the

plane of the page, so the mirror is perpendicular to the page and shows as a straight solid line instead of a solid circle. The third column in Figure 9.21 shows identical crystal faces related by the symmetry; the fourth shows more complex crystals, containing faces of different shapes, but still having the same symmetry.

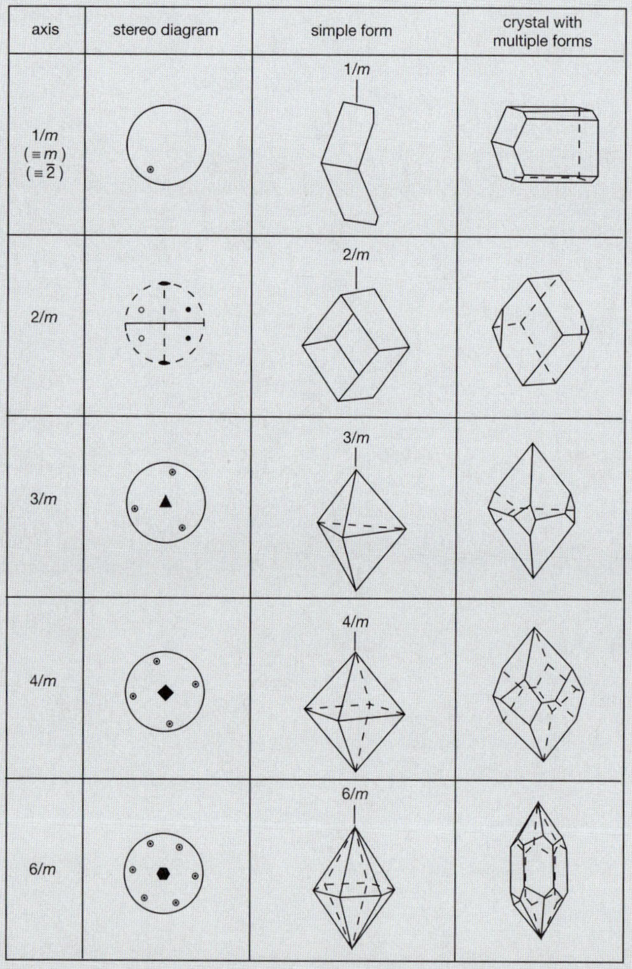

▶**FIGURE 9.21**
Stereo diagrams and drawings of faces and crystals having rotation axes with perpendicular mirror planes. Crystal faces are plotted in the same manner as in Figure 9.19.

axis	stereo diagram	simple form	crystal with multiple forms

faces) terminate prisms (open forms). Prisms, pyramids, and pedions are all open forms that, when combined, produce a closed shape. Crystals with only one form must, of necessity, have a closed form.

The number of possible forms is limited. On the basis of the relative positions of their faces, only 48 can

be distinguished (Figure 9.25). Crystallographers use various schemes to name them, but A. F. Rogers's scheme, published in 1935, is commonly used in the United States (Box 9.4, page 194). Table 9.2 gives basic names that we can expand to the entire 48 using modifiers. Some modifiers have obvious meanings. For example,

Box 9.3 *Rotoinversion*

Besides proper rotation axes (1, 2, 3, 4, and 6), many crystals contain a rotoinversion axis, symbolized by $\bar{1}, \bar{2}, \bar{3}, \bar{4}$, and $\bar{6}$. Rotoinversion involves a combination of rotation and inversion in one operation:

Symbol	Axis	Rotation Angle
	$\bar{1}$	360°
	$\bar{2}$	180°
△	$\bar{3}$	120°
◆	$\bar{4}$	90°
◉	$\bar{6}$	60°

A 1-fold rotoinversion axis involves rotating 360° and then inverting. Because a 360° rotation is equivalent to no rotation at all, a 1-axis is identical to an inversion center. A 2-fold rotoinversion

axis involves rotation of 180° followed by inversion. This has the same result as a mirror plane perpendicular to the 2-fold axis. Similarly, a 6-fold axis is equivalent to $3/m$. We normally don't talk about 1-, 2-, or 6-axes.

It is important to emphasize that rotoinversion is a combined operation. Rotation and inversion are applied together to produce an equivalent point or crystal face. It makes no difference whether we think of rotation followed by inversion, or vice versa; the results come out the same. Figure 9.22 shows the five rotoinversion axes in stereo diagrams and the forms related by them. The last column shows a complete crystal with equivalent symmetry. It is sometimes difficult to see rotoinversion symmetry in crystals.

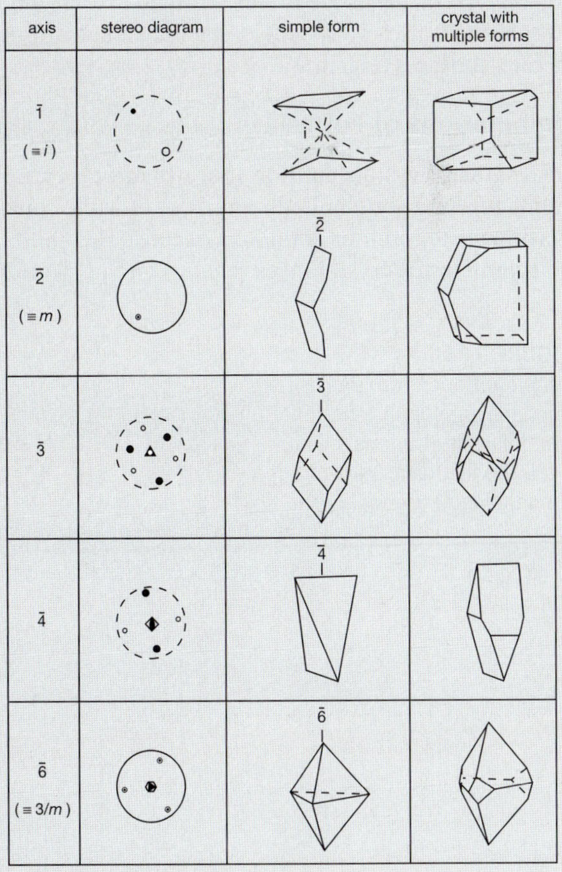

▶**FIGURE 9.22**
Stereo diagrams and drawings of faces and crystals having rotoinversion symmetry. Crystal faces are plotted in the same manner as in Figures 9.19 and 9.21.

axis	stereo diagram	simple form	crystal with multiple forms
$\bar{1}$ ($\equiv i$)		$\bar{1}$	
$\bar{2}$ ($\equiv m$)		$\bar{2}$	
$\bar{3}$		$\bar{3}$	
$\bar{4}$		$\bar{4}$	
$\bar{6}$ ($\equiv 3/m$)		$\bar{6}$	

▶**FIGURE 9.23**
Scanning electron microscope photo of small pyrite crystals (about 0.1 mm across) with multiple forms.

a hexagonal pyramid has six sides, while a tetragonal pyramid has only four. We should emphasize that although only 48 possible forms exist, they can have an infinite number of sizes and shapes. A **disphenoid,** a form consisting of four faces, may be tall and skinny or short and wide. Nevertheless, it is still a disphenoid (Figure 9.27a, b). It is also a disphenoid even if another form truncates the faces (Figure 9.27c).

Combinations of Forms

Most natural crystals contain more than one form, leading to a large but limited number of possible combinations. The number is limited because the shape and symmetry of crystal faces depend on the atomic

▶**TABLE 9.2**
Basic Names of Forms

Open Forms	
pedions	single face
pinacoids	2 parallel faces
prisms	3, 4, 6, 8, or 12 faces, all parallel to a common line
pyramids	3, 4, 6, 8, or 12 nonparallel faces that intersect at a common point
domes	2 nonparallel faces related by a mirror
sphenoids	2 nonparallel faces related by a 2-fold axis

Closed Forms	
scalenohedrons	8 or 12 scalene triangle-shaped faces
trapezohedrons	6, 8, or 12 trapezoid-shaped faces
disphenoids	4 nonequilateral triangular faces
dipyramids	two 3-, 4-, 6-, 8-, or 12-sided pyramids related by a mirror
rhombohedrons	6 rhomb-shaped faces
tetrahedrons	4 equilateral triangular faces
cubes (hexahedrons)	6 square faces
octahedrons	8 equilateral triangle-shaped faces
dodecahedrons	12 rhomb-shaped faces
pyritohedrons and tetartoids	12 five-sided faces
diploids and gyroids	24 five-sided faces

▶**FIGURE 9.24**
Examples of open forms: (a) ditetragonal pyramid; (b) tetragonal pyramid; (c) ditetragonal prism; (d) tetragonal prism; (e) pedion; (f) sphenoid; (g) trigonal prism.

(a) ditetragonal pyramid

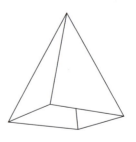

(b) tetragonal pyramid

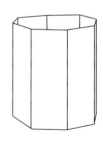

(c) ditetragonal prism

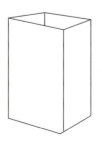

(d) tetragonal prism

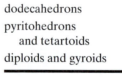

(e) pedion

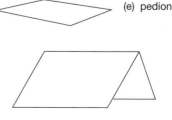

(f) sphenoid

(g) trigonal prism

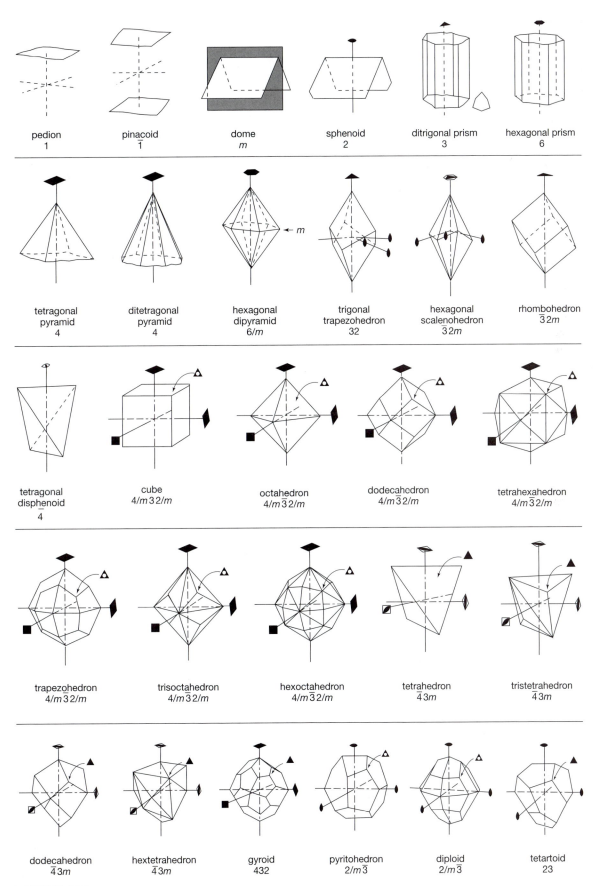

pedion 1	pinacoid $\bar{1}$	dome *m*	sphenoid 2	ditrigonal prism 3	hexagonal prism 6
tetragonal pyramid 4	ditetragonal pyramid 4	hexagonal dipyramid 6/*m*	trigonal trapezohedron 32	hexagonal scalenohedron $\bar{3}2m$	rhombohedron $\bar{3}2m$
tetragonal disphenoid $\bar{4}$	cube $4/m\bar{3}2/m$	octahedron $4/m\bar{3}2/m$	dodecahedron $4/m\bar{3}2/m$	tetrahexahedron $4/m\bar{3}2/m$	
trapezohedron $4/m\bar{3}2/m$	trisoctahedron $4/m\bar{3}2/m$	hexoctahedron $4/m\bar{3}2/m$	tetrahedron $\bar{4}3m$	tristetrahedron $\bar{4}3m$	
dodecahedron $\bar{4}3m$	hextetrahedron $\bar{4}3m$	gyroid 432	pyritohedron $2/m\bar{3}$	diploid $2/m\bar{3}$	tetartoid 23

▶**FIGURE 9.25**
Some of the 48 possible forms.

Box 9.4 Names of the Most Important Forms*

Forms have names based on geometric shapes (for example, prism or pyramid) with Greek words used as modifiers. Most of them can be translated into English with just a small Greek vocabulary.

Suffixes:

-gonal (angle)
-hedron (face)

Prefixes:

scaleno- (scalene triangle)
rhombo- (rhomb shaped)
trapezo- (trapezoid shaped)

Numerical Prefixes:

di- (two)
tris- (three)
tetra- (four)
penta- (five)
hexa- (six)
octa- (eight)
dodeca- (twelve)

In crystal drawings, forms are indicated by lower-case letters, often the first letter of the form name (for example, *o* is used to indicate octahedron faces). Some form names are single words (for example, *tetrahedron,* a four-sided closed form). Others consist of two words, the first describing the shape of the cross section of the ideal form (for example, *hexagonal pyramid,* a pyramid with a hexagonal base). The prefixes *di-, tris-, tetra-,* and *hex-* are used to indicate a doubling, tripling, and so on, of faces. If each of the six sides on a hexagonal pyramid (Figure 9.26a) is split down the middle to produce two faces (Figure 9.26c), the result is a dihexagonal pyramid. Similarly, if each of the faces on an octahedron (Figure 9.26d) is replaced by three faces, the result is a trisoctahedron (Figure 9.26e). If each of the faces on an octahedron is replaced by six faces, the result is a hexoctahedron (Figure 9.26f). A further modifying prefix can be applied to the word *pyramids.* The prefix *di-* indicates that there are two equivalent pyramids related by a mirror plane. Figure 9.26b shows a dihexagonal dipyramid.

*After Rogers, 1935.

▶**FIGURE 9.26**
Examples of forms and form names: (a) hexagonal pyramid; (b) dihexagonal dipyramid; (c) dihexagonal pyramid; (d) octahedron; (e) trisoctahedron; (f) hexoctahedron.

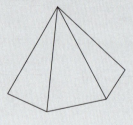

(a) hexagonal pyramid

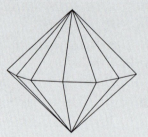

(b) dihexagonal dipyramid

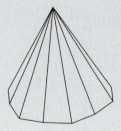

(c) dihexagonal pyramid

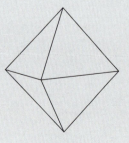

(d) octahedron

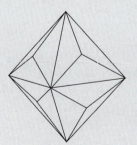

(e) trisoctahedron

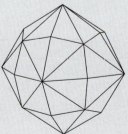

(f) hexoctahedron

►**FIGURE 9.27**
(a) and (b) are different-shaped disphenoids; (c) two disphenoids together.

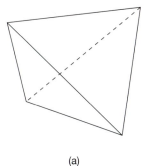
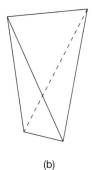
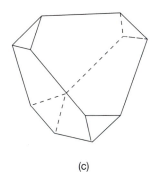

(a) (b) (c)

►**FIGURE 9.28**
Combined forms in ideal crystals and letters used to indicate various forms: (a) two disphenoids (*e* and *f*); (b) hexagonal dipyramid (*p*) and prism (*m*); (c) rhombohedron (*r*), pinacoid (*c*), and prism (*a*); (d) cube (*a*) and octahedron (*o*).

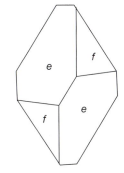
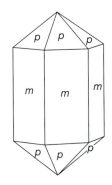

(a) two disphenoids (*e* and *f*) (b) hexagonal dipyramid (*p*) and prism (*m*)

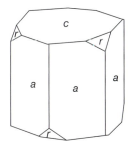
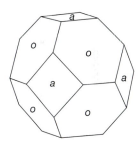

(c) rhombohedron (*r*), pinacoid (*c*), and prism (*a*) (d) cube (*a*) and octahedron (*o*)

structure of a crystal. If the atoms within a crystal are not arranged in hexagonal patterns, forms may not have hexagonal symmetry. Similarly, a crystal may not develop a cubic form unless atoms are in a cubic arrangement. Thus, certain forms never coexist in crystals, while others are often found together. Figure 9.28 shows some idealized crystals with various combinations of forms.

PLOTTING CRYSTAL FACES ON A STEREO DIAGRAM

Until now, we have used stereo diagrams to discuss symmetry in a qualitative way, and we have compared the symmetry of stereo diagrams with the symmetry of

ideal crystals. When studying the symmetry of a real crystal, being able to plot symmetry elements and crystal faces on stereo diagrams is often useful. Since three-dimensional crystals are to be shown on a flat piece of paper, such plots must involve projections. Although there are alternatives, the standard procedure used is reasonably straightforward (Box 9.5). Dwelling on the procedure is not necessary, but the implications of stereo diagrams are important. Figure 9.30 shows a hexagonal prism terminated by flat faces (pinacoids) at either end, and a corresponding stereo diagram. To show the orientations of a crystal face on a stereo diagram, we plot the orientations of a line perpendicular to the face. We call such a line a *pole*. The stereo diagram in Figure 9.30 contains eight points, corresponding to the eight face poles of the crystal. Because the six

Box 9.5 *Plotting Crystal Faces on a Stereo Diagram*

The process of plotting crystal faces on a stereo diagram is depicted in Figure 9.29. The crystal is imagined to lie at the center of a sphere. Perpendiculars, called **poles,** to each face on the crystal are extended until they intersect the sphere (Figure 9.29a). To create a two-dimensional drawing, we project the points on the surface of the sphere onto the sphere's equatorial plane, which intersects the sphere in a circle. To do this, we draw lines connecting each point in the northern hemisphere with the south pole of the sphere (Figure 9.29b), and each point in the southern hemisphere with the north pole of the sphere (Figure 9.29c). Where the lines will pass through the equatorial plane, we place an open circle to represent faces "below the equator" and solid

dots to represent faces "above the equator." Each face on the crystal is thus represented by a point on the equatorial plane and located within the circle that represents the equator. That circle and points are the stereo projection (Figure 9.29d). The stereo projection shows the same symmetry as the original crystal; in Figure 9.29 the symmetry includes a 4-fold axis and several mirrors and 2-fold axes.

Normally, we orient stereo projections so that lines of symmetry are north-south or east-west if possible (Figure 9.29e). However, we have not done that in some of the diagrams in this chapter to emphasize that orientation is only a matter of convention.

▶**FIGURE 9.29**
Constructing a stereo diagram: (a) Normals are drawn to crystal faces to project them onto a sphere. (b) and (c) The points on the sphere are projected onto the equatorial plane. (d) The result is the stereo diagram. Solid dots are points projected from on or above the equator; Open circles are points projected from below the equator; a "bull's eye" indicates both. (e) Usually we rotate the diagram so lines of symmetry are north-south or east-west, if possible.

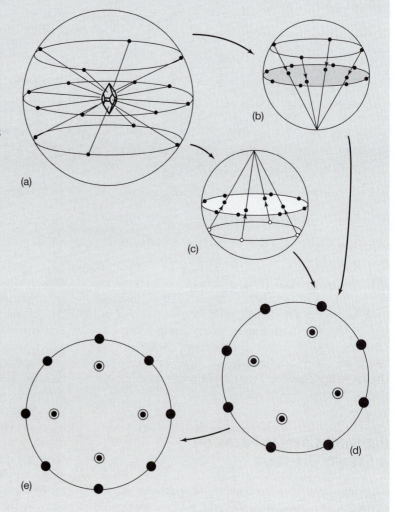

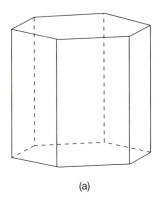

(a)

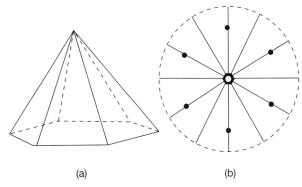

(a)   (b)

▶**FIGURE 9.31**
Hexagonal pyramid and pedion: (a) crystal drawing;
(b) stereo projection showing orientation of faces a crystal
symmetry. Note the pedion plots as an open circle at the
center of the diagram.

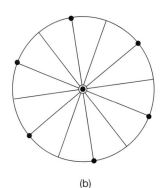

(b)

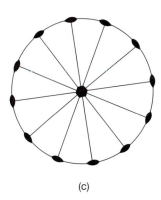

(c)

▶**FIGURE 9.30**
Hexagonal prism with terminating pinacoid: (a) crystal
drawing; (b) stereo diagram showing orientation of faces
and mirror planes; (c) stereo diagram showing the total
symmetry of the crystal. Note the pinacoid plots as a "bull's
eye" at the center of the diagram.

prism faces are vertical, their poles are horizontal and
plot as points on the circle of the stereo diagram. In
stereo projections, all points that plot on the circle represent vertical faces. Similarly, the top and bottom
faces (pinacoid) of the crystal plot as points at the center of the stereo projection. A small, open circle represents the face below the equator, while a dot represents
the one on top. Plotted on top of each other, the two
symbols resemble a bull's-eye. All points at the center
of a stereo diagram represent horizontal faces; the
faces may be either on the top or on the bottom of the

crystal. The angular relationships between the prismatic faces are well shown on the stereo diagram; the dots
representing those faces are 60° from each other. The
angle between the pinacoid and prism faces is 90° since
the end faces plot in the center of the diagram and the
prism faces on the outside. For clarity, we have plotted
the crystal faces on Figure 9.30b and the crystal symmetry on Figure 9.30c, but crystallographers usually
combine the two kinds of diagrams.

If faces are neither horizontal nor vertical, they
plot within, but not at the center of, the circle. Inclined faces plot at various distances from the center,
depending on their slopes. If nearly horizontal, they
are near the center of the diagram; if nearly vertical,
they are near the perimeter. Figure 9.31 shows a
hexagonal pyramid with a flat base (pedion) projected onto a stereo diagram. The six pyramidal faces result in solid dots, located between the center and
edge of the circle. They are solid because they represent faces above the equator. The pedion plots as an
open circle at the center of the diagram since it is a
horizontal face below the equator.

As a final example of a stereo diagram, consider a
complicated tetragonal crystal (vesuvianite) shown in
Figure 9.32a. The drawing is based on an imperfectly
formed natural sample. Figure 9.32b, a stereo diagram, shows the symmetry of the crystal. Figure 9.32c
shows the orientation of faces. The prism forms (labeled *m* and *a*) yield points on the circumference of
the stereo diagram since they are vertical. The pyramidal faces *(e)* yield bull's-eye patterns representing
four faces above and four faces below the equatorial
plane. The pinacoid *(c)* yields a bull's-eye pattern at
the center of the diagram. Note that although seeing
all symmetry elements in the crystal is difficult, they
are more easily seen on the stereo diagram. With the
aid of a diagram, we see that vesuvianite belongs to

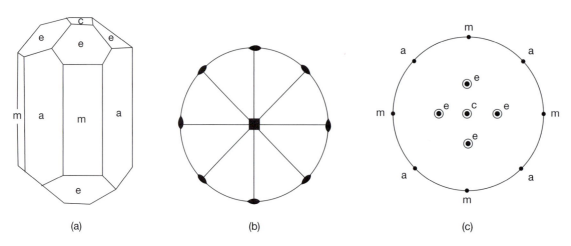

▶**FIGURE 9.32**
Vesuvianite crystal: (a) idealized crystal drawing; (b) stereo diagram showing symmetry of crystal; (c) stereo diagram showing orientation of crystal faces.

point group $4/m2/m2/m$. Its symmetry consists of five rotation axes: four 2-fold axes and a 4-fold perpendicular to them. Each rotation axis is perpendicular to a mirror.

POINT GROUPS AND CRYSTAL SYSTEMS

Table 9.1 contains 13 distinct symmetry operators. If we made a complete analysis, we would find that the number of possible ways they can combine is surprisingly small. Only 32 combinations are possible; they represent the only combinations of symmetry elements that crystals, or arrangements of atoms, can have. This leads to the division of crystals into 32 distinct point groups, also sometimes called *crystal classes,* each having their own distinct symmetry. Although the expression *point group* refers only to symmetry, and *crystal class* refers specifically to the symmetry of a crystal, the semantic difference is subtle and the two phrases are often used interchangeably by mineralogists. We can make drawings of crystal shapes with all 32 possible symmetries (Figure 9.33), but some of them are not represented by any known minerals.

Mineralogists group the 32 classes into six crystal systems based on common symmetry elements: cubic, hexagonal, orthorhombic, tetragonal, monoclinic, and triclinic:

Cubic System: This system is also called the *isometric system.* Crystals have high symmetry, all having four 3-fold or $\bar{3}$ axes. Some have three 4-fold axes as well, and some have 2-fold axes or mirror planes. Cubes and octahedra (Figure 9.8) are examples of forms belonging to the cubic system, but other forms belong to the cubic system, too. We call the general form (the form with the

most faces), having 48 faces, a hexoctahedron (see the last entry in Figure 9.33). Many complex shapes are possible, but all tend to be **equant,** meaning they are approximately equidimensional. Minerals of the cubic system are optically isotropic; they have the same optical properties in all directions.

Hexagonal System: Crystals have a single 3-fold, $\bar{3}$, or 6-fold axis. Crystals with more than one 3-fold or $\bar{3}$ axis belong to the cubic system. Hexagonal crystals may also have 2-fold axes and mirror planes. Because they have one direction that is different from others, hexagonal crystals are often prisms of three or six sides terminated by pyramids or pinacoids. Other forms, including the scalenohedron and rhombohedron, are also possible. Minerals of the hexagonal system are optically uniaxial; the single optic axes is parallel to the 3-fold, $\bar{3}$, or 6-fold axis of symmetry.

Tetragonal System: One 4-fold or $\bar{4}$ axis of symmetry characterizes all tetragonal crystals; crystals that have more than one 4-fold axis must belong to the cubic system. Tetragonal crystals may also have 2-folds and mirror planes. As with the crystals in the hexagonal system, tetragonal crystals often form prisms. Minerals of the tetragonal system are optically uniaxial; the single optic axes is parallel to the 4-fold, or $\bar{4}$ axis, of symmetry.

Orthorhombic System: Crystals have three perpendicular 2-fold axes, two perpendicular mirror planes paralleling a 2-fold axis, or three perpendicular $2/m$ axes. A shoebox shape is an excellent example, but many shapes and forms are possible. The disphenoids in Figure 9.27 have point group symmetry 222 and belong to the orthorhombic system. Sometimes orthorhombic crystals may be rather equant but they are often

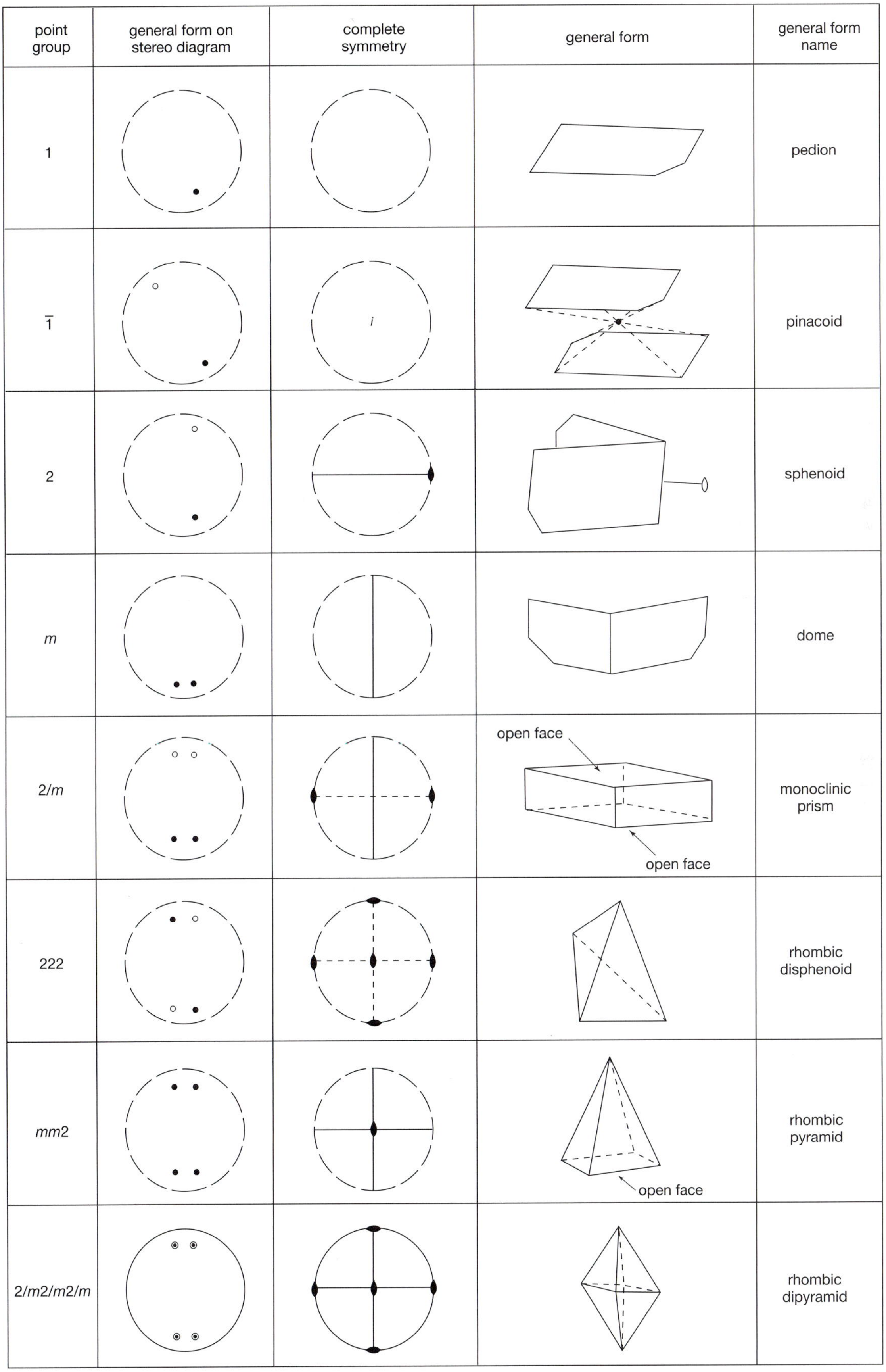
point group
general form on stereo diagram
complete symmetry
general form
general form name
1
pedion
1̄
i
pinacoid
2
sphenoid
m
dome
2/m
open face
open face
monoclinic prism
222
rhombic disphenoid
mm2
open face
rhombic pyramid
2/m2/m2/m
rhombic dipyramid

▶FIGURE 9.33
The 32 point groups.

point group	general form on stereo diagram	complete symmetry	general form	general form name
4				tetragonal pyramid
$\bar{4}$				tetragonal disphenoid
4/m				tetragonal dipyramid
422				tetragonal trapezohedron
4mm				ditetragonal pyramid
$\bar{4}2m$				tetragonal scalenohedron
4/m2/m2/m				ditetragonal dipyramid
3				trigonal pyramid

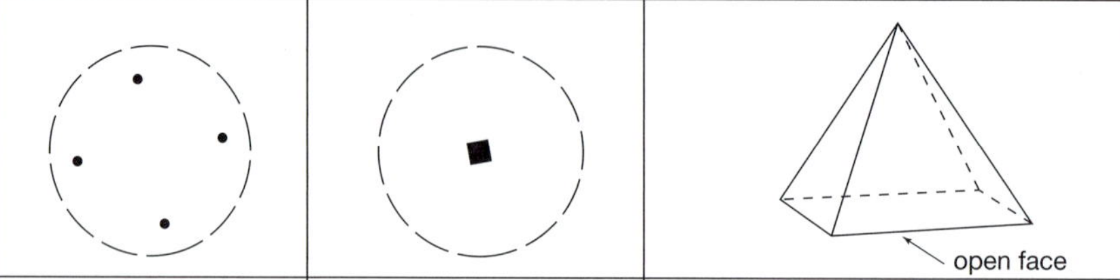

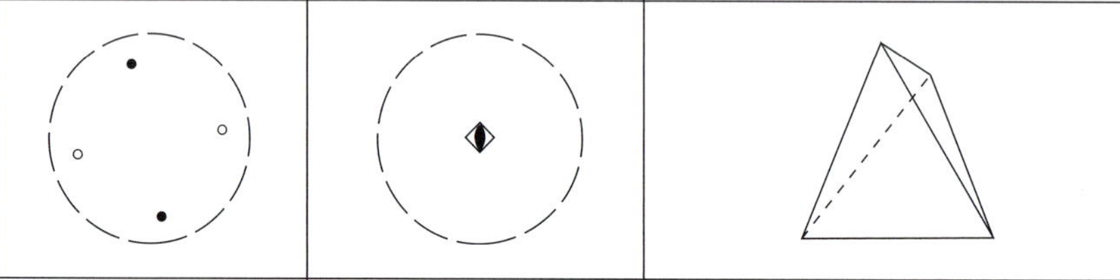

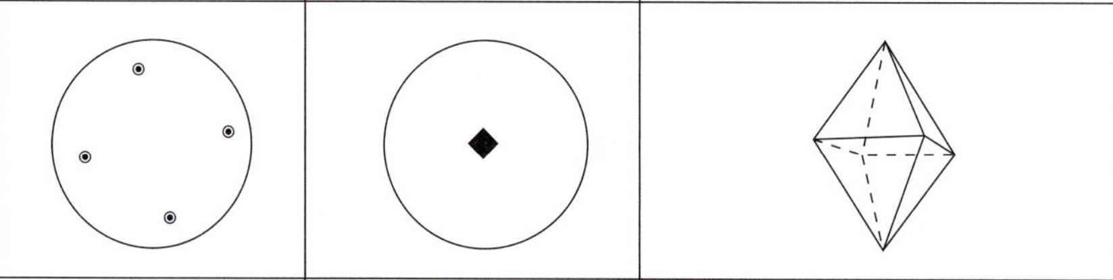

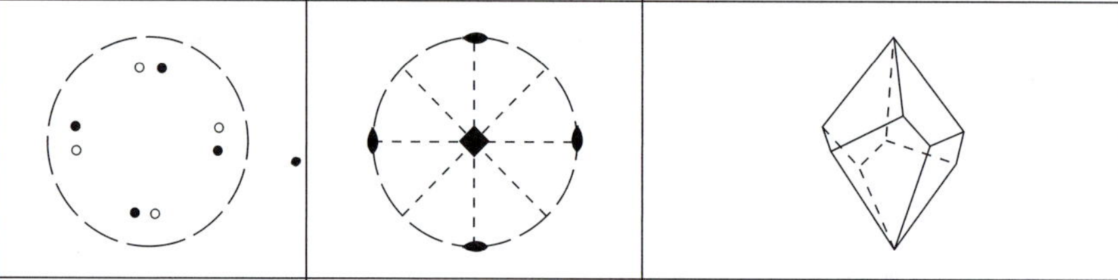

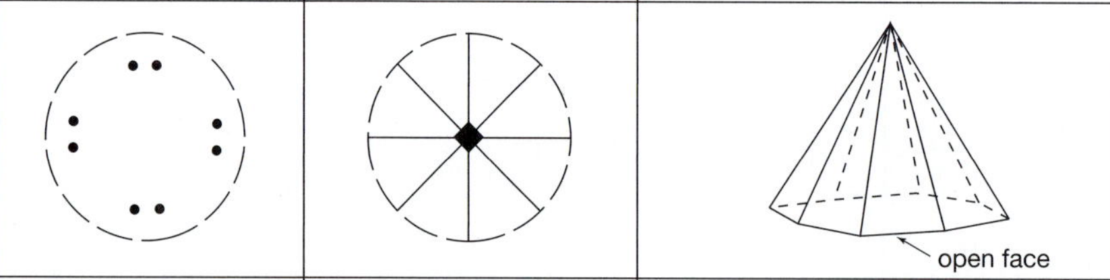

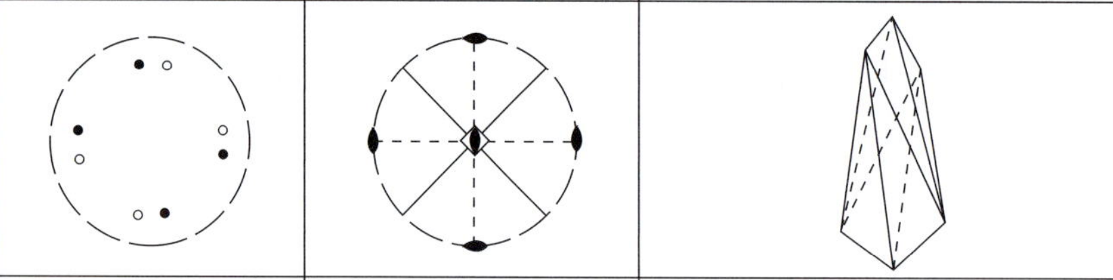

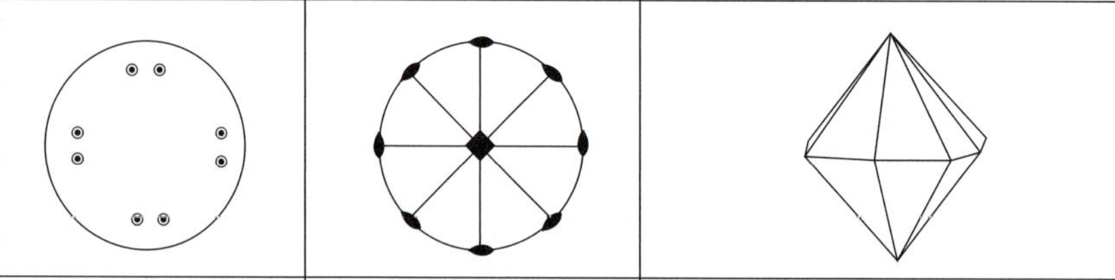

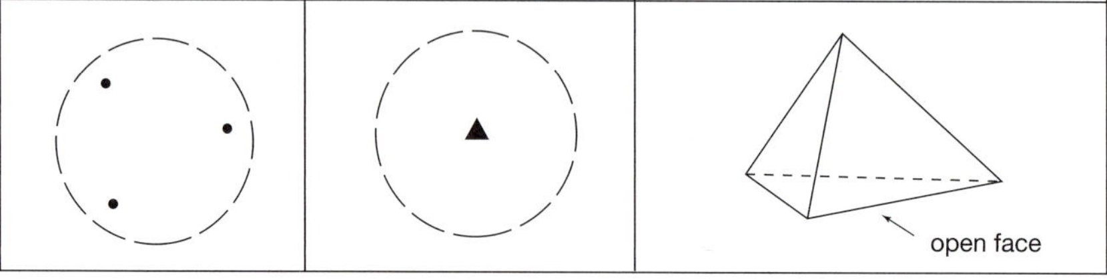

▶FIGURE 9.33 *(Continued)*

point group	general form on stereo diagram	complete symmetry	general form	general form name
$\bar{3}$	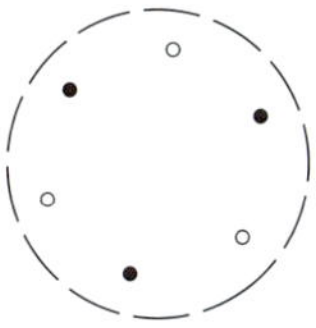	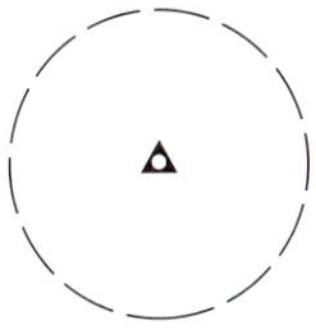	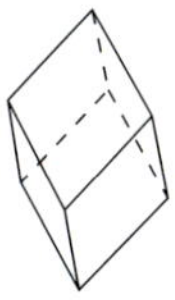	rhombohedron
32	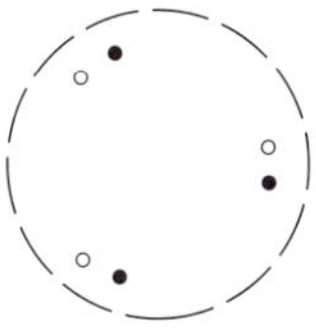	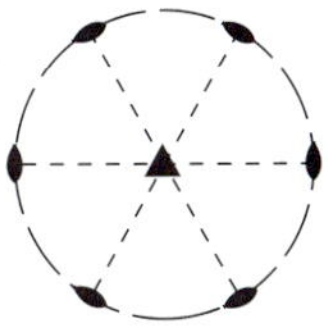	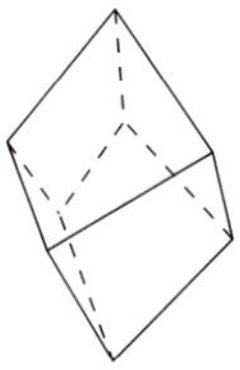	trigonal trapezohedron
3*m*	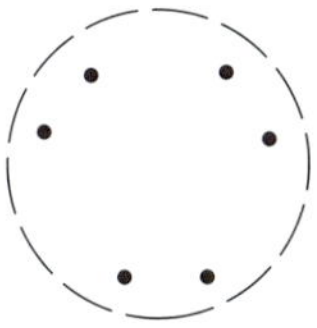	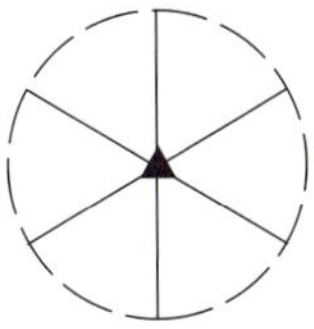	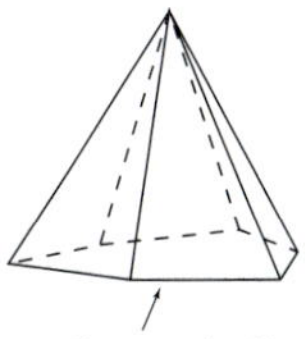 open face on bottom	ditrigonal pyramid
$\bar{3}2m$	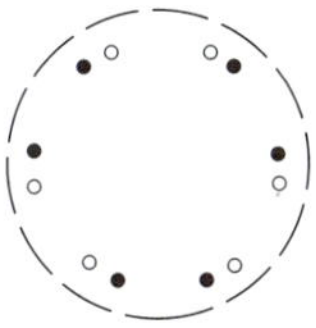	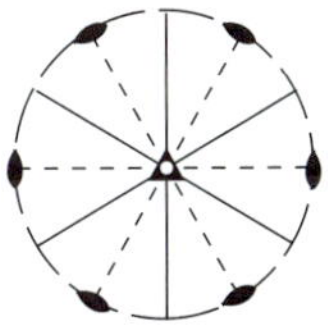	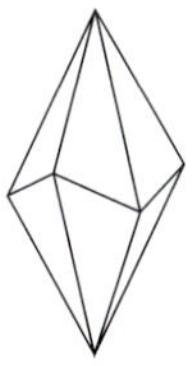	hexagonal scalenohedron
6	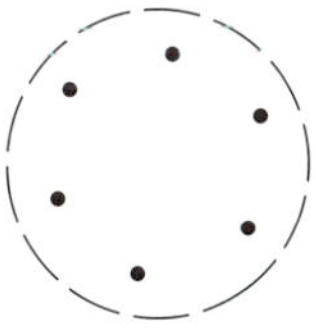	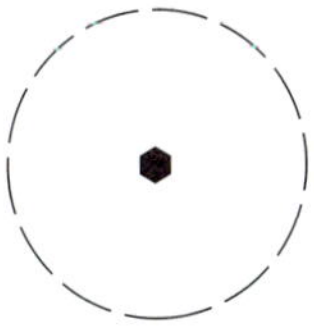	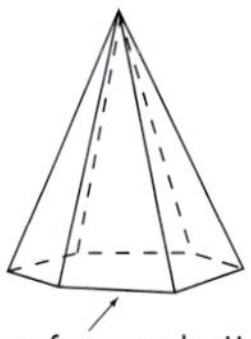 open face on bottom	hexagonal pyramid
$\bar{6}$	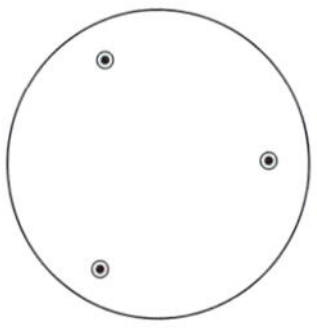		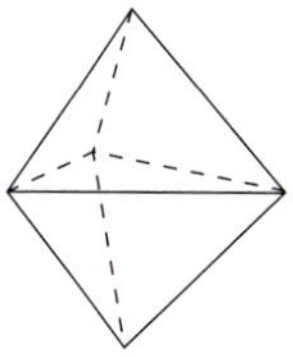	trigonal dipyramid
6/*m*	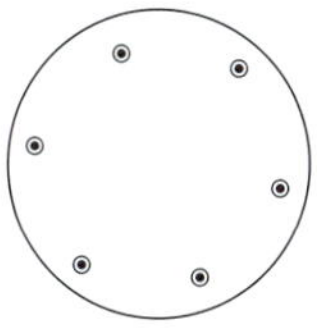	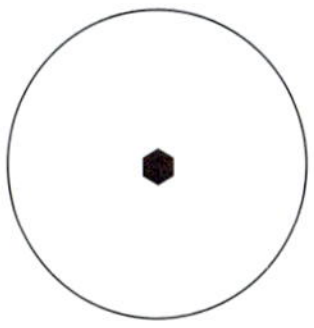	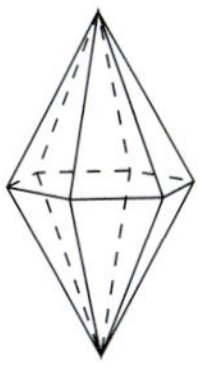	hexagonal dipyramid
622	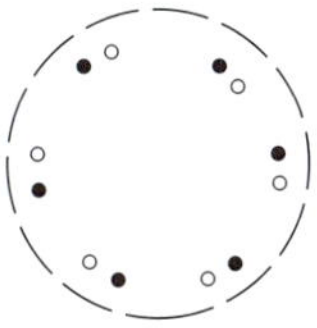	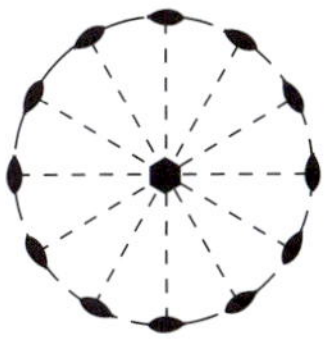	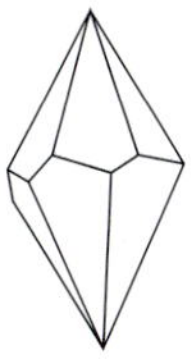	hexagonal trapezohedron

▶FIGURE 9.33 *(Continued)*

point group	general form on stereo diagram	complete symmetry	general form	general form name
6mm	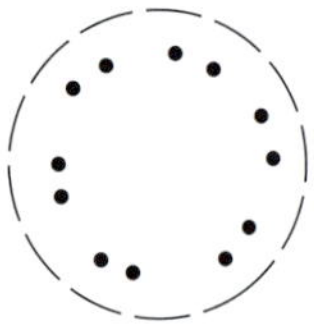	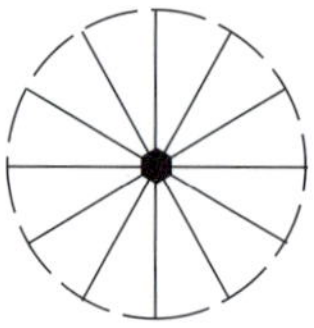	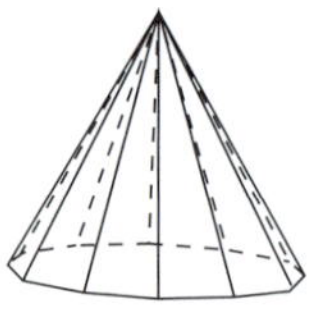 open face on bottom	dihexagonal pyramid
$\bar{6}m2$	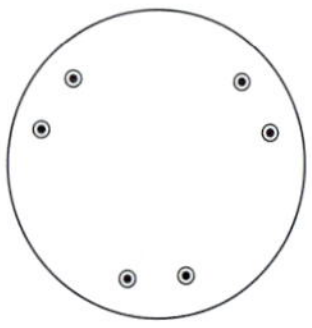	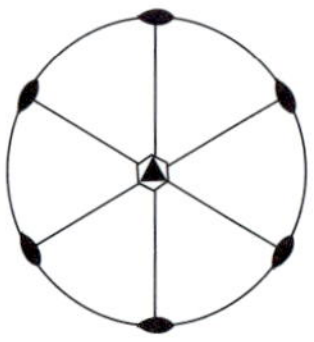	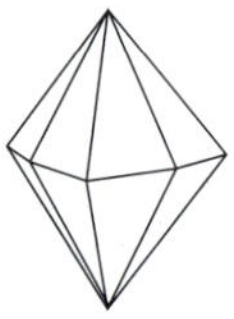	ditrigonal dipyramid
6/m2/m2/m	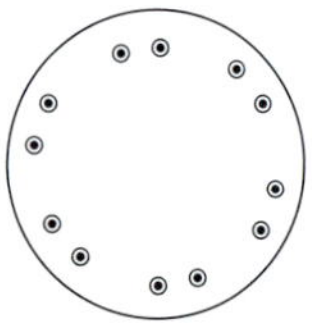	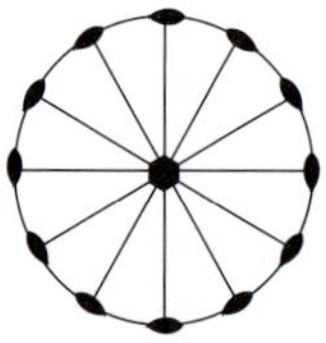	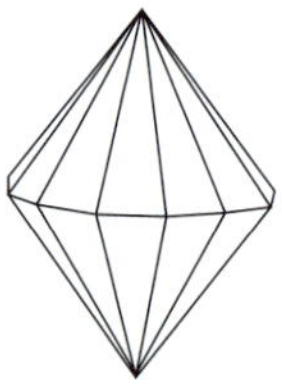	dihexagonal dipyramid
23	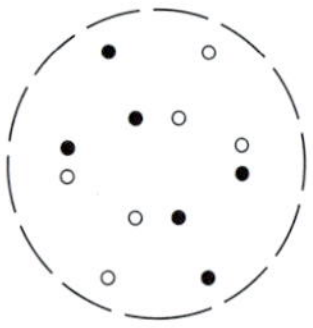	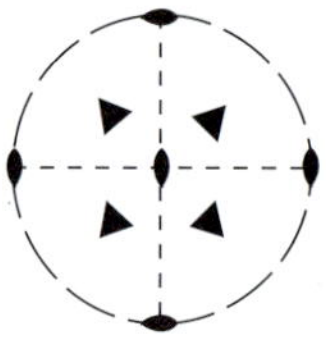	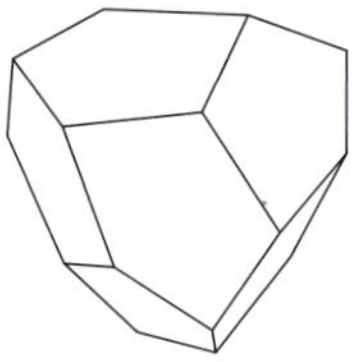	tetartoid
2/m$\bar{3}$	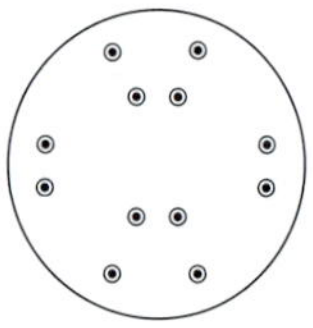	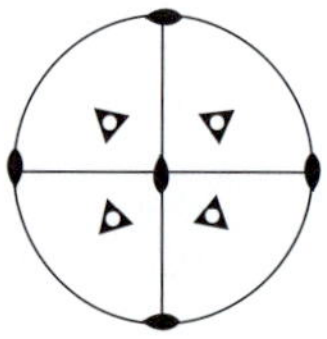	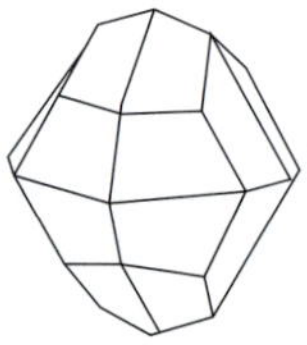	diploid
432	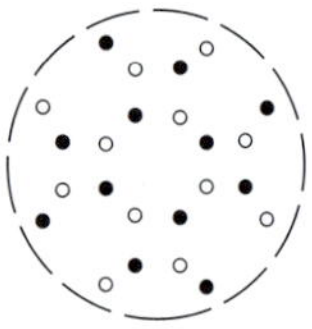		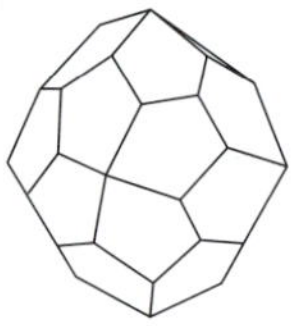	gyroid
$\bar{4}$3m				hextetrahedron
4/m$\bar{3}$2/m	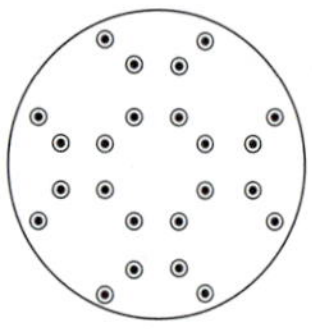		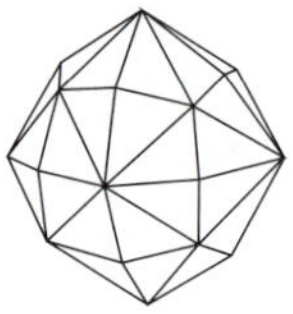	hexoctahedron

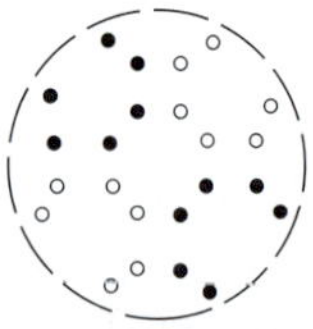

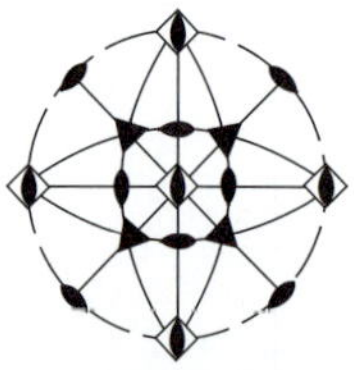

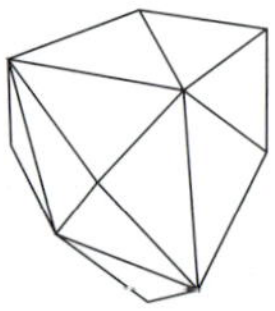

▶FIGURE 9.33 (Continued)

Box 9.6 Hermann-Mauguin Symbols

The symbols used in this book are based on notations developed by C. H. Hermann and C. V. Mauguin in the early 1900s. They have been used by most crystallographers since about 1930. Numbers refer to rotation axes of symmetry; a bar over a number indicates a rotoinversion axis. Mirrors, designated by *m*, are perpendicular to an axis if they appear as a denominator (for example, $4/m$), and parallel to an axis otherwise. When articulating the symbols, they are pronounced just as if they were typographical characters. $4/m\bar{3}2/m$, for example, is read "four over m, bar three, two over m."

Either one, two, or three symbols describe a point group; they combine in different ways for different systems (Figure 9.33):

Crystal System	First Symbol
cubic (isometric)	$4, 4/m, \bar{4}, 2, 2/m$
hexagonal	$6, 6/m, \bar{6}, 3, \bar{3}$
tetragonal	$4, 4/m, \bar{4}$
orthorhombic	$2, 2/m, m$
monoclinic	$2, 2/m, m$
triclinic	$1, \bar{1}$

Second Symbol	Third Symbol
$3, \bar{3},$	$2, 2/m, m$
$2/m, m$	$2, 2/m, m$
$2, 2/m, m$	$2, 2/m, m$
$2, 2/m, m$	$2, 2/m, m$

For cubic point groups, the first symbol describes three mutually perpendicular principal symmetry axes, oriented perpendicular to cubic faces (if cubic faces are present). The second describes four axes oriented at angles of 54°44′ to the principal axes. They correspond to the "body diagonals" of a cube, a diagonal from a corner through the center to the opposite corner. The third symbol, if present, describes six 2-fold axes or mirror planes oriented at angles of 45° to the principal axes. They correspond to "edge diagonals" of a cube, diagonals from the center of edges through the center of the cube to the opposite edge.

For hexagonal point groups, the first symbol describes the single principal axis. The second, if present, describes three secondary rotation axes oriented at 120° to each other and perpendicular to the principal axis, or three mirror planes oriented at 120° to each other and parallel to the principal axis. The third symbol, if present, represents mirror planes or 2-fold axes oriented between the secondary axes.

For tetragonal point groups, the first symbol represents the principal axis. The second, if present, represents two secondary axes perpendicular to each other and to the principal axis, or two mirror planes oriented at 90° to each other and parallel to the principal axis. The third represents axes or mirror planes between the secondary axes.

Only three orthorhombic point groups are possible. Point group 222 has three mutually perpendicular 2-fold axes. Point group *mm*2 has one 2-fold axis with two mutually perpendicular mirror planes parallel to it. Point group $2/m2/m2/m$ has three perpendicular 2-fold axes with mirror planes perpendicular to each.

For monoclinic point groups, only one symmetry element is included in the Hermann-Mauguin symbol because the only possible symmetries are a 2-fold axis, a mirror, or a 2-fold axis with a mirror perpendicular to it. Similarly, for triclinic crystals, the only possible point groups are 1 and $\bar{1}$.

---

quite tabular. Minerals belonging to the orthorhombic system are optically biaxial.

Monoclinic System: Crystals are characterized by a single 2-fold axis or mirror, or by a 2-fold axis perpendicular to a mirror. In the simplest case, they may appear as shoeboxes squashed in one direction so that one angle at some corners is not 90°. The symmetry may be hard to see because monoclinic crystals often comprise many different forms. Minerals belonging to the monoclinic system are optically biaxial.

Triclinic System: Crystals have no symmetry greater than a 1-fold axis or an inversion center. Minerals belonging to the triclinic system are optically biaxial.

▶TABLE 9.3
The Possible Forms of Each Point Group

Triclinic-Monoclinic-Orthorhombic Systems

Number of Faces	Form	1	$\bar{1}$	2	m	$\frac{2}{m}$	222	2mm	$\frac{2}{m}\frac{2}{m}\frac{2}{m}$
1	pedion	X		X	X			X	
2	pinacoid		X	X	X	X	X	X	X
2	dome				X			X	
2	sphenoid			X					
4	prism					X	X	X	X
4	disphenoid						X		
4	pyramid							X	
8	dipyramid								X

Tetragonal System

Number of Faces	Form	4	$\bar{4}$	$\frac{4}{m}$	422	4mm	$\bar{4}2m$	$\frac{4}{m}\frac{2}{m}\frac{2}{m}$
1	pedion	X				X		
2	pinacoid		X	X	X		X	X
4	tetragonal prism	X	X	X	X	X	X	X
4	tetragonal pyramid	X				X		
4	tetragonal disphenoid		X				X	
8	ditetragonal prism				X	X	X	X
8	tetragonal dipyramid			X	X		X	X
8	tetragonal trapezohedron				X			
8	tetragonal scalenohedron						X	
8	ditetragonal pyramid					X		
16	ditetragonal dipyramid							X

Hexagonal System

Number of Faces	Form	3	$\bar{3}$	32	3m	$\bar{3}\frac{2}{m}$	6	$\frac{3}{m}$	$\frac{6}{m}$	622	6mm	$\bar{6}m2$	$\frac{6}{m}\frac{2}{m}\frac{2}{m}$
1	pedion	X			X		X				X		
2	pinacoid		X	X		X		X	X	X		X	X
3	trigonal prism	X		X	X			X				X	
3	trigonal pyramid	X			X								
6	ditrigonal prism			X	X							X	
6	hexagonal prism		X	X		X	X	X	X	X	X	X	X
6	trigonal dipyramid			X				X				X	
6	rhombohedron		X	X		X							
6	trigonal trapezohedron			X									
6	ditrigonal pyramid				X								
6	hexagonal pyramid						X				X		
12	hexagonal dipyramid					X			X			X	X
12	hexagonal scalenohedron					X							
12	dihexagonal prism									X	X		X
12	ditrigonal bipyramid											X	
12	hexagonal trapezohedron									X			
12	dihexagonal pyramid										X		
24	dihexagonal dipyramid												X

Cubic System

Number of Faces	Form	23	432	$\frac{2}{m}\bar{3}$	$\bar{4}3m$	$\frac{4}{m}\bar{3}\frac{2}{m}$
4	tetrahedron	X			X	
6	cube (hexahedron)	X	X	X	X	X
8	octahedron		X	X		X
12	dodecahedron	X	X	X	X	X
12	pyritohedron	X		X		
12	tristetrahedron	X			X	
12	deltohedron	X			X	
12	tetartoid	X				
24	tetrahexahedron		X		X	X
24	trapezohedron		X	X		X
24	trisoctahedron		X	X		X
24	hextetrahedron				X	
24	diploid			X		
24	gyroid		X			
48	hexoctahedron					X

All six crystal systems are represented by common minerals. For the minerals listed in Part III of this book: 8% are cubic, 9% are tetragonal, 9% are triclinic, 17% are hexagonal, 27% are orthorhombic, and 30% are monoclinic. Within each system, the point groups have varying amounts of symmetry. Most natural crystals fall into the point group with the highest symmetry in each system. Few belong to the point groups of lowest symmetry. Figure 9.33 depicts stereo diagrams for all the 32 possible point groups and shows crystal drawings of their general forms. The point groups are designated using a conventional notation called *Hermann-Mauguin symbols* (Box 9.6).

We call the crystal shapes shown in Figure 9.33 *general forms* because they are equivalent to general points in a stereo diagram. The faces are at nonspecial angles to mirror planes and rotation axes, and the number of faces is the maximum for a single form within the point group. The 32 point groups also contain special forms. Figure 9.25 shows some of them. Faces of special forms are parallel or perpendicular to symmetry elements, or at other special angles to the symmetry operators. All special forms comprise fewer faces than general forms of the same class, and sometimes only one face. The names of the general forms in Figure 9.33 are sometimes used as names for the point groups. For example, we call the class $6/m2/m2/m$ the *dihexagonal dipyramidal class*. More commonly, however, we simply refer to the classes by their symmetry using Hermann-Mauguin symbols. Table 9.3 gives a complete list of all the forms possible in each crystal system and the number of faces comprising the form. Some forms exist in all crystal systems, some in a few, and some, like the octahedron, in only one. Table 9.3 is convenient to identify forms because the possibilities in each crystal system are limited. For example, if we note four equivalent faces on a tetragonal crystal, they must represent a prism, pyramid, or disphenoid. We can decide which by noting the relative orientations of the four faces. Prism faces are parallel, pyramid faces converge in one direction (and intersect at a point unless another face terminates them), and disphenoid faces converge in pairs (Figure 9.27).

▶QUESTIONS FOR THOUGHT

Some of the following questions have no specific correct answers; they are intended to promote thought and discussion.

1. When we think of symmetry, we usually think of some sort of repetitive motif such as a pattern on wallpaper or of a physical object such as a crystal. However, symmetry applies to other things. Give several examples of symmetry that do not involve repetition of a physical pattern or the shape of an object.
2. We can describe the symmetry of crystals using 13 different operators (Table 9.1). Why do we need 13? Why not fewer or more?
3. How can two different crystals with identical symmetry have different forms and shapes?
4. If you examine a crystal and see lots of symmetry, what can you infer about the crystal's atomic structure? On the other hand, if you examine a crystal and see no symmetry, what can you infer about the crystal's atomic structure?
5. When we list Hermann-Mauguin symbols for point groups, we never include *i*. Some point groups, however, do have inversion centers of symmetry. Why do we omit the *i* from the symbols?
6. Why can crystals have no more than three 4-fold axes of symmetry?
7. A crystal may have one 2-fold axis of symmetry. It may also have more, but no crystals have two 2-fold axes of symmetry. Why can't crystals have two 2-fold axes of symmetry? What is the maximum number of 2-fold axes of symmetry they can have?
8. Why are there only 32 point groups?
9. Although there are only 32 point groups, crystals can have many possible forms, and some forms belong to more than one point group. Why are there so many possible forms? How can they belong to more than one point group?
10. Some crystals have only one form. Others have many. What controls or limits the possible combinations of forms in crystals?
11. Can a crystal consist of only open forms?
12. To what point groups do each of the four crystals depicted in Figure 9.28 belong?

▶RESOURCES

(See also the general references listed at the end of Chapter 1.)

Bloss, F. D. *Crystallography and Crystal Chemistry, An Introduction*. New York: Mineralogical Society of America, 1994.

Buerger, M. J. *Introduction to Crystal Geometry*. New York: McGraw Hill, 1971.

McKie, D., and C. McKie. *Essentials of Crystallography*. Oxford: Blackwell, 1986.

Rogers, A. F. A tabulation of crystal forms and discussion of form names. *American Mineralogist* 20 (1935): 838–851.

Zoltai, T., and J. H. Stout. *Mineralogy: Concepts and Principles*. Minneapolis: Burgess, 1984.

Crystallography

In the last chapter we considered the external symmetry that crystals may have. In this chapter, we look at the implications of that symmetry. We find that all crystals are made of basic building blocks called *unit cells*, and that the unit cells fit together in a pattern described by a Bravais lattice. There are six possible unit cell shapes, and each defines a different crystal system. There are 14 possible Bravais lattices. We can often make inferences about unit cells and Bravais lattices on the basis of crystal habit and symmetry. The overall symmetry of an atomic structure depends on more than just unit cell and lattice; it also depends on the locations of atoms in the unit cell. There are 230 possible structure symmetries, called *space groups*. Identifying space group symmetry requires the use of X-ray techniques.

OBSERVATIONS IN SEVENTEENTH AND EIGHTEENTH CENTURIES

In 1669 Nicolaus Steno studied many quartz crystals and found angles between adjacent prism faces, termed **interfacial angles,** to be 120° no matter how the crystals had formed (Figure 10.1). Steno could not make precise measurements, and some of his contemporaries argued that he was overlooking subtle differences. A century later, in 1780, more accurate measurements became possible when Carangeot invented the **goniometer,** a protractor-like device used to measure interfacial angles on crystals. Carangeot's measurements confirmed Steno's earlier observations. Shortly after, Romé de l'Isle (1782) stated a law called the "constancy of interfacial angles," which we commonly call **Steno's law.** This law states that *angles between equivalent faces of crystals of the same mineral are always the same.* The law acknowledges that the size and shape of the crystals may vary.

In 1784 René Haüy studied calcite crystals and found that they had the same shape, no matter what their size. Haüy hypothesized the existence of basic building blocks called **integral molecules** and ar-

gued that large crystals formed when many integral molecules bonded together. Haüy erroneously concluded that integral molecules formed basic units that could not be broken down further. At the same time, Berzelius and others established that the composition of a mineral does not depend on sample size. Additionally, Joseph Proust and John Dalton proved that elements combined in proportions of small rational numbers. Scientists soon combined these crystallographic and chemical observations and came to several conclusions:

- Crystals are made of small basic building blocks.
- The blocks stack together in a regular way, creating the whole crystal.
- Each block contains a small number of atoms.
- All building blocks have the same atomic composition.
- The building block has shape and symmetry that relate to the shape and symmetry of the entire crystal.

Early in the 1800s, several researchers found that crystals of similar, but not identical, chemical composition could have identical shapes. W. H. Wol-

▶**FIGURE 10.1**
Quartz crystals. Nicolaus Steno studied quartz crystals and found that the angles between prism faces were always 120°.

▶**FIGURE 10.2**
Rhombohedral carbonates. Because calcite (left), rhodochrosite (crystals in top sample), and siderite (dark mineral in the bottom right sample) all have the same atomic arrangement, they all form the same shaped (rhombohedral) crystals.

laston (c. 1809) showed that calcite ($CaCO_3$), magnesite ($MgCO_3$), and siderite ($FeCO_3$) all formed the same distinctive rhombohedral crystals (Figure 10.2). Those who studied sulfate compounds also found that crystals of different compositions had the same crystal shape. The rhombohedral carbonates, and the sulfates, are examples of **isomorphous series.** Wollaston and others concluded that when crystal shapes in such series are truly identical, the distribution of atoms within the crystals must be identical as well, even if the compositions are not. Minerals with identical atomic distributions are termed **isostructural.** Sometimes isostructural minerals form solid solutions because they can mix to form intermediate compositions. Fayalite (Fe_2SiO_4) and forsterite (Mg_2SiO_4) are isostructural and form a complete solid solution; olivines can have any intermediate composition. In contrast, halite (NaCl) and periclase (MgO) are isostructural but do not form solid solutions. Calcite and siderite are isostructural but their mutual solubility is limited, and they form only limited (or partial) solid solutions.

In 1821 Eilhard Mitscherlich, a student of Berzelius's, discovered that the same elements may combine in different atomic structures. For instance, calcite and aragonite both have composition $CaCO_3$, but they have different crystal forms and physical properties. We call such minerals **polymorphs** because, although identical in composition, they have different crystal morphologies. Mineralogists have now studied several other $CaCO_3$ polymorphs, but none except vaterite occur naturally. In calcite and vaterite the basic building blocks are rhombohedral, while in aragonite they are orthorhombic (Figure 10.3). It soon became clear that no direct correlation exists between the shapes of building blocks and crystal composition, as Haüy had originally thought.

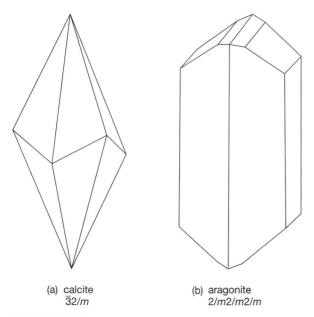

(a) calcite
$\bar{3}2/m$

(b) aragonite
$2/m2/m2/m$

▶**FIGURE 10.3**
Typical calcite and aragonite crystals. (a) Calcite $(\bar{3}2/m)$ and (b) aragonite $(2/m2/m2/m)$ have the same composition, but their atomic arrangements, crystal shapes and symmetries are different.

Despite flaws in some of his theories, however, we must credit René Haüy as one of the founders of crystallography. In later years, he pioneered the application of mathematical concepts to crystallography and established the basis for modern crystallography. Crystallographers still use much of his work today. We now accept that:

- All crystals have basic building blocks called **unit cells.**
- The unit cells are arranged in a pattern described by points in a **lattice.**
- The relative proportions of elements in a unit cell are given by the chemical formula of a mineral.
- Crystals belong to one of six crystal systems. Unit cells of distinct shape and symmetry characterize each crystal system.
- Total crystal symmetry depends on both unit cell symmetry and lattice symmetry.

In Chapter 9 we discussed symmetry due to rotation, reflection, and inversion. These are all types of **point symmetry.** The orderly repetition of patterns due to **translation** is another form of symmetry, called **space symmetry.** Space symmetry differs from point symmetry. Space symmetry repeats something an infinite number of times to fill space, while point symmetry repeats something a discrete number of times and only describes symmetry localized about a

central point. Point symmetry operators return a dot or a crystal face to its original position and orientation after 1, 2, 3, 4, or 6 repeats of the operation, but space symmetry operators do not.

We often envision translational symmetry by thinking about a lattice. Figure 10.4a shows a helicopter repeated by translation in two directions. If a dot replaces each helicopter, the resulting pattern is the lattice, a pattern of dots that repeats indefinitely (Figure 10.4b). The distances between lattice points are the translations that define the lattice. We can define a unit cell as one helicopter and the space around it (Figure 10.4c). The lattice describes how the unit cells fit together to produce the whole pattern.

UNIT CELLS AND LATTICES IN TWO DIMENSIONS

Shapes of Unit Cells in Two Dimensions

What possible shapes can unit cells have? Mineralogists begin answering this by considering only two dimensions. This is much like imagining what shapes we can use to tile a floor. Figure 10.5 shows some possible tile shapes. In the right-hand column of Figure 10.5, a single dot has replaced each tile to show how the tiles repeat.

What happens when gaps occur between unit cells? Figure 10.6 shows the two possibilities: The gaps may be either regularly (Figure 10.6a) or randomly (Figure 10.6b) distributed. If regularly, we can simply redefine the unit cells to include the gap. If randomly, the entire structure is not composed of identical building blocks that fit together in a regular way. It is not repetitive and we cannot describe it with a regular pattern. It does not, therefore, represent a crystal structure, so we need not consider unit cell shapes that do not fit together without gaps between them.

Various complex patterns can appear on tiled floors, but the tiles are usually simple shapes such as squares or rectangles. For example, to tile a floor in the L-shaped pattern shown in Figure 10.7a, we could use L-shaped tiles. However, square or rectangular tiles would do the job as well; Figure 10.7b, c, and d show some examples. In fact, we can make any repetitive two-dimensional pattern, no matter how complicated, with tiles of one of four fundamental shapes: parallelogram, rhomb, rectangle, and square. For reasons we will see later, we usually distinguish two types of rhombs: those with **nonspecial angles** and those with 60° and 120° angles. For the rest of this chapter, we will refer to the general type of rhomb as a *diamond,* and *rhomb* will be reserved

▶**FIGURE 10.4**
Crystal structures must have units that re-peat in an orderly way, just like the heli-copters in this figure: (a) a helicopter repeating in two dimensions; (b) the lattice that describes the repetition; (c) a unit cell.

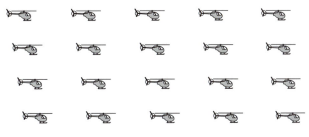

(a) pattern

(b) lattice

(c) unit cell

▶**FIGURE 10.5**
Three possible two-dimensional shapes for tiles. The dot patterns show the way the tiles fit together.

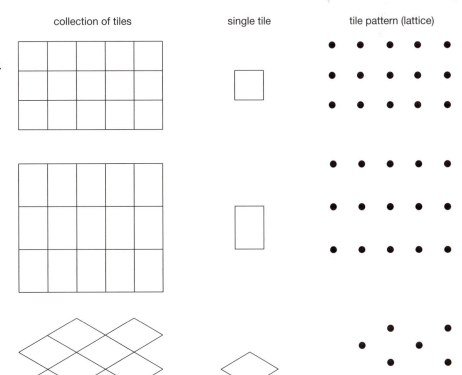

▶**FIGURE 10.6**
What happens if there are gaps in a pattern? (a) We can redefine the unit cell to include the gaps, provided the gaps are distributed in a regular manner. (b) If the gaps are randomly distributed, then there is no unit cell, or pattern, that adequately describes the pattern.

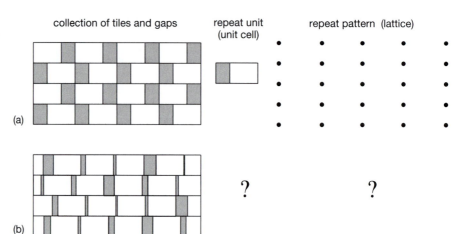

▶**FIGURE 10.7**
A floor pattern made of L-shaped polygons: (a) the overall pattern; (b) a square tile can be used to create the pattern; (c) a rectangular tile can be used to create the pattern; (d) a larger square tile can be used to create the pattern.

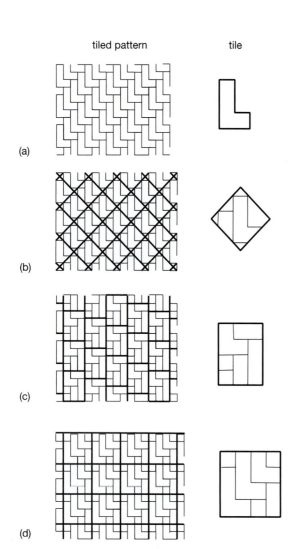

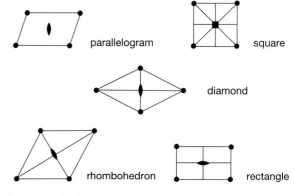

▶**FIGURE 10.8**
The five basic two-dimensional shapes and their symmetry. In this and following figures, lines indicate mirror planes. Lenses, triangles, squares, and hexagons indicate rotation axes.

for shapes with only 60° and 120° angles. Figure 10.8 shows the five basic shapes, the only ones needed to discuss two-dimensional unit cells, and their symmetries. Each has its own unique combination of symmetry elements.

Motifs and Lattices

We call designs that repeat, such as those on wallpaper, *motifs*. The term *motif* is used in an analogous way in music to refer to a sequence of notes that repeat. Interior decorators or clothing designers use the term to describe a common theme or element in their work. We can think of patterns as starting with one motif. The motif is then reproduced by translating (moving) it a certain distance and reprinting it. The distance of the translation is the distance between lattice points. If a dot replaces each motif, we get the lattice.

Figure 10.9 shows a motif (a drawing of a flag) repeated according to a rectangular lattice. When

▶**FIGURE 10.9**
Motifs and unit cells. The unit cell consists of a rectangular space including a flag. It repeats according to the pattern described by the lattice to produce the overall pattern.

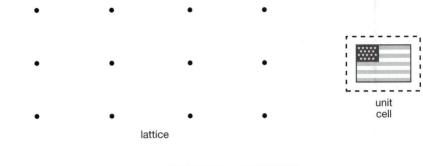

lattice

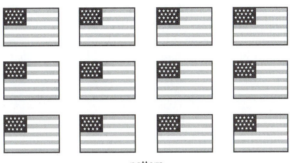

pattern

▶**FIGURE 10.10**
Pattern based on a motif containing two atoms, shown by solid and open circles; a, b, c, and d are possible choices for a unit cell.

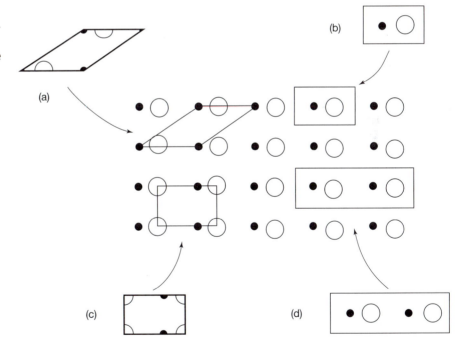

combined with surrounding empty space, motifs form unit cells. The unit cells, which are all rectangular, stack together to form the entire pattern. Unit cells may have any of the shapes in Figure 10.8.

Figure 10.10 shows a pattern made of two atoms, symbolized by solid and open circles. The entire pattern consists of motifs composed of one of each. The figure shows several possible choices of unit cells. By convention, however, we usually choose the smallest unit cell that contains the same symmetry as the en-

tire pattern or structure. Unit cells a, b, and c contain only one motif in total (one solid circle and one open circle). We call them *primitive* because they are the smallest choices possible. All primitive unit cells contain exactly one motif in total. We term unit cell d in Figure 10.10 *doubly primitive* because it contains two motifs. Unit cell b in Figure 10.10 would be the choice of most crystallographers because it is simple, primitive, and contains the same symmetry as the entire structure (Box 10.1).

Box 10.1 Choices of a Unit Cell

In two dimensions, unit cells are parallelograms, including rectangles and squares, that can be fit together to produce an entire pattern. They may be chosen in many ways, although mineralogists follow some standard conventions (see text). Consider the lattice below, called a **clinonet** (Figure 10.11a). Its symmetry is shown in Figure 10.11b: 2-fold axes (perpendicular to the page) located at each lattice point and between lattice points, and two sets of mirror planes at 90° to each other.

Figure 10.11c shows some possible choices of unit cells. Let's count the number of lattice points in the cells. If points are completely inside the cell, they count as one. If points are on a corner, they are a quarter in the unit cell because they are shared between four cells; if they are on a side, they are half in the cell. Some of the unit cells (cells 1 through 5) are **primitive,** containing only one lattice point in total. Cells 6, 7, and 8 are **doubly primitive,** containing two points. Cell 9 contains three points and cell 10 contains six.

The area of a cell depends only on the number of lattice points within it. All the primitive cells, for example, have the same area despite differences in shape and lattice point location. In addition, the doubly primitive unit cells all have exactly twice the area of the primitive unit cells. Two of the unit cells (9 and 10) are **orthogonal,** meaning that the angles at the corners are 90°. Only cell 9 is square.

Unit cells are chosen for convenience. Sometimes crystallographers choose primitive unit cells because they are simple. In this example, however, cell 9, although not primitive, is probably the choice that some crystallographers would make. Because it is square, it is easier to see. Of more importance, it contains the same symmetry elements (2-fold axis and mirror planes at 90° to each other) that the original structure does (Figure 10.11b, d).

▶**FIGURE 10.11**
A lattice, its symmetry, and some choices for a unit cell. (a) The open circles show the lattice. (b) The symmetry of the lattice includes mirror planes (dashed lines) and 2-fold axes (lens shape). (c) Some possible unit cell shapes. (d) The symmetry of cell 9, a square unit cell.

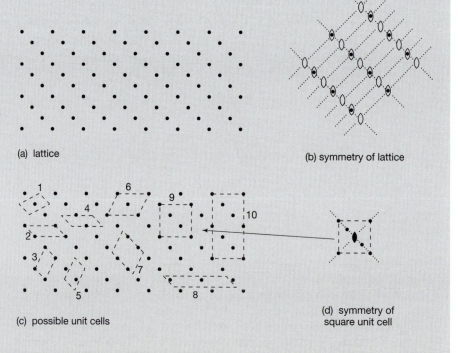

(a) lattice

(b) symmetry of lattice

(c) possible unit cells

(d) symmetry of square unit cell

Well-known Dutch artist M. C. Escher used repetitive patterns to create many complicated drawings. Figure 10.12 shows a pattern in Escher's style. A basic motif, composed of five American icons, repeats to make the entire pattern. For this pattern, we may choose a unit cell that includes one of each of the five figures and white space around them; it may have any of an infinite number of pos-

▶FIGURE 10.12
Five American icons arranged in a repetitive two-dimensional pattern. See text for explanation.

sible shapes. However, for any repetitive pattern, we can find a unit cell with one of the five shapes in Figure 10.8. For Figure 10.12, we get such a cell by drawing lines connecting four equivalent points; for example, the center of W. C. Fields's nose. Depending on the noses we choose, it is a parallelogram or has a diamond shape. If we choose four near-neighbor noses, we get a primitive unit cell; it includes one of each of the figures in total, but some figures are fragmented because they overlap the unit cell edges.

For any structure or lattice, we can always choose a primitive unit cell. Figure 10.13 shows a motif—here a running man—arranged in a diamond pattern. A primitive diamond-shaped unit cell can be chosen (Figure 10.13b, c). However, a nonprimitive rectangular unit cell, containing two running men, can also be chosen (Figure 10.13d, e). The rectangular unit cell emphasizes 90° angular relationships, so most crystallographers would choose a nonprimitive rectangular unit cell. The two rectangular unit cells shown have the same symmetry (no rotation axes or mirror planes), but the one in Figure 10.13d shows the pattern more clearly and is a better choice. Cells with an extra lattice point (motif) in the middle, such as the rectangular one in Figure 10.13d with the extra running man, are termed *centered*.

Unit cells and lattices are not real objects; they are concepts used by humans to help understand the repetitive nature of crystal structures. As emphasized in Figure 10.13, the exact outline of a unit cell is arbitrary. It is chosen for convenience, often to help visualization. Although we may think of lattice points as representing corners of unit cells, they actually do not represent corners or any other specific points within unit cells. Unit cells and lattices are just concepts that provide a convenient way to think about and discuss the repetition of structures and symmetry. Although crystallographers follow standard conventions, they are often forced to make choices. If they choose primitive unit cells, each lattice point corresponds to one unit cell and one motif. If they choose nonprimitive unit cells, lattices represent the way motifs repeat, but not the way unit cells repeat, because there is more than one lattice point per unit cell.

Figure 10.14 shows the five two-dimensional lattices, called the five **plane lattices** (corresponding to the five basic unit cell shapes) and their symmetries. Crystallographers call these lattices the *clinonet, diamond net, hexanet, orthonet,* and *square net*. Each corresponds to one of the possible unit cell shapes in Figure 10.8. To emphasize that the choice of a unit cell is arbitrary, note again that a

(a) pattern:

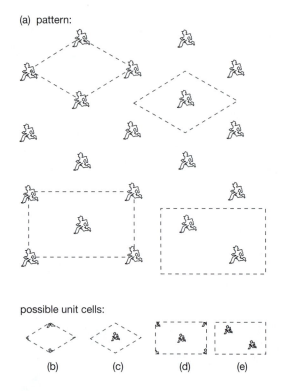

possible unit cells:

(b)   (c)   (d)   (e)

▶**FIGURE 10.13**
(a) A running man motif repeating in a diamond pattern. Figures (b) and (c) are possible choices for diamond shaped unit cells; (d) and (e) are possible choices for rectangular unit cells.

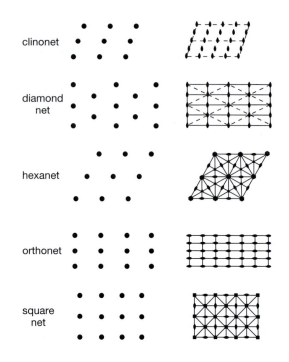

clinonet

diamond net

hexanet

orthonet

square net

▶**FIGURE 10.14**
The five plane lattices (nets) and their symmetries. Symmetry elements are shown using standard symbols.

centered rectangular unit cell (doubly primitive) can be chosen instead of a primitive diamond-shaped unit cell (Figure 10.15a), and a six-sided nonprimitive hexagonal unit cell can be chosen to replace a rhomb to emphasize the 6-fold symmetry (Figure 10.15b).

We can describe translational symmetry and lattices using vectors, and we can explain many fundamental properties of lattices using vector properties (Box 10.2). The vectors that describe the translations relating one lattice point to its near neighbors have magnitudes and orientations equivalent to edges of primitive unit cells. The five basic unit cell shapes represent all possible combinations of two vectors if the absolute magnitude of the vectors is ignored.

Figure 10.16 shows some patterns based on the flags of Canadian Provinces that have translational symmetry. They also contain rotational symmetry, mirror planes, and inversion centers, the same way plane lattices do. The same translations that relate one unit cell to its neighbors also relate the symmetry elements. In the previous chapter, we noted that rotation axes and mirror planes operate on each other; now we see that *translational symmetry also affects rotation axes and mirror planes.*

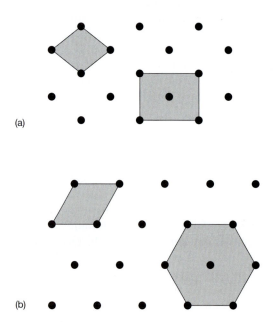

(a)

(b)

▶**FIGURE 10.15**
(a) Diamond-shaped and rectangular unit cells can both produce the same pattern but the rectangular cell is not primitive. (b) Both rhombohedral and hexagonal unit cells can create a hexanet but the hexagonal cell is not primitive.

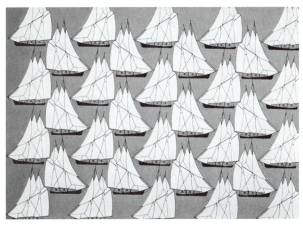

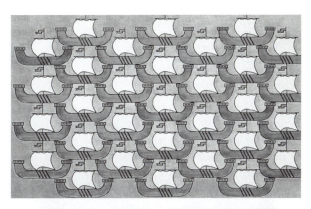

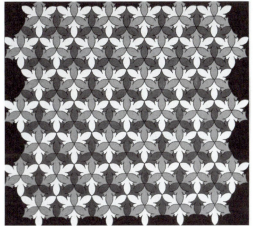

▶**FIGURE 10.16**
Flags of the Canadian Provinces. Modified from F. Brisse, La symétrie bidimensionelle et le Canada, *Canadian Mineralogist,* 19 (1981). Used by permission.

Symmetry of the Motif and the Lattice

Let's consider the motif in Figure 10.18a, in which the solid and open circles represent different atoms. The motif has a 4-fold axis of symmetry and four mirror planes at 45° to each other. In Figure 10.18b, the motif repeats according to a square net; in Figure 10.18c it repeats following an orthonet. The overall pattern formed with the square net still has 4-fold axes of symmetry and mirror planes just as the original motif does, while the pattern formed with the orthonet does not. A combination of a 4-fold rotation axis with an orthonet is an impossible combination for atomic structures. We assumed that the four open circles were identical atoms with identical bonds around them (or else 4-fold symmetry would not be present). In our final structure, therefore, the open circles must have identical surroundings. They do not with the orthonet. If the open circles all represent absolutely identical atoms, how can they pack closer together in one direction than in the other? The answer is that they cannot.

The preceding discussion points out a second important law of crystallography: *If a motif has certain symmetry, the lattice must have at least that much symmetry.* A motif with the 4-fold axis of symmetry requires a square plane lattice because it is the only plane lattice with a 4-fold axis. A motif may have less symmetry than a lattice. If a motif has a 2-fold axis of symmetry, it may be repeated according to any of the five plane lattices because they all have 2-fold axes of symmetry. Halite, NaCl, is an excellent example of a mineral in which the basic motif has less symmetry than the lattice (Figure 10.19). In two dimensions the lattice has square symmetry, but the motif, consisting of one Na and one Cl atom, does not. Notice, however, that we can choose a nonprimitive unit cell containing two motifs that does have square symmetry.

Unit Cells and Lattices in Three Dimensions

What Shapes Are Possible?

In the preceding discussion, we established that two-dimensional unit cells must have one of five shapes. What shapes are possible in *three* dimensions? We can

Box 10.2 Lattices and Vectors in Two Dimensions

The points in a lattice can be generated by starting with one point and defining two translations, described by vectors $\vec{t}_a$ and $\vec{t}_b$, which translate any point in a lattice to a new location (Figure 10.16a). The relationship between $\vec{t}_a$ and $\vec{t}_b$ determines the kind of lattice (Figure 10.17a–e).

Magnitude of $\vec{t}_a$ and $\vec{t}_b$	Angle Between $\vec{t}_a$ and $\vec{t}_b$	Unit Cell Shape	Plane Lattice
$t_a \neq t_b$	general	parallelogram	clinonet
$t_a \neq t_b$	90°	rectangle	orthonet
$t_a = t_b$	general	diamond	diamond net
$t_a = t_b$	120°	rhomb	hexanet
$t_a = t_b$	90°	square	square net

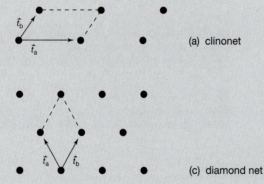

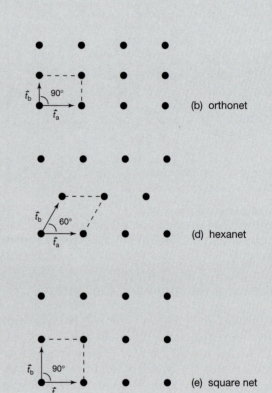

▶FIGURE 10.17
Vectors and their relationships for each of the five plane lattices.

answer this question by considering the shapes of bricks that we can stack together without leaving gaps. To begin with, the bottoms of the bricks will have one of the five two-dimensional shapes in Figure 10.8. Let's consider possible brick shapes with a square bottom (Figure 10.20). The first two, a cubic unit cell and a tetragonal unit cell, are orthogonal shapes: All angles are 90°, all faces are either squares or rectangles. The symmetries of these two shapes correspond to two of the 32 point groups: $4/m\bar{3}2/m$ and $4/m2/m2/m$, respectively. The next shape has two faces that are not square or rectangular (Figure 10.20c). At each corner,

three edges come together and form three angles. Two of the angles are 90°; one is not. This unit cell is monoclinic and has $2/m$ symmetry.

Instead of imagining a brick with a square base, we could imagine other shapes. Figure 10.21 shows some unit cells with rectangular bases. Two are simply repeats of those we derived using the square base. An important new one is the orthorhombic unit cell (Figure 10.21b), shaped like an orthorhombic prism (shoebox shape), which has symmetry $2/m2/m2/m$. Figure 10.22a shows a rhombic prism, a brick having a rhomb-shaped base. Three rhombic prisms placed to-

▶**FIGURE 10.18**
Symmetry of lattices, motifs, and patterns: (a) a motif of five atoms and its symmetry (4-fold axis and four mirrors); (b) the motif repeated according to a square net to produce a square pattern; (c) the motif repeating according to an orthonet to produce a rectangular pattern.

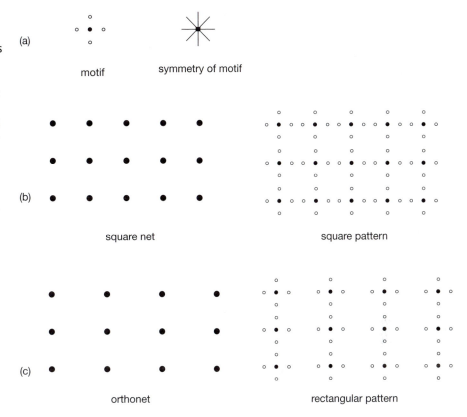

gether create a hexagonal unit cell, which has symmetry $6/m2/m2/m$ (Figure 10.22b). Finally, Figure 10.23 shows a triclinic unit cell, which is based on a parallelogram and has no edges of equal length and no angles of 90°.

Although we could use other shapes as bricks, they are equivalent to the ones just mentioned, singly or in combination. There are only six distinct possibilities, having six distinct symmetries (Figure 10.24). Each corresponds to one of the six crystal systems introduced in the last chapter (Table 10.1). This brings us to a third important law of crystallography: *The symmetries of the unit cells are the same as the point groups of greatest symmetry in each of the crystal systems.* To describe the shape of a unit cell, we give dimensions and angles. The lengths of unit cell edges and the angles between them, together called the *unit cell parameters*, are usually designated as a, b, c, α, β, and γ (Figure 10.23). For primitive unit cells, a, b, and c are the magnitudes of the vectors relating lattice points, and α, β, and γ are the angles between vectors.

In two dimensions, we observed that a plane lattice describes the way floor tiles or unit cells pack. In three dimensions, a space lattice describes the way bricks or unit cells can be packed together. We can envision space lattices as identical points that repeat indefinitely in three-dimensional space. The possible space lattices are named after the possible unit cell shapes: cubic, hexagonal, tetragonal, orthorhombic, monoclinic, and triclinic. Each has unique symmetry that is equivalent to the symmetry of the unit cell. One of the six lattices characterizes any crystal.

Figure 10.25, patterned after an Escher print, contains a motif (an egg) repeating in three directions. A primitive unit cell might include one egg and some space around it. An easier way to envisage a primitive unit cell is to consider a rectangular solid with an egg at each corner and none inside. Each egg is then shared between eight unit cells. The four largest eggs form the front face of such a solid. If we replace each egg with a dot, we get a space lattice; it shows the way unit cells repeat to give the entire three-dimensional pattern. Notice that we can think of the space lattice as plane lattices stacked on top of each other. The distance between adjacent planes is constant.

We based the preceding discussion mostly on geometry; we can also use vectors and vector properties to derive three-dimensional unit cells (Box 10.3). We need three vectors to describe the way lattice points repeat. Essentially, the first two vectors describe the way lattice points repeat to make plane lattices, while the third describes how plane lattices stack one above another to produce a space lattice.

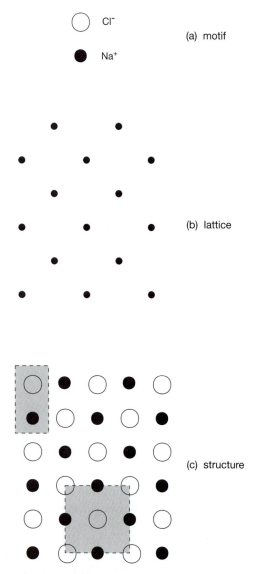

(a) motif

(b) lattice

(c) structure

▶**FIGURE 10.19**
Halite. (a) The basic motif contains one Cl and one Na atom.
(b) The motif repeats according to a square net (tilted here).
(c) The structure has the same symmetry as the lattice,
which is more than the symmetry of the motif. A primitive
unit cell does not show this symmetry, but a doubly primitive square unit cell does.

Cells With Extra Lattice Points

As pointed out earlier, we can think of three-dimensional
lattices (space lattices) as layers of two-dimensional
lattices (plane lattices) stacked on top of each other
in a regular and evenly spaced way. Figure 10.14
shows the symmetry elements of each of the plane
lattices. When we stack plane lattices, unless we take
care, symmetry elements in one layer will not correspond with those in another. The resulting space lattice will have no rotation axes or mirror planes of
symmetry.

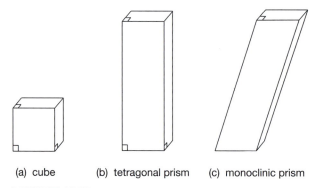

(a) cube   (b) tetragonal prism   (c) monoclinic prism

▶**FIGURE 10.20**
Bricks with square bottoms include: (a) cube; (b) tetragonal
prism; (c) monoclinic prism.

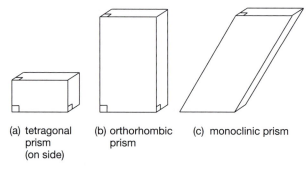

(a) tetragonal   (b) orthorhombic   (c) monoclinic prism
    prism   prism
    (on side)

▶**FIGURE 10.21**
Bricks with rectangular bottoms: (a) tetragonal prism (on its
side); (b) orthorhombic prism; (c) monoclinic prism.

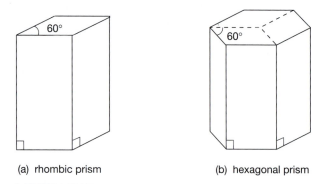

(a) rhombic prism   (b) hexagonal prism

▶**FIGURE 10.22**
Bricks having a rhomb-shaped base: (a) rhombic prism;
(b) hexagonal prism.

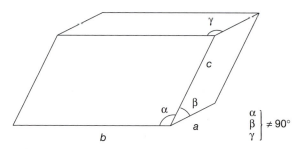

▶**FIGURE 10.23**
A triclinic unit cell with edge lengths and angles a, b, c,
α, β, and γ. All faces and edges intersect at non-90° angles.

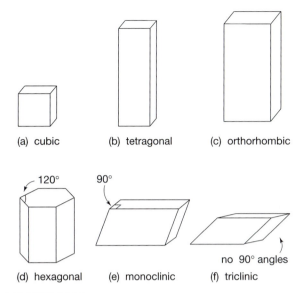

(a) cubic   (b) tetragonal   (c) orthorhombic

120°   90°

no 90° angles

(d) hexagonal   (e) monoclinic   (f) triclinic

▶**FIGURE 10.24**
The six basic three-dimensional unit cell shapes.

▶**TABLE 10.1**
Crystal Systems and Unit Cell Symmetry

Unit Cell Shape	Crystal System	Unit Cell Symmetry
cubic	cubic (isometric)	$4/m\bar{3}2/m$
tetragonal	tetragonal	$4/m2/m2/m$
hexagonal	hexagonal	$6/m2/m2/m$
orthorhombic	orthorhombic	$2/m2/m2/m$
monoclinic	monoclinic	$2/m$
triclinic	triclinic	$\bar{1}$

▶**FIGURE 10.25**
A motif (here an egg) repeating in space. If each egg is replaced with a point, we get a space lattice.

Stacking square plane lattices to produce space lattices can be done in many ways. If one plane lattice is placed randomly above another, we may preserve no symmetry. Alternatively, we may stack the layers exactly one above another to preserve maximum symmetry. If the distance between layers is the same as the distance between points in a layer, designated a, the three-dimensional lattice has a cubic arrangement of lattice points (Figure 10.27a). Each cubic unit cell contains one $(8 \times \frac{1}{8})$ lattice point in total. We designate unit cells containing only one lattice point with the letter P standing for primitive.

Figure 10.27b shows another possible way of stacking square plane lattices. As with the previous example, the distance between adjacent lattice points in a layer is a. The distance between adjacent layers is, however, $\frac{1}{2}a$. Each layer is offset from the ones above and below it. Lattice points in the first, third, fifth (and so on) layers are directly above each other, and create a cubic arrangement of lattice points. Lattice points in the second, fourth, sixth (and so on) layers lie exactly in the center of the cubes. We can think

of the cubes as nonprimitive unit cells. They are doubly primitive, having an extra lattice point at their center. We call them *body-centered* and designate them with the letter I.

We can stack diamond lattices to produce an orthorhombic unit cell (all angles 90°, but sides of different lengths) with extra lattice points in two opposite faces (Figure 10.28a). The resulting cells are called *end-centered* and designated by the letter A, B, or C, depending on which pair of faces contains the extra lattice points. Alternatively, we may stack the diamond lattices to produce a unit cell with extra lattice points in the center of each face (Figure 10.28b). We call such unit cells *face-centered*, and designate them by the letter F.

Stacking hexagonal plane lattices is more complicated than stacking other lattices because hexagonal plane lattices have more symmetry. Figure 10.29a shows the symmetry of a hexagonal plane lattice. If we stack plane lattices so that 6-fold axes in all the layers coincide, we produce a simple hexagonal space lattice. The three-dimensional result is a unit cell shaped like a rhombic prism (Figure 10.29b) or an equivalent hexagonal unit cell (Figure 10.29c).

We may also stack hexagonal plane lattices so they do not preserve the 6-fold axes of symmetry but do preserve the 3-fold axes. Figure 10.30a shows how we do this. The second layer is placed above the first so that its lattice points lie above the center of equilateral triangles in the first layer. The third layer is similarly

Box 10.3 Vectors and Space Lattices

We can generate the six possible space lattices using vectors in the same way we generated the plane lattices. Consider three vectors, $\vec{t}_a$, $\vec{t}_b$, and $\vec{t}_c$, which cause a lattice point to translate and repeat in space. The relationships between $\vec{t}_a$, $\vec{t}_b$, and $\vec{t}_c$ determine the type of lattice. The three vectors may be equal or different in magnitude, and the angles between them may be general or special. The possible combinations are summarized in the table below. Figure 10.26 depicts unit cells and vector relationships for each.

Magnitudes of $\vec{t}_a$, $\vec{t}_b$, and $\vec{t}_c$	Angles between $\vec{t}_a$, $\vec{t}_b$, and $\vec{t}_c$	Unit Cell Shape	Lattice
$\vec{t}_a \neq \vec{t}_b \neq \vec{t}_c$	all angles general	triclinic	triclinic
$\vec{t}_a \neq \vec{t}_b \neq \vec{t}_c$	one angle general, the others 90°	monoclinic	monoclinic
$\vec{t}_a = \vec{t}_b \neq \vec{t}_c$	$\vec{t}_a \wedge \vec{t}_b = 120°$ $\vec{t}_a \wedge \vec{t}_c = 90°$ $\vec{t}_b \wedge \vec{t}_c = 90°$	hexagonal prism	hexagonal
$\vec{t}_a \neq \vec{t}_b \neq \vec{t}_c$	all angles 90°	orthorhombic prism	orthorhombic
$\vec{t}_a = \vec{t}_b \neq \vec{t}_c$	all angles 90°	tetragonal prism	tetragonal
$\vec{t}_a = \vec{t}_b = \vec{t}_c$	all angles 90°	cube	cubic

▶**FIGURE 10.26**
The six unit cell shapes corresponding to the six space lattices, with vector relationships shown. *a*, *b*, and *c* are magnitudes of vectors relating lattice points at the unit cell corners. *α*, *β*, and *γ* are the angles between the vectors.

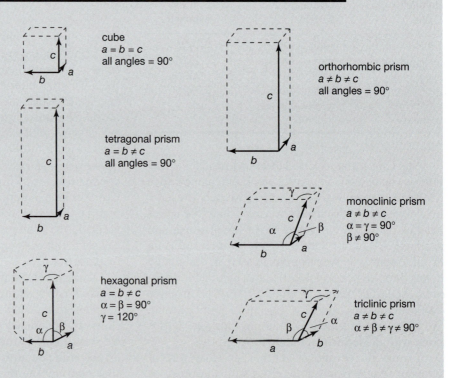

placed above the second, and the fourth layer lies directly above the first. We give this arrangement of lattice points the symbol *R*, for rhombohedral. We can connect lattice points to make a nonprimitive unit cell, shaped like a rhombic prism, with its base in the first layer and its top in the fourth layer. Lattice points are at the corners and two extra lattice points are inside the prism (Figure 10.30b). Another way of choosing a unit cell is to isolate eight lattice points from several adjacent rhombohedral prisms and connect them as

►**FIGURE 10.27**
Stacking square plane lattices to produce three-dimensional lattices: (a) primitive cubic lattice *(P)*; (b) body-centered cubic lattice *(I)*.

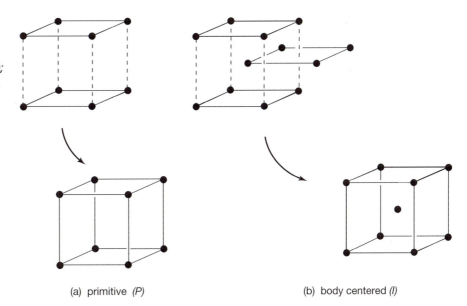

(a) primitive *(P)*   (b) body centered *(I)*

shown in Figure 10.30d, e. One point is from the first layer, three from the second, three from the third, and one from the fourth. The result is a primitive rhombohedral unit cell. It has 12 equal-length edges, and eight identically shaped faces. Lattice points are only at the corners of this unit cell.

Figure 10.31 shows several possible choices for the unit cell of calcite. If we replace each $CaCO_3$ unit (equivalent to a motif) with a point, we get a rhombohedral lattice. We can choose a unit cell in several different ways. The unit cell in Figure 10.31a is doubly primitive. We could also choose a hexagonal prism as a unit cell (Figure 10.31b). A third possibility is a face-centered rhomb (Figure 10.31c). Lattice points are at the center of each face and at the corners. Although the three possible unit cells look different, they all give the atomic structure of calcite when many unit cells are stacked together. One of the first two is the choice of most mineralogists. The third one is equivalent in shape to the standard cleavage rhombohedron that forms when calcite crystals break. Figure 10.32a shows a photo of a ball and stick model equivalent to the cleavage rhombohedron (Figure 10.32b).

BRAVAIS LATTICES

When all possibilities have been examined, we can show that only 14 distinctly different space lattices exist. We call them the 14 **Bravais lattices,** named after Auguste Bravais, a French scientist who was the first to show that there were only 14 possibilities. The 14 Bravais lattices correspond to 14 different unit cells that may have any of six different symmetries, depending on

the crystal system. Six of the possible unit cells are primitive, one in each system (Figure 10.26). For convenience, we may think of them as having lattice points only at the corners, but it is important to remember

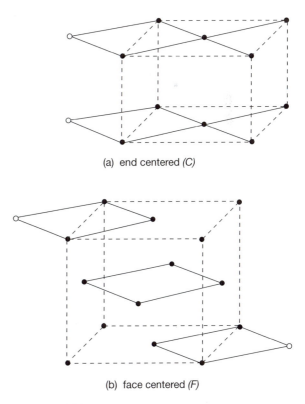

(a) end centered *(C)*

(b) face centered *(F)*

►**FIGURE 10.28**
Stacking diamond plane lattices to produce three-dimensional orthorhombic unit cells. (a) If the three-dimensional cell has extra points in two opposite faces, we designate it by *A, B,* or *C* (see text). (b) If the three-dimensional cell has extra points in the center of each face, we designate it *F*.

▶**FIGURE 10.29**
Stacking hexagonal plane lattices: (a) the symmetry of a rhomb and hexagon in the plane lattice; lattice points coincide with 6-fold rotation axes; (b) a primitive unit cell, shaped like a rhombohedral prism, that describes a hexagonal space lattice; (c) a nonprimitive unit cell, shaped like a hexagonal prism, that describes a hexagonal space lattice.

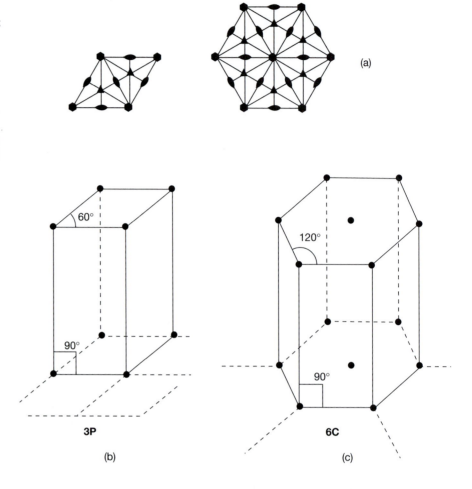

that the choice of unit cells and lattice points is arbitrary. Eight unit cells share each of the eight lattice points at the corners of a primitive unit cell, so the total number of lattice points per cell is one. The other eight Bravais lattices involve nonprimitive unit cells containing two, three, or four lattice points. Body-centered unit cells, for example, contain one extra lattice point at their center (Figure 10.27b). Face-centered unit cells have lattice points in the centers of six faces (Figure 10.28b). Each of the six points is half in the unit cell and half in an adjacent cell. The total number of lattice points is therefore 1 (at the corners) + 3 (in the faces) = 4. Table 10.2 lists the different types of space lattices and their symbols. Note that all types do not exist in all systems. For example, end-centered unit cells only exist in the orthorhombic system; they cannot exist in the other systems because of symmetry constraints.

As pointed out in the discussion of two-dimensional lattices, we sometimes make arbitrary decisions when choosing unit cells. The 14 Bravais lattices, in fact, do not represent the only 14 that we could list. For example, in the monoclinic system, the body-centered (*2I*) unit cell is equivalent to a face-centered one (*2F*), with different dimensions. Most

crystallographers consider only the body-centered cell because that is the way it has been done since the time of Bravais. The important thing to realize is that no matter what unit cells and lattices we consider, only 14 are distinct. Furthermore, the 14 in Table 10.2 are the simplest and are the ones used by most crystallographers.

UNIT CELL SYMMETRY AND CRYSTAL SYMMETRY

Consider fluorite (CaF_2), spinel ($MgAl_2O_4$), and the garnet almandine ($Fe_3Al_2Si_3O_{12}$). Fluorite and spinel have face-centered cubic unit cells; almandine has a body-centered cubic unit cell. In fluorite, Ca^{2+} is found at the corners of the unit cell and at the center of each face, while F^- occupies sites completely within the unit cell (Figure 10.33). Spinel and almandine have more complex structures, in part because they contain more than one cation.

When many unit cells stack together, we get a crystal that may or may not have the same symmetry as a single unit cell. In fluorite, spinel, and almandine

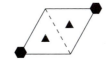

▶**FIGURE 10.30**
An alternative way to stack hexagonal plane lattices. (a) Rhomb showing locations of some 3-fold and 6-fold axes. (b) Stacking rhombs to preserve 3-fold symmetry, but not 6-fold symmetry (different layers in the stack are indicated with numbers). (c) The resulting nonprimitive three-dimensional cell has two extra lattice points within it. (d) Eight points that can be used to define an alternative three-dimensional unit cell (numbers indicate layers). (e) The result is a primitive rhombohedral unit cell.

(a) rhomb and 3-fold axes

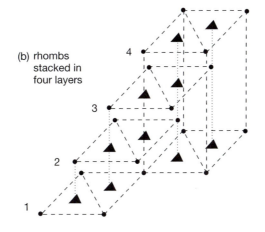

(b) rhombs stacked in four layers

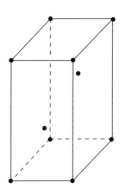

(c) nonprimitive rhombohedral unit cell

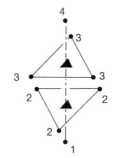

(d) lattice points needed to make a rhombohedron

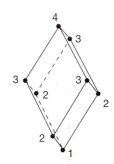

(e) primitive rhombohedral unit cell

▶**FIGURE 10.31**
The atomic structure of calcite, CaCO₃, and three choices for a unit cell. Symbols show location of Ca atoms and CO₃ radicals: (a) doubly primitive rhombohedron; (b) nonprimitive hexagonal prism; (c) nonprimitive face-centered rhombohedron. The letters a, b, c, and d show equivalent atoms in all three drawings.

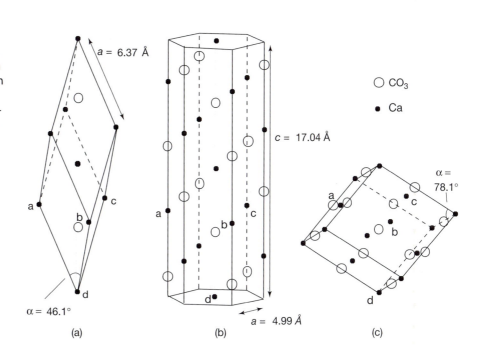

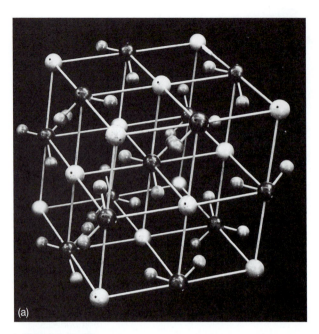

(a)

▶**FIGURE 10.32**
Ball and stick model of calcite: (a) This model does not depict the standard unit cell for calcite (see text), but does show the atomic arrangement that leads to calcite's excellent rhomboheral cleavage. (b) Calcite cleavage rhombs.

(b)

crystals, the unit cells stack together so that all symmetry elements are preserved. However, fluorite crystals are typically cubes, spinel crystals are typically octahedra, and almandine crystals are typically dodecahedra (Figure 10.34a, b, and c). Other forms are possible for these minerals, especially for almandine.

A fourth important law of crystallography is: *If a crystal has certain symmetry, the unit cell must have at least as much symmetry.* A corollary to this law is that because crystals consist of unit cells, the symmetry of a crystal can never be more than that of its unit cell. If a crystal has a 4-fold axis of symmetry, it must have a unit cell that includes a 4-fold axis of symmetry. A mineral that forms cubic crystals must have a cubic unit cell. If a crystal has certain symmetry, it is 100% certain that the unit cell and crystal's atomic structure have at least that much symmetry. They may have more.

It is important to remember that the symmetry of a crystal depends not only on the symmetry of the unit cell, but also on how the unit cells combine to make the crystal. Some minerals, such as halite and other salts, tend to develop euhedral crystals with obvious symmetry, making it easy to infer the symmetry of the unit cell. Others develop crystals with faces that are nearly identical, suggesting the presence of symmetry but leaving some uncertainty. While size and shape of faces can vary because of accidents of crystal growth, the angles between faces

vary little from the ideal (Steno's law). Consequently, crystallographers often rely on angles instead of face shapes to infer the symmetry of the unit cell. Even if a crystal is poorly formed, its internal structure is orderly and its unit cells all have the same atomic arrangement.

Some minerals appear to have more symmetry than they actually do. Euhedral biotite often appears to have hexagonal symmetry when, in fact, close examination reveals that it does not (Figure 10.35). This is a good example of **pseudosymmetry.** Biotite is a monoclinic mineral, but is so close to being hexagonal that we call it **pseudohexagonal.** Sometimes we need precise measurements, and sometimes even X-ray studies, to tell pseudosymmetry from real symmetry.

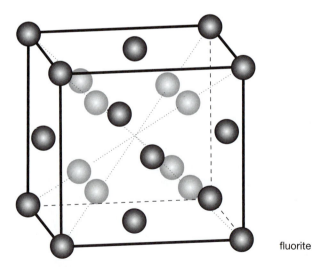

fluorite

▶**FIGURE 10.33**
The structure of fluorite showing the locations of atoms. Darker spheres are Ca; lighter spheres are F. The 8 F atoms are entirely within the unit cell. The Ca atoms are at the corners and in the centers of the faces.

▶TABLE 10.2
Space Lattices

Name	Symbol	Crystal Systems	Number of Lattice Points per Cell	Unit Cell and Lattice Points
primitive	23P	cubic	1	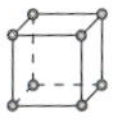
	4P	tetragonal	1	
	222P	orthorhombic	1	
	6P or 6C	hexagonal	1 or 3	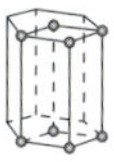
	2P	monoclinic	1	
	1P	triclinic	1	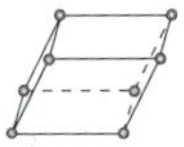
body-centered	23I	cubic	2	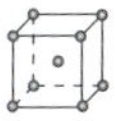
	4I	tetragonal	2	
	222I	orthorhombic	2	
	2I	monoclinic	2	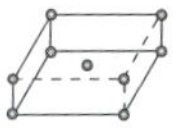
face-centered	23F	cubic	4	
	222F	orthorhombic	4	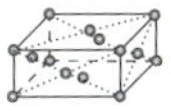
end-centered	222A or 222B or 222C	orthorhombic	2	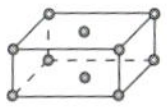
rhombohedral	3R	hexagonal	1	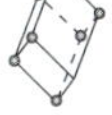

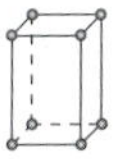

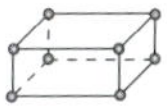

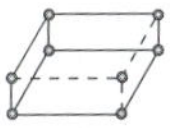

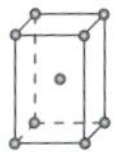

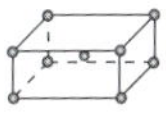

POINT GROUPS AND CRYSTAL SYSTEMS

Although we often infer unit cell shape from crystal symmetry, inferences do not work the other way. Minerals with cubic unit cells might not form crystals with cubic shape. Halite, spinel, and garnet all have cubic unit cells, but garnet and spinel do not normally grow as cubes (Figure 10.36). Garnet forms are **dodecahedrons, trapezohedrons,** and, rarely, **hexoctahedrons.** Spinel typically forms octahedra. Halite, fluorite, and garnet crystals all have $4/m\bar{3}2/m$ symmetry, the same symmetry as their unit cells. Some minerals with cubic unit cells form crystals with less symmetry. Sphalerite, for example, has a face-centered cubic unit cell, but sphalerite crystals have symmetry $\bar{4}3m$; the most common form is a tetrahedron. Sphalerite has less symmetry than a cube because the cubic unit cells do not stack in an arrangement that preserves a 4-fold axis of symmetry. Instead, a $\bar{4}$ axis is preserved.

Crystals with cubic unit cells may, in fact, belong to any of the five point groups associated with the cubic system:

$4/m\bar{3}2/m$
432
$\bar{4}3m$
$2/m\bar{3}$
23

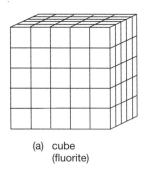

(a) cube
(fluorite)

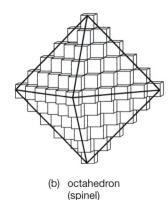

(b) octahedron
(spinel)

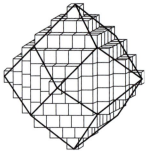

(c) dodecahedron
(garnet)

▶**FIGURE 10.34**
Cubic unit cells stacking together to make crystals. Cubic unit cells can produce crystals of many shapes. (a) Typical fluorite crystals are cubes. (b) Typical spinel crystals are octahedrons. (c) Typical garnet crystals are dodecahedrons.

▶**FIGURE 10.35**
Biotite crystals often appear to have trigonal or hexagonal symmetry but in fact do not because β, the angle between the *a* and *c* edges of the unit cell, is not 90°. (a) Photograph of a pseudorhombohedral biotite crystal. (b) Drawing of a pseudorhombohedral biotite crystal. (c) Drawing of a pseudohexagonal biotite crystal. (d) Drawing of a monoclinic biotite unit cell. Many unit cells combine to produce the crystals in a, b, and c.

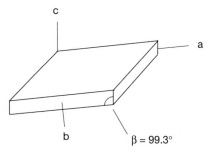

(b) pseudorhombohedral crystal

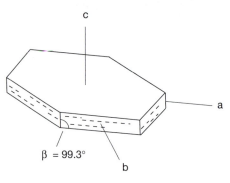

(c) pseudohexagonal crystal

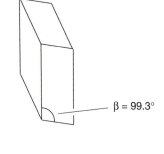

(d) monoclinic unit cell

Similarly, crystals with orthorhombic unit cells may belong to any of three point groups of the orthorhombic system:

$2/m2/m2/m$
222
$mm2$

We can now explain why we divided the 32 point groups into six crystal systems in Chapter 9. Crystals in the same crystal system all have the same-shaped unit cells, even if the crystals themselves do not look the same. The six crystal systems are defined by symmetry and named by their unit cell shape. By examining a crystal's morphology, we can often determine the point group, system, and unit cell shape. Determining whether a unit cell is primitive, face-centered, body-centered, or end-centered, however, is not possible without X-ray studies.

Table 10.3 gives a complete list of the crystal systems, point groups, and lattices. In this table we

►**FIGURE 10.36**
Cubic minerals: (a) Spinel (dark mineral, left) and halite (right); (b) garnet. The black spinel crystals are octahedra; the halite is a cube distorted by cleavage; the garnets are dodecahedra. Compare these photos to the drawings in Figure 10.34.

have divided the hexagonal system into two subdivisions: hexagonal and trigonal. This is done for convenience to distinguish point groups that contain $\bar{6}$-axes or 6-fold axes from those that do not. Some crystallographers use the term *rhombohedral* to refer to trigonal point groups that contain rhombohedral forms ($\bar{3}$ and $\bar{3}2/m$). Note that in all crystal systems the lattice symmetry is equivalent to the point group with greatest symmetry. An unfortunate and confusing shorthand notation is used to designate lattice types. The symbol 23 denotes lattices corresponding to the cubic system; 23 is used because it is the Hermann-Mauguin symbol for the cubic point group with least symmetry. Cubic lattices are therefore designated 23P, 23I, or 23F, de-

pending on whether they are primitive, body centered, or face centered. They all, however, have symmetry $4/m\bar{3}2/m$. Similar symbols are used in the other systems. Although not noted in Hermann-Mauguin symbols, all lattices (and therefore all unit cells) have an inversion center of symmetry. Point groups need not have an inversion center of symmetry. Table 10.3 lists the number and kinds of symmetry elements of each point group; only 11 contain an inversion center. Some symmetry operators are implied by the presence of others: Point groups containing $\bar{6}$ symmetry must also have 3-fold symmetry perpendicular to a mirror; point groups containing $\bar{4}$ symmetry must also have 2-fold symmetry; point groups containing $\bar{3}$ symmetry must have 3-fold

▶ **TABLE 10.3**
The Relationships Between Crystal Systems, Point Groups, Total Symmetry of Point Groups, Unit Cells, and Lattices

In the "Total Symmetry of Point Group" block, columns 6, 4, 3, $\bar{6}$, $\bar{4}$, $\bar{3}$, and 2 give the *Number of Rotation Axes*; column m gives the *Number of Mirrors*; column i indicates an *Inversion Center?*

System	Point Group	6	4	3	$\bar{6}$	$\bar{4}$	$\bar{3}$	2	m	i	Unit Cell Symmetry	Unit Cell Parameters	Lattice Symmetry	Possible Lattices
cubic	$4/m\bar{3}2/m$		3	(4)			4	6	9	i	$4/m\bar{3}2/m$	$a=b=c$ $\alpha=\beta=\gamma=90°$	$4/m\bar{3}2/m$	$23P$
	432		3	4				6						$23I$
	$\bar{4}3m$			4		3		(6)	6					$23F$
	$2/m\bar{3}$			(4)			4	3	3	i				
	23			4				3						
tetragonal	$4/m2/m2/m$		1					4	5	i	$4/m2/m2/m$	$a=b\neq c$ $\alpha=\beta=\gamma=90°$	$4/m2/m2/m$	$4P$
	422		1					4						$4I$
	$4mm$		1					(1)	4					
	$\bar{4}2m$					1		2 (3)	2					
	$4/m$		1					(1)	1	i				
	$\bar{4}$					1		(1)						
	4		1					(1)						
hexagonal (hexagonal)	$6/m2/m2/m$	1						6	7	i	$6/m2/m2/m$	$a=b\neq c$ $\alpha=\beta=90°$ $\gamma=120°$	$6/m2/m2/m$	$6P$ or $6C$
	622	1						6						
	$6mm$	1						(1)	6					
	$\bar{6}2m$				1			3	3 (4)					
	$6/m$	1						(1)	1	i				
	6	1						(1)						
	$\bar{6}$				1				1					
hexagonal (trigonal)	$\bar{3}2/m$			(1)			1	3	3	(i)	$\bar{3}2/m$	$a=b\neq c$ $\alpha=\beta=90°$ $\gamma=120°$		$3R$
	32			1				3						
	$3m$			1					3					
	$\bar{3}$			(1)			1			(i)				
	3			1										
orthorhombic	$2/m2/m2/m$							3	3	i	$2/m2/m2/m$	$a\neq b\neq c$ $\alpha=\beta=\gamma=90°$	$2/m2/m2/m$	$222P$
	222							3						$222I$
	$mm2$							1	2					$222F$ $222A/B/C$
monoclinic	$2/m$							1	1	i	$2/m$	$a\neq b\neq c$ $\alpha=\gamma=90°\neq\beta$	$2/m$	$2P$
	2							1						$2I$
	m								1					
triclinic	$\bar{1}$									i	$\bar{1}$	$a\neq b\neq c$ $\alpha\neq\beta\neq\gamma$	$\bar{1}$	$1P$
	1													

symmetry and an inversion center. In Table 10.3, the numbers in parentheses include symmetry implied by the presence of other operators, in effect counting some symmetry elements twice.

SYMMETRY OF THREE-DIMENSIONAL CRYSTAL STRUCTURES

In the preceding sections, we discussed the symmetry of crystals. We now turn our attention briefly to the symmetry of atomic structures, also called *space symmetry.* What possible symmetry can a three-dimensional atomic structure have? In Chapter 9 we showed that there are only 32 possible point groups. The 32 represent the only possible symmetries for a motif or any discrete object, such as a crystal or a collection of atoms. In this chapter we introduced translation as an additional symmetry element and discussed the 14 possible space lattices that extend motifs through three-dimensional space. An atomic structure, then, consists of groups of atoms with one of 32 symmetries being repeated an indefinite number of times according to one of 14 space lattices. Thus, the symmetry of a crystal structure depends on both the arrangement of atoms in a motif and the lattice type.

Space Group Operators

To describe space symmetry, we must consider two kinds of symmetry operators not previously discussed. Figure 10.37a shows a unit cell containing three atoms, and Figure 10.37b shows an entire line of equivalent unit cells. It is apparent from Figure 10.37b that some sort of symmetry exists that was not apparent when looking at the single unit cell in Figure 10.37a. The symmetry depicted in Figure 10.37b is an example of a **glide plane;** the dashed line shows the location of the glide plane in two dimensions. The sailing ship pattern in Figure 10.16 contains an obvious but more complicated example. Glide plane operations involve combinations of translation plus reflection. Figure 10.37d shows such a combination in three dimensions; a triangular group of atoms repeats, zigzagging back and forth from one side of the glide plane to the other.

Glide planes are one of the two important kinds of **space group operators,** symmetry elements that atomic structures can have in addition to those already discussed. **Screw axes,** the other kind of space group operator, involve combinations of translation plus rotation. Just as rotoinversion axes are not equivalent to **proper rotation axes,** glide planes are different from mirror planes, and screw axes are different from rotation axes. Glide planes and screw axes are combinations of mirror planes, rotation, and translation that describe the symmetrical placement of atoms in an infinite crystal structure.

Screw Axes

Screw axes result from the simultaneous application of translation and rotation. We can combine 2-, 3-, 4-, or 6-fold rotation operators with translation to produce screw axes. Many combinations exist; Figure 10.38 shows some examples. As with proper rotational axes (rotational axes not involving translation), each *n*-fold screw operation involves a rotation of $360°/n$. After n repeats, the screw has come full circle. Similarly, the translation associated with a screw axis must be a rational fraction of the unit cell dimension or an infinite number of atoms, all in different places in unit cells, will result.

Figure 10.39 shows the possible 4-fold screw axes, labeled with conventional symbols. A large 4 indicates a 4-fold axis, and the subscripts indicate the magnitudes of the translations. A subscript of 1 indicates the translation is ¼ of the unit cell dimension in the direction of the axis. A subscript of 2 indicates the translation is ²⁄₄ of the unit cell dimension, and a subscript of 3 indicates the translation is ¾ of the unit cell dimension. All unit cells must be identical, but the 4_2 operation shown in Figure 10.39c results in a unit cell in the second layer that differs from that in the first. The only way this operator can be made consistent is to add the extra lattice points shown in Figure 10.39d. In other words, the presence of a 4_2 axis requires the presence of a 2-fold axis.

Looking more closely at 4_1 and 4_3 axes is instructive. Consider a point at the corner of the unit cells shown in Figures 10.40a, and 40b. After four applications of either operator the total rotation is 360°, bringing the fourth point directly above the first. For the 4_1 axis the total translation is equivalent to one unit cell length, but for the 4_3 axis it is three unit cell lengths. In Figure 10.40b, showing the 4_3 axis, the three unit cells have points in different locations. This is impossible because all unit cells must be identical. The unit cells can only be identical, while also being consistent with a 4_3 axis, by adding extra points as shown in Figure 10.40c. Figure 10.39e and f shows the same result. If we compare Figure 10.40c and Figure 10.40a, we see that the 4_1 and 4_3 axes are mirror images of each other. The two axes are an **enantiomorphic pair,** sometimes called *right-handed* and *left-handed* screw axes.

When all combinations are considered, there are 20 possible rotation axes (listed in Table 10.4). As with proper rotational axes, some screw axes are

▶**FIGURE 10.37**
Glide plane: (a) single unit cell containing three atoms; (b) multiple unit cells related by a glide plane; (c) arrow showing glide operations; (d) a glide plane (gray) in three dimensions relating a triangular group of atoms.

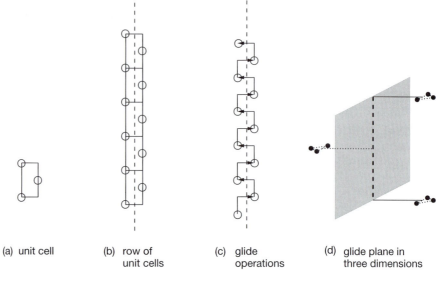

(a) unit cell   (b) row of unit cells   (c) glide operations   (d) glide plane in three dimensions

▶**FIGURE 10.38**
Four types of screw axes parallel to a unit cell edge with length t: (a) a 2_1 screw axis involves translation of $\frac{1}{2}t$ and rotation of $180°$; (b) a 3_1 screw axis involves translation of $\frac{1}{3}t$ and rotation of $120°$; (c) a 4_1 screw axis involves translation of $\frac{1}{4}t$ and rotation of $90°$; (d) a 6_1 screw axis involves translation of $\frac{1}{6}t$ and rotation of $60°$. Note that the symbols for the axes are related to the symbols for proper rotation axes.

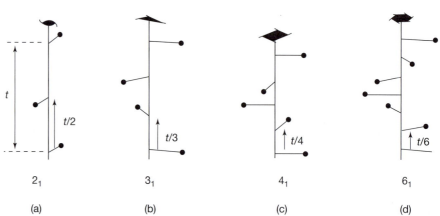

(a)   (b)   (c)   (d)

restricted to one or a few crystal systems. For example, 3_1 and 3_2, which are an enantiomorphic pair, only exist in the trigonal subdivision of the hexagonal system. Similarly, the 6_n axes only exist in the hexagonal subdivision of the hexagonal system.

Glide Planes

Glide planes result from the simultaneous application of translation and reflection (Figure 10.37). There are six types of glide planes, distinguished by having different magnitudes and directions of translation relative to crystallographic axes. Some glide planes are restricted to certain crystal systems just as some screw axes are. Glide planes may involve translation parallel to the a, b, or c axis and, for crystals in the cubic and tetragonal systems only, the translation may follow a diagonal from one corner of a unit cell to another. The letters a, b, c, n, or d symbolize these glides; Table 10.5 lists the possibilities.

Space Groups

Neither glide planes nor screw axes are present in point groups and lattices. They are only manifested when atomic motifs and lattice types combine to produce structures in three-dimensional space. For that reason we call them *space group operators*. When we combine the space group operators in Tables 10.4 and 10.5 with the 14 possible space lattices, we get 230 possible space groups. They represent all possible symmetries crystal structures can have. Deriving them all is not trivial, and crystallographers debated the exact number until the 1890s when several independent studies concluded that there could only be 230 (Box 10.4).

Crystallographers use several different notations for space groups; the least complicated is that used in the *International Tables for X-ray Crystallography (ITX)* (Hahn, 1983). *ITX* space group symbols consist of a letter indicating lattice type (P, I, F, R, A, B, or C) followed by symmetry notation similar to conventional Hermann-Mauguin symbols. An example

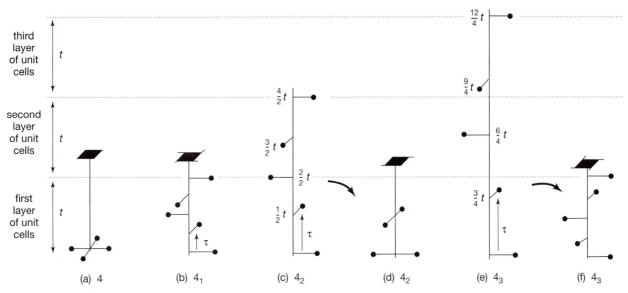

▶FIGURE 10.39
Proper 4-fold axis and all possible 4-fold screw axes parallel to a unit cell edge with length t. All involve rotation of 90°. (a) A proper 4-fold axis involves no translation. (b) A 4_1 axis involves translation of $\frac{1}{4}t$. (c) A 4_2 axis involves translation of $\frac{2}{4}t$. (d) Because all unit cells must be identical, the symmetry of a 4_2 axis requires other points to be present. (e) A 4_3 axis involves translation of $\frac{3}{4}t$. (f) Because all unit cells must be identical, the symmetry of a 4_3 axis requires other points to be present.

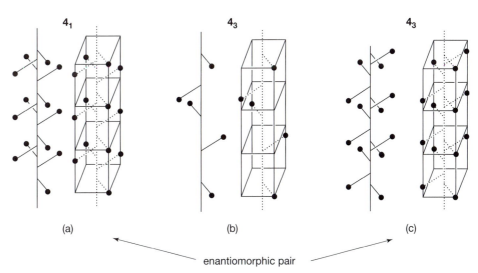

▶FIGURE 10.40
Application of 4_1 and 4_3 screw axes: (a) a 4_1 axis relating points in unit cells; (b) a 4_3 axis relating points in unit cells; the unit cells are not identical in this drawing; (c) a possible arrangement of points that satisfies 4_3 symmetry and maintains all unit cells identical.

is $P4_2/m2_1/n2/m$, the space group of rutile. Rutile has a primitive *(P)* tetragonal unit cell, a 4_2 screw axis perpendicular to a mirror plane, a 2_1 screw axis perpendicular to an n glide plane, and a proper 2-fold axis perpendicular to a mirror plane. Rutile crystals have symmetry $4/m2/m2/m$. Similarly, the space group of garnet, $I4_1/a\overline{3}2/d$, implies a body-centered

unit cell, a 4_1 screw axis perpendicular to an a glide plane, a 3-fold rotoinversion axis, and a proper 2-fold axis perpendicular to a d glide plane. Garnet crystals have point group symmetry $4/m\overline{3}2/m$. Rather than using an entire symbol, crystallographers often use abbreviations for space groups (and occasionally for point groups). Thus, they would say that

▶**TABLE 10.4**
The 20 Possible Space Symmetry Operators Involving Rotation

Operator	Type of Axis	Rotation Angle	Translation Distance*
1	identity	360°	none
$\bar{1}$	inversion center	360°	none
2	proper 2-fold	180°	none
2_1	2-fold screw	180°	½t
3	proper 3-fold	120°	none
3_1	3-fold screw	120°	⅓t
3_2	3-fold screw	120°	⅔t
$\bar{3}$	3-fold rotoinversion	120°	none
4	proper 4-fold	90°	none
4_1	4-fold screw	90°	¼t
4_2	4-fold screw	90°	²⁄₄t = ½t
4_3	4-fold screw	90°	¾t
$\bar{4}$	4-fold rotoinversion	90°	none
6	proper 6-fold	60°	none
6_1	6-fold screw	60°	⅙t
6_2	6-fold screw	60°	²⁄₆t = ⅓t
6_3	6-fold screw	60°	³⁄₆t = ½t
6_4	6-fold screw	60°	⁴⁄₆t = ⅔t
6_5	6-fold screw	60°	⅚t
$\bar{6}$	6-fold rotoinversion	60°	none

*t = unit cell dimension in direction of translation.

▶**TABLE 10.5**
The Possible Space Symmetry Operators Involving Reflection

Operator	Type of Plane	Orientation of Translation	Magnitude of Translation
m	proper mirror	none	none
a	axial glide	parallel to a	½a
b	axial glide	parallel to b	½b
c	axial glide	parallel to c	½c
n	diagonal glide	parallel to face diagonal	½t
d	diamond glide	parallel to main diagonal	¼t

*t = unit cell dimension in direction of glide.

garnet belonged to the space group $Ia3d$. A glance at Table 10.6 shows why we can use this abbreviation: Only one space group, $I4_1/a\bar{3}2/d$, includes these symmetry elements. Most mineralogy reference books use abbreviated symbols, so we often need a table, such as Table 10.6, to interpret them.

The translations associated with lattices, glide planes, and screw axes are very small, on the order of tenths of a nanometer, equivalent to a few Ångstroms. Detecting their presence by visual examination of a crystal is impossible, with or without a microscope. A crystal with symmetry $4/m2/m2/m$ could belong to the space group $I4_4/a2/c2/d$, but there are also 19

other possibilities (Table 10.6). We say that the 20 possibilities are *isogonal*, meaning that when we ignore translation they all have the same symmetry. Without detailed X-ray studies, telling one isogonal space group from another is impossible, and we are left with only the 32 distinct point groups.

CRYSTAL HABIT AND CRYSTAL FACES

Why do halite and garnet, both cubic minerals, have different crystal habits? It is not fully understood why crystals grow together in the ways they

Box 10.4 Why Are There Only 230 Space Groups?

In Chapter 9 we showed that point groups may have one of 32 symmetries; in this chapter we determined that crystal structures must have one of 14 Bravais lattices. When point groups are combined with Bravais lattices, and the space group operators in Tables 10.4 and 10.5 are considered, 230 possible space groups result (Table 10.6). The 230 space groups are the only possible symmetries that a crystal structure can have. They were tabulated in the 1890s by a Russian crystallographer, E. von Federov; a German mathematician, Artur Schoenflies; and a British amateur, William Barlow; all working independently.

Why are there only 230 space groups? The answer is that symmetry operators, as we have already seen, can only combine in certain ways. In the discussion of point group symmetry, we concluded that mirrors and rotation axes can only combine in 32 ways. Other combinations led to infinite symmetry, which is impossible. The same is true of space group operators and Bravais lattices; only certain combinations are allowed. For example, triclinic lattices ($P1$ and $P\bar{1}$) may not be combined with 2-fold axes of any sort. Similarly, 3, 3_1, and 3_2 axes are only consistent with a rhombohedral or hexagonal lattice ($3R$ or $6P$).

do. Cubic unit cells may lead to cube-shaped crystals such as halite, octahedral crystals such as spinel, dodecahedral crystals such as garnet, and many other shaped crystals. Why the differences? Crystallographers do not have complete answers, but the most important factor is the location of atoms and lattice points within the unit cell. Crystals of a particular mineral tend to have the same forms, or only a limited number of forms, no matter how they grow. Haüy and Bravais noted this and used it to infer that atomic structure controls crystal forms. In 1860 Bravais observed what we now call the *Law of Bravais:* Faces on crystals tend to be parallel to planes having a high density of lattice points. This means that, for example, crystals with hexagonal lattices often have faces related by hexagonal symmetry. Crystals with orthogonal unit cells (those in the cubic, orthorhombic, or tetragonal systems) tend to have faces at 90° to each other, etc.

The relationship between lattice symmetry and crystal habit can be seen by comparing Figure 10.31, showing the possible unit cell shapes for calcite, with Figure 10.41 showing some common forms for calcite crystals. There is a close resemblance between the unit cell shapes, which represent the lattice symmetry, and some of the crystal shapes. Thus, Bravais's Law works well. Unfortunately, some minerals, such as pyrite (FeS_2) and quartz (SiO_2), appear to violate Bravais's Law. Bravais's observations were based on considerations of the 14 Bravais lattices and their symmetries, but in the early twentieth century, P. Niggli, J. D. H. Donnay,

and D. Harker realized that space group symmetries needed to be considered as well. By extending Bravais's ideas to include glide planes and screw axes, Niggli, Donnay, and Harker explained most of the biggest inconsistencies. They concluded that crystal faces form parallel to planes of highest atom density, a slight modification of the Law of Bravais.

As a crystal grows, different faces grow at different rates. Some may dominate in the early stages of crystallization while others will dominate in the later stages. The relationship, however, is the opposite of what we might expect. Faces that grow fastest are the ones that eventually disappear. Figure 10.42 shows why this occurs. If all faces on a crystal grow at the same rate, the crystal will keep the same shape as it grows (Figure 10.42a). However, this is not true if some faces grow faster than others. In Figure 10.42b, the diagonal faces (oriented at 45° to horizontal) grew faster than those oriented vertically and horizontally. Eventually, the diagonal faces disappeared; they "grew themselves out." The final crystal has a different shape, and fewer faces, than when it started growing. We observe this phenomenon in many minerals; small crystals often have more faces than larger ones.

In summary, crystals may have any of 32 possible symmetries. corresponding to the 32 point groups. Each requires the lattice and unit cell to have one of six possible symmetries. This allows us to divide crystals into six crystal systems, each characterized by unit cells of different shape. Several lattice types are possible for most crystal systems;

▶**TABLE 10.6**
The 230 Possible Space Groups

Crystal System	Point Group	Space Groups
triclinic	1	$P1$
	$\bar{1}$	$P\bar{1}$
monoclinic	2	$P2, P2_1, C2$
	m	Pm, Pc, Cm, Cc
	$2/m$	$P2/m, P2_1/m, C2/m, P2/c, P2_1/c, C2/c$
orthorhombic	222	$P222, P222_1, P2_12_12, P2_12_12_1, C222_1, C222, F222, I222, I2_12_12_1$
	$mm2$	$Pmm2, Pmc2_1, Pcc2, Pma2, Pca2_1, Pnc2, Pmn2_1, Pba2, Pna2_1, Pnn2, Cmm2, Cmc2_1,$ $Ccc2, Amm2, Abm2, Ama2, Aba2, Fmmc, Fdd2, Imm2, Iba2, Ima2$
	$2/m2/m2/m$	$P2/m2/m2/m, P2/n2/n2/n, P2/c2/c2/m, P2/b2/a2/n, P2_1/m2/m2/a, P2/n2_1/n2/a,$ $P2/m2/n2_1/a, P2_1/c2/c2/a, P2_1/b2_1/a2/m, P2_1/c2_1/c2/n, P2/b2_1/c2_1/m, P2_1/n2_1/n2/m,$ $P2_1/m2_1/m2/n, P2_1/b2/c2_1/n, P2_1/b2_1/c2_1/a, P2_1/n2_1/m2_1/a, C2/m2/c2_1/m, C2/m2/c2_1/a,$ $C2/m2/m2/m, C2/c2/c2/m, C2/m2/m2/a, C2/c2/c2/a, F2/m2/m2/m, F2/d2/d2/d,$ $I2/m2/m2/m, I2/b2/a2/m, I2/b2/c2/a, I2/m2/m2/a$
tetragonal	4	$P4, P4_1, P4_2, P4_3, I4, I4_1$
	$\bar{4}$	$P4, I4$
	$4/m$	$P4/m, P4_2/m, P4/n, P4_2/n, I4/m, I4_1/a$
	422	$P422, P42_12, P4_122, P4_12_12, P4_222, P4_22_12, P4_322, P4_32_12, I422, I4_122$
	$4mm$	$P4mm, P4bm, P4_2cm, P4_2nm, P4cc, P4nc, P4_2mc, P4_2bc, I4mm, I4cm, I4_1md, I4_1cd$
	$\bar{4}2m$	$P\bar{4}2m, P\bar{4}2c, P\bar{4}2_1m, P\bar{4}2_1c, P\bar{4}m2, P\bar{4}c2, P\bar{4}b2, P\bar{4}n2, I\bar{4}m2, I\bar{4}c2, I\bar{4}2m, I\bar{4}2d$
	$4/m2/m2/m$	$P4/m2/m2/m, P4/m2/c2/c, P4/n2/b2/m, P4/n2/n2/c, P4/m2_1/b2/m, P4/m2_1/n2/c,$ $P4/n2_1/m2/m, P4/n2_1/c2/c, P4_1/m2/m2/c, P4_2/m2/c2/m, P4_2/n2/b2/c, P4_2/n2/n2/m,$ $P4_2/m2_1/b2/c, P4_2/m2_1/n2/m, P4_1/n2_1/m2/c, P4/n2_1/c2/m, I4/m2/m2/m, I4/m3/c2/m,$ $I4_1/a2/m2/d, I4_1/a2/c2/d$
hexagonal (hexagonal)	6	$P6, P6_1, P6_5, P6_2, P6_4, P6_3$
	$\bar{6}$	$P\bar{6}$
	$6/m$	$P6/m, P6_3/m$
	622	$P622, P6_122, P6_522, P6_222, P6_422, P6_322$
	$6mm$	$P6mm, P6cc, P6_3cm, P6_3mc$
	$\bar{6}m2$	$P\bar{6}m2, P\bar{6}c2, P\bar{6}2m, P\bar{6}2c$
	$6/m2/m2/m$	$P6/m2/m2/m, P6/m2/c2/c, P6_3/m2/c2/m, P6_2/m2/m2/c$
hexagonal (trigonal)	3	$P3, P3_1, P3_2, R3$
	$\bar{3}$	$P\bar{3}, R\bar{3}$
	32	$P312, P321, P3_112, P3_121, P3_212, P3_221, R32$
	$3m$	$P3m1, P31m, P3c1, P31c, R3m, R3c$
	$32/m$	$P31m, P31c, P3m1, P\bar{3}c1, R\bar{3}m, R\bar{3}c$
cubic	23	$P23, F23, I23, P2_13, I2_13$
	$2/m\bar{3}$	$P2/m\bar{3}, P2/n\bar{3}, F2/m\bar{3}, F2/d\bar{3}, I2/m\bar{3}, P2_1/a\bar{3}, I2_1/a\bar{3}$
	432	$P432, P4_232, F432, F4_132, I432, P4_332, P4_132, I4_132$
	$\bar{4}3/m$	$P\bar{4}3m, F\bar{4}3m, I\bar{4}3m, P\bar{4}3n, F\bar{4}3c, I\bar{4}3d$
	$4/m\bar{3}2/m$	$P4/m\bar{3}2/m, P4/n\bar{3}2/n, P4_2/m\bar{3}2/n, P4_2/n\bar{3}2/m, F4/m\bar{3}2/m, F4/m\bar{3}2/c, F4_1/d\bar{3}2/m,$ $F4_1/d\bar{3}2/c, I4/m\bar{3}2/m, I4_1/a\bar{3}2/d$

►FIGURE 10.41
Typical calcite crystals: (a) eight idealized drawings of crystals; (b) five photographs of calcite crystals.

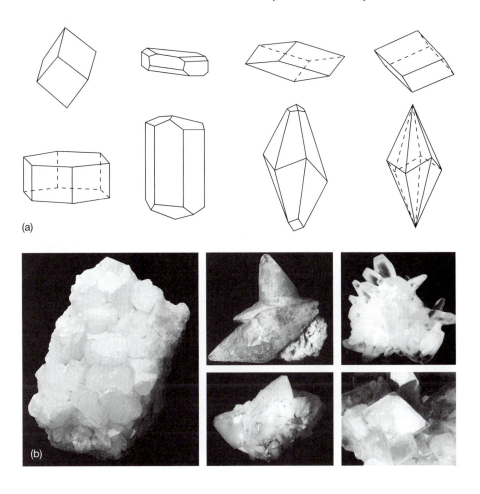

(a)

(b)

►FIGURE 10.42
Growth rates affect crystal shape. These drawings show crystals and the "growth rings" that developed as they grew to full size. (a) If all faces grow at the same rate, the crystal does not change shape as it grows. (b) If the horizontal and vertical faces grow more slowly than the diagonal faces (note thickness of growth bands), eventually the diagonal faces disappear.

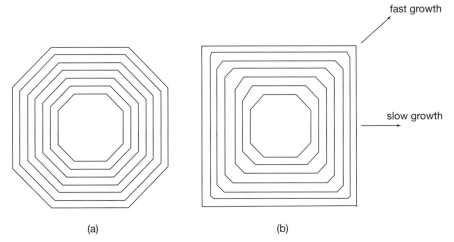

fast growth

slow growth

(a)   (b)

there are 14 different Bravais lattices in all. The combination of atomic motif symmetry, lattice type, glide planes, and screw axes defines the space group. The 230 space groups represent the possible symmetries relating atoms in crystals. Crystal morphology depends on the way unit cells are stacked together. For some crystals, we can infer crystal system and point group by examining morphology. We need X-ray studies to determine lattice type and space group.

We can now explain why we only considered 1-, 2-, 3-, 4-, and 6-fold rotation axes in Chapter 9. Crystals can have no more symmetry than their unit cells. Unit cells can have no rotation axes other than the ones we considered. Three-dimensional solids with 5-fold axes of symmetry, for example, cannot be unit cells because they cannot fit together to fill three-dimensional space. We can verify this by drawing equal-sided pentagons on a piece of paper. No matter how we fit them together, space will always be left over.

▶QUESTIONS FOR THOUGHT

Some of the following questions have no specific correct answers; they are intended to promote thought and discussion.

1. Steno's Law is one of the fundamental laws of crystallography. What is Steno's Law and what are its implications for crystal structure?
2. Today we know that crystals are made of unit cells composed of a discrete number of atoms. How do we know this?
3. In Chapter 9 we defined crystal systems and said that all crystals with four 3-fold axes of symmetry must belong to the cubic system. Why is this a good way to group crystals? What do crystals with four 3-fold axes have in common that makes grouping them appropriate?
4. Why are there only six crystal systems?
5. What is the relationship between crystal symmetry (point group) and unit cell symmetry?
6. What is the relationship between crystal symmetry (point group) and lattice symmetry?
7. What is the fundamental difference between point group operators and space group operators? Why do we need space group operators to describe the symmetry of crystal structures?
8. What is the relationship between crystal form and crystal structure? How do they both relate to crystal habit?
9. Examine the patterns in Figure 10.16. Place a piece of tracing paper over them and indicate all symmetry elements. Choose a representative unit cell for each pattern and make a drawing of the lattice.

▶RESOURCES

(See also the general references listed at the end of Chapter 1.)

Bloss, F. D. *Crystallography and Crystal Chemistry, An Introduction.* Washington, DC: Mineralogical Society of America, 1994.

Brisse, F. La Symétrie bidimensionelle et le Canada. *Canadian Mineralogist* 19 (1981): 217–224.

Buerger, M. J. *Elementary Crystallography.* New York: John Wiley and Sons, 1963.

Buerger, M. J. *Contemporary Crystallography.* New York: John Wiley and Sons, 1975.

Hahn, T., ed. *International Tables for X-Ray Crystallography, Vol. A. Space Group Symmetry.* Boston: International Union of Crystallography, D. Reidel, 1983.

MacGillavry, C. H. *Fantasy and Symmetry: The Periodic Drawings of M. C. Escher.* New York: Harry N. Abrams, 1976.

McKie, D., and C. McKie. *Essentials of Crystallography.* Oxford: Blackwell, 1986.

Phillips, F. C. *An Introduction to Crystallography,* 3rd ed. New York: John Wiley and Sons, 1963.

Unit Cells, Points, Lines, and Planes

In the last two chapters we talked about qualitative aspects of crystals and unit cells. To describe the details of crystals and crystal structures, it is necessary to be more quantitative. In this chapter we describe points, lines, and planes in crystal structures using a coordinate system based on unit cell geometry. The most important results are that we can give exact descriptions of crystal structures, and we can label and describe crystal faces and crystal properties in unambiguous ways.

In Chapter 10, we showed that a crystal is composed of many identical building blocks called *unit cells*. To describe a unit cell, we must give its shape, size, and composition. Symmetry axes, crystal faces, and other linear and planar features of crystals depend on the nature of the unit cell and on crystal growth. To describe them, we need some way to describe the orientation of lines and planes in crystals. Furthermore, for complete characterization of a crystal, we must specify the locations of atoms in its structure. We therefore need a quantitative way to describe points (for example, atom locations), lines, and planes in crystals and unit cells. To do this, it is convenient to have a coordinate system. The system used by crystallographers is similar to a **Cartesian coordinate system,** with the exception that the angles between the three axes vary according to the crystal being described, as do the unit lengths along each axis. Figure 11.1b shows this general system.

Unit Cell Parameters and Crystallographic Axes

Chapter 10 introduced the unit cell parameters a, b, c, α, β, and γ. a, b, and c are the lengths of unit cell edges; α, β, and γ are the angles between the edges,

α being the angle between b and c, β the angle between a and c, and γ the angle between a and b. Mineralogists have traditionally used angstroms ($1\text{Å} = 10^{-10}$ m) but some more recent literature uses nanometers (1 nm = 10Å) to measure a, b, and c. The angles α, β, and γ are today given as a decimal number of degrees (for example, 94.62°). Typical mineral unit cells have edges of 2–20Å; angles between edges vary greatly, although unit cells are often chosen so that angles are close to 90°.

Unit cell edges define the coordinate system used by crystallographers (Figure 11.1). Three axes, designated a, b, and c, run parallel to a, b, and c length edges of the unit cell. The angles between the axes are therefore α, β, and γ. In the cubic, tetragonal, and orthorhombic systems, α, β, and γ all equal 90° because the unit cell edges are orthogonal. Table 10.3 gives other constraints on the angles. Instead of using angstroms to give distances, we may also use unit cell dimensions as a scale. For example, we might say that a certain plane intersects the axes at distances of $3a$, $2b$, and $2c$ from the origin. It is implicit that $3a$ refers to a distance equal to three unit cell edge lengths along the a axis, $2b$ a distance equal to two unit cell lengths along the b axis, and $2c$ a distance equal to two unit cell lengths along the c axis.

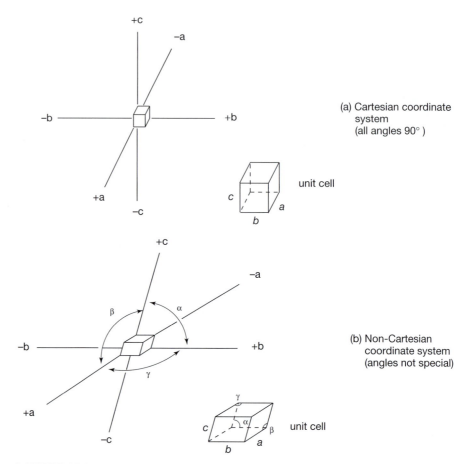

▶**FIGURE 11.1**
Coordinate system used to describe crystals: (a) a standard Cartesian coordinate system;
(b) a crystallographic coordinate system in which the angles between axes are α (the angle
between b and c), β (the angle between a and c) and γ (the angle between a and b).

▶**TABLE 11.1**
Unit Cell Parameters (a, b, c, α, β, γ), Z (number of formulas per unit cell),
and V (unit cell volume) for One Mineral from Each of the Crystal Systems*

Fluorite CaF_2	Rutile TiO_2	Beryl $Be_3Al_2Si_6O_{18}$	Enstatite $Mg_2Si_2O_6$	Sanidine $KAlSi_3O_8$	Albite $NaAlSi_3O_8$
cubic	tetragonal	hexagonal	orthorhombic	monoclinic	triclinic
$Z = 4$	$Z = 2$	$Z = 2$	$Z = 4$	$Z = 4$	$Z = 4$
$a = 5.46$	$a = 4.59$	$a = 9.23$	$a = 18.22$	$a = 8.56$	$a = 8.14$
$V = 162.77$	$c = 2.96$	$c = 9.19$	$b = 8.81$	$b = 13.03$	$b = 12.8$
	$V = 62.36$	$V = 2034.09$	$c = 5.21$	$c = 7.17$	$c = 7.16$
			$V = 836.30$	$\beta = 115.98$	$\alpha = 94.33$
				$V = 799.72$	$\beta = 116.57$
					$\gamma = 87.65$
					$V = 746.01$

*Units are Ångstroms, degrees, and cubic Ångstroms.

The symmetry of a unit cell always affects the relationships between a, b, and c. So, in the cubic system, $a = b = c$, but in the tetragonal and hexagonal
systems $a = b \neq c$ (see Table 10.3). The relationships implied by crystal systems mean that, for systems other than triclinic, we need not give six values
to describe unit cell shape. Table 11.1 gives examples
of unit cell parameters for minerals from each of the

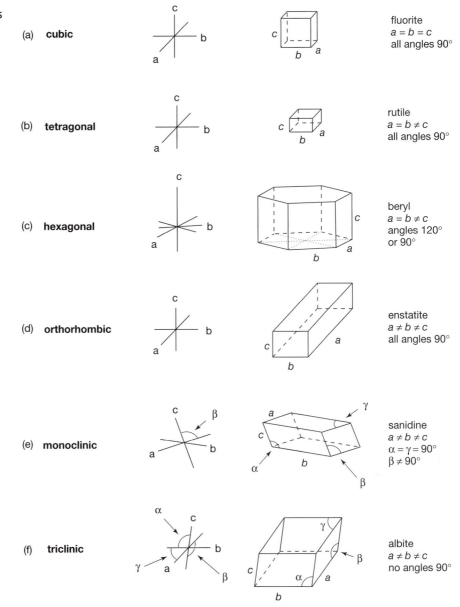

▶**FIGURE 11.2**

Examples of unit cells for minerals belonging to each of the six crystal systems: (a) fluorite (cubic); (b) rutile (tetragonal); (c) beryl (hexagonal); (d) enstatite (orthorhombic); (e) sanidine (monoclinic); (f) albite (triclinic).

(a) **cubic**

fluorite
$a = b = c$
all angles 90°

(b) **tetragonal**

rutile
$a = b \neq c$
all angles 90°

(c) **hexagonal**

beryl
$a = b \neq c$
angles 120° or 90°

(d) **orthorhombic**

enstatite
$a \neq b \neq c$
all angles 90°

(e) **monoclinic**

sanidine
$a \neq b \neq c$
$\alpha = \gamma = 90°$
$\beta \neq 90°$

(f) **triclinic**

albite
$a \neq b \neq c$
no angles 90°

crystal systems; unnecessary information has been omitted. Figure 11.2 depicts the unit cells.

Consider the triclinic mineral albite, a feldspar with composition $NaAlSi_3O_8$. The last column in Table 11.1 gives albite's cell dimensions and Figure 11.2f shows a drawing of the unit cell with all edges and angles labeled. Because none of the angles are special and none of the cell edges are equal, we need six parameters to describe the unit cell shape. In contrast, fluorite (Table 11.1, Figure 11.2a) has a cubic unit cell, so we need to give only one cell parameter, the length of the cell edge. It is implicit that all angles are 90° and all cell edges are the same length.

Although it makes no difference which edges of a unit cell we call *a*, *b*, or *c*, mineralogists normally

follow certain conventions. In triclinic minerals, none of the angles are special and *a*, *b*, and *c* are all different lengths. Although the literature contains exceptions, by modern convention edges are chosen so that $c < a < b$. In monoclinic minerals, such as sanidine, only one angle in the unit cell is not 90°. By convention, the non-90° angle is β, the angle between *a* and *c*. This convention is the **second setting** for monoclinic minerals. The value of β is the only one listed in Table 11.1 because it is implicit that the other two angles are 90°. For orthorhombic, hexagonal, tetragonal, and cubic minerals, we need not give any angles since they are all defined by the crystal system. In orthorhombic crystals, axes are most often chosen so that $c < a < b$. In tetragonal and hexagonal crystals, the *c* axis always corresponds to

▶**FIGURE 11.3**
Using subscripts to denote axes in hexagonal unit cells: (a) nonprimitive hexagonal unit cell with four axes (a_1, a_2, a_3, c); (b) primitive rhombohedral unit cell with three axes (a_1, a_2, a_3) separated by angle α

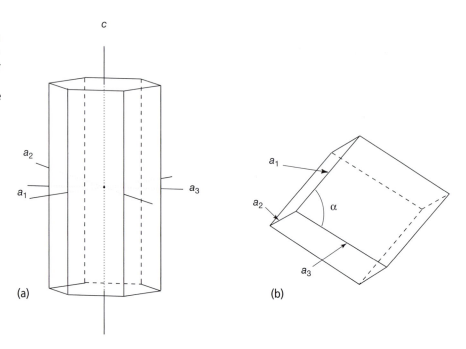

the 4-fold or 6-fold axis. The cell edge is always designated a in cubic crystals.

In this book we use a, b, and c to designate the three crystallographic axes, but other conventions are sometimes used. Some crystallographers use subscripts to indicate axes, and thus cell edges, that must be identical because of symmetry. For example, instead of a, b, and c, the three axes of cubic minerals might be designated a_1, a_2, and a_3. The axes of tetragonal crystals can be designated a_1, a_2, and c. For hexagonal crystals, crystallographers have historically used four axes: a_1, a_2, a_3, and c (Figure 11.3a). The three a axes, a_1, a_2, or a_3, are parallel to edges of a nonprimitive hexagonal unit cell (Figure 11.3a). The third axis is redundant, but has been included in the past to emphasize that there are three identical a axes perpendicular to the c axis. A further complication arises for hexagonal crystals because, as described in the last chapter, they are sometimes described by a primitive rhombohedral unit cell with three equal edges (a_1, a_2, and a_3) and three equal angles (α) between edges (Figure 11.3b). Seldom, however, are these edges used as the basis for a coordinate system. Only three axes are used in much of the modern literature, so we will only briefly mention the fourth axis in the rest of this chapter.

The Composition of Unit Cells

Physical dimensions do not completely describe a unit cell. We must also specify the nature and locations of atoms within the unit cell. To provide some of this information, Table 11.1 lists two other things: mineral formulas and Z, the number of formulas in each unit cell. For example, albite has the formula $NaAlSi_3O_8$ and $Z = 4$. This means that four $NaAlSi_3O_8$ formulas are in each unit cell. In other words, each unit cell contains 4 Na atoms, 4 Al atoms, 12 Si atoms, and 32 O atoms. Similarly, in Figure 10.31, the three possible calcite unit cells have $Z = 2$, $Z = 12$, and $Z = 4$.

Points in Unit Cells

Mineralogists describe the locations of atoms or other points in unit cells by giving their coordinates. As examples, consider the orthorhombic and monoclinic crystals shown in Figure 11.4a and b. Each has one point in the exact center. In the orthorhombic crystal the three axes intersect at 90°, in the monoclinic crystal one angle, β, is greater than 90°. Assume both crystals have cell dimensions $a = 5.20\text{Å}$, $b = 18.22\text{Å}$, and $c = 8.80\text{Å}$, and the axes' origins are at the center of the unit cells as shown. The coordinates of point P, at the lower right-hand corner in each drawing, are therefore $5.20/2 = 2.60\text{Å}$, $18.22/2 = 9.11\text{Å}$, and $-8.80/2 = -4.40\text{Å}$. The negative sign arises because the points are in a negative direction from the origin along the c axis.

We do not, however, normally report coordinates in angstroms (Å). Instead, we report distances relative to unit cell dimensions. Another way to describe the location of the points in Figure 11.4a and b would be to say they have coordinates $\frac{1}{2}a$, $\frac{1}{2}b$, and $\frac{1}{2}c$. We normalize coordinates to unit cell dimensions by dropping the a, b, and c. The coordinates then become $\frac{1}{2}$, $\frac{1}{2}$, $\frac{1}{2}$. The system to which a crys-

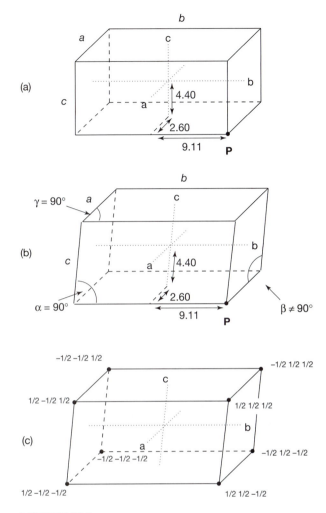

▶FIGURE 11.4
Points in orthorhombic and monoclinic crystals. (a) An orthorhombic unit cell having cell edges of *a* = 5.2, *b* = 18.22, and *c* = 8.80Å. Point P has coordinates (in Å) of 2.6, 9.11, −4.40. These values correspond to the distance along the *a*, *b*, and *c* axes from the origin to P. In (b), Point P has the same coordinates, although the unit cell is monoclinic. In both cases, we can also describe the coordinates of P as ½*a*, ½*b*, −½*c*, or simply as ½ ½ −½ and not specify the units. (c) The coordinates of all points at the corners of a cell: They all have coordinate values of ½ or −½ if the origin is at the center of the cell.

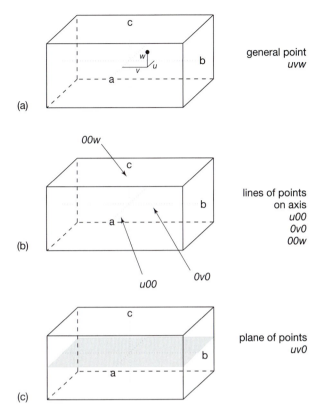

▶FIGURE 11.5
General points and special points in a unit cell. (a) A general point in a unit cell has nonspecial values for its indices *u*, *v*, and *w*. (b) Special points lying on crystallographic axes must have two indices equal to 0. (c) Special points within a plane including two axes have one index equal to 0.

tal belongs does not affect the coordinates. The points at the corners must have coordinates ±½, ±½, ±½, if the origin is at the center of the unit cell (Figure 11.4c).

Traditionally, crystallographers have used the characters *u*, *v*, and *w* to symbolize the coordinates of a point if the coordinates are rational numbers, and *x*, *y*, and *z* to symbolize the coordinates if irrational. For simplicity, in this text we will use *uvw* in all instances (Figure 11.5a). We can say that *uvw* represents a **general point** anywhere in the cell. Suppose two coordinates, for example *v* and *w*, are

zero. All the points described by the coordinates *u*00 constitute a set of **special points,** lying along the *c* axis (Figure 11.5b). Similarly, 0*v*0 and 00*w* refer to sets of special points on the *b* and *c* axes, respectively. If only one coordinate equals zero, a point lies in a plane including two axes. We might, for example, talk about special points located at *uv*0. These are points located on a plane that includes the *a* and *b* axes, as shown in Figure 11.5c. Note that no parentheses or brackets are used when giving coordinates of a point.

Figure 11.6a shows a unit cell of sphalerite, ZnS, with atom coordinates *u*, *v*, *w* given. Because three-dimensional drawings are sometimes difficult to draw and see, crystallographers often use projections, that contain the same information in a less cluttered manner (Figure 11.6b). In the projection, we need not specify *u* and *v* values because they can be estimated from the location of the atoms in the drawing. Sometimes the *w* values for atoms at the corner unit cells are omitted; by convention, this means the atoms are found on both the top and bottom of the cell as it appears in projection.

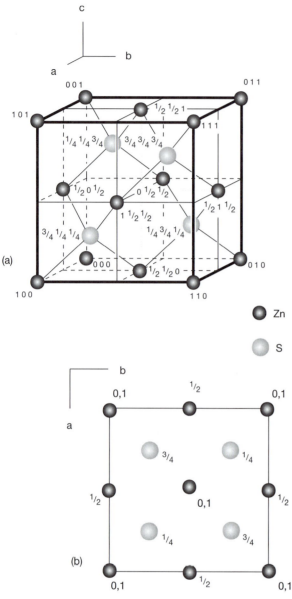

▶FIGURE 11.6
The atomic structure of sphalerite and *u*, *v*, and *w* values (atomic coordinates) of each atom: (a) the structure in three dimensions with the crystallographic origin at the corner labeled 000; (b) the structure in projection when looking down the *c* axis. *u* and *v* values can be inferred from the location of the atom symbols in the projection. *w* values are given with numbers. The atoms at the corners and center of the unit cell have two *w* values because in this projection two atoms fall on top of each other.

LINES AND DIRECTIONS IN CRYSTALS

The absolute location of a line in a crystal is not often significant, but directions do have significance because they describe the orientations of symmetry axes and other linear features. Crystallographers designate directions with three indices in square brackets, [*uvw*]. As with point locations, we give numbers

describing directions in terms of unit cell dimensions. In Figure 11.7, direction OA has indices [132]. In terms of vectors, $\overrightarrow{oa}$ is equivalent to $1\vec{a} + 3\vec{b} + 2\vec{c}$. [132] is the direction from the origin to the point 132. Notice that direction $\left(\frac{1}{3}\,1\,\frac{2}{3}\right)$ is the same as [132]. Fractions are cleared and common denominators eliminated when giving the indices of a direction. $\left(\frac{1}{3}\,1\,\frac{2}{3}\right)$ becomes [132] when we multiply all indices by 3. By convention, commas separate indices only if they have more than one digit. Although no commas are included in [132], reading sequentially, as we do for point locations, we articulate it "one-three-two."

Figure 11.8 shows several lines on a two-dimensional lattice. The direction [120] (OC) is the same as [240] (OD), which is consistent with the proportional relationship of the two sets of indices. Parallel lines have identical indices, so the choice of the origin is unimportant and OC, O'C', and O"C" have indices [120]. Line OF has indices [$\bar{2}$30], articulated "bar-two-three-zero." The bar over the 2 is equivalent to a negative sign, indicating that the line goes in the negative *a* direction.

Because all directions in Figure 11.8 lie in the a–b plane, they have indices [*uv*0]. This notation indicates that the first two indices may vary, but the third is 0. Whenever the index for one axis is 0, a line or direction must lie in a plane parallel to the other two axes (Figure 11.9). If two indices are 0, the direction is parallel to an axis: [100] is parallel to *a*, [010] parallel to *b*, and [001] parallel to *c*. Figure 11.9a, b, and c emphasize that it does not matter where lines are; orientation determines indices. We determine the indices of lines in the same way, even if the axes are not orthogonal. Note that in Figure 11.9c the line [1$\bar{1}$1] could also be designated as [$\bar{1}$1$\bar{1}$] because lines, unlike vectors, do not go in any specific direction.

PLANES IN CRYSTALS

While crystals and crystal faces vary in size and shape, Steno's Law tells us that angles between faces are characteristic for a given mineral. The absolute location and size of the faces are rarely of significance to crystallographers, while the relative orientations of faces are of fundamental importance. We can use face orientations to determine crystal systems and point groups, so having a simple method to describe the orientation crystal faces is useful. Consider a plane parallel to a crystal face. Such a plane may intersect all three axes, or it may be parallel to one or two of them (Figure 11.10). We can describe the orientation of a plane by listing its intercepts with the axes. For a face parallel to an axis, the intercept is at ∞,

▶**FIGURE 11.7**
Vector addition to describe a line.
(a) Vector $\overrightarrow{OA}$ is described by giving its components in the a, b, and c directions: 1a, 3b, 2c, given in abbreviated form as 132.
(b) Vector ⅓ 1 ⅔ parallels 132 but is shorter.

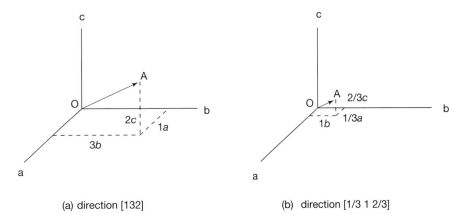

(a) direction [132]

(b) direction [1/3 1 2/3]

▶**FIGURE 11.8**
Lines and their indices on a two-dimensional lattice. All lines are assumed to lie in the a–b plane.

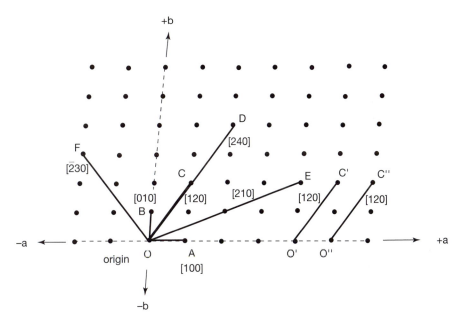

▶**FIGURE 11.9**
Special directions and indices in a unit cell: (a) directions parallel to unit cell edges; (b) directions parallel to face diagonals; (c) directions parallel to body diagonals.

(a) cell edges   (b) face diagonals   (c) body diagonals

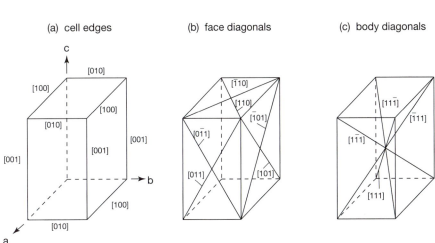

an infinite distance from the origin. In Figure 11.10a, a plane (PQR) intersects the axes at distances OQ, OP, and OR from the origin. The dots on the axes show the unit cell lengths, which may in some crystal systems be different along each axis. We could give the distances OQ, OP, and OR in Ångstroms (Å) or some other absolute units. As with points and directions, however, it is more convenient to express distances

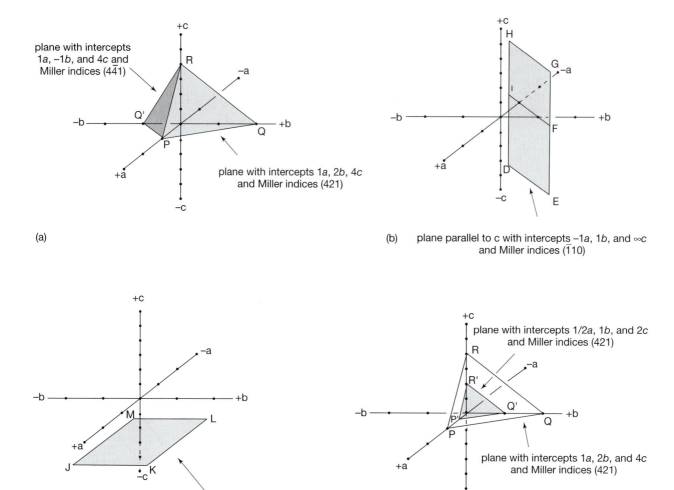

plane with intercepts
1*a*, –1*b*, and 4*c* and
Miller indices (4$\bar{4}$1)

plane with intercepts 1*a*, 2*b*, 4*c*
and Miller indices (421)

(a)

(b)   plane parallel to c with intercepts –1*a*, 1*b*, and ∞*c*
and Miller indices ($\bar{1}$10)

(c)   plane parallel to a and b with intercepts ∞*a*, ∞*b*, and –3*c*
and Miller indices (00$\bar{1}$)

(d)

plane with intercepts 1/2*a*, 1*b*, and 2*c*
and Miller indices (421)

plane with intercepts 1*a*, 2*b*, and 4*c*
and Miller indices (421)

▶**FIGURE 11.10**
Planes intersecting three, two, or one axis: (a) planes intersecting all three axes;
(b) a plane parallel to the *c* axis; (c) a plane parallel to two axes; (d) two parallel
planes intersecting all three axes.

relative to unit cell dimensions (a, b, and c). So, we would say that plane PQR (Figure 11.10a) has axial intercepts 1*a*, 2*b*, 4*c*, or just 1,2,4.

One problem with describing a plane by giving its axial intercepts is that faces often lie parallel to crystal axes. If parallel to the *c* axis, for example, a crystal face has intercepts u, v, ∞, where u and v can have any values (Figure 11.10b). A face that is parallel to two axes has two parameters that are infinity (Figure 11.10c). The use of ∞ can be confusing and awkward. A second problem with listing intercepts is that parallel planes and faces, such as those in Figure 11.10d, have different indices. Crystallographers find this inconvenient because orientation is usually the feature of importance when discussing crystal faces. To avoid these and other problems, crystallographers generally do not report axial intercepts, but instead use **Miller indices.**

Miller Indices

Miller indices were first developed in 1825 by W. Whewell, a professor of mineralogy at Cambridge University. We use them to describe the orientation of crystal faces, and also the orientations of cleavages and other planar properties. They are named after W. H. Miller, a student of Whewell's, who promoted and popularized their use in 1839. The general symbol for a Miller index is (hkl), in which the letters h, k, and l each stand for an integer. We calculate Miller indices for a plane from its axial intercepts (Figure 11.10). The procedure is as follows: Axial intercept values are inverted; fractions are cleared; ∞ becomes 0 upon inversion. Parentheses enclose the resulting Miller index. As with directions, bars show negative values and we do not include commas unless numbers have more than one digit. Conversion of the axial intercepts ∞, ∞, -3 for the plane in Figure 11.10c would go as follows: Inversion yields 0, 0, and $-\frac{1}{3}$; multiplying by 3 gives a Miller index of $(00\bar{1})$, articulated as "oh-oh-bar-one." Similarly, plane PQ'R (Figure 11.10a) has axial intercepts of $1, -1, 4$ and a Miller index $(4\bar{4}1)$.

Because crystal faces are parallel to rows of lattice points, they are parallel to planes that intercept crystal axes at an integral number of unit cells from the origin. Consequently, inversion of intercepts and clearing fractions always yields integers. This observation is known as the **Law of Rational Indices.** Another observational law, called **Haüy's Law,** says that Miller indices of faces generally contain low numbers. For example, (111) is a common face in crystals, while (972) is not. Haüy's Law is really a corollary to the Law of Bravais, already discussed, which states that faces form parallel to planes of high lattice point density; planes with low values in their Miller index have the greatest lattice point density.

Parallel planes always have the same Miller index. In Figure 11.10d, both PQR and P'Q'R' have Miller index (421). Any face lying parallel to a plane with axial intercepts at 1,2,4 has this index. Crystallographers would call it the "4-2-1 face," no matter its size or shape. The Miller index describes the orientation of a crystal face with respect to crystallographic axes, but not the absolute size or location of the face.

The replacement of unknown or variable numbers in a Miller index with $h, k,$ or l allows us to make generalizations. The index $(hk0)$ describes the family of faces with their third index equal to zero. A Miller index including a zero indicates that a face is parallel to one or more axes. The family of faces described by $(hk0)$ is parallel to the c axis; faces with the Miller index $(00l)$ are parallel to both the a axis and the b axis (Figure 11.10c).

As mentioned previously, crystallographers have in the past used four axes for crystals in the hexagonal system (Figure 11.11). The general symbols are $[hkil]$ for directions and $(hkil)$ for planes. One of the first three values $(h, k,$ or $i)$ is always redundant because we can always describe the location of a plane in three-dimensional space with three variables. In all cases:

$$h + k + i \equiv 0 \qquad (11.9)$$

Because of the redundancy, and to be consistent with other crystal systems, many crystallographers today use only three indices for hexagonal minerals.

The relationship between planes and directions in a crystal depends on the crystal system. Except in the cubic system, the direction $[uvw]$ is neither per-

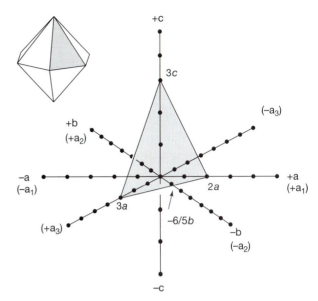

▶**FIGURE 11.11**
Indexing a crystal face on a hexagonal mineral: The shaded face is shown by itself intersecting the a and b axes at $2a$ and $-6/5b$ and the c axis at $3c$. Its Miller index is $(3\bar{5}2)$. If we consider three a axes (see text), the intercepts are $2a$, $-6/5a$, $3a$, $3c$, and the Miller index is $(3\bar{5}22)$.

pendicular nor parallel to planes with the Miller index (uvw) (Figure 11.12).

CRYSTAL FORMS AND THE MILLER INDEX

Recall that a set of faces related by symmetry comprise a form. Faces of a form always have similar Miller indices. This relationship is especially clear for crystals in the cubic system because high symmetry means that many forms may contain many identical faces. The six identical faces on the cube in Figure 11.13a have indices (001), (010), (100), $(00\bar{1})$, $(0\bar{1}0)$, and $(\bar{1}00)$. We symbolize the entire form {100}, and the { } braces indicate the form contains all faces with the numerals 1, 0, and 0 in their Miller index.

The four faces on the tetrahedron in Figure 11.13b, and the eight faces on the octahedron in Figure 11.13c are all equilateral triangles. For both, the form is {111}. As Figure 11.13b and c demonstrate, two crystals of different shapes can have the same form if they belong to different point groups. The tetrahedron in Figure 11.13b belongs to point group $\bar{4}3m$; the octahedron in Figure 11.13c belongs to point group $4/m\bar{3}2/m$. If we know the point group and the form, we can calculate the orientation of faces. If a crystal contains only one form, we then know the shape of the crystal. Note that the cube (Figure 11.13a), octahedron (Figure 11.13c), and dodecahedron (Figure 11.13d) all

Box 11.1 The Miller Indices of Planes Within a Crystal Structure

We use Miller indices to describe the orientation of crystal faces, but we also use them to describe planes within a crystal structure. We calculate the indices in the same way, with the exception that we do not clear common denominators after inversion of axial intercepts when we are talking about planes within a structure. For example, if we calculate an index of (240) for a set of planes, we do not divide by 2 to give (120), as we do when calculating a Miller index for a crystal face. We do not clear fractions because, besides orientation, the spacing and location of planes are important when we are talking about crystal structure and not crystal faces.

Figure 11.12a shows some two-dimensional unit cells cut perpendicular to the c axis. The plane passing through points A and A′, and perpendicular to the page, has axial intercepts 1, ½, ∞, so its Miller index is (120). Because all unit cells are equivalent, we know identical (120) planes exist in all the unit cells.

Figure 11.12a shows the entire family of (120) planes; when we discuss (120) planes in a crystal structure, we refer to this entire family, not just to one plane. Four (120) planes intersect each unit cell, two pass through the inside of the cell, and two through the corners in our drawing (Figure 11.12b). Note that the axial intercepts of the (120) plane closest to the origin (O) are 1, ½, ∞, equivalent to $1/h$, $1/k$, $1/l$. This relationship holds true for any family of (hkl) planes.

Different families of planes, with the same orientation but different spacings, have different indices. Figure 11.12c shows planes spaced half as far apart as those in Figure 11.12b. They have the Miller index (240). We do not clear the common denominator (divide by 2) because the families of (240) and (120) planes, although parallel, are not identical. There are twice as many (240) planes.

The consequence that the (hkl) plane closest to the origin intercepts the axes at distances $1/h$, $1/k$, $1/l$ follows from algebraic considerations. The algebraic equation of a plane cutting three axes (X, Y, and Z) at absolute distances $x°$, $y°$, and $z°$ from the origin is:

$$x/x° + y/y° + z/z° = 1 \qquad (11.3)$$

In terms of Miller indices, the corresponding equation for a plane is:

$$h\mathrm{x} + k\mathrm{y} + l\mathrm{z} = Q \qquad (11.4)$$

where Q is a rational number. In terms of crystallographic axes a, b, and c:

$$ha + kb + lc = Q \qquad (11.5)$$

where a, b, and c are coordinates of a point in the plane. This simple relationship is the equation of a plane in a-b-c space in terms of Miller indices. When $Q = 1$, the equation describes the closest plane to the origin with Miller indices (hkl). All other values of Q yield parallel planes farther from the origin.

---

belong to point group $4/m\overline{3}2/m$. The cubic form is {100}, the octahedral form is {111}, and the dodecahedral form is {110}.

Figure 11.13e shows a crystal containing three forms: cube {100}, octahedron {111}, and dodecahedron {110}. Because they all belong to point group $4/m\overline{3}2/m$, we can calculate that the faces are *oriented* as shown. However, the crystal in Figure 11.13f belongs to the same point group and contains the same forms, but the *size* and *shape* of corresponding faces are different. We do not know crystal shape if more than one form is present, unless we know some extra information.

Crystallographers sometimes label faces of the same form with the same letter (Figure 11.13a–f). For some forms, the letter is just the first letter of the form name. For example, o indicates the octahedral form and d the dodecahedral form in the cubic system. Usually, however, the symbols are less obvious (we normally designate cube faces, for example, by the letter a); they also vary from one crystal system to another.

General Forms and Special Forms

The relationship between special forms and general forms is the same as the one between special points and general points discussed in Chapter 9. Faces of a general form are neither parallel nor perpendicular to any symmetry element, while faces of special forms are. Thus, in special forms, symmetry relates fewer

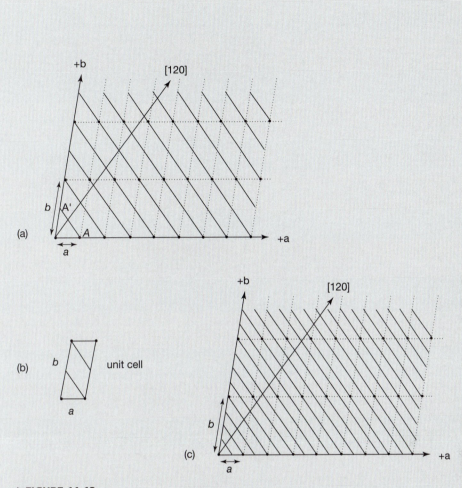

▶FIGURE 11.12
Planes and Miller indices. (a) AA' is one of a family of (120) planes. The direction [120] is not perpendicular to (120). (b) The complete family of (120) planes. (c) One unit cell and four (120) planes that intersect it. (d) The family of (240) planes.

equivalent faces. For example, the faces of a cube (Figure 11.13a) are perpendicular to the 4-fold axes of rotation, and the faces of an octahedron (Figure 11.13c) are perpendicular to the 3-fold axes of rotation. The presence of zeros, or equal values of h, k, or l, in the Miller index, suggests that these forms may be parallel or perpendicular to symmetry elements, no matter the crystal system. In the cubic system, $\{h00\}$ will always be a cube, $\{hhh\}$ an octahedron, and $\{hh0\}$ a dodecahedron, for all values of h. The general form in the cubic system is a hexoctahedron $\{hkl\}$, with h, k, and l all unequal, which has 48 faces (Box 11.2). Its standard designation, $\{123\}$, indicates that $h \neq k \neq l$ in its Miller index, and that none of the indices is zero. We can think of a hexoctahedron (Figure 11.15a) as an

octahedron (Figure 11.13c) with each of its eight faces replaced by six smaller triangular faces. In all point groups, the general form is the one with the most faces.

Zones and Zone Axes

As discussed in Chapter 9, a set of faces parallel to a common direction defines a zone. The common direction is the zone axis. Figure 11.14a, b, and c show examples of forms in the tetragonal and hexagonal systems. The prismatic faces, parallel to c in all three drawings, form [001] zones. The zone axes are [001]. Zones may contain faces from more than one form. In Figure 11.14c, the [010] zone comprises eight faces and three forms, $\{100\}$, $\{001\}$, and $\{101\}$. In all

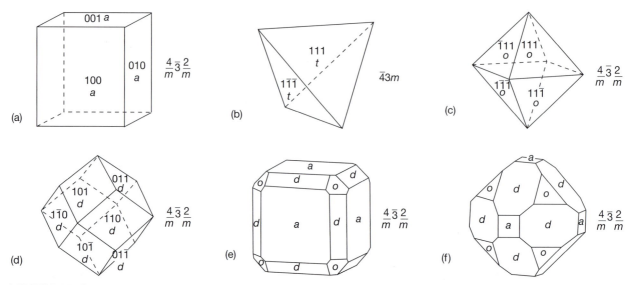

▶FIGURE 11.13
Cubic forms and Miller indices:(a) cube; (b) tetrahedron; (c) octahedron; (d) dodecahedron; (e) a combination of cube, octahedron, and dodecahedron; (f) a different combination of cube, octahedron, and dodecahedron.

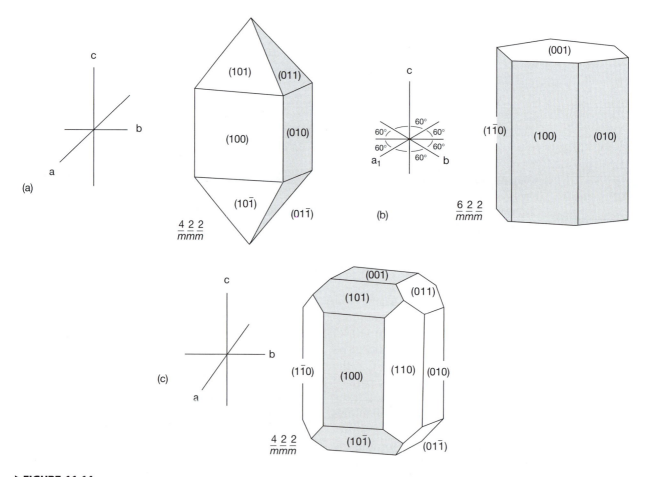

▶FIGURE 11.14
Zones in tetragonal and hexagonal crystals. The crystal in (a) is a tetragonal prism and dipyramid.The gray faces belong to the [100] zone. The crystal in (b) is a hexagonal prism and pinacoid. The gray faces belong to the [001] zone. The crystal in (c) includes a tetragonal prism, dipyramid, and pinacoids. The gray faces belong to the [010] zone.

Box 11.2 Comparison of a Hexoctahedron With Other Forms in the Cubic System

The hexoctahedron is the general form in the cubic system; its Miller index contains three different values ($h \neq k \neq l \neq 0$). All other forms are special forms and can be thought of as derivatives. Special forms have two or more identical indices in their Miller index (or have an index with value zero). If groups of faces on a hexoctahedron are systematically combined to produce one face, special forms result with no change in symmetry. Figure 11.15 shows a hexoctahedron, a trapezohedron, and a cube. The trapezohedron is formed by systematically combining pairs of faces on the hexoctahedron and replacing them with a single face. The cube is formed by systematically combining groups of eight faces and replacing them with a single face. The cube and the trapezohedron are special forms; unlike the hexoctahedron, their faces are perpendicular or parallel to symmetry elements. In the case of the cube, the faces are obviously perpendicular to 4-fold rotational axes. Less clear is that the faces of the trapezohedron are perpendicular to mirror planes. Similar relationships between general forms, symmetry elements, and special forms can be demonstrated for all point groups. The coincidence of faces with symmetry elements means that there can be fewer faces without reducing the symmetry.

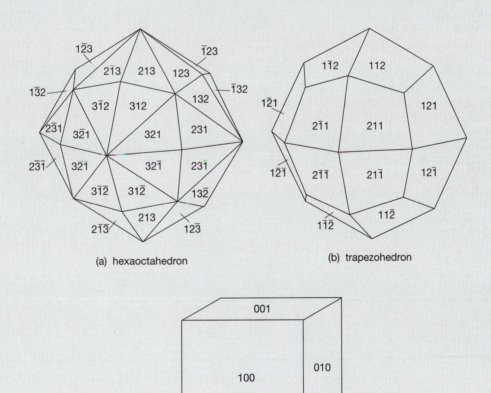

(a) hexaoctahedron

(b) trapezohedron

(c) cube

▶**FIGURE 11.15**
The general form (hexoctahedron) and some special forms (trapezohedron, cube) for point group 4/m $\bar{3}$ 2/m. Note that all faces of a form have similar Miller indices.

three drawings, additional zones are present besides [010] and [001]. The stippled faces in Figure 11.14a are part of the [100] zone; the unstippled faces are part of the [010] zone; the prismatic faces comprise the [001] zone. In this figure, the zones have simple indices because they are parallel to crystal axes; in many crystals zones are not parallel to axes, and consequently the indices contain values other than 0 and 1.

▶QUESTIONS FOR THOUGHT

Some of the following questions have no specific correct answers; they are intended to promote thought and discussion.

1. We use indices to discuss the orientations of lines and planes in crystals. Why?

2. We use unit cell parameters to describe the shape of a unit cell. Why do we need fewer parameters for some crystal systems than for others?

3. When we describe the unit cell of enstatite, we might give its formula as $Mg_2S_2iO_6$ and say that $Z = 4$ (Table 11.1). What does this mean? On the other hand, we could choose a unit cell twice as large, and then Z would be 16. Why don't we do this? Alternatively, can we choose a smaller unit cell so that $Z = 1$? Why don't we do this?

4. If you look at the mineral descriptions in Part III of this book, you will find that some cleavages are listed using parentheses (hkl), and some are listed using braces, $\{hkl\}$. For example, the cleavage of chrysoberyl is described as "good $\{011\}$, poor (010)." Why do we use two different notations?

5. Consider the crystal form with Miller index $\{111\}$. Will it be the same for crystals in all point groups? Why? Will it be the same for crystals in point groups belonging to one crystal system?

6. Consider the point group $4/m2/m2/m$. Possible forms include pinacoid, tetragonal dipyramid, tetragonal prism, ditetragonal pyramid, and ditetragonal dipyramid. They are all shown in at least one of the drawings in Figure 11.16, each containing one, two, or three forms. Which of the forms is the general form? Possible Miller indices for these forms are $\{001\}$, $\{010\}$, $\{0kl\}$, $\{hkl\}$, and $\{hk0\}$. Which forms and which of the faces in the drawings have these indices?

7. Draw stereo diagrams and plot the locations of all the faces in the crystals shown in Figure 11.16.

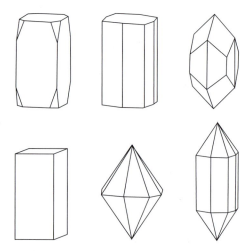

▶FIGURE 11.16
Some possible crystals belonging to point group $4/m2/m2/m$.

▶RESOURCES

(See the general references listed at the end of Chapter 1 and specific references listed for Chapters 9 and 10.)

X-ray Diffraction

X-ray diffraction is a phenomenon routinely used by mineralogists, but the discovery of X rays and X-ray diffraction are relatively modern developments. They have been understood for less than 100 years. Diffraction studies allow rapid mineral identification, even on very small samples. They also facilitate the determination of crystal structure. This chapter discusses how diffraction works and examines its historical and contemporary applications.

Mineralogists studied minerals for hundreds of years before the discovery of **X rays.** By the late nineteenth century, they believed that crystals had ordered and repetitive crystal structures. They hypothesized about atomic arrangements and the nature of crystal structures, but they lacked direct evidence. Some ideas about crystal structure were generally accepted, while others were poorly understood and hotly debated. Without a way to test hypotheses, the development of an acceptable theory of the crystal structure of minerals was stalled. William Konrad Röntgen's discovery of X rays in 1895 allowed mineralogists to proceed with their studies and eventually led to a greater understanding of crystal structures. Mineralogists quickly discarded many hypotheses disproved by X-ray studies; just as quickly, they developed and tested new ones. In less than two decades, scientists developed a firm theoretical basis for understanding the crystal structure of minerals.

Today we accept without question the idea that atoms bond together in regular arrangements to make crystals. We draw pictures, make enlarged models, and study the details of crystal structures of thousands of minerals. All this knowledge would have been unobtainable if Röntgen and his coworkers had not recognized the importance of some curious phenomena they observed while studying **cathode ray tubes** (early versions of television tubes) in 1895.

Röntgen taught and studied physics at the University of Würtzburg in Germany. He was studying the relationship between matter and force as charged particles flowed from a heated filament in an evacuated glass tube when, by chance, he observed that a nearby piece of barium platinocyanide fluoresced when he turned on the tube. Röntgen deduced that electrons interacting with the walls of the tube produced a high-energy form of radiation, and he showed that the radiation could penetrate paper and even thin metals. Because the radiation seemed to behave differently from light, he thought it a completely different phenomenon, and called it **X-radiation.** Various physicists searched for an understanding of the nature of X-radiation. X rays became inextricably involved with mineralogy when Max von Laue successfully used crystals as a tool in this search.

The diffraction of light was well understood at the time. Physicists routinely measured the wavelengths of colored light using finely spaced **diffraction gratings.** In 1911 von Laue, a physics professor in Münich, determined that lines in diffraction gratings were not spaced closely enough to diffract X rays. This partly explained Röntgen's confusion about the relationship between light and X rays. Von Laue hypothesized that the distance between atoms in crystals, being much less than the distance between lines in diffraction gratings, could lead to X-ray diffraction.

Walter Friedrich and Paul Knipping confirmed von Laue's theories in 1912 when they caused X rays to pass through crystals of copper sulfate ($CuSO_4$) and sphalerite (ZnS) and recorded diffraction patterns on film. Their studies showed that X rays are electromagnetic waves similar to light, but with much shorter wavelengths. They also confirmed that crystals must have a regular crystal structure. The impact of the work by von Laue and his students was immense. Paul Debye and Paul Scherrer soon improved X-ray techniques. They showed that all crystals have a lattice and that lattices vary in their symmetry. In 1913 William H. Bragg and his son William determined the crystal structure of sphalerite, ZnS, using data obtained with X-ray studies. For the first time, mineralogists knew the actual locations of atoms within a crystal. Other crystal structure determinations soon followed as Linus Pauling and other scientists realized the power of X-ray diffraction.

Early X-ray studies were tedious. Scientists could spend an entire career determining the crystal structure of only a few minerals. Today, with sophisticated equipment and high-speed computers, crystal structure determinations are often routine and sometimes can be done in less than a day. It is important to remember, however, that without the pioneering work of Röntgen, von Laue, the Braggs, Pauling, and others, we would have no detailed knowledge of the crystal structure of minerals. In 1901 Röntgen received the first Nobel Prize for Physics; von Laue was awarded the same prize in 1914, and both Braggs were the recipients in 1915.

WHAT ARE X RAYS?

X rays are a form of electromagnetic radiation (Figure 12.1). Whereas the wavelengths of visible light are 10^{-6} to 10^{-7} meters, X-ray wavelengths are only 10^{-8} to 10^{-12} meters. Long-wavelength X rays grade into ultraviolet light; shorter wavelengths grade into cosmic and gamma rays. Mineralogists usually give X-ray wavelengths in Angstroms (1Å equals 10^{-10} meters). The copper radiation commonly used in X-ray studies has $\lambda = 1.5418$Å.

The frequency (ν) and wavelength (λ) of electromagnetic radiation are inversely related. **Planck's law** relates them to energy:

$$E = h\nu = hc/\lambda \qquad (12.1)$$

In this equation h is Planck's constant and c is the speed of light in a vacuum. Because of their short wavelengths and high frequencies, X rays have high energy compared with visible light and most other forms of electromagnetic radiation. High energy al-

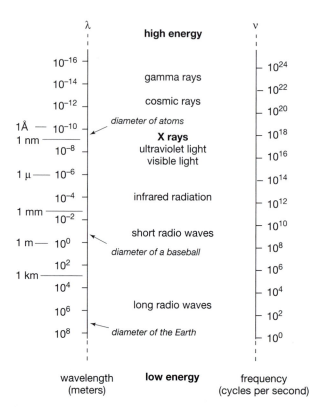

►**FIGURE 12.1**

Electromagnetic radiation covers a spectrum of wavelengths ranging over 24 orders of magnitude. Short-wavelength radiation, equivalent to gamma rays or cosmic rays, has high energy. Long-wavelength radiation has low energy. X rays, which are relatively high energy forms of radiation, have wavelengths between those of cosmic rays and visible light.

lows X rays to penetrate many natural materials, as observed by Röntgen in 1895. X rays of highest energy, called **hard radiation,** are used in many manufacturing and industrial applications, such as checking steel for flaws. X rays of relatively low energy, called **soft radiation,** are used by mineralogists and for medical diagnoses. Soft radiation is the most dangerous to people because rather than passing through tissue like many hard X rays and gamma rays, soft X rays interact with atoms in cells and tissues, causing damage. Because of the potential health hazards, crystallographers take special care to avoid exposure to the X rays in their experiments.

A beam of high-velocity electrons striking a metal target in an X-ray tube generates X rays for diffraction studies (Box 12.1). The high-velocity electrons collide with electrons orbiting atomic nuclei in the target material and bump them temporarily into higher energy levels. As the target electrons return to lower energy levels, they emit energy as X rays. The energy difference between the two levels is proportional to the energy and frequency, and inversely proportional to the wavelength, of X rays emitted. Typical X-ray tubes emit **polychromatic** radiation

Box 12.1 X-ray Tube

For diffraction studies, X rays are generated in an evacuated X-ray tube. Within the tube, a tungsten cathode filament releases electrons that accelerate and travel at high velocity to strike a metal anode. The accelerating voltage is typically tens of kilovolts, but the current is always very low. Mineralogists generally use a copper X-ray tube, but they use molybdenum, iron, or other tubes in special applications.

When the high-velocity electrons strike the target metal with sufficient energy, they cause electrons in target atoms to temporarily jump to higher energy levels. As the electrons naturally return to lower levels, the energy difference between the two levels is given off as X rays (Figure 12.2). X-ray tubes generate a tremendous amount of heat and must be constantly cooled by flowing water.

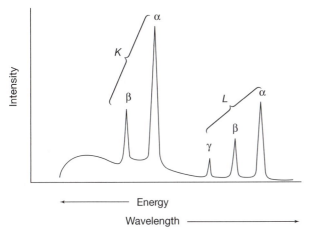

►FIGURE 12.2
Ouput from an X-ray tube. The peaks are the characteristic radiation. The wavelengths of the characteristic radiation depend on the target metal in the X-ray tube.

(Figure 12.2). The broad whaleback-shaped background (called *white*, or **continuous radiation**) is one result, while sharp peaks of high intensity (called **characteristic radiation**) occur at different wavelengths that depend on the metal in the target of the X-ray tube and the voltage. Each peak corresponds to a different electron transition, and a different X-ray energy and wavelength. We designate peaks using combinations of Roman (K, L, M) and Greek (α, β, γ) letters. The Greek letter indicates the energy level an electron reaches before it decays; the Roman letter indicates the energy level after decay. The most intense peaks are designated $K\alpha$.

We use copper $K\alpha$ radiation for most routine X-ray studies. Copper has several characteristic wavelengths, but interpretation of X-ray studies is easiest if we use one wavelength **(monochromatic radiation).** To isolate $K\alpha$ radiation from the other wavelengths, most X-ray machines have filters, monochromators, or

solid-state monochromatic detectors. X-ray tubes emit two nearly equal wavelengths of Cu-$K\alpha$ radiation: $K\alpha_1 = 1.5401\text{Å}$ and $K\alpha_2 = 1.5443\text{Å}$. They are so similar that, even if both wavelengths are present, for most applications the radiation is effectively monochromatic, and we take a weighted average of $K\alpha_1$ and $K\alpha_2$ wavelengths and assume a λ value of 1.5418Å.

INTERACTIONS OF X RAYS AND ATOMS

When an X ray strikes an atom, the wavelike character of the X ray causes electrons, protons, and neutrons to vibrate. Heavy protons and neutrons vibrate less than much lighter electrons. The oscillating electrons reemit radiation, called **secondary radiation,** at almost the same frequencies and wavelengths as the incoming beam. This process, called **scattering,** is not the same for all elements, nor is it the same in all directions. Since heavy elements have atoms with more electrons, they scatter more efficiently than light elements, and scattering by heavy elements can completely mask scattering by light ones. As a result, X-ray crystallographers often have trouble determining the location of light atoms, such as hydrogen, in crystal structures. As X rays scatter in different directions, they interact with electron clouds in various ways. Overall, those scattered at high angles to the incident beam are less intense than those scattered at low angles.

Besides being scattered, when X rays interact with atoms in a crystal, some electrons temporarily bump up to higher energy states. As the electrons return to their normal state, a release of radiation characteristic of the target atom (in the crystal) occurs. This process, called **X-ray fluorescence (XRF),** is similar to the interaction of electrons and atoms in the target metal of an X-ray tube, but results from interaction of X rays and atoms in the crystal. X-ray fluorescence, while not widely used by mineralogists, is the basis for a common analytical method used by petrologists and geochemists.

INTERFERENCE OF X-RAY WAVES

Just like visible light, X rays propagate in all directions and interact, or **interfere,** with each other while they produce **constructive** or **destructive interference.** If we could move an X-ray detector around two point sources emitting monochromatic X rays, we would find that energy is intense in some directions because the waves interact constructively. In other directions, the detector would register no X rays due to destructive interference (Figure 12.3). This channeling of energy in specific directions is *diffraction.* The directions depend on X-ray wavelength and distance between the two X-ray sources.

A narrow slit, a series of parallel grooves in a diffraction grating, regularly spaced atoms in a crystal, and many other things can cause diffraction. The main requirement is that two or more sources emit, or scatter, monochromatic waves. Diffraction resembles **reflection** in many ways, but reflection refers to the coherent scattering of energy by atoms in a two-dimensional surface such as a mirror. In principle, all electromagnetic radiation can be diffracted but unless the spacing of atoms, slits, or gratings is simi-

lar to the wavelength of the radiation, diffraction will not occur. The table on the inside front cover lists elemental atomic radii; they are all on the order of Ångstroms, about the same as the wavelength of X rays. Because atoms pack closely together in crystals, their spacing is of the same magnitude as X-ray wavelengths, and atoms in crystals can produce intense X-ray diffraction. This is the understanding that led von Laue and his coworkers to their successful experiments.

DIFFRACTION BY A ROW OF ATOMS

To discuss diffraction by a row of atoms, we can use the analogy of a group of campers standing on a lake dock. One camper throws a rock in the water, and a series of waves propagate across the lake. The front of the moving wave, the **wave front,** will have a circular or arc shape (Figure 12.4a). It loses energy and dies out as it propagates, but a canoeist might feel a small ripple as it passes by. Next, the campers roll a very long log off the dock and create a more pronounced wave that propagates across the lake. This wave front is straight, instead of curved (Figure 12.4b). Canoeists might notice the straight wave front as it passes underneath their boat.

Now, the campers on the dock all drop rocks into the water at different times, making many circular wave fronts. The wave fronts work with each other constructively in some directions and destructively in others. Most are out of phase and die out quickly. The canoeists hardly notice the waves. Finally, the campers drop their rocks simultaneously, creating a wave similar to the one created by the log (Figure 12.4c). All of the individual waves are in phase, and they add to make a straight wave front moving across the lake. The canoeists cannot determine whether a log or rocks generated the wave.

When a row of atoms scatters an X-ray beam, the atoms can produce a coherent wave front similar to the one produced when rocks hit water simultaneously. Figure 12.4c could be showing a row of atoms each emitting an X-ray wave with identical wavelength. If all the atoms emit a wave simultaneously, the waves will create a straight wave front moving in a direction 90° from the row of atoms. A situation similar to Figure 12.4c might arise if an incident monochromatic X-ray beam strikes a row of atoms at 90°. As shown in Figure 12.4d, the incident beam excites all the atoms at the same time, causing them to scatter coherent X rays. Constructive interference then yields wave fronts moving perpendicular to the row of atoms as shown in Figure 12.4c, and, as we will see shortly, in other directions.

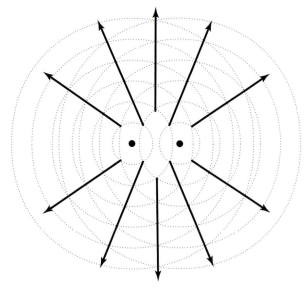

▶FIGURE 12.3
Two point sources interfere constructively in some directions (arrows) and destructively in others. The dotted-line circles show the wave fronts from both sources. Energy is channeled (diffracted) in directions (arrows) where the wave peaks coincide because the waves add constructively.

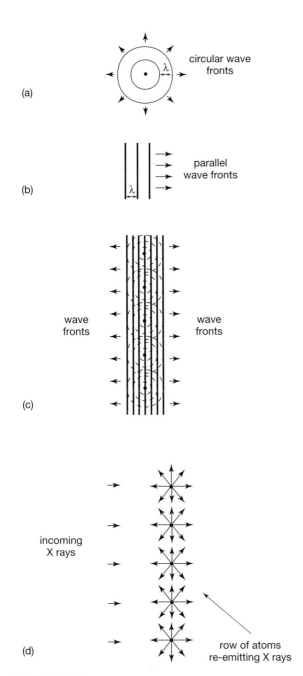

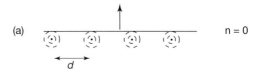

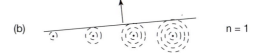

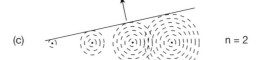

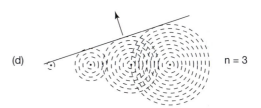

▶**FIGURE 12.5**
Wave fronts generated by a row of atoms. (a) Atoms emitting circular wave fronts that may combine to form one front moving away from the row at 90°. (b) through (d) Wave fronts moving in other directions.

▶**FIGURE 12.4**
Wave fronts in a lake. (a) A circular wave front is created by a rock dropped into water. (b) Parallel wave fronts result when a long log is dropped into water. (c) Parallel wave fronts can also be caused by multiple rocks dropped in the water. (d) When an X-ray beam strikes a row of atoms, the atoms reemit X rays to produce wave fronts similar to those in Figure 12.4c.

Figure 12.5a shows circular wave fronts emitted by atoms in a row. The waves, which have all traveled the same distance, combine to form straight wave fronts moving perpendicular to the row. Figure 12.5b, c, and d show that coherent wave fronts also travel in other directions. In Figure 12.5b, waves emitted by adjacent atoms differ in travel distances by 1λ, in c

by 2λ, and in d by 3λ. Because one cycle of a wave is the same as any other, it does not matter that they have traveled different distances if they are in phase.

The geometry in Figure 12.5 yields a simple relationship between wavelength, atomic spacing, and the angle of diffraction. If α is the angle between the diffracted wave front and the row of atoms, d is the distance between atoms, and n is the number of wavelengths that each wave is behind (or ahead of) the one next to it, then as pictured in Figure 12.6a:

$$n\lambda = d \sin \alpha \qquad (12.2)$$

n is the **order of diffraction.** When n = 0, 0th order diffraction occurs and α must be 0. When n = 1, first-order diffraction occurs; when n = 2, second-order diffraction occurs, and so on. As n increases, the angle between the diffracted wave front and the row of atoms increases. The maximum value of n corresponds to $\sin \alpha = 1$ (when $\alpha = 90°$), so n must always be less than d/λ. If $\lambda > d$, no diffraction can occur. This explains why atoms in crystals do not diffract visible light: The wavelengths are too long compared with the atomic spacings.

▶**FIGURE 12.6**
A closer look at diffraction by a row of atoms. (a) An incident wave front strikes a row of atoms at 90° and causes diffraction at angle α. (b) An incident beam strikes a row of atoms at angle α and causes diffraction at angle α'.

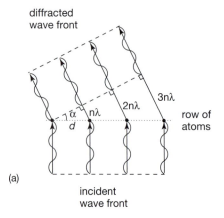
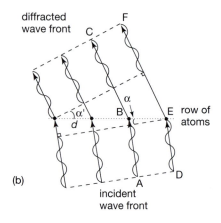

The preceding discussion assumed that an incident X-ray beam struck a row of atoms at 90°, but this is rarely the case. When the incident radiation strikes a row of atoms at another angle, coherent diffraction will occur in any direction where X rays are in phase. To analyze this situation, we need only consider two rays (Figure 12.6b). They will be in phase if the difference in their path lengths, the difference between paths ABC and DEF, is an integral number of wavelengths ($n\lambda$). For the geometry shown in Figure 12.6b, we can easily verify that:

$$d(\sin \alpha' - \sin \alpha) = n\lambda \qquad (12.3)$$

where d is the atomic spacing, α' and α are the angles of the diffracted and incident wave with the row of atoms, n is the order of diffraction, and λ is the wavelength. This is the general equation for diffraction by a row of atoms.

PLANES OF ATOMS

The Distance Between Planes in Crystals

A regularly spaced row of atoms that causes diffraction is only part of a crystal structure (Figure 12.7a). In two dimensions, adjacent unit cells combine to create many parallel rows, all causing diffraction at the same angle (Figure 12.7b). In three dimensions, an entire crystal contains many identical unit cells containing atoms with identical spacing that diffract at the same angle (Figure 12.7c). The regular spacing defines a family of planes, all separated by the same distance. Thus, diffraction involves sets of planes of atoms, diffracting at angles we can calculate with Equation 12.3.

As discussed in Chapter 11, crystallographers use Miller indices (hkl) to describe planes of atoms in crystals. We use the symbol d_{hkl} for the perpendicular distance between (hkl) planes. Figure 12.8

shows a single (120) plane; Figure 12.8b shows an entire family of (120) planes separated by the perpendicular distance d_{120}. Different families of planes, with the same orientation but different indices, have different d-values. Figure 12.8c shows the family of (240) planes; they parallel (120) but are half as far apart. For all crystal systems, d_{120} is twice d_{240}, and in general:

$$d_{hkl} = n \, d_{h'k'l'} \qquad (12.4)$$

where

$$h' = n \, h \qquad (12.5)$$
$$k' = n \, k \qquad (12.6)$$
$$l' = n \, l \qquad (12.7)$$

The relationship between Miller indices, unit cell lengths (a, b, c), and d-values depends on the crystal system. Box 12.2 gives equations relating unit cell parameters to d-values. Derivation of equations for orthogonal systems (cubic, tetragonal, and orthorhombic) is relatively simple compared with derivations for the other systems. For orthogonal systems:

$$d_{100} = a \qquad (12.11)$$
$$d_{010} = b \qquad (12.12)$$
$$d_{001} = c \qquad (12.13)$$

For the other systems, the angles α, β, and γ must also be considered. Figure 12.9 shows this by comparing d-values to unit cell lengths for square, rectangular, and monoclinic cells.

Diffraction by Planes of Atoms

Von Laue derived equations similar to Equation 12.3 that describe diffraction by three-dimensional structures. We call these the **Laue equations.** W. L. Bragg developed a simpler and more easily understood mathematical treatment. His final equations are just as valid as von Laue's, but the derivations avoided

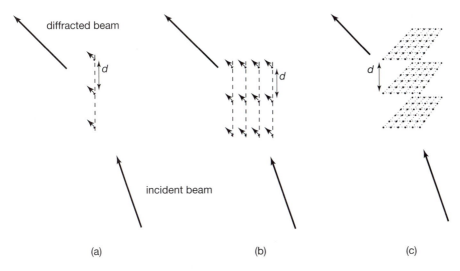

FIGURE 12.7
Diffraction by a row of atoms (a), by parallel rows of atoms (b), and by planes of atoms (c).

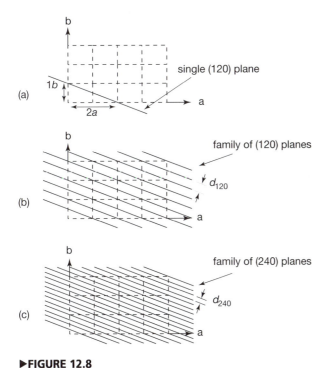

►**FIGURE 12.8**
Planes and *d*-values: (a) a single (120) plane; (b) the family of (120) planes; (c) the family of (240) planes. In these drawings, the c-axis and the planes are perpendicular to the plane of this page.

some unnecessary complexities. Although diffraction and reflection are two different processes, Bragg noted that diffracted X rays behave as if they were reflected from planes within a crystal. To model this "reflection," Bragg considered two parallel planes of atoms separated by distance d_{hkl} (Figure 12.10). ABC and DEF represent monochromatic parallel X-ray beams striking and reflecting from the planes. The angle of incidence

and the angle of reflection are both θ. Because path lengths of the beams must vary by an integral number of wavelengths if diffraction is to occur, the sum of distances GE and EH must equal nλ, so:

$$n\lambda = 2\,d_{hkl}\,\sin\theta \qquad (12.14)$$

Equation 12.14 is known as the Bragg Law. Bragg went on to prove unnecessary the consideration of situations in which incident and diffraction angles are different (Figure 12.11a). We can always describe diffraction using reflection geometry, even though true reflection is not occurring (Figure 12.11b). The angle of "reflection" θ is related to the incident angle (α) and the diffraction angle (α') by

$$\theta = (\alpha - \alpha')/2 \qquad (12.15)$$

For conventional X-ray diffraction studies, we assume the n in the Bragg Law is 1. Because first-order diffraction by planes with spacing d_{hkl} occurs at the same angle as second-order diffraction by a set of planes spaced twice as far apart, we cannot distinguish the two, and assuming first order is simplest when we are talking about X-ray diffraction (Figure 12.11c, d). In other applications of the Bragg Law, the order of diffraction is important.

In standard X-ray diffraction studies, we allow the incident X-ray beam to strike the sample at many different angles so that many different *d*-values will satisfy the Bragg Law and cause diffraction (Figure 12.12). After being diffracted, the X-ray beam travels at an angle of 2θ from the incident beam. Because a crystal contains many differently spaced planes with atoms on them, diffraction occurs at many 2θ angles. In most X-ray devices used today, X-ray source and sample are at fixed locations (although the sample may rotate),

Box 12.2 *Cell Parameters and d-values*

Each diffraction spot or line on film, or diffraction peak on a diffractogram, corresponds to a specific set of (hkl) planes having spacing d_{hkl}. d_{hkl} is a function of h, k, and l, and of the unit cell parameters $a, b, c, \alpha, \beta,$ and γ:

$$d_{hkl} = \sqrt{\frac{A}{B+C}} \qquad (12.6)$$

where

$$A = 1 - \cos^2\alpha - \cos^2\beta - \cos^2\gamma \\ + 2\cos\alpha \cos\beta \cos\gamma \qquad (12.7)$$

and

$$B = h^2 \sin^2\alpha/a^2 = k^2 \sin^2\beta/b^2 \\ + l^2 \sin^2\gamma/c^2 \qquad (12.8)$$

and

$$C = 2hk(\cos\alpha \cos\beta - \cos\gamma)/ab \\ + 2lk(\cos\gamma \cos\alpha - \cos\beta)/ca \qquad (12.9) \\ + 2kl(\cos\beta \cos\gamma - \cos\alpha)/bc$$

This equation is greatly simplified for orthogonal systems (cubic, tetragonal, and orthorhombic) because all angles are 90°, $A = 1$, and $C = 0$. For these systems, then:

$$d_{hkl} = \sqrt{\frac{1}{\left(\dfrac{h^2}{a^2} + \dfrac{k^2}{b^2} + \dfrac{l^2}{c^2}\right)}} \qquad (12.10)$$

▶FIGURE 12.9
Comparison of *d*-values to unit cell parameters: (a) in a square cell $d_{100} = d_{010} = a = b$; (b) in a rectangular cell $d_{100} = a$ and $d_{010} = b$; (c) in a monoclinic cell d_{100} and d_{010} are not equal to unit cell parameters.

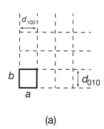
(a)

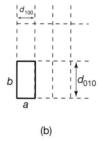
(b)

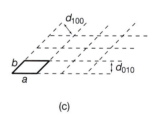
(c)

▶FIGURE 12.10
Reflection geometry. Two rays in an X-ray beam must differ in path length by an integral number of wavelengths if they are to be in phase after "reflection." This observation results in the Bragg Law.

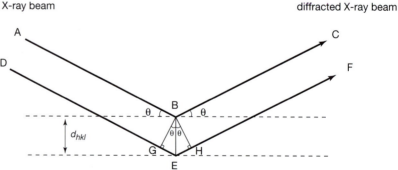

incident X-ray beam

diffracted X-ray beam

and a detector, which measures X-ray intensity, moves through an entire range of angles from near 0 to some high angle. If a diffracted beam is to hit the detector, two requirements must be met:

- A family of planes with d_{hkl} must be oriented at an angle (θ) to the incident beam so the Bragg Law is satisfied.

- The detector must be located at the correct angle (2θ) from the incident beam to intercept the diffracted X rays.

Because the angle between the detector and the X-ray beam is 2θ, mineralogists usually report X-ray data in terms of 2θ. For example, reference books list the two most intense diffraction peaks of quartz (using

▶FIGURE 12.11
Why we can use the Bragg Law. Diffraction by a row of atoms (a) can be modeled as reflection by an imaginary plane (b). (c) and (d) show that the angle of diffraction (or reflection) is the same for first-order diffraction by planes separated by d_1 and for second-order diffraction by planes spaced twice as far apart (d_2).

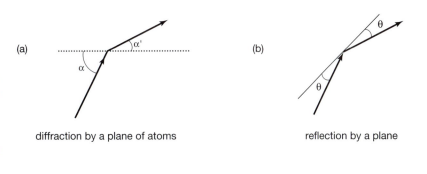

(a) diffraction by a plane of atoms

(b) reflection by a plane

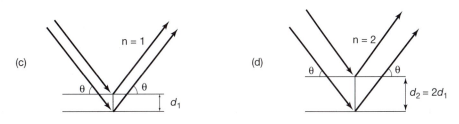

(c)

(d)

▶FIGURE 12.12
Bragg Law geometry. In most conventional X-ray diffractometers the detector rotates around the sample to measure intensity of diffracted beams over a wide range of 2θ values. Compare this with Figure 12.18.

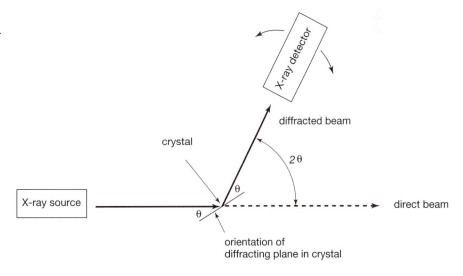

Cu-$K\alpha$ radiation) as 26.66° and 20.85° 2θ. We must remember to divide 2θ values by 2 before we use them with the Bragg Law to calculate d-values.

INTENSITY OF DIFFRACTION

Diffraction at Different Angles

The Bragg Law allows us to calculate the angle at which diffraction occurs for any particular d_{hkl} value. However, intense diffraction will only occur if many atoms occupy the (hkl) planes because without atoms no electrons are present to scatter X rays. When discussing families of planes in unit cells, h, k, and l may have any integer value, which implies the possibility of an infinite number of d_{hkl} values that could satisfy the Bragg Law. However, mineral unit cells contain a small number of atoms, which in turn limits the number of d-values corresponding to planes of high atomic density. Therefore, the number of angles at which intense diffraction occurs is limited, although weak diffraction occurs in many, sometimes more than 1,000, directions for a single crystal. An example in two dimensions will make this point.

Consider the hypothetical unit cell shown in Figure 12.13a, with all atoms on two opposite edges. It has dimensions of 3.0Å × 5.0Å. Ignoring the third dimension (or assuming that all planes are perpendicular to the page), we can say that $d_{100} = 3.0$Å and $d_{010} = 5.0$Å. If we imagine a structure made of many of these unit cells (Figure 12.13b), we see that many atoms are in rows 5.0Å apart, and diffraction corresponding to d_{010} (5.0Å) should be quite intense. We can also expect apparent diffraction by the (020) planes to be intense, although no atoms

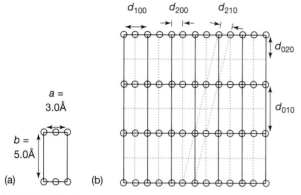

▶**FIGURE 12.13**

Prediction of X-ray peaks: (a) a unit cell with atoms only on two opposing sides; (b) a collection of unit cells showing some of the families of planes with many atoms on them. These are the planes that will produce the strongest diffraction.

occupy every other (020) plane, because second-order ($n = 2$) diffraction by (010) planes occurs at the same angle as first-order diffraction by (020) planes. In addition, (200) diffraction (d_{200} =1.5Å) will be quite intense because many atoms occupy the (200) planes. Diffraction for other d-values will be weaker.

Because the locations and natures of atoms are different in different minerals, diffraction patterns depend on mineral structure and composition. But symmetry also affects diffraction patterns. High crystal symmetry means identical spacing of atoms in multiple directions, so fewer d-values cause diffraction, and thus diffraction occurs at fewer angles with relatively uniform intensity. Low crystal symmetry means diffraction occurs in many directions, with variable intensity because atoms and atomic spacings differ in all directions. Figure 12.14 compares X-ray patterns for fluorite (a cubic mineral) and barite (an orthorhombic mineral). As predicted, the pattern for fluorite is more regular, with fewer but more intense diffraction peaks.

Extinctions

Sometimes, due to destructive interference, planes containing many atoms do not produce diffraction. We term this phenomenon a *systematic absence,* or an *extinction.* Figure 12.15a shows a two-dimensional drawing of a cubic structure in which atoms occupy the corners and the center of each unit cell. When (010) diffraction occurs at angle θ, beam X and X′ are completely in phase as shown in Figure 12.15b. If the diffraction is first order, beam X travels exactly 1λ farther than beam X′. However, the constructive interference cannot prevail when we account for the atoms between the (010) planes. X rays scattered by these

atoms travel $\frac{1}{2}\lambda$ more or less than those scattered by the adjacent (010) planes to create perfect destructive interference (Figure 12.15c). No X rays will be found at the angle θ, though it satisfies the Bragg Law for d_{010}. This effect is extinction, and we would say the (010) diffraction peak is *extinct.* Although (010) planes do not diffract, (020) planes do, but the smaller d-value results in a larger diffraction angle (Figure 12.15d). The structure shown in Figure 12.13 also produces extinctions. (100) diffraction is absent because additional atoms occur halfway between the (100) planes.

Figures 12.13 and 12.15 show two-dimensional examples of structures that result in extinctions due to end-centered and body-centered unit cells. In three dimensions, end-centered and body-centered structures also produce extinctions. We often describe extinctions using arithmetical rules. For body-centering, the rule is that an (hkl) peak will be extinct if $h + k + l$ is an odd number. Besides centering, other symmetry elements lead to extinctions. They all involve symmetry in which planes of atoms are found between other planes; screw axes and glide planes are two examples. The systematic extinction of certain X-ray peaks, then, is one way we determine space group symmetry.

SINGLE CRYSTAL DIFFRACTION

When X-raying single crystals, determining diffraction directions can be quite complicated. In the preceding discussions and figures we only considered two dimensions; both the incoming X-ray beam and the diffracted beam were in the plane of the paper because we assumed the planes causing diffraction were perpendicular to the page. Suppose the diffracting planes are not perpendicular to the page of the paper. If so, diffraction might still occur, given an appropriate angle of incidence, but the diffracted beam could go off in any direction, including straight out of the page. In an experiment where we place a stationary single crystal of a mineral in front of an X-ray beam, unless we carefully orient the crystal, the odds of the beam hitting planes of atoms at angles that satisfy the Bragg Law are very small. Diffraction is unlikely, but if it occurs, the diffracted beam could travel in any direction. The odds that it will hit a detector, even if the detector is moving in a plane, are minuscule. So how can we expect to measure diffraction directions and intensities?

Several techniques overcome this problem. When they made their pioneering studies in 1912,

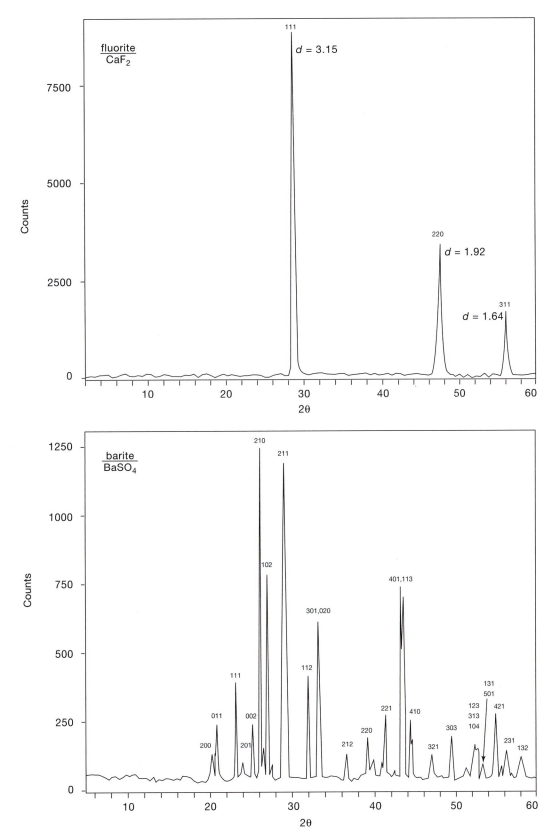

▶**FIGURE 12.14**

Comparison of X-ray patterns for fluorite and barite. The barite pattern is more complex and has more peaks because barite's atomic structure has less symmetry than fluorite's. The numbers above the peaks are *hkl* values.

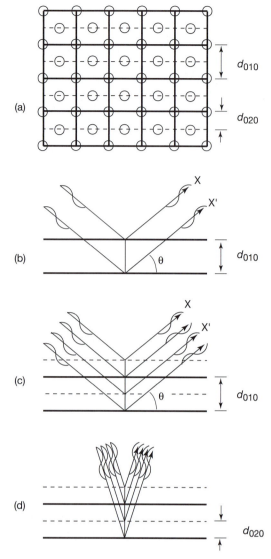

(a)

(b)

(c)

(d)

▶**FIGURE 12.15**
Extinctions occur when diffracted rays are out of phase.
(a) Some cubic unit cells with atoms (open circles) at their
corners and centers. (b) (010) diffraction produces some
rays that are in phase and diffract at an angle θ. (c) Atoms
between the (010) layers produce other rays diffracting at
the same angle but 180° out of phase, so the rays interfere
destructively and the intensity of diffraction is 0. (d) (020)
diffraction does occur, but the angle of diffraction is
greater than that for (010).

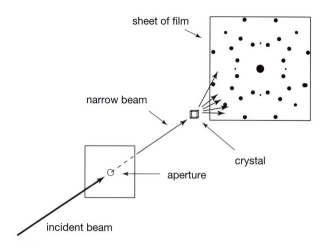

▶**FIGURE 12.16**
The Laue technique. A focused X-ray beam hits an oriented
crystal and the diffracted rays produce dark spots on a piece
of film that is shielded from light by a black envelope.

hours, the film developed black spots where hit by
diffracted beams. The size and density of each spot
varied according to the intensity of the diffracted
beam that created it.

The Laue technique reveals interesting information. For example, two isostructural minerals, if oriented the same way, yield similar patterns. Laue
patterns also reveal specific information about crystal structures. Of most importance, the symmetry of
the X-ray pattern is always greater than or equal to
that of the pattern. If a pattern reveals no symmetry,
the crystal's structure is not constrained, but if, for
example, a pattern shows 6-fold symmetry, the crystal must belong to the hexagonal system. A short-
coming of the Laue technique is that although it
reveals angles of diffraction, we cannot determine
which diffraction peaks correspond to which X-ray
wavelengths. Therefore, using the Bragg Law to
learn d-values is not possible. On the other hand, if
we use monochromatic radiation, we get few, if any,
spots on the film.

Mineralogists have made many modifications to
the Laue method. When using monochromatic radiation, one way to increase the probability of satisfying
the Bragg Law is to rotate the crystal about one or
even two axes. If we rotate through all possible orientations, many settings satisfy the Bragg Law. We can
place film behind the crystal, or we can even wrap
film around it. Alternatively, we can use an X-ray detector instead of film, although we must move it all
around the crystal in three-dimensional space to
avoid missing any diffraction peaks. All these techniques are kinds of **single crystal diffraction.**

For most of the twentieth century, the most common single crystal techniques used **Buerger precession cameras** and **Weissenberg cameras** (Figure 12.17);

Friedrich and Knipping used what we now call *the
Laue technique.* They carefully oriented a single
crystal in front of a polychromatic X-ray source so
that principal crystal faces and lattice planes were
perpendicular or parallel to the X-ray beam. Due to
the careful orientation and multiple X-ray wavelengths, many different families of planes satisfied
the Bragg Law. Behind the crystal, Friedrich and
Knipping placed a piece of film, shielded from light
in a black envelope (Figure 12.16). After several

▶FIGURE 12.17

(a) A Weissenberg camera for collecting single crystal X-ray patterns. The crystal is in the center of the cylinder with the film wrapped around it. Both the crystal and the film move as the pattern is collected. (b) A precession camera for collecting single crystal X-ray patterns. We mount the crystal, generally no larger than 1 mm in longest dimension, on a small spindle in the front-center of the apparatus. The square black envelope in the rear contains the film. We can use this same geometry with a detector instead of film to collect a diffraction pattern digitally. Photographs courtesy of Charles Supper Company, Natick, Massachusetts.

both synchronize crystal rotation with film movement to produce diffraction patterns on film. Since diffraction occurs in three dimensions, and photos are two-dimensional, we need multiple photos to record an entire spectrum of diffraction peaks. A single photograph, for example, might show all diffraction peaks corresponding to $(0kl)$ planes, but none corresponding to nonzero values of h.

Instead of a camera, an instrument equipped with an X-ray detector called a **diffractometer** is more often used today. A diffractometer rotates the crystal and moves the detector to measure intensities of all diffraction peaks. Computers controlling the diffractometer store data directly on disk. Because the process is completely automatic, the d-values and intensities of thousands of different peaks may be measured for a single crystal.

We use the data obtained from single-crystal diffraction studies to determine crystal structures. If we know the composition of a mineral, we know how many atoms of which elements are present. Computer programs can determine where atoms are in the structure, based on the diffraction data. We call this process a *crystal structure determination*. Crystal structure determination involves many complexities, but automation and high-speed computers have simplified the process considerably since the Braggs determined the first crystal structure in 1913.

ROUTINE X-RAY ANALYSES

Powder Diffraction

Single-crystal X-ray studies provide information necessary for structure determination. Simpler techniques exist for mineral identification (Figure 12.18). The most common is **powder diffraction,** the X-raying of a finely powdered sample mounted on a slide or in a holder. A powdered sample consists of a near-infinite number of small crystals in random orientations. When so many crystals in so many orientations are X-rayed, for any set of (hkl) planes there will be many crystals that satisfy the Bragg Law.

Powdered samples diffract beams in many directions, as shown in the two-dimensional drawing in Figure 12.19a. In three dimensions, they produce diffraction cones (Figure 12.19b). The angles between the cones and the direct X-ray beam are 2θ values for different (hkl) planes. Consider a powdered sample of some mineral with a set of (111) planes having a d-value of 5.0Å. For Cu-$K\alpha$, then, the Bragg Law tells us this corresponds to a 2θ value of 17.74°. When the sample is X-rayed, a sufficient number of crystals will be oriented to cause diffraction in all directions 17.74° from an X-ray beam; the result is a cone of diffraction. In most crystals, many d-values cause measurable diffraction. Each produces a cone at a different angle (2θ) to the X-ray beam

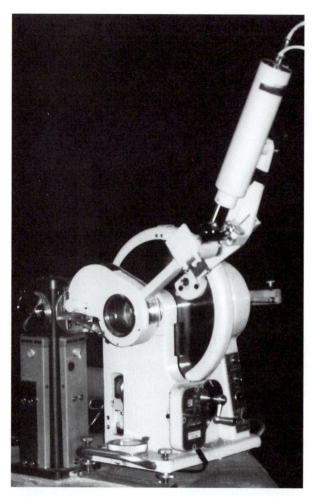

A powder diffractometer: This sort of instrument has been a mainstay for routine mineral analyses for many years; the pictured instrument is over 20 years old. Today's models are enclosed in X-ray proof boxes for safety, and in many of them the X-ray source and the detector both move. See, for example, Figure 12.22b.

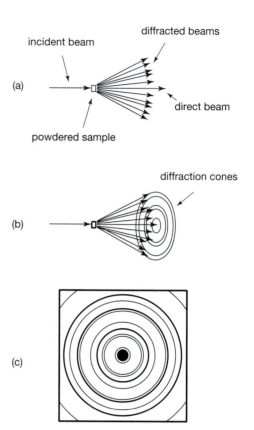

▶**FIGURE 12.19**
Geometry of powder diffraction. (a) Diffraction occurs in many directions but is symmetrical with respect to the direct X-ray beam. (b) In three dimensions this leads to cones of diffraction at many different angles to the direct X-ray beam. (c) If a flat piece of film is placed behind the crystal, it records circles corresponding to each cone.

(Figure 12.19b). If we measure all angles, we can use the Bragg Law to calculate all *d*-values corresponding to planes with atoms on them. The intensity of each diffraction peak depends on the number and kind of atoms on the planes causing the diffraction.

As with single-crystal diffraction, mineralogists use several different techniques to record powder diffraction data. Before about 1970, a photographic technique was most common. We could position a sheet of flat film to register the diffracted beams, producing circles with radii proportional to sin2θ, and darkness proportional to intensity of diffraction (Figure 12.19c). However, beams diffracted at moderate and high 2θ angles cannot strike the film.

A more often used alternative has historically been the **Debye-Scherrer camera** (Figure 12.20). We mount a small amount of powdered specimen on a

spindle positioned in the center of the camera and encircled by a piece of film (Figure 12.20). A light-proof window allows an X-ray beam to enter the camera and strike the specimen. The Debye-Scherrer camera allows diffraction at all angles, from just a few degrees to nearly 180°, to be recorded. Diffraction cones leave dark lines and arcs where they hit the film. We can determine 2θ values by measuring the distance from the center of the film to each diffraction line; we can then calculate *d*-values. Figure 12.21 shows a powder pattern recorded with a Debye-Scherrer camera.

Today most mineralogists use an X-ray powder diffractometer instead of a camera. We mount the powdered sample on a slide or in a well in a sample holder instead of on a spindle (Figure 12.22a). A focused X-ray beam strikes the sample, and the detector moves in a circular arc, usually vertically, to measure diffraction intensities for 2θ values from very low angles to more than 150°. In practice, most diffractometers cannot measure peaks at angles below 1 or 2° 2θ because the direct X-ray beam bombards the detector. The upper 2θ limit of mea-

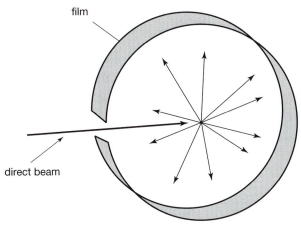

▶**FIGURE 12.20**
Drawing and photo of a Debye-Scherrer camera. Cameras of this sort are rarely used today.

▶**FIGURE 12.21**
Filmstrip showing a powder pattern. The film is 35mm film cut to the right length to fit into the Debye-Scherrer camera. The dark lines are the diffraction lines caused by the cones of diffraction striking the film.

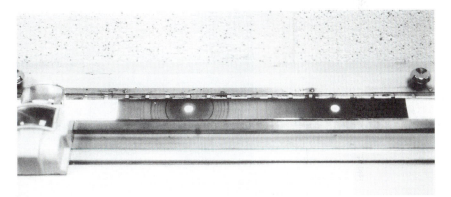

surement usually depends on the need of the mineralogist, but for most purposes, we do not need data above 60° to 70° 2θ. Figures 12.18 and 12.22b show a typical powder diffractometer used by mineralogists.

In the past, diffractometers were connected to strip chart recorders to produce diffraction patterns that typically contained 10 to 50 identifiable peaks. A mineralogist would then measure the diffraction angles on the chart and calculate *d*-values. Today we connect diffractometers to computers that can analyze data and print diffraction patterns. The diffraction patterns in Figure 12.14 were produced that way. Diffractometers connected to computers have many advantages over earlier techniques. The biggest advantages are automation, speed, and the ease with which data can be processed. Camera techniques may require hours to record diffraction patterns while, in contrast, automated diffractometers can collect and analyze a pattern in tens of minutes. Computers can produce paper copies at any scale or form wanted, whereas interpretation of strip charts is slow and problematic.

For simple mineral identification, powder diffractometers connected to computers generally require 15 to 30 minutes. For extremely precise results, however, data collection may take hours or even days.

Identifying Minerals from Powder Patterns

Powder patterns are not normally used to determine crystal structures because powder diffraction data are more difficult to interpret than single crystal data. Different X-ray peaks correspond to different crystals in the powdered sample, and the data do not reveal the orientations of planes causing diffraction. A relatively new technique for interpreting powder patterns, called the **Rietveld method,** sometimes overcomes these complications and is gaining popularity today. However, powder patterns do yield a list of *d*-values and peak intensities that mineralogists routinely use to identify minerals. Each *d*-value corresponds to sets of planes, and the intensity is a measure of how many atoms are on those planes. Because each

▶**FIGURE 12.22**
Powder diffraction is the most commonly used X-ray technique today: (a) Powdered samples may be mounted on glass slides or (less commonly) in wells on metal plates, for X-ray analysis; (b) a modern powder diffractometer.

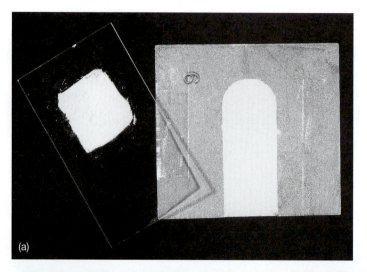

(a)

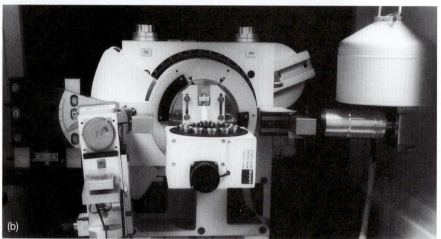

(b)

mineral has a different crystal structure, each yields a different pattern. For mineral identification, we compare measured *d*-values and intensities to reference data sets. The most complete reference, the X-ray Powder Diffraction File (PDF) compiled by the International Centre for Diffraction Data, is available in printed form or in computerized databases and contains information on almost 6,000 minerals and tens of thousands of other inorganic and organic crystalline compounds.

Box 12.3 gives an example of data that are retrievable from most X-ray databases. The data are for fluorite and barite, minerals with diffraction patterns displayed in Figure 12.14. The data include a complete list of *d*-values and relative intensities (in the right-hand column); the intensity of the strongest peak is always assigned a value of 100%. Box 12.3 follows the standard practice of listing the *d*-values corresponding to the most intense diffraction peaks, and the *d*-value corresponding to the lowest angle diffraction peak, at the top of the record. These *d*-values are the most useful for mineral identification. Basic crys-

tallographic information, including cell parameters, space group, point group, and other ancillary data are also included in most databases. Additionally, references to pertinent X-ray investigations are given.

The X-ray patterns in Figure 12.14 match the reference *d*-values in Box 12.3 quite well. The minor discrepancies probably stem from slight compositional differences between the samples X-rayed. The relative intensities of peaks do not match as well. For example, the most intense peak for fluorite in Figure 12.14 is listed as the second most intense peak in Box 12.3. X-ray peak intensities listed in databases commonly do not exactly match those obtained when we X-ray samples in the laboratory. Different samples, different methods of sample preparation, and many other things can lead to deviations. *d*-values corresponding to strong peaks, however, should always match closely, even if the intensities are not quite in the same order.

Once we learn *d*-values and intensities for an unknown mineral, it may seem an overwhelming task to find a match in a large reference file. It sometimes is.

Box 12.3 Example of Data from the Powder Diffraction File Data Set

Three strongest peaks, largest *d*-value, and relative intensities

d	1.93	3.15	1.65	3.15
I/I_{100}%	100	94	35	94

Z:	4
Crystal system:	cubic
a:	5.463Å
Space group:	$F4/m\bar{3}2/m$
Point group:	$4/m\bar{3}2/m$
Ref. index (*n*):	1.4338
Specific gravity:	3.181

Crystal structure determination: Bragg, W. L. The structure of some crystals as indicated by their diffraction of X-rays. *Proc. Roy. Soc.* 115 (1914): 705.

CaF_2 fluorite

Complete list of diffraction peaks and relative intensities (%)

d Å	I/I_{100}	*hkl*	*d* Å	I/I_{100}	*hkl*
3.153	94	111			
1.931	100	220			
1.647	35	311			
1.366	12	400			
1.253	10	331			
1.115	16	422			
1.051	16	422			
0.966	5	440			
0.923	7	531			
0.911	1	600			
0.864	9	620			
0.833	3	533			

Three strongest peaks, largest *d*-value, and relative intensities

d	3.45	3.10	2.12	5.58
I/I_{100}%	100	95	80	2

Z:	4
Crystal system:	orthorhombic
a:	8.87Å
b:	5.45Å
c:	7.14Å
Space group:	*Pnma*
Point group:	$2/m2/m2/m$
Ref. index (α, β, γ):	1.636, 1.637, 1.648
Specific gravity:	4.50

Crystal structure determinations: Basche, W., and H. Mark. Über die struktur von verbindungen des typus $MeXO_4$. *Zeit. Krist.* 64 (1926): 1; James, R. W., and W. A. Wood. The crystal structure of barites, celestine and anglesite. *Proc. Roy. Soc.* 109A (1925): 598. Rinne, F., H. Hentschel, and E. Schiebold. Zum feinbau von anhydrit und schwerspat, *Zeit. Krist.* 61 (1924): 164.

$BaSO_4$ barite

Complete list of diffraction peaks and relative intensities (%)

d Å	I/I_{100}	*hkl*	*d* Å	I/I_{100}	*hkl*
5.58	2	101	2.106	75	122,312
4.440	16	200	2.057	19	410
4.339	30	011	1.949	1	222
3.899	50	111	1.932	7	321
3.773	12	201	1.858	18	303
3.577	30	002	1.789	4	004
3.445	100	210	1.762	8	031
3.319	70	102	1.758	10	123,313
3.103	95	211	1.754	8	104
2.836	50	112	1.728	4	131
2.735	15	301	1.724	5	501
2.729	45	020	1.682	8	230
2.482	13	212	1.674	14	421
2.447	2	121,311	1.670	11	114
2.325	14	220	1.660	2	204
2.305	6	103	1.644	3	511
2.282	8	302	1.638	8	231
2.211	25	221	1.626	1	403
2.169	3	022	1.594	8	132
2.121	80	113,401	1.591	6	502

Box 12.4 Identification of Minerals from X-ray Patterns the Old Way

Before the development of computer databases to identify minerals from powder diffraction patterns, identification was done by hand with catalogs and other lists of *d*-values and *2θ* values. One of the more popular sets of search manuals is available from the International Centre for Diffraction Data. This set of manuals is still sometimes used today to supplement or replace computer searching methods.

The search manuals consist of three large books: *The Alphabetical Index—Inorganic Phases; The Search Manual (Hanawalt)—Inorganic Phases;* and *The Alphabetical Index & Search Manual— Organic Phases.* Minerals are included in the two inorganic volumes. The books also list chemical formulas, the *d*-values (to the nearest 0.01Å) of

the three most intense peaks for each compound, and the Powder Diffraction File (PDF) number of each compound, which allows users to obtain a complete list of *d*-values from PDF databases or reference cards. The alphabetized manual is most useful if you want to compare the data from your sample with a few selected minerals. If you do not know the identity of a sample, the Hanawalt search manual is more useful. The Hanawalt search manual lists the names and formulas of thousands of compounds, ordering them first by the *d*-values of their most intense peak and then by the *d*-values of their second most intense peak. The Hanawalt search manual also includes the *d*-values of the third through the eighth most intense peaks.

However, often we need only consider a few of the most intense diffraction peaks. Mineralogists have used several different schemes when manually comparing an unknown pattern to reference files; the **Hanawalt method,** developed by J. D. Hanawalt in 1936, has been most popular (Box 12.4). Today, instead of doing the matching manually, most mineralogists use computer databases to match unknown patterns with PDF data. This has two main advantages: It is rapid, and many more reference patterns can be considered. On the other hand, computer searches are "black box" procedures; they do not improve our knowledge or understanding of powder diffractometry. In addition, computers sometimes give absurd answers that a human would have rejected at a glance.

Powder diffraction is not an exact process, and many things cause X-ray patterns to deviate from those in reference files. Proper sample preparation is crucial to obtaining accurate patterns. Complications arise when we do not grind a sample properly, or when powdered crystals are not in truly random orientations. For example, because micas are hard to grind and their tiny flakes tend to pile up parallel to each other, obtaining random orientation and a good X-ray pattern is difficult. A second problem, more so for natural materials than for synthetics, is that small compositional and crystal structural variations can affect X-ray patterns. Despite these complications, the PDF data contain adequate reference patterns for most mineralogical uses.

Indexing Patterns and Determining Cell Parameters

We have labeled the peaks on the X-ray patterns in Figure 12.14 with (*hkl*) indices corresponding to the planes causing diffraction. For identifying unknown minerals, we do not need to know which (*hkl*) indices correspond to which *d*-values, but for other purposes, such as determining unit cell dimensions, we must. We call the process of matching *d*-values and (*hkl*) *indexing.* The easiest way to index a pattern is to compare it with one for a similar mineral that is already indexed. For instance, if we measured a pattern of an unknown garnet, we could compare it with the pattern for almandine (Box 12.5). This is quick and simple. Without a pattern to compare, the process becomes more complex and, except perhaps for cubic minerals, computer programs are desirable or necessary.

When indexing a cubic pattern by hand, a good general approach is to start with the largest *d*-values (peaks at lowest *2θ* angles) and most intense peaks. Large *d*-values always correspond to small indices, and high intensity usually suggests simple indices, for example (100), (110), or (111). We may not be able to detect some low-angle peaks if they are very weak or extinct. For primitive lattices, the lowest angle diffraction peak (the largest *d*-value) may correspond to (100). Due to extinctions, for body-centered and face-centered lattices, respectively, the largest possible *d*-values correspond to (110) and to (111). Once

Box 12.5 Indexing a Garnet Pattern and Determining a

To calculate unit cell dimensions from an X-ray pattern, we must first determine which *hkl* values go with which X-ray peaks. The process, called *indexing*, is best accomplished by comparing the "unknown" pattern to one that is already indexed. The table below lists the *d*-values for the five most intense peaks from a powder X-ray diffraction pattern of an unknown garnet. For comparison, the *d*- and *hkl*-values for a "known" almandine are given. By comparing the two lists, we can assign *hkl* values to the peaks in the unknown's pattern.

Garnets are cubic, so *a* is the only unit cell parameter to determine. We can derive the relation-ship between d_{hkl} and *a* from the equations in Box 12.2:

$$a^2 = (h^2 + k^2 + l^2)d^2 \qquad (12.16)$$

Once we index the peaks in the unknown's pattern, we calculate the value of *a* from each of the d_{hkl} values. The table below lists results for five peaks. The five values range from 11.82 to 11.90Å. The range is due to inaccuracies in determining 2θ from the X-ray pattern. Checking in the reference section at the back of this book, we find that our unknown garnet is probably grossular.

d-values and Miller Indices for an Unknown Garnet and for Almandine

d-values for the unknown garnet	Similar *d*-values for the "known" almandine	*hkl* values for the "known" almandine	Calculated values of *a* for the unknown garnet
1.583	1.540	642	11.82
1.652	1.599	640	11.90
1.924	1.866	611	11.84
2.653	2.569	420	11.85
2.961	2.873	400	11.84

we assume indices for some low-angle peaks, we can check to see if they are consistent with other indices for higher angle peaks. Having indexed a pattern, we can derive the unit cell parameters ($a, b, c, \alpha, \beta, \gamma$). With *d*-values for many (*hkl*) peaks, we use the equations in Box 12.2 and a least-squares approach

to derive the six cell parameters. For crystals belonging to systems with high symmetry, our task is simplified because we know some of the six parameters and others may be redundant (see Table 11.1). Box 12.5 gives an example of deriving the single unit cell parameter (*a*) for a garnet.

▶QUESTIONS FOR THOUGHT

Some of the following questions have no specific correct answers; they are intended to promote thought and discussion.

1. Why was the discovery of X rays a "key" to crystallography? What was missing before the discovery of X rays?

2. Shortly after Linus Pauling did some pioneering X-ray studies, he wrote a book on the nature of the chemical bond. What is the connection between X-ray studies and bonding theory?

3. What process does a crystallographer follow to determine a crystal's atomic structure? Are the results interpretations, or do crystal structure studies yield exact answers? Why?

4. Fluorite and sodalite are both cubic minerals, yet fluorite has a much simpler X-ray pattern. Why?

5. Do all crystals, including nonminerals, diffract X rays? Do all minerals diffract X rays? Why?

6. We use single crystal diffraction for most crystal structure studies. We normally use powder diffraction for

mineral identification. Why are different methods used for each?

7. In the diffraction pattern for barite (Figure 12.14), some of the strongest diffraction peaks are at 20.167, 25.081, 26.056, 27.063, 28.967, 33.033, 42.805, and 43.104 2θ. If you convert the angles to d-values using the Bragg Law (Equation 12.14) and compare the d-values to those listed in Box 12.3, you can index these peaks. Use the d-values for peaks with indices containing two zeros to determine a, b, and c for barite using Equation 12.10 in Box 12.2. How do your values compare to the reference values given in Box 12.3? Explain any discrepancies.

▶RESOURCES

(See also the general references listed at the end of Chapter 1.)

Azaroff, L. V. *Elements of X-ray Crystallography.* New York: McGraw-Hill, 1968.

Bish, D. L., and J. E. Post, eds. Modern powder diffraction. *Reviews in Mineralogy,* vol. 20 (1989). Washington, DC: Mineralogical Society of America.

Bragg, W. L. *The Crystalline State: A General Survey.* London: G. Bell and Sons Ltd., 1949.

13

Atomic Structure

With the development of X-ray diffraction techniques, mineralogists and chemists could directly investigate the nature of crystal structures. While they were conducting their studies, others were investigating the nature of chemical bonds. The combined results led to a new understanding of crystals and their structures. In this chapter we look at some of the details of crystal structure. We find that minerals have highly ordered atomic arrangements and that most can be described as spherical atoms held together by ionic, covalent, or metallic bonds. In some, the atoms are packed tightly together, while in others they are arranged in networks of geometric shapes. In most cases, they obey some simple and fundamental rules related to ionic size and charge.

THE IMPACT OF X-RAY CRYSTALLOGRAPHY

We cannot overstate the importance of the discoveries by Röntgen, von Laue, and the Braggs. Before their pioneering work, scientists could not test competing hypotheses for the nature of crystal structures. Within a few decades of the discovery and development of X-ray diffraction techniques, most of the basic principles of crystal structures were well known. While crystallographers were working on crystals, chemists were developing atomic theory. The Bohr model of the atom, the Schrödinger wave equation, and theories of ionic and covalent bonding were firmly established by the 1920s. Chemists such as Linus Pauling, realizing the importance of X-ray techniques, conducted X-ray studies in efforts to further understand crystal structure and bonding. In 1939 Pauling published *The Nature of the Chemical Bond;* he subsequently won the Nobel Prize in Chemistry in 1954.

IONIC CRYSTALS

Ionic crystals are those composed of cations and anions held together primarily by ionic bonds. They have overall electrical neutrality, or else electrical current would flow until they obtained charge balance, so the total number of electrons in the structure is equal to the total number of protons. Anions repel anions, cations repel cations, so ions of similar charge stay as far apart from each other as possible. Consequently, an organized and repetitive atomic arrangement, with cations packed around anions and anions packed around cations, typifies ionic crystals.

Consider the mineral halite, NaCl, which contains an equal number of Na and Cl atoms (Figure 13.1a–c). Mineralogists have determined its atomic structure through X-ray studies, finding that Na^+ and Cl^- ions pack around each other in an alternating three-dimensional structure. Each Na^+ bonds to six Cl^- and vice versa. Bonds around one ion are all equal length and at 90° to each other. Unit cells are

▶**FIGURE 13.1**
Halite and CsCl: (a) ball and stick model showing the halite structure (black spheres are Na$^+$, light spheres are Cl$^-$); (b) Na$^+$ in octahedral coordination; (c) an octahedron surrounding Na$^+$; (d) ball and stick model showing the structure of CsCl (dark spheres are Cs$^+$); (e) Cs$^+$ in cubic (8-fold) coordination; (f) a cube surrounding Cs$^+$.

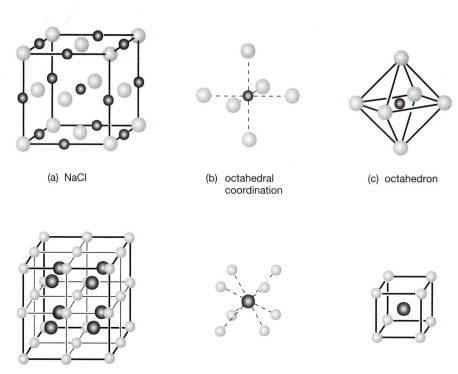

(a) NaCl

(b) octahedral coordination

(c) octahedron

(d) CsCl

(e) cubic coordination

(f) cube

therefore cubic, containing four Na$^+$ and four Cl$^-$ ions in a face-centered arrangement. Salt crystals, including the ones that come out of your salt shaker, are often perfect cubes. CsCl is also cubic, but in contrast with the halite structure, the alkali cation (Cs$^+$) is surrounded by eight anions (Figure 13.1d–f). The difference is because Cs$^+$ is larger than Na$^+$.

The ionic bonds between alkalis and Cl$^-$ are not terribly strong, and they break easily when salts dissolve in water, releasing free alkalis and Cl$^-$ ions. High solubility in water is characteristic of highly ionic crystals, especially those in which the ions only have charges of ±1. If concentrations of dissolved Na$^+$ and Cl$^-$ reach high enough levels, perhaps due to evaporation, halite may precipitate from solution. Other minerals have different atoms but have atomic structures and bonding similar to halite's; sylvite (KCl) and periclase (MgO) are both examples. In periclase, however, the ions are divalent, having a charge of ±2 (in contrast with halite and sylvite) and the bonds are 25% covalent. The stronger, more covalent bonds mean that periclase is harder and has lower solubility than sylvite and halite.

In some minerals, tightly bonded anion radicals such as carbonate $(CO_3)^{2-}$, sulfate $(SO_4)^{2-}$, or phosphate $(PO_4)^{3-}$ are present instead of simple anions. The radicals may not dissociate, even if the mineral dissolves in water, because covalent bonds hold the radicals together. Calcite, $CaCO_3$, is a good example. In the calcite structure, carbonate groups and Ca^{2+} ions alternate in three dimensions. However, the car-

bonate radicals are triangular, so the crystals cannot be isotropic and cubic like halite. When calcite dissolves, the ionic bond between calcium and the carbonate radical breaks easily, but the carbonate group itself does not dissociate into C and O. Consequently, dissolved species are Ca^{2+} and $(CO_3)^{2-}$.

IONIC RADII

Ions consist of nuclei with electron clouds around them. The electrons are constantly moving; sometimes they are farther away from the nucleus than at other times, so we can never know the exact size of the electron cloud. However, ions often behave as if they have fixed radii, and we can understand many crystal properties by thinking of crystals as collections of spherical ions packed together. The spacing between them, and the way they pack together, are directly related to their size. Knowing their **effective ionic radius** is therefore useful.

How can we determine ionic radii if ions really do not have a fixed radius? We estimate them by studying bond length in crystals. Consider the mineral periclase, MgO. Through X-ray diffraction studies we learn that the distance between the centers of the Mg^{2+} and O^{2-} ions is about 2.11Å, so the sum of the effective ionic radii of Mg^{2+} and O^{2-} is 2.11Å. Assuming for the moment that the radius of O^{2-} is 1.32Å, this yields a radius of 0.79Å for Mg^{2+} in periclase. An alternative way to find the radius of an Mg atom is to determine the distance between neighboring Mg atoms in Mg metal. Mg metal is not, however, ionically bonded, and the results do not tell us the radius of Mg^{2+}, although they do tell us that metallic Mg has a radius of about 1.60Å.

If we study many compounds, it becomes possible to learn the effective ionic radii of all individual elements. Consider the alkali oxides, which have the general formula R_2O, where R can be any alkali element. Each alkali ion is bonded to oxygen; chemists and mineralogists have determined the bond lengths through X-ray studies. Since the radius of O^{2-} is common to all alkali oxides, variations in bond length must be due to variations in the radii of the alkalis. If we assume O^{2-} has a constant radius of 1.32Å, we get the cation radii in Table 13.1. It should be no surprise that alkali radius increases with atomic number because we know elements with higher numbers have more protons and electrons, and so are larger atoms.

Radii also vary systematically across a row of the Periodic Table (Table 13.2). As we move from the margins of the periodic chart toward the center, cations get smaller because as cation charge increases, attraction between electrons and protons increases. Cation radii are always smaller than uncharged atoms of the same species. Table 13.2 also includes two anions (S^{2-} and Cl^-) from the right-hand side of the periodic chart; they are larger than the cations because they contain extra electrons in outer orbitals. Anion radii are always larger than uncharged

atoms of the same species. Note that we cannot list an ionic radius for argon because it is a noble gas and does not ionize to enter ionic structures.

We can best see the relationship between cation radius and charge by looking at elements that exist in more than one valence state. Table 13.3 gives ionic radii for manganese ions, and Table 13.4 shows similar data for vanadium. As expected, the radii decrease with charge, reflecting the greater pull nucleus protons have on outer electrons.

The radii of Na^+ given in Tables 13.1 and 13.2 are different because Table 13.1 refers to alkali oxides, whereas Table 13.2 contains average values for many different types of crystals. This distinction reminds us that the concept of ionic radius is an approximation because the effective radius of an ion depends on several things. Most significantly, radii are only constant if bond types are constant. For example, the average ionic radius of Mg^{2+} is 0.88Å, while its covalent radius is 1.36Å and its metallic radius is 1.60Å. Additionally, bond length varies with atomic structure and the number of anions surrounding a cation. A further complication may arise because in some structures ions become **polarized** (elongated) in one direction and no longer act as spheres. A final ambiguity arises because we must assume a value for the radius of O^{2-} to calculate radii for other elements. Depending on atomic structure, O^{2-} may have an effective radius between 1.27Å and 1.34Å, but the values in this book's tables are based on an assumed radius of 1.32Å. The inside front cover has a complete list of ionic radii based on this value.

▶**TABLE 13.1**
Radii of Alkali Cations in Alkali Oxides

Cation	Li^+	Na^+	K^+	Rb^+	Cs^+
Atomic Number	3	11	19	37	55
Cation Radius (Å)	0.82	1.40	1.68	1.81	1.96

▶**TABLE 13.2**
Average Radii of Ions from One Row in the Periodic Table (see inside front cover)

Ion	Na^+	Mg^{2+}	Al^{3+}	Si^{4+}	P^{5+}	S^{6+}	S^{2-}	Cl^-	Ar^0
Atomic Number	11	12	13	14	15	16	16	17	18
Ion Radius (Å)	1.18	0.79	0.55	0.41	0.25	0.20	1.72	1.70	—

▶**TABLE 13.3**
Average Radii Manganese Atom and Ions

Cation	Mn^0	Mn^{2+}	Mn^{3+}	Mn^{4+}	Mn^{6+}	Mn^{7+}
Cation Radius (Å)	1.12	0.97	0.70	0.62	0.35	0.34

▶**TABLE 13.4**
Average Radii of Vanadium Atom and Ions

Cation	V^0	V^{2+}	V^{3+}	V^{4+}	V^{+5}
Cation Radius (Å)	1.31	0.87	0.72	0.67	0.53

COORDINATION NUMBER

Most minerals, except native elements such as gold or copper, contain anions or anionic groups. Especially common are O^{2-}, S^{2-}, $(OH)^-$, and $(CO_3)^{2-}$. The large size of oxygen and other anions (and anionic groups) compared with most common cations (Figure 13.2) means that we can often think of crystal structures as large anions with small cations in **interstices** (spaces) between them. The anions are packed in a repetitive structure, with the cations at regular intervals throughout. The number of anions to which some particular cation bonds is the cation's **coordination number** (C.N.). Si^{4+}, for example, nearly always bonds to four O^{2-} in minerals, and therefore has a C.N. of 4. So we say it is "in 4-fold coordination."

The size of interstices between anions depends on how the anions are packed. In two dimensions, anions can fit together in symmetrical patterns to form hexagonal or square patterns (see Figure 13.5). In three dimensions, other possibilities including tetrahedral and cubic arrangements can exist. Figure 13.3 shows some ways that groups of identical anions may pack around a single cation. Since only one cation and one kind of anion are involved, all bond distances are the same. We give coordination arrangements geometrical names depending on the shape of the polyhedron created by connecting the centers of the anions (Figure 13.3). We call some 2-fold coordination *linear* because the ions form a line. We call 3-fold coordination *triangular* because the anions form a triangle. We call some 4-fold coordination *tetrahedral* because the four anions form tetrahedra. We call the 6-fold coordination, shown in Figure 13.3c, *octahedral* because the anions outline an octahedron (an eight-sided geometric shape). We call the 8-fold coordination, shown in Figure 13.3d, *cubic* because connecting the centers of the anions produces a cube. (Fortunately, we never call cubic coordination *hexahedral* even though a cube has six faces.) We call the 12-fold coordination, shown in

Figure 13.3e, *dodecahedral* because the coordinating polyhedron has twelve vertices. The coordination polyhedra shown in Figure 13.3 are all regular, meaning the cation to anion distance is the same for all anions. As C.N. increases, the space inside the polyhedron increases and larger cations can be accomodated. So, large cations have greater C.N.s than small cations. Cations sometimes occupy distorted sites or sites with unusual coordinations not represented by the drawings in Figure 13.3. Atoms in minerals typically have C.N. of 3, 4, 6, or 8, but 5-fold, 7-fold, 9-fold, and 10-fold coordination are possible. The atoms in some native metals are in 12-fold coordination. Most common elements have different coordinations in different minerals.

Figure 13.4 shows a typical ball and stick model for kaolinite. This model is convenient and easy to examine, but incorrect in detail since balls of similar size, separated by large distances, represent all atoms, anions, or cations. More accurate models of crystal structures could reflect variations in ionic radii. In principle we could construct very exact models, using balls of correctly proportioned sizes. In practice it is not often done because regular ball and stick models are easier to make and examine.

CLOSEST PACKING

Packing in Two Dimension

In some crystals, anions pack together in highly regular repetitive patterns. As an analogy, consider a collection of equal-sized marbles. We may arrange the marbles so rows line up and repeat at regular spacing characterized by translational symmetry. Figure 13.5 shows two alternative ways that marbles (shown as circles) can pack together in two dimensions. In Figure 13.5a groups of three marbles are arranged so that connecting their centers yields an equilateral triangle. On a slightly larger scale, each marble is surrounded by six others, and con-

▶**FIGURE 13.2**
The relative sizes and ionic charges of common cations and anions in minerals.

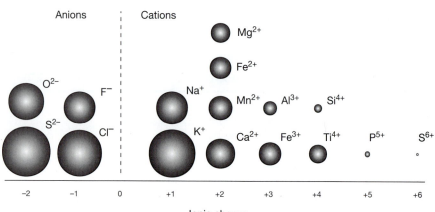

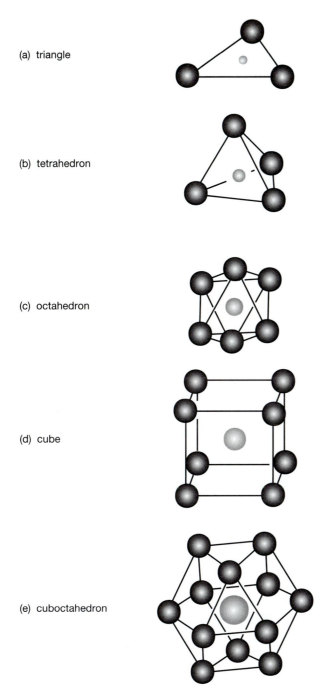

(a) triangle

(b) tetrahedron

(c) octahedron

(d) cube

(e) cuboctahedron

▶FIGURE 13.3

Packing of ions and coordination polyhedra: (a) a cation in triangular coordination; (b) a cation in tetrahedral (4-fold) coordination; (c) a cation in octahedral (6-fold coordination); (d) a cation in cubic (8-fold) coordination; (e) a cation in dodecahedral (12-fold) coordination.

necting their centers makes a hexagon. The total pattern has 2-fold, 3-fold, and 6-fold rotational symmetry, as well as translational symmetry and mirror planes (Figure 13.5b). Figure 13.5c shows an alternative arrangement of marbles in two dimensions. This arrangement has 2-fold and 4-fold axes of symmetry, but not 3-fold (Figure 13.5d). The symmetries of the two patterns in Figure

13.5 are equivalent to the symmetries of a hexanet and a square net (see Figure 10.14).

We call the arrangement of marbles in Figure 13.5a **hexagonal packing.** We call the pattern in Figure 13.5c **tetragonal packing** because of the obvious 4-fold symmetry. In hexagonal packing the marbles are closer together than in tetragonal packing. Because no other two-dimensional arrangement allows marbles to be closer together, we say the hexagonal arrangement is **closest packed.** Each marble touches six others, the maximum possible. In the tetragonal arrangement each marble only touches four others.

Packing in Three Dimensions

In three dimensions, marbles (or anions) can be closest packed in two ways. Both are equivalent to piling hexagonal packed sheets one above another. Three adjacent marbles in a hexagonal closest packed sheet make a triangle. If we put another marble on top of them, it slips into the low spot above the center of the triangle, resulting in a tetrahedral structure composed of four marbles (Figure 13.6a, b). All four marbles touch each other, so the arrangement is closest packed. Alternatively, we could put three marbles on top of the first three, as shown in Figure 13.6c, d. This arrangement of marbles, too, is closest packed. Each marble is in contact with three in the adjacent layer, but the second layer contains no marble directly above the center of the triangle in the first layer.

Now consider an entire layer of hexagonal packed marbles. If we put another hexagonal packed layer on top, its marbles naturally fall into gaps produced by groups of three in the bottom layer, so the marbles in the second layer will not be directly above those in the first (Figure 13.7). If we now place a third layer on top of the second, marbles will fill gaps as before. However, marbles in the third layer may or may not be directly above those in the first layer (Figure 13.8). If they are, we call the structure **hexagonal closest packed.** If they are not, we call it **cubic closest packed.** (Box 13.1) Although it may not be immediately obvious, if we add a fourth layer, its marbles must lie directly above those of one of the other layers. Hexagonal closest packed (HCP) structures are often described as having ABABAB layering because alternate layers are directly above each. Cubic closest packed (CCP) structures have an ABCABCABC packing sequence; it takes three layers before they repeat. In both HCP and CCP, every marble (or anion) is in contact with 12 others.

Exceptions to Closest Packing

Johannes Kepler first broached the idea of atoms as touching spheres in 1611. William Barlow described the systematics of closest packed structures more

▶**FIGURE 13.4**
Photograph of a ball and stick model of kaolinite. The white marbles represent Al, the black marbles Si, and the gray marbles are O and (OH).

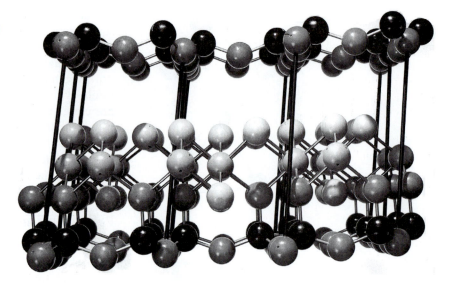

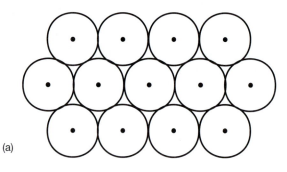

(a)

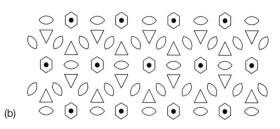

(b)

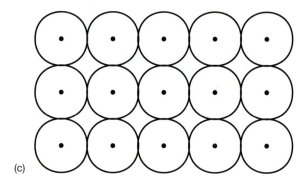

(c)

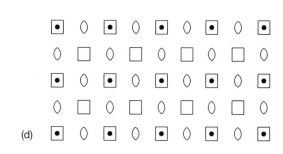

(d)

▶**FIGURE 13.5**
Circles arranged in hexagonal and cubic patterns: (a) hexagonal packing; (b) the rotational symmetry elements of hexagonal packing; (c) tetragonal packing; (d) the rotational symmetry elements of tetragonal packing. For comparison, the small dots show equivalent points in the packing diagrams and the symmetry diagrams. Mirror planes of symmetry have been ommitted for clarity.

than 250 years later in 1883. For a long time, all structures were thought to have simple repetitive closest packing. It was not long after the Braggs' X-ray studies led to the first crystal structure determination that scientists found exceptions. Anions in closest packed structures are arranged so that only tetrahedral and octahedral sites exist between them (Figure 13.6). We can use closest packing to describe metals, sulfides,

halides, some oxides, and other structures in which all cations are in tetrahedral or octahedral coordination. Sphalerite (ZnS), halite (NaCl), and native metals gold, silver, platinum, and copper are all examples of cubic closest packed minerals. Wurtzite (ZnS), magnesium metal, and zinc metal are hexagonal closest packed. Other mineral structures are, however, not truly closest packed. In general, dense minerals with

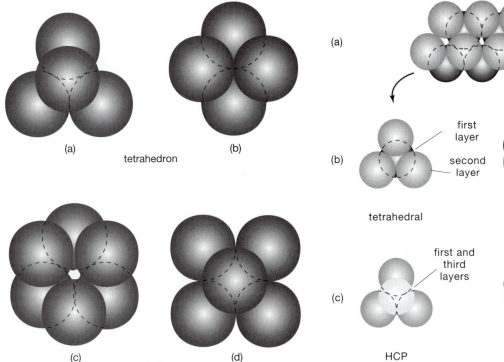

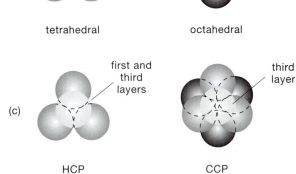

▶FIGURE 13.8
Cubic and hexagonal closest packing. (a) One closest packed layer on top of another. (b) View of tetrahedral and octahedral sites between closest packed layers. (c) In hexagonal closest packing (HCP), a third layer is directly above the first. In cubic closest packing (CCP), marbles in a third layer are not above marbles in either of the first two layers.

▶FIGURE 13.6
Tetrahedron and octahedron formed by closest packing spheres: (a) and (b) Two different views of four marbles in a tetrahedron; (c) and (d) two different views of an octahedron made by six marbles. In (d), one marble does not show because it is behind the others.

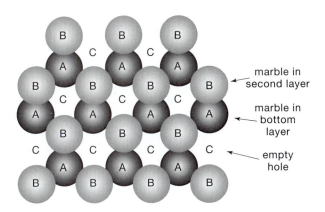

▶FIGURE 13.7
Stacking of two layers of marbles in hexagonal arrangements (for clarity, the size of the circles has been reduced so that both layers can be seen). The bottom layer is designated by the letter A and the top layer by the letter B. Open spaces, designated by C, might contain marbles in a third layer.

few large cations are most closely packed. Today we know that some minerals have complicated structures with mixed stacking sequences, and that many minerals have polyhedral frameworks that are not closest packed at all. Minerals containing alkali or al-kaline earth elements cannot be closest packed be-cause alkalis and alkaline earths are too large to fit in tetrahedral or octahedral sites. The closest packed model also fails for other minerals, such as fluorite, CaF_2, in which small anions are between large cations, and for metals that have a body-centered cubic structure in which each atom contacts eight others.

PAULING'S RULES

In 1929, Linus Pauling summarized five general rules that apply to ionic structures. The rules, now called **Pauling's rules,** provide a convenient framework for examining some details of ionic structures.

Pauling's Rule 1

Pauling's rule 1, sometimes called the *Radius Ratio Principle,* is that the distance between cations and an-ions can be calculated from their effective ionic radii, and cation coordination depends on the relative radii of the cation and surrounding anions. In essence, this

Box 13.1 *Why Are They Called Hexagonal Closest Packed (HCP) and Cubic Closest Packed (CCP)?*

In closest packed structures, each atom is surrounded by 12 others. Standard unit cells are usually chosen: For HCP, atoms from three layers combine to form a unit cell with the shape of a hexagonal prism; for CCP, atoms from four layers combine to form a unit cell with the shape of a cube. The difference between the two is that HCP involves ABABAB stacking, while CCP involves ABCABC stacking. The sphere, and ball and stick, views in Figure 13.9 show both unit cells and, for comparison, a body-centered cubic unit cell (not closest packed) in which all atoms contact eight neighbors. Metals, alloys, and minerals of all three structures are known.

FIGURE 13.9
Atoms packed together as spheres and as balls connected by sticks (in a and b, different closest packed layers are shown with different shades of gray): (a) cubic closest packing; (b) hexagonal closest packing; (c) a body-centered cubic arrangement of atoms (not closest packed).

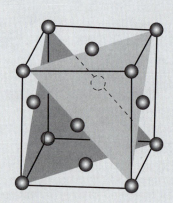

(a) cubic closest packing

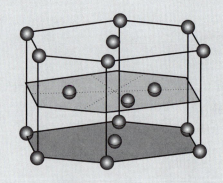

(b) hegagonal closest packing

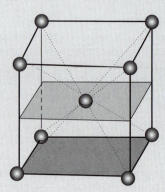

(c) body-centered cube

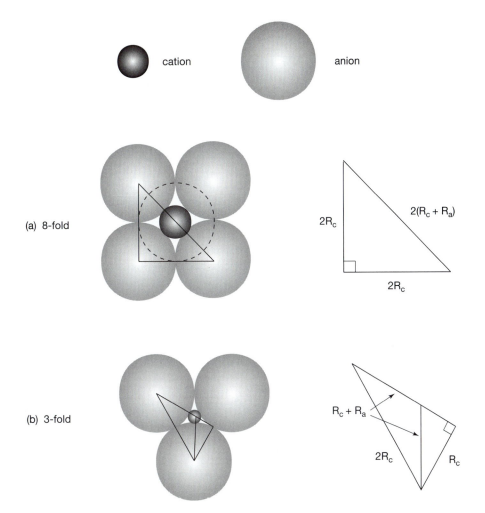

▶**FIGURE 13.10**
Two examples of how geometry can be used to calculate the limits on cation size in different coordinations: (a) octahedral coordination (additional cations above and below the anion are shown with dashed lines); (b) triangular coordination.

Box 13.2 Pauling's Rules

Rule 1. Radius Ratio Principle: Cation-anion distances are equal to the sum of their effective ionic radii, and cation coordination numbers are determined by the ratio of cation to anion radii.

Rule 2. Electrostatic Valency Principle: The strength of an ionic bond is equal to ionic charge divided by coordination number.

Rule 3. Sharing of edges or faces by coordinating polyhedra is inherently unstable.

Rule 4. Cations of high valence and small coordination number tend not to share anions with other cations.

Rule 5. Principle of Parsimony: The number of different components in a crystal tends to be small.

rule says that very small cations will bond to only a few anions, while very large cations may bond to many anions. In other words, as the radius of the cation increases, so too will C.N. Figure 13.10a shows

the limiting case when a cation just fits into the opening between four touching anions. If additional anions are directly above and below the cation, the cation is in perfect octahedral coordination

Box 13.3 Who Was Linus Pauling?

Linus Carl Pauling was a prolific American chemist and, in his later years, a peace and health activist. His success as a scientist stemmed from his ability to cross traditional discipline boundaries, an uncanny ability to identify key questions, and courage to put forth new, although sometimes incorrect, ideas. Born in Portland, Oregon, on February 28, 1901, Pauling received a B.S. in chemical engineering from Oregon State University in 1922 and a Ph.D. from California Institute of Technology (Cal Tech) in 1925. For several years he was a postdoctoral fellow in Europe, where he worked with such renowned scientists as Niels Bohr, Erwin Schrödinger, and Sir William Henry Bragg. He returned to the United States and began his career as a professor of chemistry at Cal Tech in 1927. In 1963 he left Cal Tech to join the Center for the Study of Democratic Institutions at Santa Barbara, where he spent his time working for world peace. In the late 1960s he worked for a brief period of time at the University of California–Santa Barbara before moving to Stanford University. He died in Big Sur, California, on August 29, 1994, at the age of 93.

Pauling's chemical studies covered many fields, including both organic and inorganic chemistry. One of the first to interpret crystal structures using quantum mechanics, he was also one of the pioneers of X-ray diffraction. His studies of chemical bonding resulted in the publication of his book *The Nature of the Chemical Bond, and the Structure of Molecules and Crystals* in 1939. In the 1930s and 1940s, Pauling turned his attention to molecular chemistry, producing significant papers concerning blood, proteins, and sickle cell anemia. In 1954 he was awarded the Nobel Prize in Chemistry "for research into the nature of the chemical bond and its application to the elucidation of the structure of complex substances." During his later years, Pauling received acclaim for his investigations of vitamin C. His vitamin C studies resulted in many publications including the books *Cancer and Vitamin C* and *Vitamin C and the Common Cold*, published in the 1970s.

With the advent of nuclear weapons, Pauling became concerned about potential world destruction. In 1958 he published *No More War!*, and during the same year he delivered a petition urging the end of nuclear testing, signed by more than 11,000 scientists, to the United Nations. In 1962 he was awarded the Nobel Peace Prize, making him one of only a few individuals ever to win two Nobel prizes. During his long career he received many other honors, including the Mineralogical Society of America's Roebling Medal in 1967.

▶**FIGURE 13.11**
Linus Pauling.

(compare with Figure 13.6d). Application of the Pythagorean Theorem to the right triangle shown in Figure 13.10a reveals the ratio of cation radius to anion radius (R_c/R_a) to be 0.414. Figure 13.10b shows how we can use geometry to calculate R_c/R_a for cations in perfect triangular coordination; R_c/R_a is 0.155. We can make similar, though more complicated, calculations for cations in tetrahedral (4-fold), cubic (8-fold), or dodecahedral (12-fold)

coordination (Table 13.5). As C.N. increases, space between anions increases, and the size of the cation that fits increases. Pauling argued, therefore, that as R_c/R_a increases, cations will move from 2- or 3-fold to higher coordinations in atomic structures. He further argued that stretching a polyhedron to hold a cation slightly larger than ideal might be possible. However, it was unlikely, he argued, that a polyhedron would be stable if cations were smaller than

▶TABLE 13.5
R_c/R_a and Coordination of Cations

R_c/R_a	Expected Coordination of Cation	C.N.
<0.15	2-fold coordination	2
0.15	ideal triangular coordination	3
0.15–0.22	triangular coordination	
0.22	ideal tetrahedral coordination	4
0.22–0.41	tetrahedral coordination	
0.41	ideal octahedral coordination	6
0.41–0.73	octahedral coordination	
0.73	ideal cubic coordination	8
0.73–1.0	cubic coordination	
1.0	ideal dodecahedral coordination	12
>1.0	dodecahedral coordination	

ideal. In nature, the upper limits given for various coordinations are sometimes stretched; the lower ones are rarely violated.

As an example of application of Pauling's first rule, let's take another look at halite. The radii of Na^+ and Cl^- are 1.10Å and 1.72Å. The radius ratio, R_c/R_a, is 1.10/1.72 = 0.64. Thus we can expect the cation Na^+ to be in octahedral (6-fold) coordination, consistent with the model shown in Figure 13.1a. If Na^+ is in 6-fold coordination, Cl^- must be as well, since the structure contains an equal number of both.

Pauling's Rule 2

Pauling's second rule, sometimes called the *Electrostatic Valency Principle,* says that we can calculate the strength of a bond (its electrostatic valence) by dividing an ion's valence by its C.N. Consequently, the sum of all bonds to an ion must be equal to the charge on the ion. In halite, six anions bond to each Na^+ and the strength of each bond is $1/6$, total charge/number of bonds (Figure 13.12a). The strength of each bond around Cl^- is $1/6$ as well. Six bonds of charge $1/6$ add up to 1, the charge on each ion.

We can use Pauling's first two rules to analyze a more complicated mineral, fluorite (CaF_2), in which Ca-F bonds are the only bonds present (Figure 13.12b). The radii of Ca^{2+} and F^- are 1.12Å and 1.31Å. R_c/R_a is 0.85 and, as predicted by Rule 1, Ca^{2+} is in 8-fold (cubic) coordination (Table 13.5). Each bond has a strength of $2/8$ (total charge ÷ number of bonds) = $1/4$. Since each F^- has a total charge of -1, it must be bonded to four Ca^{2+} to satisfy Rule 2, so F^- is in tetrahedral coordination.

Although Pauling's first two rules are useful guides to crystal structures, they have shortcomings. First, ionic radii vary with C.N. and valence, among other things. Sometimes radius-ratio calculations may be ambiguous because they require choosing a C.N. before we may make calculations. Second, bonds in minerals are rarely completely ionic, and ionic radius varies somewhat with the nature of the bond. Third, in cases where R_c/R_a is near a limiting value, we

▶FIGURE 13.12
Pauling's second rule: (a) bond strengths in halite (NaCl); (b) bond strengths in fluorite (CaF_2).

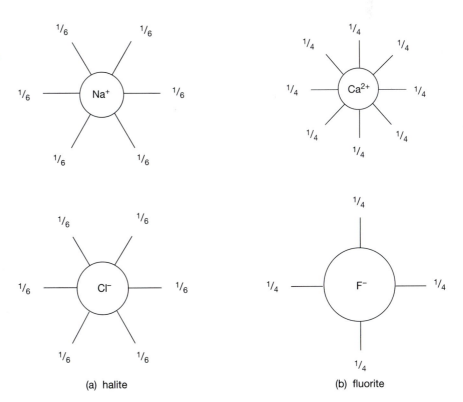

(a) halite

(b) fluorite

cannot be certain whether the higher or lower C.N. will prevail. Fourth, some coordination polyhedra, especially for high C.N.s, may be irregular in shape.

Halite and fluorite are relatively simple minerals; they each contain one cation and one anion and, therefore, one kind of bond. Most minerals contain more than two elements and may have many kinds of bonds. Spinel, $MgAl_2O_4$, contains Mg-O and Al-O bonds. X-ray studies reveal that Mg^{2+} is in tetrahedral coordination and Al^{3+} is in octahedral coordination. Consequently, the strength of the bonds around Mg^{2+} is $2 \div 4 = \frac{1}{2}$, and the strength of the bonds around Al^{3+} is $3 \div 6 = \frac{1}{2}$. Compounds such as spinel, in which all bonds have the same strength, are termed **isodesmic.**

Sulfates such as anhydrite, $CaSO_4$, are examples of **anisodesmic** compounds. In sulfates, S^{6+} is in 4-fold coordination with O^{2-}. The strength of a sulfur-oxygen (S-O) bond is therefore $6 \div 4 = 1\frac{1}{2}$. Because the strength of an S-O bond is greater than half the charge on a coordinating oxygen, oxygen is bonded to S^{6+} more tightly than to other cations. Consequently, $(SO_4)^{2-}$ radicals are tight units within the crystal structure. All sulfates, carbonates, nitrates, and other anisodesmic compounds have tightly bonded radicals. So, we often think of $(SO_4)^{2-}$, $(CO_3)^{2-}$, and $(NO_3)^{-}$ as single anionic units within crystals.

Silicate minerals belong to a special group of compounds called **mesodesmic** compounds. If a bond distribution is mesodesmic, cation-anion bond strength equals exactly half the charge on the anion. In silicates, Si^{4+} is in tetrahedral coordination and each Si-O bond has a strength of 1, exactly half the charge of O^{2-}. Consequently, the oxygen may coordinate to another cation just as strongly as to its coordinating Si^{4+}. In some silicates, the "other" cation is another Si^{4+}, so two silica tetrahedra may share an oxygen. This is why silica tetrahedra can polymerize to form pairs, chains, sheets, or networks. It also helps explain why many silicate minerals are quite hard.

Pauling's Rule 3

Pauling's third rule is that coordinating polyhedra become less stable when they share edges and are extremely unstable if they share faces. In all crystal structures, anions bond to more than one cation, which holds structures together. Rule 3 states that it is unlikely that two cations will share *two* anions (polyhedral edge) and extremely unlikely that they will share *three* (polyhedral face). Instability results because if polyhedra share edges or faces, cations in the centers of the polyhedra are too close together and will repel. The instability is especially important for cations of high charge, high C.N., or in cases when R_c/R_a is near limiting values. Figure 13.13 shows several ways silica tetrahedra might be associated in an

atomic structure. In Figure 13.13a, tetrahedra do not share any common oxygen. In 13.13b, pairs of tetrahedra share one oxygen, called a **bridging oxygen.** In Figure 13.13c, adjacent tetrahedra share an edge (two oxygen). In Figure 13.13d, they share a face (three oxygen). We can see that as the structure progresses from Figure 13.13a to 13.13d, Si^{4+} in the centers of the tetrahedra get closer together. Yet the Si^{4+} are all positively charged, so we expect them to repel each other. This is the essence of Pauling's third rule.

Pauling's Rule 4

Pauling's fourth rule is an extension of his third, stating that small cations with high charge do not share anions easily with other cations. This is another consequence of the fact that highly charged cations will repel each other. For example, all silicate minerals contain Si^{4+} tetrahedra. Yet in all the many known silicate minerals, none contains $(SiO_4)^{4-}$ polyhedra that shares edges or faces because that would bring Si^{4+} cations too close together.

Pauling's Rule 5

We call Pauling's last rule the **principle of parsimony.** It states that atomic structures tend to be composed of only a few distinct components. This means that atomic structures tend to be simple and ordered. They normally have few types of bonds and only a few types of cation or anion sites. While a mix of ions on a particular site is possible, the mix is limited and controlled.

OXYGEN AND OTHER COMMON ELEMENTS

Oxygen is the most abundant element in the Earth's crust, accounting for about 60 wt. %. It is not surprising, then, that O^{2-} is the most common anion in min-

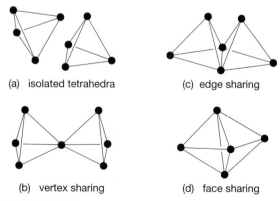

(a) isolated tetrahedra   (c) edge sharing

(b) vertex sharing   (d) face sharing

▶**FIGURE 13.13**

Sharing of oxygen by SiO_4 tetrahedra: (a) no sharing (isolated); (b) vertex sharing; (c) edge sharing; (d) face sharing. For clarity, the Si^{4+} ions at the centers of the tetrahedra have been omitted.

▶**TABLE 13.6**
Ionic Radii and Typical Coordination Number With Oxygen

Ion	Ionic Radius (Å)	Typical C.N. With Oxygen	Examples in minerals	
			C.N.	Minerals and Formulas
K^+	1.59–1.68	8–12 (cubic or dodecahedral)	9	nepheline, $(Na,K)AlSiO_4$; orthoclase, $KAlSi_3O_8$
			12	leucite, $KAlSi_2O_6$; muscovite, $KAl_2(AlSi_3O_{10})(OH)_2$
Na^+	1.10–1.24	6–8 (octahedral or cubic)	6	pectolite, $NaCa_2(SiO_3)_3H$; albite, $NaAlSi_3O_8$
			7	sodalite, $Na_3Al_3Si_3O_{12} \cdot NaCl$
			8	nepheline, $(Na,K)AlSiO_4$
Ca^{2+}	1.08–1.20	6–8 (octahedral or cubic)	6	wollastonite, $CaSiO_3$; pectolite, $NaCa_2(SiO_3)_3H$
			7	plagioclase, $(Ca,Na)(Al,Si)_4O_8$; titanite, $CaTiSiO_5$
			8	diopside, $CaMgSi_2O_6$; garnet, $(Mg,Fe,Ca,Mn)_3Al_2Si_3O_{12}$
Mn^{2+}	0.83–1.01	6–8 (octahedral or cubic)	6	rhodonite, $MnSiO_3$
			8	garnet, $(Mg,Fe,Ca,Mn)_3Al_2Si_3O_{12}$
Mg^{2+}	0.80–0.97	6–8 (octahedral or cubic)	6	diopside, $CaMgSi_2O_6$; olivine, $(Mg,Fe)_2SiO_4$
			8	garnet, $(Mg,Fe,Ca,Mn)_3Al_2Si_3O_{12}$
Fe^{2+}	0.71–0.77	6–8 (octahedral or cubic)	4	staurolite, $Fe_2Al_9Si_4O_{23}(OH)$
			6	biotite, $K(Mg,Fe)_3(AlSi_3O_{10})(OH)_2$; olivine, $(Mg,Fe)_2SiO_4$
			8	garnet, $(Ca,Mn,Fe,Mg)_3Al_2Si_3O_{12}$
Ti^{4+}	0.69	6 (octahedral)	6	titanite, $CaTiSiO_5$
Fe^{3+}	0.57–0.68	4–6 (tetrahedral or octahedral)	6	epidote, $Ca_2(Al,Fe)_3Si_3O_{12}(OH)$
Al^{3+}	0.47–0.61	4–6 (tetrahedral or octahedral)	4	muscovite, $KAl_2(AlSi_3O_{10})(OH)_2$; orthoclase, $KAlSi_3O_8$
			5	andalusite, Al_2SiO_5
			6	muscovite, $KAl_2(AlSi_3O_{10})(OH)_2$; beryl, $Be_3Al_2Si_6O_{18}$
Si^{4+}	0.34–0.48	4 (tetrahedral)	4	quartz, SiO_2; tremolite, $Ca_2Mg_5Si_8O_{22}(OH)_2$
			6	stishovite, SiO_2
P^{5+}	0.025	4 (tetrahedral)	4	apatite, $Ca_5(PO_4)_3(OH,F,Cl)$
C^{+4}	—	3 (triangular)	3	calcite, $CaCO_3$; malachite, $Cu_2(CO_3)(OH)_2$

erals. The ionic radius of oxygen varies from about 1.27Å to 1.34Å, depending on its C.N. The abundant crustal cations include Si^{4+}, Al^{3+}, Fe^{2+}, Fe^{3+}, Mg^{2+}, Mn^{2+}, Na^+, Ca^{2+}, and K^+. Table 13.6 lists R_c/R_a values for the common cations and O^{2-}. A range of radius values has been considered for each cation because cation radius varies slightly with structure and coordination. The predicted C.N.s in Table 13.6 are consistent with most known mineral structures. Small cations such as C^{4+} and B^{3+} can have triangular coordination. Si^{4+} is nearly exclusively tetrahedral, while Al^{3+} may be either tetrahedral or octahedral. Although radius ratios predict only 6-fold coordination for Fe^{3+}, in nature Fe^{3+} can be either tetrahedral or octahedral. Other elements can be in octahedral coordi-

nation as well. The alkalis and the alkaline earths are the only elements that normally can be in cubic or dodecahedral coordination. When contradictions between nature and Table 13.6 occur, it is usually for cations in highly irregularly shaped sites. For example, the aluminum in andalusite is in both 5-fold and 6-fold coordination, some magnesium in anthophyllite is in 7-fold coordination, and the potassium in microcline is in 10-fold coordination. We do not include 5-fold and 7-fold coordination in Tables 13.5 and 13.6 because no regular polyhedra have five or seven vertices, so application of Pauling's first rule is problematic.

Sometimes we wish to show cation coordination in a mineral formula. Traditionally, this has been done using superscript Roman numerals. For example, we may write andalusite's formula as $Al^{V}Al^{VI}SiO_5$ to remind us of the unusual aluminum coordination. We may write magnetite's as $Fe^{IV}Fe_2^{VI}O_4$ to show that iron occupies two differently coordinated sites. Some chemists today use Arabic numerals in square brackets instead of Roman numerals.

SILICATE STRUCTURES IN GENERAL

In Chapter 5 we briefly discussed silicate mineral structures. Here we take another look at silicate structures in light of Pauling's rules and other crystal-chemical principles. Because oxygen and silicon are the two most abundant elements in the Earth's crust, and because the (SiO_4) tetrahedron is such a stable radical, silicate minerals are extremely stable and abundant in crustal rocks and sediments. They dominate igneous and metamorphic rocks, as well as many sedimentary rocks. An individual silica tetrahedron has a charge of -4. Because minerals must be charge balanced, silicon tetrahedra must share oxygen ions, or must be bonded to other cations. The sharing of oxygen between silica tetrahedra is a form of **polymerization.** Quartz and tridymite (SiO_2), for example, are highly **polymerized.** In most of the SiO_2 minerals, two $(SiO_4)^{4-}$ tetrahedra share each oxygen (see Figure 5.6). The strength of each Si-O bond is 1; each Si^{4+} bonds to four oxygen, and each O^{2-} to two silicon, so charge balance is maintained and the overall formula is SiO_2.

Polymerization is absent in some silicates, such as olivine, $(Mg,Fe)_2SiO_4$ (see Figure 5.25). Instead, cations link individual silica tetrahedra. In many silicates, a combination of oxygen sharing between silicon tetrahedra, and the presence of additional cations leads to charge balance. The more oxygen sharing, the fewer additional cations needed. In still other silicates, Al^{3+} replaces some tetrahedral Si^{4+}. Consequently, more additional cations must be present to maintain charge balance. In albite, for example, Al^{3+} replaces one-fourth of the tetrahedral Si^{4+}. Na^+ ions between tetrahedra maintain charge balance. Albite's formula is $NaAlSi_3O_8$, which we may write $Na(AlSi_3)O_8$ to emphasize that both Al^{3+} and Si^{4+} occupy the same structural sites. In anorthite, another feldspar, Al^{3+} replaces half the Si^{4+}, resulting in the formula $Ca(Al_2Si_2)O_8$. Besides feldspars, tetrahedral aluminum is common in micas, amphiboles, and, to a lesser extent, in pyroxenes.

The orderly way silica (or alumina) tetrahedra polymerize leads naturally to the division of silicate minerals into the subclasses introduced in Chapter 2 (Table 13.7). We call silicates such as olivine, in which tetrahedra share no O^{2-}, **isolated tetrahedral silicates** (also called **nesosilicates** or **orthosilicates**). Silicates in

which pairs of tetrahedra share oxygen are **paired tetrahedral silicates (sorosilicates).** If two oxygen on each tetrahedron link to other tetrahedra, we get **single-chain silicates (inosilicates)** or **ring silicates (cyclosilicates).** If some oxygen are shared between two tetrahedra, and some between three, we get **double-chain silicates** (also considered **inosilicates**). If three oxygen on each tetrahedron link to other tetrahedra to form tetrahedral planes, we get **sheet silicates** (also called **layer silicates** or **phyllosilicates**), and if all oxygen are shared between tetrahedra we get **framework silicates** (also called **network silicates** or **tectosilicates**). The ratio of Si : O, then, indicates silicate subclass because different ratios result from different amounts of oxygen sharing (Table 13.7). In minerals containing tetrahedral aluminum, the ratio of $(Al^{IV}+Si^{IV})$: O, which we can abbreviate T : O, reflects the silicate subclass. If the only tetrahedral cation is silicon, isolated tetrahedral silicates are often characterized by SiO_4 in their formulas, paired tetrahedral silicates by Si_2O_7, single-chain silicates by SiO_3 or Si_2O_6, ring silicates by Si_6O_{18}, double-chain silicates by Si_4O_{11}, sheet silicates by Si_2O_5 or Si_4O_{10}, and framework silicates by SiO_2.

The chemistries of silicates correlate, in a general way, with the subclass to which they belong (Table 13.7). This correlation reflects silicon : oxygen ratios, and it also reflects the way in which silica polymerization controls atomic structures. There are many variables, but we can make some generalizations. Isolated tetrahedral silicates and chain silicates include minerals rich in Fe^{2+} and Mg^{2+}, but framework silicates do not. The three-dimensional polymerization of framework silicates generally lacks sufficient anionic charge and the small crystallographic sites necessary for small highly charged cations. For opposite reasons, Na^+ and K^+ enjoy highly polymerized structures, which have large sites that easily accommodate monovalent cations.

In Chapter 5, we talked about the melting temperatures of minerals. Figure 5.4 (Bowen's Reaction Series) compares the melting temperatures of most common igneous minerals. Quartz has the lowest melting temperature, K-feldspar the second lowest, followed by muscovite, biotite, amphiboles, pyroxenes, and finally olivine. Bowen's reaction series is based on melting temperatures determined by laboratory experiments, but it mirrors the degree to which silicate minerals are polymerized (which is also a reflection of how much silica they contain). Quartz, feldspars, and other framework silicates are highly polymerized, and they melt at the lowest temperatures. Muscovite, biotite, and other sheet silicates are less polymerized, and melt at higher temperatures. Amphiboles (double chain silicates) and pyroxenes (single chain silicates) are still less polymerized, and melt at even higher temperatures. Olivine and other isolated tetrahedral silicates

▶**TABLE 13.7**
Silicate Structures

Silicate Subclass	Example Minerals	Mineral Formula	Cation Coordination	Si : O or (Si,Al^{IV}) : O	Number of Oxygen Shared by Tetrahedra
isolated tetrahedral silicates	olivine almandine	$(Mg,Fe)_2SiO_4$ $Fe_3Al_2Si_3O_{12}$	Fe^{VI},Mg^{VI}, Si^{IV} $Fe^{VIII}, Al^{VI}, Si^{VI}$	1:4	0
paired tetrahedral silicates	lawsonite åkermanite	$CaAl_2(Si_2O_7)(OH)_2 \cdot H_2O$ $Ca_2MgSi_2O_7$	$Ca^{VIII}, Al^{VI}, Si^{IV}$ $Ca^{VIII},Mg^{IV},Si^{IV}$	2:7(1:3.5)	1
single-chain silicates	diopside wollastonite	$CaMgSi_2O_6$ $CaSiO_3$	$Ca^{VIII}, Mg^{VI},Si^{IV}$ Ca^{VI},SI^{IV}	2:6 or 1:3	2
ring silicates	tourmaline beryl	$(Na,Ca)(Fe,Mg,Al,Li)_3Al_6(BO_3)_3Si_6O_{18}(OH)_4$ $Be_3Al_2Si_6O_{18}$	$Na^{VI},Ca^{VI},Li^{VI},Fe^{VI},Mg^{VI},Al^{VI},B^{III},Si^{IV}$ Be^{IV},Al^{VI},Si^{IV}	1:3	2
double-chain silicates	anthophyllite tremolite	$(Mg,Fe)_7Si_8O_{22}(OH)_2$ $Ca_2Mg_5Si_8O_{22}(OH)_2$	Mg^{VI-VII},Si^{IV} $Ca^{VIII},Mg^{VI},Si^{IV}$	4:11 (1:2.75)	2 or 3
sheet silicates	talc phlogopite kaolinite	$Mg_3Si_4O_{10}OH_2$ $KMg_3(AlSi_3O_{10})(OH)_2$ $Al_2Si_2O_5(OH)_4$	Mg^{VI},Si^{IV} $K^{VI},Mg^{VI},Al^{IV},Si^{IV}$	4:10 (1:2.5)	3
framework silicates	quartz microcline	SiO_2 $KAlSi_3O_8$	Si^{IV} K^X,Al^{IV},Si^{IV}	1:2	4

are not polymerized at all, and melt at the highest temperatures. Why does the polymerization affect melting temperatures? The answer lies not so much with the nature of the minerals, but with the nature of the melt they create. Magmas, just like minerals, become polymerized when silicon and oxygen form complexes in the melt. Magmas richest in silicon and oxygen are more polymerized, and have lower Gibbs free energy. In a sense, magmas that are highly polymerized form at lower temperatures than those that are less polymerized because fewer bonds need to be broken to create the melt. So, silica-rich magmas, and silica-rich minerals, melt at lower temperatures than those that are silica-poor.

In Chapter 6, we pointed out that the order in which silicate minerals weather is opposite the order in which they melt. Those minerals that melt at lowest temperature are most resistant to weathering (Figure 6.4). This phenomenon, too, is partly a result of the amount of silica in the different silicate minerals. Minerals rich in silica are more tightly bonded (and thus more resistant to weathering) because they contain more $(SiO_4)^{4-}$ radicals and because the valence of ionic bonds is generally greater than in minerals poorer in silica. Because of stronger bonds, they are less easily attacked by water and other weathering agents.

Silicate crystal structures may be complex. Many silicates contain anions or anionic groups other than O^{2-}. Muscovite, for example, contains $(OH)^-$ and has the formula $KAl_2(AlSi_3O_{10})(OH)_2$. Other silicates, such as kyanite, $Al_2(SiO_4)O$, and titanite, $CaTi(SiO_4)O$, contain O^{2-} ions unassociated with the $(SiO_4)^{4-}$ tetrahedra. In muscovite and many other minerals, aluminum is in both tetrahedral and octahedral coordination. Still other silicates do not fit neatly into a subclass. Zoisite, $Ca_2Al_3O(SiO_4)(Si_2O_7)(OH)$, contains both isolated tetrahedra and paired tetrahedra. Some mineralogy texts and reference books separate elements and include extra parentheses in mineral formulas (as has been done in this paragraph) to emphasize crystal structure, but often we omit such niceties for brevity. In shorter form, we can write muscovite's formula as $KAl_3Si_3O_{10}(OH)_2$; kyanite's becomes Al_2SiO_5; titanite's becomes $CaTiSiO_5$; and zoisite's becomes $Ca_2Al_3Si_3O_{12}(OH)$. While being shorter and, perhaps, easier to write, these formulas give little hint of crystal structure.

ELEMENTAL SUBSTITUTIONS IN SILICATES

While quartz is usually >99.9% SiO_2, most minerals have variable chemistries due to elemental substitutions. Consistent with Pauling's rules, the nature and extent of substitutions depend primarily on ionic size and charge and on the nature of atomic bonding in a mineral's structure. Because silicate minerals are dominantly ionic, the nature of bonding is relatively constant; size and ionic charge control most substitutions. Figure 13.2 shows the relative sizes and charges of the most common elements in silicate minerals. Elements of similar size and charge may occupy similar sites in crystal structures without causing distortion or charge imbalance. For example, Ca^{2+}, Mn^{2+}, Fe^{2+}, and Mg^{2+} substitute for each other in many silicates (and other minerals), including garnets and pyroxenes.

The extent to which elements may substitute for each other is often limited. Natural garnets may have any composition described by the formula $(Ca,Mn,Fe,Mg)_3Al_2Si_3O_{12}$. In contrast, the substitution of Ca^{2+} for Mn^{2+}, Fe^{2+}, or Mg^{2+} in pyroxene is limited at all but the highest temperatures, due to the large size of Ca^{2+} compared with the other ions. Consequently, there is a solvus between orthopyroxene and clinopyroxene, and a similar solvus exists in the Ca-Mn-Fe-Mg carbonate system.

Similarity in size and charge allows K^+ and Na^+ to substitute for each other in feldspars, amphiboles, and other minerals. Fe^{3+} and Al^{3+} replace each other in minerals such as garnet and spinel. These are both examples of **simple substitutions.** In a simple substitution the substituting ion has the same charge as the one it replaces. Sometimes simple substitutions are described using equations such as $Fe^{2+} = Mg^{2+}$ or $Fe^{3+} = Al^{3+}$. Other elemental substitutions are more complex. For example, Ca^{2+} may replace Na^+ in feldspar. To maintain charge balance, Al^{3+} replaces Si^{4+} at the same time, and we describe the **coupled substitution** as $Ca^{2+}Al^{3+} = Na^+Si^{4+}$ or just $CaAl = NaSi$. No common substitutions involve ions with charge difference of greater than 1.

Anions, too, may substitute for each other in minerals. In some micas and amphiboles, for example, F^- or Cl^- may replace $(OH)^-$. More complex substitutions in micas involve the replacement of $(OH)^-$ by O^{2-}, which requires some compensatory substitution to maintain charge balance. In scapolite and a few other minerals, $(CO_3)^{2-}$ or $(SO_4)^{2-}$ may replace Cl^-. To add further complexity, in some minerals substitutions involve vacancies. For example, in hornblende $\square Si = KAl$ is a common substitution. The $\square$ symbol indicates a vacancy. Some of the more common and most important elemental substitutions are tabulated in Table 13.8.

While the elemental substitutions listed in Table 13.8 occur in many minerals, including many nonsilicates, they do not necessarily occur in all. For example, there is only very minor substitution of Fe^{3+}

▶TABLE 13.8
Some Typical Elemental Substitutions in Silicates

Substitution	Example Minerals
$Na^+ = K^+ = Li^+$	alkali feldspar, hornblende, micas
$Ca^{2+} = Mg^{2+} = Fe^{2+} = Mn^{2+}$	pyroxene, amphiboles, micas, garnet, carbonates
$F^- = Cl^- = OH^-$	amphiboles, micas, apatite
$Fe^{3+} = Al^{3+}$	garnet, spinels
$Ca^{2+}Al^{3+} = Na^+Si^{4+}$	plagioclase, hornblende
$\square Si^{4+} = K^+Al^{3+}$	hornblende
$O^{2-} = (OH)^-$	biotite, titanite

for Al^{3+} in corundum, even though complete solid solution exists between andradite ($Ca_3Fe_2Si_3O_{12}$) and grossular ($Ca_3Al_2Si_3O_{12}$). Similarly, periclase is always close to end member MgO, rarely forming significant solid solution with FeO or with MnO. Nonetheless, the substitutions listed in Table 13.8 are common, occur in more than one mineral class, and explain most of the mineral end members discussed in Chapters 5, 6, and 7. A glance back to Chapters 5 and 6, for example, will show that elemental substitutions, including the nature and extent of solvi, are about the same in pyroxenes, amphiboles, and carbonates. Major substitutions for both are $Ca^{+2} = Mg^{+2} = Fe^{2+} = Mn^{2+}$.

STRUCTURES OF THE BASIC SILICATE SUBCLASSES

Framework Silicates

Framework silicates consist of a three-dimensional polymerized network of Si or (Al,Si) tetrahedra. Quartz is the most common framework silicate. In quartz and the other SiO_2 polymorphs (except stishovite), oxygen links each silica tetrahedron to four others. Each SiO_2 polymorph has a different arrangement of tetrahedra; some contain 4-, 6-, or 8-membered loops, and some contain channels. In some framework silicates, Al replaces some Si. Three-dimensional polymerization leaves large holes, cages, loops, or channels that can hold large cations such as K^+, Na^+, or Ca^{2+}. Figure 13.14 shows such a mineral, orthoclase. In orthoclase and other feldspars, the sites occupied by K^+, Na^+, or Ca^{2+} are distorted; coordination is 6 to 9, depending on the feldspar. In other framework silicates, including some feldspathoids (for example, analcime) and zeolites, the openings between silica tetrahedra are large enough to hold molecular water. Framework

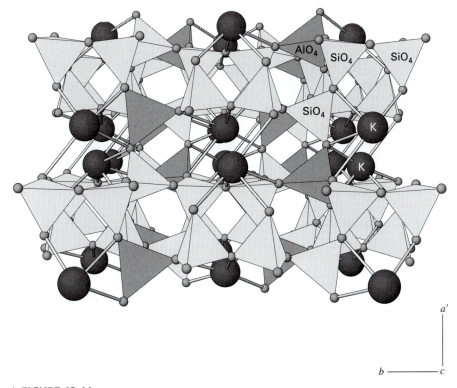

▶FIGURE 13.14
Atomic arrangement in orthoclase: The large spheres represent K^+; the tetrahedra are $(AlO_4)^{5-}$ and $(SiO_4)^{4-}$ groups. This view is down the c-axis.

silicates are often twinned. Quartz is typically twinned according to the **Dauphiné** or **Brazil laws,** although we cannot normally see the twins with the naked eye. Feldspars and other framework silicates twin by many different laws.

Sheet Silicates

In sheet silicates, sheet-like tetrahedral and octahedral layers combine to produce a planar structure. The tetrahedral layers are stacked with other layers containing octahedrally coordinated cations, and sometimes with layers containing alkalis. We call the octahedral layers **gibbsite layers** if they contain Al^{3+} and **brucite layers** if they contain Mg^{2+} or Fe^{2+} because the structures of the layers resemble these minerals. Some sheet silicates contain three divalent octahedral cations for every four tetrahedra; others contain two trivalent octahedral cations instead. So, we call muscovite, $KAl_2(AlSi_3O_{10})(OH)_2$, and other aluminous sheet silicates **dioctahedral.** Biotite, $K(Mg,Fe)_3(AlSi_3O_{10})(OH)_2$, and other Mg-Fe sheet silicates are **trioctahedral.** Figure 13.15 shows the atomic structure of serpentine, a dioctahedral sheet silicate. Units composed of one tetrahedral layer and one dioctahedral layer stack on top of each other but are slightly offset. Compare Figure 5.15 (muscovite) with Figure 13.15. Both show sheet silicates.

Tetrahedral and octahedral layers may stack in various ways. In kaolinite and serpentine they alternate. In pyrophyllite, talc, and micas, tetrahedral layers surround octahedral layers to produce T-O-T sandwiches. K^+, and more rarely Na^+ or Ca^{2+}, occupy sites between the sandwiches in micas. In chlorites (Figure 13.16), additional octahedral layers separate the sandwiches, and in other sheet silicates the stacking may be more complex. Clay minerals, in particular, have complex layered structures involving interlayer H_2O. Micas have several different polymorphs, which are related by the way in which octahedral and tetrahedral layers are stacked with respect to each other. They also commonly twin, but the subtle distinctions between polymorphs and the twinning are difficult to detect without detailed X-ray or transmission electron microscope studies.

Chain Silicates

In Chapter 5 we pointed out that the chain silicates, including the single-chain silicates (pyroxenes and pyroxenoids) and the double-chain silicates (amphiboles), all have similar structures based on chains of silica tetrahedra (see Figure 5.17). The chains, which run parallel to the c axis, alternate with bands of octahedrally coordinated cations (Figure 13.17). Tetrahedra in adjacent chains point in opposite directions. Two oxygen in each

▶**FIGURE 13.15**
Atomic arrangement in serpentine: The tetrahedra are $(SiO_4)^{4-}$; the spheres represent O^{2-}, OH^-, and Mg^{2+}. The orientation of the c-axis is shown.

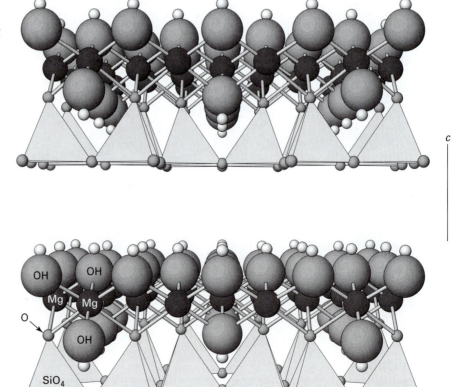

tetrahedron are shared with adjacent tetrahedra. The other two provide links to the octahedral cations. In pyroxenes and pyroxenoids, four octahedral sites are found between pairs of chains. Seven octahedral sites characterize amphiboles. In both pyroxenes and amphiboles, two octahedral sites are larger than the others and may contain Ca^{2+} or Na^+. Any of the octahedral sites may contain Mg^{2+} or Fe^{2+}, and some may contain Al^{3+}. K^+ is present in a large interlayer site in hornblende and some other amphiboles. Twinning, both simple and complex, is common in many pyroxenes and amphiboles.

Ring Silicates

Tourmaline is the only common mineral in which all tetrahedra are linked to form independent rings (Figure 13.18). Other minerals, including beryl, $Be_3Al_2Si_6O_{18}$, and cordierite, $(Mg,Fe)_2Al_4Si_5O_{18}$, contain rings, but they also contain tetrahedra joined in other ways. In tourmaline, 6-membered tetrahedral rings are connected to octahedral Fe^{2+}, Mg^{2+}, or Al^{3+}, and to triangular $(BO_3)^-$ groups. Ca^{2+}, Na^+, and K^+ occupy large sites centered in the tetrahedral rings and coordinated to the borate groups and silica tetrahedra.

Paired Tetrahedral Silicates

Lawsonite, a rare mineral found in blueschists, and the melilites åkermanite and gehlenite (also rare) are perhaps the best examples of paired tetrahedral silicate (Figure 13.19). The paired silicon tetrahedra result in Si_2O_7 groups. Other minerals contain some paired and some unpaired tetrahedra. Vesuvianite and epidote, for example, contain both SiO_4 and

▶**FIGURE 13.16**
Atomic arrangement in chlorite: The tetrahedra are $(SiO_4)^{4-}$; the spheres represent O^{2-}, OH^-, and Mg^{2+}. This drawing shows two "talc" layers with a "brucite" layer between. This view is down the *a*-axis.

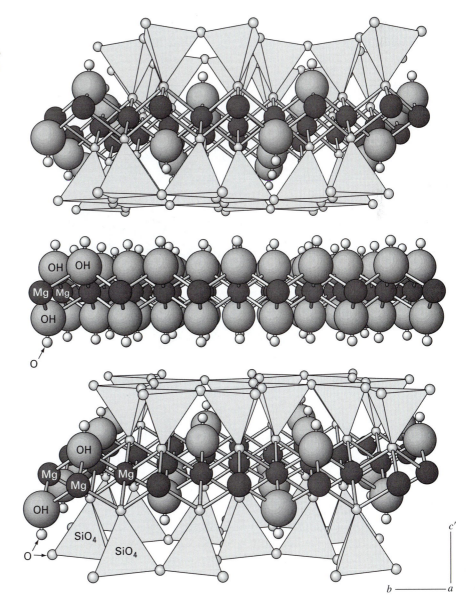

▶FIGURE 13.17
Atomic arrangement in (a) diopside (pyroxene) and (b) tremolite (amphibole). The tetrahedra are $(SiO_4)^{4-}$ with small spheres at the corners representing the oxygen. The larger spheres represent Ca^{2+} and Mg^{2+}, and $(OH)^-$ in tremolite. Both views are down the c-axis.

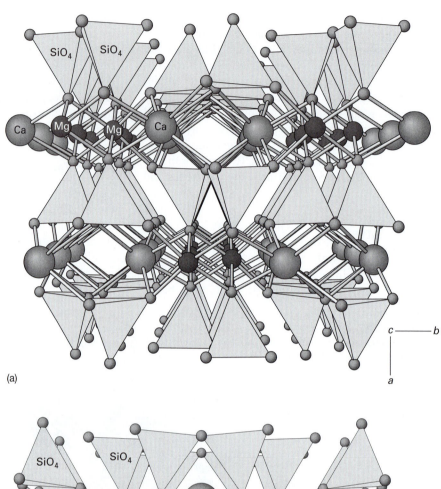

(a)

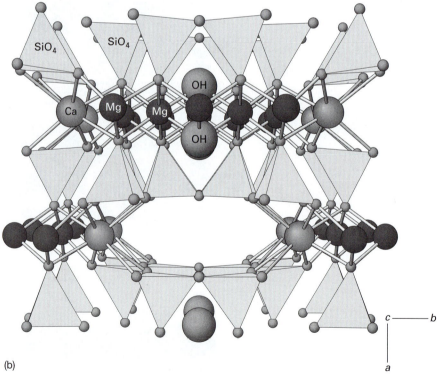

(b)

Si_2O_7 groups and are often grouped with lawsonite and the melilites. In lawsonite, Al^{3+} links pairs of silica tetrahedra. In vesuvianite and the melilites, octahedral Al^{3+}, Fe^{2+}, or Mg^{2+} links tetrahedral groups, and Ca^{2+} occupies large sites between layers of octahedra.

Isolated Tetrahedral Silicates

Mineralogists often call isolated tetrahedral silicates **island silicates** because tetrahedra do not share oxygen. Olivine, $(Mg,Fe)_2SiO_4$, and garnet, $(Ca,Mn,Fe,Mg)_3$ $Al_2Si_3O_{12}$ are examples. In olivine, divalent octahe-

▶**FIGURE 13.18**
Atomic arrangement in tourmaline: The smallest spheres are O^{2-}. The largest spheres are OH^- and Na^+ (or Ca^{2+}). The medium sized spheres are Al^{3+} (or Li^+, Fe^{2+}, Mg^{2+}). This view is looking down the *c*-axis.

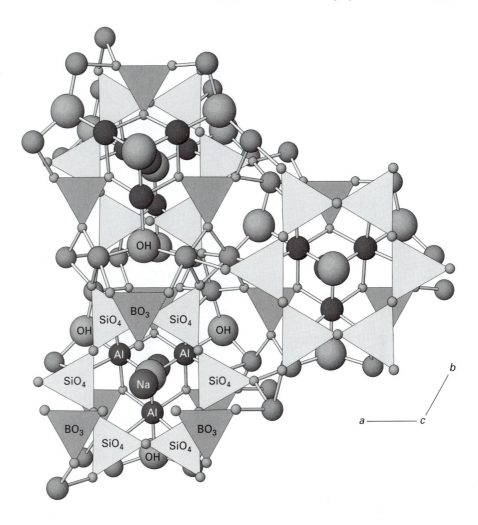

dral cations occupying two slightly different sites link independent silica tetrahedra (Figure 13.20; see also Figure 5.25). The larger site contains Ca^{2+} in monticellite, $CaMgSiO_4$, which may be considered a calcic olivine. In the garnet structure, isolated tetrahedra share oxygen with two kinds of cationic sites: an octahedral site and a highly distorted site with 8-fold coordination. The octahedral site usually contains Al^{3+} or Fe^{3+}; the other contains Ca^{2+}, Mg^{2+}, or Fe^{2+}.

STRUCTURES AND CHEMISTRY OF NONSILICATES

Having discussed silicate structures, we could go on and discuss other mineral groups. However, unless we wanted to go into great detail, such a discussion would not be particularly fruitful, for several reasons. First, in our preceding discussion of silicate structures, we ignored or glossed over some complications that become apparent when we examine the finer details of crystal structures. Second, because silicate structures are largely ionic, they are simpler and more regular than those of most other

mineral groups. Third, for many mineral groups we cannot make meaningful generalizations or categorize structures in a useful way. For example, the sulfide minerals involve structures that are covalent and metallic, or both. Sulfur may have any of a number of valences, and sulfide structures involve many different coordination polyhedra, layers, clusters, and other complex structural units. Generalizations made about structure types will inevitably be too detailed or not detailed enough for different purposes. Instead of worrying about the details of structure and chemistry of all mineral groups, we note that the same principles that apply to silicates also apply to other minerals. To find the details, we must go to mineralogical literature. Some excellent references are the five-volume series *Rock Forming Minerals* by Deer, Howie, and Zussman (Longman, 1972–1977), *Mineralogy: Concepts and Principles* by Zoltai and Stout (Burgess, 1984), *Manual of Mineralogy* by Klein and Hurlbut (John Wiley and Sons, 1995), and the *Reviews in Mineralogy* series (Mineralogical Society of America, 1974–1995).

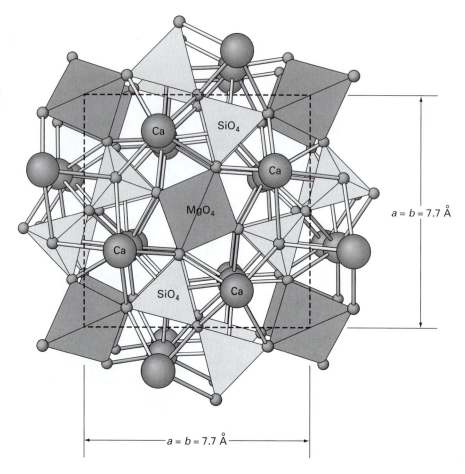

►**FIGURE 13.19**
Atomic arrangement in åkermanite, $Ca_2MgSi_2O_7$, a rare mineral that is a member of the melilite group. The tetrahedra are $(SiO_4)^{4-}$ and $(MgO_4)^{6-}$. The $(MgO_4)^{6-}$ do not look like tetrahedra because they are viewed edge-on. The spheres are Ca^{2+}. Melilites are tetragonal; the dashed lines outline a unit cell.

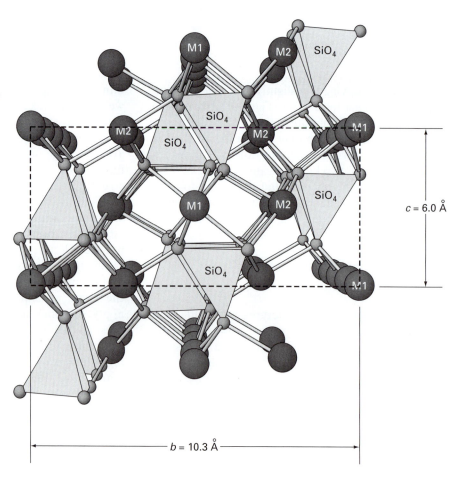

►**FIGURE 13.20**
Atomic arrangement in forstertite, Mg_2SiO_4: The tetrahedra are $(SiO_4)^{4-}$ and the larger spheres are Mg^{2+} in two sites (M1 and M2). Forsterite and other olivines are orthorhombic; the dashed lines outline a unit cell.

▶QUESTIONS FOR THOUGHT

Some of the following questions have no specific correct answers; they are intended to promote thought and discussion.

1. Crystals are not all ionic, but we have concentrated on ionic crystals in this book. Why?
2. What are the factors that control the ability of an element to substitute for another one in a mineral?
3. Ca^{2+} and Fe^{2+} mix freely in garnet, so we have solid solutions between grossular ($Ca_3Al_2Si_3O_{12}$) and almandine ($Fe_3Al_2Si_3O_{12}$). On the other hand, Ca^{2+} and Fe^{2+} do not mix freely in carbonates. Only limited solid solution is possible between calcite ($CaCO_3$) and siderite ($FeCO_3$), and an intermediate compound called *ankerite* is stable. What is ankerite? Why is there only limited substitution of Ca^{2+} for Fe^{2+} in carbonates?
4. Silicate radicals, $(SiO_4)^{4-}$, polymerize to form chains, sheets, or networks. Carbonate radicals, $(CO_3)^{2-}$, do not. Why?
5. Why are silicates, especially alkali alumino silicates, so abundant in the Earth's crust?
6. We divide silicates into subgroups, such as sheet silicates, based on the way silica tetrahedra polymerize. Can all silicates be classified unambiguously into one of the groups?
7. Titanium can be found in biotite, amphibole, and a few other common minerals, but typically most of the titanium in rocks is in titanium-rich minerals such as ilmenite ($FeTiO_3$) or rutile (TiO_2). Why?
8. Consider the crystal structures of pyroxenes. Why do many mineralogists prefer to write the formula of enstatite as $Mg_2Si_2O_6$ rather than $MgSiO_3$?

▶RESOURCES

(See also the general references listed at the end of Chapter 1.)

Bloss, F. D. *Crystallography and Crystal Chemistry.* Washington, DC: Mineralogical Society of America, 1994.

Deer, W. A., R. A. Howie, and J. Zussman. *Rock Forming Minerals* series, 5 vols. London: Longman, 1972–1977.

Evans, R. C. *An Introduction to Crystal Chemistry.* 3rd ed. London: Cambridge University Press, 1964.

Klein, C., and C. Hurlbut. *Manual of Mineralogy,* 27 vols. New York: John Wiley and Sons, 1995.

Pauling, L. C. *The Nature of the Chemical Bond.* Ithaca, NY: Cornell University, 1960.

Pauling, L. C. *No More War!* New York: Dodd and Mead, 1958.

Ribbe, P. H., ed. *Reviews in Mineralogy.* Washington, DC: Mineralogical Society of America, 1974–1995.

Whittaker, E. J. W., and R. Muntus. Ionic radii for use in geochemistry. *Geochimica et Cosmochimica Acta,* 34 (1970), 945–956.

Zoltai, T., and J. H. Stout. *Mineralogy: Concepts and Principles.* Minneapolis: Burgess, 1984.

Mineral Descriptions

This medley of crystals includes 4 cm long beryl crystals, doubly terminated quartz crystals (Herkimer diamonds), and octahedra of spinel and magnetite.

Descriptions
of Minerals

Geologists and mineralogists have described more than 3,000 minerals; most are exceedingly rare, so it is unnecessary and impractical to try to describe them all in this book. The most common minerals are listed here, as well as those of the greatest economic importance. Other species are listed if they have unique structures or chemistries or demonstrate principles or properties not well represented by the common or economic minerals. Still others are included if they are useful indicators of geological environments and processes or if they can be used for practical purposes, such as age determinations. There are descriptions of about 200 minerals, which is more than sufficient for most purposes. Intended for students of mineralogy, emphasis has been placed on those properties that best aid in practical mineral identification: hand specimen characteristics and, to a lesser extent, occurrences, associations, and optical properties. The mineral descriptions contain only brief discussions of atomic structure and crystal chemistry.

The descriptions are arranged in order based on the classification scheme presented in Chapter 2. A brief tabulation of mineral species introduces each of the classes, subclasses, or groups listed in bold type below.

I. SILICATES

FRAMEWORK SILICATES

Silica Group Minerals

quartz	SiO_2
cristobalite	SiO_2
tridymite	SiO_2
coesite	SiO_2
stishovite	SiO_2

The silica group minerals have composition SiO_2, and all except stishovite have structures based on SiO_4 tetrahedra linked at their vertices by "bridging" oxygen. Quartz, tridymite, and cristobalite have high-temperature and low-temperature polymorphs; coesite and stishovite do not. Scientists have synthesized several additional SiO_2 polymorphs in the laboratory. Although mineralogists have described many polymorphs, only common quartz, properly called *low quartz*, exists in substantial amounts; it is the second most abundant mineral in the Earth's crust.

The different polymorphs vary in the way SiO_4 tetrahedra join to form a three-dimensional framework. Consequently, they vary in symmetry: High quartz is hexagonal, low quartz is trigonal, low tridymite is orthorhombic, low cristobalite is tetragonal, and coesite is monoclinic.

Structural variations among the SiO_2 polymorphs reflect the different conditions under which they form. Low quartz is the only stable SiO_2 polymorph under normal Earth surface conditions, but some rocks contain metastable stishovite, coesite, cristobalite, or tridymite.

For more general information about the silica group, see Chapter 5.

Quartz (α-quartz) SiO_2

Origin of Name
From German, *quartz,* of unknown origin.

Hand Specimen Identification
Hardness, lack of cleavage, conchoidal fracture, and vitreous luster usually serve to identify quartz. When euhedral, its pseudohexagonal prismatic habit can be distinctive. Quartz is sometimes confused with calcite, beryl, cordierite, or feldspars. Plates 2.1 through 2.8 show color pictures of quartz. Figures 2.2, 2.4, 2.10, 2.11, 3.4, 3.6e, 6.13b, 8.8, 8.12, 8.16, 9.6b, 10.1, 14.1, and 14.2 contain additional photographs and drawings.

Physical Properties

hardness	7
specific gravity	2.65
cleavage/fracture	no cleavage or parting; brittle/conchoidal fracture
luster/transparency	vitreous/transparent
color	colorless, white, milky; less commonly purple, pink, yellow, brown, or black
streak	white

Optical Properties
In thin section, quartz is distinguished by low relief, low birefringence (maximum interference colors are gray), lack of color, lack of cleavage, lack of visible twinning, lack of alteration, usually anhedral character, and undulatory extinction. Uniaxial $(+)$; $\omega = 1.544$, $\epsilon = 1.553$, $\delta = 0.009$. Plates 5.5, 5.6, 5.7, and 5.8 show quartz in thin section.

Crystallography
Hexagonal (rhombohedral), $a = 4.913$, $c = 5.405$, $Z = 3$; space group $R3_12$ or $R3_22$; point group 32.

Habit
Crystals are prismatic or massive. They belong to class 32, but may appear to have true 6-fold symmetry. Common crystals are six-sided prisms terminated by rhombohedrons, sometimes appearing to be hexagonal dipyramids (Figures 14.1 and 14.2). Prism faces often show horizontal growth striations. Rare forms include trapezohedra. Most, if not all, quartz is twinned. Two

▶**FIGURE 14.1**
Quartz crystals from Minas Gerais, Brazil (left), and Middleville, New York (right).

▶**FIGURE 14.2**
Citrine, a yellow to brown variety of quartz.

kinds of twins, Dauphinè and Brazil, are common, but normally cannot be seen with the naked eye.

Structure and Composition
Quartz is always essentially pure SiO_2 but may contain trace amounts of other elements. It consists of a three-dimensional framework of SiO_4 tetrahedra, with all oxygens shared by two tetrahedra. At one atmosphere, upon heating to 573 °C, minor changes in bond angles cause low quartz to change into high quartz (β-quartz) with crystal symmetry 622; it inverts back to low quartz when cooled.

Occurrence and Associations
Quartz is a common and essential ingredient in many sedimentary, metamorphic, and igneous rocks. It dominates in sandstone and quartzite and occurs in all silicic metamorphic and igneous rocks. It also dominates in beach sands, many soils, and other sediments.

Varieties
Coarsely crystalline varieties of quartz include citrine (yellow), amethyst (purple), rose quartz (pink), smoky quartz (yellow-brown to black), and milky quartz (milky white). Fibrous microcrystalline varieties include many types of chalcedony, such as carnelian (red), sard (brown), chrysoprase (apple green), agate (banded or variegated), onyx (white and gray bands). Jasper (iron red), chert (light gray), and flint (dull dark color) are granular microcrystalline varieties of quartz.

Related Minerals
More than a half dozen SiO_2 polymorphs exist, the principal ones being low quartz (α-quartz); high quartz (β-quartz); coesite; stishovite; low and high cristobalite; and low, middle, and high tridymite. Low quartz is the only common one. Opal is an amorphous variety of SiO_2 that contains some H_2O (Figures 1.1 and 8.17c).

Cristobalite   SiO₂

Origin of Name
Named after occurrence at Cerro San Cristóbal, Mexico.

Hand Specimen Identification
Cristobalite is difficult to identify in hand specimen. X-ray or optical techniques are required. Cristobalite is sometimes confused with zeolites in hand specimen. It may appear as white "snowflakes" in obsidian (Figure 2.2).

Physical Properties

hardness	$6\frac{1}{2}$
specific gravity	2.33
cleavage/fracture	none/conchoidal
luster/transparency	vitreous/translucent to transparent
color	colorless
streak	white

Optical Properties
Crystal habit, low birefringence, and moderate negative relief characterizes cristobalite in thin section. Low cristobalite: uniaxial ($-$), $\omega = 1.489$, $\epsilon = 1.482$, $\delta = 0.007$.

Crystallography
Low cristobalite is tetragonal, $a = 4.97$, $c = 6.93$, $Z = 4$; space group $P4_32_12$; point group 422. High cristobalite is cubic, $a = 7.13$, $Z = 8$; space group $F4_1/d\bar{3}2/m$; point group $4/m\bar{3}2/m$.

Habit
Cubic, octahedral, or coarse aggregates typify cristobalite. Although low cristobalite crystals belong to

crystal class 422, they often appear as small octahedra or globby aggregates.

Structure and Composition
Cristobalite, always nearly pure SiO_2, may contain minor amounts of Al^{3+} and alkalis. As with quartz, the structure is based on a three-dimensional framework of SiO_4 tetrahedra. The difference between cristobalite, quartz, and other SiO_2 polymorphs is the way in which the tetrahedra are linked. Mineralogists have described two cristobalite polymorphs: low cristobalite (α-cristobalite) is tetragonal, high cristobalite (β-cristobalite) is cubic.

Occurrence and Associations
Cristobalite is only found in high-temperature silicic extrusive igneous rocks. Rapid cooling may keep it from changing into the more stable, low quartz. Typically it occurs as small spherical grains or aggregates in vugs, as misty inclusions in volcanic glass, or as principal components in fine-grained ground mass. It is associated with other high-temperature minerals including sanidine and tridymite.

Related Minerals
See α-**quartz.**

Tridymite (low tridymite)   SiO$_2$

Origin of Name
From Greek for "threefold," a reference to its habit of forming compound crystals of three individuals or triangular wedge-shaped crystals.

Hand Specimen Identification
Tridymite is usually sufficiently fine grained that X-ray or optical measurements are needed for identification. It is sometimes confused with zeolites.

Physical Properties

hardness	6–7
specific gravity	2.28
cleavage/fracture	none/conchoidal
luster/transparency	vitreous/transparent to translucent
color	colorless
streak	white

Optical Properties
In thin section, tridymite can be distinguished by its low birefringence, low R.I., moderate relief, and typical habit often showing wedge-shaped twins. Low tridymite: biaxial (+); $\alpha = 1.478$, $\beta = 1.479$, $\gamma = 1.481$, $\delta = 0.003$, $2V = 70°$.

Crystallography
Low tridymite is orthorhombic, $a = 9.9$, $b = 17.1$, $c = 16.3$, $Z = 64$; space group $P222$; point group 222.

Habit
Characteristic habit includes wedge-shaped crystals in vesicles or on the walls of cavities of volcanic rocks. Crystals belong to crystal class 222, but often appear as twinned pseudomorphs after high tridymite ($6/m2/m2/m$).

Structure and Composition
Tridymite may contain minute amounts of Al^{3+} and alkalis. Its structure consists of sheets of SiO_4 tetrahedra joined together by bridging oxygens (see Figure 5.6). Three tridymite polymorphs are known (low, middle, and high tridymite).

Occurrence and Associations
Tridymite is found in high-temperature silicic igneous rocks, where it commonly associates with other high-temperature minerals, including sanidine and cristobalite. It is also found in some stony meteorites and lunar basalts.

Related Minerals
See α-**quartz.**

Coesite   SiO$_2$

Origin of Name
Named after L. Coes, who first described it in detail.

Hand Specimen Identification
Coesite cannot be identified in hand specimen.

Physical Properties

hardness	7–8
specific gravity	2.93
cleavage/fracture	none/conchoidal
luster/transparency	vitreous/transparent
color	colorless
streak	white

Optical Properties
Biaxial (+), $\alpha = 1.59$, $\beta = 1.60$, $\gamma = 1.60$, $\delta = 0.01$, $2V = 64°$.

Crystallography
Monoclinic, $a = 7.17$, $b = 12.33$, $c = 7.17$, $\beta = 120.0°$, $Z = 16$; space group $C2/c$; point group $2/m$.

Habit
Crystals are usually thin tabs, invisible without a microscope.

Structure and Composition
Coesite's structure consists of a very dense three-dimensional network of SiO_4 tetrahedra. In some ways, the structure is similar to that of feldspars.

Occurrence and Associations

Coesite, a rare high-pressure polymorph of SiO_2, is known only from meteor impact craters and some rare xenoliths and other rocks of deep crust or mantle origin.

Related Minerals
See **α-quartz**.

Stishovite   SiO₂

Origin of Name
Named after S. M. Stishov, an early investigator of high-pressure SiO_2 polymorphs.

Physical Properties

hardness	7–8
specific gravity	4.30
cleavage/fracture	{110}/conchoidal
luster/transparency	vitreous/transparent
color	colorless
streak	white

Optical Properties
Uniaxial $(+)$, $\omega = 1.799$, $\epsilon = 1.826$, $\delta = 0.027$.

Crystallography
Tetragonal, $a = 4.18$, $c = 2.66$, $Z = 2$; space group $P4_2/m2_1/n2/m$; point group $4/m2/m/m$.

Habit
Stishovite has a prismatic habit.

Structure and Composition
Stishovite is always nearly pure SiO_2. The structure is extremely dense and resembles the structure of rutile. Si^{4+} is in octahedral coordination, in contrast with the other SiO_2 polymorphs.

Occurrence and Associations
Stishovite, a rare, high-pressure polymorph of SiO_2, is known only from meteor impact craters.

Related Minerals
See **α-quartz**.

Feldspar Group Minerals

Alkali Feldspar Series

orthoclase	$KAlSi_3O_8$
sanidine	$(K,Na)AlSi_3O_8$
microcline	$KAlSi_3O_8$
albite	$NaAlSi_3O_8$

Plagioclase Feldspar Series

albite	$NaAlSi_3O_8$
anorthite	$CaAl_2Si_2O_8$

Feldspars are the most abundant minerals in the Earth's crust. Their compositions can be described with the general formula $(Ca,Na,K)(Si,Al)_4O_8$. Feldspar structures are based on SiO_4 and AlO_4 tetrahedra linked to form a three-dimensional framework. They form two series that share one end member: the alkali feldspar series (orthoclase-albite) and the plagioclase feldspar (albite-anorthite) series.

The three $KAlSi_3O_8$ polymorphs differ in the way SiO_4 and AlO_4 tetrahedra are distributed in their structures. Sanidine, the high-temperature polymorph, is most disordered; microcline, the low-temperature polymorph, is most ordered. Orthoclase has intermediate and somewhat variable ordering. Albite and anorthite also exhibit variable ordering, depending on temperature. Figure 5.8 shows some of the more common feldspars.

Ordering complicates phase relations among the feldspars. Figures 5.11 and 5.13 are simplified diagrams showing phase relationships for the two binary series. At high temperatures alkali feldspars form complete solid solutions between monalbite (a high-temperature albite polymorph) and sanidine. Intermediate compositions are often termed *anorthoclase*. At intermediate and low temperatures, however, anorthoclase is unstable because a solvus limits solid solutions between orthoclase/microcline and albite (Figure 5.14).

At high temperatures plagioclase feldspar may have any composition between albite and anorthite. At low temperatures variable ordering and several small solvi lead to complications not reflected in Figure 5.13. Most alkali feldspars contain minor Ca, and most plagioclase feldspars contain small amounts of K. The extent to which the two series can mix is limited and varies with temperature (Figure 5.14).

Feldspars commonly form simple or polysynthetic twins, or both. Both contact and penetration twins are possible. Sometimes twinning is visible with the naked eye, but often must be seen with a microscope. Twin laws include Carlsbad, Baveno, Mannebach, albite, and pericline.

For more general information about the feldspar group, see Chapter 5.

Orthoclase   KAlSi₃O₈

Origin of Name
From Greek *orthos* (right angle) and *klasis* (to break), referring to this mineral's perpendicular cleavages.

Hand Specimen Identification
Color, hardness, cleavage, and association usually serve to identify orthoclase. Near 90° angle between cleavages and lack of polysynthetic twin striations

help distinguish it from other feldspars. It is sometimes confused with calcite or corundum but can be distinguished by its hardness. Plate 1.5 is a color photo of twinned orthoclase crystals. Figures 2.4, 2.7, and 5.8b contain additional photographs.

Physical Properties

hardness	6
specific gravity	2.56
cleavage/fracture	90° cleavage angle; perfect (001), good (010), poor {110}/uneven
luster/transparency	pearly, vitreous/translucent
color	white, pink, turbid
streak	white

Optical Properties
In thin section, orthoclase has low birefringence, moderate relief, and resembles quartz. However, it has negative relief, is biaxial, and is often clouded by fine-grained alteration. It is distinguished from sanidine by $2V$ and from microcline by its lack of plaid twinning. Biaxial, $\alpha = 1.521$, $\beta = 1.525$, $\gamma = 1.528$, $\delta = 0.007$, $2V = 60°-65°$.

Crystallography
Monoclinic, $a = 8.56$, $b = 12.99$, $c = 7.19$, $\beta = 116.01°$, $Z = 4$; space group $C2/m$; point group $2/m$.

Habit
Crystals are prismatic, stubby to elongate, and may be flattened or doubly terminated. Penetration twins and contact twins are common.

Structure and Composition
The structure of orthoclase consists of a three-dimensional framework of SiO_4 and AlO_4 tetrahedra (Figure 13.14). K^+ ions occupy available holes between the tetrahedra. Most orthoclase contains some Na replacing K; complete solid solution between orthoclase and albite ($NaAlSi_3O_8$) is possible only at high temperature. Some orthoclase contains small amounts of CaAl replacing NaSi.

Occurrence and Associations
Orthoclase is common in many kinds of silicic igneous rocks, sediments such as arkoses, and a variety of metamorphic rocks. Quartz and micas are typically associated minerals.

Related Minerals
The principal K-feldspar polymorphs are sanidine (high-temperature form), orthoclase (moderate-temperature form), and microcline (low-temperature form). They differ in the way SiO_4 and AlO_4 tetrahe-

▶**FIGURE 14.3**
Amazonite, a variety of microcline, from Teller Co., Colorado.

dra are arranged in their structure. Several other related minerals are known: Adularia is a colorless, transparent form of K-feldspar that forms prismatic crystals. If it shows opalescence, we call it *moonstone*. Perthite is a form of K-feldspar containing exsolved patches or lamellae of albitic feldspar (Plate 6.7).

Sanidine   (K,Na)AlSi₃O₈

Origin of Name
From Greek *sanis* (tablet) and *idios* (appearance), referring to this mineral's typical habit.

Hand Specimen Identification
Sanidine may be difficult to tell from other feldspars, but its restricted occurrence and associations are helpful diagnostic tools. Certain identification requires X-ray analysis or thin sections. Plagioclase and microcline have different kinds of twins than sanidine.

Physical Properties

hardness	6
specific gravity	2.56
cleavage/fracture	perfect (001), good (010)
luster/transparency	vitreous/transparent to translucent
color	white, variable
streak	white

Optical Properties
Sanidine is similar to orthoclase but greater $2V$. Carlsbad twins may divide crystals into halves. Manebach and Baveno twins may also be present.

Biaxial $(-)$, $\alpha = 1.521$, $\beta = 1.525$, $\gamma = 1.528$, $\delta = .007$, $2V$ varies depending on structure.

Crystallography
Monoclinic, $a = 8.56$, $b = 13.03$, $c = 7.17$, $\beta = 116.58°$, $Z = 4$; space group $C2/m$; point group $2/m$.

Habit
Crystals are prismatic, may be tabular or elongate, and often have a square cross section. Carlsbad twins are common.

Structure and Composition
Similar to orthoclase, the structure of sanidine consists of a three-dimensional framework of SiO_4 and AlO_4 tetrahedra. In sanidine the two kinds of tetrahedra are randomly distributed in the structure, while in orthoclase they are partially ordered. K^+ ions occupy available holes between the tetrahedra. Na may replace K; complete solid solution between sanidine and albite $(NaAlSi_3O_8)$ is possible at high temperature.

Occurrence and Associations
Sanidine occurs in silicic igneous rocks but is restricted to rocks that have cooled quickly. If cooling is slow, orthoclase will be present instead. Typical occurrences are as phenocrysts in rocks such as trachyte or rhyolite.

Related Minerals
Related minerals include the other $KAlSi_3O_8$ polymorphs, orthoclase and microcline, and the plagioclase feldspar series. See **orthoclase.**

Microcline   KAlSi₃O₈

Origin of Name
From Greek *micros* (small) and *klinein* (to lean), referring to this mineral's cleavage angles being close to 90°.

Hand Specimen Identification
Hardness, luster, cleavage, habit, and association help identify microcline, but it is easily confused with other feldspars, especially orthoclase, its polymorph. Twinning or a deep green color are diagnostic but may not be present or visible (Figures 3.6 and 5.8d). X-ray or optical measurements are needed for certain identification.

Physical Properties
hardness	6
specific gravity	2.56
cleavage/fracture	perfect (001), good (010)
luster/transparency	pearly, vitreous/translucent
color	white, green, salmon-pink
streak	white

Optical Properties
Microcline's "tartan plaid" twinning (a combination of albite and pericline twinning) is its most diagnostic characteristic in thin section. Plagioclase may show two sets of twins at about 90°, but in plagioclase the twins have sharp parallel boundaries, while in microcline they pinch and swell. Biaxial $(-)$, $\alpha = 1.518$, $\beta = 1.524$, $\gamma = 1.528$, $\delta = 0.010$, $2V = 77°–84°$.

Crystallography
Triclinic, $a = 8.58$, $b = 12.96$, $c = 7.21$, $\alpha = 89.7°$, $\beta = 115.97°$, $\gamma = 90.87°$, $Z = 4$; $P1$; point group 1.

Habit
Prismatic, stubby to elongate crystals are typical. It is also common as cleavable masses or irregular grains. Twins, both contact and penetration, may be present. Combinations of polysynthetic albite and pericline twinning results in the "tartan plaid" appearance, normally only visible under a microscope.

Structure and Composition
The structure of microcline is similar to orthoclase and sanidine, but the SiO_4 and AlO_4 tetrahedra are more regularly ordered, leading to less symmetry. Ordering decreases with increasing temperature of formation: Low microcline has complete tetrahedral ordering. Intermediate and high microcline are less well ordered. A solvus limits solid solutions between microcline and albite, $NaAlSi_3O_8$, at low temperatures.

Occurrence and Associations
Microcline is common in many kinds of silicic igneous rocks, sediments such as arkoses, and a variety of metamorphic rocks. It easily forms from sanidine or orthoclase as rocks cool from high temperatures. Quartz and micas are typically associated minerals.

Varieties
When colored a deep green, microcline is given the name *amazonite* (Plate 8.3).

Related Minerals
Related minerals include the other $KAlSi_3O_8$ polymorphs, orthoclase, microcline, adularia, and perthite (see **orthoclase**), and the plagioclase feldspar series.

Albite   NaAlSi₃O₈

Origin of Name
From Latin *albus,* meaning "white."

Hand Specimen Identification
Cleavage, hardness, luster, association, and fine polysynthetic twinning help identify albite and other plagioclase feldspars. If not twinned, it may be difficult to tell from K-feldspar. Distinguishing albite from

other plagioclase feldspars cannot be done precisely without detailed X-ray or optical data. Plate 6.8 shows rare euhedral albite. Figures 2.7 and 5.8a also show plagioclase.

Physical Properties

hardness	6
specific gravity	2.62
cleavage/fracture	perfect (001), good (010), poor {110}/uneven
luster/transparency	pearly, vitreous/translucent
color	white, gray, green
streak	white

Optical Properties
In thin section, plagioclase shows no color, has low relief, and exhibits gray interference colors. It is similar to K-feldspar and superficially similar to quartz. However, cleavage, biaxial character, and "zebra stripes" caused by polysynthetic twinning usually serve to identify it. Biaxial $(+)$, $\alpha = 1.527$, $\beta = 1.531$, $\gamma = 1.538$, $\delta = 0.011$, $2V = 77°$. Plates 5.7 and 5.8 and Figures 5.5a, 5.8c, and 7.7b show plagioclase in thin section.

Crystallography
Triclinic, $a = 8.14$, $b = 12.79$, $c = 7.16$, $\alpha = 93.17°$, $\beta = 115.85°$, $\gamma = 87.65°$, $Z = 4$; space group $P\bar{1}$; point group 1.

Habit
Masses or subhedral grains are common. Rare euhedral crystals are prismatic, tabular, or bladed. Most crystals are twinned according to the pericline law, and some are twinned by the albite law. Albite twins give plagioclase the characteristic polysynthetic twinning that is often visible as fine striations in hand specimen and as stripes in thin section. Figure 2.7g is a drawing of plagioclase twinning.

Structure and Composition
Albite is an end member of both the plagioclase feldspar and the alkali felspar series. As with K-feldspar, ordering of AlO_4 and SiO_4 tetrahedra decreases with increasing temperature, leading to minor changes in structure. Low albite's structure is similar to that of low microcline; high albite's structure is more disordered. At very high temperature a completely disordered albite, called *monalbite* because it is monoclinic, is stable. At all but the lowest temperatures, complete solid solution exists between albite and the other plagioclase end member, anorthite, $CaAl_2Si_2O_8$. Albite, and other plagioclase feldspars, also form limited solid solutions with orthoclase, $KAlSi_3O_8$.

Occurrence and Associations
The most abundant mineral of the Earth's crust, plagioclase feldspars are found in a wide variety of igneous, metamorphic, and, less commonly, sedimentary rocks. Most are intermediate between albite and anorthite, but compositions approaching end members are known. Albite, defined as plagioclase with greater than 90% $NaAlSi_3O_8$, is found in silicic igneous rocks such as granite, syenite, trachyte, or rhyolite, where it associates with quartz and orthoclase.

Varieties
Clevelandite is a form of albite, typified by curved plates, found in pegmatites. Opalescent varieties of albite or other plagioclase feldspars are called *moonstone*.

Related Minerals
Albite is closely related to the other, more calcic plagioclase feldspars, and to the other alkali feldspars (orthoclase, sanidine, and microcline).

Anorthite   CaAl₂Si₂O₈

Origin of Name
From the Greek word meaning "oblique," in reference to anorthite's crystal shape.

Hand Specimen Identification
Cleavage, hardness, luster, and association help identify anorthite, but it can be extremely difficult to tell from other feldspars. See plagioclase in Figures 2.7 and 5.8a. Several kinds of twinning are common (see albite). Albite twins, if present, may be difficult to see in hand specimen.

Physical Properties

hardness	$6-6\frac{1}{2}$
specific gravity	2.76
cleavage/fracture	perfect (001), good (010), poor {110}/uneven
luster/transparency	pearly, vitreous/translucent
color	white, gray
streak	white

Optical Properties
In thin section, anorthite is similar to other plagioclase feldspars. Biaxial $(-)$, $\alpha = 1.577$, $\beta = 1.585$, $\gamma = 1.590$, $\delta = 0.013$, $2V = 78°$. See Figures 5.5a, 5.8c, and 7.7b.

Crystallography
Triclinic, $a = 8.17$, $b = 12.88$, $c = 14.16$, $\alpha = 93.33°$, $\beta = 115.60°$, $\gamma = 91.22°$, $Z = 8$; space group $P\bar{1}$; point group $\bar{1}$.

Habit
Anorthite is common as cleavable masses or irregular grains. Euhedral crystals are rare. They may be prismatic, tabular, or bladed and are frequently twinned according to the same laws as albite. When

present, albite twins give calcic plagioclase characteristic polysynthetic twinning, but the width of the twins is usually greater than is common for albitic plagioclase.

Structure and Composition
Anorthite is the calcic end member of the plagioclase feldspar series, but the name is also used for any plagioclase containing >90% $CaAl_2Si_2O_8$. Its structure is similar to those of albite and orthoclase. As with the other feldspars, the ordering of AlO_4 and SiO_4 tetrahedra decreases with increasing temperature, leading to minor changes in structure. Complete solid solution with albite, $NaAlSi_3O_8$, is possible at all but very low temperatures. Minor solid solution with orthoclase, $KAlSi_3O_8$, is common.

Occurrence and Associations
Anorthite, found primarily in mafic igneous rocks, is rarer than other plagioclase feldspars. In igneous rocks, it associates with amphibole, pyroxene, or olivine. It is occasionally found in metamorphosed carbonates.

Related Minerals
Anorthite is closely related to the other, more sodic plagioclase feldspars, and to the potassic feldspars (orthoclase, sanidine, and microcline).

Feldspathoid Group Minerals

analcime	$NaAlSi_2O_6 \cdot H_2O$
leucite	$KAlSi_2O_6$
nepheline	$(Na,K)AlSiO_4$

Feldspathoid minerals are similar to feldspars in many ways, but they contain more Al and less Si. They have structures based on three-dimensional frameworks of AlO_4 and SiO_4 tetrahedra; alkalis occupy holes between the tetrahedra. Al:Si ratios vary from 1:1 in nepheline, $NaAlSiO_4$, to 1:4 in the rare feldspathoid petalite, $LiAlSi_4O_{10}$. Feldspathoids are closely related to zeolites; the distinction between the two groups is hazy: The major difference is that zeolite structures contain large cavities or open channels and usually contain H_2O.

The most important feldspathoids are leucite, nepheline, and analcime, although analcime is often considered a zeolite because it contains molecular H_2O. Sodalite, $Na_3Al_3Si_3O_{12} \cdot NaCl$, and haüyne, $Na_3CaAl_3Si_3O_{12}(SO_4)$, are sometimes grouped with the feldspathoids but have a cage structure more similar to that of zeolites.

For more general information about the feldspathoid group, see Chapter 5.

Analcime $NaAlSi_2O_6 \cdot H_2O$

Origin of Name
From Greek *analkimos* (weak), referring to its weak pyroelectric character.

Hand Specimen Identification
Crystal form and association help identify analcime, but it may be difficult to tell from leucite. Because of its typical form, it is occasionally confused with garnet.

Physical Properties
hardness	$5-5\frac{1}{2}$
specific gravity	2.26
cleavage/fracture	poor cubic {100}/uneven
luster/transparency	vitreous/transparent to translucent
color	white, gray, pink
streak	white

Optical Properties
In thin section, analcime is colorless and exhibits low negative relief and low birefringence. It may be confused with leucite, which has higher indices of refraction, or with sodalite, which is usually slightly bluish. Isotropic, $n = 1.482$.

Crystallography
Cubic, $a = 13.71$, $Z = 16$; space group $I4_1/a\bar{3}2/d$; point group $4/m\bar{3}2/m$.

Habit
Analcime typically forms distinct euhedral crystals. Trapezohedrons and cubes are common forms, often in combination. Massive and granular aggregates are also known.

Structure and Composition
The framework structure of analcime consists of AlO_4 and SiO_4 groups joined to make rings of four, six, or eight tetrahedra. Rings align, producing channels that hold H_2O molecules. The channels can hold additional absorbed ions or groups. Na^+ ions occupy nonchannel sites between rings. Minor substitution of K or Cs for Ca and of Al for Si are common. Analcime forms a limited solid solution with pollucite, $(Cs,Na)_2(AlSi_2O_6)_2 \cdot H_2O$.

Occurrence and Associations
Analcime is found in cavities in basalt and as a primary mineral in alkalic igneous rocks such as Na-rich basalt or syenite. In cavities, it is associated with zeolites, calcite, or prehnite.

Related Minerals
Analcime is similar in structure to zeolites such as wairakite, $Ca(Al_2Si_4O_{12}) \cdot 2H_2O$, and to leucite, $KAlSi_2O_6$.

Leucite $KAlSi_2O_6$

Origin of Name
From the Greek *leukos,* meaning "white," in reference to leucite's color.

Hand Specimen Identification
Association, crystal form, and color help identify leucite. It may be confused with analcime, but leucite typically forms as a matrix mineral whereas analcime forms in cavities.

Physical Properties
hardness	$5\frac{1}{2}$–6
specific gravity	2.48
cleavage/fracture	poor {100}, poor (001)/ conchoidal
luster/transparency	vitreous/transparent to translucent
color	white, gray
streak	white

Optical Properties
Low relief, gray interference colors, and lamellar or concentric twins help identify leucite. Uniaxial (+), $\omega = 1.508$, $\epsilon = 1.509$, $\delta = 0.001$.

Crystallography
Tetragonal, $a = 13.04$, $c = 13.85$, $Z = 16$; space group $I4_1/a$; point group $4/m$.

Habit
Although tetragonal at low temperatures, leucite normally has the form of its high-temperature cubic polymorph. Trapezohedral crystals are typical. Polysynthetic twinning may give faces fine striations.

Structure and Composition
Leucite's structure consists of a framework of 4-, 6-, and 8-membered rings of AlO_4 and SiO_4 tetrahedra. K^+ ions occupy half the available sites between the rings. Leucite is generally close to end member composition, although small amounts of Fe, Na, and other alkalis may be present.

Occurrence and Associations
Leucite is a rare mineral found in Si-poor, K-rich volcanic rocks. It is never found with quartz.

Related Minerals
Leucite is isostructural with pollucite, $(Cs,Na)_2$ $(AlSi_2O_6)_2 \cdot H_2O$. Chemically, it is closely related to orthoclase, $KAlSi_3O_8$; kaliophilite, $KAlSiO_4$; and to analcime, $NaAlSi_2O_6 \cdot H_2O$. A cubic polymorph of leucite exists above 605 °C.

Nepheline $(Na,K)AlSiO_4$

Origin of Name
From the Greek *nephele,* meaning "cloud," because crystals turn cloudy when immersed in acid.

Hand Specimen Identification
Association and greasy luster are diagnostic. It may be confused with feldspar or quartz but is softer. Occasionally it is confused with apatite.

Physical Properties
hardness	$5\frac{1}{2}$–6
specific gravity	2.60
cleavage/fracture	poor {100}, poor (001)/ subconchoidal
luster/transparency	vitreous, greasy/transparent to translucent
color	colorless, turbid
streak	white

Optical Properties
Low birefringence and relief and common alteration identify nepheline. It is distinguished from the feldspars by its uniaxial nature and from quartz by its optic sign. Uniaxial (−), $\omega = 1.540$, $\epsilon = 1.536$, $\delta = 0.004$.

Crystallography
Hexagonal, $a = 10.01$, $c = 8.41$, $Z = 8$; space group $P6_3$; point group 6.

Habit
Massive, compact, and embedded grains are common. Crystals are short and prismatic with six or twelve sides.

Structure and Composition
The structure of nepheline derives from that of tridymite: Every other Si is replaced by Al, and Na occupies large sites between Al and Si tetrahedra. All natural nepheline contains some K substituting for Na, but a solvus exists between nepheline and kalsilite, $KAlSiO_4$, at temperatures below 1,000 °C. Exsolution, similar to perthite (see orthoclase), is common.

Occurrence and Associations
Nepheline is characteristic of some Si-poor igneous rocks, such as syenite. It is found with feldspars, apatite, cancrinite, sodalite, zircon, and biotite.

Related Minerals
Nepheline is isostructural with tridymite. It has a high-temperature polymorph above 900 °C. It forms

a solid solution with kalsilite, $KAlSiO_4$, and is chemically related to kaliophilite, $KAlSiO_4$. Cancrinite, $(Na_3Ca_2)_2CO_3(Si_3Al_3O_{12}) \cdot 2H_2O$ is similar to nepheline in many ways and occurs in the same type of rocks.

Scapolite Series Minerals

marialite $Na_4(AlSi_3O_8)_3Cl$
meionite $Ca_4(Al_2Si_2O_8)_3(CO_3,SO_4)$

Scapolites are metamorphic minerals with formulas related to the feldspars, but with structures more closely related to nepheline. The two principal end members are marialite, $Na_4(AlSi_3O_8)_3Cl$, equivalent in composition to albite plus halite; and meionite, $Ca_4(Al_2Si_2O_8)_3(CO_3,SO_4)$ equivalent in composition to anorthite plus calcite/anhydrite. Complete solid solution between the two end members is possible; F and OH may replace Cl and CO_3. Figure 9.6a shows scapolite crystals.

Scapolite   **Solid solutions of marialite, $Na_4(AlSi_3O_8)_3Cl$, and meionite, $Ca_4(Al_2Si_2O_8)_3(CO_3,SO_4)$**

Origin of Name
From the Greek *skapos,* meaning "stalk," in reference to its common, woody appearance.

Hand Specimen Identification
The scapolites are characterized by prismatic crystals with square cross sections and 45° cleavage (Figure 9.6a). Massive samples often have a distinct woody or fibrous appearance. Scapolite may be confused with feldspar.

Physical Properties

hardness	5–6
specific gravity	2.55
cleavage/fracture	good {100}, poor {110}/ conchoidal
luster/transparency	vitreous/transparent to translucent
color	colorless, variable
streak	white

Optical Properties
Scapolite may appear similar to feldspars but is uniaxial and has different cleavage. Uniaxial $(-)$, $\omega = 1.540$, $\epsilon = 1.536$, $\delta = 0.004$.

Crystallography
Tetragonal, $a = 12.11$, $c = 7.56$, $Z = 2$; space group $I4/m$; point group $4/m$.

Habit
Crystals are typically tetragonal prisms, often woody looking. Massive varieties are common.

Structure and Composition
The structure, closely related to that of nepheline, consists of AlO_4 and SiO_4 joined in a three-dimensional framework. The structure contains two types of holes for anions and anionic groups: One holds Na and Ca; the other Cl or CO_3. CaAl substitutes for NaSi freely; SO_4 and F may substitute for Cl and CO_3. The two important end members are marialite, $Na_4(AlSi_3O_8)_3Cl$, and meionite, $Ca_4(Al_2Si_2O_8)_3(CO_3,SO_4)$. Natural scapolites have intermediate compositions. Chemically, scapolites are equivalent to plagioclase feldspar plus $CaCO_3$, NaCl, or $CaSO_4$. Minor K, OH, and F may be present.

Occurrence and Associations
Most scapolite occurrences are in calcic metamorphic rocks: marbles, marls or mafic gneisses, and amphibolites. Rare occurrences are reported from igneous rocks. Associated minerals include plagioclase, clinopyroxene, hornblende, apatite, garnet, and sphene.

Zeolite Group Minerals

natrolite	$Na_2Al_2Si_3O_{10} \cdot 2H_2O$
chabazite	$CaAl_2Si_4O_{12} \cdot 6H_2O$
heulandite	$CaAl_2Si_7O_{18} \cdot 6H_2O$
stilbite	$CaAl_2Si_7O_{18} \cdot 7H_2O$
sodalite	$Na_3Al_3Si_3O_{12} \cdot NaCl$

Zeolites are a large and important group of minerals. Only five are considered in detail here, but more than 40 natural species are known, and many equivalent phases have been synthesized. All are framework silicates, containing a three-dimensional network of SiO_4 and AlO_4 tetrahedra linked to form channels, cages, rings, or loops. Alkalis or alkaline earths occupy large sites in the structures. Some zeolites have 4-, 6-, 8-, or 10-member tetrahedral rings. Others contain complex polyhedra resulting in cage-like openings. Unlike other framework silicates, zeolites contain large open cavities and channels in their structures that permit cations and H_2O molecules to pass in and out without disruption. Consequently, zeolites are used as molecular sieves in water softeners and other industrial applications.

Crystal symmetry and morphology vary between species. Different zeolites are cubic, tetragonal, orthorhombic, hexagonal, or monoclinic, and habits may be fibrous, tabular or platy, prismatic, or equant. Figure 14.4 shows two examples: chabazite and stilbite.

Although sodalite is sometimes grouped with the feldspathoids, it is here grouped with the zeolites because it has a cagelike structure with 4- and 6-member silica rings.

Natrolite $Na_2Al_2Si_3O_{10} \cdot 2H_2O$

Origin of Name
From the Greek *natron,* meaning "soda."

Hand Specimen Identification
Natrolite typically occurs as white radiating needles or as radial aggregates. Its perfect prismatic cleavage and uneven fracture and association also aid identification. Natrolite may be confused with aragonite or pectolite.

Physical Properties

hardness	$5–5\frac{1}{2}$
specific gravity	2.23
cleavage/fracture	perfect prismatic {110}/ uneven
luster/transparency	vitreous/transparent to translucent
color	colorless, gray
streak	white

Optical Properties
Natrolite may be difficult to distinguish from other zeolites. It has two perfect cleavages, is of parallel extinction, is length slow, and has a moderate *2V*. Biaxial $(+)$, $\alpha = 1.48$, $\beta = 1.48$, $\gamma = 1.49$, $\delta = 0.012$, $2V = 38°–62°$.

Crystallography
Orthorhombic, $a = 18.30$, $b = 18.63$, $c = 6.60$; space group *Fd2d*; point group *mm2*.

Habit
Acicular crystals, often radiating, are typical. Crystals may show vertical striations. Radiating rounded masses are also common. Less commonly, it is fibrous, massive, or granular.

Structure and Composition
Natrolite and all zeolites are framework structures built of AlO_4 and SiO_4 tetrahedra. The tetrahedra are linked to form chains, cages, rings, or loops. In contrast with some other framework silicates, the zeolite structure contains many large holes and channels, which hold weakly bonded H_2O. Slightly smaller holes contain Na, Ca, or K. In natrolite, scolecite, and some other zeolites, the tetrahedral framework has a strong linear fabric parallel to *c*. Crystal habit is therefore fibrous or acicular. Natrolite always contains minor amounts of Ca and K and has variable H_2O content.

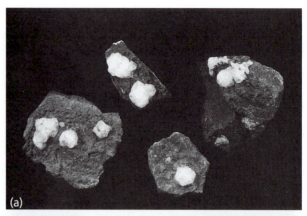

(a)

(b)

▶**FIGURE 14.4**
(a) Chabazite, $CaAl_2Si_4O_{12} \cdot 6H_2O$, is one of the more common zeolites. Here it is shown as euhedral crystals on basalt from Melbourne, Australia. (b) Stilbite, $CaAl_2Si_7O_{18} \cdot 7H_2O$, is another relatively common zeolite. This specimen containing stilbite needles is from Golden, Colorado.

Occurrence and Associations
Natrolite and most other zeolites are secondary minerals that form in cracks or on cavity walls in mafic igneous rocks. It is found with calcite and other zeolites.

Related Minerals
Scolecite, $CaAl_2Si_3O_{10} \cdot 3H_2O$, and thomsonite, $NaCa_2 Al_5Si_5O_{20} \cdot 6H_2O$, are closely related to natrolite.

Chabazite $CaAl_2Si_4O_{12} \cdot 6H_2O$

Origin of Name
From the Greek *chabazios,* meaning "hail."

Hand Specimen Identification
Form, color, and association help identify chabazite. Crystals are transparent to translucent pseudocubic rhombs, similar to calcite. Chabazite is distinguished from calcite by its poorer cleavage and lack of reaction to HCl. Figure 14.4a shows chabazite crystals on basalt; Figure 3.3a illustrates twinned chabazite.

Physical Properties

hardness	4–5
specific gravity	2.1
cleavage/fracture	poor rhombohedral {101}/uneven
luster/transparency	vitreous/transparent to translucent
color	colorless, red
streak	white

Optical Properties
Uniaxial $(-)$, $\omega = 1.484$, $\epsilon = 1.481$, $\delta = 0.003$.

Crystallography
Trigonal, $a = 13.17$, $c = 15.06$, $Z = 6$; space group $R\bar{3}2/m$; point group $\bar{3}2/m$.

Habit
Chabazite usually forms simple rhombohedral crystals that may, at first glance, appear cubic. Some crystals are more complicated, showing more than one rhombohedral form or exhibiting penetration twins.

Structure and Composition
Chabazite is similar in structure to natrolite and other zeolites. Tetrahedra form large cagelike openings, which can hold a variety of loosely bonded ions and molecules. The large openings allow diffusion of some small molecules through the structure. Considerable substitution of Na and K for Ca is common.

Occurrence and Associations
Chabazite is a secondary minerals that form in cracks or on cavity walls in mafic igneous rocks and as an alteration product in silicic igneous rocks. It is found with calcite and other zeolites.

Related Minerals
Several other rare zeolites have structures similar to chabazite's.

Heulandite $\qquad$ CaAl$_2$Si$_7$O$_{18}$ · 6H$_2$O

Origin of Name
Named for H. Heuland, a British mineralogist.

Hand Specimen Identification
Occurrence, association, crystal form, perfect one-direction cleavage, and luster help identify heulandite.

Physical Properties

hardness	$3\frac{1}{2}$–4
specific gravity	2.15
cleavage/fracture	one perfect (010)/subconchoidal
luster/transparency	vitreous/transparent to translucent
color	white, variable
streak	white

Optical Properties
Biaxial $(+)$, $\alpha = 1.49$, $\beta = 1.50$, $\gamma = 1.50$, $\delta = 0.005$, $2V = 35°$.

Crystallography
Monoclinic, $a = 17.73$, $b = 17.82$, $c = 7.43$, $\beta = 116.3°$, $Z = 4$; space group Cm; point group m.

Habit
Crystals are typically platy with a diamond or modified diamond shape.

Structure and Composition
The structure of heulandite is similar to that of other zeolites, except that tetrahedra are linked in 6-member rings that align to give a more planar structure than most others. End member calcic heulandite, called *mordenite*, is rare. Considerable solid solution towards clinoptiloite, $(Na,K)Si_7Al_2O_{18} · 6H_2O$, is typical.

Occurrence and Associations
Heulandite is one of the more common zeolites. It is a secondary mineral, found with calcite and other zeolites, that forms in cracks or on cavity walls in mafic igneous rocks, and in some metamorphic rocks.

Related Minerals
Other zeolites, including clinoptiloite, $(Na,K)Si_7Al_2O_{18} · 6H_2O$ and stilbite, $CaAl_2Si_7O_{18} · 7H_2O$, have structures similar to heulandite's.

Stilbite $\qquad$ CaAl$_2$Si$_7$O$_{18}$ · 7H$_2$O

Origin of Name
From the Greek *stilbein*, meaning "to shine," referring to this mineral's luster.

Hand Specimen Identification
Sheaflike aggregates of twinned crystals, one excellent cleavage, pearly luster, color, and association identify stilbite. See Figure 14.4b.

Physical Properties

hardness	$3\frac{1}{2}$–4
specific gravity	2.15
cleavage/fracture	perfect (010)/subconchoidal
luster/transparency	pearly/transparent to translucent
color	gray
streak	gray

Optical Properties
Cruciform twins and parallel or near-parallel extinction help identify stilbite in thin section. Biaxial $(-)$, $\alpha = 1.49$, $\beta = 1.50$, $\gamma = 1.50$, $\delta = 0.010$, $2V = 30°$–$50°$.

Crystallography
Monoclinic, $a = 13.64$, $b = 18.24$, $c = 11.27$, $\beta = 129.16°$, $Z = 8$; space group $C2/m$; point group $2/m$.

Habit
Simple crystals are extremely rare. Aggregates of twinned crystals, having the appearance of sheaves of grain, are common. Sometimes aggregates appear fibrous. More rarely, stilbite forms crystals displaying cruciform twinning.

Structure and Composition
Stilbite's structure is similar to heulandite's. Considerable substitution of Na and K for Ca is common.

Occurrence and Associations
Stilbite is one of the more common zeolites. It is a secondary mineral, found with calcite and other zeolites, that forms in cracks or on cavity walls in igneous rocks and in some schists associated with hydrothermal ore bodies.

Related Minerals
Stilbite is grouped with heulandite, $CaAl_2Si_7O_{18}$ $\cdot 6H_2O$, and several others because of its chemical and structural similarity.

Sodalite $Na_3Al_3Si_3O_{12} \cdot NaCl$

Origin of Name
This mineral's name refers to its sodium content.

Hand Specimen Identification
Color and association are usually diagnostic for sodalite. If not blue, identification may require chemical tests to tell it from zeolites or analcime. It is sometimes confused with lazulite but has different associations.

Physical Properties

hardness	$5\frac{1}{2}$–6
specific gravity	2.3
cleavage/fracture	six poor {110} at 60° angles/conchoidal
luster/transparency	vitreous/transparent to translucent
color	blue, white
streak	white

Optical Properties
In thin section, sodalite is distinguished by being isotropic and having a low index of refraction and, sometimes, a hexagonal outline. Isotropic, $n = 1.485$.

Crystallography
Cubic, $a = 8.87$, $Z = 2$; space group $P\bar{4}3m$; point group $\bar{4}3m$.

Habit
Often massive or in embedded grains, sodalite forms rare dodecahedral crystals.

Structure and Composition
The structure of sodalite is similar to many zeolites, containing 4- and 6-member tetrahedral rings but, unlike true zeolites, it contains no molecular water that can be driven off easily. The Al and Si tetrahedral rings are linked to form a framework with cage-like openings that hold Cl and sometimes S or SO_4. Sodalite is usually close to end member composition. Small amounts of K or Ca may also be present.

Occurrence and Associations
Sodalite is associated with nepheline, $(Na,K)AlSiO_4$, cancrinite, $(Na_3Ca)_2(Al_3Si_3O_{12})CO_3 \cdot 2H_2O$, leucite, $KAlSi_2O_6$, and with feldspars in Si-poor, alkali-rich igneous rocks.

Varieties
Hackmanite is a sulfurous form of sodalite.

Related Minerals
Sodalite is structurally and chemically related to other zeolites and to cancrinite, $(Na_3Ca_2)_2(Al_3Si_3O_{12})$ $CO_3 \cdot 2H_2O$.

Other Framework Silicates

beryl	$Be_3Al_2Si_6O_{18}$
cordierite	$(Mg,Fe)_2Al_4Si_5O_{18}$

Beryl and cordierite, which contain 6-membered tetrahedral rings, are sometimes considered to be ring silicates. They are, however, more properly classified as framework silicates because they have an overall three-dimensional framework of connecting tetrahedra. Open channels in their structures almost always contain some H_2O.

Beryl $Be_3Al_2Si_6O_{18}$

Origin of Name
From the Greek *beryllos*, meaning "a blue-green gem."

Hand Specimen Identification
Association, crystal form, very poor cleavage, color, and hardness identify beryl. See Figures 1.2b, 9.9b, and 14.5.

Physical Properties

hardness	$7\frac{1}{2}$–8
specific gravity	2.7–2.9
cleavage/fracture	poor (001)/even
luster/transparency	vitreous/transparent to translucent

▶FIGURE 14.5
Beryl crystals, from Minas Gerais, Brazil: These 4 cm long crystals are light blue and nearly clear.

color	colorless, variable
streak	white

Optical Properties
In thin section, beryl may appear similar to quartz, apatite, or topaz, but quartz is $(+)$, apatite has higher relief, and topaz is biaxial. Uniaxial $(-)$, $\omega = 1.568$, $\epsilon = 1.562$, $\delta = 0.006$.

Crystallography
Hexagonal, $a = 9.23$, $c = 9.19$, $Z = 2$; space group $P6/m2/c2/c$; point group $6/m2/m2/m$.

Habit
Beryl typically forms hexagonal prisms. When present, terminating faces are pinacoids or, more rarely, pyramids. Single crystals are common; columnar aggregates are less so.

Structure and Composition
In the structure, 6-membered tetrahedral rings form sheets that are linked by tetrahedral Be and octahedral Al. Some classification schemes group beryl with true ring silicates such as tourmaline, but in beryl the framework structure is not completely planar; there is much cross linking. Small amounts of Na, Rb, and Li may substitute for Be; minor H_2O and CO_2 may occupy spaces within the rings.

Occurrence and Associations
Beryl is found in granitic rocks, notably in pegmatites. It may also be found in schists and in rare ore deposits.

Varieties
Beryl can be many different colors due to small amounts of trace elements: emerald is vivid green; aquamarine is pale greenish blue (Plates 3.3 and 8.5); morganite is rose colored; heliodor is gold.

Related Minerals
Cordierite, $(Mg,Fe)_2Al_4Si_5O_{18}$, is similar in structure to beryl. Euclase, $BeAlSiO_4(OH)$, is another of the rare beryllium silicates.

Cordierite   (Mg,Fe)$_2$Al$_4$Si$_5$O$_{18}$

Origin of Name
Named after P. L. A. Cordier (1777–1861), the French mineralogist who first described the mineral.

Hand Specimen Identification
Prismatic cleavage, color, and association aid identification, but if it is not blue, this mineral is often difficult to distinguish without a microscope. It may be easily confused with quartz in hand specimen and with plagioclase in thin section.

Physical Properties

hardness	$7–7\frac{1}{2}$
specific gravity	2.5–2.8
cleavage/fracture	fair prismatic (010), poor (100)
luster/transparency	vitreous/transparent to translucent
color	indigo to gray-blue
streak	white

Optical Properties
In thin section, cordierite may be clear or pale blue-violet. Lamellar twinning is common. Pleochroic halos around zircon inclusions are diagnostic. Cordierite often alters to a fine-grained mass called *pinnite*. It may be confused with quartz, feldspar, or nepheline. Uniaxial $(+$ or $-)$, $\alpha = 1.54$, $\beta = 1.55$, $\gamma = 1.56$, $\delta = 0.02$, $2V = 65°–105°$.

Crystallography
Orthorhombic, $a = 17.13$, $b = 9.80$, $c = 9.35$, $Z = 4$; space group $C2/c2/c2/m$; point group $2/m2/m2/m$.

Habit
Rare, euhedral crystals are short and prismatic and may appear pseudohexagonal. Twins are common. Cordierite is more typically granular, massive, or compact.

Structure and Composition
Cordierite, like beryl, consists of 6-membered tetrahedral rings joined in a three-dimensional framework. Mg^{2+}, Fe^{2+}, and Al^{3+} link the rings to each other. Hollow channels, sometimes occupied by H_2O or alkalis, run parallel to the c-axis. Fe:Mg ratio is variable.

Occurrence and Associations
Cordierite is found as a product of contact or regional metamorphism in high-grade metamorphosed aluminous rocks. Rare occurrences in igneous rocks

have been reported. Associated minerals include garnet, sillimanite, spinel, plagioclase, anthophyllite, and orthopyroxene.

Varieties
A high-temperature polymorph of cordierite, indialite, is isostructural with beryl.

Related Minerals
Cordierite is structurally similar to beryl, and is chemically and structurally similar to osumilite, $(K,Na)(Fe,Mg)_2(Al,Fe)_3(Si,Al)_{12}O_{30} \cdot H_2O$.

SHEET SILICATES

Serpentine Group Minerals

antigorite	$Mg_6Si_4O_{10}(OH)_8$
chrysotile	$Mg_6Si_4O_{10}(OH)_8$
lizardite	$Mg_6Si_4O_{10}(OH)_8$

Antigorite, chrysotile, and lizardite comprise the serpentine group. Antigorite and lizardite are typically massive and fine grained; chrysotile is fibrous and is one of the few asbestiform minerals. Most of the world's asbestos is chrysotile; crocidolite and amosite, both amphibole varieties, account for the rest. Chrysotile and lizardite are true polymorphs, but antigorite has a slightly different composition not reflected in its formula. Figures 3.5c and 14.6 show typical massive serpentine.

Antigorite $Mg_6Si_4O_{10}(OH)_8$

Origin of Name
Named after Valle Antigorio, Italy, the first reported locality in which this mineral was found.

Hand Specimen Identification
Green color, greasy luster, and lack of cleavage and association identify antigorite. Chrysotile is a more fibrous variety of serpentine.

Physical Properties

hardness	3–4
specific gravity	2.6
cleavage/fracture	perfect (001)/flexible
luster/transparency	resinous, silky/translucent
color	green, yellow-green
streak	white

Optical Properties
In thin section, serpentine typically has a netlike pattern and exhibits wavy extinction. Very low birefringence results in anomalous interference colors. It may be confused with chlorite or brucite. Biaxial $(-)$, $\alpha = 1.56$, $\beta = 1.57$, $\gamma = 1.57$, $\delta = 0.007$, $2V = 20°–60°$.

Crystallography
Monoclinic, $a = 5.32$, $b = 9.50$, $c = 14.9$, $\beta = 101.9°$, $Z = 4$; space group $C2/m$; point group $2/m$.

Habit
Antigorite is commonly fine grained, massive, or platy, in contrast with fibrous chrysotile.

Structure and Composition
The layered structure consists of paired sheets of SiO_4 tetrahedra and $Mg(O,OH)_6$ octahedra stacked on top of each other (Figure 13.15). Antigorite has a slightly different composition from that indicated by its formula because the ratio of brucite layers to tetrahedral layers is slightly greater than 1. Because atoms in the tetrahedral and octahedral layers have slightly mismatched spacings, layers curve slightly. In antigorite the sheets curve in both directions, alternating on a fine scale, so the overall structure retains sheetlike properties. Small amounts of Ni, Mn, Al, Ti, and Fe typically substitute for Mg.

Occurrence and Associations
Antigorite is a common secondary mineral in mafic and ultramafic igneous rocks. It is also found in some marbles. In some serpentinites it is the only mineral present. Associated minerals include magnesium silicates, carbonates, hydroxides, and oxides, as well as olivine, pyroxene, amphibole, magnesite, spinel, chromite, magnetite, brucite, and talc.

Varieties
Garnierite is a Ni-rich variety of serpentine associated with Ni-peridotites.

Related Minerals
Antigorite is closely related to lizardite and chrysotile (common asbestos). Greenalite, $Fe_3Si_2O_5(OH)_4$, the Fe equivalent of serpentine, has a different structure.

Chrysotile $Mg_6Si_4O_{10}(OH)_8$

Origin of Name
From the Greek *chrysos* and *tilos,* meaning "golden" and "fiber."

Hand Specimen Identification
Fiberlike, asbestiform appearance is diagnostic (Figures 3.5c and 14.6). Chrysotile may be distinguished from most fibrous amphiboles by its greenish white color; compare Plates 3.4 and 3.7.

Physical Properties

hardness	3–5
specific gravity	2.5–2.6
cleavage/fracture	none/uneven
luster/transparency	greasy, waxy/translucent
color	variable white, greenish white
streak	white

▶FIGURE 14.6
Chrysotile, $Mg_6Si_4O_{10}(OH)_8$, from Waldheim, Germany. Chrysotile is the asbestiform variety of serpentine.

Crystallography
Monoclinic, $a = 5.34$, $b = 9.25$, $c = 14.65$, $\beta = 93°$, $Z = 8$; or orthorhombic, depending on polymorph.

Habit
Fibrous, asbestiform habit typifies chrysotile (Figure 14.6).

Structure and Composition
Composition and structure are similar to those of antigorite (figure 13.15), except that the mismatch in spacing of the octahedral and tetrahedral layers causes them to curl in one direction only, forming fibers.

Occurrence and Associations
Occurrence and associations are the same as for antigorite. In many serpentine samples, chrysotile layers are separated by a fine-grained platy polymorph called *lizardite*.

Related Minerals
Chrysotile has two polymorphs, lizardite and antigorite. Greenalite, $Fe_3Si_2O_5(OH)_4$, the Fe equivalent of serpentine, has a different structure. Amosite and crocidolite, varieties of amphibole, are also asbestiform.

Clay Mineral Group

montmorillonite	$(Ca,Na)_{0.2-0.4}(Al,Mg,Fe)_2$ $(Si,Al)_4O_{10}(OH)_2 \cdot nH_2O$
illite	micalike clay mineral
kaolinite	$Al_2Si_2O_5(OH)_4$
pyrophyllite	$Al_2Si_4O_{10}(OH)_2$
talc	$Mg_3Si_4O_{10}(OH)_2$

To a petrologist, the term *clay* refers to a kind of rock or sediment, usually made of a number of minerals referred to as *clay minerals*. To a mineralogist, clays are a group of sheet silicates with related atomic structures. Most are hydrated aluminum or magnesium silicates that form as products of weathering. Clay mineral compositions are highly variable, in part due to poor crystallinity and mixed structures. The most common clay minerals are kaolinite, illite, and montmorillonite. Montmorillonite is the most common clay in bentonite, altered volcanic ash. *Smectite* is a general term for a number of clay minerals similar to montmorillonite. Figure 6.5 shows some typical clay minerals.

Clay minerals fall into three main subgroups: smectite, illite, and kaolinite. For many clays, crystal structure and chemistry are variable or poorly determined. Illite and smectite are basically "three-layer" structures, while kaolinite clays are "two-layer" structures. Illite group clays, including illite, are transitional from smectites to true micas.

Similar in structure to serpentine, kaolinite is composed of alternating layers of Al octahedra and Si tetrahedra. Other clays have more complex structures. Illite is close to mica in structure but lacks essential alkalis. Pyrophyllite and talc, sometimes grouped with serpentine or considered to belong to a separate group, are here grouped with the other clay minerals because of similar chemistry.

For more general information about clays, see Chapter 6.

Montmorillonite $(Ca,Na)_{0.2-0.4}(Al,Mg,Fe)_2(Si,Al)_4O_{10}(OH)_2$

Origin of Name
From its original discovery at Montmorillon, near Limoges, France.

Hand Specimen Identification
Clays are difficult to tell apart without detailed X-ray study. When massive, they are unctuous and earthy if wet, and appear as soft, very fine-grained aggregates if dry. See Figure 6.5b.

Physical Properties

hardness	$1-1\frac{1}{2}$
specific gravity	2.0–2.7
cleavage/fracture	perfect {001} rarely visible/ irregular
luster/transparency	dull/translucent
color	white, gray, variable

Optical Properties
Optical properties are highly variable due to variable chemistry and crystallinity.

Crystallography
Monoclinic, $a = 5.17$, $b = 8.94$, $c = 15.20$, $\beta \simeq 90°$ $Z = 2$.

Habit
Earthy masses are typical.

Structure and Composition
The structure is based on groups of three layers: Single sheets of $(Al,Mg)(O,OH)_6$ octahedra are sandwiched between two sheets of SiO_4 tetrahedra. Montmorillonite is a member of the smectite group and forms solid solutions with beidellite, $(Na,Ca)Al_2(Si,Al)_4O_{10}(OH)_2 \cdot nH_2O$. Other minor solid solutions are common, and the amount of H_2O is variable.

Occurrence and Associations
Clays are secondary minerals, often residual, formed by alteration of Al-rich silicates.

Kaolinite $Al_2Si_2O_5(OH)_4$

Origin of Name
From the Chinese *Kao-ling,* the name of the hill that was the first source of kaolinite sent to Europe for ceramics.

Hand Specimen Identification
Claylike properties, including softness, habit, feel, and earthy smell, help identify kaolinite, but without X-ray data it cannot be distinguished from other light-colored clays. See Figure 6.5c.

Physical Properties

hardness	$2–2\frac{1}{2}$
specific gravity	2.6
cleavage/fracture	perfect (001) but rarely seen
luster/transparency	dull/translucent
color	white
streak	white

Optical Properties
Optical identification of kaolinite is very difficult. Biaxial $(-)$, $\alpha = 1.556$, $\beta = 1.563$, $\gamma = 1.565$, $\delta = 0.007$, $2V = 40°$.

Crystallography
Triclinic, $a = 5.15$, $b = 8.92$, $c = 7.38$, $\alpha = 91.8°$, $\beta = 104.8°$, $\gamma = 90.0°$, $Z = 2$; space group $P\bar{1}$; point group $\bar{1}$.

Habit
Kaolinite is usually massive or a fine-grained aggregate; rare platy, pseudohexagonal crystals have been found.

Structure and Composition
Kaolinite has a two-layer structure: Layers of $Al(O,OH)_6$ octahedra alternate with sheets of SiO_4 tetrahedra. Several minor substitutions are possible: Alkalis or alkaline earths may be present, as well as excess H_2O. Figures 2.16a and 13.4 show the structure of kaolinite.

Occurrence and Associations
Kaolinite is a common secondary mineral, forming after aluminous silicates. It is a rock-forming mineral, a component of soils, and replaces feldspar in rocks undergoing weathering. Associated minerals include quartz and other minerals resistant to alteration.

Related Minerals
In composition and structure, kaolinite is equivalent to an aluminous serpentine. It has two polymorphs, dickite and nacrite.

Pyrophyllite $Al_2Si_4O_{10}(OH)_2$

Origin of Name
From the Greek *pyro* and *phyllon,* meaning "fire" and "leaf," in reference to this mineral's behavior when heated.

Hand Specimen Identification
Association, greasy feel, and cleavage help identify pyrophyllite, but it cannot be told from other clays without X-ray data. It is easily confused with talc. See Figure 14.7.

▶**FIGURE 14.7**
Radiating splays of pyrophyllite, $Al_2Si_4O_{10}(OH)_2$, from Mariposa County, California. Mineralogists often group pyrophyllite with clay minerals although it has less water and a more ordered structure than true clays.

Physical Properties

hardness	1–2
specific gravity	2.8
cleavage/fracture	perfect basal (001)
luster/transparency	pearly/translucent
color	white
streak	white

Optical Properties

High birefringence, perfect cleavage, bird's-eye maple appearance and lack of color identify pyrophyllite. Talc and muscovite have smaller $2V$s. Biaxial $(-)$, $\alpha = 1.553$, $\beta = 1.588$, $\gamma = 1.600$, $\delta = 0.047$, $2V = 52°–62°$.

Crystallography

Triclinic, $a = 5.16$, $b = 8.96$, $c = 9.35$, $\alpha = 90.03°$, $\beta = 100.37°$, $\gamma = 89.75°$, $Z = 2$; space group $P1$; point group 1.

Habit

Individual crystals are unknown. Pyrophyllite is usually massive and foliated, sometimes forming platy or radiating masses, such as in Figure 14.7.

Structure and Composition

The three-layered structure consists of individual sheets of $Al(O,OH)_6$ octahedra sandwiched between sheets of SiO_4 tetrahedra. See Figure 2.16d. Fe may replace some of the Al; minor Mg, Ca, Na, or K may also be present.

Occurrence and Associations

Pyrophyllite is found in low- and medium-grade metamorphosed shales. Associated minerals include kyanite, feldspar, and quartz.

Related Minerals

Pyrophyllite is isostructural with talc, $Mg_3Si_4O_{10}(OH)_2$, and structurally similar to minnesotaite, $Fe_3Si_4O_{10}(OH)_2$.

Talc $\quad\quad$ Mg₃Si₄O₁₀(OH)₂

Origin of Name

Unknown, perhaps from the Arabic *talq,* meaning "pure."

Hand Specimen Identification

Greasy feel, massive habit, cleavage, and association help identify talc. It may be confused with pyrophyllite, serpentine, or chlorite.

Physical Properties

hardness	1
specific gravity	2.8
cleavage/fracture	perfect basal (001)/flexible
luster/transparency	resinous, silky/translucent
color	gray, white
streak	white

Optical Properties

Talc is similar to muscovite, chlorite, and pyrophyllite, but has a smaller $2V$ and often appears smeared or poorly defined when viewed under crossed polars. Biaxial $(-)$, $\alpha = 1.54$, $\beta = 1.58$, $\gamma = 1.58$, $\delta = 0.05$, $2V = 6°–30°$.

Crystallography

Monoclinic, $a = 5.29$, $b = 9.10$, $c = 18.81$, $\beta = 100.00°$, $Z = 4$; space group Cc; point group m.

Habit

Rare tabular pseudohexagonal crystals have been found, but talc is usually very fine grained and massive.

Structure and Composition

Talc is isostructural with pyrophyllite, being composed of layers of $Mg(O,OH)_6$ octahedra sandwiched between layers of SiO_4 tetrahedra. Talc may contain some Ti, Ni, Fe, or Mn but is generally quite pure.

Occurrence and Associations

Talc is a primary mineral in some low-grade metamorphic rocks, including marbles and ultramafic rocks, and less commonly a secondary mineral in mafic igneous rocks.

Varieties

When massive, talc is sometimes called *steatite* or *soapstone.*

Related Minerals

Talc is isostructural with pyrophyllite, $Al_2Si_4O_{10}(OH)_2$, and structurally similar to minnesotaite, $Fe_3Si_4O_{10}(OH)_2$.

Mica Group Minerals

Biotite Series

phlogopite	$KMg_3(AlSi_3O_{10})(OH)_2$
annite	$KFe_3(AlSi_3O_{10})(OH)_2$

Other Micas

muscovite	$KAl_2(AlSi_3O_{10})(OH)_2$
margarite	$CaAl_2(Al_2Si_2O_{10})(OH)_2$
lepidolite	$K(Li,Al)_{2-3}(AlSi_3O_{10})(OH)_2$

The mica group consists of a number of minerals, the principal ones being biotite and muscovite. The name *biotite* refers to a series dominated by end members phlogopite and annite. Mg-rich biotite (phlogopite) and Fe-rich biotite (annite) are often difficult to distinguish without detailed X-ray or chemical analyses. In common use, mineralogists call any black mica *biotite* and reserve the term *phlogopite* for brown micas. Muscovite, sometimes referred to as a "white mica," is in sharp contrast to biotite. However, it may be confused with margarite, lepidolite, and some other rarer micas. Margarite is one of the "brittle micas," a group that includes rarer clintonite and xanthophyllite. Lepidolite is a Li-rich mica found in pegmatites. Although annite and phlogopite enjoy complete solid solution, only limited solution is possible among other end members.

For more general information about the mica group, see Chapter 5.

Phlogopite $KMg_3(AlSi_3O_{10})(OH)_2$

Origin of Name
From the Greek *phlogopos,* meaning "fiery," in reference to this mineral's brown color.

Hand Specimen Identification
Phlogopite, an end member biotite, is identified by cleavage, color, and association. If not brown, it is indistinguishable from other biotites without additional analytical data.

Physical Properties

hardness	$2\frac{1}{2}$–3
specific gravity	2.8
cleavage/fracture	perfect basal (001)/elastic
luster/transparency	pearly/transparent
color	yellow-brown
streak	white

Optical Properties
Phlogopite is similar to other micas in thin section. It has a smaller $2V$ than muscovite and is often colored light brown. It may be pleochroic, but not as much as the Fe-rich biotites. Biaxial $(-)$, $\alpha = 1.56$, $\beta = 1.60$, $\gamma = 1.60$, $\delta = 0.04$, $2V = 0°$–$20°$.

Crystallography
Monoclinic, $a = 5.31$, $\beta = 9.19$, $c = 10.15$, $\beta = 95.18°$, $Z = 2$; space group $C2/m$; point group $2/m$.

Habit
Coarse books of pseudohexagonal crystals are distinctive but uncommon. More frequently, phlogopite is disseminated as irregular grains or flakes or foliated masses.

Structure and Composition
The basic phlogopite structure is similar to that of talc and pyrophyllite: Two tetrahedral SiO_4 layers surround an octahedral $Mg(O,OH)_6$ layer. Unlike talc and pyrophyllite, however, the three-layer sandwiches are linked by K^+ ions occupying large sites be-

tween them. Fe often substitutes for Mg, leading to complete solid solution with annite, $KFe_3(AlSi_3O_{10})$ $(OH)_2$. Al substitutes for both Mg and Si, creating solid solutions with siderophilite, $K(Fe,Al)_3(Si,Al)_4$ $O_{10}(OH)_2$. Limited solid solutions with muscovite, $KAl_2(AlSi_3O_{10})(OH)_2$, are also possible. Mn, Ti, and a number of alkalis and alkaline earths may also be present, and F may replace some OH.

Occurrence and Associations
Phlogopite is a common mineral in marbles where it associates with calcite, dolomite, quartz, diopside, and tremolite; less commonly, it is found in highly magnesium-rich igneous rocks.

Related Minerals
Phlogopite has several different polymorphs, which are difficult to tell apart without detailed X-ray studies. It is isostructural with muscovite, $KAl_2(AlSi_3O_{10})$ $(OH)_2$, and isotypical with other micas, forming complete or limited solutions with most.

Biotite $K(Mg,Fe)_3(AlSi_3O_{10})(OH)_2$

Origin of Name
Named after J. B. Biot (1774–1862), a French scientist who conducted detailed studies of micas.

Hand Specimen Identification
Biotite is distinguished by its association, micaceous nature, and dark color. It may be confused with other micas, especially chlorite. Composition is variable and cannot be determined without detailed analytical work. See Figures 2.4, 3.6c, and 10.35.

Physical Properties

hardness	$2\frac{1}{2}$–3
specific gravity	2.9–3.1
cleavage/fracture	perfect basal {001}/ragged
luster/transparency	vitreous/transparent to opaque
color	black, greenish black, brown-black
streak	white

Optical Properties
Brown, red, or green in thin section, biotite exhibits strong pleochroism, perfect cleavage, bird's-eye extinction, and parallel or near-parallel extinction (Figures 7.4d and 7.7b). Biaxial $(-)$, $\alpha = 1.57$, $\beta = 1.60$, $\gamma = 1.61$, $\delta = 0.04$, $2V = 0°$–$32°$. Plates 5.3, 5.4, 5.7, and 5.8 show biotite in thin section. Stilpnomelane, a rarer related sheet silicate, looks much like biotite but lacks "bird's eye" extinction.

Crystallography
Monoclinic, pseudohexagonal, $a = 5.33$, $\beta = 9.31$, $c = 10.16$, $\beta = 99.3°$, $Z = 2$; space group $C2/m$; point group $2/m$.

Habit

Rare, foliated books of pseudohexagonal crystals are distinctive but uncommon (Plate 6.6). More frequently, biotite is disseminated as irregular grains or flakes or foliated masses.

Structure and Composition

The basic biotite structure is identical to that of phlogopite: Two tetrahedral layers surround an octahedral layer. The three-layer sandwiches are linked by K^+ ions occupying large sites between them. Fe mixes freely with Mg in octahedral sites, leading to complete solid solution between the two principal biotite end members: annite, $KFe_3(AlSi_3O_{10})(OH)_2$, and phlogopite, $KMg_3(AlSi_3)O_{10}(OH)_2$. Coupled substitution of Al for both Mg and Si creates limited solid solutions with siderophilite, $K(Fe,Al)_3(Si,Al)_4O_{10}(OH)_2$. Minor solid solution with muscovite, $KAl_2(AlSi_3O_{10})(OH)_2$, is also possible: Two Al replace three (Fe,Mg), leaving every third octahedral site vacant. Mn, Ti, and a number of alkalis and alkaline earths may also be present, and F may replace some OH.

Occurrence and Associations

Biotite is common in a wide variety of igneous and metamorphic rocks and in immature sediments. Associated minerals include other micas, amphiboles, quartz, and feldspars.

Varieties

Annite, phlogopite, and siderophilite are names given to end member Fe-, Mg-, and Al-biotite.

Related Minerals

Several biotite polymorphs differ in the ways the tetrahedral and octahedral sheets are stacked. Biotite is isostructural or isotypical with other micas, and similar in many ways to other sheet silicates.

Muscovite $KAl_2(AlSi_3O_{10})(OH)_2$

Origin of Name

Named after the Muscovy principality of thirteenth-century Russia, which produced muscovite for use in window panes.

Hand Specimen Identification

Micaceous character (Figure 14.8a) and silver color distinguish muscovite. It may be confused with other white micas, such as paragonite or margarite. Certain identification requires optical, X-ray, or chemical analysis.

Physical Properties

hardness	$2-2\frac{1}{2}$
specific gravity	2.8
cleavage/fracture	perfect basal (001)/elastic
luster/transparency	vitreous/transparent
color	white, gray-brown
streak	white

Optical Properties

In thin section, muscovite appears clear, has moderate to high birefringence, sometimes has bird's-eye extinction, a high index of refraction, and one perfect cleavage. It may be confused with colorless phlogopite or with other white micas. Biaxial $(-)$, $\alpha = 1.565$, $\beta = 1.596$, $\gamma = 1.600$, $\delta = 0.035$, $2V = 30°-40°$.

Crystallography

Monoclinic, $a = 5.19$, $\beta = 9.04$, $c = 20.08$, $\beta = 95.5°$, $Z = 4$; space group $C2/c$; point group $2/m$.

Habit

Individual crystals are rare. Muscovite typically forms books of mica, often in massive aggregates (Figure 14.8a), with or without a pseudohexagonal outline, or is found as disseminated grains within a

▶**FIGURE 14.8**
Muscovite and vermiculite: (a) Books of muscovite that give a hint of hexagonal symmetry; (b) vermiculite, a clay-like sheet silicate related to micas, puffs up and expands when heated above 150 °C.

quartz-feldspar matrix. Penetration twins may be present.

Structure and Composition

The muscovite structure is identical to that of biotite except that Al has replaced two out of every three (Fe,Mg). See Figure 2.16c and 5.15a. Very limited solid solution with annite, $KFe_3(AlSi_3O_{10})(OH)_2$, and with phlogopite, $KMg_3(AlSi_3O_{10})(OH)_2$, are possible. Na may be present, resulting in solid solution toward paragonite, $NaAl_2(AlSi_3O_{10})(OH)_2$. Other alkalis and alkaline earths may also replace K. Li may replace some Al, and F may replace some OH.

Occurrence and Associations

Muscovite, the most common mica, is found in many silicic to intermediate igneous rocks, in a wide variety of metamorphic rocks, and in some immature sediments. Associated minerals include other micas, K-feldspar, and quartz.

Varieties

Sericite is a fine-grained form of muscovite created by alteration of feldspars and other alkali-aluminum silicates.

Related Minerals

Several different muscovite polymorphs are known. Muscovite is isotypical with biotite and other micas, such as lepidolite, $K(Li,Al)_{2-3}(AlSi_3O_{10})(OH)_2$. It shares many properties with paragonite, $NaAl_2(AlSi_3O_{10})(OH)_2$, and with margarite, $CaAl_2(Al_2Si_2O_{10})(OH)_2$.

Margarite $CaAl_2(Al_2Si_2O_{10})(OH)_2$

Origin of Name

From the Greek *margarites,* meaning "pearl," a reference to margarite's color and luster.

Hand Specimen Identification

Margarite is characterized by its micaceous nature, brittleness, and pearly luster. Association and brittleness usually distinguish it from other white micas.

Physical Properties

hardness	$3\frac{1}{2}$–5
specific gravity	3.1
cleavage/fracture	perfect basal (001)/uneven
luster/transparency	vitreous/transparent to translucent
color	gray-yellow, pinkish silver
streak	white

Optical Properties

Margarite is colorless in thin section, resembling muscovite and other white micas. It is distinguished by a higher index of refraction, lower birefringence,

and a 6°–8° extinction angle. Biaxial $(-)$, $\alpha = 1.635$, $\beta = 1.645$, $\gamma = 1.648$, $\delta = 0.013$, $2V = 45°$.

Crystallography

Monoclinic, $a = 5.14$, $\beta = 9.00$, $c = 19.81$, $\beta = 100.8°$, $Z = 4$; space group $C2/c$; point group $2/m$.

Habit

Margarite is typically found in massive, micaceous books and aggregates; individual crystals are rare.

Structure and Composition

Margarite has a structure transitional between muscovite and chlorite. Because the interlayer site is occupied by Ca^{2+} rather than K^+, bonds between layers are stronger; thus, margarite is more brittle. Minor Na^+ and K^+ may replace Ca^{2+}; charge balance is maintained by replacement of Al^{3+} by Si^{4+}. Be, Ba, Sr, K, Mn, Fe, Mg, and excess OH may also be present.

Occurrence and Associations

Margarite, typically found with corundum and diaspore, is an alteration product of corundum.

Related Minerals

Margarite is a member of the brittle mica group. Other members include clintonite and xanthophyllite, both having compositions $Ca(Mg,Al)_{2-3}(Al_2Si_2)O_{10}(OH)_2$. Margarite also shares similarities with stilpnomelane, $K(Fe,Al)_{10}Si_{12}O_{30}(OH)_{12}$.

Lepidolite $K(Li,Al)_{2-3}(AlSi_3O_{10})(OH)_2$

Origin of Name

From the Greek *lepid,* meaning "scale," referring to this mineral's usual habit.

Hand Specimen Identification

Distinctive color and association usually serve to identify lepidolite.

Physical Properties

hardness	2.5–4
specific gravity	2.9
cleavage/fracture	perfect basal {001}/uneven
luster/transparency	pearly/translucent
color	lilac to rose-red; less commonly, yellow, gray, white
streak	white

Optical Properties

Lepidolite is generally colorless in thin section. It resembles muscovite but has lower relief and lower birefringence. Biaxial $(-)$, $\alpha = 1.53$–1.55, $\beta = 1.55$–1.59, $\gamma = 1.55$–1.59, $\delta = 0.02$–0.04, $2V = 0°$–60°.

Crystallography

Monoclinic, $a = 5.21$, $\beta = 8.97$, $c = 20.16$, $\beta = 100.8°$, $Z = 4$; space group $C2/m$; point group $2/m$.

Habit

Lepidolite usually forms coarse- to fine-grained scaly aggregates. It is less commonly disseminated as fine flakes.

Structure and Composition

Lepidolite is actually a complex solid solution series with a structure similar to muscovite. Chemically, it is equivalent to muscovite with half or more of the octahedral Al replaced by Li, and perhaps Fe or Mg. In addition, other alkalis may substitute for K, and O or F may replace OH.

Occurrence and Associations

Lepidolite is restricted to Li-rich pegmatites. Associated minerals include other Li minerals such as tourmaline, amblygonite, and spodumene, as well as the more common muscovite, feldspar, and quartz.

Related Minerals

Lepidolite has several polymorphs that cannot be differentiated without detailed X-ray study. It is isotypical with other micas.

Chlorite Group Minerals

Ideal End Members

clinochlore	$(Mg_5Al)(AlSi_3)O_{10}(OH)_8$
chamosite	$(Fe_5Al)(AlSi_3)O_{10}(OH)_8$
nimite	$(Ni_5Al)(AlSi_3)O_{10}(OH)_8$
pennantite	$(Mn,Al)_6(Si,Al)_4O_{10}(OH)_8$

Chlorite is the general name given to a number of Mg-rich sheet silicates with similar chemistry and structure. Compositions vary widely, especially Mg, Fe, and Al contents, but cannot be determined without careful X-ray, chemical, or optical studies. The complex chemical variations and similarity of all chlorites make identifying individual species problematic, but a few ideal end member compositions have been given names. Structurally, all chlorites consist of alternating talc-like and brucite-like layers. In both kinds of layers, Fe and Al may substitute for Mg; other elements, such as Ni, may also be present.

Chlorite **Composition variable; common chlorites are close to $(Mg,Fe,Al)_6(Si,Al)_4O_{10}(OH)_8$**

Origin of Name

From the Greek *chloros,* meaning "green."

Hand Specimen Identification

Deep green color and micaceous habit and cleavage are usually adequate to identify chlorite. The many individual species and varieties are difficult to tell apart. Chlorite is sometimes confused with talc.

Physical Properties

hardness	$2-2\frac{1}{2}$
specific gravity	3.0
cleavage/fracture	perfect basal (001)/flexible
luster/transparency	vitreous/transparent to translucent
color	green, variable
streak	white, green

Optical Properties

Chlorite is generally green and sometimes green-brown in thin section. It usually exhibits yellow-green-brown pleochroism and has moderate to moderately high relief. Birefringence is low. Interference colors normally are between anomalous blue, brown, or purple and first-order yellow. Biaxial $(-)$, $\alpha = 1.56-1.60$, $\beta = 1.57-1.61$, $\gamma = 1.58-1.61$, $\delta = 0.006-0.020$, $2V = 0°-40°$.

Crystallography

Monoclinic, $a = 5.37$, $b = 9.30$, $c = 14.25$, $\beta = 97.4°$, $Z = 2$; space group $C2/m$; point group $2/m$.

Habit

Chlorite has habits similar to the other micas: Foliated books, scaly aggregates, and individual flakes in a quartz-feldspar matrix are common. Rare pseudohexagonal crystals are known.

Structure and Composition

The structure of chlorite consists of stacked talc, $Mg_3Si_4O_{10}(OH)_2$ and brucite, $Mg(OH)_2$, layers (Figures 2.16b and 13.16). The stacking order is variable, leading to a great deal of variety. In both kinds of layers, Al replaces some of the Mg, and in most chlorites Fe is present as well. A general formula, $(Mg,Fe,Al)_6(Si,Al)_4O_{10}(OH)_8$, does not describe all of the compositional variability.

Occurrence and Associations

Chlorite is a common mineral in low- to intermediate-grade metamorphic rocks, diagnostic of the greenschist facies. It is also a common secondary mineral after biotite, muscovite, and other mafic silicates in igneous and metamorphic rocks, and is sometimes found in sediments. Many greenish rocks owe their color to the presence of chlorite. Associated minerals include quartz and feldspars, epidote, muscovite, actinolite, albite, and a number of ferromagnesian silicates.

Varieties

Names given to some idealized chlorite compositions are clinochlore, $(Mg_5Al)(AlSi_3O_{10})(OH)_8$; chamosite, $(Fe_5Al)(AlSi_3O_{10})(OH)_8$; nimite, $(Ni_5Al)(AlSi_3O_{10})(OH)_8$; and pennantite, $(Mn,Al)_6(Si,Al)_4O_{10}(OH)_8$.

Related Minerals
Chlorite has many varieties and polymorphs. Other similar minerals include cookeite, $LiAl_4(AlSi_3)O_{10}(OH)_8$; sudoite, $Mg_2(Al,Fe)_3(AlSi_3O_{10})(OH)_8$; and a number of other hydrated aluminosilicates.

Other Sheet Silicates

prehnite	$Ca_2Al(AlSi_3O_{10})(OH)_2$
apophyllite	$KCa_4Si_8O_{20}F \cdot 8H_2O$
sepiolite	$Mg_4Si_6O_{15}(OH)_2 \cdot 6H_2O$
chrysocolla	$Cu_4Si_4O_{10}H_4(OH)_8 \cdot nH_2O$
glauconite	$\sim(K,Na)(Fe,Mg,Al)_2(Si,Al)_4O_{10}(OH)_2$

A number of other minerals may be considered sheet silicates but do not fit into any of the groups previously discussed; five examples are listed above. Prehnite and apophyllite are secondary minerals, similar in occurrence and association to zeolites. Due to highly variable properties, chemistry, and structure, glauconite, sepiolite, and chrysocolla are not discussed in detail. Glauconite and sepiolite are both claylike minerals, and chrysocolla is a secondary copper mineral of highly variable chemistry.

Prehnite $Ca_2Al(AlSi_3O_{10})(OH)_2$

Origin of Name
Named after H. van Prehn, who discovered this mineral in 1774.

Hand Specimen Identification
Habit, color, and association often serve to identify prehnite. It may be confused with hemimorphite or smithsonite.

Physical Properties

hardness	6–6½
specific gravity	2.9
cleavage/fracture	good basal (001)/uneven
luster/transparency	vitreous/transparent to translucent
color	pale green
streak	white

Optical Properties
Prehnite is colorless in thin section. It has moderate birefringence but often exhibits anomalous interference figures, sometimes with an "hourglass" or "bow tie" structure. It may be confused with lawsonite, pumpellyite, epidote, datolite, and a number of zeolites but has higher birefringence than all of them. Biaxial (+), $\alpha = 1.625$, $\beta = 1.635$, $\gamma = 1.655$, $\delta = 0.03$, $2V = 65°–70°$.

Crystallography
Orthorhombic, $a = 4.65$, $b = 5.48$, $c = 18.49$, $Z = 2$; space group $P2/n2_1/c2/m$; point group $2/m2/m2/m$.

Habit
Botryoidal, globular, barrel-shaped, and reniform aggregates are typical. Rarely, prehnite is found as individual tabular or prismatic crystals.

Structure and Composition
Prehnite consists of tetrahedral $(Si,Al)O_4$ sheets connected by $Al(O,OH)_6$ octahedra. Fe may replace some of the Al.

Occurrence and Associations
Prehnite may be a product of low-grade metamorphism but is more commonly a secondary mineral that forms as crusts or fillings in basalt and other mafic igneous rocks. Associated minerals include pumpellyite, zeolites, datolite, pectolite, and calcite.

Related Minerals
Prehnite is sometimes grouped with the zeolites because of its similar occurrences. Its structure, however, is significantly different.

Apophyllite $KCa_4Si_8O_{20}F \cdot 8H_2O$

Origin of Name
From the Greek words for "form" and "leaf," because it becomes flaky on heating.

Hand Specimen Identification
Perfect one direction of cleavage, form, color, and luster help identify apophyllite.

Physical Properties

hardness	4½–5
specific gravity	2.3
cleavage/fracture	one perfect (001), poor {110}/ uneven
luster/transparency	pearly/transparent to translucent
color	colorless, white, gray
streak	white

Optical Properties
Apophyllite has negative relief, is colorless, and has very low birefringence. It has perfect (001) cleavage, and often exhibits anomalous interference colors. Uniaxial (+), $\omega = 1.535$, $\epsilon = 1.537$, $\delta = 0.002$.

Crystallography
Tetragonal, $a = 8.96$, $c = 15.78$, $Z = 2$; space group $P4/m2_1/n2/c$; point group $4/m2/m2/m$.

Habit

Apophyllite has a varied habit, including short prisms, pinacoids, and bipyramids. It often has pseudosymmetry, appearing to be combinations of cubes and octahedra.

Structure and Composition

Sheets in the apophyllite structure are composed of 4- and 8-member rings of SiO_4 tetrahedra. Interlayer Ca, K, and F link the tetrahedral sheets.

Occurrence and Associations

Apophyllite is a secondary mineral, which sometimes lines openings in basalt and other mafic igneous rocks. Associated minerals include zeolites, datolite, calcite, and pectolite.

CHAIN SILICATES

Pyroxene Group Minerals

Orthopyroxene Series

enstatite	$Mg_2Si_2O_6$
ferrosilite	$Fe_2Si_2O_6$

Clinopyroxene (Diopside) Series

diopside	$CaMgSi_2O_6$
hedenbergite	$CaFeSi_2O_6$

Pyroxene Solid Solutions

hypersthene	$(Mg,Fe)_2Si_2O_6$
pigeonite	$(Ca,Mg,Fe)_2Si_2O_6$
augite	$(Ca,Mg,Fe,Na)(Mg,Fe,Al)(Si,Al)_2O_6$
omphacite	$(Ca,Na)(Fe,Mg,Al)(Si,Al)_2O_6$
aegirine	$Na(Al,Fe^{3+})Si_2O_6$

Na- and Li-pyroxenes

jadeite	$NaAlSi_2O_6$
spodumene	$LiAlSi_2O_6$

All pyroxenes have a structure based on chains of SiO_4 tetrahedra linked by shared (bridging) oxygen. Octahedral cations (Ca,Na,Mg,Fe,Al) occupy sites between nonbridging oxygens of adjacent chains (Figures 5.17 and 13.17). Small amounts of Al sometimes replace tetrahedral Si.

Mineralogists divide pyroxenes into two groups based on their crystal symmetry (see Figure 5.18). Orthopyroxenes (orthorhombic) have two principal end members: enstatite, $Mg_2Si_2O_6$, and ferrosilite, $Fe_2Si_2O_6$. Compositions between are generally called hypersthene but more specific names can be found in the literature. Clinopyroxenes (monoclinic) include diopside-hedenbergite solid solutions, high-temperature polymorphs of hypersthene, and a number of other Ca- and Na-bearing species. The most important end member is diopside, $CaMgSi_2O_6$; many natural clinopyroxenes are close to diopside composition, so the name *diopside* is often used to refer to any green or greenish black clinopyroxene. Petrologists use the names *hypersthene, pigeonite, augite, omphacite,* and *aegirine* to describe solid solution clinopyroxenes having specific physical properties or compositions.

Except at very high temperature, a large solvus exists between monoclinic calcic pyroxenes and orthorhombic Ca-free pyroxenes. Consequently, many mafic rocks contain both clinopyroxene and orthopyroxene.

For more general information about the pyroxene group, see Chapter 5.

Enstatite $Mg_2Si_2O_6$

Origin of Name

From the Greek *enstates,* meaning "adversary," which refers to this mineral's resistance to melting.

Hand Specimen Identification

Association, luster, color, and two cleavages at about 90° to each other identify enstatite. Exact composition cannot be determined without additional analytical data. It may be confused with other pyroxenes or with amphibole, but the latter has two cleavages at about 60° to each other.

Physical Properties

hardness	5–6
specific gravity	3.2–3.5
cleavage/fracture	two perfect prismatic {210}/uneven
luster/transparency	vitreous, pearly/translucent
color	gray, green
streak	white, gray

Optical Properties

Enstatite is nearly colorless in thin section. If Fe has replaced some of the Mg, it may exhibit slight to pronounced pink to green pleochroism. An 86° cleavage angle, parallel extinction in prismatic section, and relatively low birefringence distinguish it from clinopyroxene. Biaxial (+), $\alpha = 1.657$, $\beta = 1.659$, $\gamma = 1.665$, $\delta = 0.008$, $2V = 54°$.

Crystallography

Orthorhombic, $a = 18.22$, $b = 8.81$, $c = 5.21$, $Z = 4$; space group $P2_1/b2_1/c2_1/a$; point group $2/m2/m2/m$.

Habit

Enstatite is usually massive, blocky, fibrous, or lamellar. Individual crystals may be prismatic or acicular.

Structure and Composition

Enstatite is one of two principal orthopyroxene end members; the other is ferrosilite, $Fe_2Si_2O_6$. Enstatite and other pyroxenes contain chains of zigzagging SiO_4 tetrahedra running parallel to the *c*-axis. Each tetrahedron shares two oxygens with neighbors in its chain and has one unshared oxygen at an apex pointing perpendicular to *c*. Pairs of chains face each other; Mg is located in four adjacent octahedral sites between unshared apices of tetrahedra. Complete solid solution exists between enstatite and ferrosilite. Except at high temperature, only limited solid solution exists with the clinopyroxene end members diopside, $CaMgSi_2O_6$, and hedenbergite, $CaFeSi_2O_6$. Mn, Cr, Al, and Ti may also be present in small amounts. See Chapter 5 for more details.

Occurrence and Associations

Enstatite is common in mafic igneous rocks, including gabbro, basalt, and norite, commonly associating with plagioclase and clinopyroxene. It is also found in some high-grade metamorphic rocks and is considered diagnostic for the granulite facies.

Varieties

Bronzite (Mg $\gg$ Fe) and hypersthene (Mg > Fe) are varietal names given to Mg-Fe orthopyroxene.

Related Minerals

Enstatite is isostructural or isotypical with other pyroxenes. It is closely related to ferrosilite, $Fe_2Si_2O_6$, and donpeacorite, $(Mn,Mg)_2Si_2O_6$. Most natural orthopyroxenes are Mg-Fe solid solutions with Mg dominant.

▶FIGURE 14.9
This sample of a marble, from near Chenaux, Quebec, contains diopside (dark) and calcite (light). Diopside is common in marbles, but usually is not as darkly colored as in this sample.

Diopside $CaMgSi_2O_6$

Origin of Name

From the Greek *dis* and *opsis*, meaning "two" and "appearance," in reference to the fact that diopside appears different when viewed in different ways.

Hand Specimen Identification

Association, form, two cleavages at 86° to each other, and color identify diopside (Figures 7.8c and 14.9). It may be confused with other pyroxenes or with hornblende, but the latter has two cleavages at 60° to each other. Diopside may contain substantial Fe, Al, or other impurities; exact composition cannot be determined without additional analytical data. Diopside is often white to green or pale green; augite tends to be dark green or black. Plates 6.2 and 6.3 contain color photos of diopside.

Physical Properties

hardness	$5\frac{1}{2}-6\frac{1}{2}$
specific gravity	3.2–3.5
cleavage/fracture	two perfect prismatic {110}/uneven
luster/transparency	vitreous/transparent to translucent
color	whitish green, variable
streak	white, gray

Optical Properties

Diopside is colorless in thin section when Fe-free. With increasing iron content, it may become pleochroic in greens or browns. Higher birefringence and inclined extinction distinguish clinopyroxene from orthopyroxene. Grains are prismatic or blocky, depending on orientation. Extinction angle, optic sign, and *2V* help distinguish the different clinopyroxenes, but telling them apart may be difficult. Biaxial (+), $\alpha = 1.665$, $\beta = 1.672$, $\gamma = 1.695$, $\delta = 0.030$, $2V = 56°-62°$.

Crystallography

Monoclinic, $a = 9.7$, $b = 8.9$, $c = 5.25$, $\beta = 105.83°$, $Z = 4$; space group $C2/c$; point group $2/m$.

Habit

Prismatic crystals often have a square or octahedral cross section. Diopside may also be massive or finely disseminated. Polysynthetic twins are common but may not be visible without a microscope.

Structure and Composition

The structure of diopside is similar to that of enstatite (see **enstatite**). Chains of SiO_4 tetrahedra run parallel to the *c*-axis, with octahedral Ca and Mg connecting opposing chains to each other. Ca and Mg occupy different structural sites; Ca:Mg ratios are always $\leq 1:1$ (Figure 13.17a). A complete solid solution series exists between diopside, hedenbergite,

$CaFeSi_2O_6$, and johannsenite, $CaMn(Si_2O_6)$. Small amounts of Al, Mn, Na, Ti, and Cr may be present.

Occurrence and Associations
Diopside is the most common pyroxene. It is found in mafic and ultramafic igneous rocks, associated with plagioclase, hornblende, and olivine. It is found in marbles (Plate 6.3) associated with calcite, quartz or forsterite, tremolite, scapolite, and garnet. It is also found in medium- and high-grade metamorphosed mafic rocks.

Varieties
Chrome diopside is a chromium-rich variety known for its vivid green color (Plate 6.2).

Related Minerals
All pyroxenes are closely related. Hedenbergite, $CaFeSi_2O_6$, is the iron end member of the diopside series. Augite, $(Ca,Mg,Fe,Na)(Mg,Fe,Al)Si_2O_6$, is a related pyroxene with slightly different structure. Pigeonite is a high-temperature, subcalcic, pyroxene with compositions that approach those of Fe-Mg diopside.

Pigeonite $(Ca,Mg,Fe)_2Si_2O_6$

Origin of Name
Named after the locality where this mineral was originally found, Pigeon Cove, Minnesota.

Hand Specimen Identification
Association, form, two cleavages at 86°, and color identify pyroxene, but telling pigeonite from the others is difficult. Only X-ray studies can make the distinction.

Physical Properties

hardness	6
specific gravity	3.40
cleavage/fracture	two perfect prismatic {110}/uneven
luster/transparency	vitreous/transparent to translucent
color	brown, green
streak	white

Optical Properties
Pigeonite is similar to other clinopyroxenes in thin section (see diopside) but has a lower $2V$. Biaxial $(+)$, $\alpha = 1.69$, $\beta = 1.69$, $\gamma = 1.72$, $\delta = 0.025$, $2V = 0°–32°$.

Crystallography
Monoclinic, $a = 9.73$, $b = 8.95$, $c = 5.26$, $\beta = 108.55°$, $Z = 4$; space group $C2/c$; point group $2/m$.

Habit
Rare euhedral crystals are prismatic. Pigeonite is most commonly in granular masses or in columnar aggregates or is found as fine grains or zones associated with other pyroxenes.

Structure and Composition
Pigeonite is similar in structure to other pyroxenes (enstatite and diopside), and may contain the same impurities. It contains more Ca than enstatite and less than diopside, resulting in a slightly different structure.

Occurrence and Associations
Pigeonite is only found in high-temperature igneous rocks that have cooled rapidly. In many cases, it inverts to lower temperature pyroxenes with cooling, but its former presence may be inferred from textural features or from the presence of exsolved grains of augite and orthopyroxene.

Related Minerals
Pigeonite is closely related to other pyroxenes, especially augite. See also **enstatite** and **diopside.**

Augite $(Ca,Mg,Fe,Na)(Mg,Fe,Al)(Si,Al)_2O_6$

Origin of Name
From the Greek word *augites,* meaning "brightness," referring to its shiny cleavage surfaces.

Hand Specimen Identification
Association, form, two cleavages at 86° to 87°, and color identify pyroxene, but differentiating augite is problematic without X-ray data. White to green or pale green pyroxenes are often diopside; augite tends to be dark green or black (Figure 5.19a). Augite is occasionally confused with hornblende. Plate 6.1 is a color photo of massive augite.

Physical Properties

hardness	5–6
specific gravity	3.2–3.4
cleavage/fracture	two perfect {110}/uneven
luster/transparency	vitreous/transparent to translucent
color	black, dark green
streak	white

Optical Properties
Augite is similar to other clinopyroxenes in thin section. Grains are prismatic or blocky, depending on orientation. Color may be various shades of light brown, yellow-brown, or green. Augite may exhibit weak pleochroism. Extinction angle, optic sign, and $2V$ help distinguish the different clinopyroxenes, but telling them apart may be difficult. Biaxial $(+)$, $\alpha = 1.671–1.735$, $\beta = 1.672–1.741$, $\gamma = 1.703–1.761$, $\delta = 0.030$, $2V = 25°–60°$.

Crystallography
Monoclinic, $a = 9.8$, $b = 9.0$, $c = 5.25$, $\beta = 105°$, $Z = 4$; space group $C2/c$; point group $2/m$.

Habit
Individual crystals are typically poorly formed stubby black or green prisms with octagonal or square cross sections. Simple and polysynthetic twins may be present.

Structure and Composition
The structure of augite is similar to that of the other clinopyroxenes (see diopside). Its chemistry is, however, more complex and variable. Ca : Mg : Fe ratios vary. Significant solid solution occurs with jadeite, $NaAlSi_2O_6$, aegirine, $NaFeSi_2O_6$, and Ca-Tschermaks pyroxene, $CaAl_2SiO_6$. Ti, Li, Mn, and a number of other elements may also be present in small amounts.

Occurrence and Associations
Augite is the most common pyroxene found in mafic to intermediate igneous rocks, both plutonic and volcanic. Associated minerals include hornblende and plagioclase.

Related Minerals
Augite is equivalent in composition to diopside with many impurities, especially Na. It is structurally and chemically closely related to other pyroxenes, especially pigeonite, and to pyroxenoids. Omphacite is a bright green variety of augite rich in Na and Al.

Jadeite $NaAlSi_2O_6$

Origin of Name
Name of unknown origin. The term *jade* refers to either jadeite or to the amphibole, nephrite.

Hand Specimen Identification
Association, form, two (rarely seen) cleavages at near 90°, green color, and tenacity identify jadeite. It is distinguished from nephrite by its luster. An optical microscope may be needed to confirm identification.

Physical Properties

hardness	$6\frac{1}{2}$–7
specific gravity	3.30
cleavage/fracture	two perfect {110}/uneven
luster/transparency	vitreous, greasy/translucent
color	variable shades of green to white
streak	white

Optical Properties
Jadeite is colorless to very pale green in thin section. Birefringence is low; anomalous blue interference colors are common, maximum interference colors are first-order red or yellow. Jadeite exhibits typical clinopyroxene shape and cleavage but has a higher $2V$ than most. Biaxial (+), $\alpha = 1.65$, $\beta = 1.66$, $\gamma = 1.67$, $\delta = 0.02$, $2V = 70°–75°$.

Crystallography
Monoclinic, $a = 9.50$, $b = 8.61$, $c = 5.24$, $\beta = 110.46°$, $Z = 4$; space group $C2/c$; point group $2/m$.

Habit
Jadeite is usually granular, forming tenacious masses; less commonly, in prismatic or tabular crystals.

Structure and Composition
Jadeite is similar in structure to other clinopyroxenes and may contain some of the same impurities (see diopside). It forms limited solid solutions with aegirine, $NaFeSi_2O_6$, with omphacite, $(Ca,Na)(Fe,Mg,Al)Si_2O_6$, and with other pyroxene end members.

Occurrence and Associations
Jadeite is a high-pressure pyroxene found in metamorphic rocks of the blueschist facies. It is associated with other high-pressure minerals such as glaucophane, lawsonite, or aragonite, and with quartz and epidote. A bright green variety of jadeite, omphacite, occurs in eclogites with pyrope-rich garnet. Omphacite is also found in kimberlites.

Related Minerals
Jadeite is closely related to other Na-pyroxenes: aegirine, $NaFeSi_2O_6$; omphacite, $(Ca,Na)(Fe,Mg,Al)Si_2O_6$; and aegirine-augite, $(Ca,Na)(Fe^{2+},Fe^{3+},Mg)Si_2O_6$. It is chemically similar to nepheline, $(Na,K)AlSiO_4$, and to albite, $NaAlSi_3O_8$.

Spodumene $LiAlSi_2O_6$

Origin of Name
From the Greek *spodoumenos*, meaning "ashes."

Hand Specimen Identification
Prismatic cleavage, hardness, color, and association help identify spodumene, but it may be difficult to tell from feldspar or scapolite. Spodumene often has the appearance of a log or of petrified wood.

Physical Properties

hardness	$6\frac{1}{2}$–7
specific gravity	3.15
cleavage/fracture	two perfect prismatic {110}/uneven
luster/transparency	vitreous/transparent to translucent
color	highly variable: colorless, white, gray, pink, green, yellow, purple, or tan
streak	white

Optical Properties
Spodumene is similar to other clinopyroxenes in thin section. Its occurrence in pegmatites and its small extinction angle (20° to 26°) help identify it. Biaxial (+), $\alpha = 1.65$, $\beta = 1.66$, $\gamma = 1.67$, $\delta = 0.02$, $2V = 60°–80°$.

Crystallography
Monoclinic, $a = 9.52$, $b = 8.32$, $c = 5.25$, $\beta = 110.46°$, $Z = 4$; space group $C2/c$; point group $2/m$.

Habit
Prismatic crystals with well-developed {100} faces showing vertical striations are common. Crystals are often polysynthetically twinned and may be very large. In some pegmatites they are tens of meters long. Spodumene also occurs as cleavable masses.

Structure and Composition
Spodumene is a pyroxene with structure similar to that of diopside. Minor Na substitutes for Li. Fe, Ca, Mn, Mg, and rare earths are also present in small amounts.

Occurrence and Associations
Spodumene is found in granitic pegmatites, where it associates with K-feldspar, muscovite, quartz, tourmaline, beryl, and lepidolite.

Varieties
Hiddenite is a name given to emerald-green spodumene; kunzite to lilac/pink spodumene; and triphane to colorless or yellow spodumene.

Related Minerals
Spodumene is similar to other pyroxenes (see **diopside**). It is similar in composition to eucryptite, $LiAlSiO_4$.

Amphibole Group Minerals

Monoclinic Amphiboles
cummingtonite series
    cummingtonite $(Mg,Fe)_7Si_8O_{22}(OH)_2$
    grunerite $Fe_7Si_8O_{22}(OH)_2$
actinolite series
    tremolite $Ca_2Mg_5Si_8O_{22}(OH)_2$
    actinolite $Ca_2(Mg,Fe)_5Si_8O_{22}(OH)_2$
hornblende $(K,Na)_{0-1}(Ca,Na,Fe,Mg)_2$
                      $(Mg,Fe,Al)_5(Si,Al)_8O_{22}(OH)_2$
Na-amphiboles
    glaucophane $Na_2Mg_3Al_2Si_8O_{22}(OH)_2$
    riebeckite $Na_2(Fe,Mg)_3(Fe,Al)_2Si_8O_{22}(OH)_2$

Orthorhombic Amphiboles
anthophyllite $(Mg,Fe)_7Si_8O_{22}(OH)_2$
gedrite $(Mg,Fe,Al)_7(Al,Si)_8O_{22}(OH)_2$

Amphiboles are double-chain silicates, sharing many physical and chemical properties with pyroxenes (see Figures 5.17, 5.23, and 13.17b). Major chemical variations mirror those of the pyroxene group. Calcic amphiboles are monoclinic; Ca-free amphiboles are generally orthorhombic. A solvus between calcic and noncalcic amphiboles results in many rocks containing two, or in some cases three, different amphibole species.

Many end member amphiboles have specific names, the most important are listed above (see **hornblende** entry for additional names). The detailed entries that follow do not include gedrite and riebeckite because they are similar to the more common anthophyllite and glaucophane.

The name *hornblende* is used by petrologists to refer to the common black amphibole found in many igneous and metamorphic rocks. Hornblendes have complex and highly variable chemistry; the formula above only partially reflects the variations.

For more general information about the amphibole group, see Chapter 5.

Cummingtonite $(Mg,Fe)_7Si_8O_{22}(OH)_2$

Origin of Name
Named after Cummington, Massachusetts, its type locality.

Hand Specimen Identification
Prismatic habit, two perfect cleavages intersecting at near 60° when viewed in basal section, habit, color, and association identify cummingtonite. It may be confused with other amphiboles.

Physical Properties

hardness	$5\frac{1}{2}–6$
specific gravity	2.9–3.2
cleavage/fracture	two perfect prismatic {110}/uneven
luster/transparency	vitreous, silky, fibrous/ transparent to translucent
color	white, green
streak	white

Optical Properties
Cummingtonite is colorless to pale green in thin section and exhibits weak pleochroism. Interference colors may be up to second order. Basal sections show typical amphibole cleavage displaying 56° and 124° angles. Extinction is inclined (15°–21°) to prismatic cleavage, polysynthetic twinning is common, and birefringence is greater than for anthophyllite-gedrite. Biaxial (+), $\alpha = 1.644$, $\beta = 1.657$, $\gamma = 1.674$, $\delta = 0.030$, $2V = 80°–90°$.

Crystallography
Monoclinic, $a = 9.51$, $b = 18.19$, $c = 5.33$, $\beta = 101.83°$, $Z = 2$; space group $C2/m$; point group $2/m$.

Habit
Cummingtonite forms prismatic, fibrous crystals; aggregates of radiating fibers or blades are common.

Structure and Composition

Cummingtonite, like other amphiboles, has a double-chain structure. SiO_4 tetrahedra are linked to make double chains that run parallel to the *c*-axis. Each tetrahedron shares two or three oxygen with neighbors, and has an unshared oxygen at the vertex pointing perpendicular to *c*. Chains are paired; unshared oxygens point toward each other and are bonded to the five octahedral cations occupying sites between them. A complete solid solution series exists between Mg-cummingtonite, $Mg_7Si_8O_{22}(OH)_2$, and grunerite, $(Fe_7Si_8O_{22}(OH)_2$. The name *cummingtonite* is given to intermediate compositions, $(Mg,Fe)_7Si_8O_{22}(OH)_2$, with Mg > Fe. Substantial Mn may replace Mg; Al and Ca may be present in small amounts.

Occurrence and Associations

Cummingtonite occurs in mafic or marly medium-grade metamorphic rocks. Common associated minerals include other amphiboles (hornblende, actinolite, or anthophyllite), garnet, plagioclase, and cordierite. Cummingtonite also occurs in a few rare kinds of igneous rocks.

Varieties

Amosite is an asbestiform amphibole similar to Fe-rich cummingtonite.

Related Minerals

Cummingtonite is closely related to the other amphiboles and is polymorphic with members of the anthophyllite series.

Grunerite $Fe_7Si_8O_{22}(OH)_2$

Origin of Name

Named after L. E. Grüner, a nineteenth-century mineralogist who first analyzed grunerite.

Hand Specimen Identification

Habit, color, two prominent cleavages at 56° to each other, and association help identify grunerite, but it cannot be distinguished from other members of the cummingtonite series without chemical or X-ray analysis.

Physical Properties

hardness	6
specific gravity	3.1–3.6
cleavage/fracture	two perfect prismatic {110}/uneven
luster/transparency	silky, vitreous/transparent to translucent
color	dark green or brown
streak	white

Optical Properties

Grunerite is similar to other members of the cummingtonite-grunerite series (see **cummingtonite**), but exhibits less pleochroism than cummingtonite, has an extinction angle of 10° to 15° to prismatic cleavage, and may show interference colors up to third order. Biaxial (−), $\alpha = 1.69$, $\beta = 1.71$, $\gamma = 1.73$, $\delta = 0.040$, $2V = 80°–90°$.

Crystallography

Monoclinic, $a = 9.6$, $b = 18.3$, $c = 5.3$, $\beta = 101.8°$, $Z = 2$; space group $C2/m$; point group $2/m$.

Habit

Grunerite typically forms fibrous, bladed, or columnar crystals, often radiating.

Structure and Composition

Grunerite is an end member of the cummingtonite-grunerite series. Structure and composition are analogous to cummingtonite. The name *grunerite* is by definition restricted to compositions close to end member $Fe_7Si_8O_{22}(OH)_2$.

Occurrence and Associations

Grunerite is found with Fe-rich minerals such as magnetite, hematite, minnesotaite, hedenbergite, fayalite, or garnet in metamorphosed Fe-rich sediments.

Related Minerals

See **cummingtonite.**

Tremolite $Ca_2Mg_5Si_8O_{22}(OH)_2$

Origin of Name

Named after Val Tremola, Switzerland, where it was first found.

Hand Specimen Identification

Association, perfect prismatic cleavages, and 56° cleavage angle when viewed in basal section, fibrous/bladed or thin columnar crystals, and generally very light color identify tremolite (Figure 3.6). A 56° cleavage angle distinguishes it from pyroxenes and pyroxenoids; light color distinguishes it from hornblende. It may also be confused with vesuvianite or wollastonite.

Physical Properties

hardness	5–6
specific gravity	3.0–3.3
cleavage/fracture	two perfect prismatic {110}/uneven
luster/transparency	transparent to translucent
color	white, green
streak	white

Optical Properties

Generally colorless when Fe-free, tremolite may be green and pleochroic when Fe is present. Amphibole cleavage angles (56° and 124°), 10° to 21° extinction angle in prismatic section, large $2V$, and upper first- to second-order interference colors identify tremolite. Biaxial $(-)$, $\alpha = 1.608$, $\beta = 1.618$, $\gamma = 1.630$, $\delta = 0.022$, $2V = 85°$.

Crystallography

Monoclinic, $a = 9.86$, $b = 18.11$, $c = 5.34$, $\beta = 105.00°$, $Z = 2$; space group $C2/m$; point group $2/m$.

Habit

Tremolite is typically prismatic. It may be in radiating or parallel blades, fibrous, asbestiform, or columnar. It is commonly twinned parallel to {100}.

Structure and Composition

Tremolite, $Ca_2Mg_5Si_8O_{22}(OH)_2$, is the Mg end member of the calcic amphibole series. Complete solid solution exists between tremolite and Fe-actinolite, $Ca_2Fe_5Si_8O_{22}(OH)_2$. Intermediate compositions are simply termed *actinolite*. Like other amphiboles, tremolite has a double-chain structure. SiO_4 tetrahedra are linked to make double chains that run parallel to the c-axis (Figure 13.17). Each tetrahedron shares two or three oxygen with neighbors and has an unshared oxygen at the vertex pointing perpendicular to c. Chains are paired; unshared oxygens point toward each other and are bonded to the five octahedral cations occupying sites between them. The two larger octahedral sites are occupied by Ca. Other alkalis and alkaline earths may substitute in small amounts for Ca, and some Al may be present in either the octahedral or tetrahedral sites. If impurities are present in sufficient quantities, the amphibole becomes dark and, in the absence of analytical data, we call it *hornblende*.

Occurrence and Associations

Tremolite is one of the first minerals to form when impure carbonates are metamorphosed. It is associated with calcite, dolomite, talc, quartz or forsterite, diopside and phlogopite.

Related Minerals

All amphiboles are structurally similar. Tremolite is closely related to Fe-actinolite $Ca_2Fe_5Si_8O_{22}(OH)_2$, the other principal calcic amphibole end member.

Actinolite $Ca_2(Fe,Mg)_5Si_8O_{22}(OH)_2$

Origin of Name

From the Greek *actis* (ray), referring to its common habit of radiating needles (see Plate 3.2).

Hand Specimen Identification

A needlelike or columnar habit, prismatic cleavages, 56° and 124° cleavage angles, and distinctive green color usually serve to identify actinolite. It is sometimes confused with epidote because of its green color. Mg:Fe ratios may vary; exact composition cannot be determined in hand specimen. Plate 3.2 shows a color photo of massive actinolite.

Physical Properties

hardness	5–6
specific gravity	3.0–3.3
cleavage/fracture	two perfect prismatic {110} /uneven
luster/transparency	vitreous/transparent to translucent
color	dark green
streak	white

Optical Properties

Actinolite is similar to tremolite in thin section (see **tremolite**), but is generally more strongly colored and pleochroic. Biaxial $(-)$, $\alpha = 1.66–1.67$, $\beta = 1.62–1.68$, $\gamma = 1.63–1.69$, $\delta = 0.03$, $2V = 70°–80°$.

Crystallography

Monoclinic, $a = 9.84$, $b = 18.05$, $c = 5.27$, $\beta = 104.7°$, $Z = 2$; space group $C2/m$; point group $2/m$.

Habit

Actinolite typically forms needles—either radiating or in parallel aggregates—or columnar masses.

Structure and Composition

Actinolite is the name given to green amphiboles with compositions intermediate between tremolite, $Ca_2Mg_5Si_8O_{22}(OH)_2$, and Fe-actinolite, $Ca_2Fe_5Si_8O_{22}(OH)_2$. Mn, Al, F, and Cr are sometimes present in minor amounts. Actinolite has the same structure as other calcic amphiboles (see **tremolite**).

Occurrence and Associations

Actinolite is characteristic of medium-grade metamorphosed mafic rocks. It is one of the minerals that gives greenschists their characteristic color. Associated minerals typically include albite, epidote, chlorite, and quartz.

Varieties

Nephrite is a Na-Al variety of actinolite.

Related Minerals

All amphiboles are structurally similar. Actinolite is closely related to tremolite, $Ca_2Mg_5Si_8O_{22}(OH)_2$, the other calcic amphibole end member.

Hornblende $(K,Na)_{0-1}(Ca,Na,Fe,Mg)_2(Mg, Fe,Al)_5(Si,Al)_8O_{22}(OH)_2$

Origin of Name
From the German *horn* (horn) and *blenden* (blind), referring to its luster and its lack of value.

Hand Specimen Identification
Habit, 56° angle between two prominent cleavages, and dark color usually serve to identify hornblende. It is occasionally confused with augite, but augite has a near 90° cleavage angle. Figure 5.23a shows typical hornblende. In the absence of compositional data, the name hornblende is often used for any black amphibole.

Physical Properties

hardness	5–6
specific gravity	3.0–3.5
cleavage/fracture	two perfect prismatic {110}/uneven
luster/transparency	vitreous/translucent
color	black or dark green
streak	white

Optical Properties
Hornblende may be various shades of brown, green, blue-green, or yellow-brown in thin section. Moderate to strong pleochroism is typical. Cross sections may be pseudohexagonal or diamond shaped. It may appear superficially like biotite, but has two good cleavages, and generally higher birefringence. 56° and 124° cleavage angles distinguish it from pyroxenes (Figures 5.24b and 7.7b). Biaxial $(-)$, $\alpha = 1.65$, $\beta = 1.66$, $\gamma = 1.67$, $\delta = 0.02$, $2V = 50°-80°$. Plates 5.3, 5.4, 5.5, and 5.6 show hornblende in thin section.

Crystallography
Monoclinic, $a = 8.97$, $b = 18.01$, $c = 5.33$, $\beta = 105.75°$, $Z = 2$; space group $C2/m$; point group $2/m$.

Habit
Hornblende may be massive or prismatic and is sometimes bladed, columnar, or fibrous. Euhedral crystals are often prismatic with a pseudohexagonal cross section. {100} contact twins are common.

Structure and Composition
Hornblende structure is similar to other amphiboles (see **cummingtonite** and **tremolite**), except that a large site, vacant in most of them, is partly occupied by Na or K. Thus, hornblende contains close to eight octahedral cations instead of seven. Hornblende composition varies greatly. Many end members have names; some of the more commonly used ones are

edenite	$Ca_2NaMg_5(AlSi_7)O_{22}(OH)_2$
ferro-edenite	$Ca_2NaFe_5(AlSi_7)O_{22}(OH)_2$
pargasite	$Ca_2NaMg_4Al(Al_2Si_6)O_{22}(OH)_2$
ferro-pargasite	$Ca_2NaFe_4Al(Al_2Si_6)O_{22}(OH)_2$
tschermakite	$Ca_2Mg_3Al_2(Al_2Si_6)O_{22}(OH)_2$
ferro-tschermakite	$Ca_2Fe_3Al_2(Al_2Si_6)O_{22}(OH)_2$
tremolite	$Ca_2(Mg,Fe_5)Si_8O_{22}(OH)_2$
ferro-actinolite	$Ca_2(Fe,Mg_5)Si_8O_{22}(OH)_2$
glaucophane	$Na_2Mg_3Al_2Si_8O_{22}(OH)_2$
kaersutite	$Ca_2Na(Mg,Fe)_4Ti(Al_2Si_6)O_{22}(OH)_2$

Besides compositional variations described by the end members just listed, some hornblende varieties include F^- or O^{2-} substituting for $(OH)^-$, or Fe^{3+} substituting for Fe^{2+}.

Occurrence and Associations
Hornblende is common in many kinds of igneous rocks covering a wide range of composition. It is usually associated with plagioclase and may coexist with quartz or with mafic minerals such as pyroxene or olivine. It is an essential mineral in those of intermediate composition such as syenite or diorite. Hornblende is also found in metamorphosed mafic rocks, especially in amphibolites that have hornblende and plagioclase as dominant minerals.

Related Minerals
All of the amphiboles are closely related in composition and structure. Hornblende has a more variable composition than most of the others.

Glaucophane $Na_2Mg_3Al_2Si_8O_{22}(OH)_2$

Origin of Name
From Greek words meaning "to appear bluish."

Hand Specimen Identification
Association, fibrous habit, near 60° cleavage angle, and blue color are distinctive of glaucophane and related Na-amphiboles crossite and riebeckite. See Figure 7.11.

Physical Properties

hardness	6–6½
specific gravity	3.1–3.2
cleavage/fracture	two perfect prismatic {110}/uneven
luster/transparency	vitreous/transparent to translucent
color	blue, gray
streak	white to very light blue

Optical Properties
Glaucophane may be difficult to tell from other blue amphiboles. It is colorless to blue or violet in thin section and often strongly pleochroic. Interference colors may range up to low second order, but are sometimes masked by mineral color. It exhibits typical amphibole cleavage and often forms fine prisms or needles with diamond-shaped cross sections. Biaxial $(-)$, $\alpha = 1.66$, $\beta = 1.67$, $\gamma = 1.65$, $\delta = 0.01-0.02$, $2V = 0°-50°$.

Crystallography
Monoclinic, $a = 9.78$, $b = 17.80$, $c = 5.30$, $\beta = 103.76°$, $Z = 2$; space group $C2/m$; point group $2/m$.

Habit
Acicular, asbestiform, or fibrous habit characterizes glaucophane.

Structure and Composition
Glaucophane has a structure similar to the calcic amphiboles (see **tremolite**). Although glaucophane has end member composition $Na_2Mg_3Al_2Si_8O_{22}(OH)_2$, most natural samples contain substantial Fe: Fe^{2+} replaces Mg^{2+} and Fe^{3+} replaces Al^{3+}. If Fe^{2+} and Fe^{3+} replace most of the Mg^{2+} and Al^{3+}, the amphibole becomes riebeckite. In riebeckite, some of the Na enters a normally vacant interlayer site. Compositions intermediate between glaucophane and riebeckite are called *crossite*.

Occurrence and Associations
Glaucophane is a high-pressure metamorphic mineral characteristic of the blueschist facies. Other blueschist minerals include jadeite, lawsonite, and aragonite.

Related Minerals
Glaucophane, $Na_2Mg_3Al_2Si_8O_{22}(OH)_2$, is similar in structure and chemistry to all amphiboles, but in particular the sodic amphiboles: riebeckite, $Na_2Fe_3Fe_2Si_8O_{22}(OH)_2$; eckermannite, $NaNa_2Mg_4AlSi_8O_{22}(OH)_2$; and arfvedsonite, $NaNa_2Fe_5Si_8O_{22}(OH)_2$. In the latter two, substantial Na occupies a normally unoccupied interlayer site.

Anthophyllite $\qquad$ $(Mg,Fe)_7Si_8O_{22}(OH)_2$

Origin of Name
From the Latin *anthophyllum*, meaning "clove leaf," referring to this mineral's color.

Hand Specimen Identification
Anthophyllite is characterized by its clove-brown color, usual prismatic habit, and prismatic cleavages with a 54° to 55° cleavage angle, but it is difficult to distinguish from other amphiboles such as grunerite or cummingtonite. Some samples of anthophyllite are fibrous (Plate 3.4).

Physical Properties

hardness	$5\frac{1}{2}$–6
specific gravity	2.9–3.2
cleavage/fracture	two perfect prismatic {210}, poor (100)/uneven
luster/transparency	vitreous/transparent to translucent
color	brown to green
streak	white

Optical Properties
In thin section, anthophyllite is colorless to pale brown or green and may be weakly pleochroic. It shows typical amphibole cleavage angles (56° and 124°) and up to second-order interference colors. It is difficult to tell from gedrite, but parallel extinction distinguishes it from clinoamphiboles. Biaxial (+ or −), $\alpha = 1.60$, $\beta = 1.62$, $\gamma = 1.63$, $\delta = 0.03$, $2V = 65°–90°$.

Crystallography
Orthorhombic, $a = 18.56$, $b = 18.01$, $c = 5.28$, $Z = 4$; space group $P2/n2/m2/a$; point group $2/m2/m2/m$.

Habit
Crystals are prismatic, fibrous, bladed, or columnar with diamond-shaped cross sections.

Structure and Composition
Anthophyllite is part of a solid solution series extending from $Mg_7Si_8O_{22}(OH)_2$ toward $Fe_7Si_8O_{22}(OH)_2$. Although compositionally identical to the monoclinic cummingtonite-grunerite amphiboles, anthophyllite is orthorhombic. Anthophyllite is usually Mg-rich; Fe-rich compositions yield cummingtonite. Al and Na may be present in anthophyllite; if Al content is great enough, the amphibole is called *gedrite*. The structure of anthophyllite is similar to that of cummingtonite and other amphiboles (see **cummingtonite**).

Occurrence and Associations
Anthophyllite is found in low-grade Mg-rich metamorphic rocks where it may be associated with cordierite. It is sometimes secondary after high-temperature minerals such as pyroxene and olivine, and is common in some serpentines.

Varieties
Amosite is asbestiform anthophyllite.

Related Minerals
Anthophyllite is similar to all the other amphiboles, especially gedrite, $(Mg,Fe,Mn)_2(Mg,Fe,Al)_5(Si,Al)_8O_{22}(OH)_2$, and holmquistite, $Li_2(Mg,Fe)_3Al_2Si_8O_{22}(OH)_2$. It is polymorphic with cummingtonite.

Pyroxenoid Group Minerals

wollastonite	$CaSiO_3$
rhodonite	$MnSiO_3$
pectolite	$NaCa_2(SiO_3)_3H$

Pyroxenoids are chain silicates having structures similar, but not identical, to pyroxenes. In pyroxene chains, SiO_4 tetrahedra zigzag back and forth (Figure 5.17).

Every other tetrahedron has the same orientation; the chains have a repeat distance of two tetrahedra, approximately 5.2Å. In pyroxenoids, repeat distances involve three or more tetrahedra, sometimes in complex arrangements. Ca^{2+}, Mn^{2+}, and other cations, bonded to unshared chain oxygens, occupy distorted octahedral sites between chains. The less regular structures of pyroxenoids have less symmetry; pyroxenoids are triclinic, whereas pyroxenes are monoclinic or orthorhombic.

Wollastonite   CaSiO₃

Origin of Name
Named after W. H. Wollaston (1766–1828), who discovered palladium and rhodium, invented the reflecting goniometer, and developed the camera lucida.

Hand Specimen Identification
Wollastonite's restricted occurrence, two perfect cleavages, and habit make it distinctive (Figure 7.8d). Tremolite has many features in common with wollastonite, but wollastonite can be distinguished by its two perfect cleavages about 84° apart (c.f., 56° in tremolite). Pectolite may be confused with wollastonite.

Physical Properties

hardness	$5-5\frac{1}{2}$
specific gravity	3.1
cleavage/fracture	perfect (100) and (001), good (102)/uneven
luster/transparency	silky/transparent to translucent
color	white
streak	white

Optical Properties
In thin section, wollastonite is clear and has low birefringence. It has two good-perfect cleavages in cross section and one in prismatic section, and it has near-parallel extinction. Wollastonite is distinguished from tremolite, pectolite, and diopside by its lower birefringence. Biaxial $(-)$, $\alpha = 1.620$, $\beta = 1.632$, $\gamma = 1.634$, $\delta = 0.014$, $2V = 39°$.

Crystallography
Triclinic, $a = 7.94$, $b = 7.32$, $c = 7.07$, $\alpha = 90.03°$, $\beta = 95.37°$, $\gamma = 103.43°$, $Z = 4$; space group $P\bar{1}$; point group $\bar{1}$.

Habit
Wollastonite typically forms cleavable masses or fibrous aggregates. Occasionally, tabular or prismatic crystals may be found.

Structure and Composition
Wollastonite, generally close to 100% $CaSiO_3$, may contain minor Mn substituting for Ca. Its structure is similar to other pyroxenoids and to pyroxenes (see the introduction to pyroxenoid group minerals and **enstatite**).

Occurrence and Associations
Wollastonite is common in high-grade marbles and other calcareous metamorphic rocks, especially contact metamorphic rocks. Common associated minerals are calcite, dolomite, tremolite, epidote, garnet, diopside, and vesuvianite.

Related Minerals
Wollastonite and other pyroxenoids are related to pyroxenes in both chemistry and structure. Other pyroxenoids include rhodonite, $MnSiO_3$; pectolite, $NaCa_2(SiO_3)_3H$; and bustamite, $(Mn,Ca,Fe)SiO_3$. Pseudowollastonite, a high-temperature polymorph, is similar to wollastonite, and several other wollastonite polymorphs are known.

Rhodonite   MnSiO₃

Origin of Name
From the Greek *rhodon*, meaning "rose," in reference to rhodonite's color.

Hand Specimen Identification
Rhodonite is one of the few pink minerals and has a nearly perfect 90° cleavage angle. It is occasionally confused with rhodochrosite but is harder and has a different habit.

Physical Properties

hardness	$5\frac{1}{2}-6$
specific gravity	3.5–3.7
cleavage/fracture	perfect {110}, perfect {1$\bar{1}$0}/conchoidal
luster/transparency	vitreous/transparent to translucent
color	pink, occasionally red; weathers to dark-colored Mn-oxide
streak	white

Optical Properties
Rhodonite is weakly pleochroic, colorless to light pink in thin section; maximum interference color is first-order yellow. Inclined extinction, high index of refraction, and low birefringence help identify it. Biaxial $(+)$, $\alpha = 1.717$, $\beta = 1.720$, $\gamma = 1.730$, $\delta = 0.013$, $2V = 63°-76°$.

Crystallography
Triclinic, $a = 7.68$, $b = 11.82$, $c = 6.71$, $\alpha = 92.35°$, $\beta = 93.95°$, $\gamma = 105.67°$, $Z = 2$; space group $P\bar{1}$; point group $\bar{1}$.

Habit
Cleavable masses or discrete grains are typical of rhodonite; large, as are irregular tabular crystals.

Structure and Composition

The rhodonite structure is similar to that of other pyroxenoids (see the introduction to pyroxene group minerals on the previous page). Rhodonite always contains some Ca substituting for Mn. If Ca content is great enough, it becomes bustamite. Fe and Zn may be present as well.

Occurrence and Associations

Rhodonite is found in manganese deposits and some iron formations. It is often found with Zn-minerals. Other associated minerals include the Mn-minerals rhodocrosite, bustamite, pyrolusite, tephroite, zincite, willemite, calcite, and quartz.

Related Minerals

Rhodonite is similar in composition and structure to other pyroxenoids, especially pyroxmangite, $(Mn,Fe)SiO_3$, and bustamite, $(Mn,Ca,Fe)SiO_3$. Pyroxmangite, however, contains less Ca and more Fe, and bustamite contains significantly more Ca.

Pectolite NaCa₂(SiO₃)₃H

Origin of Name
From the Greek *pectos,* meaning "well put together."

Hand Specimen Identification
Association, habit, two cleavages, and opacity help identify pectolite. It breaks into sharp acicular fragments when cleaved. It is occasionally confused with wollastonite or zeolites. Plate 3.1 shows a color photo of pectolite.

Physical Properties

hardness	5
specific gravity	2.9
cleavage/fracture	two perfect prismatic {100}, perfect {001}
luster/transparency	silky/translucent
color	white
streak	white

Optical Properties
Pectolite is colorless in thin section, has moderate birefringence and relief, and shows two perfect cleavages with parallel extinction. Biaxial $(+)$, $\alpha = 1.59$, $\beta = 1.61$, $\gamma = 1.63$, $\delta = 0.04$, $2V = 35°–63°$.

Crystallography
Triclinic, $a = 7.99$, $b = 7.04$, $c = 7.02$, $\alpha = 90.05°$, $\beta = 95.28°$, $\gamma = 102.47°$, $Z = 2$; space group $P\bar{1}$; point group $\bar{1}$.

Habit
Pectolite is typically fibrous or acicular. Acicular radiating crystals forming compact masses are common.

Structure and Composition

Pectolite is similar in structure to wollastonite. Twisted chains of SiO_4 tetrahedra run parallel to the *b*-axis and are connected by Na and Ca in octahedral coordination. Pectolite may contain small amounts of Fe, K, or Al.

Occurrence and Associations

Pectolite is usually a secondary mineral, resembling zeolites in appearance and occurrence. It is typically found as crusts, in cavities, or along joints in basalt. Associated minerals include zeolites, calcite, and prehnite. It is occasionally found as a primary mineral in alkalic igneous rocks or in calcic metamorphic rocks.

Related Minerals

Pectolite is structurally and chemically related to other pyroxenoids and pyroxenes.

RING SILICATES

Ring silicates have structures consisting of planar rings of silica tetrahedra not connected to each other by other tetrahedra. Using this definition, tourmaline is the only common example. Mineralogists sometimes group beryl and cordierite with tourmaline, but in beryl and cordierite, tetrahedral rings are connected by additional silica tetrahedra. The very rare minerals dioptase, $CuSiO_2(OH)_2$, and benitoite, $BaTiSi_3O_9$, are also ring silicates.

Tourmaline
(Na,Ca)(Fe,Mg,Al,Li)₃Al₆(BO₃)₃Si₆O₁₈(OH)₄

Origin of Name
From the Sinhalese *toramalli,* meaning "brown," the color of some tourmaline gemstones from Ceylon.

Hand Specimen Identification
Crystal habit, deformed triangular cross section, vitreous luster, hardness, and conchoidal fracture are characteristic. Black tourmaline may be confused with hornblende; small crystals may superficially resemble staurolite. Plates 1.1, 3.3, and 8.5 show color photographs of tourmaline; see also Figures 5.3, 8.17a, and 14.10.

Physical Properties

hardness	$7–7\frac{1}{2}$
specific gravity	2.9
cleavage/fracture	poor {101}, poor {110}/ subconchoidal
luster/transparency	resinous/translucent

color	variable black, blue, green, red, colorless; often zoned
streak	white

Optical Properties

Color and pleochroism are strong and variable: black, brown, green, blue, yellow, red, or pink. The color usually masks the birefringence. Pseudohexagonal or triangular cross sections are typical. Uniaxial $(-)$, typically $\omega = 1.645–1.670$, $\epsilon = 1.625–1.640$, $\delta = 0.020–0.030$.

Crystallography

Hexagonal (rhombohedral), $a = 15.84–1.603$, $c = 7.10–0.722$, $Z = 3$; space group $R3m$; point group $3m$.

Habit

Elongate trigonal prisms with vertical striations are common. Cross sections appear hexagonal or ditrigonal, with or without some rounded angles between faces (Figure 14.10). Tourmaline may occur as parallel or radiating crystal aggregates or may be massive and compact.

Structure and Composition

One of the few common boron minerals, tourmaline's composition is highly variable, leading to many different colored varieties. Fe-rich varieties are black; Mg-rich varieties are often brown or yellow; Li-rich varieties may be blue or green. The basic structure consists of rings of six $(SiO_4)^{4-}$ tetrahedra. The rings are connected to borate groups $(BO_3)^-$ and to (O^{2-}, OH^-), which form octahedra around $(Al^{3+}, Mg^{2+}, Fe^{2+})$. Cations such as Na^+ occupy positions in the center of the rings (Figure 13.18).

Occurrence and Associations

Tourmaline is a common accessory mineral in many granitic igneous rocks and in some metamorphic rocks. Rarely, it is found as detrital grains in sediments. It commonly associates with quartz and K-feldspar. It may be a major mineral in pegmatites, where it is often associated with lepidolite, beryl, apatite, spodumene, or fluorite.

Varieties

The most common varieties are
Black-blue: schorl $NaFe_3Al_6B_3O_9Si_6O_{18}(OH)_4$
Brown-yellow: dravite $NaMg_3Al_6B_3O_9Si_6O_{18}(OH)_4$
Blue-variable: elbaite $Na(Al,Li)_3Al_6B_3O_9Si_6O_{18}(OH)_4$

Related Minerals

Tourmaline is structurally related to beryl, $Be_3Al_2Si_6O_{18}$; cordierite, $(Mg,Fe)_2Al_4Si_5O_{18}$; dioptase, $CuSiO_2$; and benitoite, $BaTiSi_3O_9$.

(a)

(b)

►**FIGURE 14.10**
(a) Black tourmaline (schorl) from Minas Gerais, Brazil (8 cm tall). (b) Rubellite, a pink variety of tourmaline, in quartz from the Black Hills, South Dakota, USA (6 cm tall).

ISOLATED TETRAHEDRAL SILICATES

Garnet Group Minerals

Pyralspite Series

pyrope	$Mg_3Al_2Si_3O_{12}$
almandine	$Fe_3Al_2Si_3O_{12}$
spessartine	$Mn_3Al_2Si_3O_{12}$

Ugrandite Series

grossular	$Ca_3Al_2Si_3O_{12}$
andradite	$Ca_3Fe_2Si_3O_{12}$
uvarovite	$Ca_3Cr_2Si_3O_{12}$

The garnets all have chemistries $A_3B_2Si_3O_{12}$. Their structure consists of isolated $(SiO_4)^{4-}$ tetrahedra linked to distorted octahedrons containing (Al^{3+}, Fe^{3+}) and to distorted dodecahedrons containing $(Ca^{2+}, Mg^{2+}, Fe^{2+}, Mn^{2+})$. Mineralogists conveniently divide garnets into two series, the pyralspites (*pyrope-al*mandine-*spe*ssartine) and the ugrandites (*u*varovite-*gross*ular-*and*radite). Complete solid solution exists within each series, but only limited solid solution occurs between the two. Figures 2.1, 2.12, 7.2, 7.4, 7.8, and 10.36 are all photographs of garnet.

Pyrope $Mg_3Al_2Si_3O_{12}$

Origin of Name
From the Greek *pyropos,* meaning "fiery," a reference to this mineral's luster.

Hand Specimen Identification
Crystal form, lack of cleavage, luster, and color identify garnets. The different species can sometimes be distinguished by color or association. Determining exact composition requires analytical data.

Physical Properties

hardness	7
specific gravity	3.54
cleavage/fracture	none/subconchoidal
luster/transparency	resinous/transparent to translucent
color	red; occasionally black
streak	white

Optical Properties
Pyrope is generally clear or very pale in thin section, has high relief, and exhibits no cleavage. Euhedral and subhedral crystals are common. Isotropic, $n = 1.71$.

Crystallography
Cubic, $a = 11.46$, $Z = 8$; space group $I4_1/a\bar{3}2/d$; point group $4/m\bar{3}2/m$.

Habit
Equant grains, sometimes displaying dodecahedral or trapezohedral faces, characterize pyrope and other garnets. Massive occurrences are rare.

Structure and Composition
The name *pyrope* refers to garnets close in composition to the Mg end member of the pyralspite series. The structure of pyrope is similar to those of other garnets.

Occurrence and Associations
Pyrope is only stable in high-pressure rocks and is found in eclogites and other mafic and ultramafic rocks from deep in the Earth. Commonly associated minerals include olivine, pyroxene, spinel, and, occasionally, diamond.

Almandine $Fe_3Al_2Si_3O_{12}$

Origin of Name
From Alabanda, a Middle Eastern trade center where garnets were cut and polished in the first century A.D.

Hand Specimen Identification
Crystal form, lack of cleavage, luster, and color identify garnets, and the different species can sometimes be inferred by color or association. Determining exact composition requires analytical data. Plate 8.4 and Figures 2.1, 2.12b, 7.2b, and 7.4c show almandine.

Physical Properties

hardness	7
specific gravity	4.33
cleavage/fracture	none/subconchoidal
luster/transparency	resinous/transparent to translucent
color	deep red
streak	white

Optical Properties
Almandine is generally clear or very pale in thin section, has high relief, and exhibits no cleavage. Euhedral and subhedral crystals are common. Isotropic, $n = 1.83$. See Figures 7.2d and 7.4d.

Crystallography
Cubic, $a = 11.53$, $Z = 8$; space group $I4_1/a\bar{3}2/d$; point group $4/m\bar{3}2/m$.

Habit
Almandine is characterized by euhedral to subhedral equant grains displaying dodecahedral or trapezohedral faces. Massive occurrences are rare.

Composition and Structure
Almandine, the Fe end member garnet of the pyralspite series, may contain appreciable amounts of Ca, Mg, or Mn replacing Fe. It has the same structure as other garnets.

Occurrence and Associations
Almandine is a common mineral in medium- and high-grade metamorphic rocks. It is often found with quartz, feldspar, micas, staurolite, cordierite, chloritoid, tourmaline, and kyanite or sillimanite.

Spessartine   Mn₃Al₂Si₃O₁₂

Origin of Name
From Spessart, a district in Germany.

Hand Specimen Identification
Crystal form, lack of cleavage, luster, and color identify garnets. The different species can sometimes be
inferred by color or association. Determining exact
composition requires analytical data.

Physical Properties
hardness   7
specific gravity   4.19
cleavage/fracture   none/subconchoidal
luster/transparency   resinous/transparent to
                        translucent
color   pink to violet
streak   white

Optical Properties
Spessartine is generally clear or very pale in thin section, has high relief, and exhibits no cleavage. Euhedral and subhedral crystals are common. Isotropic,
$n = 1.80$.

Crystallography
Cubic, $a = 11.62$, $Z = 8$; space group $I4_1/a\bar{3}2/d$;
point group $4/m\bar{3}2/m$.

Habit
Euhedral to subhedral equant grains displaying dodecahedral or trapezohedral faces characterize spessartine. Massive occurrences are rare.

Structure and Composition
Spessartine is the Mn end member of the pyralspite
series. Natural spessartine may contain appreciable
amounts of Ca, Mg, or Fe replacing Mn. Spessartine
has the same structure as other garnets.

Occurrence and Associations
Spessartine is found with other Mn minerals in Mn-
rich skarns, low-grade metamorphic rocks, some rare
granites and rhyolites, and, occasionally, in pegmatites. Common associated minerals in igneous
rocks are quartz, feldspar, and micas.

Grossular   Ca₃Al₂Si₃O₁₂

Origin of Name
From *grossularia*, the Latin name for the pale green
gooseberry, which is the same color as some grossular.

Hand Specimen Identification
Crystal form, lack of cleavage, luster, and color identify grossular. The different species can sometimes be
inferred by color or association. Determining exact

▶FIGURE 14.11
Subhedral and euhedral grossular from Chihuahua, Mexico.

composition requires analytical data. See Figures 7.8d,
10.36b, and 14.11.

Physical Properties
hardness $6\frac{1}{2}$
specific gravity   3.56
cleavage/fracture   none/subconchoidal
luster/transparency   resinous/transparent to
                        translucent
color   pink, brown, yellow, or green
streak   white

Optical Properties
Grossular is generally clear or very pale in thin section, has high relief, and exhibits no cleavage. Euhedral and subhedral crystals are common. Isotropic,
$n = 1.75$.

Crystallography
Cubic, $a = 11.85$, $Z = 8$; space group $I4_1/a\bar{3}2/d$;
point group $4/m\bar{3}2/m$.

Habit
Grossular is characterized by euhedral to subhedral
equant grains displaying dodecahedral or trapezohedral faces (Figure 14.11). Massive occurrences are rare.

Structure and Composition
Grossular is the Al end member of the ugrandite se-
ries, $Ca_3(Fe^{3+}, Al^{3+}, Cr^{3+})_2Si_3O_{12}$, but natural grossular may contain appreciable amounts of Fe^{3+} or Cr^{3+}
in solid solution. The structure is the same as other
garnets.

Occurrence and Associations
Grossular is found in marbles where it may be associated with calcite, dolomite, quartz, tremolite, diopside, and wollastonite.

Silicates 335

Varieties
Hydrogrossular (hibschite) is the name given to grossular in which a substantial amount of Si^{4+} has been replaced by $4H^+$.

Andradite $\qquad$ Ca₃Fe₂Si₃O₁₂

Origin of Name
Named after J. B. d'Andrada e Silva (1763–1838), a Portuguese mineralogist.

Hand Specimen Identification
Crystal form, lack of cleavage, luster, and color identify garnets. The different species can sometimes be inferred by color or association. Determining exact composition requires analytical data.

Physical Properties
hardness — 7
specific gravity — 3.86
cleavage/fracture — none/subconchoidal
luster/transparency — resinous/transparent to translucent
color — yellow, brown, green
streak — white

Optical Properties
Andradite is generally clear or very pale in thin section, has high relief, and exhibits no cleavage. Euhedral and subhedral crystals are common. Isotropic, $n = 1.87$.

Crystallography
Cubic, $a = 12.05$, $Z = 8$; space group $I4_1/a\bar{3}2/d$; point group $4/m\bar{3}2/m$.

Habit
Andradite is characterized by euhedral to subhedral equant grains displaying dodecahedral or trapezohedral faces. Massive occurrences are rare.

Structure and Composition
Andradite, the Fe^{3+} end member of the garnet ugrandite series, may contain appreciable amounts of Al^{3+} or Cr^{3+} replacing Fe^{3+}. It has the same structure as other garnets.

Occurrence and Associations
Andradite is found in marbles and occasionally as an accessory mineral in igneous rocks. Typical associated minerals include hedenbergite and magnetite.

Varieties
Melanite is a black variety of andradite.

Uvarovite $\qquad$ Ca₃Cr₂Si₃O₁₂

Origin of Name
Named after Count S. S. Uvarov (1785–1855), president of the St. Petersburg Academy in Russia.

Hand Specimen Identification
Crystal form, lack of cleavage, luster, and color identify garnets. The different species can sometimes be inferred by color or association. Determining exact composition requires analytical data.

Physical Properties
hardness — $7\frac{1}{2}$
specific gravity — 3.80
cleavage/fracture — none/subconchoidal
luster/transparency — resinous/transparent to translucent
color — emerald green
streak — white

Optical Properties
Uvarovite is generally clear or very pale in thin section, has high relief, and exhibits no cleavage. Euhedral and subhedral crystals are common. Isotropic, $n = 1.85$.

Crystallography
Cubic, $a = 12.00$, $Z = 8$; space group $I4_1/a\bar{3}2/d$; point group $4/m\bar{3}2/m$.

Habit
Uvarovite is characterized by euhedral to subhedral equant grains displaying dodecahedral or trapezohedral faces. Massive occurrences are rare.

Structure and Composition
Uvarovite, the Cr^{3+} end member of the garnet ugrandite series, may contain appreciable amounts of Fe^{3+}, or Al^{3+} replacing Cr^{3+}. It has the same structure as other garnets.

Occurrence and Associations
Uvarovite is a rare mineral found primarily in peridotites and often associated with chrome ore. Associated minerals include chromite, olivine, pyroxene, and serpentine. It is more rarely found in metamorphic rocks.

Olivine Group Minerals

forsterite	Mg_2SiO_4
fayalite	Fe_2SiO_4
tephroite	Mn_2SiO_4
monticellite	$CaMgSiO_4$

Olivine, an abundant mineral in mafic and ultramafic igneous rocks, has the general formula $(Mg,Fe,Mn)SiO_4$. Its structure is similar to that of garnet: Isolated SiO_4 tetrahedra are linked by divalent cations in octahedral coordination (Figures 5.25 and 13.20). Complete solid solution exists between important end members forsterite and fayalite, and, less important, tephroite. Limited solid solution toward a Ca_2SiO_4

end member is also possible, but the rare mineral larnite, with composition Ca_2SiO_4, does not have the olivine structure. Monticellite, $CaMgSiO_4$, is grouped with the other olivines, but because of its highly distorted structure is not considered a true olivine.

For more general information about the olivine group, see Chapter 5.

Forsterite   Mg₂SiO₄

Origin of Name
Named after J. Forster, a scientist and founder of Heuland Cabinet.

Hand Specimen Identification
Olivine is distinguished by its glassy luster, conchoidal fracture, and usually olive-green color. Association and alteration to serpentine help identification. Olivines are sometimes confused with epidote or pyroxene. Telling forsterite from other olivines requires detailed optical or X-ray data. Plate 6.5 shows some typical forsterite in a basalt from San Carlos, Arizona; Figure 5.31a shows forsterite in an igneous rock called *dunite*.

Physical Properties

hardness	$6\frac{1}{2}$
specific gravity	3.2
cleavage/fracture	poor (010) and (100)/ conchoidal
luster/transparency	vitreous/transparent to translucent
color	colorless or green
streak	white

Optical Properties
Mg-rich olivines are colorless in thin section. Index of refraction and birefringence are high. Poor cleavage, irregular fracture, often equant grains, relatively high birefringence, and alteration to serpentine or chlorite help identification (Figures 5.31b and 14.12). Biaxial (+), $\alpha = 1.635$, $\beta = 1.651$, $\gamma = 1.670$, $\delta = 0.035$, $2V = 85°-90°$.

Crystallography
Orthorhombic, $a = 4.78$, $b = 10.28$, $c = 6.00$, $Z = 4$; space group $P2_1/b2/n2/m$; point group $2/m2/m2/m$.

Habit
Rare crystals are combinations of prisms and dipyramids, often having a tabular or lozenge shape. Granular forms that resemble green sand, or embedded grains, are common.

Structure and Composition
Isolated SiO_4 tetrahedra are linked by octahedral Mg (Figures 5.25 and 13.20). Complete solid solution exists between forsterite, Mg_2SiO_4; fayalite, Fe_2SiO_4;

▶**FIGURE 14.12**
Gemmy olivine (peridot) crystals in a basalt from near the Hawaiian Volcanoes National Park. The crystals are hard to pick out in the rock sample; some individual crystals (about 1 cm across) are also shown.

and tephroite, Mn_2SiO_4. Minor Ca or Ni may also be present as replacement for Mg.

Occurrence and Associations
Mg-rich olivine, a primary mineral in many mafic and ultramafic rocks, is typically associated with pyroxenes, plagioclase, spinel, garnet, and serpentine. It is common as both an igneous and a metamorphic mineral (in marbles) and more rarely is found in sediments.

Varieties
Peridot is a gemmy green transparent variety of forsterite.

Related Minerals
The principal olivine end members are forsterite, Mg_2SiO_4; fayalite, Fe_2SiO_4; and tephroite, Mn_2SiO_4. Olivine is isostructural with chrysoberyl, $BeAl_2O_4$. Minerals with similar but not identical structures include monticellite, $CaMgSiO_4$; sinhalite, $MgAl(BO_4)$; larnite, Ca_2SiO_4; and kirschsteinite, $CaFe(SiO_4)$.

Fayalite   Fe₂SiO₄

Origin of Name
Named after Fayal Island of the Azores, where fayalite was once found.

Hand Specimen Identification
Common olivine is distinguished by its glassy luster, conchoidal fracture, and usually olive-green color. Fayalite, Fe-rich olivine, however, may be various shades of brown or yellow and difficult to identify.

Exact composition cannot be determined without additional X-ray or optical data.

Physical Properties

hardness	$6\frac{1}{2}$
specific gravity	3.4
cleavage/fracture	poor (010) and (100)/ conchoidal
luster/transparency	vitreous/transparent to translucent
color	green to yellow
streak	white yellow

Optical Properties
Fe-rich olivines are pale yellow or green in thin section and may be weakly pleochroic. Index of refraction and birefringence are high. Poor cleavage, often equant grains, and alteration to serpentine or chlorite help identification. Biaxial $(-)$, $\alpha = 1.827$, $\beta = 1.877$, $\gamma = 1.880$, $\delta = 0.053$, $2V = 47°–54°$.

Crystallography
Orthorhombic, $a = 4.81$, $b = 10.61$, $c = 6.11$, $Z = 4$; space group $P2_1/b2/n2/m$; point group $2/m2/m2/m$.

Habit
Rare crystals are combinations of prisms and dipyramids, often having a tabular or lozenge shape. Embedded grains are common.

Structure and Composition
Fayalite structure is the same as that of other olivines: Isolated SiO_4 tetrahedra are linked by octahedral Fe. Complete solid solution exists between fayalite, Fe_2SiO_4, forsterite, Mg_2SiO_4, and tephroite, Mn_2SiO_4. Minor Ca or Ni may also be present as replacement for Fe.

Occurrence and Associations
Fayalite, less common than Mg-rich olivine, is found in some Fe-rich granitic igneous or metamorphic rocks.

Related Minerals
The principal olivine end members are forsterite, Mg_2SiO_4; fayalite, Fe_2SiO_4; and tephroite, Mn_2SiO_4. Many other minerals have related structures (see **forsterite**).

Monticellite   CaMgSiO₄

Origin of Name
Named after Italian mineralogist T. Monticelli.

Hand Specimen Identification
Association, color, conchoidal fracture, and habit help identify monticellite. It is difficult to distinguish from other olivine minerals without optical or X-ray data.

Physical Properties

hardness	$5\frac{1}{2}$
specific gravity	3.15
cleavage/fracture	poor (010) and (100)/ conchoidal
luster/transparency	vitreous/transparent to translucent
color	colorless, gray, or green
streak	white

Optical Properties
Monticellite is similar to olivine but has greater $2V$. Biaxial $(-)$, $\alpha = 1.645$, $\beta = 1.655$, $\gamma = 1.665$, $\delta = 0.020$, $2V = 72°–82°$.

Crystallography
Orthorhombic, $a = 4.82$, $b = 11.08$, $c = 6.38$, $Z = 4$; space group $P2_1/b2/n2/m$; point group $2/m2/m2/m$.

Habit
Crystals tend to be subequant combinations of prisms and dipyramids. Monticellite is usually embedded grains or massive patches in a carbonate-rich host.

Structure and Composition
The structure of monticellite is similar to that of olivine, but the mismatch in size between Ca and Mg leads to some slight differences (see **forsterite**). Fe may substitute for Mg, leading to solid solutions with kirschsteinite, $CaFeSiO_4$. Minor Al and Mn may also be present.

Occurrence and Associations
A rare mineral, monticellite is found in skarns and, less commonly, in regionally metamorphosed rocks. Associated minerals include calcite, forsterite, åkermanite, merwinite, and tremolite. Very minor occurrences have been reported from ultramafic igneous rocks.

Related Minerals
The true olivines (forsterite, fayalite, tephroite) are closely related to monticellite. Kirschsteinite, $CaFe(SiO_4)$, is isostructural with monticellite. Minerals with similar but not identical structures include sinhalite, $MgAl(BO_4)$, and larnite, Ca_2SiO_4.

Humite Group Minerals

norbergite	$Mg_3SiO_4(OH,F)_2$
chondrodite	$Mg_5(SiO_4)_2(OH,F)_2$
clinohumite	$Mg_9(SiO_4)_4(OH,F)_2$

Chondrodite is the most common of the humite minerals. All are isolated tetrahedral silicates with structural similarity to olivine. Their general formula is $nMg_2SiO_4 \cdot Mg(OH,F)_2$, where n is 1, 2, 3, and 4, respectively, for norbergite, chondrodite, humite, and clinohumite.

Norbergite — Mg₃SiO₄(OH,F)₂

Origin of Name
Named after the type locality at Norberg, Sweden.

Hand Specimen Identification
The members of the humite group (norbergite, chondrodite, and clinohumite) cannot be distinguished without optical or X-ray data. They are usually identified by association, light color, and form, but may be difficult to tell from olivine.

Physical Properties

hardness	$6\frac{1}{2}$
specific gravity	3.16
cleavage/fracture	none/subconchoidal
luster/transparency	vitreous/transparent to translucent
color	white, yellow, brown, red
streak	white

Optical Properties
Norbergite and other humite minerals resemble olivine in thin section, but most olivines are biaxial $(-)$, and humites have lower birefringence. Biaxial $(+)$, $\alpha = 1.561$, $\beta = 1.570$, $\gamma = 1.587$, $\delta = 0.026$, $2V = 44°–50°$.

Crystallography
Orthorhombic, $a = 4.70$, $b = 10.22$, $c = 8.72$, $Z = 4$; space group $P2_1/b2_1/n2_1/m$; point group $2/m2/m2/m$.

Habit
Crystals, usually found as isolated grains, are variable and display many forms. Highly modified orthorhombic or pseudo-orthorhombic crystals are common.

Structure and Composition
The structure consists of alternating layers of forsterite and brucite structure. Norbergite is usually close to end member composition, although F:OH ratios are variable. Some Fe may replace Mg.

Occurrence and Associations
Norbergite is a rare mineral found in metamorphosed carbonate rocks. Associated minerals include calcite, dolomite, phlogopite, diopside, spinel, wollastonite, grossular, forsterite, and monticellite.

Chondrodite — Mg₅(SiO₄)₂(OH,F)₂

Origin of Name
From the Greek word meaning "grain," referring to this mineral's occurrence as isolated grains.

Hand Specimen Identification
The members of the humite group (norbergite, chondrodite, and clinohumite) are usually identified by association, light color, and form. Chondrodite is difficult to distinguish from other humite minerals and from olivine, even under an optical microscope.

Physical Properties

hardness	$6–6\frac{1}{2}$
specific gravity	3.16–3.26
cleavage/fracture	poor (100)/subconchoidal
luster/transparency	vitreous/transparent to translucent
color	white to yellow
streak	white

Optical Properties
Chondrodite and other humite minerals resemble olivines in thin section, but most olivines are biaxial $(-)$, and humites have lower birefringence. Biaxial $(+)$, $\alpha = 1.60$, $\beta = 1.62$, $\gamma = 1.63$, $\delta = 0.03$, $2V = 60°–90°$.

Crystallography
Monoclinic, $a = 4.73$, $b = 10.27$, $c = 7.87$, $\beta = 109.1°$, $Z = 2$; space group $P2_1/b$; point group $2/m$.

Habit
Usually found as isolated grains, crystals are variable and display many forms. Highly modified orthorhombic or pseudo-orthorhombic crystals, with or without {001} twinning, are common.

Structure and Composition
The structure consists of layers of forsterite and brucite structure. Chondrodite contains two forsterite layers for each brucite layer. F:OH ratios are variable. Some Fe may replace Mg, and Ti content can be substantial.

Occurrence and Associations
Like norbergite, chondrodite is a rare mineral found in metamorphosed carbonate rocks. Associated minerals include calcite, dolomite, phlogopite, diopside, spinel, wollastonite, grossular, forsterite, and monticellite. Chondrodite has also been found in a few rare carbonatites.

Clinohumite — Mg₉(SiO₄)₄(OH,F)₂

Origin of Name
Named after English mineralogist Sir Abraham Hume (1749–1839).

Hand Specimen Identification
The members of the humite group (norbergite, chondrodite, and clinohumite) cannot be distinguished from each other, and sometimes from olivine, without

very detailed optical or X-ray data. They are usually identified by association, light color, and form.

Physical Properties

hardness	6
specific gravity	3.21–3.35
cleavage/fracture	poor (100)/subconchoidal
luster/transparency	vitreous/transparent to translucent
color	white to yellow
streak	white

Optical Properties
Clinohumite and other humite minerals resemble olivines in thin section, but most olivines are biaxial (−), and humites have lower birefringence. Biaxial (+), $\alpha = 1.63$, $\beta = 1.64$, $\gamma = 1.59$, $\delta = 0.03$–0.04, $2V = 73°$–$76°$.

Crystallography
Monoclinic, $a = 4.75$, $b = 10.27$, $c = 13.68$, $\beta = 100.8°$, $Z = 2$; space group $P2_1/b$; point group $2/m$.

Habit
Usually found as isolated grains, crystals are variable and display many forms. Highly modified pseudo-orthorhombic crystals with or without {001} twinning are common.

Structure and Composition
Structure is similar to the other humite minerals (see **chondrodite**) except that the ratio of forsterite : brucite is 4:1. Some Fe may replace Mg, and F : OH ratios are variable. Ti is almost always present in small amounts.

Occurrence and Associations
The occurrences of clinohumite are primarily in metamorphosed carbonate rocks, similar to other humite minerals.

Varieties
Titanoclinhumite, an especially Ti-rich variety, has been reported from a few rare serpentinites and gabbros.

Aluminosilicate Group Minerals

kyanite	Al_2SiO_5
andalusite	Al_2SiO_5
sillimanite	Al_2SiO_5

The aluminosilicate polymorphs vary little from end member Al_2SiO_5 composition. All are isolated tetrahedral silicates but have distinctly different structures: In kyanite all the Al is in 6-fold coordination, in andalusite half is in 5-fold coordination and half is in 6-fold coordination, and in sillimanite half is in 4-fold coordination and half is in 6-fold coordination. The aluminosilicates are important minerals in pelitic metamorphic rocks. The presence of a particular polymorph indicates a general range of pressure-temperature at which the rock must have formed (see Figure 7.12). For this reason, and because they are relatively common, the aluminosilicates are key metamorphic index minerals.

Kyanite Al_2SiO_5

Origin of Name
From the Greek *kyanos*, meaning "blue."

Hand Specimen Identification
Kyanite forms blue bladed crystals and is easily cleaved into acicular fragments (Plate 3.6, Figures 2.12c and 7.4a).

Physical Properties

hardness	5–7
specific gravity	3.60
cleavage/fracture	two prominent: perfect (100), good (010)/uneven
luster/transparency	pearly, vitreous/transparent to translucent
color	blue to white
streak	white

Optical Properties
Kyanite is typically colorless in thin section, but may be weakly blue and pleochroic. High relief, low birefringence, and excellent cleavage aid identification. It may be confused with sillimanite or andalusite, but sillimanite has a small *2V*, and andalusite has parallel extinction. Biaxial (−), $\alpha = 1.712$, $\beta = 1.720$, $\gamma = 1.728$, $\delta = 0.016$, $2V = 82°$–$83°$.

Crystallography
Triclinic, $a = 7.10$, $b = 7.74$, $c = 5.57$, $\alpha = 90.08°$, $\beta = 101.03°$, $\gamma = 105.73°$, $Z = 4$; space group $P\bar{1}$; point group $\bar{1}$.

Habit
Kyanite is usually in long blade-shaped or tabular crystals, sometimes forming parallel or radiating aggregates.

Structure and Composition
Chains of AlO_6 octahedra are cross linked by additional AlO_6 octahedra and by SiO_4 tetrahedra. Kyanite is always near to end member composition; it may contain very minor Fe, Mn, or Cr.

Occurrence and Associations
Kyanite is primarily a metamorphic mineral found in medium- to high-pressure schists and gneisses. Typical associated minerals are quartz, feldspar, mica, garnet,

corundum, and staurolite. It is known from a few aluminous eclogites and other rocks of deep origin.

Related Minerals
Kyanite has two polymorphs, andalusite and sillimanite.

Andalusite Al_2SiO_5

Origin of Name
Named for Andalusia, a province of Spain.

Hand Specimen Identification
Andalusite is recognized primarily by crystal form and association. Hardness and nearly square (diamond) cross sections help identify andalusite. A variety called *chiastolite* displays a "maltese cross" pattern that is diagnostic (Figure 7.4b). Andalusite is sometimes confused with scapolite.

Physical Properties

hardness	$7\frac{1}{2}$
specific gravity	3.18
cleavage/fracture	rarely seen, good {110}, poor (100)/subconchoidal
luster/transparency	vitreous/transparent to translucent
color	brown to red
streak	white

Optical Properties
Usually clear in thin section, andalusite may be weakly colored and pleochroic. Euhedral crystals with a square outline, sometimes showing penetration twins or a maltese cross pattern, are diagnostic. Andalusite is length fast and has a high *2V*. Biaxial $(-)$, $\alpha = 1.632$, $\beta = 1.640$, $\gamma = 1.642$, $\delta = 0.010$, $2V = 75°-85°$.

Crystallography
Orthorhombic, $a = 7.78$, $b = 7.92$, $c = 5.57$, $Z = 4$; space group $P2_1/n2_1/n2/m$; point group $2/m2/m2/m$.

Habit
Stubby to elongate square prisms characterize andalusite. Individual crystals may be rounded. Massive and granular forms are also known.

Structure and Composition
Andalusite consists of chains of AlO_6 octahedra parallel to *c*, cross linked by SiO_4 tetrahedra and by AlO_5 polyhedra. Andalusite is usually close to Al_2SiO_5 in composition. Small amounts of Mn, Fe, Cr, and Ti may be present.

Occurrence and Associations
Andalusite is a metamorphic mineral characteristic of relatively low pressures. It occurs in pelitic rocks, often associated with cordierite, sillimanite, kyanite, garnet, micas, and quartz.

Varieties
Chiastolite is a variety of andalusite that has a square cross section (001) displaying a "maltese cross" pattern. The pattern results from carbonaceous impurities included during crystal growth.

Related Minerals
Andalusite has two polymorphs, sillimanite and kyanite.

Sillimanite Al_2SiO_5

Origin of Name
Named after Benjamin Silliman, a chemistry professor at Yale University.

Hand Specimen Identification
Sillimanite is found in high-grade pelites, often forming thin, acicular crystals with square cross sections. Aggregates may be masses or splays. Very fine-grained sillimanite mats, often having a satiny appearance under the microscope, are termed *fibrolite*. Sillimanite is occasionally confused with anthophyllite.

Physical Properties

hardness	6–7
specific gravity	3.23
cleavage/fracture	perfect but rarely seen (010)/uneven
luster/transparency	vitreous/transparent to translucent
color	white to brown
streak	white

Optical Properties
Sillimanite typically forms needles, often in masses or mats, with square cross sections showing one good diagonal cleavage (Figure 7.4d). It has high relief, small *2V*, $(+)$ optic sign, and is length slow. Biaxial $(+)$, $\alpha = 1.658$, $\beta = 1.662$, $\gamma = 1.680$, $\delta = 0.022$, $2V = 20°-30°$.

Crystallography
Orthorhombic, $a = 7.44$, $b = 7.60$, $c = 5.75$, $Z = 4$; space group $P2_1/b2/m2/n$; point group $2/m2/m2/m$.

Habit
Long, slender prisms, needles, or fibers are common. Subparallel aggregates and splays are typical. Fine-grained fibrous mats are also common.

Structure and Composition
The structure consists of chains of AlO_6 octahedra, parallel to *c*, cross linked by SiO_4 and AlO_4 tetrahedra. Composition is always close to end member Al_2SiO_5; small amounts of Fe may be present.

Occurrence and Associations

Sillimanite is the high-temperature Al_2SiO_5 polymorph, found in high-grade pelites associated with garnet, cordierite, spinel, hypersthene, orthoclase, biotite, and quartz.

Varieties

Fibrolite is the name given to fine-grained fibrous masses of sillimanite.

Related Minerals

Sillimanite has two polymorphs: andalusite and kyanite.

Other Isolated Tetrahedral Silicates

staurolite	$Fe_2Al_9Si_4O_{23}(OH)$
chloritoid	$(Fe,Mg)Al_2SiO_5(OH)_2$
titanite	$CaTiSiO_5$
topaz	$Al_2SiO_4(F,OH)_2$
zircon	$ZrSiO_4$

Besides those already listed, a number of other isolated tetrahedral silicates are common and important minerals. Although they have structures based on isolated SiO_4 tetrahedra, they do not fit into any of the previously discussed structural groups. Staurolite and chloritoid are important metamorphic minerals in rocks rich in Fe and Al. Titanite and zircon are common accessory minerals in silicic igneous rocks and in many metamorphic rocks. Topaz is most commonly found in pegmatites and hydrothermal veins associated with granites and other silicic igneous rocks.

Staurolite $\qquad Fe_2Al_9Si_4O_{23}(OH)$

Origin of Name

From the Greek *stauros,* meaning "cross," in reference to its cruciform twins.

Hand Specimen Identification

Staurolite is often easily recognized by its characteristic twins and crystals (Figures 2.7d, 2.12a, and 7.4c). It is sometimes confused with andalusite but has different habit. Pyroxene, tourmaline, titanite, and amphibole may look superficially like staurolite.

Physical Properties

hardness	$7–7\frac{1}{2}$
specific gravity	3.75
cleavage/fracture	poor {010}/subconchoidal
luster/transparency	vitreous, resinous/translucent
color	brown to black
streak	white or gray

Optical Properties

Staurolite may be clear to yellow or light brown in thin section, often pleochroic. Birefringence is low; maximum colors are first-order yellow. Anhedral to euhedral porphyroblasts, often exhibiting a "sieve" structure due to quartz inclusions, or penetration twins, are common. Biaxial (+), $\alpha = 1.740$, $\beta = 1.744$, $\gamma = 1.753$, $\delta = 0.013$, $2V = 80°–88°$.

Crystallography

Monoclinic, $a = 7.82$, $b = 16.52$, $c = 5.63$, $\beta = 90.0°$, $Z = 2$; space group $C2/m$; point group $2/m$.

Habit

Staurolite is usually found as prismatic crystals, often flattened in one direction and having several terminating forms. Massive varieties are rare. Penetration twins are common, sometimes resulting in perfect "cruciform" crosses.

Structure and Composition

Staurolite structure is closely related to that of kyanite. Layers of Al_2SiO_5, including AlO_6 octahedra in chains, alternate with layers of $Fe(OH)_2$. Pure end member Fe-staurolite does not exist in nature; Mg is always present, replacing up to 35% of the Fe. Small amounts of Ti and Mn are generally present as well. Water content is slightly variable.

Occurrence and Associations

Staurolite is a metamorphic mineral common in medium- to high-grade metamorphic rocks. Associated minerals include kyanite, garnet, chloritoid, micas, and tourmaline.

Chloritoid $\qquad (Fe,Mg)Al_2SiO_5(OH)_2$

Origin of Name

Named for its resemblance to chlorite.

Hand Specimen Identification

Color, cleavage, and association help identify chloritoid. Thin sections may be required to distinguish it from chlorite, biotite, or stilpnomelane.

Physical Properties

hardness	$6\frac{1}{2}$
specific gravity	3.50
cleavage/fracture	poor{110}/uneven
luster/transparency	pearly/transparent
color	dark green
streak	gray

Optical Properties

Chloritoid is typically colorless to green in thin section and may exhibit pleochroism in various shades of green, yellow, or blue. It has high relief and anomalous interference colors and is frequently twinned. Chlorite looks superficially like chloritoid but has lower R.I. and relief and a smaller $2V$. Biaxial (+), $\alpha = 1.715$, $\beta = 1.720$, $\gamma = 1.725$, $\delta = 0.010$, $2V = 45°–65°$.

Crystallography
Monoclinic, $a = 9.52$, $b = 5.47$, $c = 18.19$, $\beta = 101.65°$, $Z = 6$; space group $C2/c$; point group $2/m$.

Habit
Coarse masses or thin scales are typical; individual crystals are rare. Tabular crystals are platy and foliated with common {001} twinning.

Structure and Composition
The chloritoid structure is layered. Alternating brucite-like and corundum-like layers, perpendicular to the *c*-axis, are linked by SiO_4 tetrahedra and hydrogen bonds. Chloritoid is not a layer silicate, like chlorite, because the SiO_4 tetrahedra do not share oxygen. Chloritoid is generally Fe-rich, but Fe:Mg ratios are variable; end members are not found in nature. Some Mn may be present.

Occurrence and Associations
Chloritoid is common in low- or medium-grade Fe- and Al-rich schists. Associated minerals include quartz, feldspars, muscovite, chlorite, staurolite, garnet, andalusite, and kyanite. In some rare high-pressure metamorphic rocks, it occurs with glaucophane and other blueschist minerals.

Varieties
Ottrelite is Mn-rich chloritoid; carboirite is Ge-containing chloritoid.

Related Minerals
Several different polytypes and polymorphs have been described.

Titanite (Sphene) $CaTiSiO_5$

Origin of Name
This mineral's name refers to its titanium content. Its older name, *sphene,* refers to its crystal shape.

Hand Specimen Identification
Luster, color, and wedge-shaped crystals help identify titanite. It may be confused with staurolite and zircon, but is softer; or with sphalerite, but is harder.

Physical Properties

hardness	$5–5\frac{1}{2}$
specific gravity	3.50
cleavage/fracture	good prismatic {110}, poor (100)/uneven
luster/transparency	adamantine/transparent to translucent
color	gray, black, brown, greenish yellow
streak	white

Optical Properties
Very high relief and birefringence and crystal form characterize titanite. Biaxial $(+)$, $\alpha = 1.86$, $\beta = 1.93$, $\gamma = 2.10$, $\delta = 0.15$, $2V = 23°–50°$.

Crystallography
Monoclinic, $a = 6.56$, $b = 8.72$, $c = 7.44$, $\beta = 119.72°$, $Z = 4$; space group $C2/c$; point group $2/m$.

Habit
Sphenoidal crystals, tabular with a wedge or diamond shape in cross section, are typical. Less commonly, titanite is massive or lamellar. Titanite is normally fine grained but occasionally occurs as large crystals (Figure 14.13).

Structure and Composition
The structure contains TiO_6 polyhedra and SiO_4 tetrahedra that share corners, forming distorted chains parallel to *a*. Ca is in 7-fold coordination, in large holes between the Ti- and Si-polyhedra. Many elements may substitute in titanite; especially important are the rare earths.

Occurrence and Associations
Titanite is an often overlooked common and widespread accessory mineral. In many rocks it is the only important Ti mineral present. It is common in many igneous rocks, especially silicic to intermediate ones, and many metamorphic rocks. It also has been found in some limestones and a few rare clastic sediments. Associated minerals include just about all the important rock forming minerals, including pyroxene, amphibole, feldspar, and quartz.

▶**FIGURE 14.13**
Large black crystals of titanite surrounded by quartz and mica. This sample is from Eganville, Ontario, Canada.

Varieties

Greenovite is the name given to red or pink titanite.

Related Minerals

Titanite is isostructural with tilasite, $CaMg(AsO_4)F$; malayaite, $CaSn(SiO_4)O$; and with fersmantite, $(Ca, Na)_4(Ti,Nb)_2Si_2O_{11}(F,OH)_2$. Other related minerals include perovskite, $CaTiO_3$, benitoite, $BaTiSi_3O_9$, and neptunite, $KNa_2Li(Fe,Mn)_2Ti_2O(Si_4O_{11})_2$.

Topaz   (F,OH)$_2$

Origin of Name

Named after an island in the Red Sea, Topazion.

Hand Specimen Identification

Orthorhombic form, hardness, one good cleavage, color, and luster identify topaz. It may be confused with quartz but is not hexagonal. The very rare mineral danburite, $Ca(B_2Si_2O_8)$, is similar to topaz in form and properties, and cannot be distinguished without chemical analysis. Gem topaz is shown in Figure 8.17b. See also Figure 14.14.

Physical Properties

hardness	8
specific gravity	3.5–3.6
cleavage/fracture	one perfect (001)/ subconchoidal
luster/transparency	vitreous/transparent to translucent
color	colorless or variable
streak	white

Optical Properties

Topaz is typically colorless in thin section and has low birefringence. It resembles quartz and apatite but has one perfect cleavage and higher relief than quartz, is biaxial, and is length slow. Biaxial $(+)$, $\alpha = 1.61$, $\beta = 1.61$, $\gamma = 1.62$, $\delta = 0.01$, $2V = 48°-65°$.

Crystallography

Orthorhombic, $a = 4.65$, $b = 8.80$, $c = 8.40$, $Z = 4$; space group $P2_1/b2/n2/m$; point group $2/m2/m2/m$.

Habit

Typical crystals are orthorhombic prisms, terminated by dipyramids and pinacoids in combination (Figure 14.14). Prism faces show striations. Cross sections may be square, rectangular, diamond shaped, or octagonal. Coarse- or fine-grained masses are also common.

Structure and Composition

The structure consists of chains, parallel to c, containing pairs of edge-sharing $Al(OH,F)_6$ octahedra alternating with SiO_4 tetrahedra. F and OH content are variable, but F:OH ratio is usually in excess of 6:1. No other significant substitutions are known.

►**FIGURE 14.14**
Clear topaz crystals look superficially like quartz but have orthorhombic instead of hexagonal symmetry. These samples are from Topaz Mountain, Utah.

Occurrence and Associations

Topaz is a late-stage igneous or hydrothermal mineral. It is an accessory in granite, rhyolite, and granitic pegmatites and may be found in contact aureoles adjacent to silicic plutons. It is often associated with Li and Sn mineralization. Associated minerals include quartz, feldspar, muscovite, tourmaline, fluorite, cassiterite, apatite, and beryl.

Related Minerals

Euclase, $BeAl(SiO_4)(OH)$, is isotypical with topaz.

Zircon   ZrSiO$_4$

Origin of Name

From the Persian *zar* ("gold") and *gun* ("color").

Hand Specimen Identification

Zircon is usually identified by its distinctive crystal shape, color, luster, hardness, and density.

Physical Properties

hardness	$7\frac{1}{2}$
specific gravity	4.68
cleavage/fracture	poor (100), poor {101}/conchoidal
luster/transparency	adamantine/transparent to translucent
color	brown to green, also gray, red, or colorless
streak	colorless to white

Optical Properties

Zircon is normally colorless but may be pale yellow or brown and faintly pleochroic. Grains are typically small, exhibiting very high relief and birefringence. Pleochroic halos around grains are due to decay of radioactive elements. Uniaxial $(+)$, $\omega = 1.99$, $\epsilon = 1.93$, $\delta = 0.06$.

Crystallography
Tetragonal, $a = 6.59$, $c = 5.99$, $Z = 4$; space group $I4_1/a2/m2/d$; point group $4/m2/m2/m$.

Habit
Zircon typically is found as square prisms, pyramids, or as combinations of the two. Rounded grains are also common.

Structure and Composition
The zircon structure contains SiO_4 tetrahedra sharing corners or edges with distorted cubic polyhedra containing Zr. Zircon is generally close to end member composition, but frequently contains small amounts of Al, Fe, Mg, Ca, rare earths, and water.

Occurrence and Associations
Zircon is a common and widespread accessory mineral in igneous rocks, especially silicic ones. It is found in many metamorphic rocks and is common in sediments and sedimentary rocks. It is an important mineral for some kinds of radiometric dating.

Varieties
Zircon is often metamict (structurally damaged by the decay of radioactive elements in its structure), causing variation in color and optical properties.

Related Minerals
Zircon has a number of isotypes, including thorite, $(Th,U)(SiO_4)$; xenotime, $Y(PO_4)$; and huttonite, $ThSiO_4$. Baddeleyite, ZrO_2, is another Zr-rich mineral.

PAIRED TETRAHEDRAL SILICATES

lawsonite	$CaAl_2Si_2O_7(OH)_2 \cdot H_2O$
epidote	$Ca_2(Al,Fe)_3Si_3O_{12}(OH)$
clinozoisite	$Ca_2Al_3Si_3O_{12}(OH)$
vesuvianite	$Ca_{10}(Mg,Fe)_2Al_4Si_9O_{34}(OH)_4$

Mineralogists often group these four minerals together because they all contain pairs of SiO_4 tetrahedra sharing a single bridging oxygen. In lawsonite all silica tetrahedra are paired. However, in epidote, clinozoisite, and vesuvianite, structures are more complicated because some SiO_4 tetrahedra are unpaired. For this reason, these three minerals are sometimes not considered to be paired tetrahedral silicates. Åkermanite and gehlenite, rare minerals belonging to the melilite group, are also paired tetrahedral silicates (see Figure 13.19) but are not considered here.

Lawsonite $CaAl_2Si_2O_7(OH)_2 \cdot H_2O$

Origin of Name
Named after A. C. Lawson, a professor at the University of California.

Hand Specimen Identification
Hardness, bladed character, color, and association help identify lawsonite, but it is often very fine grained.

Physical Properties

hardness	8
specific gravity	3.1
cleavage/fracture	perfect (100) and (010), fair {101}/uneven
luster/transparency	greasy, vitreous/transparent
color	gray, white, colorless
streak	white

Optical Properties
Lawsonite is usually colorless and exhibits high relief and interference colors no higher than first-order red. Biaxial $(+)$, $\alpha = 1.665$, $\beta = 1.674$, $\gamma = 1.685$, $\delta = 0.020$, $2V = 76°-86°$.

Crystallography
Orthorhombic, $a = 8.90$, $b = 5.76$, $c = 13.33$, $Z = 4$; space group $C2/c2/m2/m$; point group $2/m2/m2/m$.

Habit
Lawsonite may form tabular or prismatic crystals. Simple or lamellar twins are common.

Structure and Composition
Lawsonite is the only relatively common silicate in which all $(SiO_4)^{4-}$ tetrahedra are paired, thus producing Si_2O_7 groups. The Si_2O_7 groups link $Al(O,OH)$ octahedra; Ca occupies holes between the Si_2O_7 groups and the octahedra. Small amounts of Ti, Fe, Mg, Na, and K may be present, but no major solid solutions are known.

Occurrence and Associations
Lawsonite is a metamorphic mineral typical of the blueschist facies. Associated high-pressure minerals include glaucophane, jadeite, pumpellyite, or aragonite. Other associated minerals include chlorite, plagioclase, titanite, quartz, and epidote.

Related Minerals
Hemimorphite, $Zn_4(Si_2O_7)(OH)_2 \cdot H_2O$, and ilvaite, $CaFe_3O(Si_2O_7)(OH)$, are other sorosilicates in which all SiO_4 tetrahedra are paired. Other minerals usually considered sorosilicates are epidote, $Ca_2(Al,Fe)_3Si_3O_{12}(OH)$; clinozoisite, $Ca_2Al_3Si_3O_{12}(OH)$ piemontite, $Ca_2(Al,Mn)_3Si_3O_{12}(OH)$; allanite, $(Ca,Ce)_2(Al,Fe,Mg)_3Si_3O_{12}(OH)$; and vesuvianite, $Ca_{10}(Mg,Fe)_2Al_4Si_9O_{34}(OH,F)_4$.

Epidote   Ca₂(Al,Fe)₃Si₃O₁₂(OH)

Origin of Name
From the Greek word *epididonai*, meaning "increase," referring to the base of an epidote prism, one side of which is longer than the other.

Hand Specimen Identification
Epidote is characterized by its pistachio-green color, habit, one perfect cleavage, and association. It is often fine grained and massive (Figure 14.15).

Physical Properties

hardness	6–7
specific gravity	3.4–3.5
cleavage/fracture	perfect (001), poor (100)/ uneven
luster/transparency	vitreous/transparent to translucent
color	pistachio-green, yellow-green to black
streak	white

Optical Properties
Epidote has high relief and is usually colorless, but may be light green or pink and pleochroic, depending on composition. Interference colors range up to third order as Fe content increases. Football-shaped grains with concentric interference color rings are diagnostic of epidote. Biaxial $(-)$, $\alpha = 1.71$–1.75, $\beta = 1.72$ – 1.78, $\gamma = 1.73$–1.80, $\delta = 0.01$–0.05, $2V = 90°$–115°.

Crystallography
Monoclinic, $a = 8.98$, $b = 5.64$, $c = 10.22$, $\beta = 115.4°$, $Z = 2$; space group $P2_1/m$; point group $2/m$.

Habit
Crystals are prismatic, fibrous, or acicular, usually elongated parallel to b, with faces showing striations. Granular, massive, and fibrous aggregates are common.

Structure and Composition
Epidote is usually considered a paired tetrahedral silicate (sorosilicate), but contains both paired and unpaired $(SiO_4)^{4-}$ tetrahedra (see discussion under **lawsonite**). Chains of edge-sharing $Al(O,OH)_6$ octahedra, cross linked by Si_2O_7 and SiO_4 groups, are parallel to b. Ca occupies large sites between the various groups. A complete solid solution exists between Fe-epidote, $Ca_2Fe_3Si_3O_{12}(OH)$, and clinozoisite, $Ca_2Al_3Si_3O_{12}$ (OH). Limited substitution exists between epidote and piemontite, $Ca_2(Mn,Al)_3Si_3O_{12}(OH)$. Cr, Pb, V, Sr, Sn, and rare earths may be present in small amounts.

Occurrence and Associations
Epidote is a common and widespread mineral, characteristic of low- to medium-grade metabasites and marbles. Associated minerals include actinolite, chlorite, and albite in mafic rocks and diopside, grossular, and vesuvianite in marbles. Epidote is also produced by alteration of feldspar, pyroxene, and amphibole.

▶**FIGURE 14.15**
A vein of epidote in a granite, probably formed by alteration of feldspar.

Varieties
Tanzanite is a blue epidote gem; thulite is a rose-colored variety; allanite is epidote rich in rare earth elements. *Sausserite* is a name given to fine-grained epidote produced by alteration of plagioclase.

Related Minerals
Piemontite, a rare-earth mineral, and about 12 other minerals belong to the epidote group.

Clinozoisite   Ca₂Al₃Si₃O₁₂(OH)

Origin of Name
Named after Baron von Zois (1747–1819), an Austrian who financed mineralogists.

Hand Specimen Identification
Clinozoisite is difficult to identify in hand specimen. If green, it is mistaken for epidote; if uncolored or lightly colored, it is often overlooked. It cannot be distinguished from zoisite (orthorhombic) without X-ray analysis.

Physical Properties

hardness	6–6½
specific gravity	3.1–3.4
cleavage/fracture	perfect (001)/uneven
luster/transparency	vitreous/transparent to translucent
color	light green to yellow or gray
streak	white

Optical Properties

Clinozoisite has high relief, and is usually colorless. Interference colors are often anomalous; birefringence is low. Biaxial (+), $\alpha = 1.67–1.72$, $\beta = 1.67–1.72$, $\gamma = 1.69–1.73$, $\delta = 0.005–0.015$, $2V = 14°–90°$.

Crystallography

Monoclinic, $a = 8.94$, $b = 5.61$, $c = 10.23$, $\beta = 115.0°$, $Z = 2$; space group $P2_1/m$; point group $2/m$.

Habit

Crystals are prismatic, fibrous, or acicular, usually elongated parallel to b, with faces showing striations. Granular, massive, and fibrous aggregates are common.

Structure and Composition

The clinozoisite structure is similar to epidote's, containing both paired and unpaired $(SiO_4)^{4-}$ tetrahedra (see **epidote**).

Occurrence and Associations

Clinozoisite, like epidote, is a product of metamorphism of Ca-rich rocks. It forms instead of epidote in relatively Fe-poor rocks.

Related Minerals

A polymorph of clinozoisite, named simply *zoisite,* is orthorhombic. Clinozoisite is structurally and chemically related to all the other sorosilicates (see **lawsonite**).

Vesuvianite (Idocrase)

$Ca_{10}(Mg,Fe)_2Al_4Si_9O_{34}(OH)_4$

Origin of Name

From the Italian Mount Vesuvius locality where this mineral was found.

Hand Specimen Identification

Association, tetragonal or columnar habit, and brown color help identify vesuvianite. It is sometimes confused with epidote, tourmaline, or garnet.

Physical Properties

hardness	$6\frac{1}{2}$
specific gravity	3.4
cleavage/fracture	poor (001), (100), {110}/ subcochoidal
luster/transparency	vitreous, resinous/transparent
color	brown, also yellow, green, blue, or red
streak	white

Optical Properties

Vesuvianite is usually colorless but may be pleochroic in light green, brown, or yellow. High relief, low birefringence, and anomalous interference colors are typical. It may be confused with zoisite and clinozoisite but is uniaxial and lacks a good prismatic cleavage. Uniaxial (−), $\omega = 1.706$, $\epsilon = 1.701$, $\delta = 0.005$.

Crystallography

Tetragonal, $a = 15.66$, $c = 11.85$, $Z = 4$; space group $P4/n2/n2/c$; point group $4/m2/m2/m$.

Habit

Typical vesuvianite occurs as coarse, prismatic, brown tetragonal prisms (Figures 3.3 and 9.32). Faces may be striated and crystals may be terminated by pyramids. Crystals may combine to form striated columnar masses, fibrous sheaves, or granular aggregates.

Structure and Composition

Vesuvianite, usually considered a sorosilicate, contains both paired and unpaired $(SiO_4)^{4-}$ tetrahedra (see **lawsonite**). Its structural similarity to grossular leads to some misidentification. Composition is highly variable. Mn, Na, and K may replace Ca. Ti and Al may replace (Mg,Fe). Other elements, including B, Be, Cr, Cu, Li, Zn, and rare earths, may be present.

Occurrence and Associations

Vesuvianite is found primarily in contact aureoles associated with impure limestones or dolomites. Associated minerals include garnet, wollastonite, epidote, diopside, and carbonates. It is also found in altered mafic rocks, including serpentinites.

Varieties

Cyprine is a blue variety of vesuvianite.

Related Minerals

Vesuvianite is structurally and chemically similar to the garnet grossular; it is less similar to epidote and other sorosilicates.

II. NATIVE ELEMENTS

Metals

gold	Au
silver	Ag
platinum	Pt
copper	Cu

Semimetals

arsenic	As
bismuth	Bi
antimony	Sb

Nonmetals

diamond	C
graphite	C
sulfur	S

Gold, silver, platinum, and copper are the most common of the native metals. Iron, zinc, nickel, lead, and indium have occasionally been reported from meteorites or altered igneous rocks. All native metals have similar properties: metallic luster, high thermal and electrical conductivity, malleability, and opaqueness to visible light. Complex solid solutions are possible; many natural alloys have been given their own names. Kamacite and taenite, for example, are Fe-Ni alloys.

The native semimetals, all rare, are found in hydrothermal deposits but rarely have economic importance. The native nonmetals are diverse in occurrence and properties. Graphite is common as an accessory mineral in many metamorphic rocks, sulfur exists in massive beds or as encrustations associated with fumaroles, and diamond is primarily restricted to kimberlite pipes and mantle nodules.

For more general information about native elements, see Chapter 8.

Gold
Au

Origin of Name
The name of this mineral refers to its color.

Hand Specimen Identification
Gold is metallic and yellow. High specific gravity, sectile nature, and slightly different color and luster distinguish gold from the yellow sulfides pyrite and chalcopyrite. Plate 8.7 shows native gold.

Physical Properties
hardness	$2\frac{1}{2}$–3
specific gravity	15.6–19.3
cleavage/fracture	none/hackly
luster/transparency	metallic/opaque
color	gold-yellow
streak	gold-yellow

Crystallography
Cubic, $a = 4.0783$, $Z = 4$; space group $F4/m\bar{3}2/m$; point group $4/m\bar{3}2/m$.

Habit
Crystals are octahedral, rarely showing other forms. More typically, gold is arborescent, fills fractures, or is found as nuggets, grains, or wire scales.

Structure and Composition
Gold's face-centered cubic structure (Figure 8.9a) is the same as those of platinum and copper. Its composition is often close to pure Au, but substantial Ag may be present in solid solution. *Electrum* is the name given to intermediate Ag-Au solutions. Small amounts of other elements, such as Cu and Fe, may be present.

Occurrence and Associations
Gold is most often found in quartz veins associated with altered silicic igneous rocks. It is also concentrated in placer deposits. Associated minerals include quartz, pyrite, chalcopyrite, galena, stibnite, sphalerite, arsenopyrite, tourmaline, and molybdenite.

Varieties
Most natural gold contains up to 10% alloyed metals, thus giving rise to a number of slightly different colors and properties.

Related Minerals
Other gold minerals include calaverite, $AuTe_2$; sylvanite, $(Au,Ag)Te_2$; petzite, Ag_3AuTe_2; maldonite, Au_2Bi; and uytenbogaardtite, Ag_3AuS_3.

Silver
Ag

Origin of Name
From the Old English word for this metal, *seolfor*.

Hand Specimen Identification
Silver is distinguished by its habit, color, high specific gravity, and malleability (Figures 8.8 and 14.16). It is occasionally confused with the platinum group minerals.

Physical Properties
hardness	$2\frac{1}{2}$–3
specific gravity	10.1–10.5
cleavage/fracture	none/hackly
luster/transparency	metallic/opaque
color	silver-white
streak	silver-white

Crystallography
Cubic, $a = 4.0856$, $Z = 4$; space group $F4/m\bar{3}2/m$; point group $4/m\bar{3}2/m$.

Habit
Distorted cubes, octahedra, or dodecahedra are known, but silver is typically acicular. Flakes, plates, scales, and filiform or arborescent masses are common (Figures 8.8 and 14.16).

Structure and Composition
Silver has a face-centered cubic structure (Figure 8.9a) that is isostructural with copper. It may contain substantial amount of Au, Hg, Cu, As, Sb, Bi, Pt, or Fe in solid solution.

Occurrence and Associations
Silver is found with sulfides and arsenides in oxidized zones of ore deposits, or in hydrothermal deposits. The many associated minerals include, most significantly, species containing Co, Ni, and As.

Varieties
Amalgam is a solid solution of Ag and Hg. Electrum is a solid solution of Ag and Au.

▶FIGURE 14.16
Native silver from Batopilas, Mexico. This sample is about 10 cm across.

Related Minerals
Silver is isostructural with copper. Other Ag minerals include dyscrasite, Ag_3Sb; argentite, Ag_2S; proustite, Ag_3AsS_3; and pyrargyrite, Ag_3SbS_3.

Platinum   Pt

Origin of Name
From the Spanish *platina,* meaning "silver."

Hand Specimen Identification
Platinum is easily identified by its malleability, color, and high specific gravity.

Physical Properties

hardness	$4-4^1/_2$
specific gravity	21.47
cleavage/fracture	none/hackly
luster/transparency	metallic/opaque
color	gray-silver, steel-gray
streak	gray-silver, steel-gray

Crystallography
Cubic, $a = 3.9237$, $Z = 4$; space group $F4/m\overline{3}2/m$; point group $4/m\overline{3}2/m$.

Habit
Crystals, generally poorly formed, are rare. Masses, nuggets, or small grains are typical.

Structure and Composition
Platinum has a cubic closest packed structure (Figure 8.9a) similar to gold's. It forms alloys with other elements, notably Fe, Cu, Pd, Rh, and Ir.

Occurrence and Associations
Primary platinum is found with chromite, spinel, and olivine in ultramafic rocks. It is also found in some placer deposits.

Related Minerals
Platinum is isotypical with copper.

Copper   Cu

Origin of Name
From the Greek *kyprios,* referring to Cyprus, one of the earliest places where copper was mined.

Hand Specimen Identification
Color, hackly fracture, malleability, and specific gravity identify copper. Plate 7.7 and Figure 14.17 show native copper from Michigan.

Physical Properties

hardness	$2^1/_2-3$
specific gravity	8.7–8.9
cleavage/fracture	none/hackly
luster/transparency	metallic/opaque
color	copper-red
streak	copper-red

Crystallography
Cubic, $a = 3.6153$, $Z = 4$; space group $F4/m\overline{3}2/m$; point group $4/m\overline{3}2/m$.

Habit
Copper may form cubes, octahedra, or dodecahedra. Contact or penetration twins are common. Crystals are usually malformed, dendritic, arborescent, or in irregular plates, scales, or masses (Figure 14.17).

Structure and Composition
Copper has a cubic closest packed structure similar to gold and platinum (Figure 8.9a). It often contains solid solutions of Ag, Fe, As, or other elements.

Occurrence and Associations
Copper is found in the oxidized zones of many copper deposits, and as primary mineralization from hydrothermal fluids passing through mafic lavas. Copper is often deposited in voids or cracks. Associated

▶FIGURE 14.17
Native copper from Michigan's upper peninsula.

minerals include silver, sulfides, calcite, chlorite, zeolites, cuprite, malachite, and azurite.

Related Minerals
Gold, silver, platinum, and lead are isotypical with copper.

Diamond   C

Origin of Name
From the Greek *adamas,* meaning "invincible."

Hand Specimen Identification
Diamond is distinguished from minerals that resemble it by its hardness, octahedral cleavage, and luster (Figures 1.2 and 8.11b).

Physical Properties

hardness	10
specific gravity	3.5
cleavage/fracture	perfect octahedral {111}/conchoidal
luster/transparency	adamantine/transparent
color	typically colorless but rare, colored varieties may be valuable
streak	white

Optical Properties
Isotropic, $n = 2.419$.

Crystallography
Cubic, $a = 3.5668$, $Z = 8$; space group $F4/d\bar{3}2/m$; point group $4/m\bar{3}2/m$.

Habit
Crystals are usually octahedral and often distorted or twinned. More rarely, diamond forms cubes or dodecahedra. Curved faces are common.

Structure and Composition
Diamond is essentially pure carbon but may contain inclusions of other material.

Occurrence and Associations
Diamond is found in altered ultramafic rock of mantle origin or in placer deposits. Associated minerals include pyrope, olivine, kyanite, and zircon.

Related Minerals
Graphite is a polymorph of diamond.

Graphite   C

Origin of Name
The name comes from the Greek *graphein,* meaning "to write," in correlation to its use in pencils.

Hand Specimen Identification
Easily recognized by its greasy feel, softness, color, and foliated nature (Figures 3.2 and 8.11a).

Physical Properties

hardness	1–2
specific gravity	2.1–2.2
cleavage/fracture	perfect basal (001)/elastic, flexible
luster/transparency	submetallic/opaque
color	lead-gray, black
streak	black

Crystallography
Hexagonal (rhombohedral), $a = 2.46$, $c = 10.06$, $Z = 6$: space group $R\bar{3}2/m$; point group $\bar{3}2/m$.

Habit
Individual crystals are hexagonal tablets. Foliated and scaly masses are common; radiating or granular aggregates are less common.

Structure and Composition
The structure is composed of stacked planes of covalently bonded C atoms arranged in a hexagonal pattern. Graphite is essentially pure carbon.

Occurrence and Associations
Graphite is common in a wide variety of metamorphic rocks including schists, marbles, and gneisses. It is a rare mineral in some igneous rocks. Graphite is usually disseminated as fine flakes, but may form large books.

Related Minerals
Diamond is a polymorph of graphite.

Sulfur   S

Origin of Name
From the Middle English *sulphur,* meaning "brimstone."

Hand Specimen Identification
Sulfur can be easily identified by its yellow color, hardness, and density. It is occasionally confused with orpiment or sphalerite.

Physical Properties

hardness	$1\frac{1}{2}$–$2\frac{1}{2}$
specific gravity	2.1
cleavage/fracture	poor {101} and {110}/conchoidal
luster/transparency	resinous/transparent to translucent
color	bright yellow
streak	white

Optical Properties
Pale yellow in thin section; often pleochroic. Sulfur is characterized by extreme relief and birefringence. Biaxial (+), $\alpha = 1.958$, $\beta = 2.038$, $\gamma = 2.245$, $\delta = 0.29$, $2V = 69°$.

Crystallography
Orthorhombic, $a = 10.44$, $b = 12.84$, $c = 24.37$, $Z = 128$; space group $F2/d2/d2/d$; point group $2/m2/m2/m$.

Habit
Tabular crystals often display combinations of dipyramids and pinacoids. Typically, sulfur is massive, colloform, or stalactitic.

Structure and Composition
The sulfur structure consists of covalently bonded S_8 groups, stacked parallel to c, and weakly connected to each other. Sulfur is essentially pure S but may contain small amounts of Se in solid solution.

Occurrence and Associations
Sulfur is found as a deposit associated with volcanic fumaroles, in veins where it forms from sulfides, or in sediments where it forms by the reduction of sulfates by bacterial action. The most substantial occurrences are thick beds in sedimentary sequences. Associated minerals include celestite, gypsum, anhydrite, and carbonates.

Related Minerals
Sulfur has several different polymorphs.

III. SULFIDES

Tetrahedral Sulfide Group Minerals

sphalerite	ZnS
wurtzite	ZnS
chalcopyrite	$CuFeS_2$
bornite	Cu_5FeS_4
enargite	Cu_3AsS_4

In the tetrahedral sulfides, sulfur and arsenic are nearly closest packed and all metal atoms are in tetrahedral coordination. Complex solid solutions are possible, especially at high temperature. At low temperature, many sulfide minerals exhibit exsolution resulting from unmixing to form more stable phases.

For more general information about sulfides, see Chapter 8.

Sphalerite　　　　　　　　　　　　　ZnS

Origin of Name
This mineral's name comes from the Greek *sphaleros,* meaning "treacherous," alluding to problems identifying the mineral.

Hand Specimen Identification
Sphalerite is generally recognized by its luster, density, and perfect cleavage. Black varieties (Figure 2.11) may be confused with galena but yield a brown

streak. Sphalerite is also sometimes confused with siderite, sulfur, or enargite. Plates 2.7 and 7.4 show color photos of sphalerite.

Physical Properties

hardness	$3\frac{1}{2}$–4
specific gravity	4.0
cleavage/fracture	perfect dodecahedral {110}/conchoidal
luster/transparency	metallic, resinous/ transparent to opaque
color	brown-orange-red, colorless when pure, also brown to black and yellow
streak	white to brown or yellow

Optical Properties
Sphalerite is colorless to pale brown or yellow in thin section. It has extremely high relief and good cleavage. Isotropic, $n = 2.42$.

Crystallography
Cubic, $a = 5.43$, $Z = 4$; space group $F\bar{4}3m$; point group $\bar{4}3m$.

Habit
Crystals may be distorted or rounded. Crystals show combinations of tetrahedra, dodecahedra, and cubes. Polysynthetic twinning is common. Sphalerite commonly forms cleavable masses. It is typically brown and resinous.

Structure and Composition
Sphalerite is isostructural with diamond, with S arranged in a face-centered cubic pattern and Zn occupying tetrahedral sites within the cube (Figure 11.6). Many solid solutions are possible. Fe and, to a lesser extent, Mn and Cd are nearly always present.

Occurrence and Associations
Sphalerite is a common mineral, found in several different kinds of deposits. It is found with galena, chalcopyrite, pyrite, barite, fluorite, carbonates, and quartz in voids and fracture fillings of carbonate hosts. It is found in hydrothermal veins with pyrrhotite, pyrite, and magnetite, and is also found in contact metamorphic aureoles.

Related Minerals
Wurtzite is a high-temperature hexagonal polymorph of sphalerite. Greenockite, CdS, is isostructural with sphalerite at low temperature and with wurtzite at high temperature.

Chalcopyrite　　　　　　　　　　　$CuFeS_2$

Origin of Name
From the Greek *chalkos,* meaning "copper," and *pyrites,* meaning "to ignite."

Hand Specimen Identification
Luster, brass-yellow color, and greenish-black streak identify chalcopyrite. It may be confused with pyrite, but is softer, and with gold, but is more brittle. Plate 7.8 shows chalcopyrite.

Physical Properties
hardness · $3\frac{1}{2}$–4
specific gravity · 4.2
cleavage/fracture · poor {011}/uneven
luster/transparency · metallic/opaque
color · brass-yellow
streak · greenish black

Crystallography
Tetragonal, $a = 5.25$, $c = 10.32$, $Z = 4$; space group $I\bar{4}2d$; point group $\bar{4}2m$.

Habit
Crystals are usually pseudotetrahedral, displaying disphenoidal faces, sometimes in combination with prisms. Polysynthetic and penetration twins are common. Massive aggregates are common.

Structure and Composition
The structure is similar to that of sphalerite. Cu and Fe alternate in tetrahedral sites between S arranged in a face-centered cubic pattern. Small amounts of Ag, Au, Zn, and other elements are commonly present.

Occurrence and Associations
Chalcopyrite, the most important Cu ore mineral, is widespread and common. It is present in most sulfide deposits, but the most significant ores are formed by hydrothermal veins or by replacement. Common associated minerals include pyrite, sphalerite, bornite, galena, and chalcocite. Chalcopyrite is also found as magmatic segregations associated with pyrrhotite and pentlandite and in black shales.

Related Minerals
A number of other sulfides are isotypical with chalcopyrite, including stannite, Cu_2FeSnS_4; gallite, $CuGaS_2$; and roquesite, $CuInS_2$. Other similar minerals include talnakhite, $Cu_9(Fe,Ni)_8S_{16}$; mooihoekite, $Cu_9Fe_9S_{16}$; and haycockite, $Cu_4Fe_5S_8$.

Bornite · Cu_5FeS_4

Origin of Name
Named after I. von Born (1742–1791), a German mineralogist.

Hand Specimen Identification
Density, luster, color, and purplish blue (peacock) tarnish usually identify bornite. It may be confused with niccolite, pyrrhotite, chalcocite, and covellite.

Physical Properties
hardness · 3
specific gravity · 6
cleavage/fracture · poor {111}/conchoidal
luster/transparency · metallic/opaque
color · bronze
streak · grayish black

Crystallography
Tetragonal, $a = 10.94$, $c = 21.88$, $Z = 16$; space group $P\bar{4}2_1c$; point group $\bar{4}2m$.

Habit
Crystals are often pseudocubic, less commonly pseudododecahedral and pseudo-octahedral. Tetragonal crystals with distorted or curved faces are rarer. Massive aggregates are common.

Structure and Composition
The structure is complex. S is distributed in a modified face-centered cubic arrangement. Cu and Fe are in tetrahedral sites, each coordinated to four S. The structure contains many defects. At high temperatures bornite forms solid solutions with chalcopyrite, $CuFeS_2$. Consequently, Cu:Fe ratios are somewhat variable. If cooling is slow, exsolution occurs. Small amounts of Pb, Au, Ag, and other elements may also be present.

Occurrence and Associations
The most important bornite occurrences are in sulfide veins and as a secondary mineral in enriched zones of sulfide deposits. Typical associated minerals include chalcopyrite, chalcocite, covellite, pyrrhotite, pyrite, and quartz.

Related Minerals
A cubic polymorph of bornite exists above 228°C. Related minerals include chalcopyrite, $CuFeS_2$, and pentlandite, $(Ni,Fe)_9S_8$.

Enargite · Cu_3AsS_4

Origin of Name
From the Greek word *enarges,* meaning "distinct," referring to its cleavage.

Hand Specimen Identification
Enargite is identified by its density, softness, prismatic cleavage, and color. It may be confused with stibnite or sphalerite.

Physical Properties
hardness · 3
specific gravity · 4.5
cleavage/fracture · perfect prismatic {110}, good (100), good (010), poor (001)/uneven
luster/transparency · metallic/opaque

| color | bronze, grayish black, iron black |
| streak | dark gray, gray-black |

Crystallography
Orthorhombic, $a = 6.47$, $b = 7.44$, $c = 6.19$, $Z = 1$; space group $Pnm2$; point group $mm2$.

Habit
Crystals are tabular or columnar, and striated. Massive, columnar, or granular aggregates are common.

Structure and Composition
The closest packed structure is closely related to that of wurtzite, the hexagonal polymorph of sphalerite. Sb commonly substitutes for some As. Some Fe, Zn, and Ge replace Cu.

Occurrence and Associations
Enargite is found in vein or replacement sulfide deposits with other sulfides such as chalcocite, covellite, galena, bornite, sphalerite, and pyrite.

Related Minerals
Enargite is the most common of the sulfosalts, a group of minerals similar to sulfides, but have S, As, Sb, or Bi in chains or sheets. Other related sulfosalts are pyrargyrite, Ag_3SbS_3; tetrahedrite, $Cu_{12}Sb_4S_{13}$; and tennantite, $Cu_{12}As_4S_{13}$. Enargite has a rare very low temperature tetragonal polymorph, luzonite. Luzonite is isostructural with famatinite, Cu_3SbS_4, with which enargite forms a partial solid solution. Other related minerals are sulvanite, Cu_3VS_4, and germantite, Cu_3GeS_4.

Octahedral Sulfide Group Minerals

galena	PbS
pyrrhotite	$Fe_{1-x}S$
niccolite	NiAs

Galena and pyrrhotite are the two most important octahedral sulfides; niccolite, although not containing sulfur, is included in this group because of structural similarities. In the octahedral sulfide structure, S and As are closest packed and metal atoms occupy octahedral sites. Pyrrhotite is often slightly deficient in Fe, so its formula is written as $Fe_{1-x}S$, and the name *troilite* is given to end member FeS.

For more general information about sulfides, see Chapter 8.

Galena   PbS

Origin of Name
From the Latin *galene*, a name originally given to lead ore.

Hand Specimen Identification
Galena is recognized by its density, color, habit, good cleavage, and softness. Plate 7.5, Figures 3.5a, 8.5, and 14.18 show galena.

Physical Properties

hardness	$2\frac{1}{2}$
specific gravity	7.6
cleavage/fracture	perfect {100}/subconchoidal
luster/transparency	metallic/opaque
color	lead-gray
streak	lead-gray

Crystallography
Cubic, $a = 5.94$, $Z = 4$; space group $F4/m\bar{3}2/m$; point group $4/m\bar{3}2/m$.

Habit
Crystals are typically cubes or cubes modified by octahedra. Penetration and contact twins are common. Lamellar twins are less common. Aggregates are massive, fine granular, or plumose.

Structure and Composition
The structure of galena is similar to that of halite. Alternating Pb and S are arranged in a face-centered cubic pattern. Galena often contains small amounts of Fe, As, or Sb, and even smaller amounts of Zn, Cd, Bi, and Se. Other impurities exist in trace amounts.

Occurrence and Associations
Galena is common in many types of sulfide deposits. Associated minerals include sphalerite, chalcopyrite, pyrite, fluorite, barite, marcasite, cerussite, anglesite,

▶**FIGURE 14.18**
Galena, PbS, often forms cubic crystals and exhibits cubic cleavage.

calcite, dolomite, and quartz. Galena is often associated with silver minerals such as silver, acanthite, or pyrargyrite.

Related Minerals
Besides halite, NaCl, other minerals isostructural with galena are periclase, MgO; wustite, FeO; alabandite, MnS; altaite, PbTe; and clausthalite, PbSe.

Pyrrhotite $Fe_{1-x}S$

Origin of Name
From the Greek *pyrrhos,* meaning "flame colored."

Hand Specimen Identification
Pyrrhotite is recognized by its luster, color, and magnetism. It may be confused with pentlandite, bornite, and pyrite.

Physical Properties
hardness	4
specific gravity	4.6
cleavage/fracture	poor (001)/uneven
luster/transparency	metallic/opaque
color	brownish bronze
streak	gray to black

Crystallography
Hexagonal, $a = 5.69$, $c = 11.75$, $Z = 6$; space group $P\bar{6}2c$; point group $\bar{6}2m$.

Habit
Pyrrhotite may be massive or disseminated. Rare crystals are hexagonal plates or tabs, often twinned.

Structure and Composition
The complex structure is similar to that of niccolite, NiAs. Fe occupies sites between hexagonally closest packed S. The amount and distribution of Fe are complex functions of composition and crystallization history, so composition is variable. Most pyrrhotite has less Fe than S. Ni, Co, Mn, and Cu are often present in small amounts.

Occurrence and Associations
Pyrrhotite is typically found in mafic igneous rocks. Associated minerals include pyrite, pentlandite, galena, magnetite, and chalcopyrite. Other pyrrhotite occurrences are in pegmatites, contact aureoles, and vein deposits.

Related Minerals
Pyrrhotite has a hexagonal polymorph stable at high temperature. Troilite is end member FeS. Isotypical minerals include troilite, niccolite, NiAs; and breithauptite, NiSb.

Niccolite NiAs

Origin of Name
The name refers to this mineral's nickel content.

Hand Specimen Identification
Niccolite is easily recognized by its pale copper-red color and its alteration to green nickel bloom (annabergite).

Physical Properties
hardness	$5–5\frac{1}{2}$
specific gravity	4.6
cleavage/fracture	poor (001)/uneven
luster/transparency	metallic/opaque
color	copper-red
streak	brownish black

Crystallography
Hexagonal, $a = 3.58$, $c = 5.11$, $Z = 2$; space group $P6_3/m2/m2/c$; point group $6/m2/m2/m$.

Habit
Rare crystals are tabular with pyramidal faces and sometimes cyclic twins. Niccolite is usually massive and sometimes colloform or columnar.

Structure and Composition
The structure involves hexagonal closest packed As with Ni between. Sb usually replaces some of the As; Fe, Co, and S are also present in small amounts.

Occurrence and Associations
Niccolite is found in veins with Co and Ag minerals and in sulfide deposits hosted by mafic igneous rocks. Associated minerals include pyrrhotite, chalcopyrite, skutterudite, silver, and a variety of other sulfosalts.

Related Minerals
Breithauptite, NiSb; freboldite, CoSe; and kotulskite, Pd(Te,Bi) are isostructural with niccolite. Other related minerals include millerite, NiS; pentlandite, $(Ni,Fe)_9S_8$; and langisite, (Co,Ni)As. Annabergite, $Ni_3As_2O_8 \cdot 8H_2O$, also called *nickel bloom,* is a common green alteration product of niccolite.

Other Sulfide Minerals

The eight sulfides just discussed and several other less important ones have relatively simple structures based on closest packing and metal ions occupying either tetrahedral or octahedral sites, but not both. Pentlandite, the most important ore mineral of nickel, is the only common sulfide containing metal atoms in both coordinations. Many other sulfides have more complex structures. Metal ions may be in 5-fold or other unusual coordinations, they may occupy several different sites in the structures, and the sites may be highly polarized or distorted.

For more general information about sulfides, see Chapter 8.

Pentlandite   (Ni,Fe)$_9$S$_8$

Origin of Name
Named after J. B. Pentland, the geologist working in Sudbury, Ontario, Canada, who first described the mineral.

Hand Specimen Identification
Luster, color, and association help identify pentlandite. It resembles pyrrhotite in appearance but is not magnetic. Plate 8.1 shows pentlandite.

Physical Properties

hardness	$3^1/_2$–4
specific gravity	5.0
cleavage/fracture	perfect {100}, good octahedral {111}/uneven
luster/transparency	metallic/opaque
color	light bronze to yellow-bronze
streak	light bronze to brown

Crystallography
Cubic, $a = 10.05$, $Z = 4$; space group $F4/m\overline{3}2/m$; point group $4/m\overline{3}2/m$.

Habit
Crystals are very rare. Pentlandite is usually massive or in granular aggregates. Sometimes {111} parting develops.

Structure and Composition
Pentlandite has a complicated face-centered cubic structure. The basic structure consists of (Ni,Fe)S$_6$ octahedra sharing corners. Additional Ni and Fe occupy distorted tetrahedral sites between the octahedra. Co commonly substitutes for (Ni,Fe). Mn and Cu are other common impurities.

Occurrence and Associations
Pentlandite, the most important nickel ore mineral, is found in late-stage sulfide deposits with other nickel minerals (millerite, niccolite), pyrrhotite, and chalcopyrite. Pentlandite often occurs as exsolved blebs and lamellae within pyrrhotite.

Related Minerals
Pentlandite forms solid solutions with cobalt pentlandite, Co$_9$S$_8$. It is isostructural with a number of minerals, including argentopentlandite, Ag(Fe,Ni)$_8$S$_8$. Other related minerals are bornite, Cu$_5$FeS, and niccolite, NiAs.

Molybdenite   MoS$_2$

Origin of Name
From the Greek *molybdos,* meaning "lead," which refers to a misidentification by early mineralogists.

Hand Specimen Identification
Luster, color, flexibility, basal cleavage, and hardness identify molybdenite. It superficially resembles graphite. Plate 8.2 shows molybdenite in quartz.

Physical Properties

hardness	1–$1^1/_2$
specific gravity	4.7
cleavage/fracture	perfect basal (001)/flexible
luster/transparency	metallic/opaque
color	silver, lead-gray
streak	green-gray

Crystallography
Hexagonal, $a = 3.16$, $c = 12.32$, $Z = 2$; space group $P6_3/m2/m2/m$; point group $6/m2/m2/m$.

Habit
Crystals form hexagonal plates or stubby prisms. Foliated or scaly aggregates are flexible, but not elastic.

Structure and Composition
The molybdenite structure involves two sheets of S, arranged in a hexagonal pattern, sandwiching a sheet of Mo atoms. Each Mo atom is bonded to three S in each of the two sheets. The three-layer units are stacked up to produce the entire structure. Molybdenite is usually quite close to end member but may contain traces of Au, Ag, Re, and Se.

Occurrence and Associations
Molybdenite occurs as an accessory mineral in some granitic rocks, including pegmatites. It also is found in porphyry copper deposits; in vein deposits with scheelite, cassiterite, wolframite, and fluorite; and in some contact aureoles.

Millerite   NiS

Origin of Name
Named after W. H. Miller, who was the first to study the crystals.

Hand Specimen Identification
Millerite is easily recognized by its luster, color, radiating acicular crystals, and rhombohedral cleavage.

Physical Properties

hardness	3–$3^1/_2$
specific gravity	5.5
cleavage/fracture	perfect {101} and {012}/uneven
luster/transparency	metallic/opaque
color	brass yellow
streak	greenish black, green-gray

Crystallography
Hexagonal (rhombohedral), $a = 9.62$, $c = 3.15$, $Z = 9$; space group $R3m$; point group 3m.

Habit
Crystals are typically acicular or filiform. Millerite may form radiating splays or velvety crusts.

Structure and Composition
The structure of millerite is a complex derivative of niccolite. Both Ni and S are in 5-fold coordination. Co, Fe, and As are minor impurities.

Occurrence and Associations
Millerite is a low-temperature mineral that forms as a replacement for other nickel minerals or in cavities. It is associated with calcite, fluorite, dolomite, hematite, siderite, pyrrhotite, and chalcopyrite.

Related Minerals
Millerite has structural similarity with niccolite, NiAs, and with pyrrhotite, $Fe_{1-x}S$. A high-temperature polymorph exists above 379 °C.

Cinnabar   HgS

Origin of Name
From the Persian *zinjifrah,* perhaps referring to red resin.

Hand Specimen Identification
High density, red color, and streak identify cinnabar. It may be confused with hematite, cuprite, or realgar.

Physical Properties

hardness	$2\frac{1}{2}$
specific gravity	8.1
cleavage/fracture	perfect prismatic {100}/ subconchoidal
luster/transparency	adamantine/transparent to translucent
color	bright red to brownish red
streak	scarlet

Optical Properties
Uniaxial $(+)$, $\omega = 2.90$, $\epsilon = 3.25$, $\delta = 0.35$.

Crystallography
Hexagonal (trigonal), $a = 4.15$, $c = 9.50$, $Z = 3$; space group $P3_121$; point group 32.

Habit
Rare crystals are rhombohedral, thick tabs or prisms; less commonly acicular. Most occurrences are granular or earthy masses; often they are crusty, sometimes disseminated.

Structure and Composition
Cinnabar's structure consists of Hg-S-Hg chains spiraling parallel to the *c*-axis. It is usually close to HgS in composition; only traces of other elements are present.

Occurrence and Associations
Cinnabar is the most significant Hg ore mineral. It is found as masses in volcanic or sedimentary rocks, in veins, or as disseminated grains. Associated minerals include native mercury, realgar, stibnite, pyrite, marcasite, calcite, quartz, and opal.

Related Minerals
Metacinnabar (cubic) and hypercinnabar (hexagonal) are polymorphs of cinnabar. Other related minerals include coloradoite, HgTe, and tiemannite, HgSe.

Covellite   CuS

Origin of Name
Named after N. I. Covelli (1790–1829), who discovered Vesuvian covellite.

Hand Specimen Identification
Density, luster, indigo-blue color, and association identify covellite. Plate 7.3 shows covellite forming on pyrite.

Physical Properties

hardness	$1\frac{1}{2}$–2
specific gravity	4.6
cleavage/fracture	perfect basal (001)/ conchoidal
luster/transparency	metallic/opaque
color	indigo-blue; purplish tarnish
streak	dark gray to black

Crystallography
Hexagonal, $a = 3.80$, $c = 16.36$, $Z = 6$; space group $P6_3/m2/m2/c$; point group $6/m2/m2/m$.

Habit
Rare hexagonal crystals are tabular or platy; covellite is usually in massive or foliated aggregates or in overgrowths and coatings on other copper minerals.

Structure and Composition
Covalent sulfur bonds link layers of CuS_4 tetrahedra to layers of CuS_3 triads. Weak bonds between the layers result in excellent planar cleavage. Fe often replaces some Cu; Se replaces some S.

Occurrence and Associations
Primarily a secondary (supergene) mineral, covellite occurs with other Cu sulfides in veins or disseminated deposits. Associated minerals include bornite, chalcopyrite, chalcocite, and enargite.

Related Minerals
Covellite is similar in some ways to klockmannite, CuSe, with which it forms solid solutions.

Chalcocite   Cu₂S

Origin of Name
From the Greek *chalkos,* meaning "copper."

Hand Specimen Identification
Gray, often sooty color with blue tarnish, luster, hardness, sectile nature, and density identify chalcocite.

Physical Properties

hardness	$2\frac{1}{2}$–3
specific gravity	5.8
cleavage/fracture	poor prismatic {110}/ conchoidal
luster/transparency	metallic/opaque
color	blue-white, shining lead-gray, dull sooty gray
streak	grayish black

Crystallography
Monoclinic, a = 15.25, b = 11.88, c = 13.49, β = 116.35°, Z = 48; space group $P2_1/c$; point group $2/m$.

Habit
Chalcocite is usually fine grained and massive with conchoidal fracture. Squat prisms or tabular crystals—sometimes with a hexagonal outline, sometimes displaying striations—are rare.

Structure and Composition
Hexagonal close packed S atoms characterize the structure. Two-thirds of the Cu atoms occupy trigonal sites in the planes of S atoms; the other third is in octahedral coordination between planes. Fe and Ag are common replacements for Cu; Se may replace some S.

Occurrence and Associations
Chalcocite is a common primary or secondary copper ore mineral. It occurs both in veins and in altered zones. Associated primary minerals include bornite, chalcopyrite, enargite, galena, tetrahedrite, cuprite, and pyrite. Covellite, malachite, or azurite are common alteration products.

Related Minerals
Normal chalcocite is monoclinic, but a hexagonal polymorph exists at elevated temperatures. Solid solution with berzelianite, Cu_2Se, is common. Similar minerals include stromeyerite, $AgCuS$, and digenite, $Cu_{2-x}S$.

Argentite (Acanthite)   Ag₂S

Origin of Name
From the Latin *argentum,* which means "silver."

Hand Specimen Identification
Density, luster, color, and sectility identify argentite. It may be confused with chalcocite and tetrahedrite.

Physical Properties

hardness	2–$2\frac{1}{2}$
specific gravity	7.1
cleavage/fracture	poor cubic {100}/ subconchoidal
luster/transparency	metallic/opaque
color	lead-gray to black
streak	black or shiny black

Crystallography
Cubic, a = 4.89, Z = 2; space group $I4/m\overline{3}2/m$; point group $4/m\overline{3}2/m$.

Habit
Wiry, branching, columnar, and massive habit are common. Crystals are cubes, octahedra, dodecahedra, or combinations. Penetration twins are common.

Structure and Composition
Sulfur atoms are arranged in a distorted body-centered arrangement. Ag atoms occupy 2-fold and 3-fold sites between S.

Occurrence and Associations
Argentite, an important Ag ore mineral, is found in veins associated with other silver minerals, galena, sphalerite, tetrahedrite, and Co-Ni sulfides.

Related Minerals
Argentite has both high-temperature and low-temperature polymorphs. Argentite is the proper name of the high-temperature cubic polymorph, and acanthite is the name of the low-temperature monoclinic polymorph. However, because low-temperature Ag_2S is usually twinned, appearing pseudocubic, it is also commonly referred to as argentite. Other related minerals include hessite, Ag_2Te; petzite, Ag_3AuTe; fischesserite, Ag_3AuSe_2; naumannite, Ag_2Se; eucairite, $CuAgSe$; jalpaite, Ag_3CuS_2; and aguilarite, Ag_4SeS.

Pyrite   FeS₂

Origin of Name
From the Greek *pyr,* meaning "fire," because it sparks when struck by steel.

Hand Specimen Identification
Density, metallic luster, brass-yellow color, and hardness identify pyrite. It is sometimes confused with chalcopyrite and marcasite, both of which have slightly different colors. Plates 1.3 and 2.7 show color photos of pyrite. Figures 2.11, 3.1, 8.10, and 8.12 contain additional photos.

Physical Properties

hardness	6–$6\frac{1}{2}$
specific gravity	5.1

cleavage/fracture	poor {100}/subconchoidal
luster/transparency	metallic/opaque
color	brass-yellow
streak	greenish black to green-gray

Crystallography
Cubic, $a = 5.42$, $Z = 4$; space group $P2_1/a\overline{3}$; point group $2/m\overline{3}$.

Habit
Crystals are typically cubes, pyritohedra, or octahedra. Striated faces, combinations of forms, and penetration twinning ("iron cross twins") are common.

Structure and Composition
The structure is closely related to that of NaCl; Fe and S_2 alternate in a three-dimensional cubic array. Pyrite commonly contains some Ni and Co as replacements for Fe. Cu, V, Mo, Cr, W, Au, or Tl may also be present.

Occurrence and Associations
Pyrite is the most common and widespread sulfide and is often called "fool's gold." It is an accessory mineral in many igneous, sedimentary, and metamorphic rocks. It is common in all sulfide deposits, associated with a wide variety of ore minerals. It also replaces organic material in coal, wood, or shells.

Related Minerals
Marcasite is an orthorhombic polymorph of pyrite. Pyrite forms solid solutions with vaesite, NiS_2, and cattierite, CoS_2. Cobaltite, CoAsS, and hauerite, MnS_2, are isostructural with pyrite. Many other minerals are isotypical with pyrite. Other similar minerals include arsenoferrite, $FeAs_2$; pyrrhotite and mackinawite, $Fe_{1-x}S$; greigite, Fe_3S_4; and smythite, $(Fe,Ni)_9S_{11}$.

Cobaltite (Co,Fe)AsS

Origin of Name
From the German *kobold*, meaning "goblin," because early miners found it difficult to mine.

Hand Specimen Identification
Density, luster, color, cleavage, and habit identify cobaltite. It is sometimes confused with skutterudite.

Physical Properties
hardness	$5\frac{1}{2}$
specific gravity	6.3
cleavage/fracture	good cubic {100}/uneven
luster/transparency	metallic/opaque
color	tin-white or silver-white
streak	gray-black

Crystallography
Orthorhombic, $a = 5.58$, $b = 5.58$, $c = 5.58$, $Z = 4$; space group $Pa2c$; point group $mm2$.

Habit
Cobaltite is usually massive. Aggregates may be granular or compact. Rare individual crystals are pseudocubic, similar in form to pyrite.

Structure and Composition
Cobaltite is isostructural with pyrite. As and S occupy the S_2 sites in pyrite. Fe and Ni commonly substitute for Co; Sb substitutes for As.

Occurrence and Associations
Cobaltite is commonly found with other cobalt and nickel sulfides, arsenides, and related minerals. Pyrrhotite, chalcopyrite, galena, and magnetite are also associated minerals. It may be veined or disseminated. Cobaltite is also found in a few rare metamorphic rocks.

Related Minerals
Cobaltite forms solid solutions with gersdorffite, NiAsS; ullmanite, NiSbS; and willyamite, (Co,Ni)SbS. Other similar minerals include hollingworthite, (Rh,Pt,Pd)AsS; irarsite, (Ir,Ru,Rh,Pt)AsS; platarsite, (Pt,Rh,Ru)AsS; and tolovkite, IrSbS.

Marcasite FeS$_2$

Origin of Name
From Markashitu, an ancient province of Persia.

Hand Specimen Identification
Density, luster, color, and orthorhombic habit identify marcasite. Cockscomb groups are diagnostic. Marcasite may be confused with its cubic polymorph, pyrite, if crystals are not distinct.

Physical Properties
hardness	$6-6\frac{1}{2}$
specific gravity	4.9
cleavage/fracture	poor {101}/uneven
luster/transparency	metallic/opaque
color	white-green to pale bronze yellow, often slightly tarnished
streak	grayish black

Crystallography
Orthorhombic, $a = 4.44$, $b = 5.41$, $c = 3.38$, $Z = 2$; space group $P2_1/n2_1/n2/m$; point group $2/m2/m2/m$.

Habit
Crystals are typically tabular, often with curved faces, showing orthorhombic symmetry. They combine to form needlelike groups, sometimes radiating,

colloform, globular, reniform, or stalactitic. Twinning often produces cockscomb aggregates.

Structure and Composition
Zigzagging FeS_2 chains run parallel to the *c*-axis; FeS_6 octahedra share corners and edges. Marcasite is nearly pure FeS_2; traces of Cu may be present.

Occurrence and Associations
Marcasite is a low-temperature mineral found in sulfide veins with lead and zinc minerals or as a replacement mineral in limestones or shale. Common associates are galena, pyrite, chalcopyrite, calcite, and dolomite.

Related Minerals
Pyrite is a more stable polymorph of marcasite. Isostructural minerals include hastite, $CoSe_2$; ferroselite, $FeSe_2$; frohbergite, $FeTe_2$; kullerudite, $NiSe_2$; and mattagamite, $CoTe_2$.

Arsenopyrite   FeAsS

Origin of Name
Named for its composition.

Hand Specimen Identification
Density, luster, color, and crystal form help identify arsenopyrite. It may be confused with marcasite, pyrite, or skutterudite, but color distinguishes it from the first two and form from the latter. Arsenopyrite sometimes smells like garlic when struck. Plate 7.2 shows a color photo of arsenopyrite.

Physical Properties
hardness $5^1/_2$–6
specific gravity   6.1
cleavage/fracture   good (101)/uneven
luster/transparency   metallic/opaque
color   silver-white
streak   black

Crystallography
Monoclinic, $a = 5.76$, $b = 5.69$, $c = 5.785$, $\beta = 112.2°$, $Z = 4$; space group $P2_1/c$; point group $2/m$.

Habit
Prismatic, striated crystals are typical. Penetration, contact, and cyclic twins are common. It may be disseminated, massive, or granular.

Structure and Composition
The structure is similar to that of marcasite with half the S replaced by As. $FeAs_3S_3$ octahedra share vertices and edges. As:S ratios vary slightly, but arsenopyrite is always close to FeAsS in composition. Minor Co and Bi may replace Fe and other elements may be present in trace amounts.

Occurrence and Associations
Arsenopyrite, the most abundant and widespread arsenic mineral, is found in Fe, Cu, Sn, Co, Ni, Ag, Au, and Pb ores. It occurs in veins, pegmatites, contact aureoles, or as disseminations in low- to medium-grade metamorphic rocks. Common associated minerals include chalcopyrite, pyrite, sphalerite, cassiterite, and gold and silver minerals.

Related Minerals
Arsenopyrite forms solid solutions with glaucodot, $(Co,Fe)AsS$. It is isotypical with marcasite, FeS_2, and with gudmundite, FeSbS. Other related minerals include lautite, CuAsS; osarsite, $(Os,Ru)AsS$; and ruarsite, RuAsS.

Skutterudite   (Co,Ni)As₃

Origin of Name
From the type locality at Skutterude, Norway.

Hand Specimen Identification
The skutterudite series contains a number of related minerals; they have similar properties and are difficult to tell apart. Density, luster, and color help with identification, but chemical or X-ray analysis may be needed. Skutterudite is sometimes confused with arsenopyrite or cobaltite.

Physical Properties
hardness $5^1/_2$–6
specific gravity   6.1–6.8
cleavage/fracture   good (101)/uneven
luster/transparency   metallic/opaque
color   tin-white to silver-gray
streak   black

Crystallography
Cubic, $a = 8.20$, $Z = 8$; space group $I2/m\overline{3}$; point group $2/m\overline{3}$.

Habit
Cubes and octahedra are common forms. Skutterudite usually forms dense to granular aggregates.

Structure and Composition
(Co,Ni) in octahedral coordination are linked to square AsS_4 groups. Fe and Bi may substitute for (Co,Ni). Some As may be missing or replaced by S or Sb.

Occurrence and Associations
Skutterudite is a vein mineral associated with other Co and Ni minerals such as cobaltite or niccolite. Other associated minerals include arsenopyrite, native silver, silver sulfosalts, native bismuth, calcite, siderite, barite, and quartz.

Related Minerals

The three important members of the skutterudite series are skutterudite, $(Co,Ni)As_3$; smaltite, $(Co,Ni)As_{3-x}$; and chloanthite, $(Ni,Co)As_{3-x}$. Linnaeite, Co_3S_4, is closely related.

Stibnite   Sb₂S₃

Origin of Name
From the Greek *stibi,* a name originally used by Pliny, a first-century Greek encyclopedist and scientist.

Hand Specimen Identification
Stibnite is characterized by its perfect cleavage in one direction, soft black streak, bladed habit, and lead-gray color.

Physical Properties
hardness	2
specific gravity	4.6
cleavage/fracture	one perfect (010)/subconchoidal
luster/transparency	metallic/opaque
color	lead-gray
streak	lead-gray to black

Crystallography
Orthorhombic, $a = 11.12$, $b = 11.30$, $c = 3.84$, $Z = 4$; space group $P2_1/b2_1/n2_1/m$; point group $2/m2/m2/m$.

Habit
Prismatic striated crystals, often slender, long, and curved, are common. Faces show striations; terminating faces may be steep. Typically, stibnite is found in aggregates containing granular to coarse columns or needles.

Structure and Composition
The structure has zigzagging chains of Sb_2S_3 parallel to c. Small amounts of other metals may replace Sb. Fe, Pb, and Cu are the most common impurities. Ag, Au, Zn, and Co may also be present in trace amounts.

Occurrence and Associations
Stibnite is found in hydrothermal veins, in replacement deposits, and more rarely, in hot spring deposits. Typical associates include orpiment, realgar, cinnabar, galena, sphalerite, pyrite, barite, and sometimes gold.

Related Minerals
Bismuthinite, Bi_2S_3, and guanajuatite, Bi_2Se_3, are isostructural with stibnite.

Tetrahedrite   Cu₁₂Sb₄S₁₃

Origin of Name
Named for its typical crystal form.

Hand Specimen Identification
Tetrahedral crystals, lack of cleavage, and color help identify members of the tetrahedrite series, $Cu_{12}(Sb,As)_4S_{13}$. They cannot be told apart without chemical analysis or X-ray study.

Physical Properties
hardness	$3–4\frac{1}{2}$
specific gravity	4.5–5.1
cleavage/fracture	none/subconchoidal
luster/transparency	metallic/opaque
color	silver or grayish black to black
streak	brown to black

Crystallography
Cubic, $a = 10.34$, $Z = 2$; space group $I\bar{4}3m$; point group $\bar{4}3m$.

Habit
Crystals are typically tetrahedra, sometimes with modifying faces. Penetration twins are common. Crystal aggregates may be massive or granular.

Structure and Composition
The structure is similar to that of sodalite. CuS_4 tetrahedra share corners. Sb or As occupy openings between the tetrahedra. Fe, Zn, Pb, Ag, and Hg may replace Cu. Complete solid solution exists between end members tetrahedrite, $Cu_{12}Sb_4S_{13}$, and tennantite, $Cu_{12}As_4S_{13}$.

Occurrence and Associations
Tetrahedrite, one of the most common sulfosalts, is found in veins and in replacement deposits. Associated minerals include chalcopyrite, sphalerite, galena, pyrite, argentite, and many other minerals.

Varieties
Freibergite is a Ag-rich variety. Schwatzite is an Hg-rich variety.

Related Minerals
Tennantite, $Cu_{12}As_4S_{13}$, is isostructural with tetrahedrite. Other related minerals include germanite, $Cu_3(Ge,Fe)S_4$; colusite, $Cu_3(As,Sn,Fe)S_4$; and sulvanite, Cu_3VS_4.

Pyrargyrite (Ruby Silver)   Ag₃SbS₃

Origin of Name
From the Greek words meaning "fire" and "silver," relating to its color and composition.

Hand Specimen Identification
Color, translucence, and density identify pyrargyrite. It may be confused with proustite but generally has a ruby-red color, while proustite is more vermillion. It is occasionally mistaken for cuprite.

Physical Properties

hardness	2
specific gravity	5.85
cleavage/fracture	good {101}/subconchoidal
luster/transparency	adamantine/translucent
color	ruby-red
streak	red to purple

Optical Properties

Uniaxial $(-)$, $\omega = 3.08$, $\epsilon = 2.88$, $\delta = 0.20$.

Crystallography

Hexagonal (trigonal), $a = 11.06$, $c = 8.73$, $Z = 6$; space group $R3c$; point group $3m$.

Habit

Trigonal striated prisms are typical. Simple, repeated, and cyclic twins are common. Aggregates may be massive or granular.

Structure and Composition

The structure is composed of S_3 bonded to Sb to form pyramids. Ag occupies large sites between the pyramids. Cu may replace Ag; As and Bi may replace Sb.

Occurrence and Associations

Pyrargyrite is found in low-temperature veins and in replacement deposits. Associated minerals include native silver, argentite, tetrahedrite, galena, sphalerite, carbonates, and quartz.

Related Minerals

Pyrargyrite has two polymorphs: pyrostilpnite and xanthoconite. Proustite, Ag_3AsS_3, is isostructural with pyrargyrite, but solid solutions are limited. Both pyrargyrite and proustite are called *ruby silvers*.

Orpiment   As₂S₃

Origin of Name

From the Latin *aurum* and *pigmentum,* meaning "golden paint," referring to its color.

Hand Specimen Identification

Color, foliated structure, and hardness identify orpiment. When visible, perfect cleavage distinguishes it from sulfur. Plate 7.6 shows a color photo of orpiment.

Physical Properties

hardness	$1\frac{1}{2}$–2
specific gravity	3.49
cleavage/fracture	one perfect (010)/even or sectile
luster/transparency	resinous, also pearly on cleavage face/translucent
color	lemon-yellow to orange
streak	pale yellow to yellow

Optical Properties

Biaxial $(+)$, $\alpha = 2.40$, $\beta = 2.81$, $\delta = 3.02$, $2V = 76°$.

Crystallography

Monoclinic, $a = 11.49$, $b = 9.59$, $c = 4.25$, $\beta = 90.45°$, $Z = 4$; space group $P2_1/n$; point group $2/m$.

Habit

Rare crystals are small, tabular, or prismatic, often poorly formed. Columnar or foliated aggregates are common.

Structure and Composition

AsS_3 pyramids share edges, producing 6-member rings. Crumpled layers of rings are stacked on top of each other.

Occurrence and Associations

Orpiment, a rare mineral found in hot spring deposits and some gold deposits, is commonly associated with realgar. Other associated minerals include stibnite, native arsenic, calcite, barite, and gypsum.

Related Minerals

Getchellite, $AsSbS_3$, is similar in structure to orpiment. Realgar, AsS, is closely related in composition.

Realgar   AsS

Origin of Name

From the Arabic *rahj al-ghar,* meaning "powder of the mine."

Hand Specimen Identification

Association with orpiment, resinous luster, orange-red streak, and red color identify realgar. It is sometimes confused with cinnabar, cuprite, or hematite. Plate 7.6 shows orpiment and realgar.

Physical Properties

hardness	$1\frac{1}{2}$–2
specific gravity	3.56
cleavage/fracture	good (010)/conchoidal
luster/transparency	resinous/transparent to translucent
color	red to orange
streak	red to orange

Optical Properties

Biaxial $(-)$, $\alpha = 2.538$, $\beta = 2.864$, $\gamma = 2.704$, $\delta = 0.166$, $2V = 41°$.

Crystallography

Monoclinic, $a = 9.29$, $b = 13.53$, $c = 6.57$, $\beta = 106.55°$, $Z = 4$; space group $P2_1/n$; point group $2/m$.

Habit

Realgar may form short prismatic crystals having vertical striations. It is often massive granular or forms as earthy crusts.

Structure and Composition
Uneven rings of As_4S_4 form layers in the structure. The As atoms, lying alternately above and below the plane of the S atoms, are covalently bonded to As in adjacent layers.

Occurrence and Associations
Realgar is associated with lead, gold, and silver ores. Associated minerals include orpiment, other arsenic minerals, and stibnite.

Related Minerals
Orpiment, As_2S_3, is closely related in composition.

IV. HALIDES

halite	NaCl
sylvite	KCl
chlorargerite	AgCl
atacamite	$Cu_2Cl(OH)_3$
fluorite	CaF_2
cryolite	Na_3AlF_6

Fluorite and other halogens generally form nearly pure ionic bonds of moderate strength. Consequently, they bond with alkali and alkali earth elements to make chlorides, fluorides, bromides, and other salts referred to collectively as *halides*. Due to the generally large cation size and the nature of the bonding, halide structures tend to be simple with high symmetry. At high temperatures, some halides exhibit mutual solubility, but under normal Earth surface conditions most are usually close to end member composition. Halite and fluorite are the most common halide minerals, but others may be locally abundant.

For more general information about halides, see Chapter 6.

Halite   NaCl

Origin of Name
From the Greek *halos,* meaning "salt."

Hand Specimen Identification
Halite is characterized by its color, transparency, salty taste, and cubic cleavage. It may be confused with sylvite, but it is distinguished by its less bitter taste. See Figures 3.6e, 6.11, 6.15, 9.1b, 10.36a, and 14.19 for photographs and drawings.

Physical Properties
hardness	$2\frac{1}{2}$
specific gravity	2.16
cleavage/fracture	perfect cubic {100}/ conchoidal

luster/transparency	vitreous/transparent to translucent
color	colorless, white, or if impure may contain shades of red, blue, purple, or yellow
streak	white

Optical Properties
Halite has low relief, perfect cubic cleavage, and is colorless in thin section. Isotropic, $n = 1.5446$.

Crystallography
Cubic, $a = 5.6404$, $Z = 4$; space group $F4/m\bar{3}2/m$; point group $4/m\bar{3}2/m$.

Habit
Cubic crystals, often massive or granular, characterize halite. Cubic cleavage is pronounced (Figure 14.19).

Structure and Composition
Closest packed Na^+ and Cl^- alternate in a face-centered cubic arrangement. Both ions are in perfect octahedral coordination (Figures 6.12 and 13.1a).

Occurrence and Associations
Halite, a rock-forming mineral, occurs in salt flats, in sedimentary beds, in salt domes, and as deposits from volcanic gasses. It is the most common evaporite mineral. Associated minerals include many other salts, gypsum, calcite, sylvite, anhydrite, sulfur, and clay.

Related Minerals
Galena, PbS; alabandite, MnS; periclase, MgO; sylvite, KCl; carobbiite, KF; and chlorargyrite, AgCl are all isostructural with halite.

▶**FIGURE 14.19**
Halite, NaCl, showing cubic cleavage.

Sylvite   KCl

Origin of Name
From the old Latin name for the mineral.

Hand Specimen Identification
Sylvite is characterized by its color, transparency, taste, and cubic cleavage. It is distinguished from halite by its more bitter taste.

Physical Properties
hardness	2
specific gravity	1.99
cleavage/fracture	perfect cubic {100}/uneven
luster/transparency	vitreous/transparent to translucent
color	colorless, white, or shades of yellow, blue, or red caused by impurities
streak	white

Optical Properties
Sylvite has low relief, perfect cubic cleavage, and is colorless in thin section. Isotropic, $n = 1.490$.

Crystallography
Cubic, $a = 6.29$, $Z = 4$; space group $F4/m\overline{3}2/m$; point group $4/m\overline{3}2/m$.

Habit
Crystals form cubes, often with modifying octahedra. Massive or granular aggregates are typical.

Structure and Composition
Sylvite is isostructural with halite. K^+ and Cl^- are arranged in a face centered cubic manner (Figure 2.3). Only minor solid solution exists between the two salts.

Occurrence and Associations
Sylvite is rarer than halite but has the same origin, associates, and occurrences (see **halite**).

Related Minerals
Sylvite and halite are isostructural with a number of other minerals (see **halite**). Associated potassium salts include carnallite, $KMgCl_3 \cdot 6H_2O$; kainite, $KMg(Cl,SO_4) \cdot nH_2O$; and polyhalite, $K_2Ca_2Mg(SO_4)_4 \cdot 2H_2O$.

Chlorargerite   AgCl

Origin of Name
From the Greek *chlor* and *argyros,* meaning "green" and "silver."

Hand Specimen Identification
Chlorargerite is characterized by its waxlike appearance and heft.

Physical Properties
hardness	$2\frac{1}{2}$
specific gravity	5.55
cleavage/fracture	poor {100}/subconchoidal
luster/transparency	resinous/transparent to translucent
color	colorless, pale green, pearl-gray
streak	white

Optical Properties
Isotropic, $n = 2.071$.

Crystallography
Cubic, $a = 5.55$, $Z = 4$; space group $F4/m\overline{3}2/m$; point group $4/m\overline{3}2/m$.

Habit
Rare crystals are cubic. Chlorargyrite is usually massive or columnar.

Structure and Composition
Chlorargyrite is isostructural with halite. Br, F, and I may substitute for Cl. Hg or Fe may be present in small amounts.

Occurrence and Associations
Chlorargyrite is a secondary silver mineral found in the oxidized zones of silver deposits. Associated minerals are many, including native silver.

Varieties
Embolite is a Br-rich variety.

Related Minerals
Chlorargyrite forms complete solid solutions with bromargyrite, AgBr. It is isostructural with a number of other minerals (see **halite**). Other related minerals include iodargyrite, AgI.

Atacamite $Cu_2Cl(OH)_3$

Origin of Name
From Atacama, a province in Chile.

Hand Specimen Identification
Atacamite is characterized by its granular crystalline aggregates and its bright green to blackish green color.

Physical Properties
hardness	$3-3\frac{1}{2}$
specific gravity	3.76
cleavage/fracture	perfect basal (010), good {101}/conchoidal
luster/transparency	adamantine/transparent to translucent
color	various shades of green
streak	green

Optical Properties
Biaxial $(-)$, $\alpha = 1.831$, $\beta = 1.861$, $\gamma = 1.880$, $\delta = 0.049$, $2V = 75°$.

Crystallography
Orthorhombic, $a = 6.02$, $b = 9.15$, $c = 6.85$, $Z = 4$; space group $P2_1/n2/a2/m$; point group $2/m2/m2/m$.

Habit
Slender, striated prisms are typical. Massive, granular, or fibrous aggregates are common.

Structure and Composition
$CuCl(OH)_5$ and $CuCl_2(OH)_4$ octahedra characterize the structure. Mn commonly substitutes for Cu.

Occurrence and Associations
Atacamite occurs in the oxidized zones of copper deposits and in sands. It is associated with malachite, cuprite, and other secondary copper minerals.

Fluorite   CaF₂

Origin of Name
From the Latin *fluere*, meaning "to flow," referring to the ease with which it melts.

Hand Specimen Identification
Crystal form, cleavage, hardness, and color characterize fluorite. When uncolored, fluorite is sometimes confused with calcite or quartz, but may be distinguished by hardness and habit. See Figures 2.7, 3.6f, 6.11, and 14.20.

Physical Properties
hardness	4
specific gravity	3.18
cleavage/fracture	perfect octahedral {111}/ conchoidal and splintery
luster/transparency	vitreous/transparent to translucent

▶FIGURE 14.20
Fluorite typical forms cubic crystals, although cleavage fragments may be octahedra. This fluorite sample from near Thunder Bay, Ontario, has cubes up to 1 cm on edge.

color	colorless to light green, blue-green, yellow, or purple; more rarely can also be white, brown, or rose
streak	white

Optical Properties
Fluorite is usually colorless in thin section; occasionally, it is light purple or green. Octahedral cleavage and low R.I. distinguish it from other clear isotropic minerals. Isotropic, $n = 1.434$.

Crystallography
Cubic, $a = 5.463$, $Z = 4$; space group $F4/m\bar{3}2/m$; point group $4/m\bar{3}2/m$.

Habit
Cubic crystals are common; octahedral cleavage fragments may appear to be crystals (Figure 14.20). Penetration twins are common. Fluorite may be massive or granular.

Structure and Composition
The cubic unit cell has Ca^{2+} ions in 8-fold coordination at the corners and in the middle of faces. F^- ions are located in tetrahedral coordination within the cell just inside each of the four corners (Figures 6.12 and 10.33). Fluorite is generally close to end member composition. Minor Y, Ce, and other rare earths may substitute for Ca. Cl, Sr, Ba, Fe, and Na may be present in small amounts.

Occurrence and Associations
Fluorite is common and widespread. It is found in veins and associated with quartz, calcite, galena, barite, and a number of other minerals. In some carbonate hosted ore deposits, it is a replacement or fracture filling mineral associated with pyrrhotite, galena, and pyrite. It is also found as an accessory mineral in limestones and in igneous and metamorphic rocks.

Related Minerals
Fluorite is isostructural with thorianite, ThO_2; cerianite, $(Ce,Th)O_2$; and uraninite, UO_2. It is closely related to sellaite, MgF_2, and frankdicksonite, BaF_2.

Cryolite   Na₃AlF₆

Origin of Name
From the Greek words for "ice" and "stone," referring to its icy appearance.

Hand Specimen Identification
Peculiar luster, color, habit, and hardness help identify cryolite.

Physical Properties
hardness	$2\frac{1}{2}$
specific gravity	2.97

cleavage/fracture	none/uneven; cubic parting
luster/transparency	pearly, greasy, vitreous/
	transparent to translucent
color	colorless to snow-white
streak	white

Optical Properties
Biaxial (+), $\alpha = 1.3385$, $\beta = 1.3389$, $\gamma = 1.3396$, $\delta = 0.0011$, $2V = 43°$.

Crystallography
Monoclinic, $a = 5.46$, $b = 5.60$, $c = 7.80$, $\beta = 90.18°$, $Z = 2$; space group $P2_1/n$; point group $2/m$.

Habit
Individual pseudocubic crystals are rare. Aggregates are massive, lamellar, or columnar, often exhibiting pseudocubic parting.

Structure and Composition
In cryolite, both Na and Al are coordinated to six F. Na octahedra are distorted. F occupies a tetrahedral site, coordinated to three Na and one Al.

Occurrence and Associations
Cryolite is a rare mineral. The most significant samples are from Greenland, where it is in ore deposits hosted by granitic rocks. Associated minerals include quartz, K-feldspar, siderite, galena, sphalerite, and chalcopyrite, and less commonly other sulfides, wolframite, cassiterite, fluorite, and columbite.

Related Minerals
Cryolite has a high-temperature cubic polymorph.

V. OXIDES

Tetrahedral and Octahedral Oxides

Tetrahedral Oxides

| zincite | ZnO |

Octahedral Oxides

rutile	TiO_2
periclase	MgO
hematite	Fe_2O_3
corundum	Al_2O_3
ilmenite	$FeTiO_3$
cassiterite	SnO_2
pyrolusite	MnO_2
columbite	$(Fe,Mn)Nb_2O_6$
tantalite	$(Fe,Mn)Ta_2O_6$

Similar to the sulfides, mineralogists divide oxide minerals into those having metal ions only in tetrahedral or only in octahedral coordination, and those in which the ions occupy sites with mixed or unusual coordinations. Zincite, a rare mineral, is the only known example of a purely tetrahedral oxide, but more than a dozen octahedral oxides are known. At high temperatures, Mg-, Fe-, and Ti-oxides form stable solid solutions. At low temperatures, most intermediate compositions are unstable; exsolution is common.

For more general information about oxides, see Chapter 8.

Zincite   ZnO

Origin of Name
Zincite is named for its composition.

Hand Specimen Identification
Zincite is characterized by its association with willemite and franklinite, its orange-yellow streak, and its red color.

Physical Properties

hardness	$4-4\frac{1}{2}$
specific gravity	5.4–5.7
cleavage/fracture	perfect basal (001)/
	subconchoidal
luster/transparency	subadamantine/translucent
color	orange, yellow to deep red
streak	orange-yellow

Optical Properties
Uniaxial (+), $\omega = 2.013$, $\epsilon = 2.029$, $\delta = 0.016$.

Crystallography
Hexagonal, $a = 3.25$, $c = 5.19$, $Z = 2$; space group $P6_3mc$; point group $6mm$.

Habit
Zincite is commonly massive, platy, or granular. The rare crystals are hexagonal prisms terminated by pyramids and pedions.

Structure and Composition
The structure is identical to that of wurtzite; Zn is hexagonal closest packed. Mn and minor Fe may substitute for Zn.

Occurrence and Associations
Zincite is a rare mineral, primarily found at Franklin, New Jersey. Associated minerals include calcite, dolomite, franklinite, and willemite.

Related Minerals
Zincite is isostructural with wurtzite, ZnS; enargite, Cu_3AsS_4; and greenockite, CdS.

Rutile   TiO₂

Origin of Name
From the Latin *rutilas,* meaning "red."

Hand Specimen Identification
Rutile is characterized by its color, adamantine luster, crystal form, and twinning. It sometimes occurs as needles in quartz (Plate 2.1).

Physical Properties

hardness	6–$6\frac{1}{2}$
specific gravity	4.24
cleavage/fracture	good prismatic {100} and {110}/subconchoidal
luster/transparency	adamantine, submetallic/transparent to translucent
color	red, red-brown to black
streak	pale or light brown

Optical Properties
Rutile appears deep red, red-brown, or yellow-brown in thin section. The strong color may mask its extreme birefringence. Relief is very high. Uniaxial (+), $\omega = 2.61$, $\epsilon = 2.90$, $\delta = 0.29$.

Crystallography
Tetragonal, $a = 4.59$, $c = 2.96$, $Z = 2$; space group $P4_2/m2_1/n2/m$; point group $4/m2/m2/m$.

Habit
Rutile may be massive but more commonly forms stubby to acicular tetragonal crystals. Striated prismatic crystals, terminated by prisms and often twinned, are common.

Structure and Composition
Distorted TiO₆ octahedra share edges to form chains. Chains are connected by corner-sharing octahedra. Each O is in triangular coordination, bonded to three Ti. Fe, Ta, Nb, V, Sn, and Cr may be present as substitutions.

Occurrence and Associations
Rutile, although not particularly abundant, is widespread. It is found typically as small grains in intermediate to mafic igneous rocks, in some metamorphic rocks, in veins, in pegmatites, and in some sediments. Associated minerals include quartz, feldspar, ilmenite, and hematite.

Varieties
Sagenite is the name given to rutile that exists as needled patches within other minerals such as quartz and pyroxene.

Related Minerals
Rutile has several polymorphs; most important are anatase and brookite. Minerals with similar structures to rutile include cassiterite, SnO₂; pyrolusite, MnO₂; plattnerite, PbO₂; and stishovite, SiO₂. Other similar minerals include baddeleyite, ZrO₂, and paratellurite, TeO₂.

Periclase   MgO

Origin of Name
From the Greek *peri* and *klasis,* meaning "even" and "fracture," referring to its perfect cubic cleavage.

Hand Specimen Identification
Color, crystal form, cubic cleavage, and association help identify periclase.

Physical Properties

hardness	$5\frac{1}{2}$
specific gravity	3.56
cleavage/fracture	perfect {100}, poor {111}/uneven
luster/transparency	vitreous/transparent to translucent
color	colorless or gray
streak	orange-yellow

Optical Properties
Periclase is colorless in thin section, has high relief and cubic cleavage. Isotropic, $n - 1.736$.

Crystallography
Cubic, $a = 4.21$, $Z = 4$; space group $F4/m\bar{3}2/m$; point group $4/m\bar{3}2/m$.

Habit
Typically periclase crystals are cubes or octahedra. Coarse or granular masses are common.

Structure and Composition
Mg and O alternate in a three-dimensional cubic framework. Fe, Zn, and minor Mn may substitute for Mg.

Occurrence and Associations
Periclase is a high-temperature mineral found in metamorphosed carbonates. It is typically in contact aureoles, associated with calcite, forsterite, diopside, and a number of other Ca- and Ca-Mg-silicates.

Related Minerals
Periclase is isostructural many other minerals (see **halite**).

Hematite   Fe₂O₃

Origin of Name
From the Greek *haimatos,* meaning "blood," a reference to its color when powdered.

Hand Specimen Identification
Density and a characteristic red streak identify most hematite. It may be confused with cinnabar. See Figure 3.5e.

Physical Properties

hardness	$5\frac{1}{2}-6\frac{1}{2}$
specific gravity	4.9–5.3
cleavage/fracture	none/subconchoidal
luster/transparency	submetallic/translucent to opaque
color	steel-gray, red-brown to black
streak	red

Optical Properties
Hematite is usually opaque; when very thin, it may have a deep red color. Uniaxial $(-)$, $\omega = 3.22$, $\epsilon = 2.96$, $\delta = 0.28$.

Crystallography
Hexagonal (rhombohedral), $a = 5.04$, $c = 13.76$, $Z = 6$; space group $R\bar{3}2/c$; point group $\bar{3}2/m$.

Habit
Hematite exhibits many habits. Aggregates may be massive, in rosettes, botryoidal, reniform, micaceous and foliated, or earthy. Individual crystals are tabular, with many forms making up the edges. A metallic variety, specularite, may be a massive or fine-grained aggregate. Twins are common.

Structure and Composition
Hematite structure is similar to that of corundum. Closest packed O^{2-} forms hexagonal layers; Fe^{3+} occupies $\frac{2}{3}$ of the available interlayer octahedral sites. FeO_6 octahedra, linked by edge sharing, form 6-sided rings. Ti, Fe, Al, and Mn may replace Fe in small amounts.

Occurrence and Associations
Hematite is a common mineral in many kinds of rocks. It is found in red sandstones, iron formations, and its metamorphic equivalent, as an accessory in igneous rocks, as coatings and nodules, in hydrothermal veins, and associated with the altered zone of ore deposits. Usually hematite is secondary, forming after magnetite, Fe-sulfides, or Fe-silicates.

Varieties
Specularite (specular hematite) is the name given to micaceous hematite exhibiting a splendent metallic luster. Ocher is a red, earthy form of hematite. *Martite* is the name given to hematite pseudomorphs after magnetite.

Related Minerals
Hematite has a rare polymorph, maghemite. Isostructural minerals include corundum, Al_2O_3; eskolaite, Cr_2O_3; and karelianite, V_2O_3. Turgite, $2Fe_2O_3 \cdot nH_2O$, and limonite, $Fe_2O_3 \cdot nH_2O$, are equivalent to hydrated hematite with variable water content. Other related minerals include goethite and lepidocrocite, both $FeO(OH)$, and bixbyite, Mn_2O_3.

Corundum Al_2O_3

Origin of Name
Derived from *kauruntaka*, the Indian name for the mineral.

Hand Specimen Identification
Corundum is identified primarily by its hardness. Luster, specific gravity, and habit aid identification. Figure 8.17d shows gem corundum (sapphire).

Physical Properties

hardness	9
specific gravity	3.9–4.1
cleavage/fracture	none/uneven; rectangular parting
luster/transparency	adamantine/transparent to translucent
color	shades of brown, blue, or pink, colorless, variable
streak	white

Optical Properties
In thin section, corundum is typically colorless. Some varieties are pale pink, green, blue, or yellow. It has extremely high relief and is weakly birefringent. Uniaxial $(-)$, $\omega = 1.768$, $\epsilon = 1.760$, $\delta = 0.008$.

Crystallography
Hexagonal (rhombohedral), $a = 4.95$, $c = 13.78$, $Z = 6$; space group $R\bar{3}2/c$; point group $\bar{3}2/m$.

Habit
Hexagonal crystals may be tabular or prismatic. Multiple pyramidal forms combined with pinacoids give a tapering or barrel-shaped appearance (Figure 3.3). Euhedral individual crystals are rare; corundum is usually massive or granular.

Structure and Composition
The corundum structure is identical to that of hematite (see **hematite**). Minor amounts of Fe, Ti, Cr, Ni, and Mn replace Al.

Occurrence and Associations
Corundum is an accessory mineral in metamorphosed carbonates and sediments, in some Al-rich igneous rocks, and in placers. Massive corundum, forming emery deposits, is found in carbonate skarns.

Varieties
Sapphire (typically blue) and ruby (red) are gem varieties of corundum. Emery refers to hard compact corundum-magnetite-hematite ore.

Related Minerals
Corundum is isostructural with hematite, Fe_2O_3; ilmenite, $FeTiO_3$; eskolaite, Cr_2O_3; and karelianite, V_2O_3.

Ilmenite   FeTiO₃

Origin of Name
Named after the Ilmen Mountains in Russia.

Hand Specimen Identification
Density and color help identify ilmenite. It is distinguished from magnetite by its lack of strong magnetism and from hematite by its streak. Ilmenite is occasionally confused with chromite.

Physical Properties

hardness	$5\frac{1}{2}$–6
specific gravity	4.5–5
cleavage/fracture	none/subconchoidal
luster/transparency	metallic/opaque
color	iron-black
streak	brownish red to black

Crystallography
Hexagonal (trigonal), $a = 5.08$, $c = 14.08$, $Z = 6$; space group $R3$; point group 3.

Habit
Crystals are tabular or prismatic, often showing rhombohedral forms, and commonly twinned. Most occurrences are massive, granular, compact, scaly, or appearing as skeletal crystals.

Structure and Composition
The structure of ilmenite is the same as that of corundum and hematite (see **hematite**). Ilmenite may contain excess Fe replacing Ti, forming limited solid solutions with hematite. Mg and Mn may replace Fe.

Occurrence and Associations
Ilmenite, a common vein mineral, is found as masses in igneous rocks, in pegmatites, as an accessory in high-grade metamorphic rocks, and is present in black sands where associated minerals include quartz, hematite, magnetite, rutile, zircon, monazite, and other dense minerals.

Related Minerals
Geikielite, $MgTiO_3$, and pyrophanite, $MnTiO_3$, form complete solid solutions with ilmenite. Several minerals are isostructural with ilmenite (see **hematite**).

Cassiterite   SnO₂

Origin of Name
From the Greek *kassiteros*, meaning "tin."

Hand Specimen Identification
Cassiterite is recognized by crystal form, streak, high specific gravity, indistinct prismatic cleavage, and adamantine luster. It may be confused with rutile. See Figure 2.7.

Physical Properties

hardness	6–7
specific gravity	7.0
cleavage/fracture	good {100}, poor {111}/ subconchoidal
luster/transparency	adamantine/transparent to translucent
color	brown or black
streak	white

Optical Properties
Cassiterite is yellow, brown, red, or colorless in thin section. Relief and birefringence are high. Uniaxial (+), $\omega = 2.006$, $\epsilon = 2.097$, $\delta = 0.091$.

Crystallography
Tetragonal, $a = 4.74$, $c = 3.19$, $Z = 2$; space group $P4_2/m2_1/n2/m$; point group $4/m2/m2/m$.

Habit
Crystals are pyramidal, stubby prismatic, or acicular. Faces may be striated; contact and penetration twins are common; complex multiple twinning is less common. Cassiterite may be massive, colloform, reniform, or fibrous.

Structure and Composition
Distorted SnO_6 octahedra share edges to form chains. Chains are connected by corner-sharing octahedra. Each O is in triangular coordination, bonded to three Sn. Cassiterite may contain minor Fe and Ta substituting for Sn. Mn, W, Nb, and Sc also may be present in trace amounts.

Occurrence and Associations
SnO_2 is widespread but rarely concentrated as tin ore. It occurs in pegmatites, veins, contact aureoles, altered zones of ore deposits, and placers. Associated minerals include quartz, topaz, tourmaline, fluorite, muscovite, lepidolite, wolframite, scheelite, and others.

Related Minerals
Cassiterite is isostructural with rutile, TiO_2; pyrolusite, MnO_2; plattnerite, PbO_2; and paratellurite, TeO_2.

Pyrolusite   MnO₂

Origin of Name
From the Greek *pyr* and *louein,* meaning "fire" and "to wash," referring to its use in glass manufacturing.

Hand Specimen Identification
Pyrolusite is characterized by softness and black streak.

Physical Properties

hardness	1–2
specific gravity	4.5–5.0
cleavage/fracture	perfect but rarely seen prismatic {110}/uneven
luster/transparency	metallic/opaque
color	black
streak	black

Crystallography
Tetragonal, $a = 4.40$, $c = 2.87$, $Z = 2$; space group $P4_2/m2_1/n2/m$; point group $4/m2/m2/m$.

Habit
Rare crystals form perfect tetragonal prisms. More commonly pyrolusite forms orthorhombic pseudomorphs after manganite, or it is dendritic, fibrous, reniform or columnar (Figure 14.21).

Structure and Composition
Pyrolusite has the rutile structure (see **rutile**). It is commonly close to pure Mn-oxide, but Mn valence may be slightly variable. Small amounts of H_2O may be present.

Occurrence and Associations
Pyrolusite is a secondary mineral found as coatings, nodules, dendrites, and in beds. Associated minerals include barite, limonite, romanechite (psilomelane), hematite, magnetite, and other Mn- and Fe-oxides.

Varieties
Wad is the name given to mixtures of Mn-oxides.

Related Minerals
Pyrolusite is isostructural with a number of minerals (see **rutile**). It has several polymorphs, including nsutite, ramsdellite, and vernadite. Other related minerals include

manganite, $MnO(OH)$; romanechite (psilomelane), $BaMn_9O_{16}(OH)_4$; and birnessite, $Na_4Mn_{14}O_{27} \cdot 9H_2O$.

Columbite-Tantalite   (Fe,Mn)(Nb,Ta)$_2$O$_6$

Origin of Name
Columbium, an early name for the element tantalum, is named after Columbia, where the original samples of this mineral were found. Tantalite refers to the Greek myth of Tantalus.

Hand Specimen Identification
Association, color, luster, streak, and density help identify members of the columbite-tantalite series. Crystals often display heart-shaped twins. These minerals may be confused with wolframite or uraninite.

Physical Properties

hardness	6
specific gravity	6.0
cleavage/fracture	good (010)/subconchoidal
luster/transparency	submetallic/translucent to opaque
color	iron-black to brown
streak	brown, dark red to black

Optical Properties
Biaxial $(+)$, $\alpha = 2.44$, $\beta = 2.32$, $\gamma = 2.38$, $\delta = 0.12$, $2V = 75°$.

Crystallography
Orthorhombic, $a = 5.10$, $b = 14.27$, $c = 5.74$, $Z = 4$; space group $P2_1/b2/c2_1/n$; point group $2/m2/m2/m$.

Habit
Short prisms or tabular crystals, sometimes with heart-shaped twins, are typical. Crystal aggregates or masses are common.

Structure and Composition
The structure consists of chains of $(Fe,Mn)O_6$ and $(Nb,Ta)O_6$ octahedra. Edge sharing joins them together. A complete solid solution exists between columbite, $(Fe,Mn)Nb_2O_6$, and tantalite, $(Fe,Mn)Ta_2O_6$. Mg may substitute for (Fe,Mn). Sn and W may be present in small amounts.

Occurrence and Associations
Members of the columbite-tantalite series are uncommon. They occur in pegmatites with quartz, feldspar, mica, Li-minerals, phosphates, and other typical pegmatite minerals; in carbonatites; and in placer deposits with other dense minerals.

Varieties
Tapiolite is a polymorph of columbite-tantalite. Other related minerals include microlite, $Ca_2Ta_2O_6(O,OH,F)$; pyrochlore, $(Ca,Na)_2(Nb,Ta)_2O_6(O,OH,F)$; and fergusonite, $(REE)NbO_4$.

▶**FIGURE 14.21**
Dendritic pyrolusite, MnO_2, on a slab of sandstone from Spearfish, South Dakota.

Spinels and Other Oxides with Mixed or Unusual Coordinations

"Inverse" Spinels
spinel	$MgAl_2O_4$
magnetite	Fe_3O_4

"Normal" Spinels
chromite	$FeCr_2O_4$
franklinite	$ZnFe_2O_4$

Others
perovskite	$CaTiO_3$
chrysoberyl	$BeAl_2O_4$
uraninite	UO_2
thorianite	ThO_2
cuprite	Cu_2O

In spinel group minerals both tetrahedral and octahedral sites are occupied by metal ions. In "normal" spinel minerals each metal species is found in either tetrahedral or octahedral coordination but not both; their formulas may be written as XY_2O_4, with X representing the tetrahedral cation and Y the octahedral cation. In "inverse" spinels one metal species occupies both coordinations. A general formula might be more appropriately written as $Y[XY]O_4$, with the brackets identifying metal ions in octahedral sites.

Mineralogists have identified other oxides that do not fit into the tetrahedral, octahedral, or spinel groups. In uraninite and thorianite, metal atoms occupy cubic sites. In perovskite, Ti and Ca are in 6-fold and 12-fold coordinations, respectively. In cuprite, Cu is in 2-fold coordination and chrysoberyl is isostructural with olivine.

For more general information about oxides, see Chapter 8.

Spinel $MgAl_2O_4$

Origin of Name
From the Latin *spina,* meaning "thorn," a reference to the sharp crystals.

Hand Specimen Identification
The term *spinel* is used in a generic sense to describe any of the numerous minerals with spinel structure. The specific mineral called *spinel,* $MgAl_2O_4$, is recognized by association, octahedral crystals, hardness, and sometimes a vitreous luster. See Figures 2.7, 10.36a, and 14.22.

▶FIGURE 14.22
These spinel crystals from Sri Lanka vary from poor to well-developed octahedra. They are about 1 cm across.

Physical Properties
hardness	8
specific gravity	3.5–4.0
cleavage/fracture	none/conchoidal
luster/transparency	vitreous/transparent to translucent
color	variable, red, lavender, blue, green, brown, white, or black
streak	white

Optical Properties
Pure Mg spinel is colorless in thin section but is pleochroic green or blue-green if Fe substitutes for Mg. Octahedral shape and high index of refraction aid identification. Isotropic, $n = 1.74$.

Crystallography
Cubic, $a = 8.09$, $Z = 8$; space group $F4/d32/m$; point group $4/m32/m$.

Habit
Spinel typically forms octahedral crystals (Figures 10.34b and 10.36); twinning and modifying faces are common. Massive forms and irregular grains are also known.

Structure and Composition
Spinel minerals are relatively simple cubic structures. MgO_6 octahedra and AlO_4 tetrahedra share edges and are closest packed. Fe^{2+}, Zn, and Mn may substitute for Mg. Fe^{3+} and Cr may substitute for Al.

Occurrence and Associations
Spinel is a high-temperature mineral found in metamorphosed carbonates or schists, as an accessory in mafic igneous rocks, and in placers. Associated minerals include calcite, dolomite, garnet, and Ca-Mg silicates in marbles; garnet, corundum, sillimanite, andalusite and cordierite in highly aluminous rocks; diopside, olivine, chondrodite, in mafic ones; and other dense minerals in placers.

Varieties
Ruby spinel is the name given to gemmy-red spinel; various other gem names are used to a lesser extent. *Pleonaste* is the name given to intermediate Fe-Mg spinels.

Related Minerals
Spinel is isostructural with other members of the spinel group and with bornhardite and linnaeite, both Co_3Se_4; polydymite, Ni_3S_4; indite, $FeIn_2S_4$; and greigite, Fe_3S_4. Spinel forms solid solutions with other members of the spinel group, including hercynite, $FeAl_2O_4$; gahnite, $ZnAl_2O_4$; galaxite, $MnAl_2O_4$; zincochromite, $ZnCr_2O_4$; and magnesiochromite, $MgCr_2O_4$.

Magnetite Fe_3O_4

Origin of Name
Named after Magnesia, near Macedonia in Thessaly, where the Greeks found this mineral.

Hand Specimen Identification
Magnetite, a member of the spinel group, is characterized by its strong magnetism, hardness, and black color. Magnetism distinguishes it from ilmenite and chromite.

Physical Properties

hardness	6
specific gravity	5.20
cleavage/fracture	none/subconchoidal
luster/transparency	metallic/opaque
color	black
streak	black

Crystallography
Cubic, $a = 8.397$, $Z = 8$; space group $F4/d\bar{3}2/m$; point group $4/m\bar{3}2/m$.

Habit
Crystals are typically octahedra, sometimes displaying contact or lamellar twins. Magnetite is also common as massive or granular aggregates or disseminated as fine grains.

Structure and Composition
Magnetite has the spinel structure (see **spinel**). Ti is usually present; at high temperature a complete solid solution to Fe_2TiO_4 is possible. Minor amounts of Mg, Mn, Ni, Al, Cr, and V may be present.

Occurrence and Associations
Magnetite is common and widespread. It is found as an accessory in many types of igneous, metamorphic, and sedimentary rocks and in unconsolidated sediments. It may be concentrated to form ore bodies by magmatic, metamorphic, or sedimentary processes.

Related Minerals
Magnetite forms solid solutions with ulvöspinel, Fe_2TiO_4; magnesioferrite, $MgFe_2O_4$; and jacobsite, $MnFe_2O_4$; and to a lesser extent with maghemite, Fe_2O_3.

Chromite $FeCr_2O_4$

Origin of Name
The name *chromite* refers to this mineral's composition.

Hand Specimen Identification
Color, density, streak, luster, and association distinguish chromite (Figure 8.3). It may be slightly magnetic and is sometimes confused with magnetite or ilmenite.

Physical Properties

hardness	$5\frac{1}{2}$
specific gravity	5.10
cleavage/fracture	none/conchoidal
luster/transparency	metallic/subtranslucent; opaque
color	black, brownish black
streak	brown, dark brown

Crystallography
Cubic, $a = 8.37$, $Z = 8$; space group $F4/d\bar{3}2/m$; point group $4/m\bar{3}2/m$.

Habit
Rare euhedral crystals are octahedral; chromite is generally massive or granular.

Structure and Composition
The structure of chromite is the same as that of all the spinel minerals (see **spinel**). Mg, Fe, Al, and Zn are typical impurities.

Occurrence and Associations
Primary chromite is found with olivine, pyroxene, spinel, magnetite, and sulfides in ultramafic rocks. It is also found in placers and black sands. Chromite is common as an accessory mineral but may be concentrated by gravity or magmatic processes.

Related Minerals
Chromite has one polymorph, donathite. It forms solid solutions with magnesiochromite, $MgCr_2O_4$, and hercynite, $FeAl_2O_4$, and to lesser extent with other spinel minerals.

Franklinite $ZnFe_2O_4$

Origin of Name
Named after Franklin, New Jersey, a classic locality.

Hand Specimen Identification
Franklinite resembles other dark-colored spinels and Fe-Ti oxides but has a dark brown streak and is only slightly magnetic.

Physical Properties

hardness	6
specific gravity	5.32

cleavage/fracture	none/conchoidal
luster/transparency	metallic/opaque
color	black, iron-black
streak	black, reddish brown to dark brown

Crystallography
Cubic, $a = 8.43$, $Z = 8$; space group $F4/d\bar{3}2/m$; point group $4/m\bar{3}2/m$.

Habit
Octahedral crystals, often with modifying faces, are common in Franklin, New Jersey, the only place where it is found in large quantities. It is also found as discrete rounded grains, as granular masses, or in massive lenses.

Structure and Composition
The structure is the same as that of other spinel minerals (see **spinel**). Franklinite normally contains substantial Mn substituting for Zn. Mn^{3+} may also substitute for Fe^{3+}. Mg, Cr, and V also may be present.

Occurrence and Associations
Franklinite is associated with zincite and willemite in zinc ore deposits at Franklin, New Jersey. The host rock is a coarse-grained limestone.

Related Minerals
Franklinite is similar in many ways to other dark-colored spinel minerals. It forms minor solid solutions with most of them.

Chrysoberyl BeAl$_2$O$_4$

Origin of Name
From the Greek words meaning "golden beryl."

Hand Specimen Identification
Color, luster, hardness, and common twinning characterize chrysoberyl.

Physical Properties
hardness	$8\frac{1}{2}$
specific gravity	3.7–3.8
cleavage/fracture	good but indistinct prismatic {011}, poor (010)/ subconchoidal
luster/transparency	vitreous/transparent to translucent
color	yellow, green, or brown
streak	white

Optical Properties
Biaxial (+), $\alpha = 1.747$, $\beta = 1.748$, $\gamma = 1.757$, $\delta = 0.010$, $2V = 45°$.

Crystallography
Orthorhombic, $a = 4.24$, $b = 9.39$, $c = 5.47$, $Z = 4$; space group $P2_1/b2_1/n2_1/m$; point group $2/m2/m2/m$.

Habit
Chrysoberyl is gene[...] shaped or pseudohexa[...] Faces are often striated.

Structure and Composition
The structure, similar to that of o[...] hexagonal closest packed oxygens with [...] hedral sites and Al in octahedral sites.

Occurrence and Associations
Chrysoberyl is a rare mineral occurring in granites, pegmatites, mica schists, and some placers.

Varieties
Cat's eye (cymophane) is a green chatoyant gem variety. Alexandrite is an emerald-green gem variety that appears red under artificial light. Both are very valuable.

Related Minerals
Chrysoberyl is isostructural with olivine minerals. It is chemically similar to spinels but has a different structure.

Uraninite UO$_2$

Origin of Name
The name *uraninite* refers to this mineral's composition.

Hand Specimen Identification
Uraninite is characterized by its radioactivity, association, high specific gravity, streak, color, and luster.

Physical Properties
hardness	$5\frac{1}{2}$
specific gravity	7–9.5
cleavage/fracture	none/conchoidal
luster/transparency	pitchy dull to submetallic/opaque
color	black
streak	brown to black

Crystallography
Cubic, $a = 5.4682$, $Z = 4$; space group $F4/m\bar{3}2/m$; point group $4/m\bar{3}2/m$.

Habit
Individual crystals are rare; they form cubes, octahedra, or combinations. Massive, colloform, or botryoidal forms are typical.

Structure and Composition
Uraninite is isostructural with fluorite (see **fluorite**). U valence and U:O ratios are somewhat variable; uraninite is really a mixture of UO_2 and U_3O_8. Th may substitute for U; N, Ar, Fe, Ca, Zr, and rare

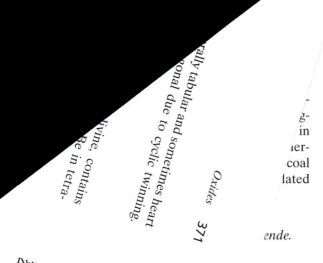

ende.

Uraninite is ... $_2$, and cerianite, $(Ce,Th)O_2$. ... ...id solution
with thorianite, ThO_2. Other ... minerals include baddeleyite, ZrO_2.

Cuprite Cu_2O

Origin of Name
From the Latin *cuprum*, meaning "copper."

Hand Specimen Identification
Red color, form, luster, streak, and association all
help identify cuprite. It may be confused with cassiterite, hematite, and cinnabar.

Physical Properties
hardness	$3\frac{1}{2}$–4
specific gravity	5.9–6.1
cleavage/fracture	poor {111}/conchoidal
luster/transparency	submetallic, metallic
color	various shades of red
streak	brownish red

Crystallography
Cubic, $a = 4.27$, $Z = 2$; space group $P4_2/n\bar{3}2/m$;
point group $4/m\bar{3}2/m$.

Habit
Octahedral, cubic, and dodecahedral forms, often in
combination, are common. Cuprite may occur as
elongated capillary crystals called *chalcotrichite*.

Structure and Composition
Oxygen, in tetrahedral groups, is arranged in a body
centered cubic array. Each Cu is bonded to two O.
Cuprite is generally close to end member composition; Fe is a common minor impurity.

Occurrence and Associations
Cuprite is a secondary mineral found in the oxidized
zones of copper deposits. Native copper, limonite, and
secondary copper minerals such as malachite, azurite,
and chrysocolla are typically associated minerals.

Related Minerals
Tenorite, CuO, is similar in composition and occurrence.

VI. HYDROXIDES

gibbsite	$Al(OH)_3$
brucite	$Mg(OH)_2$
manganite	$MnO(OH)$
goethite	$FeO(OH)$
diaspore	$AlO(OH)$
romanechite	
(psilomelane)	$BaMn_9O_{16}(OH)_4$

The hydroxide minerals all contain OH^- as an essential anion. Some, such as diaspore, also contain O^{2-}.
The structures are all generally simple; brucite and
gibbsite have layered structures equivalent to the trioctahedral and dioctahedral layers in micas. Romanechite has a structure related to rutile and spinel.
Only limited solid solution occurs between the various hydroxides, but they are often found in intimate
mixtures with each other and with oxide minerals.
Bauxite refers to a mixture of Al oxides and hydroxides, *limonite* to a mixture of Fe oxides and hydroxides, and *wad* to a mixture of Mn oxides and
hydroxides.

For more general information about hydroxides,
see Chapter 8.

Gibbsite $Al(OH)_3$

Origin of Name
Named after Colonel G. S. Gibbs (1777–1834), a mineral collector.

Hand Specimen Identification
Earthy smell and appearance and color help identify
gibbsite. It may be confused with kaolinite, talc, or
brucite.

Physical Properties
hardness	$2\frac{1}{2}$–$3\frac{1}{2}$
specific gravity	2.40
cleavage/fracture	perfect (001)/uneven
luster/transparency	vitreous, pearly, earthy/trans-
parent to translucent
color	white or gray to greenish
streak	white

Optical Properties
Gibbsite is usually colorless in thin section; maximum interference colors are upper first order. It is
difficult to distinguish from clay minerals. Biax-
ial (+), $\alpha = 1.57$, $\beta = 1.57$, $\gamma = 1.59$, $\delta = 0.02$,
$2V = 0°$–$40°$.

Crystallography
Monoclinic, $a = 8.641$, $b = 5.07$, $c = 9.719$, $\beta = 94.57°$, $Z = 8$; space group $P2_1/n$; point group $2/m$.

Habit
Foliated or tabular crystals are typically very small with a pseudohexagonal shape. Granular aggregates, colloform or radiating masses, and coatings are most common.

Structure and Composition
In the layered gibbsite structure, octahedral Al^{3+} occupies $2/3$ of the available sites between OH sheets. Small amounts of Fe may replace Al.

Occurrence and Associations
Gibbsite is a secondary mineral associated with aluminum deposits, bauxites, and laterites (Figure 14.23). Diaspore and böhmite, other aluminum hydroxides, are typically intimate associates.

Related Minerals
Gibbsite is similar in structure to brucite, $Mg(OH)_2$. It has several polymorphs, including bayerite, doyleite, and nordstrandite. Other related minerals include diaspore and böhmite, both $AlO(OH)$, and bauxite, a mixture of gibbsite, böhmite, and diaspore.

Brucite $\qquad$ Mg(OH)₂

Origin of Name
A. Bruce (1777–1818), an early American mineralogist, was the inspiration for this mineral's name.

Hand Specimen Identification
Good platy or micaceous cleavage, flexible sheets, color, and luster help identify brucite. It may be confused with gypsum or gibbsite.

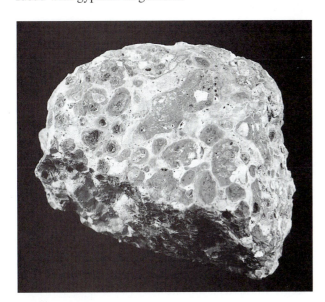

▶**FIGURE 14.23**
Bauxite, a mixture of gibbsite and other aluminum hydroxide and oxide minerals, is a product of weathering and leaching.

Physical Properties

hardness	$2\frac{1}{2}$
specific gravity	2.4–2.5
cleavage/fracture	perfect basal (001)/sectile
luster/transparency	vitreous, pearly/transparent to translucent
color	white, light green, or gray
streak	white

Optical Properties
Brucite is colorless in thin section and displays anomalous second-order red or blue interference colors. Brucite may be mistaken for talc, white mica, or gypsum, but they are all biaxial. Uniaxial $(+)$, $\omega = 1.57$, $\epsilon = 1.58$, $\delta = 0.02$.

Crystallography
Hexagonal (rhombohedral), $a = 3.147$, $c = 4.769$, $Z = 1$; space group $P\bar{3}2/m1$; point group $\bar{3}2/m$.

Habit
Broad platy, foliated, or tabular crystals are typical. Massive or fibrous aggregates and foliated masses are common.

Structure and Composition
The structure is similar to that of gibbsite: Sheets of OH sandwich Mg in octahedral coordination. Fe, Mn, and Zn may substitute in minor amounts for Mg.

Occurrence and Associations
Brucite occurs in veins of mafic rocks, serpentinite, or talc-chlorite schists, and in metamorphosed carbonates or marls. Associated minerals include chlorite and other secondary magnesium minerals in mafic rocks and calcite, dolomite, talc, magnesite, and periclase in carbonates.

Related Minerals
Brucite is isostructural with gibbsite, $Al(OH)_3$. It is isotypical with pyrochroite, $Mn(OH)_2$; amakinite, $(Fe,Mg)(OH)_2$; portlandite, $Ca(OH)_2$; and theophrastite, $Ni(OH)_2$.

Manganite $\qquad$ MnO(OH)

Origin of Name
The name *manganite* refers to this mineral's composition.

Hand Specimen Identification
Manganite is recognized by its prismatic crystals and black color. It may be confused with pyrolusite, with which it is frequently found, but is harder and has a brown streak.

Physical Properties

hardness	4
specific gravity	4.2–4.4
cleavage/fracture	perfect basal (010), good {110}/sectile
luster/transparency	submetallic/opaque
color	gray to black
streak	dark to red-brown

Crystallography
Monoclinic, $a = 8.84$, $b = 5.23$, $c = 5.74$, $\beta = 90.0°$, $Z = 8$; space group $B2_1/d$; point group $2/m$.

Habit
Prismatic crystals, sometimes twinned, often show striations and complicated terminations. Bundled, stalactitic, columnar, bladed, or fibrous aggregates are common.

Structure and Composition
O and OH are nearly hexagonal closest packed. $Mn(O,OH)_6$ octahedra share corners to make a three-dimensional framework. Fe and Mg may replace Mn in small amounts.

Occurrence and Associations
Manganite is an uncommon secondary mineral found in veins with other Mn oxides or hydroxides, carbonates, limonite, and barite.

Related Minerals
A number of manganese oxides and hydroxides are closely related, including pyrolusite, MnO_2; partridgite and bixbyite, both Mn_2O_3; hausmannite, Mn_3O_4; hollandite, $Ba_2Mn_8O_{16}$; romanechite, $BaMn_9O_{16}(OH)_4$; pyrochroite, $Mn(OH)_2$; vernadite, $Mn(OH)_4$; and takanelite, $Mn_5O_9 \cdot 3H_2O$. Wad is a mixture of these manganese minerals.

Goethite $\qquad$ FeO(OH)

Origin of Name
Named after J. W. Goethe (1749–1832), a German poet and scientist.

Hand Specimen Identification
Association, habit, and streak identify goethite. It is sometimes confused with hematite but has a brownish yellow streak, in contrast with hematite's red. Figure 14.24 and Plate 8.6 show goethite crystals that grew in vugs in iron formation. See also Figure 3.5d.

Physical Properties

hardness	5–5½
specific gravity	4.3
cleavage/fracture	perfect but rarely seen (010)/uneven

▶**FIGURE 14.24**
Goethite crystals in iron formation from Atikokan, Ontario, Canada. See also Plate 8.6.

luster/transparency	subadamantine to earthy/subtranslucent to opaque
color	yellow-brown to dark brown or black
streak	brown-yellow

Optical Properties
Goethite is pleochroic yellow, orange-red, or various shades of brown in thin section. It has high positive relief and extremely high birefringence that may be masked by its color. Goethite is difficult to distinguish from other Fe oxides. Biaxial $(-)$, $\alpha = 2.26–2.27$, $\beta = 2.39–2.41$, $\gamma = 2.40–2.52$, $\delta = 0.15$, $2V = 0°–27°$.

Crystallography
Orthorhombic, $a = 4.65$, $b = 10.02$, $c = 3.04$, $Z = 4$; space group $P2_1/b2_1/n2_1/m$; point group $2/m2/m2/m$.

Habit
Euhedral goethite crystals are rare; it is usually platy, prismatic, fibrous, botryoidal, or mammilary. Concentric growth bands are common.

Structure and Composition
Goethite is isostructural with diaspore. Edge-sharing $Fe(O,OH)$ octahedra form bands running parallel to the c-axis. Perpendicular to c, the bands form a checkerboard pattern, leaving empty channels between them. Up to several weight percent of Mn and absorbed water are often present.

Occurrence and Associations
Goethite is a widely distributed secondary mineral formed by the weathering of Fe-rich compounds (Figure 14.24). It concentrates in sediments, gossans,

and laterites. Common associated minerals include siderite, pyrite, magnetite, and many residual weathering products.

Varieties
Bog ore is a porous, poorly consolidated form of goethite. *Limonite* refers to a mixture of hydrous iron oxides of variable chemistry and crystallinity (Figure 3.5f).

Related Minerals
Lepidocrocite, akaganeite, and feroxyhyte are all rare polymorphs of goethite. Diaspore has the same structure except that Al replaces $\frac{2}{3}$ of the Fe. Other related minerals are manganite, $MnO(OH)$; heterogenite, $CoO(OH)$; and montroseite, $(V,Fe)O(OH)$.

Diaspore   AlO(OH)

Origin of Name
From the Greek *diaspora*, meaning "to scatter," referring to its decrepitation when heated.

Hand Specimen Identification
Bladed habit, hardness, good cleavage, and luster identify diaspore. It is occasionally confused with brucite, but brucite is much softer.

Physical Properties
hardness	$6\frac{1}{2}$–7
specific gravity	3.2–3.5
cleavage/fracture	one perfect (010), poor 210/conchoidal
luster/transparency	pearly, vitreous/translucent
color	colorless, yellow, gray, white or green
streak	white-yellow

Optical Properties
Diaspore has high relief, is colorless or pale, displays parallel extinction, and shows up to third-order interference colors. It may be confused with gibbsite or sillimanite, but gibbsite has lower relief and inclined extinction, and sillimanite has lower relief and birefringence. Biaxial (+), $\alpha = 1.68$–1.71, $\beta = 1.71$–1.72, $\gamma = 1.73$–1.75, $\delta = 0.04$, $2V = 85°$.

Crystallography
Orthorhombic, $a = 4.42$, $b = 9.40$, $c = 2.84$, $Z = 4$; space group $P2_1/b2_1/n2_1/m$; point group $2/m2/m2/m$.

Habit
Platy, tabular, or acicular crystals are common. Massive forms or foliated aggregates are common.

Structure and Composition
Diaspore structure is the same as th[at] [except] that Al replaces $\frac{2}{3}$ of the Fe (se[e] generally close to end member comp[osition] Fe or Mn may replace Al.

Occurrence and Associations
Diaspore is found in emery deposits with c[orundum,] magnetite, spinel, and chlorite; in bauxites w[ith] aluminum oxides and hydroxides; as a rare mi[neral in] some pegmatites.

Related Minerals
Böhmite is a polymorph of diaspore. Diaspore [is] isostructural with goethite and close in compositio[n] to gibbsite, $Al(OH)_3$. Bauxite is a mixture of gibbsite, böhmite, and diaspore.

Romanechite (Psilomelane) BaMn₉O₁₆(OH)₄

Origin of Name
From the original locality in Romaneche, France.

Hand Specimen Identification
Color, luster, and habit identify romanechite. It is sometimes confused with pyrolusite, but is harder. A brown-black streak separates it from limonite and other hydrous iron oxides.

Physical Properties
hardness	5–6
specific gravity	3.5–4.7
luster/transparency	submetallic, dull/opaque
color	black
streak	brown-black

Crystallography
Monoclinic, $a = 9.56$, $b = 2.88$, $c = 13.85$, $\beta = 90.5°$, $Z = 2$; space group $A2/m$; point group $2/m$.

Habit
Reniform, botryoidal, or dendritic masses are typical. Distinct crystals are not known.

Structure and Composition
Structure is related to those of rutile and spinel. $Mn(O,OH)_4$ octahedra form distorted chains; Ba and H_2O occupy holes between the chains. Many other elements, including As, V, W, Co, Cu, Ni, Mg, Ca, and alkalis may be present in small or trace amounts.

Occurrence and Associations
Romanechite is a rare secondary mineral associated with pyrolusite, manganite, calcite, and hematite.

Related Minerals
Related minerals include pyrolusite, MnO_2; manganite, $MnO(OH)$; cryptomelane, KMn_8O_{16}; hollandite, $BaMn_8O_{16}$; and wad, a mixture of manganese oxides and hydroxides.

...at of goethite, ex-
... **goethite**). It is
...osition. Minor

...orundum,
...ith other
...eral in

...is
...n

...g(CO$_3$)$_2$
...aFe(CO$_3$)$_2$
CaMn(CO$_3$)$_2$

...roup

... CaCO$_3$
...ite BaCO$_3$
...ontianite SrCO$_3$
cerussite PbCO$_3$

Other Carbonates
malachite Cu$_2$CO$_3$(OH)$_2$
azurite Cu$_3$(CO$_3$)$_2$(OH)$_2$

Nitrate Group
nitratite NaNO$_3$
niter (saltpeter) KNO$_3$

Mineralogists divide carbonate minerals into three main groups based on atomic structure: the calcite group, dolomite group, and aragonite group. Several other species that have more complex structures and chemistries are classified separately. Calcite group minerals have structures analogous to those of halite and galena: six Ca, Mg, Fe, Mn, or Zn surround anionic (CO$_3$)$^{2-}$ groups and each metal atom is surrounded by six carbonate groups. Dolomite group minerals have structures similar to calcite's, but Ca and Mg, Fe, or Mn occupy alternate layers. Aragonite group minerals are orthorhombic. Solid solutions are common within a structural group, although some carbonates, such as aragonite, are almost always close to end member composition.

   Due to very high solubility in water, nitrate minerals are rare. They have structures similar to carbonates, but contain monovalent rather than divalent cations because the NO$_3^-$ anionic group is monovalent. Over half a dozen nitrates are known, but nitratite and niter are the only common ones in more than just a few localities.

   For more general information about carbonates and nitrates, see Chapter 6.

Calcite CaCO$_3$

Origin of Name
From the Latin *calx*, meaning "burnt lime."

Hand Specimen Identification
Calcite is identified by its hardness, rhombohedral cleavage, and effervescence in cold dilute HCl. It may be confused with dolomite or aragonite. Plate 3.8 shows euhedral calcite crystals and Figure 14.25 shows "sandy" calcite. Other photographs are Figures 3.6e, 3.12, 4.14, 5.2, 7.8b, 10.2, 10.32b, and 10.41.

Physical Properties

hardness	3
specific gravity	2.71
cleavage/fracture	perfect rhombohedral {101}/conchoidal
luster/transparency	vitreous/transparent to translucent
color	colorless to white; may also be tinted gray, red, blue, yellow, or green; brown to black when impure
streak	white

Optical Properties
Calcite is colorless in thin section and has extremely high birefringence, resulting in pale, washed out, or white interference colors. Polysynthetic twinning is nearly always visible. It shows variable relief upon stage rotation. Calcite may be confused with other hexagonal carbonates. Orthorhombic carbonates have parallel extinction and are biaxial. Uniaxial ($-$), $\omega = 1.658$, $\epsilon = 1.486$, $\delta = 0.172$. Plates 5.1, 5.2, and Figures 6.7b, c, and 7.8b show calcite in thin section.

Crystallography
Hexagonal (rhombohedral), $a = 4.99$, $c = 17.04$, $Z = 6$; space group $R\bar{3}2/c$; point group $\bar{3}2/m$.

▶**FIGURE 14.25**
Calcite crystals encrusted with sand. The field of view is about 20 cm across.

Habit

Calcite has many habits. The most common are hexagonal prisms with simple to complex terminations; scalenohedra, often with combinations of other forms; rhombohedra, either acute or flattened; and tabs with well-developed basal faces. Polysynthetic twinning is common but usually requires a microscope to detect. Calcite is also found as a massive rock-forming mineral, as nodules or crusts, in speleothems, and as fine to coarse granular aggregates.

Structure and Composition

In calcite, Ca^{2+} ions alternate with $(CO_3)^{2-}$ groups in a three-dimensional array (Figures 6.6, 10.31, and 10.32a). The structure is similar to that of cubic salts, such as halite or periclase, but is not cubic because the structure has been squashed along the equivalent of a main diagonal of the cube. The shortened direction is the c-axis in calcite; planar (CO_3) groups are perpendicular to c, giving the structure a 3-fold axis of symmetry in that direction only. Mg, Fe, Mn, Zn, and a number of others may substitute for some of the Ca; except for Mn, most solid solutions are quite limited.

Occurrence and Associations

Calcite is a common and widespread mineral. It is an essential and major mineral in limestones and marbles, occurs in cave deposits, and occurs as a vein mineral with other carbonates, sulfides, barite, fluorite, and quartz. Calcite also occurs in some rare carbonate-rich igneous rocks and is a common cement in some sandstones. Calcite is common as a weathering product. Organic calcite is common in shells and skeletal material.

Varieties

Iceland spar refers to clear calcite, usually in rhombohedral cleavage fragments; *dogtooth spar* refers to crystals with steep scalenohedral forms; *nailhead spar* refers to flat rhombs or stubby prismatic crystals.

Related Minerals

Calcite has two polymorphs, aragonite and vaterite. It is isostructural with magnesite, $MgCO_3$; siderite, $FeCO_3$; sphaerocobaltite, $CoCO_3$; smithsonite, $ZnCO_3$; nitratite, $Na(NO_3)$; dolomite, $CaMg(CO_3)_2$; and gaspeite, $(Ni,Mg,Fe)(CO_3)$. Calcite and rhodocrosite form extensive solid solutions at room temperature and a complete solid solution above about 550 °C. Calcite forms limited solid solutions with ankerite, $CaFe(CO_3)_2$; dolomite, $CaMg(CO_3)_2$; and kutnohorite, $CaMn(CO_3)_2$, at all temperatures.

Magnesite   MgCO₃

Origin of Name

The name refers to its composition.

Hand Specimen Identification

Massive forms may be chalky or porcelain-like. They are occasionally confused with chert but have inferior hardness. Coarse crystals of magnesite may be difficult to tell from other carbonates, but magnesite is denser than dolomite and does not react to cold HCl like calcite.

Physical Properties

hardness	$3\frac{1}{2}$–5
specific gravity	3.00
cleavage/fracture	perfect rhombohedral {101}/conchoidal
luster/transparency	porcelainous/transparent to translucent
color	white, gray, brown, or yellow
streak	white

Optical Properties

Magnesite is similar to calcite in thin section but has higher index of refraction (see **calcite**). Uniaxial $(-)$, $\omega = 1.700$, $\epsilon = 1.509$, $\delta = 0.191$.

Crystallography

Hexagonal (rhombohedral), $a = 4.59$, $c = 14.87$, $Z = 6$; space group $R\overline{3}2/c$; point group $\overline{3}2/m$.

Habit

Crystals are rare; magnesite is usually massive, granular, fibrous, or earthy.

Structure and Composition

Magnesite is isostructural with calcite (see **calcite**). Large amounts of Fe commonly substitute for Mg. Mn, Ca, Ni, and Zn may also be present in small amounts.

Occurrence and Associations

Magnesite is most common in veins or masses as an alteration product of mafic minerals. It also occurs in some Mg-rich schists and as a primary mineral in some rare chemical sediments and is found as a replacement for calcite or dolomite in limestone.

Varieties

Breunnerite is a Fe-rich variety of magnesite; hoshiite is an Ni-rich variety.

Related Minerals

Magnesite is isostructural with calcite and many other minerals (see **calcite**). It forms complete solid solutions with siderite, $FeCO_3$, and with gaspeite,

$(Ni,Mg,Fe)CO_3$. Related minerals include hydromagnesite, $Mg_5(CO_3)_4(OH)_2 \cdot 4H_2O$.

Siderite $FeCO_3$

Origin of Name
From the Greek *sideros,* meaning "iron."

Hand Specimen Identification
Siderite is distinguished from other carbonates by its high specific gravity and color. It effervesces in warm HCl. It may be confused with sphalerite. See Figure 10.2.

Physical Properties

hardness	$3\frac{1}{2}$–4
specific gravity	3.96
cleavage/fracture	perfect rhombohedral {101}/subconchoidal
luster/transparency	vitreous/translucent
color	light to dark brown
streak	white

Optical Properties
Siderite is colorless to pale yellow-brown in thin section. It is similar to calcite and other carbonates, but they have lower index of refraction and lack color (see **calcite**). Uniaxial $(-)$, $\omega = 1.875$, $\epsilon = 1.633$, $\delta = 0.242$.

Crystallography
Hexagonal (rhombohedral), $a = 4.72$, $c = 15.46$, $Z = 6$; space group $R\bar{3}2/c$; point group $\bar{3}2/m$.

Habit
Crystals are typically rhombohedra, often with curved faces. Fine- to coarse-grained aggregates, colloform, globular, botryoidal, fibrous, and earthy forms are common.

Structure and Composition
Siderite is isostructural with calcite (see **calcite**). Mn and Mg often substitute for Fe. Small amounts of Ca, Zn, and Co may be present.

Occurrence and Associations
Siderite is a relatively common mineral found in veins with galena, pyrite, chalcopyrite, and tetrahedrite; as a rock-forming mineral associated with limestone, clay, shale, coal or ironstone; as a replacement mineral in limestone; and less commonly in metamorphic rocks.

Related Minerals
Siderite is isostructural with calcite and a number of other minerals (see **calcite**). It forms complete solid solutions with rhodochrosite, $MnCO_3$, and magnesite, $MgCO_3$.

Rhodochrosite $MnCO_3$

Origin of Name
From the Greek words meaning "rose" and "color," referring to its rose-pink color.

Hand Specimen Identification
Pink color and rhombohedral carbonate morphology identify rhodochrosite. It may be confused with rhodonite, but rhodonite is harder and does not effervesce in HCl. See Figure 10.2.

Physical Properties

hardness	$3\frac{1}{2}$–4
specific gravity	3.70
cleavage/fracture	perfect rhombohedral {101}/uneven
luster/transparency	vitreous, pearly/transparent to translucent
color	rose-red, light pink to dark brown
streak	white

Optical Properties
Rhodochrosite is colorless or pale pink in thin section, has extremely high birefringence, and three perfect cleavage directions. It may be confused with calcite and other carbonates (see **calcite**). Uniaxial $(-)$, $\omega = 1.816$, $\epsilon = 1.597$, $\delta = 0.219$.

Crystallography
Hexagonal (rhombohedral), $a = 4.74$, $c = 15.51$, $Z = 6$; space group $R\bar{3}2/m$; point group $\bar{3}2/m$.

Habit
Rhodochrosite forms rare rhombohedral crystals. It is usually massive, sometimes granular, botryoidal, columnar, or crusty.

Structure and Composition
Rhodochrosite is isostructural with calcite (see **calcite**). Zn commonly replaces some Mn; Ca, Mg, Cd, and Co may be present in limited amounts.

Occurrence and Associations
Rhodochrosite is uncommon. It is found with other manganese minerals in Mn-rich metamorphic rocks, as a primary mineral in sulfide veins and some replacement bodies, and as a secondary mineral in residual deposits.

Related Minerals
Rhodochrosite has the same structure as calcite and number of other minerals (see **calcite**). It forms solid solutions with calcite, $CaCO_3$; siderite, $FeCO_3$; and kutnohorite, $CaMn(CO_3)_2$.

Smithsonite ZnCO₃

Origin of Name
Named after J. Smithson (1754–1829), founder of the Smithsonian Institute.

Hand Specimen Identification
Rhombohedral carbonate habit, color, density, and association identify smithsonite. If not distinctly colored, it may be difficult to tell from other dense carbonates. It is occasionally confused with hemimorphite. Plates 1.2 and 1.8 show color photos of smithsonite. See also Figures 2.9 and 6.14.

Physical Properties
hardness	4–4½
specific gravity	4.43
cleavage/fracture	perfect rhombohedral {101}/subconchoidal
luster/transparency	pearly, vitreous/transparent to translucent
color	typically green; also gray, dirty brown, white, colorless, blue, or pink
streak	white

Optical Properties
Uniaxial $(-)$, $\omega = 1.850$, $\epsilon = 1.625$, $\delta = 0.225$.

Crystallography
Hexagonal (rhombohedral), $a = 4.61$, $c = 14.88$, $Z - 6$; space group $R\bar{3}2/m$; point group $\bar{3}2/m$.

Habit
Crystals, when they are euhedral or subhedral, show rhombohedral form and cleavage. More typical, smithsonite is massive, colloform, earthy, stalactitic, or forms crusts.

Structure and Composition
Smithsonite is isostructural with calcite (see **calcite**). It typically contains substantial amounts of Fe; smaller amounts of Ca, Co, Cu, Cd, Mg, or Mn; and traces of Ge or Pb.

Occurrence and Associations
Smithsonite is a secondary mineral found in zinc deposits. Associated minerals include sphalerite, hemimorphite, cerussite, malachite, azurite, and anglesite.

Related Minerals
Smithsonite is isostructural with many other minerals (see **calcite**). It forms limited solid solutions with most other carbonates, including otavite, CdCO₃.

Dolomite CaMg(CO₃)₂

Origin of Name
Named after D. de Dolomieu (1750–1801), a French chemist and geologist.

Hand Specimen Identification
Dolomite is characterized by typical rhombohedral carbonate habit and cleavage and effervescence by cold dilute HCl only when powdered. It is sometimes confused with calcite or ankerite. See Figures 6.7a, 7.8a and 14.26.

Physical Properties
hardness	3½–4
specific gravity	2.85
cleavage/fracture	perfect rhombohedral {101}/subconchoidal
luster/transparency	vitreous/transparent to translucent
color	usually a shade of pink, can be white, colorless, brown, black, green, or gray
streak	white

Optical Properties
Dolomite is similar to calcite in thin section, but has lower R.I. and two possible orientations of polysynthetic twins (see **calcite**). Uniaxial $(-)$, $\omega = 1.679$, $\epsilon = 1.500$, $\delta = 0.179$.

▶ FIGURE 14.26
(a) Dolomite crystals from West Virginia; (b) dolomite crystals dusted with drusy chalcopyrite from Joplin, Missouri.

Crystallography
Hexagonal (rhombohedral), $a = 4.84$, $c = 15.96$, $Z = 3$; space group $R\bar{3}$; point group $\bar{3}$.

Habit
Crystals are typically rhombohedral, having the shape of cleavage fragments, often with curved faces (Figure 14.26). Less commonly they are prismatic or steep rhombohedra. Lamellar twinning is nearly always present but may be hard to see. Massive dolomite, showing rhombohedral cleavage, is common.

Structure and Composition
Dolomite is isostructural with calcite (see **calcite**). Fe and Mn may substitute for Mg in substantial amounts. Co, Pb, Zn, Ce, or excess Ca may also be present.

Occurrence and Associations
Dolomite is a common mineral, found in massive carbonate sediments and in marbles, often with calcite. It also occurs in hydrothermal veins with fluorite, barite, other carbonates, and quartz, and as a secondary mineral or alteration product in limestone.

Related Minerals
Dolomite is isostructural with calcite, $CaCO_3$, nordenskiöldine, $CaSnB_2O_6$, and a number of other minerals (see **calcite**). Huntite, $CaMg_3(CO_3)_4$, is quite similar. Dolomite forms solid solutions with ankerite, $CaFe(CO_3)_2$; kutnohorite, $CaMn(CO_3)_2$; minrecordite, $CaZn(CO_3)_2$; and norsethite, $BaMg(CO_3)_2$.

Ankerite $CaFe(CO_3)_2$

Origin of Name
Named after M. J. Anker (1772–1843), an Austrian mineralogist.

Hand Specimen Identification
Typical rhombohedral carbonate habit and cleavage, effervescence by cold dilute HCl when powdered, and yellow-brown to brown color usually identify ankerite. It is sometimes confused with calcite or dolomite.

Physical Properties

hardness	$3\frac{1}{2}$
specific gravity	3.10
cleavage/fracture	perfect rhombohedral {101}/subconchoidal
luster/transparency	vitreous/transparent to translucent
color	white, yellow-brown
streak	white

Optical Properties
Ankerite is similar to dolomite in thin section (see **dolomite**), but tends to be stained red or brown by iron oxidation. Uniaxial $(-)$, $\omega = 1.750$, $\epsilon = 1.548$, $\delta = 0.202$.

Crystallography
Hexagonal (rhombohedral), $a = 4.82$, $c = 16.14$, $Z = 3$; space group $R\bar{3}$; point group $\bar{3}$.

Habit
Rare crystals are usually rhombohedral, often with curved faces, or prisms. Lamellar twinning is nearly always present but may be hard to see. Granular ankerite is sometimes found. Massive ankerite showing rhombohedral cleavage is most typical.

Structure and Composition
Ankerite is isostructural with calcite, dolomite, and many other minerals (see **calcite**). It may contain substantial Mg replacing Fe but rarely significant Mn. Co, Pb, Zn, Ce, or excess Ca may also be present.

Occurrence and Associations
Ankerite is most common in Precambrian iron formations. It is also found in veins and as replacements in limestones.

Related Minerals
Ankerite is isostructural with calcite, $CaCO_3$, dolomite, $CaMg(CO_3)_2$, and a number of other minerals (see **calcite** and **dolomite**).

Aragonite $CaCO_3$

Origin of Name
Named after the original locality in Aragon, Spain.

Hand Specimen Identification
Aragonite's softness, color, and association help identify it. Like calcite, it effervesces in cold dilute HCl. However, it does not have rhombohedral cleavage. If not well crystallized or showing cleavage, it may be difficult to distinguish from calcite (Figure 10.3). Aragonite is also sometimes confused with strontianite.

Physical Properties

hardness	$3\frac{1}{2}$–4
specific gravity	2.94
cleavage/fracture	good (010), poor {110}/subconchoidal
luster/transparency	vitreous/transparent to translucent
color	colorless to white, and pale yellow
streak	white

Optical Properties

Aragonite is similar to calcite (see **calcite**) in thin section but displays parallel extinction and is biaxial $(-)$ with a small *2V*. Biaxial $(-)$, $\alpha = 1.530$, $\beta = 1.681$, $\gamma = 1.685$, $\delta = 0.155$, $2V = 18°$.

Crystallography

Orthorhombic, $a = 4.95$, $b = 7.96$, $c = 5.73$, $Z = 4$; space group $P2_1/m2_1/c2_1/n$; point group $2/m2/m2/m$.

Habit

Aragonite crystals are acicular, tabular, prismatic, or fibrous. Crystals may form radiating splays, crusts, or masses of many different morphologies. Contact and cyclic twins are common, sometimes giving it a pseudohexagonal appearance.

Structure and Composition

In aragonite, triangular $(CO_3)^{2-}$ groups are in layers perpendicular to *c*, as in calcite. Alternate layers, however, have their $(CO_3)^{2-}$ groups pointing in opposite directions. Ca is coordinated to nine oxygens in six surrounding $(CO_3)^{2-}$ groups, giving orthorhombic (pseudohexagonal) symmetry. Solid solutions are much more restricted than for calcite. Aragonite is usually near end member composition, with only minor amounts of Sr, Pb, or Zn substituting for Ca.

Occurrence and Associations

Aragonite is found as disseminated carbonate in gypsum beds, as hot spring deposits, as precipitates from Ca-oversaturated waters, associated with sedimentary iron ores, in oxidized zones of ore deposits, in some cave formations, and in blueschist facies metamorphic rocks. It also occurs in shells and other organic carbonate material. Associated minerals typically include gypsum, siderite, celestite, sulphur, limonite, calcite, malachite, azurite, smithsonite, and cerussite.

Varieties

Flos ferri is a coral-like form of aragonite associated with iron deposits.

Related Minerals

Aragonite has two significant polymorphs, calcite and vaterite. Strontianite, $SrCO_3$; witherite, $BaCO_3$; cerussite, $PbCO_3$; and niter, KNO_3, are all isostructural with aragonite.

Witherite   BaCO₃

Origin of Name

Named after D. W. Withering (1741–1799), who first demonstrated that witherite was different from barite.

Hand Specimen Identification

Density, effervescence with dilute HCl, and hardness identify witherite. It is occasionally confused with barite, but barite does not react with HCl.

Physical Properties

hardness	$3\frac{1}{2}$
specific gravity	4.29
cleavage/fracture	good basal (010), poor {110} and {012}/uneven
luster/transparency	resinous/transparent to translucent
color	gray, white, or colorless
streak	white

Optical Properties

Witherite is similar to aragonite and other orthorhombic carbonates (see **aragonite**) in thin section. It may be confused with strontianite, but the latter has lower R.I. and two good cleavages. Biaxial $(-)$, $\alpha = 1.529$, $\beta = 1.676$, $\gamma = 1.677$, $\delta = 0.148$, $2V = 16°$.

Crystallography

Orthorhombic, $a = 5.26$, $b = 8.85$, $c = 6.55$, $Z = 4$; space group $P2_1/m2_1/c2_1/n$; point group $2/m2/m2/m$.

Habit

Crystals are orthorhombic, but multiple twinning yields pseudohexagonal pyramids. Columnar, globular, and botryoidal aggregates are common.

Structure and Composition

The structure of witherite is the same as that of aragonite. Minor Sr, Mg, and Ca may substitute for Ba.

Occurrence and Associations

Witherite is a rare low-temperature vein mineral, usually associated with galena and barite.

Related Minerals

Witherite has two high-temperature polymorphs. Strontianite, $SrCO_3$; aragonite, $CaCO_3$; cerussite, $PbCO_3$; and niter, KNO_3, are all isostructural with witherite.

Strontianite   SrCO₃

Origin of Name

Named after the first known locality at Strontian, Scotland.

Hand Specimen Identification

Form, hardness, color, density, and reaction to cold dilute HCl help identify strontianite. It may be confused with aragonite but has different cleavage.

Physical Properties

hardness	$3\frac{1}{2}$–4
specific gravity	3.72
cleavage/fracture	good prismatic {110}, poor (010)/uneven
luster/transparency	vitreous/transparent to translucent
color	white, pale green, yellow, gray
streak	white

Optical Properties

Strontianite is similar to aragonite and other orthorhombic carbonates in thin section. Aragonite and witherite both have higher indices of refraction, aragonite has only one well-developed cleavage, and witherite has a larger *2V*. Biaxial $(-)$, $\alpha = 1.520$, $\beta = 1.667$, $\gamma = 1.668$, $\delta = 0.148$, $2V = 7°$.

Crystallography

Orthorhombic, $a = 5.13$, $b = 8.42$, $c = 6.09$, $Z = 4$; space group $P2_1/m2_1/c2_1/n$; point group $2/m2/m2/m$.

Habit

Strontianite is typically acicular but may be prismatic, fibrous, granular, or massive. Twins, creating pseudohexagonal or lamellar habits, are common.

Structure and Composition

Strontianite has the same structure as aragonite and a number of other minerals (see **aragonite**). Substantial replacement of Sr by Ca or Ba is common; Pb may also be present.

Occurrence and Associations

Strontianite is an uncommon mineral. It occurs in hydrothermal veins with barite, celestite, and calcite. Hosts include limestones, sulfide veins, vugs, and concretions.

Related Minerals

Strontianite has one high-temperature polymorph. Aragonite, $CaCO_3$; witherite, $BaCO_3$; cerussite, $PbCO_3$; and niter, KNO_3, are all isostructural with strontianite.

Cerussite $PbCO_3$

Origin of Name

From the Latin *cerussa,* meaning "white lead."

Hand Specimen Identification

Density, color, and luster identify cerussite. It may be confused with anglesite, but it has a different habit and effervesces in cold dilute HCl, while anglesite does not.

Physical Properties

hardness	3–$3\frac{1}{2}$
specific gravity	6.55
cleavage/fracture	good prismatic {110}, poor {021}/conchoidal
luster/transparency	adamantine/transparent to translucent
color	colorless, gray, or white
streak	white

Optical Properties

Biaxial $(-)$, $\alpha = 1.804$, $\beta = 2.076$, $\gamma = 2.078$, $\delta = 0.274$, $2V = 9°$.

Crystallography

Orthorhombic, $a = 5.15$, $b = 8.47$, $c = 6.11$, $Z = 4$; space group $P2_1/m2_1/c2_1/n$; point group $2/m2/m2/m$.

Habit

Crystals are variable but most commonly tabular, with twinning giving a pseudohexagonal appearance. Cerussite may also appear prismatic, acicular, granular, and massive. Coarse intergrowths with platy fabric are typical (Figure 14.27).

Structure and Composition

Cerussite has the same structure as aragonite and a number of other minerals (see **aragonite**). It is generally quite close to end member composition, although minor amounts of Ba, Sr, Ag, or Zn are sometimes present.

Occurrence and Associations

Cerussite is a common secondary lead mineral found in altered ore deposits. Typical associated minerals include galena, anglesite, limonite, and pyromorphite. It occurs in both veins and bedded deposits.

▶**FIGURE 14.27**

Cerussite, $PbCO_3$, is a rare carbonate mineral. This sample is from Whim Creek, Australia.

Related Minerals
Strontianite, $SrCO_3$; aragonite, $CaCO_3$; witherite, $BaCO_3$; and niter, KNO_3, are all isostructural with cerussite. Hydrocerussite, $Pb_3(CO_3)_2(OH)_2$, is a closely related mineral.

Malachite $Cu_2CO_3(OH)_2$

Origin of Name
From the Greek *moloche*, meaning "mallows," referring to the green color of mallow leaves.

Hand Specimen Identification
Habit, color, and association help identify malachite. It may be confused with other secondary copper minerals but effervesces in cold dilute HCl.

Physical Properties

hardness	$3\frac{1}{2}$–4
specific gravity	3.7–4.0
cleavage/fracture	perfect (201)/subconchoidal
luster/transparency	adamantine/transparent to translucent
color	bright green
streak	pale green

Optical Properties
Biaxial (−), $\alpha = 1.655$, $\beta = 1.875$, $\gamma = 1.909$, $\delta = 0.254$, $2V = 43°$.

Crystallography
Monoclinic, $a = 9.48$, $b = 12.03$, $c = 3.21$, $\beta = 98.0°$, $Z = 4$; space group $P2_1/a$; point group $2/m$.

Habit
Crystals are rare. Malachite is usually massive; frequently colloform or banded, and often intergrown with other secondary copper minerals.

Structure and Composition
The structure resembles those of other carbonates. Triangular $(CO_3)^{2-}$ groups are surrounded by three $Cu(O,OH)_6$ octahedra. The octahedra share edges to form chains. Zn and Co are commonly present in small amounts.

Occurrence and Associations
Malachite is a secondary copper mineral typically found in carbonate rocks with azurite, cuprite, native copper, limonite, and chrysocolla (Figure 14.28).

Related Minerals
Similar minerals include azurite, $Cu_3(CO_3)_2(OH)_2$; hydrozincite, $Zn_5(CO_3)_2(OH)_6$; aurichalcite, $(Zn,Cu)_5(CO_3)_2(OH)_6$; and a number of other rare hydrated Cu and Zn carbonates.

▶**FIGURE 14.28**
Malachite, $Cu_2CO_3(OH)_2$, and limonite (Fe-hydroxide) are often found together. This sample is from Tooele County, Utah.

Azurite $Cu_3(CO_3)_2(OH)_2$

Origin of Name
From the French *azur*, meaning "sky-blue color."

Hand Specimen Identification
Azurite's distinctive blue color, association with malachite, effervescence in cold HCl, and habit are distinctive.

Physical Properties

hardness	$3\frac{1}{2}$–4
specific gravity	3.77
cleavage/fracture	brittle, perfect {011}, good {100}/conchoidal
luster/transparency	normally dull and earthy/transparent to translucent
color	blue
streak	blue

Optical Properties
Biaxial (+), $\alpha = 1.730$, $\beta = 1.756$, $\gamma = 1.836$, $\delta = 0.106$, $2V = 68°$.

Crystallography
Monoclinic, $a = 4.97$, $b = 5.84$, $c = 10.29$, $\beta = 92.4°$, $Z = 2$; space group $P2_1/c$; point group $2/m$.

Habit
Azurite is typically massive and earthy. It may form as a crust on other copper minerals. Rarer individual crystals are tabular or prismatic.

Structure and Composition

Usually nearly pure copper carbonate, azurite has a complex structure consisting of Cu^{2+} ions coordinated to two oxygens of adjacent CO_3 groups and to two OH radicals, making a square planar group. The square groups are linked to form chains.

Occurrence and Associations

Azurite, like malachite, is a secondary copper mineral formed by alteration of copper oxides and sulfides. It is less common than malachite.

Related Minerals

Chemically similar minerals include malachite, $Cu_2CO_3(OH)_2$; hydrocerussite, $Pb_3(CO_3)_2(OH)_2$; hydromagnesite, $Mg_5(CO_3)_4(OH)_2$; and aurichalcite, $(Zn,Cu)_5(CO_3)_2(OH)_3$.

Nitratite (Soda Niter) $NaNO_3$

Origin of Name

Nitratite is one of the two more common nitrate minerals.

Hand Specimen Identification

Hardness, distinctive "cooling" taste, habit, and solubility help identify nitratite.

Physical Properties

hardness	1–2
specific gravity	2.29
cleavage/fracture	perfect but rarely seen rhombohedral {104}/conchoidal
luster/transparency	vitreous/transparent to translucent
color	colorless
streak	white

Optical Properties

Uniaxial $(-)$, $\omega = 1.587$, $\epsilon = 1.336$, $\delta = 0.251$.

Crystallography

Hexagonal (rhombohedral), $a = 5.07$, $c = 16.82$, $Z = 6$; space group $R\bar{3}2/c$; point group $\bar{3}2/m$.

Habit

Individual crystals have a rhombohedral habit, but nitratite is often too fine grained or too massive for crystals to be easily seen. It sometimes forms as a crust.

Structure and Composition

Nitratite is isostructural with calcite; Na^+ and NO_3^- take the place of Ca^{2+} and CO_3^{2-}, respectively. It is always nearly pure, forming very limited solid solution with niter (saltpeter).

Occurrence and Associations

Because nitratite is highly soluble in water, it is only found in arid regions where it may be associated with other evaporite minerals.

Related Minerals

The only other common nitrate is niter (saltpeter).

Niter (Saltpeter) KNO_3

Origin of Name

Niter is the second most common nitrate mineral.

Hand Specimen Identification

Hardness, salty taste, habit, and high solubility in water help identify this mineral.

Physical Properties

hardness	2
specific gravity	2.10
cleavage/fracture	perfect but rarely seen rhombohedral {011}/uneven
luster/transparency	vitreous/translucent
color	white
streak	white

Optical Properties

Biaxial $(-)$, $\alpha = 1.333$, $\beta = 1.505$, $\gamma = 1.505$, $\delta = 0.172$, $2V = 7°$.

Crystallography

Orthorhombic, $a = 5.43$, $b = 9.19$, $c = 6.46$, $Z = 4$; space group $P2_1/c2_1/m2_1/n$; point group $2/m2/m2/m$.

Habit

When visible, crystals are typically acicular. Niter is common as crusts and coatings or fine dusty aggregates.

Structure and Composition

Niter is isostructural with aragonite. It forms minor solid solutions with nitratite.

Occurrence and Associations

Niter is found in arid-region soils and unconsolidated sediments in caves.

Related Minerals

Eight or nine nitrate minerals are known, but all except niter and nitratite are extremely rare.

VIII. BORATES

Anhydrous Borate Group

boracite	$Mg_3ClB_7O_{13}$
sinhalite	$MgAlBO_4$

Hydrous Borate Group

borax	$Na_2B_4O_5(OH)_4 \cdot 8H_2O$
kernite	$Na_2B_4O_6(OH)_2 \cdot 3H_2O$
ulexite	$NaCaB_5O_6(OH)_6 \cdot 5H_2O$
colemanite	$CaB_3O_4(OH)_3 \cdot H_2O$
dumortierite	$Al_{6\frac{1}{2}-7}BSi_3O_{15}(O,OH)_3$

Borate minerals have complex structures and chemistries, due in large part to the small size and trivalent nature of ionic boron. They have structural similarity to carbonates and nitrates because boron combines with oxygen to form anionic groups: $(BO_3)^{3-}$ or $(BO_4)^{5-}$. Mineralogists have identified many borate minerals, but most are very rare.

Borax $Na_2B_4O_5(OH)_4 \cdot 8H_2O$

Origin of Name
From the Persian *burah,* meaning "white."

Hand Specimen Identification
Low specific gravity, softness, prismatic habit, solubility in water, and association help identify borax. Borax can be confused with other borates, especially kernite, $Na_2B_4O_6(OH)_2 \cdot 3H_2O$.

Physical Properties

hardness	$2-2\frac{1}{2}$
specific gravity	1.7–1.9
cleavage/fracture	perfect {100}, good {110}/conchoidal
luster/transparency	vitreous, resinous/ translucent
color	white, gray, rarely light blue or green
streak	white

Optical Properties
Borax is colorless in thin section, has three distinct cleavages, displays second-order interference colors, and shows anomalous interference colors in some orientations. Biaxial $(-)$, $\alpha = 1.447$, $\beta = 1.469$, $\gamma = 1.472$, $\delta = 0.025$, $2V = 40°$.

Crystallography
Monoclinic, $a = 11.84$, $b = 10.63$, $c = 12.32$, $\beta = 106.58°$, $Z = 4$; space group $C2/c$; point group $2/m$.

Habit
Euhedral crystals are stubby prisms with complex combinations of terminating faces. Borax is common in massive or granular aggregates.

Structure and Composition
Although the Na:B ratio is fixed, the amounts of $(OH)^-$ and H_2O in borax are variable. The structure consists of chains of $Na(H_2O)_6$ octahedra connected to isolated groups of boron tetrahedra and double boron triangles. Weak van der Waals and hydrogen bonds link the octahedral chains to the boron groups, resulting in perfect prismatic cleavage.

Occurrence and Associations
Borax, associated with evaporite deposits in volcanic terranes, is the most common of the hydrous borate minerals. It is found in thick beds, similar to other salts, and as crusts and surface coatings. Common associated minerals are halite, NaCl; colemanite, $CaB_3O_4(OH)_3 \cdot H_2O$; ulexite, $NaCaB_5O_6(OH)_6 \cdot 5H_2O$; and gypsum, $CaSO_4 \cdot 2H_2O$.

Related Minerals
Borax dehydrates easily to tincalconite, $Na_2B_4O_5(OH)_4 \cdot 3H_2O$.

Kernite $Na_2B_4O_6(OH)_2 \cdot 3H_2O$

Origin of Name
Named after its only major occurrence, in Kern County, California.

Hand Specimen Identification
Kernite resembles borax but is characterized by long, splintery cleavage fragments.

Physical Properties

hardness	3
specific gravity	1.90
cleavage/fracture	prismatic, perfect (100) and (001), poor (010)/uneven
luster/transparency	vitreous, pearly/transparent
color	colorless, white
streak	white

Optical Properties
Kernite is colorless in thin section and has negative relief and two perfect cleavages. Interference colors range up to second-order red or orange. Borax has a moderate $2V$ and lower birefringence; ulexite and colemanite have only one cleavage and are biaxial $(+)$. Biaxial $(-)$, $\alpha = 1.454$, $\beta = 1.472$, $\gamma = 1.488$, $\delta = 0.034$, $2V = 80°$.

Crystallography
Monoclinic, $a = 15.68$, $b = 9.09$, $c = 7.02$, $\beta = 108.87°$, $Z = 4$; space group $P2/a$; point group $2/m$.

Habit
Kernite typically is in massive or coarse aggregates that cleave into long splintery fragments.

Structure and Composition
The structure is complex, consisting of mixed chains of $(BO_4)^{5-}$ tetrahedra and $(BO_3)^{3-}$ triangles. The chains are linked by bonds to Na^+ ions.

Occurrence and Associations
The only major deposit known is in Kern County, California, where it occurs with borax and ulexite.

Related Minerals
Kernite is similar to other borates in composition but, being identified by its distinctive cleavage, is rarely misidentified.

Ulexite   NaCaB₅O₆(OH)₆ · 5H₂O

Origin of Name
Named for German chemist G. L. Ulex (1811–1883), who discovered it.

Hand Specimen Identification
Soft, rounded masses with a loose, "cottonball" appearance are common. Ulexite is similar to, and may be confused with, other borates.

Physical Properties

hardness	1–2½
specific gravity	1.96
cleavage/fracture	perfect but rarely seen {010}/uneven
luster/transparency	silky/transparent to translucent
color	white
streak	white

Optical Properties
Biaxial (+), $\alpha = 1.491$, $\beta = 1.505$, $\gamma = 1.520$, $\delta = 0.029$, $2V = 73°$.

Crystallography
Triclinic, $a = 8.73$, $b = 12.75$, $c = 6.70$, $\alpha = 90.27°$, $\beta = 109.13°$, $\gamma = 105.12°$, $Z = 2$; space group $P\bar{1}$; point group $\bar{1}$.

Habit
Acicular and fibrous crystals are typical; crystal aggregates may form rounded masses with a "cottonball" appearance. A variety called *television rock* has massive, closely packed fibers, resulting in fiberoptic-like properties (Figure 14.29).

Structure and Composition
The structure is complex, consisting of large B(O,OH)₃ and B(O,OH)₄ anionic groups, Ca^{2+} in 8- to 10-fold coordination, and Na^+ in 6-fold coordination.

Occurrence and Associations
Similar to other borate minerals, ulexite forms in arid regions from evaporating water. It is commonly associated with borax, kernite, and colemanite.

►**FIGURE 14.29**
Ulexite, a fibrous borate mineral, from Boron, California. Some varieties of ulexite are called *television rock* because of their fiberoptic-like properties.

Colemanite   CaB₃O₄(OH)₃ · H₂O

Origin of Name
Named after W. T. Coleman (1824–1893), the Californian who founded the California borax industry.

Hand Specimen Identification
Colemanite is similar in chemistry and properties to other borates. Excellent cleavage in one direction, color, transparency, and association help identify it.

Physical Properties

hardness	4–4½
specific gravity	2.42
cleavage/fracture	one perfect (010)/ subconchoidal
luster/transparency	vitreous/transparent to translucent
color	colorless, white, gray
streak	white

Optical Properties
Colemanite is colorless in thin section, appearing similar to other borates (see **borax**) but having a higher index of refraction. Biaxial (+), $\alpha = 1.586$, $\beta = 1.592$, $\gamma = 1.614$, $\delta = 0.028$, $2V = 56°$.

Crystallography
Monoclinic, $a = 8.74$, $b = 11.26$, $c = 6.10$, $\beta = 110.12°$, $Z = 4$; space group $P2_1/a$; point group $2/m$.

Habit
Crystals vary, usually being short and prismatic but sometimes massive or granular.

Structure and Composition
The structure consists of uneven sheets containing rings of $(BO_4)^{5-}$ tetrahedra and $(BO_3)^{3-}$ triangles.

Occurrence and Associations
Usually associated with ulexite, kernite, and borax, colemanite deposits form thick layers in ancient lake beds.

IX. SULFATES

Anhydrous Sulfate Group

anhydrite	$CaSO_4$
barite	$BaSO_4$
celestite	$SrSO_4$
anglesite	$PbSO_4$

Hydrous Sulfate Group

gypsum	$CaSO_4 \cdot 2H_2O$
chalcanthite	$CuSO_4 \cdot 5H_2O$
epsomite	$MgSO_4 \cdot 7H_2O$
antlerite	$Cu_3SO_4(OH)_4$
alunite	$KAl_3(SO_4)_2(OH)_6$

Mineralogists divide sulfate minerals into two groups: the anhydrous sulfates and the hydrous sulfates. Chemistries and structures of the anhydrous sulfates are related to the carbonates, with SO_4 replacing CO_3. More than 100 sulfate minerals are known, and most are rare. Gypsum and anhydrite are the only rock-forming sulfates.

For more general information about sulfates, see Chapter 6.

Anhydrite   CaSO₄

Origin of Name
From the Greek *anhydros,* because it lacks water (compared to gypsum).

Hand Specimen Identification
Anhydrite has three cleavages at 90° to each other. It is distinguished from calcite by its higher specific gravity and from gypsum by its hardness.

Physical Properties

hardness	$3-3\frac{1}{2}$
specific gravity	2.98
cleavage/fracture	perfect cubic (010), good (100) and (001)/uneven, splintery

luster/transparency	vitreous, pearly/transparent
color	colorless
streak	white

Optical Properties
Anhydrite is colorless in thin section, displays up to third-order green interference colors, has pseudocubic cleavage, and displays parallel extinction. Gypsum has lower relief and birefringence; barite and celestite have higher indices of refraction and low birefringence. Biaxial (+), $\alpha = 1.570$, $\beta = 1.575$, $\gamma = 1.614$, $\delta = 0.044$, $2V = 44°$.

Crystallography
Orthorhombic, $a = 6.22$, $b = 6.97$, $c = 6.96$, $Z = 4$; space group $C2/c2/m2_1/m$; point group $2/m2/m2/m$.

Habit
Anhydrite is usually massive, granular, or fibrous; individual crystals, typically tabular or prismatic, are rare.

Structure and Composition
The structure is similar to zircon's; $(SO_4)^{2-}$ tetrahedra share edges and are linked by (CaO_8) polyhedra. Anhydrite is generally close to $CaSO_4$ in composition but may be partially hydrated (tending toward gypsum).

Occurrence and Associations
Anhydrite is typically an evaporite mineral associated with gypsum, sulfur, halite, calcite, or dolomite. Thick anhydrite beds are well known. It is also found in amygdules or cracks in basalt, as a gangue mineral in hydrothermal ore deposits, as a component of soils, or as a hot spring deposit.

Related Minerals
Anhydrite is chemically related to other anhydrous sulfates, including barite $(BaSO_4)$; celestite $(SrSO_4)$; and anglesite $(PbSO_4)$, but has a different structure. A polymorph of anhydrite, γ-$CaSO_4$, forms when gypsum $(CaSO_4 \cdot 2H_2O)$ is dehydrated.

Barite   BaSO₄

Origin of Name
From the Greek *barys,* meaning "heavy."

Hand Specimen Identification
High specific gravity, white or light color, distinctive crystal habit, and two principle cleavages at 90° all help identify barite (Figures 6.10a and 14.30).

Physical Properties

hardness	$3-3\frac{1}{2}$
specific gravity	4.5
cleavage/fracture	perfect (001), good (010) and {210}/uneven

luster/transparency	vitreous, pearly/transparent to translucent
color	white, gray, or colorless
streak	white

Optical Properties

Barite is colorless in thin section and displays up to second-order yellow interference colors. Biaxial (+), $\alpha = 1.636$, $\beta = 1.637$, $\gamma = 1.648$, $\delta = 0.012$, $2V = 37°$.

Crystallography

Orthorhombic, $a = 8.87$, $b = 5.45$, $c = 7.14$, $Z = 4$; space group $P2_1/n2_1/m2_1/a$; point group $2/m2/m2/m$.

Habit

Barite's crystal habit is complex and variable. Tabular crystals may combine to form cockscomb aggregates called "barite roses" or "crested barite." Individual rosettes are common (Figure 14.30). Barite is also common as massive concretions, veins, or beds.

Structure and Composition

Barite contains $(SO_4)^{2-}$ tetrahedra linked by (BaO_{12}) polyhedra. Each (BaO_{12}) group is bonded to seven individual $(SO_4)^{2-}$ tetrahedra. The structure differs from anhydrite's.

Occurrence and Associations

Barite is a common gangue mineral in hydrothermal veins, associated with fluorite, galena, quartz, calcite, or dolomite. It is also found in veins in limestone, and as residual masses in clays.

Related Minerals

Barite is chemically and structurally similar to celestite, $SrSO_4$, and anglesite, $PbSO_4$, although solid solutions between the three are limited in nature.

▶FIGURE 14.30
Typical rosettes of barite, about 3 cm across.

Celestite   SrSO₄

Origin of Name

From the Latin *caelestis*, meaning "celestial," in reference to the sky-blue color of some celestite.

Hand Specimen Identification

The light blue color is distinctive, if present. Celestite may resemble barite, but has a lower specific gravity. Crystal form and 90 ° cleavage angle also aid identification. Plate 1.6 shows a color photograph of celestite with a hint of light blue color. Figure 6.10b contains an additional photograph.

Physical Properties

hardness	$3–3\frac{1}{2}$
specific gravity	3.97
cleavage/fracture	perfect basal (001), good prismatic {210}, poor (011)/uneven
luster/transparency	vitreous, pearly/transparent to translucent
color	colorless, blue, rarely light red
streak	white

Optical Properties

Celestite is similar to barite in thin section but may have a light blue color. Biaxial (+), $\alpha = 1.622$, $\beta = 1.624$, $\gamma = 1.631$, $\delta = 0.009$, $2V = 51°$.

Crystallography

Orthorhombic, $a = 8.38$, $b = 5.37$, $c = 6.85$, $Z = 4$; space group $P2_1/n2_1/m2_1/a$; point group $2/m2/m2/m$.

Habit

Tabular crystals, similar to those of barite, are typical. Acicular, fibrous, reniform, or granular crystals are also common.

Structure and Composition

Although in principal a complete solid solution exists between celestite $(SrSO_4)$ and barite $(BaSO_4)$, most celestite is close to end member $SrSO_4$. Small amounts of Pb may substitute for Sr.

Occurrence and Associations

Celestite is a rare mineral found in sedimentary rocks and in veins. Associated minerals often include dolomite, gypsum, halite, calcite, fluorite, or barite.

Related Minerals

Celestite is isostructural with barite, $BaSO_4$, and anglesite, $PbSO_4$. It may alter to strontianite, $SrCO_3$, the only other Sr end member mineral of significance.

Anglesite   PbSO₄

Origin of Name
Named after the Welsh island of Anglesey, where it was discovered.

Hand Specimen Identification
Adamantine luster, high specific gravity, and association with galena distinguish anglesite.

Physical Properties
hardness	$2\frac{1}{2}$–3
specific gravity	6.38
cleavage/fracture	perfect (001), good {210}/conchoidal
luster/transparency	adamantine/translucent
color	white
streak	white

Optical Properties
Biaxial (+), $\alpha = 1.877$, $\beta = 1.883$, $\gamma = 1.894$, $\delta = 0.017$, $2V = 75°$.

Crystallography
Orthorhombic, $a = 8.47$, $b = 5.39$, $c = 6.94$, $Z = 4$; space group $P2_1/n2_1/m2_1/a$; point group $2/m2/m2/m$.

Habit
Anglesite is normally blocky massive or in granular aggregates; individual crystals may be tabular, prismatic, bipyramidal, or nearly equant.

Structure and Composition
Anglesite, being isostructural with barite ($BaSO_4$) and celestite ($SrSO_4$), may contain significant amounts of Ba or Sr.

Occurrence and Associations
Anglesite, a common alteration product of galena, PbS, is found in oxidized portions of Pb deposits. Associated minerals include cerussite, $SrSO_4$; wulfenite ($PbMoO_4$); smithsonite, $ZnCO_3$; hemimorphite, $Zn_4(Si_2O_7)(OH)_2 \cdot H_2O$; and pyromorphite, $Pb_5(PO_4)_3Cl$.

Related Minerals
Anglesite is isostructural with barite, $BaSO_4$, and celestite, $SrSO_4$, and is one of only a few common Pb minerals.

Gypsum   CaSO₄ · 2H₂O

Origin of Name
From the Arabic *jibs,* meaning "plaster."

Hand Specimen Identification
Softness and three cleavages, tabular or platy crystals, and gray or white color distinguish gypsum. It is sometimes confused with anhydrite. See Plate 3.5 and Figures 2.7, 3.5b, 3.8b, 6.10c, and 14.31.

▶**FIGURE 14.31**
Gypsum, $CaSO_4 \cdot 2H_2O$, is the most common of the sulfate minerals. Here it is showing asymmetrical "swallow tail" twinning.

Physical Properties
hardness	2
specific gravity	2.32
cleavage/fracture	perfect basal (010), good (100) and {011}/conchoidal
luster/transparency	vitreous, pearly/transparent to translucent
color	colorless, white, variable
streak	white

Optical Properties
Gypsum is colorless in thin section and displays up to first-order yellow interference colors. It may be confused with anhydrite or barite, but anhydrite has higher relief and birefringence, and barite has higher relief and parallel extinction. Biaxial (+), $\alpha = 1.520$, $\beta = 1.523$, $\gamma = 1.529$, $\delta = 0.009$, $2V = 58°$.

Crystallography
Monoclinic, $a = 5.68$, $b = 15.518$, $c = 6.29$, $\beta = 113.83°$, $Z = 4$; space group $A2/n$; point group $2/m$.

Habit
Typical crystals are tabular, thick to thin, often forming as elongated masses or rosettes. Gypsum also forms acicular splays and may be massive or granular. Twinning and crystal intergrowths are common (Figure 14.31).

Structure and Composition
In gypsum, layers of H_2O alternate with layers containing Ca^{2+} and SO_4. Ca^{2+} is bonded to six O and two H_2O. The Ca^{2+} polyhedra link isolated SO_4 tetrahedra. Gypsum rarely contains significant impurities.

Occurrence and Associations
Gypsum, the most common sulfate mineral, is a rock-forming mineral of many evaporite deposits where it may be associated with other bedded salts. It is also

found interlayered with limestones or shales and may be found in fractures or cracks in a variety of sedimentary rocks. It is a gangue mineral or alteration product in some ore deposits and is occasionally found around fumaroles.

Varieties
Well-known varieties of gypsum include selenite (clear, often needle-like, crystals), alabaster (compact white masses), and satin spar (fibrous deposits in veins).

Related Minerals
Other hydrous sulfates include chalcanthite, $CuSO_4 \cdot 5H_2O$; epsomite, $MgSO_4 \cdot 7H_2O$; antlerite, $Cu_3SO_4(OH)_4$; and alunite, $KAl_3(SO_4)_2(OH)_6$. Gypsum forms from, or alters to, anhydrite, $CaSO_4$.

Epsomite $MgSO_4 \cdot 7H_2O$

Origin of Name
Derived from Epsom, England, where epsom salts were first precipitated from mineral waters.

Hand Specimen Identification
Crystal habit, low specific gravity, and salty taste are characteristic of epsomite.

Physical Properties

hardness	$2-2\frac{1}{2}$
specific gravity	1.68
cleavage/fracture	perfect (010), good {011}/conchoidal
luster/transparency	vitreous/transparent to translucent
color	colorless
streak	white

Optical Properties
Biaxial (+), $\alpha = 1.433$, $\beta = 1.455$, $\gamma = 1.461$, $\delta = 0.028$, $2V = 52°$.

Crystallography
Orthorhombic, $a = 11.96$, $b = 12.05$, $c = 6.88$, $Z = 4$; space group $P2_12_12_1$; point group 222.

Habit
Epsomite may be botryoidal, fibrous, or colloform, and typically forms as crusts. Large crystals are rare.

Structure and Composition
Epsomite contains two types of H_2O molecules; some are coordinated with Mg and some are not.

Occurrence and Associations
Epsomite is uncommon but occurs in caves or mine adits as encrustations, as precipitates on carbonate or mafic igneous rocks, as an evaporite mineral, or as gangue in ore deposits. It is usually associated with other sulfates.

Antlerite $Cu_3SO_4(OH)_4$

Origin of Name
Derived from the Antler Mine, Arizona, where it was first described.

Hand Specimen Identification
Antlerite has a characteristic green color and one good cleavage. It is frequently best identified by association with other Cu minerals.

Physical Properties

hardness	$3\frac{1}{2}-4$
specific gravity	3.9
cleavage/fracture	perfect (010)/uneven
luster/transparency	vitreous/transparent to translucent
color	green, white, gray
streak	green, gray

Optical Properties
Biaxial (+), $\alpha = 1.726$, $\beta = 1.738$, $\gamma = 1.789$, $\delta = 0.063$, $2V = 53°$.

Crystallography
Orthorhombic, $a = 8.24$, $b = 11.99$, $c = 6.03$, $Z = 4$; space group $P2_1/a2_1/a2_1/m$; point group $2/m2/m2/m$.

Habit
Individual prismatic crystals, often showing striations, are common. Massive forms or aggregates are also typical.

Structure and Composition
$CuO(OH)_5$ and $CuO_3(OH)_3$ octahedra are linked by $(SO_4)^{2-}$ tetrahedra.

Occurrence and Associations
Antlerite is a rare mineral associated with Cu mineralization, forming either as a primary or a secondary mineral. It may be locally abundant, and is sometimes mined as a Cu ore mineral.

Related Minerals
Chalcanthite, $CuSO_4 \cdot 5H_2O$, is another rare hydrous copper sulfate.

Alunite $KAl_3(SO_4)_2(OH)_6$

Origin of Name
From the Latin *alumen*, meaning "alum." The element aluminum was named for its presence in this mineral.

Hand Specimen Identification
Alunite's association helps identify it, but it can be difficult to tell from other massive minerals or from limestone and dolomite.

Physical Properties

hardness	4
specific gravity	2.6–2.9
cleavage/fracture	good (001), poor (101)/ conchoidal
luster/transparency	vitreous, pearly/transparent to translucent
color	white, gray, red
streak	white, gray

Optical Properties

In thin section, alunite is clear and displays up to second-order interference colors. It may be confused with brucite but has different associations. Uniaxial (+), $\omega = 1.572$, $\epsilon = 1.592$, $\delta = 0.020$.

Crystallography

Hexagonal (rhombohedral), $a = 6.97$, $c = 17.38$, $Z = 3$; space group $R3m$; point group $3m$.

Habit

Alunite forms as crusts and coatings and is usually massive. Fibrous, columnar, and granular occurrences are known.

Structure and Composition

Alunite consists of layers of isolated $(SO_4)^{2-}$ tetrahedra alternating with $Al(OH)_6$ octahedra. The two polyhedra share oxygens at their corners. Distorted $K(O,OH)_6$ octahedra also share oxygen with the $(SO_4)^{2-}$ groups. Na may substitute for K; minor amounts of Fe and P may substitute for Al and S.

Occurrence and Associations

A rare mineral, alunite is associated with fumaroles or with zones of hydrothermal alteration in K-rich igneous rocks. It is found with quartz, kaolinite, and other clay minerals.

Related Minerals

Alunite is isostructural with jarosite, $KFe_3(SO_4)_2(OH)_6$, a secondary mineral associated with some iron ores. It forms solid solutions with natroalunite, $NaAl_3(SO_4)_2(OH)_6$.

X. TUNGSTATES, MOLYBDATES, AND CHROMATES

Tungstate Group
wolframite series
   huebnerite $MnWO_4$
   ferberite $FeWO_4$
scheelite $CaWO_4$

Molybdate Group
wulfenite $PbMoO_4$

Chromate Group
crocoite $PbCrO_4$

Tungstates, molybdates, and chromates are chemically and structurally related to the anhydrous sulfates. Generally rare minerals, they may be locally concentrated in ore deposits.

Wolframite   (Fe,Mn)WO$_4$

Origin of Name

The name's origin is unclear; perhaps it comes from the German for "wolf" and *rahm*, meaning "soot," in reference to its color.

Hand Specimen Identification

High specific gravity, dark color, and one good cleavage help identify wolframite (Figure 8.4c). It is sometimes confused with hornblende and occasionally with scheelite.

Physical Properties

hardness	$4-4\frac{1}{2}$
specific gravity	7.25–7.60
cleavage/fracture	one perfect (010)/uneven
luster/transparency	submetallic/opaque unless very thin
color	black, red-brown
streak	black, red-brown

Optical Properties

Generally opaque. Biaxial (+), $\alpha = 2.17-2.31$, $\beta = 2.22-2.40$, $\gamma = 2.30-2.46$, $\delta = 0.13-0.15$, $2V = 73°-79°$.

Crystallography

Monoclinic, $a = 4.71-4.85$, $b = 5.70-5.77$, $c = 4.94-4.98$, $\beta = 90-91°$, $Z = 2$; space group $P2/c$; point group $2/m$.

Habit

Crystals are short to long prisms or tabs, often showing vertical striations. Wolframite may form bladed, subparallel crystal groups.

Structure and Composition

A complete solid solution exists between ferberite ($FeWO_4$) and huebnerite ($MnWO_4$), the two principal wolframite end members. The basic structure consists of layers of distorted $(WO_4)^{2-}$ tetrahedra joined by octahedral Fe or Mn.

Occurrence and Associations

Wolframite is a rare mineral. Usually found in high-temperature quartz veins associated with granitic igneous rocks, it is, however, the most important tungsten ore mineral. It is also found in sulfide-rich veins, associated with scheelite, cassiterite, pyrite, or galena. It may contain minor amounts of Ca, Mg, or rare earth elements.

Related Minerals

Similar minerals include the rare tungstates sanmartinite, $(Zn,Fe)WO_4$, and raspite, $Pb(WO_4)$.

Scheelite   CaWO₄

Origin of Name

Named after K. W. Scheele (1742–1786), a Swedish chemist.

Hand Specimen Identification

Crystal form, distinct cleavage, high specific gravity, and fluorescence under ultraviolet light are characteristic. It may be confused with quartz or feldspar when uncolored.

Physical Properties

hardness	$4\frac{1}{2}$–5
specific gravity	6.11
cleavage/fracture	good pyramidal {101}, poor {112}/uneven
luster/transparency	subadamantine/transparent to translucent
color	colorless, white, yellow, brown
streak	white

Optical Properties

Uniaxial (+), $\omega = 1.920$, $\epsilon = 1.934$, $\delta = 0.014$.

Crystallography

Tetragonal, $a = 5.25$, $c = 11.40$, $Z = 4$; space group $I4_1/a$; point group $4/m$.

Habit

Scheelite is found in massive, columnar, or granular aggregates and as individual crystals, usually dipyramids.

Structure and Composition

Scheelite is usually close to end member composition, but a limited solid solution exists with powellite, $CaMoO_4$. It may also incorporate small amounts of Cu or Mn. In the structure, isolated $(WO_4)^{2-}$ tetrahedra are linked by 8-coordinated Ca.

Occurrence and Associations

Scheelite is a high-temperature mineral found in metamorphic aureoles, in granites and pegmatites, and in some hydrothermal veins. It may be found with cassiterite, SnO_2; topaz, $Al_2SiO_4(F,OH)_2$; fluorite, CaF_2; apatite, $Ca_5(PO_4)_3(OH,F,Cl)$; and with other tungstates or molybdates.

Related Minerals

Scheelite is isostructural with wulfenite, $PbMoO_4$.

Wulfenite   PbMoO₄

Origin of Name

Named after Austrian mineralogist F. X. von Wulfen (1728–1805).

Hand Specimen Identification

Crystal habit and orange-yellow color are distinctive. Wulfenite is occasionally confused with native sulfur.

Physical Properties

hardness	3
specific gravity	6.7–7.0
cleavage/fracture	good but rarely seen pyramidal {011}, poor (001)/subconchoidal
luster/transparency	adamantine, silvery/transparent to translucent
color	yellow, orange, red, brown, green
streak	white

Optical Properties

Uniaxial (−), $\omega = 2.404$, $\epsilon = 2.283$, $\delta = 0.121$.

Crystallography

Tetragonal, $a = 5.42$, $c = 12.10$, $Z = 4$; space group $I4_1/a$; point group $4/m$.

Habit

Crystals are square tablets, often thin or pyramidal.

Occurrence and Associations

Wulfenite is a rare secondary mineral found in the oxidized portions of Pb deposits. It may be associated with galena, PbS; cerussite, $PbCO_3$; vanadinite, $Pb_5(VO_4)_3Cl$; or pyromorphite, $Pb_5Cl(PO_4)_3$. Wulfenite is isostructural with scheelite, $CaWO_4$, with which it forms limited solid solutions. It also forms limited solid solutions with powellite, $CaMoO_4$, and raspite, $Pb(WO_4)$.

Crocoite   PbCrO₄

Origin of Name

From the Greek *krokos,* meaning "saffron," referring to its color.

Hand Specimen Identification

Distinctive color and luster, its specific gravity and habit, and association usually make crocoite easy to identify. It is occasionally confused with wulfenite, $PbMoO_4$, or with vanadinite, $Pb_5(VO_4)_3Cl$.

Physical Properties

hardness	$2\frac{1}{2}$–3
specific gravity	6.0
cleavage/fracture	perfect {110}, poor (001)/subconchoidal
luster/transparency	adamantine/translucent
color	reddish orange
streak	orange

Optical Properties

Biaxial (+), $\alpha = 2.31$, $\beta = 2.37$, $\gamma = 2.66$, $\delta = 0.35$, $2V = 54°$.

Crystallography

Monoclinic, $a = 7.11$, $b = 7.41$, $c = 6.81$, $\beta = 102.55°$, $Z = 4$; space group $P2_1/n$; point group $2/m$.

Habit

Crystals are acicular or columnar. The thin prismatic crystals often have striations parallel to prism faces. More rarely, crocoite forms granular aggregates or patches.

Structure and Composition

Crocoite is isostructural with monazite, $(Ce,La,Th,Y)PO_4$. It consists of distorted $(CrO_4)^{2-}$ tetrahedra alternating with Pb in 9-fold coordination.

Occurrence and Associations

Crocoite is a rare secondary Pb mineral associated with veined lead deposits. It may be found with cerussite, $PbCO_3$; pyromorphite, $Pb(PO_4)_3Cl$; or wulfenite, $PbMoO_4$.

Related Minerals

Crocoite is isostructural with monazite, $(Ce,La,Th,Y)PO_4$, and with several other rare earth silicates and phosphates. It is closely related to xenotime, $Y(PO_4)$, and pucherite, $Bi(VO_4)$.

XI. PHOSPHATES, ARSENATES, AND VANADATES

Phosphate Group Minerals

monazite	$(Ce,La,Th,Y)PO_4$
triphylite	$Li(Fe,Mn)PO_4$
apatite	$Ca_5(PO_4)_3(OH,F,Cl)$
pyromorphite	$Pb_5(PO_4)_3Cl$
amblygonite	$LiAl(PO_4)F$
lazulite	$(Mg,Fe)Al_2(PO_4)_2(OH)_2$
wavellite	$Al_3(PO_4)_2(OH)_3 \cdot 5H_2O$
turquoise	$CuAl_6(PO_4)_4(OH)_8 \cdot 4H_2O$
autunite	$Ca(UO_2)_2(PO_4)_2 \cdot 10H_2O$

Vanadate Group Minerals

vanadinite	$Pb_5(VO_4)_3Cl$
carnotite	$K_2(UO_2)_2(VO_4)_2 \cdot 3H_2O$

Arsenate Group Minerals

erythrite	$Co_3(AsO_4)_2 \cdot 8H_2O$

The phosphate group contains many minerals, but most are extremely rare. Apatite is the only common example. Vanadates and arsenates, which are closely related to the phosphates in chemistry and structure, are also rare.

Monazite $(Ce,La,Th,Y)PO_4$

Origin of Name

From the Greek *monazein*, meaning "to live alone," referring to its rare occurrences in the outcrops where it was first found.

Hand Specimen Identification

Radioactivity, color, crystal habit, and associations help identify monazite. It may be confused with zircon, $ZrSiO_4$, but is not as hard and has different forms. It can be distinguished from titanite (sphene), $CaTiSiO_5$, by its crystal shape and high density.

Physical Properties

hardness	5–$5\frac{1}{2}$
specific gravity	4.9–5.2
cleavage/fracture	perfect (001), good (100)/subconchoidal
luster/transparency	variable, subresinous/translucent
color	red, brown, yellowish
streak	white

Optical Properties

Monazite is colorless, gray, or yellow-brown in thin section. It has high positive relief and displays up to third or fourth order interference colors. Biaxial (+), $\alpha = 1.785$–1.800, $\beta = 1.786$–1.801, $\gamma = 1.838$–1.850, $\delta = 0.005$, $2V = 10°$–$20°$.

Crystallography

Monoclinic, $a = 6.79$, $b = 7.01$, $c = 6.46$, $\beta = 104.4°$, $Z = 4$; space group $P2_1/n$; point group $2/m$.

Habit

Crystals are usually small, tabular, or prismatic, often forming granular masses or individual grains in sand.

Structure and Composition

Monazite is isostructural with crocoite. In its structure, distorted $(PO_4)^{3-}$ polyhedra are bonded to rare earth elements in 9-fold coordination. All the rare earths may be present, but Ce, La, and Th are usually the dominant large cations. Small amounts of Si may substitute for P in the tetrahedral sites.

Occurrence and Associations
Monazite is a rare secondary mineral is silicic igneous rocks. It is also found in unconsolidated beach or stream sediments, where it is associated with other heavy minerals such as magnetite and ilmenite.

Varieties
Rare earth content varies, so the names monazite-(Ce), monazite-(La), and so on are sometimes used to designate the dominant rare earth.

Related Minerals
Monazite is isostructural with crocoite, $PbCrO_4$, and forms solid solutions with huttonite, $ThSiO_4$. It is chemically related to xenotime, $Y(PO_4)$, with which it forms a minor solid solution.

Triphylite   Li(Fe,Mn)PO₄

Origin of Name
From the Greek words for "three" and "family," in reference to its three cations.

Hand Specimen Identification
Association, 90° cleavage, and resinous luster help identify triphylite.

Physical Properties
hardness   5–5½
specific gravity   3.5–5.5
cleavage/fracture   perfect (001), good (010)/subconchoidal
luster/transparency   vitreous, resinous/ transparent to translucent
color   variable, blue, green, brown
streak   white, gray

Optical Properties
Biaxial $(-)$, $\alpha = 1.68$, $\beta = 1.68$, $\gamma = 1.69$, $\delta = 0.01$, $2V = 0°–56°$.

Crystallography
Orthorhombic, $a = 6.01$, $b = 4.68$, $c = 10.36$, $Z = 4$; space group $P2_1/m2_1/c2_1/n$; point group $2/m2/m2/m$.

Habit
Euhedral crystals are rare; triphylite is typically fine grained and massive.

Structure and Composition
The cations occupy octahedra forming zigzag chains between $(PO_4)^{3-}$ tetrahedra. A complete solid solution exists between the Fe and Mn end members.

Compositions near the Mn end member are given the name *lithiophilite*.

Occurrence and Associations
Triphylite, typically a pegmatite mineral, is found with other phosphates, quartz, feldspar, spodumene, and beryl.

Apatite   Ca₅(PO₄)₃(OH,F,Cl)

Origin of Name
From the Greek *apate,* meaning "deceit," because it is often difficult to distinguish it from other minerals.

Hand Specimen Identification
Crystal habit, color, hardness, and lack of distinct cleavage are most diagnostic of apatite. It is occasionally confused with beryl or with epidote. (Figure 14.32)

Physical Properties
hardness   5
specific gravity   3.2
cleavage/fracture   good (001), poor {100}/ conchoidal
luster/transparency   subresinous/transparent to translucent
color   green, yellow, variable
streak   white

►**FIGURE 14.32**
Large (8 cm tall) apatite crystal from Minas Gerias, Brazil

Optical Properties

Apatite appears similar to quartz in thin section. It is colorless, does not develop good cleavage, and has very low birefringence. Quartz, however, has lower relief and is uniaxial $(+)$. Uniaxial $(-)$, $\omega = 1.633$, $\epsilon = 1.630$, $\delta = 0.003$.

Crystallography

Hexagonal, $a = 9.38$, $c = 6.86$, $Z = 2$; space group $P6_3/m$; point group $6/m$.

Habit

Apatite typically forms prismatic crystals, but may be colloform, massive, or granular.

Structure and Composition

Complete solid solution exists between hydroxylapatite (OH end member), fluorapatite (F end member), and chlorapatite (Cl end member). In addition, transition metals or Sr may replace Ca; and $(CO_3)^{2-}$, OH^-, or $(SO_4)^{2-}$ may replace some $(PO_4)^{3-}$. In the apatite structure, Ca-PO_4 chains run parallel to the c-axis. Ca^{2+} is located around channels occupied by (F,Cl,OH).

Occurrence and Associations

Apatite is a common accessory mineral but only rarely a major rock former. It is common in all igneous rocks, including pegmatites and hydrothermal veins, in metamorphic rocks, and in marine sediments.

Varieties

Collophane is a massive cryptocrystalline form of apatite that comprises some phosphate rocks and bones.

Related Minerals

A large number of phosphates, sulfates, arsenates, vanadanates, and silicates are isostructural with apatite, but none are common.

Pyromorphite   Pb₅ (PO₄)₃Cl

Origin of Name

From the Greek words meaning "fire" and "form," because it typically develops large faces when crystallizing from a magma.

Hand Specimen Identification

Form, density, color, and luster identify pyromorphite. It is sometimes confused with apatite.

Physical Properties

hardness	$3\frac{1}{2}$–4
specific gravity	7.0
cleavage/fracture	poor {100}, poor {101}
luster/transparency	resinous/transparent to translucent
color	green, yellow, variable
streak	white, yellow

Optical Properties

Uniaxial $(-)$, $\omega = 2.058$, $\epsilon = 2.048$, $\delta = 0.010$.

Crystallography

Hexagonal, $a = 9.97$, $c = 7.32$, $Z = 2$; space group $P6_3/m$; point group $6/m$.

Habit

Pyromorphite is typically prismatic, having a barrel shape; it is less commonly granular, fibrous, cavernous (hollow prisms), globular, or reniform.

Structure and Composition

Pyromorphite is isostructural with apatite (see **apatite**). Some Ca may substitute for Pb. P may be replaced by As.

Occurrence and Associations

Pyromorphite is a secondary mineral, found with other oxidized Pb or Zn minerals, in oxidized zones associated with Pb deposits. It is isostructural with apatite and with mimetite, $Pb(AsO_4)_3Cl$, with which it forms a complete solid solution.

Amblygonite   LiAl(PO₄)F

Origin of Name

From the Greek *amblygonios*, referring to its cleavage angle.

Hand Specimen Identification

Single perfect cleavage and conchoidal fracture help distinguish amblygonite from feldspar.

Physical Properties

hardness	6
specific gravity	3.0
cleavage/fracture	perfect (100), good (110), poor (011)/subconchoidal
luster/transparency	vitreous, greasy/transparent to translucent
color	white, green
streak	white

Optical Properties

Biaxial $(-)$, $\alpha = 1.59$, $\beta = 1.60$, $\gamma = 1.62$, $\delta = 0.03$, $2V = 52°$–$90°$.

Crystallography

Triclinic, $a = 5.19$, $b = 7.12$, $c = 5.04$, $\alpha = 112.02°$, $\beta = 97.82°$, $\gamma = 68.12°$, $Z = 2$; space group $P\bar{1}$; point group $\bar{1}$.

Habit

Rare crystals may be equant or columnar. Amblygonite is more commonly found as rough masses or in irregular aggregates.

Structure and Composition
Amblygonite is composed of alternating $(PO_4)^{3-}$ tetrahedra and AlO_5F octahedra linked by Li^+ ions. F may be replaced by OH. Minor Na may be present.

Occurrence and Associations
Amblygonite is a rare mineral found in pegmatites rich in Li and P. Typical associated minerals include lepidolite, spodumene, apatite, and tourmaline.

Related Minerals
Several other phosphate minerals, some occurring as gems, are similar to amblygonite and are found in pegmatites: herderite, $CaBe(PO_4)(F,OH)$; beryllonite, $NaBePO_4$; and brazilianite, $NaAl_3(PO_4)_2(OH)_4$. Complete solid solution exists between amblygonite and a hydroxy end member, montebrasite, $LiAl(PO_4)(OH)$.

Lazulite $(Mg,Fe)Al_2(PO_4)_2(OH)_2$

Origin of Name
From an Arabic word meaning "heaven," referring to its sky-blue color.

Hand Specimen Identification
Lazulite's color is distinctive. When crystalline, the pyramidal form distinguishes lazulite from other blue minerals. When massive, identification is problematic.

Physical Properties

hardness	$5-5\frac{1}{2}$
specific gravity	3.0
cleavage/fracture	poor (011)/uneven
luster/transparency	vitreous/translucent
color	azure-blue
streak	white

Optical Properties
Biaxial $(-)$, $\alpha = 1.612$, $\beta = 1.634$, $\gamma = 1.643$, $\delta = 0.031$, $2V = 70°$.

Crystallography
Monoclinic, $a = 7.16$, $b = 7.26$, $c = 7.24$, $\beta = 120.67°$, $Z = 2$; space group $P2_1/c$; point group $2/m$.

Habit
Typically massive, granular, or compact; rare crystals are prismatic or pyramidal with steep faces.

Structure and Composition
Octahedral Mg and Fe are linked to octahedral Al by sharing of O^{2-} and OH^-. The (Mg,Fe,Al) octahedra bond to $(PO_4)^{3-}$ tetrahedra. Complete solid solution exists between the Mg and Fe end members. *Scorzalite* is the name given to Fe-rich members of the series.

Occurrence and Associations
Lazulite and related minerals are rare, found only in some pegmatites and high-grade metamorphic rocks.

They may be associated with rutile, kyanite, corundum, and sillimanite.

Wavellite $Al_3(PO_4)_2(OH)_3 \cdot 5H_2O$

Origin of Name
Named after W. Wavel, who discovered it.

Hand Specimen Identification
Globular aggregates containing radiating crystals, light color, and associations distinguish wavellite (Figure 14.33).

Physical Properties

hardness	$3\frac{1}{2}-4$
specific gravity	2.36
cleavage/fracture	perfect prismatic {101}, perfect (010)/subconchoidal
luster/transparency	vitreous/translucent
color	white, greenish yellow, gray, brown
streak	white

Optical Properties
Biaxial $(+)$, $\alpha = 1.525$, $\beta = 1.534$, $\gamma = 1.552$, $\delta = 0.027$, $2V = 72°$.

Crystallography
Orthorhombic, $a = 9.62$, $b = 17.36$, $c = 6.99$, $Z = 4$; space group $P2_1/c2_1/m2_1/n$; point group $2/m2/m2/m$.

Habit
Globular aggregates and radiating crystals are most common (Figure 14.33).

▶FIGURE 14.33
Wavellite, $Al_3(PO_4)_2(OH)_3 \cdot 5H_2O$, from Montgomery County, Arkansas, showing typical radiating crystals.

Structure and Composition
Wavellite's structure is incompletely known. It may contain small amounts of Fe and Mg.

Occurrence and Associations
Wavellite is a secondary mineral found in rock cavities or on joint surfaces, in low-grade aluminous metamorphic rocks, and in phosphorite deposits. It is commonly associated with limonite and with other phosphate minerals.

Turquoise $CuAl_6(PO_4)_4(OH)_8 \cdot 4H_2O$

Origin of Name
From the French *turquoise,* meaning "Turkish," in reference to the source of the original stones imported into Europe.

Hand Specimen Identification
Turquoise has a distinctive blue color; it is distinguished from chrysocolla, hydrated Cu-silicate, by its superior hardness.

Physical Properties
hardness	6
specific gravity	2.7
cleavage/fracture	perfect but rarely seen (001), good (010)/subconchoidal, brittle
luster/transparency	resinous, waxy/translucent
color	blue, green
streak	white, green

Optical Properties
Biaxial (+), $\alpha = 1.61$, $\beta = 1.62$, $\gamma = 1.65$, $\delta = 0.04$, $2V = 40°$.

Crystallography
Triclinic, $a = 7.48$, $b = 9.95$, $c = 7.69$, $\alpha = 111.65°$, $\beta = 115.38°$, $\gamma = 69.43°$, $Z = 1$; space group $P\bar{1}$; point group $\bar{1}$.

Habit
Rare small crystals are known; reniform, massive, or granular varieties are more typical.

Structure and Composition
The structure consists of a framework of $(PO_4)^{3-}$ tetrahedra and Al octahedra. Holes in the structure contain Cu, which bonds to the polyhedra, OH^- and to H_2O. Fe^{3+} may substitute for Al.

Occurrence and Associations
Turquoise occurs as a secondary mineral associated with Al-rich volcanic rocks. It forms in small seams, veins, stringers, and crusts. Associated minerals include kaolinite, $Al_2Si_2O_5(OH)_4$; limonite, $Fe(O,OH)_n$; and chalcedony, SiO_2.

Varieties
Blue-green varieties of turquoise are called *faustite.*

Related Minerals
Complete solid solution exists between turquoise and chalcosiderite, $CuFe_6(PO_4)_4(OH)_8 \cdot 4H_2O$.

Autunite $Ca(UO_2)_2(PO_4)_2 \cdot 10H_2O$

Origin of Name
Named for Autun, France, where it is found.

Hand Specimen Identification
Radioactivity, fluorescence under ultraviolet light, and typical yellow-green tetragonal plates typify autinite.

Physical Properties
hardness	$2-2\frac{1}{2}$
specific gravity	3.15
cleavage/fracture	perfect basal (001), good prismatic (100), (010) and {110}/uneven
luster/transparency	pearly, adamantine/transparent to translucent
color	yellow
streak	yellow

Optical Properties
Uniaxial (−), $\omega = 1.577$, $\epsilon = 1.553$, $\delta = 0.024$.

Crystallography
Tetragonal, $a = 7.00$, $c = 20.67$, $Z = 4$; space group $I4/m2/m2/m$; point group $4/m2/m2/m$.

Habit
Thin tabular or flaky crystals, often square, are typical (Figure 14.34).

▶**FIGURE 14.34**
Flakes of autunite, a rare uranium phosphate mineral, from Spokane, Washington.

Structure and Composition
$(PO_4)^{3-}$ tetrahedra link to U octahedra to form uneven layers, which are joined by weakly bonded H_2O molecules. Other alkaline earths may substitute for Ca in small amounts; the amount of water in the structure is somewhat variable.

Occurrence and Associations
Autunite is a secondary uranium mineral, often forming after uraninite, UO_2. Torbernite, $Cu(UO_2)_2$ $(PO_4)_2 \cdot nH_2O$, and uraninite are common associated minerals.

Related Minerals
Torbernite is isostructural with autunite and has similar properties.

Vanadinite $Pb_5(VO_4)_3Cl$

Origin of Name
Named in reference to its vanadium content.

Hand Specimen Identification
Hexagonal crystal form, density, luster, and color help identify vanadinite. Color helps distinguish it from apatite, pyromorphite, or mimetite.

Physical Properties
hardness	3
specific gravity	6.9
cleavage/fracture	none/subconchoidal
luster/transparency	resinous/translucent
color	ruby red, orange-red, brown, yellow
streak	white, yellow

Optical Properties
Uniaxial $(-)$, $\omega = 2.416$, $\epsilon = 2.350$, $\delta = 0.066$.

Crystallography
Hexagonal, $a = 10.33$, $c = 7.35$, $Z = 2$; space group $P6_3/m$; point group $6/m$.

Habit
Vanadinite often forms prisms, sometimes hollow, with or without pyramidal faces. It may also be rounded or globular.

Structure and Composition
Vanadinite is isostructural with apatite. P or As may substitute for V in small amounts. Minor amounts of Ca, Zn, and Cu may also be present.

Occurrence and Associations
Vanadinite is a rare mineral found in the oxidized portions of Pb deposits where it is often associated with galena, cerussite, or limonite.

Related Minerals
Vanadinite is isostructural with apatite, $Ca_5(PO_4)_3$ (OH,F,Cl), and with a number of other arsenates, vanadates, and phosphates. It forms solid solutions with mimetite, $Pb_5(AsO_4)_3Cl$, and intermediate compositions are called *endlichite*.

Carnotite $K_2(UO_2)_2(VO_4)_2 \cdot 3H_2O$

Origin of Name
Named after M-A. Carnot (1839–1920), a French mining engineer and Inspector General of Mines.

Hand Specimen Identification
Radioactivity and yellow color, often with a greenish stain, characterize carnotite. It is sometimes confused with other secondary uranium minerals.

Physical Properties
hardness	1
specific gravity	4.5
cleavage/fracture	perfect but rarely seen (001)/uneven
luster/transparency	dull, earthy/translucent
color	yellow, yellow-green
streak	yellow

Optical Properties
Biaxial $(-)$, $\alpha = 1.75$, $\beta = 1.92$, $\gamma = 1.95$, $\delta = 0.20$, $2V = 38°–44°$.

Crystallography
Monoclinic, $a = 10.47$, $b = 8.41$, $c = 6.91$, $\beta = 103.67°$, $Z = 1$; space group $P2_1/a$; point group $2/m$.

Habit
Fine powder or crumbly aggregates characterize carnotite. It may also be disseminated.

Structure and Composition
The structure contains layers of edge-sharing uranium and vanadium polyhedra. The layers are joined by weak bonds to interlayer K and H_2O.

Occurrence and Associations
Carnotite is a secondary mineral found in sandstones or conglomerates that have been altered by circulation of meteoric waters.

Related Minerals
A number of other hydrated uranium oxides are known, including tyuyamunite, $Ca(UO_2)_2(VO_4)_2 \cdot nH_2O$; torbernite, $Cu(UO_2)_2(PO_4)_2 \cdot nH_2O$; and autunite, $Ca(UO_2)_2(PO_4)_2 \cdot 10H_2O$.

Erythrite $Co_3(AsO_4)_2 \cdot 8H_2O$

Origin of Name
From the Greek *erythros*, meaning "red."

Hand Specimen Identification
Association and distinctive color identify erythrite.

Physical Properties

hardness	$1\frac{1}{2}$–$2\frac{1}{2}$
specific gravity	3.06
cleavage/fracture	perfect basal (010)/sectile
luster/transparency	adamantine/transparent to translucent
color	crimson, pink, purple-red
streak	pale purple

Optical Properties
Biaxial (−), $\alpha = 1.626$, $\beta = 1.661$, $\gamma = 1.699$, $\delta = 0.073$, $2V = 90°$.

Crystallography
Monoclinic, $a = 10.26$, $b = 13.37$, $c = 4.74$, $\beta = 105.1°$, $Z = 2$; space group $C2/m$; point group $2/m$.

Habit
Erythrite may be prismatic, acicular, reniform, or globular. Erythrite typically forms as coatings and crusts.

Structure and Composition
The atomic structure is layered, with the vertex sharing by As tetrahedra and Co octahedra. Ni may substitute for Co.

Occurrence and Associations
Erythrite may form as a pink powdery coating, called *cobalt bloom,* on other cobalt minerals such as cobaltite, $(Co,Fe)AsS$, or skutterudite, $(Co,Ni)As_3$.

Related Minerals
Annabergite, $Ni_3(AsO_4)_2 \cdot 8H_2O$, also called *nickel bloom,* is isostructural with erythrite, but has an apple-green color. A complete solid solution exists between erythrite and annabergite.

Appendix

Classified List of Minerals

Class (Subclass)	Group	Series or Subgroup	Species	Chemical Formula
Silicate (Framework Silicates)	Silica Group		quartz	SiO_2
			cristobalite	SiO_2
			tridymite	SiO_2
			coesite	SiO_2
			stishovite	SiO_2
	Feldspar Group	Potassium Feldspars	orthoclase	$KAlSi_3O_8$
			sanidine	$(K,Na)AlSi_3O_8$
			microcline	$KAlSi_3O_8$
		Plagioclase Feldspar Series	albite	$NaAlSi_3O_8$
			anorthite	$CaAl_2Si_2O_8$
	Feldspathoid Group		analcime	$NaAlSi_2O_6 \cdot H_2O$
			leucite	$KAlSi_2O_6$
			nepheline	$(Na,K)AlSiO_4$
	Scapolite Group	Scapolite Series	marialite	$Na_4(AlSi_3O_8)_3Cl$
			meionite	$Ca_4(Al_2Si_2O_8)_3(CO_3,SO_4)$
	Zeolite Group		natrolite	$Na_2Al_2Si_3O_{10} \cdot 2H_2O$
			chabazite	$CaAl_2Si_4O_{12} \cdot 6H_2O$
			heulandite	$CaAl_2Si_7O_{18} \cdot 6H_2O$
			stilbite	$CaAl_2Si_7O_{18} \cdot 7H_2O$
			sodalite	$Na_3Al_3Si_3O_{12} \cdot NaCl$
	Other		beryl	$Be_3Al_2Si_6O_{18}$
			cordierite	$(Mg,Fe)_2Al_4Si_5O_{18}$
Silicate (Sheet Silicates)	Serpentine Group	Serpentine Minerals	antigorite	$Mg_6Si_4O_{10}(OH)_8$
			chrysotile	$Mg_6Si_4O_{10}(OH)_8$
			lizardite	$Mg_6Si_4O_{10}(OH)_8$
	Clay Mineral Group	Smectite Subgroup	montmorillonite	$(Ca,Na)_{0.2-0.4}(Al,Mg,Fe)_2(Si,Al)_4O_{10}(OH)_2 \cdot nH_2O$
		Illite Subgroup	illite	micalike clay of variable composition
		Other Clays	kaolinite	$Al_2Si_2O_5(OH)_4$
			pyrophyllite	$Al_2Si_4O_{10}(OH)_2$
			talc	$Mg_3Si_4O_{10}(OH)_2$
	Mica Group	Biotite Series	phlogopite	$KMg_3(AlSi_3O_{10})(OH)_2$
			annite	$KFe_3(AlSi_3O_{10})(OH)_2$

Class (Subclass)	Group	Series or Subgroup	Species	Chemical Formula
		Other Micas	muscovite	$KAl_2(AlSi_3O_{10})(OH)_2$
			margarite	$CaAl_2(Al_2Si_2O_{10})(OH)_2$
			paragonite	$NaAl_2(AlSi_3O_{10})(OH)_2$
			lepidolite	$K(Li,Al)_{2-3}(AlSi_3O_{10})(OH)_2$
	Chlorite Group		chlorite	variable combinations of talc + brucite
	Other Sheet Silicates		prehnite	$Ca_2Al(AlSi_3O_{10})(OH)_2$
			sepiolite	$Mg_4Si_6O_{15}(OH)_2 \cdot 6H_2O$
			apophyllite	$KCa_4Si_8O_{20}F \cdot 8H_2O$
Silicate (Chain Silicates)	Pyroxene Group	Orthopyroxene (Hypersthene Series)	enstatite	$Mg_2Si_2O_6$
			ferrosilite	$Fe_2Si_2O_6$
		Clinopyroxene (Diopside-Hedenbergite Series)	diopside	$CaMgSi_2O_6$
			hedenbergite	$CaFeSi_2O_6$
		Other Clinopyroxenes	pigeonite	$(Ca,Mg,Fe)_2Si_2O_6$
			augite	$(Ca,Mg,Fe,Na)(Mg,Fe,Al)(Si,Al)_2O_6$
			jadeite	$NaAlSi_2O_6$
			spodumene	$LiAlSi_2O_6$
	Amphibole Group	Cummingtonite Series	cummingtonite	$(Mg,Fe)_7Si_8O_{22}(OH)_2$
			grunerite	$Fe_7Si_8O_{22}(OH)_2$
		Tremolite Series	tremolite	$Ca_2Mg_5Si_8O_{22}(OH)_2$
			actinolite	$Ca_2(Fe,Mg)_5Si_8O_{22}(OH)_2$
		Other Amphiboles	anthophyllite	$(Mg,Fe)_7Si_8O_{22}(OH)_2$
			glaucophane	$Na_2Mg_3Al_2Si_8O_{22}(OH)_2$
			hornblende	$(K,Na)_{0-1}(Ca,Na,Fe,Mg)_2(Mg,Fe,Al)_5(Si,Al)_8O_{22}(OH)_2$
			kaersutite	Ti-rich hornblende
	Pyroxenoid Group		wollastonite	$CaSiO_3$
			rhodonite	$MnSiO_3$
			pectolite	$NaCa_2(SiO_3)_3H$
Silicate (Ring Silicates)	Melilite Group		gehlenite	$Ca_2Al_2SiO_7$
			åkermanite	$Ca_2MgSi_2O_7$
	Other		tourmaline	$(Na,Ca)(Fe,Mg,Al,Li)_3Al_6(BO_3)_3Si_6O_{18}(OH)_4$
Silicate (Isolated Tetrahedral Silicates)	Garnet Group	Pyralspite Series	pyrope	$Mg_3Al_2Si_3O_{12}$
			almandine	$Fe_3Al_2Si_3O_{12}$
			spessartine	$Mn_3Al_2Si_3O_{12}$
		Ugrandite Series	uvarovite	$Ca_3Cr_2Si_3O_{12}$
			grossular	$Ca_3Al_2Si_3O_{12}$
			andradite	$Ca_3Fe_2Si_3O_{12}$
	Olivine Group	Olivine Series	forsterite	Mg_2SiO_4
			fayalite	Fe_2SiO_4
		Other Olivines	monticellite	$CaMgSiO_4$
	Humite Group		norbergite	$Mg_3SiO_4(OH,F)_2$
			chondrodite	$Mg_5(SiO_4)_2(OH,F)_2$
			humite	$Mg_7(SiO_4)_3(OH,F)_2$
			clinohumite	$Mg_9(SiO_4)_4(OH,F)_2$
	Aluminosilicate Group		kyanite	Al_2SiO_5
			andalusite	Al_2SiO_5

Class (Subclass)	Group	Series or Subgroup	Species	Chemical Formula
			sillimanite	Al_2SiO_5
	Other Isolated Tetrahedral Silicates		staurolite	$Fe_2Al_9Si_4O_{23}(OH)$
			chloritoid	$(Fe,Mg)Al_2SiO_5(OH)_2$
			titanite (sphene)	$CaTiSiO_5$
			topaz	$Al_2SiO_4(F,OH)_2$
			zircon	$ZrSiO_4$
Silicate (Paired Tetrahedral Silicates)	Lawsonite Group		lawsonite	$CaAl_2Si_2O_7(OH)_2 \cdot H_2O$
	Epidote Group		epidote	$Ca_2(Al,Fe)_3Si_3O_{12}(OH)$
			clinozoisite	$Ca_2Al_3Si_3O_{12}(OH)$
	Other Paired Tetrahedral Silicates		vesuvianite	$Ca_{10}(Mg,Fe)_2Al_4Si_9O_{34}(OH)_4$
Native Element	Native Metal Group		gold	Au
			silver	Ag
			platinum	Pt
			copper	Cu
	Native Semimetal Group		arsenic	As
			antimony	Sb
			bismuth	Bi
	Native Nonmetal Group		diamond	C
			graphite	C
			sulfur	S
Sulfide	Tetrahedral Sulfide Group		chalcopyrite	$CuFeS_2$
			bornite	Cu_5FeS_4
			enargite	Cu_3AsS_4
			wurtzite	ZnS
			sphalerite	ZnS
	Octahedral Sulfide Group		galena	PbS
			pyrrhotite	$Fe_{1-x}S$
			niccolite	NiAs
	Mixed Sulfide Group		pentlandite	$(Ni,Fe)_9S_8$
	Other Sulfides, Arsenides, and Sulfosalts	Sulfides	molybdenite	MoS_2
			millerite	NiS
			cinnabar	HgS
			covellite	CuS
			chalcocite	Cu_2S
			argentite	Ag_2S
			pyrite	FeS_2
			cobaltite	$(Co,Fe)AsS$
			marcasite	FeS_2
		Arsenides and Sulfosalts	arsenopyrite	FeAsS
			skutterudite	$(Co,Ni)As_3$
			stibnite	Sb_2S_3
			boulangerite	$Pb_5Sb_4S_{11}$
			tetrahedrite	$Cu_{12}Sb_4S_{13}$
			proustite	Ag_3AsS_3
			pyrargyrite	Ag_3SbS_3
			orpiment	As_2S_3
			realgar	AsS

Class (Subclass)	Group	Series or Subgroup	Species	Chemical Formula
Halide			halite	NaCl
			sylvite	KCl
			chlorargerite	AgCl
			atacamite	$Cu_2Cl(OH)_3$
			carnallite	$KMgCl_3 \cdot 6H_2O$
			fluorite	CaF_2
			cryolite	Na_3AlF_6
Oxide	Tetrahedral Oxide Group		zincite	ZnO
	Octahedral Oxide Group	TiO_2 Minerals	anatase	TiO_2
			rutile	TiO_2
			brookite	TiO_2
		Other Octahedral Oxides	periclase	MgO
			hematite	Fe_2O_3
			corundum	Al_2O_3
			ilmenite	$FeTiO_3$
			cassiterite	SnO_2
			pyrolusite	MnO_2
			columbite	$(Fe,Mn)Nb_2O_6$
			tantalite	$(Fe,Mn)Ta_2O_6$
	Spinel Group	"Inverse" Spinels	spinel	$MgAl_2O_4$
			magnetite	Fe_3O_4
		"Normal" Spinels	chromite	$FeCr_2O_4$
			franklinite	$ZnFe_2O_4$
	Cubic Oxides		uraninite	UO_2
	Other Oxides		chrysoberyl	$BeAl_2O_4$
			perovskite	$CaTiO_3$
			cuprite	Cu_2O
Hydroxide			gibbsite	$Al(OH)_3$
			brucite	$Mg(OH)_2$
			manganite	$MnO(OH)$
			goethite	$FeO(OH)$
			diaspore	$AlO(OH)$
			lepidocrocite	$FeO(OH)$
			romanechite	$BaMn_9O_{16}(OH)_4$
Carbonate and Nitrate	Calcite Group		calcite	$CaCO_3$
			magnesite	$MgCO_3$
			siderite	$FeCO_3$
			rhodochrosite	$MnCO_3$
			smithsonite	$ZnCO_3$
	Dolomite Group		dolomite	$CaMg(CO_3)_2$
			ankerite	$CaFe(CO_3)_2$
	Aragonite Group		aragonite	$CaCO_3$
			witherite	$BaCO_3$
			strontianite	$SrCO_3$
			cerussite	$PbCO_3$
	Other Carbonates		malachite	$Cu_2CO_3(OH)_2$
			azurite	$Cu_3(CO_3)_2(OH)_2$
			trona	$Na_3H(CO_3)_2 \cdot 2H_2O$

Class (Subclass)	Group	Series or Subgroup	Species	Chemical Formula
	Nitrate Group		nitratite	$NaNO_3$
			niter	KNO_3
Borate	Anyhdrous Borate Group		boracite	$Mg_3ClB_7O_{13}$
			sinhalite	$MgAlBO_4$
	Hydrous Borate Group		borax	$Na_2B_4O_5(OH)_4 \cdot 8H_2O$
			kernite	$Na_2B_4O_6(OH)_2 \cdot 3H_2O$
			ulexite	$NaCaB_5O_6(OH)_6 \cdot 5H_2O$
			colemanite	$CaB_3O_4(OH)_3 \cdot H_2O$
Sulfate	Anhydrous Sulfate Group		anhydrite	$CaSO_4$
			barite	$BaSO_4$
			celestite	$SrSO_4$
			anglesite	$PbSO_4$
	Hydrous Sulfate Group		gypsum	$CaSO_4 \cdot 2H_2O$
			chalcanthite	$CuSO_4 \cdot 5H_2O$
			epsomite	$MgSO_4 \cdot 7H_2O$
			antlerite	$Cu_3SO_4(OH)_4$
			alunite	$KAl_3(SO_4)_2(OH)_6$
Chromate, Tungstate, and Molybdate	Tungstate Group	Wolframite Series	huebnerite	$MnWO_4$
			ferberite	$FeWO_4$
		Other Tungstates	scheelite	$CaWO_4$
	Molybdate Group		wulfenite	$PbMoO_4$
			powellite	$Ca(Mo,W)O_4$
	Chromate Group		crocoite	$PbCrO_4$
Phosphate, Arsenate, and Vanadate	Phosphate Group		monazite	$(Ce,La,Th,Y)PO_4$
			triphylite	$Li(Fe,Mn)PO_4$
			apatite	$Ca_5(PO_4)_3(OH,F,Cl)$
			pyromorphite	$Pb_5(PO_4)_3Cl$
			amblygonite	$LiAl(PO_4)F$
			lazulite	$(Mg,Fe)Al_2(PO_4)_2(OH)_2$
			wavellite	$Al_3(PO_4)_2(OH)_3 \cdot 5H_2O$
			vivianite	$Fe_3(PO_4)_2 \cdot 8H_2O$
			variscite	$AlPO_4 \cdot 2H_2O$
			turquoise	$CuAl_6(PO_4)_4(OH)_8 \cdot 4H_2O$
			autunite	$Ca(UO_2)_2(PO_4)_2 \cdot 10H_2O$
	Vanadate Group		vanadinite	$Pb_5(VO_4)_3Cl$
			carnotite	$K_2(UO_2)_2(VO_4)_2 \cdot 3H_2O$
	Arsenate Group		erythrite	$Co_3(AsO_4)_2 \cdot 8H_2O$
			scorodite	$FeAsO_4 \cdot 4H_2O$

Appendix

Mineral Identification Tables

This appendix contains tables that can be used to help identify unknown minerals in hand specimen. For the sake of being complete, we have included a few commonly used unaccredited mineral names, and some common nonminerals that have mineral-like properties. The names of the most common minerals are in bold print. The tables do not include information on many extremely rare or uncommon minerals. To identify one of those, you will have to use other reference books or rely on optical, X-ray diffraction, or other sophisticated methods to make identification.

The identification process is keyed to two physical properties that normally can be determined with minimal equipment: luster and hardness. First you must decide whether the unknown mineral has metallic or nonmetallic luster. This question is not always answered unambiguously. For example, graphite is not considered to be a metal. However, most mineralogists consider graphite to have a metallic luster. Other minerals, such as muscovite, have pearly or vitreous lusters, which may be misinterpreted as metallic. Because some submetallic minerals are difficult to classify as metallic or nonmetallic, these minerals have been placed on both the metallic and nonmetallic lists.

After determining the luster of an unknown mineral, you reduce the list of possible minerals by comparing the hardness of the unknown mineral to three common reference materials: an American copper penny (Mohs' hardness scale of 3), a glass slide (5.5), and quartz (7). Relative hardness may be difficult to determine if the unknown mineral has a value that is close to one of the reference materials or if it cleaves or fractures easily. In addition, some minerals have a range of possible hardness values. For these reasons, some minerals appear in more than one of the tables.

Tables B.1 to B.7 consist of six or seven columns. The data given for a mineral include hardness, name, chemical formula, color, the mineral's environment (where it is found), and miscellaneous remarks that describe its properties. Metallic minerals also have a column for their streak color. The streak color of nonmetallic minerals is often white or nonexistent, especially for the harder minerals, and so is omitted. The hardness of the minerals is based on the traditional 1–10 Mohs' scale. Several minerals have a range of possible hardness values. The names of the minerals may include British spellings or other common variations, which are listed in parentheses.

The environment column lists the possible rock and sediment types where a particular mineral *may* be found. Typically, however, only a small minority of a particular rock or sediment type will contain the mineral in question. Even if they do contain the mineral, it may be present as a **minor mineral,** or an **accessory mineral.** For example, gold is listed as occurring in placer deposits (river and stream sediments), but the vast majority of river and stream sediments do not contain gold. Even when it is present, gold is rarely visible until the ore has been concentrated. Some minerals are **essential minerals,** and therefore **major minerals;** these are, in particular, rock or sediment types, by definition. For example, pyroxenites mostly consist of pyroxenes, and granites always contain major amounts of quartz and K-feldspar.

The remarks section contains information on the properties of the mineral and other comments that may be useful for identification. Properties listed in this column include: luster, magnetism, radioactivity, density, cleavage, crystal form, flame color, fluorescence, and solubility in water and acids.

The remarks sections of the tables also contain words of caution about radioactive or potentially poisonous minerals. In most cases, radioactive minerals are not harmful, but storing large volumes of uraninite or other radioactive minerals in small, un-

To Identify a Mineral, Start Here:

Does the mineral have a nonmetallic (dull, resinous, pearly, vitreous, or otherwise) or a metallic (shiny) luster?

Nonmetallic

If the mineral fails to scratch a penny (is softer than about 3), go to the list of minerals in Table B.1.

If the mineral scratches a penny but fails to scratch glass (has hardness of $3-5\frac{1}{2}$), go to the list of minerals in Table B.2.

If the mineral scratches a penny and glass, but fails to scratch quartz (has hardness of $5\frac{1}{2}-7$), go to the list of minerals in Table B.3.

If the mineral scratches quartz (hardness greater than 7), go to the list of minerals in Table B.4.

Metallic

If the mineral fails to scratch a penny (is softer than about 3), go to the list of minerals in Table B.5.

If the mineral scratches a penny but fails to scratch glass (has hardness of $3-5\frac{1}{2}$), go to the list of minerals in Table B.6.

If the mineral scratches a penny and glass, but fails to scratch quartz (has hardness of $5\frac{1}{2}-7$), go to the list of minerals in Table B.7.

If the mineral appears to scratch quartz, you may have made a mistake. Recheck hardness and luster.

ventilated areas may generate significant amounts of radiation and radon gas, which may be undesirable.

Although taste is often useful in identifying many salts and other minerals, *avoid tasting minerals or breathing the fumes of minerals that have been heated or dissolved in acids.* Besides possibly being toxic or radioactive, minerals may also have germs or parasites that could be harmful if ingested.

▶RESOURCES

(See also the general references listed at the end of Chapter 1.)

Blatt, H., G. Middleton, and R. Murray. *Origin of Sedimentary Rocks,* 2nd ed., Englewood Cliffs, N.J.: Prentice-Hall, 1986.

Heinrich, E. W. *Microscopic Identification of Minerals.* New York: McGraw-Hill, 1965.

Prinz, M., G. Harlow, and J. Peters. *Guide to Rocks and Minerals,* New York: Simon and Schuster, 1978.

►**TABLE B.1**
Minerals With a Nonmetallic Luster and Hardness of Less Than 3

Name	Formula	Hardness	Color	Environment
Annabergite	$Ni_3(AsO_4)_2 \cdot 8H_2O$	$2\frac{1}{2}$–3	Green	Alteration of nickel deposits
Remarks: Possibly poisonous; do not taste, heat, or place in acids; vitreous to opaque luster; often radiating crystals				
Aurichalcite	$(Zn,Cu)_5(CO_3)_2(OH)_3$	2	Green or blue	Oxidation of copper and zinc deposits
Remarks: Pearly luster; effervesces in dilute hydrochloric acid; very light blue-green streak				
Autunite	$Ca(UO_2)_2(PO_4)_2 \cdot 12H_2O$	2–$2\frac{1}{2}$	Yellowish green	Oxidation and weathering of uranium deposits
Remarks: Radioactive; vitreous luster; yellow streak; strongly fluoresces yellowish green in ultraviolet light; soluble in most acids				
Biotite	$K(Mg,Fe)_3(AlSi_3O_{10})(OH)_2$	$2\frac{1}{2}$–3	Black-brown	Schists; intrusive granitic rocks; some felsic volcanics; sometimes in intermediate and mafic intrusives
Remarks: Mica; platy mineral that peels off in "sheets"; pearly or vitreous luster				
Borax	$Na_2B_4O_5(OH)_4 \cdot 8H_2O$	2–$2\frac{1}{2}$	White to colorless	Arid lake deposits
Remarks: Vitreous luster; soluble in water; white streak				
Brucite	$Mg(OH)_2$	$2\frac{1}{2}$	White, gray, or light green	Serpentinite; talc schists; chlorite schists; some marbles
Remarks: Pearly, waxy, or vitreous luster; dissolves in many cold dilute acids without effervescence				
Carnallite	$KMgCl_3 \cdot 6H_2O$	1	Colorless or red	Evaporites
Remarks: Deliquescent; conchoidal fracture; vitreous or greasy luster; phosphorescent; very water soluble; lightweight.				
Carnotite	$K_2(UO_2)_2(VO_4)_2 \cdot 3H_2O$	1	Bright yellow	Groundwater alteration of uranium and vanadium deposits; sometimes found in sandstones and petrified "wood"
Remarks: Radioactive; dull or earthy luster				
Cerargyrite	AgCl	2–3	Gray to colorless	Silver deposits
Remarks: Transparent to translucent; hornlike shape; fresh surface rapidly darkens to violet-brown when exposed to light				
Chalcanthite	$CuSO_4 \cdot 5H_2O$	2.5	Sky blue	Oxidized zones of copper deposits
Remarks: Water soluble; conchoidal fracture; vitreous luster; transparent to translucent				
Chlorite	variable combinations of talc+brucite	2–$2\frac{1}{2}$	Greenish	Low-grade metamorphics, including metabasalts and many schists; shales
Remarks: Mica; vitreous or pearly luster; platy cleavage like other micas				
Chrysocolla	approx. $Cu_4H_4Si_4O_{10}(OH)_8$	2–4	Bluish green	Oxidation of copper deposits
Remarks: Vitreous to greasy luster; decomposes in hydrochloric acid; produces green flame				
Cinnabar	HgS	$2\frac{1}{2}$	Red	Mercury deposits associated with volcanics and hot springs
Remarks: Do not heat, will release poisonous mercury vapors; also do not taste or dissolve in acids in unventilated areas; scarlet streak; high specific gravity (8.1)				
Crocoite	$PbCrO_4$	$2\frac{1}{2}$–3	Orange red	Oxidized deposits containing lead and chromium
Remarks: Possibly poisonous; do not taste, heat, or dissolve in acids; adamantine luster; orange yellow streak				
Cryolite	Na_3AlF_6	$2\frac{1}{2}$	White to colorless	Granitic pegmatites
Remarks: Vitreous to pearly luster; because of refractive index, almost disappears in water; white streak				
Epsomite	$MgSO_4 \cdot 7H_2O$	2–$2\frac{1}{2}$	White to colorless	Hydrothermal deposits
Remarks: "Epsom salts"; effloresces; forms crusts on walls; vitreous to silky appearance; dissolves in water; lightweight				
Erythrite	$Co_3(AsO_4)_2 \cdot 8H_2O$	$1\frac{1}{2}$–$2\frac{1}{2}$	Pink	Oxidation of cobalt arsenide deposits
Remarks: Possibly poisonous; do not taste or heat; adamantine to vitreous luster; acicular crystals; red streak; dissolves and produces red solution in hydrochloric acid; *do not breathe any vapors!*				
Gibbsite	$Al(OH)_3$	$2\frac{1}{2}$–$3\frac{1}{2}$	Colorless to gray	Bauxite; laterite (a tropical soil)
Remarks: Sometimes red; radial or fine-grained texture; earthy luster				

►**TABLE B.1** *(continued)*

Minerals With a Nonmetallic Luster and Hardness of Less Than 3

Name	Formula	Hardness	Color	Environment
Glauconite	$(K,Na,Ca)_{0.5-1}(Fe^{3+},Al,Fe^{2+},Mg)_2(Si,Al)_4$ $O_{10}(OH)_2 \cdot nH_2O$	2	Green	Marine sands and sandstones

Remarks: Often spherical; earthy or greasy luster

Gypsum	$CaSO_4 \cdot 2H_2O$	2	White, gray, or colorless	Evaporites; hydrothermal deposits

Remarks: Vitreous or silky luster; transparent to translucent; as hardness suggests, can be scratched with a fingernail; clear variety is called *selenite*, which should not be confused with the selenium ion with the same name

Halite	NaCl	$2\frac{1}{2}$	Typically white	Evaporites

Remarks: "Table salt"; cubic crystals; translucent to transparent; dissolves in water; avoid tasting any mineral, including this one

Hydrozincite	$Zn_5(CO_3)_2(OH)_6$	$2-2\frac{1}{2}$	Gray or white	Oxidation of zinc deposits

Remarks: Earthy or vitreous masses or crusts; sometimes fibrous or stalactites; white streak; soluble in hydrochloric acid

Kaolinite	$Al_2Si_2O_5(OH)_4$	2	Usually white	Common clay in humid climates

Remarks: Clay; earthy luster; plastic when wet; insoluble in water; often difficult to distinguish from other clays without sophisticated analytical techniques

Lepidolite	$K(Li,Al)_{2-3}(AlSi_3O_{10})(O,OH,F)_2$	$2\frac{1}{2}-4$	Pink to grayish white	Lithium pegmatites

Remarks: Lithium mica; platy cleavage typical of micas; produces a crimson flame because of lithium; insoluble in acids

Montmorillonite	$(Na,Ca)(Al,Mg)_2(Si_4O_{10})(OH)_2 \cdot nH_2O$	$1-1\frac{1}{2}$	White or gray	Weathered volcanic ash; tropical soils

Remarks: Clay; tends to expand or "swell" in water; earthy luster; crumbles easily; greasy feel

Muscovite	$KAl_2(AlSi_3O_{10})(OH)_2$	$2-2\frac{1}{2}$	White or yellow	Schists; phyllites; granites; granitic pegmatites

Remarks: Mica; platy mineral that peels off in "sheets"; pearly or vitreous luster; transparent to translucent; difficult to distinguish from paragonite, the sodium analog of muscovite

Niter (saltpeter)	KNO_3	2	White	Some soils and rock crusts

Remarks: Saltpeter; crusts of fine acicular crystals; vitreous luster; usually translucent; very soluble in water; produces violet flame because of potassium; nondeliquescent, unlike soda niter

Orpiment	As_2S_3	$1\frac{1}{2}-2$	Lemon yellow	Hydrothermal deposits

Remarks: Poisonous; do not taste, heat, or place in acids; resinous luster; pale yellow streak; translucent; perfect cleavage in one direction, unlike sulfur

Phlogopite	$KMg_3(AlSi_3O_{10})(OH)_2$	$2\frac{1}{2}-3$	Often yellowish brown	Metamorphosed dolostones; marbles; some metamorphosed ultramafic rocks; kimberlites

Remarks: Mica; platy mineral that peels off in "sheets"; pearly or vitreous luster; generally lighter color than biotite, but darker than muscovite; unlike muscovite, will decompose in sulfuric acid

Proustite	Ag_3AsS_3	$2-2\frac{1}{2}$	Scarlet red	Low-temperature silver deposits

Remarks: Very similar to pyrargyrite; red streak; adamantine luster

Pyrargyrite	Ag_3SbS_3	$2-2\frac{1}{2}$	Red	Low-temperature silver deposits

Remarks: Very similar to proustite; red streak, darker than proustite; adamantine luster

Pyrophyllite	$Al_2Si_4O_{10}(OH)_2$	$1-2$	White, gray, or brown	Hydrothermally altered rhyolites; some kyanite schists; some sulfide deposits

Remarks: Pearly to greasy luster; translucent; perfect cleavage in one direction; may be difficult to distinguish from talc

Realgar	AsS	$1\frac{1}{2}-2$	Red	Lead, silver, or gold deposits; hot springs; some volcanic deposits

Remarks: Possibly poisonous; do not taste, heat, or dissolve in acids; resinous luster; orange-red streak; good cleavage in one direction; closely associated with orpiment; long exposure to light breaks realgar down into an orange powder

Sepiolite	$Mg_4Si_6O_{15}(OH)_2 \cdot 6H_2O$	$2-2\frac{1}{2}$	White or pale yellow	Alteration of serpentinite; some lake sediments

Remarks: Earthy luster; even fine grains are opaque; porous samples may float in water; becomes plastic when mixed with water; insoluble in water and most acids; lightweight

Nitratite (Soda Niter)	$NaNO_3$	$1-2$	Usually white or colorless	Arid deposits

Remarks: Vitreous luster; very water soluble; transparent to translucent; deliquescent; produces yellow flame; usually in crusts; crystals are rare

▶**TABLE B.1** *(continued)*
Minerals With a Nonmetallic Luster and Hardness of Less Than 3

Name	Formula	Hardness	Color	Environment
Sulfur	S	$1\frac{1}{2}-2\frac{1}{2}$	Yellow	Volcanic deposits; some evaporites

Remarks: Produces blue flame (only heat sulfur in a well-ventilated hood); resinous luster; no cleavage; unpleasant smell even at room temperature

Sylvite	KCl	2	Usually white to colorless	Evaporites

Remarks: Soluble in water; cubic cleavage; produces violet flame, whereas sodium in halite produces a yellow flame; taste is more bitter than halite, however tasting minerals is not recommended; less common than halite; lightweight

Talc	$Mg_3Si_4O_{10}(OH)_2$	1	Greenish white or gray	Schists; low-temperature metamorphism of mafic rocks and dolostones

Remarks: Greasy luster and feel, which is why talc is often called "soap stone"; insoluble in most acids

Torbernite	$Cu(UO_2)_2(PO_4)_28-12H_2O$	$2-2\frac{1}{2}$	Emerald green	Alteration of uranium deposits

Remarks: Radioactive; pearly luster; perfect cleavage; dissolves in some strong acids; nonfluorescent in ultraviolet light

Ulexite	$NaCaB_5O_6(OH)_6 \cdot 5H_2O$	$1-2\frac{1}{2}$	White	Evaporites in arid lakes

Remarks: Silky luster; may have "cottonball" appearance; produces yellow flame; soluble in hot water; lightweight

Vermiculite	$(Mg,Ca)_{0.3}(Mg,Fe,Al)_3(Al,Si)_4O_{10}-(OH)_4 \cdot 8H_2O$	$1\frac{1}{2}$	Green or yellowish brown	Alteration of biotite and phlogopite

Remarks: Often used as cat litter; platy appearance; vitreous to pearly luster; pale yellow streak; slightly soluble in some strong acids; expands when heated to more than 300 °C to yellowish brown layers that may be used as packing material

Vivianite	$Fe_3(PO_4)_2 \cdot 8H_2O$	$1\frac{1}{2}-2$	Blue to green	Oxidation of sulfide deposits; alteration of iron and manganese phosphates in pegmatites; sometimes associated with fossilized bones and shells

Remarks: Radiating crystals; transparent to translucent; vitreous to pearly luster, sometimes earthy; white streak; soluble in strong acids

►**TABLE B.2**
Minerals With a Nonmetallic Luster and Hardness of $3-5\frac{1}{2}$

Name	Formula	Hardness	Color	Environment
Actinolite	$Ca_2(Mg,Fe)_5Si_8O_{22}(OH)_2$	5–6	Green	Metamorphosed mafic rocks

Remarks: Amphibole; green; vitreous luster; needlelike crystals with 56° angle; crystals tend to radiate; transparent to translucent; perfect cleavage in one direction; a solid solution exists between tremolite and ferroactinolite, with actinolite as an intermediate

Alunite	$KAl_3(SO_4)_2(OH)_6$	4	White, gray, or red	Alteration of felsic volcanics, syenites, and granitic intrusives by natural sulfuric acid solutions

Remarks: White, gray, or red; vitreous, pearly, or earthy luster; often massive; dissolves in sulfuric acid; produces an acidic solution in water; produces violet potassium flame; poor cleavage in one direction; conchoidal fracture; white streak

Analcime	$NaAlSi_2O_6 \cdot H_2O$	$5-5\frac{1}{2}$	Usually white or colorless	Hydrothermal deposits in basalts and syenites

Remarks: Usually white or colorless; vitreous luster; transparent to translucent; produces yellow flame from sodium; no cleavage; dissolves in most acids

Anglesite	$PbSO_4$	3	White, gray, or other colors	Oxidized lead deposits

Remarks: White, gray, or other colors; dense (specific gravity of about 6.3); transparent to translucent; vitreous or adamantine luster; unlike cerussite, will not effervesce in nitric acid; good cleavage in one direction; conchoidal fracture; when rubbed, gives off slight electrical charge

Anhydrite	$CaSO_4$	$3-3\frac{1}{2}$	Usually colorless	Evaporites; fills cavities of some basalts

Remarks: Usually colorless; vitreous luster; good to perfect cleavages in three directions; unlike calcite, largely insoluble in hydrochloric acid and does not effervesce

Ankerite	$CaFe(CO_3)_2$	$3\frac{1}{2}$	Yellowish white or yellowish brown	Carbonate layers within banded iron formations; dolostones

Remarks: Yellowish white or yellowish brown; crystals are rare; properties are similar to dolomite; powdered form will effervesce in hydrochloric acid

Antigorite	$Mg_6Si_4O_{10}(OH)_8$	~4	Green	Serpentinites; altered ultramafic rocks

Remarks: Platy serpentine; green; greasy luster; translucent

Antlerite	$Cu_3SO_4(OH)_4$	$3\frac{1}{2}-4$	Green to dark green	Copper deposits

Remarks: Green to dark green; vitreous luster; perfect cleavage in one direction; light green streak; striated prismatic crystals sometimes occur; transparent to translucent; unlike malachite, does not effervesce in hydrochloric acid

Apatite	$Ca_5(PO_4)_3(OH,F,Cl)$	5	Usually green, blue, or brown	Teeth and bones; pegmatites; common in trace amounts in carbonate rocks and many metamorphic rocks

Remarks: Usually green, blue, or brown; prismatic hexagonal crystals are common; vitreous luster; soluble in most acids; poor cleavage in one direction; common varieties include fluorapatite, chlorapatite, and hydroxylapatite

Apophyllite	$KCa_4(Si_4O_{10})_2F \cdot 8H_2O$	$4\frac{1}{2}-5$	Usually white, gray, or colorless	Hydrothermal deposits in basalts

Remarks: Usually white, gray, or colorless; vitreous luster; tetragonal prismatic crystals; produces violet potassium flame; low solubility in most acids

Aragonite	$CaCO_3$	$3\frac{1}{2}-4$	Usually white or colorless	Mollusk shells; hydrothermal deposits; blueschists; cavities of some basalts

Remarks: Usually white or colorless; vitreous luster; like calcite, effervesces in hydrochloric acid; harder than calcite, although calcite is far more abundant at surface conditions; pearls consist of aragonite

Arsenic	As	$3\frac{1}{2}$	White when fresh	Silver, cobalt, and silver deposits

Remarks: Poisonous; do not taste, heat, or dissolve in acids; white when fresh; gray tarnish; brittle; perfect cleavage; opaque; dull luster

Astrophyllite	$(K,Na)_3(Fe,Mn)_7(Ti,Zr)_2Si_8(O,OH)_{31}$	3	Yellow or yellowish brown	Nepheline syenites and other alkaline rocks

Remarks: Yellow or yellowish brown; perfect platy cleavage; vitreous luster, almost metallic; yellow streak; partially soluble in some acids

Atacamite	$Cu_2Cl(OH)_3$	$3-3\frac{1}{2}$	Green	Oxidized copper deposits, rarer than most other copper minerals

Remarks: Green; adamantine to vitreous luster; transparent to translucent; prismatic striated crystals; fibrous; unlike malachite, does not effervesce in hydrochloric acid

Augite	$(Ca,Mg, Fe,Na)(Mg,Fe,Al)(Si,Al)_2O_6$	5–6	Dark green to black	Basalts; andesites; gabbros; peridotites; high-temperature metamorphic rocks (for example, granulites)

Remarks: Very common pyroxene; dark green to black; cleavage at 87° and 93°, but imperfect; short prismatic crystals; vitreous to resinous luster; insoluble in most acids; augite is an intermediate member of a solid solution between diopside and hedenbergite

▶**TABLE B.2** (*continued*)
Minerals With a Nonmetallic Luster and Hardness of 3–5$\frac{1}{2}$

Name	Formula	Hardness	Color	Environment
Azurite	$Cu_3(CO_3)_2(OH)_2$	3$\frac{1}{2}$–4	Bright blue	Copper deposits

Remarks: Bright blue; usually vitreous luster; striated crystals common; crystals may radiate; perfect cleavage in one direction; light blue streak; effervesces in hydrochloric acid; not as common as malachite

Barite	$BaSO_4$	3–3$\frac{1}{2}$	Usually white or colorless	Limestones; siltstones; clays; hydrothermal deposits

Remarks: Usually white or colorless; dense compared to most other nonmetallic minerals (specific gravity of 4.5); vitreous luster; perfect cleavage in one direction; insoluble in water and many acids; may be fluorescent in ultraviolet light

Brochantite	$Cu_4SO_4(OH)_6$	3$\frac{1}{2}$–4	Green	Oxidation of copper deposits

Remarks: Green; acicular or prismatic crystals; may radiate; crystals may have striations; vitreous luster; pale green streak; soluble in acids; produces green flame

Calcite	$CaCO_3$	3	Usually white or colorless	Limestones; many marine fossils; chalk; marbles; hot springs; carbonatites; veins in some metamorphic and igneous rocks

Remarks: Very common mineral; usually white or colorless; easily effervesces with dilute hydrochloric acid; rhombohedral crystals; transparent to translucent; vitreous to dull luster; exhibits double refraction: Words on paper may appear double when viewed through large transparent crystals

Celestite	$SrSO_4$	3–3$\frac{1}{2}$	Usually colorless or white	Limestones; evaporites; lead and hydrothermal deposits

Remarks: Usually colorless or white; vitreous luster; transparent to translucent; prismatic crystals; perfect cleavage parallel to base; may be fluorescent in ultraviolet light; partially soluble in water and acids; produces bright red flame because of strontium; lower density than barite

Cerussite	$PbCO_3$	3–3$\frac{1}{2}$	Usually white, gray, or colorless	Lead deposits

Remarks: Usually white, gray, or colorless; very dense (specific gravity of about 6.5); good cleavage in one direction; adamantine luster; insoluble in hydrochloric acid; effervesces and dissolves in warm dilute nitric acid *(Note: do not pour any acidic solutions containing lead down the drain; they should be disposed of properly and legally);* may produce bluish green fluorescence in ultraviolet light

Chabazite	$Ca_2Al_2Si_4O_{12} \cdot 6H_2O$	4–5	Variable	Hydrothermal deposits in cavities of basalts and other igneous rocks

Remarks: White, brown, or variety of other colors; vitreous luster; transparent to translucent; rhombohedral (pseudocubic) crystals are common; poor cleavage in one direction; soluble in hydrochloric acid, but does not effervesce

Chrysocolla	approx. $Cu_4H_4Si_4O_{10}(OH)_8$	2–4	Green or bluish green	Oxidized copper deposits

Remarks: Green or bluish green; black when impure; typically amorphous; vitreous, greasy, or earthy luster; conchoidal fracture; decomposes in hydrochloric acid to produce silica gel; produces green flame; similar appearance to turquoise, but not as hard

Chrysotile	$Mg_6Si_4O_{10}(OH)_8$	4	Usually green to white	Serpentinites; altered ultramafic rocks

Remarks: Asbestos; *do not breathe fibers, possible carcinogen;* usually green to white; fibrous serpentine; dissolves in strong acids

Colemanite	$CaB_3O_4(OH)_3 \cdot H_2O$	4–4$\frac{1}{2}$	Usually white or colorless	Arid lake deposits

Remarks: Usually white or colorless; short prismatic crystals; vitreous luster; transparent to translucent; perfect cleavage in one direction; insoluble in water; acid solutions produce green flame because of boron

Crocidolite	$NaFe_3^{2+}Fe_2^{3+}Si_8O_{22}(OH)_2$	4	Dark blue to nearly black	Granites; syenites; pegmatites

Remarks: Blue asbestos (fibrous riebeckite); dark blue to nearly black; white to light blue streak; vitreous to silky luster; produces yellow flame because of sodium; insoluble in most acids; crossite is an intermediate in the partial solid solution series between glaucophane and riebeckite

Cuprite	Cu_2O	3$\frac{1}{2}$–4	Ruby red	Oxidized copper deposits

Remarks: Ruby red; adamantine to almost metallic luster; translucent to nearly opaque; reddish brown streak; cubic, octahedral, or dodecahedral crystals; may be fibrous; produces green flame; dissolves in concentrated hydrochloric acid

Datolite	$CaB(SiO_4)(OH)$	5–5$\frac{1}{2}$	White or light green	Cavities in basalts; serpentinites; hydrothermal deposits

Remarks: White or light green; vitreous luster; transparent to translucent; multifaceted monoclinic crystals; conchoidal fracture; dissolves in most acids

Diopside	$CaMgSi_2O_6$	5–6	Typically light green	Marbles; metamorphosed siliceous limestones and dolostones

Remarks: Pyroxene; white, light green, or other colors depending on impurities such as chromium, manganese, or vanadium; cleavage at 87° and 93°, but imperfect; prismatic crystals; vitreous to resinous luster; insoluble in most acids; forms solid solution with hedenbergite

▶**TABLE B.2** *(continued)*
Minerals With a Nonmetallic Luster and Hardness of 3–5$\frac{1}{2}$

Name	Formula	Hardness	Color	Environment
Dioptase	$Cu_6(Si_6O_{18}) \cdot 6H_2O$	5	Bright green	Oxidized copper deposits

Remarks: Bright green; rhombohedral crystals are often noticeable; green streak; vitreous luster; transparent to translucent; decomposes in most strong acids to yield silica gel

Dolomite	$CaMg(CO_3)_2$	$3\frac{1}{2}$–4	Commonly white-pink	Dolostones; limestones; marbles; talc schists; serpentinites; hydrothermal deposits

Remarks: Very common mineral; white, pink, colorless, or a variety of other colors depending on presence of impurities; vitreous luster; transparent to translucent; white streak; tends not to effervesce in hydrochloric acid unless the mineral is powdered; perfect cleavage in one direction; rhombohedral crystals; forms complete solid solution series with ankerite

Fluorite	CaF_2	4	Variable	Dolostones; limestones; hydrothermal deposits; some felsic volcanics; some granitic and intermediate intrusives

Remarks: Purple, colorless, or a variety of other colors; perfect cleavage; vitreous luster; transparent to translucent; cubic or octahedral cleavage; insoluble in water and hydrochloric acid; strongly fluorescent in ultraviolet light

Goethite	$FeO(OH)$	5–5$\frac{1}{2}$	Yellow to brown	Weathering product of serpentine, hematite, and other iron-bearing minerals

Remarks: Yellow to brown; adamantine, greasy, or earthy luster; often fibrous, acicular, or tabular; yellowish brown streak; nonmagnetic, unless heated; perfect cleavage in one direction; partially soluble in hydrochloric acid

Greenockite	CdS	3–3$\frac{1}{2}$	Yellow	Hydrothermal deposits with sphalerite

Remarks: Possibly poisonous; do not heat, taste, or dissolve in acids; yellow; usually coatings rather than distinct crystals; perfect cleavage; adamantine to resinous luster; orange or red streak; when zinc is present, greenockite fluoresces yellowish orange

Harmotome	$Ba(Al_2Si_6O_{16}) \cdot 6H_2O$	4$\frac{1}{2}$	White to gray	Hydrothermal deposits; basalts; phonolites; trachytes

Remarks: Zeolite; white, gray, or a variety of other colors; distinct cleavage; vitreous luster; transparent to translucent; decomposes in hydrochloric acid

Hedenbergite	$CaFeSi_2O_6$	5–6	Black	Iron-rich metamorphic rocks, including skarns; iron-rich igneous intrusions

Remarks: Pyroxene; black; short prismatic crystals; often radiating; nearly opaque; greenish brown streak; cleavage at 87° and 93°, but imperfect; insoluble in most acids

Hemimorphite	$Zn_4(Si_2O_7)(OH)_2 \cdot H_2O$	4$\frac{1}{2}$–5	Usually white	Oxidized zinc deposits

Remarks: Usually white; hemimorphic orthorhombic crystals; often fibrous; transparent to translucent. Vitreous luster; strongly piezoelectric and pyroelectric; dissolves in most acids to produce a silica gel; specific gravity of about 3.4, which is higher than prehnite

Heulandite	$CaAl_2Si_7O_{18} \cdot 6H_2O$	$3\frac{1}{2}$–4	White, orange, colorless, red, or yellow	Cavities of basalts

Remarks: Usually white, orange, colorless, red, or yellow; vitreous luster; perfect cleavage in one direction; decomposes in hydrochloric acid to produce a silica gel; white streak

Hornblende	$(K,Na)_{0-1}(Ca,Na,Fe,Mg)_2(Mg,Fe,Al)_5$-$(Si,Al)_8O_{22}(OH)_2$	5–6	Dark green to black	Amphibolites; schists; granodiorites; tonalites; monzonites; hornblendites; quartz porphyryes; dacites; latites; andesites

Remarks: Very common amphibole; dark green to black; fibrous to prismatic crystals with 56° and 124° cleavage angles; vitreous luster; insoluble in most acids

Hypersthene	$(Mg,Fe)SiO_3$	5–6	Dark green to brown	Peridotites; gabbros; pyroxenites; basalts; granulites; gneisses

Remarks: Orthopyroxene in enstatite-orthoferrosilite solid solution series (that is, hypersthene is 50%–80% enstatite end member); dark green to brown; vitreous luster, may be pearly to submetallic; translucent to opaque; insoluble in most acids; may be difficult to distinguish from augite without optical methods

Kyanite	Al_2SiO_5	5–7	Usually blue	High-pressure Al-rich metamorphics, including schists, some eclogites, and kimberlites

Remarks: Polymorph of sillimanite and andalusite; usually blue; hardness of about 5 parallel to length of crystals and about 7 perpendicular; bladed crystals with perfect cleavage in one direction; vitreous luster; insoluble in most acids

Laumontite	$Ca(Al_2Si_4O_{12}) \cdot 4H_2O$	4	White, pink, or light yellow	Cavities and veins in variety of intrusive igneous and volcanic rocks; hydrothermal deposits

Remarks: Usually white, pink, or light yellow; prismatic or fibrous crystals; perfect cleavage; vitreous luster; transparent to translucent; decomposes in hydrochloric acid to form silica gel; becomes powdery when exposed to dry air and light

▶**TABLE B.2** *(continued)*
Minerals With a Nonmetallic Luster and Hardness of 3–5$\frac{1}{2}$

Name	Formula	Hardness	Color	Environment
Lazulite	$(Mg,Fe)Al_2(PO_4)_2(OH)_2$	5–5$\frac{1}{2}$	Blue	Pegmatites; high-grade metamorphics, including certain quartzites and schists

Remarks: Blue; vitreous luster; translucent; insoluble in most acids; falls apart when heated; usually massive, rarely pseudodipyramidal crystals

| Lazurite | $(Na,Ca)_8(AlSiO_4)_6(SO_4,S,Cl)_2$ | 5–5$\frac{1}{2}$ | Blue, greenish blue, or violet blue | Contact metamorphosed limestones; usually associated with pyrite |

Remarks: Lapis lazuli gems; blue, greenish blue, or violet blue; usually massive; vitreous luster; translucent; produces yellow flame from sodium and dissolves in hydrochloric acid, but *beware of release of poisonous hydrogen sulfide ("rotten egg") gases!*

| Lepidochrosite | $FeO(OH)$ | 5 | Red to brown | Laterites (soils found in warm and humid climates); associated with its polymorph goethite |

Remarks: Red to brown; platy to fibrous crystals; orange streak; translucent; adamantine luster; soluble in nitric acid, less soluble in hydrochloric acid; perfect platy cleavage; nonmagnetic unless heated

| Lepidolite | $K(Li,Al)_{2-3}(AlSi_3O_{10})(O,OH,F)_2$ | 2$\frac{1}{2}$–4 | Pink or light gray | Lithium-rich pegmatites |

Remarks: Lithium mica; pink or light gray; vitreous to pearly luster; platy micaceous cleavage in one direction; translucent; produces red flame from the presence of lithium; insoluble in most acids

| Limonite | $FeO(OH) \cdot nH_2O$ | 5–5$\frac{1}{2}$ | Yellow to brown | Laterites (soils found in warm and humid climates); weathering product of many iron minerals |

Remarks: Mineraloid, since it is mostly amorphous (noncrystalline); yellow to brown; light brown streak; often earthy coatings on iron minerals; sometimes vitreous; dissolves slowly in most strong acids; variable composition and properties; may contain some crystalline iron phases, such as goethite

| Lithiophilite-Triphylite | $Li(Mn,Fe)PO_4$ to $Li(Fe,Mn)PO_4$ (lithiophilite has more manganese than iron, while triphylite has more iron than manganese) | 4$\frac{1}{2}$–5 | Blue-pink | Pegmatites |

Remarks: Solid solution series exists between lithiophilite and triphylite; lithiophilite is usually brown or pink, while triphylite tends to be bluish gray; vitreous to resinous luster; may be stained with black manganese oxide stain; produces red flame from lithium; usually massive with nearly perfect cleavage in one direction and imperfect cleavage in another

| **Magnesite** | $MgCO_3$ | 3$\frac{1}{2}$–5 | Usually white, gray, or yellow | Hydrothermal deposits in peridotites and other ultramafic igneous rocks; pegmatites; serpentinites; talc schists; limestones; dolostones |

Remarks: Solid solution exists between magnesite and siderite; usually white, gray, or yellow; vitreous luster; transparent to translucent; higher density than dolomite or calcite (specific gravity of about 3.1); white streak; sometimes fluorescent; drops of cold dilute hydrochloric acid will not effervesce on the mineral

| **Malachite** | $Cu_2CO_3(OH)_2$ | 3$\frac{1}{2}$–4 | Green | Oxidized copper deposits; can be seen on weathered pennies and copper statues as a green stain |

Remarks: Green; often earthy; sometimes occurs as crystals with adamantine or vitreous luster; may occur as agate-like concretions; light green streak; effervesces in hydrochloric acid; copper in the mineral produces a green flame

| Margarite | $CaAl_2(Al_2Si_2O_{10})(OH)_2$ | 3$\frac{1}{2}$–5 | White, pink, or gray | Chlorite, staurolite, and other schists; weathering product of corundum and other aluminum minerals |

Remarks: White, pink, or gray; translucent; brittle; micaceous (platy) cleavage; insoluble in cold hydrochloric acid, but partially soluble in hot hydrochloric acid

| Mimetite | $Pb_5(AsO_4)_3Cl$ | 3$\frac{1}{2}$ | Yellowish brown, colorless, or orange | Oxidation of lead deposits |

Remarks: Possibly poisonous; do not heat, taste, or dissolve in acids; usually yellowish brown, colorless, or orange; prismatic hexagonal crystals; resinous or adamantine luster; translucent; white streak

| Monazite | $(Ce,La,Y,Th)PO_4$ | 5–5$\frac{1}{2}$ | Yellowish to reddish brown | Usually, but not always, as microscopic crystals in granites, syenites, gneisses, pegmatites, and placer deposits |

Remarks: Thorium varieties are *radioactive;* yellowish to reddish brown; conchoidal fracture; translucent; insoluble in hydrochloric acid; fairly high density (specific gravity of about 5.0); resinous luster

| Natrolite | $Na_2Al_2Si_3O_{10} \cdot 2H_2O$ | 5–5$\frac{1}{2}$ | Usually white or colorless | Cavities in basalts; hydrothermal deposits in nepheline syenites, phonolites, and serpentinites |

Remarks: Usually white or colorless; transparent to translucent; vitreous luster; perfect cleavage in one direction; decomposes in hydrochloric acid to produce silica gel; presence of sodium produces a yellow flame; prismatic to fibrous crystals; sometimes fluoresces in ultraviolet light

| Neptunite | $KNa_2Li(Fe,Mn)_2TiO_2(Si_4O_{11})_2$ | 5–6 | Usually black | Nepheline pegmatites and syenites; serpentinites |

Remarks: Usually black; prismatic monoclinic crystals; vitreous luster; perfect cleavage; translucent to opaque; red streak; insoluble in hydrochloric acid

►**TABLE B.2** *(continued)*
Minerals With a Nonmetallic Luster and Hardness of 3–5$\frac{1}{2}$

Name	Formula	Hardness	Color	Environment
Omphacite	$(Ca,Na)(Mg,Fe,Al)Si_2O_6$	5–6	Green	Eclogites

Remarks: Sodium-rich pyroxene; green; prismatic crystals; cleavage at 87° and 93°, but imperfect; vitreous luster; translucent; white to light green streak; insoluble in most acids

Name	Formula	Hardness	Color	Environment
Opal	$SiO_2 \cdot nH_2O$	5–6	Variable	Hydrothermal deposits; volcanic tuffs; petrified wood; marine sediments

Remarks: Mineraloid; white, colorless, yellow, black, red, or a variety of other colors depending on the presence of impurities; vitreous to resinous luster; transparent to translucent; may fluoresce in ultraviolet light; conchoidal fracture; insoluble in most acids

Name	Formula	Hardness	Color	Environment
Pectolite	$NaCa_2(SiO_3)_3H$	5	Usually white, colorless, or gray	Cavities in basalts; hydrothermal deposits; serpentinites; contact metamorphosed limestones

Remarks: Pyroxenoid; usually white, colorless, or gray; vitreous luster; white streak; decomposes in hydrochloric acid and forms silica gel; produces yellow flame; transparent; perfect cleavage in two directions, which may produce needlelike crystals

Name	Formula	Hardness	Color	Environment
Phillipsite	$KCa(Al_3Si_5O_{16}) \cdot 6H_2O$	4$\frac{1}{2}$–5	Usually white or colorless	Cavities in basalts; hydrothermal alterations of volcanic ashes and feldspar-bearing rocks

Remarks: Usually white or colorless; small prismatic crystals; distinct cleavage; vitreous luster; translucent and transparent; dissolves in acids

Name	Formula	Hardness	Color	Environment
Romanechite (Psilomelane)	$BaMn_9O_{16}(OH)_4$	5–6	Black	Associated with pyrolusite (MnO_2) in marine sedimentary rocks (for example, limestones)

Remarks: Black; submetallic luster; greasy feel; brownish black streak; opaque; dissolves in hydrochloric acid; branchlike forms on rock surfaces

Name	Formula	Hardness	Color	Environment
Pyromorphite	$Pb_5(PO_4)_3Cl$	3$\frac{1}{2}$–4	Usually green, brown, or colorless	Oxidized lead deposits

Remarks: Usually green, brown, or colorless; translucent; dense (specific gravity of about 7); prismatic hexagonal crystals; sometimes crystals are hollow; resinous to adamantine luster; white streak; soluble in most acids

Name	Formula	Hardness	Color	Environment
Rhodochrosite	$MnCO_3$	3$\frac{1}{2}$–4	Usually pink	Hydrothermal deposits of silver, lead, or copper; manganese deposits

Remarks: Usually pink; vitreous luster; white or light pink streak; transparent or translucent; rhombohedral cleavage; insoluble in cold hydrochloric acid, but effervesces and is soluble in hot concentrated hydrochloric acid; softer than rhodonite

Name	Formula	Hardness	Color	Environment
Riebeckite	$Na_2Fe_3^{2+}Fe_2^{3+}Si_8O_{22}(OH)_2$	5	Dark blue to nearly black	Granites; syenites; pegmatites

Remarks: Amphibole; if coarse shows typical 60°–120° amphibole cleavage angles; dark blue to nearly black; white to light blue streak; vitreous luster; translucent; produces yellow flame because of sodium; insoluble in most acids; crossite is an intermediate in the partial solid solution series between glaucophane and riebeckite; optical or other sophisticated analytical methods may be needed to distinguish this amphibole from other bluish amphiboles

Name	Formula	Hardness	Color	Environment
Scapolite	$(Na,Ca,K)_4Al_3(Al,Si)_3Si_6O_{24}(Cl,SO_4,CO_3)$	5–6	White, blue, gray, or pink	Schists; amphibolites; gneisses; granulites; skarns; pegmatites

Remarks: Group of metamorphic minerals, including marialite-meionite solid solution series with mizzonite as an intermediate member; usually white, blue, gray, or pink; poor cleavage in two directions; vitreous luster; transparent or translucent; usually fluoresces in ultraviolet light; decomposes in hydrochloric acid, leaving a silica gel

Name	Formula	Hardness	Color	Environment
Scheelite	$CaWO_4$	4$\frac{1}{2}$–5	Usually yellow, brown, or green	Hydrothermal deposits in granitic rocks; granitic pegmatites; contact metamorphic rocks

Remarks: Partial solid solution exists with powellite ($CaMoO_4$); usually yellow, brown, or green; very dense (specific gravity of about 6); vitreous to adamantine luster; usually translucent; usually fluoresces in ultraviolet light; good cleavage in one direction; white streak

Name	Formula	Hardness	Color	Environment
Scolecite	$CaAl_2Si_3O_{10} \cdot 3H_2O$	5–5$\frac{1}{2}$	Usually white or colorless	Cavities in basalt; schists; contact metamorphosed limestones

Remarks: Usually white or colorless; radiating and striated prismatic crystals; perfect cleavage; decomposes in hydrochloric acid and leaves a silica gel; vitreous luster; usually transparent

Name	Formula	Hardness	Color	Environment
Serpentine	$Mg_3Si_2O_5(OH)_4$	3–5	Usually green	Serpentinites; weathered mafic and ultramafic rocks

Remarks: Two major serpentine minerals: antigorite (platy variety) and chrysotile (asbestiform variety); usually green; translucent to opaque; massive to fibrous; antigorite is greasy, while chrysotile tends to have silky luster

Name	Formula	Hardness	Color	Environment
Siderite	$FeCO_3$	3$\frac{1}{2}$–4	Light to dark brown	Shales; coal; hydrothermal deposits

Remarks: Light to dark brown; vitreous luster; transparent to translucent; rhombohedral crystals; fairly dense (specific gravity of about 4); white streak; often in concretions; insoluble in cold hydrochloric acid; soluble and effervescent in hot hydrochloric acid

Name	Formula	Hardness	Color	Environment
Smithsonite	$ZnCO_3$	4–4$\frac{1}{2}$	Variable	Zinc deposits, especially, in carbonate rocks

Remarks: Usually brown or blue, but may be colorless, white, green, violet, or pink

►**TABLE B.2** *(continued)*
Minerals With a Nonmetallic Luster and Hardness of $3-5\frac{1}{2}$

Name	Formula	Hardness	Color	Environment
Sphalerite	ZnS	$3\frac{1}{2}-4$	Variable	Zinc deposits; contact metamorphic deposits

Remarks: Usually brown to black, sometimes yellow, red, colorless, pink, or green; resinous to submetallic luster; white, yellow, red, or brown streak; coarser crystals show perfect cleavage in one direction; *releases poisonous* hydrogen sulfide gas when heated or dissolved in strong acids; *use ventilation!*

Titanite (Sphene)	$CaTiSiO_5$	$5-5\frac{1}{2}$	Variable	Usually as microscopic crystals in granites, granodiorites, diorites, and syenites; felsic and intermediate volcanics; gneisses; schists; amphibolites; hydrothermal deposits; contact metamorphosed carbonate rocks

Remarks: Green, black, yellow, or a variety of other colors; resinous to adamantine luster; often wedge-shaped crystals; transparent to translucent; white streak; soluble in sulfuric acid and partially soluble in hydrochloric acid

Stilbite	$CaAl_2Si_7O_{18} \cdot 7H_2O$	$3\frac{1}{2}-4$	Usually white	Cavities in basalt

Remarks: Usually white; vitreous luster; transparent to translucent; tabular or radiating fibrous crystals, often in a sheaf-like mass (like a bundle of wheat); perfect cleavage in one direction; decomposes in hydrochloric acid, but the silica in the mineral does not dissolve

Strontianite	$SrCO_3$	$3\frac{1}{2}-4$	Usually colorless, white, or gray	Low-temperature hydrothermal deposits, especially in carbonate rocks

Remarks: Usually colorless, white, or gray; adamantine luster; transparent to translucent; very dense (specific gravity near 6.6); effervesces and dissolves in warm nitric acid; soluble in hydrochloric acid; crystals may be fibrous or in prismatic aggregates; white streak; may fluoresce in ultraviolet light; strontium in the mineral produces a bright red flame

Thomsonite	$NaCa_2(Al_5Si_5O_{20}) \cdot 6H_2O$	$3-4\frac{1}{2}$	Usually white	Cavities in basalts; schists; contact metamorphosed limestones

Remarks: Usually white or sometimes brown; often prismatic crystals in radiating spheres; vitreous to pearly luster; decomposes in hydrochloric acid to produce a silica gel

Tremolite	$Ca_2Mg_5Si_8O_{22}(OH)_2$	$5-6$	Usually white	Dolomitic marbles; schists (often associated with talc); serpentinites

Remarks: Amphibole, often displaying typical $60°-120°$ cleavage angle; usually white; forms complete solid solution series with ferroactinolite, with actinolite as an intermediate; vitreous luster; usually prismatic or radiating needlelike crystals; transparent to nearly opaque; insoluble in most acids

Trona	$Na_3H(CO_3)_2 \cdot 2H_2O$	3	Usually colorless, white, or yellowish white	Evaporites in arid lake deposits

Remarks: Usually colorless, white, or yellowish white; may occur as microcrystalline earthy crusts or as radiating prismatic or acicular crystals; perfect cleavage; white streak; soluble in water and most acids

Vanadinite	$Pb_5(VO_4)_3Cl$	3	Usually red, orange-red, yellow, or brown	Oxidized lead deposits

Remarks: Usually red, orange-red, yellow, or brown; resinous to adamantine luster; very dense (specific gravity of about 6.9); transparent to translucent; often hexagonal prismatic crystals; may have fibrous or radiating crystals; no cleavage; light yellow streak; soluble in most acids

Variscite	$Al(PO_4) \cdot 2H_2O$	$3\frac{1}{2}-4\frac{1}{2}$	Bluish green	Hydrothermal alteration of feldspar-rich igneous rocks

Remarks: Bluish green; usually occurs as turquoiselike crusts or veins, but softer than turquoise; conchoidal fracture; vitreous to greasy luster; translucent

Wavellite	$Al_3(PO_4)_2(OH)_3 \cdot 5H_2O$	$3\frac{1}{2}-4$	Usually white, yellow, gray, green, or brown	Hydrothermal deposits in pegmatites and pelitic metamorphic rocks

Remarks: Usually white, yellow, gray, green, or brown; often as fibrous radiating crystals; translucent

Witherite	$BaCO_3$	$3\frac{1}{2}$	Usually white, colorless, or gray	Lead deposits

Remarks: Usually white, colorless, or gray; vitreous luster; translucent; often occurs as pseudohexagonal striated dipyramids; effervesces and dissolves in hydrochloric acid; when a few drops of sulfuric acid are added to witherite dissolved in dilute hydrochloric acid, a white barium sulfate precipitate will form; produces yellowish green flame

Wolframite	$(Fe,Mn)WO_4$	$4-4\frac{1}{2}$	Brown to black	Pegmatites; granites; placer deposits

Remarks: Solid solution series exists between ferberite ($FeWO_4$) and heubnerite ($MnWO_4$) with wolframite as an intermediate; brown (heubnerite) to black (ferberite); submetallic to resinous luster; translucent to opaque; very dense (specific gravity of 7–7.5); brown to black streak; perfect cleavage in one direction; insoluble in most acids

Wollastonite	$CaSiO_3$	$5-5\frac{1}{2}$	Usually white	Contact metamorphic limestones

Remarks: Pyroxenoid; usually white; vitreous luster; prismatic or fibrous crystals, sometimes radiating; perfect to good cleavage in three directions; two perfect cleavages are separated by approximately 84°; decomposes in hydrochloric acid to produce insoluble silica

►**TABLE B.2** *(continued)*
Minerals With a Nonmetallic Luster and Hardness of $3-5^1/_2$

Name	Formula	Hardness	Color	Environment
Wulfenite	$PbMoO_4$	3	Usually orange red	Oxidized lead deposits

Remarks: Usually orange red; vitreous, resinous, or adamantine luster; white streak; very dense (specific gravity of about 6.8); transparent to translucent; tabular tetragonal crystals; dissolves slowly in most strong acids

Name	Formula	Hardness	Color	Environment
Zincite	ZnO	4	Usually red or yellowish orange	Some zinc deposits, very rare

Remarks: Usually red or yellowish orange; adamantine luster; usually massive, crystals are very uncommon; yellowish orange streak; translucent; soluble in hydrochloric acid

▶**TABLE B.3**
Minerals With a Nonmetallic Luster and Hardness of $5\frac{1}{2}-6\frac{1}{2}$

Name	Formula	Hardness	Color	Environment
Actinolite	$Ca_2(Mg,Fe)_5Si_8O_{22}(OH)_2$	5–6	Green	Metamorphosed mafic rocks

Remarks: Amphibole; a solid solution exists between tremolite and ferroactinolite with actinolite as an intermediate; green; vitreous luster; needlelike crystals with 56° and 124° cleavage angle; crystals tend to radiate; transparent to translucent; perfect cleavage in one direction

Aegirine (Acmite)	$NaFe^{3+}Si_2O_6$	$6-6\frac{1}{2}$	Green or brown	Nepheline syenites; phonolites

Remarks: Pyroxene; green or brown; vitreous luster; translucent; prismatic crystals with angles of 87° and 93°; optical tests may be required to distinguish from other pyroxenes

Albite	$NaAlSi_3O_8$	6	Usually white or gray	Granites; syenites; rhyolites; trachytes; granitic pegmatites; schists; gneisses

Remarks: Plagioclase feldspar; usually white or gray; bladed crystals; vitreous luster; good basal cleavage; white streak; striations (twinning) are often visible; insoluble in hydrochloric acid; optical or other sophisticated analytical methods are probably needed to distinguish from other feldspars

Allanite	$(Ca,Ce)_2FeAl_2Si_3O_{12}(OH)$	$5\frac{1}{2}-6$	Brown or black	Often as microscopic crystals in granites, pegmatites, syenites, and diorites

Remarks: Brown or black; weakly radioactive; resinous to submetallic luster; may be coated with yellowish brown oxide layer; usually massive; translucent to opaque; decomposes in hydrochloric acid to produce a silica gel

Amblygonite	$LiAlFPO_4$	6	Variable	Granitic pegmatites

Remarks: White, light green, or variety of other colors; vitreous luster; translucent; usually massive; produces red flame from presence of lithium; white streak; insoluble in most acids; perfect cleavage in one direction; good to distinct cleavages in two other directions

Anatase	TiO_2	$5\frac{1}{2}-6$	Usually black, yellow, or blue	Hydrothermal deposits in granites and granitic pegmatites; placer deposits

Remarks: Usually black, yellow, or blue; perfect cleavage; usually dipyramidal crystals; adamantine to submetallic luster; transparent to translucent; pale yellow streak; insoluble in most acids

Andesine	$Ab_{70}An_{30}$ to $Ab_{50}An_{50}$	6	Usually white or gray	Diorites; andesites

Remarks: Plagioclase feldspar; usually white or gray; bladed crystals; vitreous luster; perfect basal cleavage; white streak; striations (twinning) are often visible along with a "play of colors" on the faces of the mineral; insoluble in hydrochloric acid; optical methods are usually required to distinguish from other plagioclase minerals

Anorthite	$CaAl_2Si_2O_8$	6	Usually white	Gabbros; peridotites; contact metamorphic limestones

Remarks: Plagioclase feldspar; white or a variety of other colors; tabular crystals; vitreous luster; white streak; unlike albite, decomposes in hydrochloric acid to form a silica gel; optical methods are usually required to distinguish from other plagioclase minerals

Anthophyllite	$(Mg,Fe)_7Si_8O_{22}(OH)_2$	$5\frac{1}{2}-6$	Gray, green, or brown	Cordierite, talc, or other schists; amphibolites; serpentinites; metamorphosed ultramafic rocks

Remarks: Amphibole, polymorph of cummingtonite; gray, green, or brown; vitreous luster; often fibrous; translucent; insoluble in most acids; optical methods are usually needed to distinguish from other amphiboles

Arfvedsonite	$Na_3Fe_4^{2+}Fe^{3+}Si_8O_{22}(OH)_2$	6	Green to black	Nepheline syenites and pegmatites

Remarks: Amphibole; green to black; prismatic, tabular crystals; opaque; vitreous luster; dark blue streak; insoluble in acids; usually requires optical or other sophisticated analytical methods to distinguish from other amphiboles

Augite	$(Ca,Mg,Fe,Na)(Mg,Fe,Al)(Si,Al)_2O_6$	5–6	Dark green to black	Basalts; andesites; gabbros; peridotites; high-temperature metamorphic rocks (for example, granulites)

Remarks: Pyroxene; dark green to black; cleavage at 87° and 93°, but imperfect; short prismatic crystals; vitreous to resinous luster; insoluble in most acids; augite is an intermediate member of a solid solution between diopside and hedenbergite

Axinite	$(Ca,Fe,Mn)_3Al_2(BO_3)Si_4O_{12})(OH)$	$6\frac{1}{2}-7$	Variable	Cavities in granites

Remarks: Brown, gray, violet, or other colors are possible; vitreous luster; transparent to translucent; triclinic crystals with very acute angles; distinct cleavage in one direction

Benitoite	$BaTiSi_3O_9$	$6\frac{1}{2}$	Blue	Blueschist and serpentinite complexes

Remarks: Blue; vitreous to adamantine luster; transparent to translucent; prismatic dipyramidal crystals; pleochroic; fluoresces in ultraviolet light; crystals are frequently zoned

Bronzite	$(Mg,Fe)_2Si_2O_6$	$5\frac{1}{2}$	Brownish green to black	Some gabbros and basalts; some ultramafic plutons

Remarks: Orthopyroxene; bronzite is an intermediate member in the enstatite-ferrosilite solid solution series, bronzite is 80%–88.5% enstatite end member; brownish green to black; submetallic bronzelike luster; good cleavage; short prismatic crystals; insoluble in acids; may need optical or other sophisticated methods to distinguish from other pyroxenes

▶**TABLE B.3** *(continued)*
Minerals With a Nonmetallic Luster and Hardness of $5\frac{1}{2}$–$6\frac{1}{2}$

Name	Formula	Hardness	Color	Environment
Brookite	TiO_2	$5\frac{1}{2}$–6	Brown to black	Granites; gneisses; contact metamorphic rocks

Remarks: Polymorph of rutile; brown to black; tabular crystals, often striated; adamantine to submetallic luster; poor cleavage; transparent to translucent; light yellow or light brown streak; insoluble in most acids

Bytownite	$Ab_{30}An_{70}$ to $Ab_{10}An_{90}$	6	Typically white to gray	Some gabbros, diabases, basalts, anorthosites, and diorites

Remarks: Plagioclase feldspar; white or a variety of other colors; tabular crystals; vitreous luster; white streak; optical or other sophisticated analytical methods are usually required to distinguish from other plagioclase minerals

Cancrinite	$Na_6Ca(CO_3)(AlSiO_4)_6 \cdot 2H_2O$	5–6	Violet, red, or other	Syenites

Remarks: Violet, red, or other colors; usually massive or fine grained; perfect cleavage; vitreous to greasy luster; transparent to translucent; white streak; effervesces in concentrated hydrochloric acid and produces a silica gel

Cassiterite	SnO_2	6–7	Brown to black	Hydrothermal deposits in granites and granitic pegmatites; placer deposits

Remarks: Usually brown to black; submetallic to adamantine luster; translucent; very dense (specific gravity near 7); white streak; may have elbow-shaped twins; imperfect cleavage in one direction; conchoidal fracture; insoluble in most acids

Chloritoid	$(Fe,Mg)Al_2SiO_5(OH)_2$	$6\frac{1}{2}$	Usually green to black	Iron-rich phyllites and pelitic schists; some quartzites; some contact metamorphic marbles; some hydrothermal deposits

Remarks: Chloritoid is not a mica—do not confuse with chlorite, which is softer; usually green to black; often massive; sometimes pseudohexagonal crystals; good cleavage in one direction, but not as good as micas; brittle; translucent; vitreous to pearly luster; light green to colorless streak; soluble in concentrated sulfuric acid, but insoluble in hydrochloric acid

Chondrodite	$Mg_5(SiO_4)_2(F,OH)_2$	6–$6\frac{1}{2}$	Pale yellow, reddish brown, or red	Dolomitic marbles

Remarks: Pale yellow, reddish brown, or red; vitreous to resinous luster; translucent; usually massive; soluble in hot hydrochloric acid; optical and other sophisticated analytical methods may be needed to distinguish from other humites

Clinozoisite	$Ca_2Al_3(Si_3O_{12})(OH)$	6–$6\frac{1}{2}$	Light green to gray	Amphibolites; grossular schists; slates; quartzites; contact metamorphic rocks; hydrothermal alteration of anorthitic (calcic) plagioclases in intrusives

Remarks: Light green to gray; vitreous luster; striated crystals; perfect cleavage in one direction and imperfect in another; transparent to translucent; insoluble in most acids; optical and sophisticated analytical methods may be needed to positively identify

Columbite-Tantalite solid solution series	$(Fe,Mn)Nb_2O_6$ to $(Fe,Mn)Ta_2O_6$, often with tin (Sn) and tungsten (W) impurities	6	Black	Granitic intrusives, including pegmatites

Remarks: Black; dark red or black streak; submetallic luster; may be iridescent; good cleavage in one direction; fairly dense, specific gravity up to 7.9; prismatic or tabular crystals; insoluble in acids

Cristobalite	SiO_2	$6\frac{1}{2}$	Colorless, white, or gray	Felsic volcanics (that is, rhyolites, andesites, trachytes, and obsidian), especially cavities in these rocks; low (tetragonal) and high (isometric) forms; high cristobalite is stable at 1,470° to the melting point of SiO_2 at 1,728°, although the mineral may metastably exist at lower temperatures; low cristobalite is stable below 268°

Remarks: Colorless, white, or gray; vitreous luster; usually translucent; usually microcrystalline massive deposits; no cleavage; conchoidal fracture; insoluble in hydrochloric and nitric acids

Cummingtonite	$(Mg,Fe)_7Si_8O_{22}(OH)_2$	$5\frac{1}{2}$–6	White to brown	Amphibolites; schists; some gneisses

Remarks: Amphibole; usually white to brown (darker specimens have more iron); usually fibrous, sometimes used as asbestos; perfect prismatic cleavage in one direction; translucent; vitreous to silky luster; insoluble in most acids; forms a solid solution with grunerite, the iron end member; optical and sophisticated analytical techniques are often needed to distinguish from other amphiboles

Diaspore	$AlO(OH)$	$6\frac{1}{2}$–7	Usually white, gray, or green	Bauxites; aluminum-rich schists and other metamorphic rocks

Remarks: Usually white, gray, or green; vitreous to pearly luster; transparent to translucent; perfect cleavage in one direction; massive or bladed crystals; insoluble in most acids, but may partially dissolve in very alkaline solutions

Diopside	$CaMgSi_2O_6$	5–6	White-green	Marbles; metamorphosed siliceous limestones and dolostones

Remarks: Pyroxene; white, light green, or other colors depending on impurities, such as chromium, manganese, or vanadium; cleavage at 87° and 93°, but imperfect; prismatic crystals; vitreous to resinous luster; insoluble in most acids; forms solid solution with hedenbergite

▶**TABLE B.3** *(continued)*
Minerals With a Nonmetallic Luster and Hardness of $5\frac{1}{2}$–$6\frac{1}{2}$

Name	Formula	Hardness	Color	Environment
Enstatite	$MgSiO_3$	$5\frac{1}{2}$	Usually green	Peridotites; pyroxenites; gabbros; basalts; granulites; meteorites

Remarks: Orthopyroxene; usually green; short prismatic crystals; vitreous luster; translucent; good cleavage in one direction; forms a solid solution series with the uncommon iron end member, ferrosilite ($FeSiO_3$), and with bronzite and hypersthene as intermediate members; optical and other sophisticated analytical methods are required to distinguish from other pyroxenes

Epidote	$Ca_2(Al,Fe)Al_2Si_3O_{12}(OH)$	6–7	Usually yellowish green	Hydrothermal veins in granitic plutons; amphibolites; basalts

Remarks: Usually yellowish green; vitreous to dull luster; often as coatings on other minerals; sometimes as prismatic crystals; insoluble in most acids

Fayalite	$(Fe,Mg)_2SiO_4$ with $(Fe \gg Mg)$	$6\frac{1}{2}$	Dark green to brown	Highly metamorphosed banded iron formations; pegmatites; some granitic plutons and felsic volcanics

Remarks: Iron-rich olivine, forms a solid solution series with forsterite, magnesium-rich olivine; rarer than forsterite; dark green to brown; vitreous luster; short crystals; conchoidal fracture; slowly decomposes in hydrochloric acid to produce reddish brown silica gel; pure fayalite is practically non-existent; optical or analytical methods are required to determine the amount of iron and magnesium in this olivine

Fergusonite	$(Y,Er,Ce,Fe)NbO_4$	$5\frac{1}{2}$–6	Usually gray, yellow, or brown	Granitic pegmatites

Remarks: Usually gray, yellow, or brown; usually massive with dark brown surface coating; light brown to black streak; indistinct cleavage; conchoidal fracture; dull luster, although finer grains may be vitreous; partially soluble in hydrochloric acid

Forsterite	$(Mg,Fe)_2SiO_4$ with $Mg \gg Fe$	$6\frac{1}{2}$	Green	Gabbros; peridotites; basalts; other mafic and ultramafic igneous rocks; dolomitic marbles

Remarks: Magnesium olivine; green; vitreous luster; transparent to translucent; conchoidal fracture; slowly decomposes in hydrochloric acid to produce a silica gel; optical or other sophisticated analytical methods are needed to determine the magnesium or iron contents of an olivine specimen

Gadolinite	$YFeBe_2Si_2O_{10}$	$6\frac{1}{2}$–7	Usually green or brown	Syenitic or granitic pegmatites

Remarks: Radioactive; usually green or brown; vitreous luster; no cleavage; conchoidal or bladed fracture; usually massive; usually transparent; decomposes in most strong acids to produce silica gel

Garnet	$(Mg,Fe^{2+},Mn^{2+},Ca)_3(Al,Cr,Fe^{3+})_2Si_3O_{12}$	$6\frac{1}{2}$–$7\frac{1}{2}$	Variable	Schists; gneisses; amphibolites; peridotites; kimberlites; eclogites; contact metamorphosed limestones; rarer in granitic and alkaline igneous intrusives

Remarks: Red, brown, green, or other colors, depending on composition; no cleavage; adamantine luster; transparent to translucent; dodecahedrons are common; insoluble in most acids; common types: almandine, andradite, grossular, pyrope, spessartine, and mixtures of these end members; hydrogrossularite $(Ca_3Al_2Si_2O_8(SiO_4)_{1-n}(OH)_{4n})$, hydrated garnet, may contain up to 8.5% water

Glaucophane	$Na_2Mg_3Al_2Si_8O_{22}(OH)_2$	6–$6\frac{1}{2}$	Usually blue to violet-blue	Blueschists; eclogites

Remarks: Amphibole, rarely coarse enough to show 60°–120° cleavage angle; usually blue to violet-blue; vitreous luster; translucent; often fibrous; perfect cleavage; insoluble in most acids; optical or other analytical methods are needed to distinguish glaucophane from riebeckite and other amphiboles

Grossular	$Ca_3Al_2Si_3O_{12}$	$6\frac{1}{2}$	Usually white, green, yellow, or brown	Marbles; contact metamorphosed limestones

Remarks: Garnet; usually white, green, yellow, or brown; vitreous luster; usually transparent; often dodecahedra; no cleavage; fluorescent in ultraviolet light; insoluble in most acids

Grunerite	$Fe_7Si_8O_{22}(OH)_2$	6	Brown	Iron-rich schists

Remarks: Amphibole, often showing 60°–120° cleavage angle; forms solid solution series with cummingtonite; brown; translucent; silky luster; often radiating and fibrous; optical or other analytical methods are needed to distinguish grunerite from other amphiboles

Hauynite	$(Na,Ca)_{4-8}(AlSiO_4)_6(SO_4)_{1-2}$	$5\frac{1}{2}$–6	Usually blue, white, or green	Nepheline syenites; phonolites

Remarks: Usually blue, white, or green; usually occurs as small round grains rather than euhedral dodecahedral crystals; vitreous luster; transparent to translucent; decomposes in strong acids to form a silica gel; optical or other sophisticated analytical techniques may be needed to positively identify

Hedenbergite	$CaFeSi_2O_6$	5–6	Black	Iron-rich metamorphic rocks, including skarns; iron-rich igneous intrusives

Remarks: Pyroxene; black; short prismatic crystals; often radiating; nearly opaque; greenish brown streak; cleavage at 87° and 93°, but imperfect; insoluble in most acids; optical or other analytical methods may be needed to distinguish from other pyroxenes

▶**TABLE B.3** *(continued)*
Minerals With a Nonmetallic Luster and Hardness of $5\frac{1}{2}$–$6\frac{1}{2}$

Name	Formula	Hardness	Color	Environment
Hematite	Fe_2O_3	$5\frac{1}{2}$–$6\frac{1}{2}$	Red, brown, or black	Sandstones; banded iron formations; granitic intrusives; felsic volcanics; various metamorphic rocks; common in at least trace amounts in many other lithologies

Remarks: Red, brown, or black; red streak; metallic to earthy luster; slowly dissolves in hydrochloric acid; may be iridescent

Name	Formula	Hardness	Color	Environment
Hornblende	$(K,Na)_{0-1}(Ca,Na,Fe,Mg)_2(Mg,Fe,Al)_3$-$(Si,Al)_8O_{22}(OH)_2$	5–6	Dark green to black	Amphibolites; schists; granodiorites; tonalites; monzonites; hornblendites; quartz porphyries; dacites; latites; andesites

Remarks: Very common amphibole; dark green to black; fibrous to prismatic crystals with 56° and 124° angles; translucent; vitreous luster; insoluble in most acids

Name	Formula	Hardness	Color	Environment
Hypersthene	$(Mg,Fe)SiO_3$	5–6	Usually dark green to brown	Peridotites; gabbros; pyroxenites; basalts; granulites; gneisses

Remarks: Orthopyroxene, intermediate in the enstatite-orthoferrosilite solid solution series; hypersthene is 50%–80% enstatite end member; usually dark green to brown; vitreous luster, may be pearly to submetallic; translucent to opaque; insoluble in most acids; may be difficult to distinguish from augite without optical methods

Name	Formula	Hardness	Color	Environment
Vesuvianite (Idocrase)	$Ca_{10}(Mg,Fe)_2Al_4Si_9O_{34}(OH)_4$	$6\frac{1}{2}$	Usually brown, yellow, or green	Contact metamorphosed limestones

Remarks: Usually brown, yellow, or green; vitreous to resinous luster; usually translucent; often occurs as striated tetragonal prisms; poor cleavage in one direction; conchoidal fracture; insoluble in most acids

Name	Formula	Hardness	Color	Environment
Jadeite	$NaAlSi_2O_6$	$6\frac{1}{2}$–7	Usually light or emerald green	Blueschists; serpentinites; other high-pressure metamorphic rocks

Remarks: Sodium-rich pyroxene; usually light or emerald green; vitreous luster; like other pyroxenes, cleavage angles of 87° and 93°; usually fibrous; insoluble in most acids; optical or other sophisticated analytical methods may be needed to distinguish from other pyroxenes

Name	Formula	Hardness	Color	Environment
Johannsenite	$CaMnSi_2O_6$	6	Usually dark green or gray	Marbles; skarns; hydrothermal deposits in rhyolites

Remarks: Pyroxene; usually dark green or gray; vitreous or greasy luster; good cleavage with angles of 87° and 93°; translucent to opaque; may have black manganese oxide coating; insoluble in hydrochloric acid unless the acid is heated; optical or other sophisticated analytical methods may be needed to distinguish from other pyroxenes

Name	Formula	Hardness	Color	Environment
Labradorite	$Ab_{50}An_{50}$ to $Ab_{30}An_{70}$	6	Typically blue-gray	Gabbros; basalts; anorthosites

Remarks: Plagioclase feldspar; white or a variety of other colors; often twinned; usually massive; vitreous luster; white streak; often shows iridescence (play of colors); optical methods are usually required to distinguish from other plagioclase minerals

Name	Formula	Hardness	Color	Environment
Lazulite-Scorzalite	$(Mg,Fe)Al_2(PO_4)_2(OH)_2$ (Lazulite has $Mg > Fe$, while scorzalite has $Fe > Mg$)	5–$5\frac{1}{2}$	Usually blue	Pegmatites; pelitic schists

Remarks: Usually blue; vitreous luster; usually translucent; usually massive; indistinct prismatic cleavage; fragments when heated; insoluble in most acids; optical and other sophisticated analytical methods may be needed to distinguish from other blue minerals. Associated with pyrite

Name	Formula	Hardness	Color	Environment
Lechatelierite	SiO_2	6–7	Variable	Lightning strikes on sand deposits; recent meteorite impacts; sand deposits in contact with recently erupted volcanics

Remarks: Natural silica glass; mineraloid; amorphous; colors vary depending on impurities; glassy luster; natural glasses tend to weather away rapidly over time

Name	Formula	Hardness	Color	Environment
Leucite	$KAlSi_2O_6$	$5\frac{1}{2}$–6	Usually white or gray	Recent silica-poor volcanics (for example, basalts and phonolites), often as phenocrysts

Remarks: Usually white or gray; translucent to opaque; vitreous luster; tetragonal form often consists of trapezohedrons; no cleavage; conchoidal fracture; soluble in most strong acids; tetragonal form is stable below 605 °C, while isometric form occurs above this temperature

Name	Formula	Hardness	Color	Environment
Marialite-Meionite	$Na_4(AlSi_3O_8)_3(Cl_2,CO_3,SO_4)$ to $Ca_4(Al_2Si_2O_8)_3(Cl_2,CO_3,SO_4)$	$5\frac{1}{2}$–6	Usually white or gray	Schists; gneisses; amphibolites; contact metamorphosed limestones; pegmatites

Remarks: Scapolite minerals; marialite is the sodium end member of a solid solution series with meionite, the calcium end member; usually white or gray, but may be a variety of colors; vitreous luster; transparent to translucent; often prismatic crystals with vertical striations; distinct cleavage in two directions; often fluoresces in ultraviolet light; decomposes in hydrochloric acid to produce a silica gel

Name	Formula	Hardness	Color	Environment
Microcline	$KAlSi_3O_8$	6	Typically white	Granitic intrusives; syenites; arkosic sandstones and conglomerates; gneisses

Remarks: Potassium feldspar; very common mineral; usually white, sometimes bluish green ("amazonstone"); perfect cleavage in one direction and good in another; vitreous luster; may be twinned, especially carlsbad (penetration twins); usually translucent; prismatic crystals; insoluble in hydrochloric and nitric acids; optical and other sophisticated analytical techniques may be needed to distinguish from other feldspars

▶**TABLE B.3** *(continued)*
Minerals With a Nonmetallic Luster and Hardness of $5\frac{1}{2}-6\frac{1}{2}$

Name	Formula	Hardness	Color	Environment
Microlite	$Ca_2Ta_2O_6(O,OH,F)$	$5\frac{1}{2}$	Usually brown	Granitic pegmatites

Remarks: Usually brown, but may be yellow or colorless; usually massive; vitreous to resinous luster; usually translucent to opaque; yellow to brown streak; partially soluble in hydrochloric acid

| Nepheline | $(Na,K)AlSiO_4$ | $5\frac{1}{2}-6$ | White, yellow-gray | Syenites; syenitic pegmatites; trachytes |

Remarks: Feldspathoid; white, gray, or a variety of other colors; often massive with greasy luster, but may exist as transparent vitreous crystals; distinct cleavage in one direction; unlike feldspars, decomposes in hydrochloric acid to produce silica gels

| Neptunite | $KNa_2Li(Fe,Mn)_2TiO_2(Si_4O_{11})_2$ | 5–6 | Usually black | Nepheline pegmatites and syenites; serpentinites |

Remarks: Usually black; prismatic monoclinic crystals; vitreous luster; perfect cleavage; translucent to opaque; red streak; insoluble in hydrochloric acid

| **Oligoclase** | $Ab_{90}An_{10}$ to $Ab_{70}An_{30}$ | 6 | Usually white or gray | Granodiorites; monzonites; dacites; latites |

Remarks: Plagioclase feldspar; usually white or gray; bladed crystals; vitreous luster; mostly insoluble in hydrochloric acid; optical or other sophisticated analytical methods are probably needed to distinguish from other feldspars

| **Olivine** | $(Mg,Fe)_2SiO_4$ | $6\frac{1}{2}-7$ | Various shades of green | Gabbros; peridotites; basalts; other mafic and ultramafic igneous rocks; dolomitic marbles; highly metamorphosed banded iron formations |

Remarks: Forsterite-fayalite solid solution series dominates the composition of olivines; most are Mg-rich in various shades of green; vitreous luster; transparent to translucent; conchoidal fracture; slowly decomposes in hydrochloric acid to produce a silica gel; optical or other sophisticated analytical methods are needed to determine the magnesium or iron contents of an olivine specimen

| Omphacite | $(Ca,Na)(Mg,Fe,Al)Si_2O_6$ | 5–6 | Green | Eclogites |

Remarks: Sodium-rich pyroxene; green; prismatic crystals; cleavage at 87° and 93°, but imperfect; vitreous luster; translucent; white to light green streak; insoluble in most acids

| **Opal** | $SiO_2 \cdot nH_2O$ | 5–6 | Variable | Hydrothermal deposits; volcanic tuffs; "petrified wood"; marine sediments |

Remarks: Mineraloid; amorphous; white, colorless, yellow, black, red, or a variety of other colors, depending on the presence of impurities; vitreous to resinous luster; transparent to translucent; may fluoresce in ultraviolet light; conchoidal fracture; insoluble in most acids

| **Orthoclase** | $KAlSi_3O_8$ | 6 | Usually white, gray, or pink | Granites; granodiorites; syenites; granitic pegmatites; gneisses; arkosic sandstones and conglomerates |

Remarks: Potassium feldspar; usually white, gray, or pink; vitreous luster; often carlsbad (penetration) twins; prismatic or tabular crystals, but may be massive; imperfect to perfect cleavage in three directions, two cleavages at almost 90°; insoluble in most acids; optical or other sophisticated analytical methods may be needed to distinguish from other feldspars

| Periclase | MgO | $5\frac{1}{2}$ | White | Dolomitic marbles |

Remarks: Rare; usually microscopic; octahedral sheets or cubic crystals

| Perovskite | $CaTiO_3$ | $5\frac{1}{2}$ | Reddish brown, yellow, or black | Nepheline syenites; carbonatites |

Remarks: Reddish brown, yellow, or black; usually pseudocubic crystals or massive; adamantine to submetallic luster; no cleavage; conchoidal fracture; light yellow streak

| Petalite | $Li(AlSi_4O_{10})$ | $6-6\frac{1}{2}$ | Usually white, gray, or colorless | Lithium-bearing pegmatites |

Remarks: Usually white, gray, or colorless; vitreous luster; brittle; perfect cleavage in one direction and good cleavage in another; often columnar crystals; transparent to translucent; produces red flame from the presence of lithium; insoluble in most acids

| Piemontite | $Ca_2MnAl_2(Si_3O_{12})(OH)$ | $6\frac{1}{2}$ | Reddish or purplish brown | Manganese deposits |

Remarks: Manganese epidote; reddish or purplish brown; bright red streak; often massive, sometimes in long slender prisms; vitreous luster; translucent; insoluble in acids; optical or other analytical methods may be needed to distinguish from other epidotes

| **Pigeonite** | $(Ca,Mg,Fe)_2Si_2O_6$ | 6 | Greenish brown, brown, or black | Basalts; diabases |

Remarks: Pyroxene; greenish brown, brown, or black; good cleavage in one direction; optical or other analytical methods are required to distinguish pigeonite from other pyroxenes, especially augite

▶**TABLE B.3** *(continued)*
Minerals With a Nonmetallic Luster and Hardness of $5\frac{1}{2}$–$6\frac{1}{2}$

Name	Formula	Hardness	Color	Environment
Plagioclase	$Ab_{100}An_0$ to Ab_0An_{100}	6	Usually white or gray	Intrusive igneous rocks of various types (for example, granites, syenites, and gabbros); felsic to mafic volcanics; pegmatites; schists; gneisses; contact metamorphic limestones

Remarks: Very common feldspar; usually white or gray; bladed crystals; vitreous luster; usually translucent; white streak; twinning is common; some intermediate varieties may show a "play of colors"; albite is insoluble in hydrochloric acid, while anorthite decomposes to form a silica gel; optical and other sophisticated analytical methods are needed to quantify the calcium and sodium compositions of most plagioclases

Prehnite	$Ca_2Al(AlSi_3)O_{10}(OH)_2$	6–$6\frac{1}{2}$	Usually white or light green	Cavities in basalts

Remarks: Usually white or light green; vitreous luster; usually translucent; often massive, radiating crystals; optical or other sophisticated analytical techniques may be needed to positively identify

Psilomelane	$(Ba,Mn)_3(O,OH)_6Mn_8O_{16}$ (Romanechite)	5–6	Black	Associated with pyrolusite (MnO_2) in marine sedimentary rocks (for example, limestones)

Remarks: Black; submetallic luster; greasy feel; brownish black streak; opaque; dissolves in hydrochloric acid; branchlike forms on rock surfaces

Rhodonite	$MnSiO_3$	$5\frac{1}{2}$–6	Usually pink	Manganese deposits; skarns

Remarks: Pyroxenoid; usually pink; vitreous luster; transparent to translucent; may have brown or black surface coatings; crystals may be fibrous or tabular; nearly 90° cleavages; insoluble in hydrochloric acid, unlike rhodochrosite; also harder than rhodochrosite

Rutile	TiO_2	6–$6\frac{1}{2}$	Usually red, brown, or black	Usually as microscopic crystals in granites, granitic pegmatites, gneisses, schists, marbles, and quartz grains

Remarks: Usually red, reddish brown, or black; adamantine to metallic luster; light brown streak; transparent to translucent; often striated prismatic crystals; sometimes elbowlike twinning; frequently found as hairlike crystals inside of quartz crystals; fair cleavage in one direction; insoluble in most acids

Sanidine	$(K,Na)AlSi_3O_8$	6	Usually colorless or white	Phenocrysts in rhyolites and trachytes

Remarks: Potassium feldspar; high-temperature solid solution series exists between sanidine and high albite with anorthoclase as an intermediate; usually colorless or white; vitreous luster; often transparent; white streak; nearly 90° cleavage; insoluble in most acids; optical and analytical methods are needed to distinguish sanidine from other feldspars

Scapolite	$(Na,Ca)_4(Al_2Si_2O_8)_3(CO_3,SO_4,Cl)$	5–6	Usually white, blue, gray, or pink	Schists; amphibolites; gneisses; granulites; skarns; pegmatites

Remarks: Group of metamorphic minerals, including marialite-meionite solid solution series with mizzonite as an intermediate member; usually white, blue, gray, or pink; poor cleavage in two directions; vitreous luster; transparent or translucent; usually fluoresces in ultraviolet light; decomposes in hydrochloric acid leaving a silica gel

Sillimanite	Al_2SiO_5	6–7	Usually white, gray, or brown	High-temperature pelitic schists and gneisses

Remarks: Polymorph of kyanite and andalusite; usually brown, gray, or white; vitreous luster; perfect cleavage in one direction; usually translucent; elongated crystals or fibrous; insoluble in most strong acids

Sodalite	$Na_3Al_3Si_3O_{12} \cdot NaCl$	$5\frac{1}{2}$–6	Typically blue	Nepheline syenites; phonolites; trachytes

Remarks: Usually blue, sometimes white, gray, or light green; vitreous luster; usually translucent; poor cleavage in one direction; usually massive; produces yellow flame from the presence of sodium; decomposes in hydrochloric acid to produce a silica gel; optical or sophisticated analytical methods may be necessary to distinguish from other blue minerals

Spodumene	$LiAlSi_2O_6$	$6\frac{1}{2}$–7	Typically white	Lithium-rich pegmatites

Remarks: Lithium pyroxene; usually white, but may be gray, yellow, green, or pink, depending on the presence of impurities; vitreous luster; transparent to translucent; prismatic vertically striated crystals; cleavage at 87° and 93°, like most pyroxenes; lithium in the mineral produces a red flame; insoluble in most acids

Thorianite	ThO_2	$6\frac{1}{2}$	Black to dark gray	Granitic pegmatites; placer deposits

Remarks: Radioactive; black to dark gray; dark green streak; often cubic or octahedral crystals; submetallic luster; opaque; very dense (specific gravity near 9.8); poor cleavage; conchoidal fracture; insoluble in hydrochloric acid, but partially soluble in nitric acid

Tremolite	$Ca_2Mg_5Si_8O_{22}(OH)_2$	5–6	Usually white	Dolomitic marbles; schists (often associated with talc); serpentinites

Remarks: Amphibole; usually white; forms complete solid solution series with ferroactinolite, with actinolite as an intermediate; vitreous luster; usually prismatic or radiating needlelike crystals; transparent to nearly opaque; insoluble in most acids

▶**TABLE B.3** (continued)
Minerals With a Nonmetallic Luster and Hardness of $5\frac{1}{2}-6\frac{1}{2}$

Name	Formula	Hardness	Color	Environment
Turquoise	$CuAl_6(PO_4)_4(OH)_8 \cdot 4H_2O$	6	Light blue to greenish blue	Altered aluminum-rich volcanics in arid environments

Remarks: Light blue to greenish blue; usually massive; perfect cleavage in one direction; conchoidal fracture; greasy luster; opaque; harder than chrysocolla; white to light green streak; insoluble in most acids, but soluble in hot hydrochloric acid

Name	Formula	Hardness	Color	Environment
Uraninite	UO_2	$5\frac{1}{2}$	Black	Granitic intrusives, including pegmatites; hydrothermal deposits associated with copper minerals, silver, or sulfides; sandstones

Remarks: Radioactive; black; brownish black streak; pitchblende is a massive variety, where some of the uranium is hexavalent rather than tetravalent; submetallic to dull luster; usually massive (pitchblende variety), but uraninite consists of octahedral, cubic, or dodecahedral crystals; very dense (specific gravity of 7.5 to 9.7 for uraninite and 6.5 to 9 for pitchblende variety); uraninite, including pitchblende, may contain substantial amounts of rare earth elements, which affects physical properties; generally insoluble in hydrochloric acid, but soluble in nitric and sulfuric acids

Name	Formula	Hardness	Color	Environment
Willemite	Zn_2SiO_4	$5\frac{1}{2}$	Black-brown	Zinc deposits in marbles

Remarks: Black, brown, sometimes colorless, or a variety of other colors; vitreous to resinous luster; transparent to translucent; usually massive; may fluoresce in ultraviolet light; good cleavage in one direction; powder decomposes in hydrochloric acid to produce a silica gel; sophisticated analytical methods may be needed to positively identify

Name	Formula	Hardness	Color	Environment
Zoisite	$Ca_2Al_3Si_3O_{12}(OH)$	6	Pink-white	Gneisses; granulites; eclogites; hydrothermal sulfide deposits

Remarks: Orthorhombic polymorph of clinozoisite; white, pink, or a variety of other colors depending on the presence of impurities; vitreous luster; usually transparent or translucent; often striated prismatic crystals or massive; insoluble in most acids; optical or sophisticated analytical methods may be needed to positively identify

▶**Table B.4**
Minerals With a Nonmetallic Luster and Hardness of 7 or Greater

Name	Formula	Hardness	Color	Environment
Almandine	$Fe_3Al_2Si_3O_{12}$	7	Red	Schists

Remarks: Red garnet; no cleavage; adamantine luster; often dodecahedral form; insoluble in most acids

Andalusite	Al_2SiO_5	$7\frac{1}{2}$	Red, brown, or green	Low pressure pelitic (aluminum-rich) metamorphic rocks, especially schists

Remarks: Polymorph of kyanite and sillimanite; red, brown, or green; vitreous luster; usually translucent to opaque; crosses may occur within prismatic crystals; insoluble in most acids

Andradite	$Ca_3Fe_2Si_3O_{12}$	7	Brown, red, black, green, or yellow	Contact metamorphic limestones; skarns

Remarks: Garnet; brown, red, black (melanite), green, or yellow; no cleavage; adamantine luster; often dodecahedral form; insoluble in most acids

Beryl	$Be_3Al_2Si_6O_{18}$	$7\frac{1}{2}$–8	Typically green to white	Granitic intrusives; pegmatites; biotite schists

Remarks: Green (emeralds), greenish blue (aquamarine variety), yellow, white, pink, or colorless; transparent to translucent; vitreous luster; hexagonal crystals; insoluble in most acids

Chert	SiO_2	7	Variable	Carbonate rocks, including chalk; banded iron formations; marine sediments of various types

Remarks: Light colored; darker varieties called *flint;* jasper is the red variety; fine grained (microcrystalline); produces sparks when struck; other properties similar to quartz

Chrysoberyl	$BeAl_2O_4$	$8\frac{1}{2}$	Variable	Granitic rocks; pegmatites; schists; placer deposits

Remarks: Green, brown, yellow, gray, or colorless; vitreous luster; insoluble in most acids; often in pseudohexagonal form; fair cleavage in one direction; transparent to translucent

Cordierite	$(Mg,Fe)_2Al_4Si_5O_{18} \cdot nH_2O$	7–$7\frac{1}{2}$	Bluish gray, black, or blue	Schists; gneisses; contact metamorphic rocks

Remarks: Bluish gray, black, or blue; pleochroic; prismatic crystals; vitreous luster; poor cleavage in one direction; conchoidal fracture; often translucent; insoluble in most acids

Corundum	Al_2O_3	9	Variable	Schists; gneisses; syenites; peridotites; pegmatites; placer deposits

Remarks: White, red (rubies), blue (sapphires), gray, green, and other colors; adamantine to vitreous luster; transparent to translucent; extremely hard; insoluble in most acids; some fluoresce yellow; others display asterism, or an image of a star parallel to the c-axis of the crystal

Diamond	C	10	Usually colorless or yellow	Kimberlites; placer deposits

Remarks: Hardest natural substance; usually colorless or yellow; adamantine luster; perfect cleavage in one direction; will burn at high temperatures; sometimes fluorescent in ultraviolet light

Dumortierite	$Al_7BSi_3O_{18}$	7	Blue, violet, or reddish brown	Aluminum-rich metamorphic rocks, including contact metamorphic rocks; pegmatites

Remarks: Blue, violet, or reddish brown; vitreous luster; translucent; poor cleavage; insoluble in most acids; columnar or fibrous and radiating crystals

Euclase	$BeAlSiO_4(OH)$	$7\frac{1}{2}$	Variable	Granitic pegmatites; placer deposits

Remarks: Colorless, white, blue, green, or other colors; prismatic crystals; often striated; perfect cleavage; conchoidal fracture; transparent to translucent; vitreous luster; insoluble in most acids

Flint	SiO_2	7	Dark	Carbonate rocks, including chalk; banded iron formations; marine sediments of various types

Remarks: Dark colored; red variety is jasper, while lighter varieties are called *chert;* fine grained (microcrystalline); produces sparks when struck; other properties similar to quartz

Gahnite	$ZnAl_2O_4$	$7\frac{1}{2}$–8	Dark green	Granitic pegmatites; zinc deposits; contact metamorphosed limestones

Remarks: Zinc spinel; dark green; vitreous luster; commonly forms octahedral crystals

Garnet	$(Mg,Fe^{2+},Mn^{2+},Ca)_3$-$(Al,Cr,Fe^{3+})_2Si_3O_{12}$	$6\frac{1}{2}$–$7\frac{1}{2}$	Variable	Schists; gneisses; amphibolites; peridotites; kimberlites; contact metamorphosed limestones; rarer in granitic and alkaline igneous intrusives

Remarks: Red, brown, green, or other colors, depending on composition; no cleavage; adamantine luster; often dodecahedral form; insoluble in most acids; common types: almandine, andradite, grossular, pyrope, spessartine, and mixtures of these end members

▶TABLE B.4 *(continued)*

Minerals With a Nonmetallic Luster and Hardness of 7 or Greater

Name	Formula	Hardness	Color	Environment
Hercynite	$FeAl_2O_4$	$7\frac{1}{2}$–8	Dark green	Gneisses
Remarks: An iron spinel; dark green; octahedral crystals; nonmagnetic; no streak				
Kyanite	Al_2SiO_5	5–7	Usually blue	High-pressure aluminum-rich schists; kimberlites; some eclogites
Remarks: Polymorph of andalusite and sillimanite; usually blue; hardness of about 5 parallel to length of crystals and about 7 perpendicular; bladed crystals with perfect cleavage in one direction; vitreous luster; insoluble in most acids				
Lawsonite	$CaAl_2Si_2O_7(OH)_2 \cdot H_2O$	8	Blue to gray or colorless	Blueschists
Remarks: Blue, bluish gray or colorless; vitreous luster; translucent; good cleavage in two directions				
Melanite	$Ca_3Fe_2Si_3O_{12}$	7	Black	Alkaline igneous intrusives
Remarks: Black andradite garnet; no cleavage; adamantine luster; often dodecahedral form; insoluble in most acids				
Phenacite (Phenakite)	Be_2SiO_4	$7\frac{1}{2}$–8	Colorless or white	Pegmatites; mica schists
Remarks: Colorless or white; vitreous luster; transparent to translucent; rhombohedral crystals, often twinned; poor cleavage in one direction; insoluble in most acids				
Pyrope	$Mg_3Al_2Si_3O_{12}$	7	Purple, red, or black	Ultramafic rocks, including peridotites, serpentinites, and kimberlites; eclogites
Remarks: Magnesium garnet; purple, red, or black; often transparent; no cleavage; adamantine luster; often dodecahedral form; insoluble in most acids				
Quartz	SiO_2	7	Variable	Very common; igneous rocks, especially felsic volcanics and granitic intrusives and pegmatites; schists, gneisses, and most other metamorphic rocks; sandstones and most sedimentary rocks, even many carbonate rocks
Remarks: Often white, sometimes colorless, purple, red, blue, yellow, or other colors; vitreous luster; conchoidal fracture; may contain fibrous needles of rutile; only soluble in hydrofluoric acid. *Warning! Hydrofluoric acid is very dangerous, use with great care!*				
Spessartine	$Mn_3Al_2Si_3O_{12}$	7	Brown to red	Skarns; some manganese deposits
Remarks: Manganese garnet; brown to red; no cleavage; adamantine luster; often dodecahedral form; insoluble in most acids				
Spinel	$MgAl_2O_4$	8	Variable	High-temperature silica-poor metamorphic rocks, including gneisses and some marbles; high-pressure ultramafics
Remarks: Red or various colors, depending on the amount of iron and other impurities; pure magnesium spinels are red and are easily confused with rubies; often contains some iron (hercynite); octahedral crystals; no cleavage; vitreous luster; transparent, translucent, or opaque; nonmagnetic; insoluble in most acids				
Staurolite	$Fe_2Al_9Si_4O_{23}(OH)_2$	7–$7\frac{1}{2}$	Brown, reddish brown, or black	Medium-temperature aluminum-rich schists; placer deposits
Remarks: Brown, reddish brown, or black; pseudo-orthorhombic crystals; twinning is common and twins often form crosses or X shapes; usually resinous or dull luster; translucent to opaque; insoluble in most acids				
Topaz	$Al_2SiO_4(F,OH)_2$	8	Typically colorless	Pegmatites; granitic intrusives; felsic volcanics
Remarks: Very hard; colorless or various colors depending on impurities; vitreous luster; transparent to translucent; perfect basal cleavage; insoluble in most acids				
Tourmaline	$Na(Fe,Mg,Al,Li)_3Al_6(BO_3)_3Si_6O_{18}(OH)_4$	7–$7\frac{1}{2}$	Variable	Granitic pegmatites and other granitic intrusives; schists; some metamorphosed carbonate rocks
Remarks: Black, blue, brown, green, yellow, pink, and other colors; vitreous to resinous luster; usually prismatic crystals with characteristic triangular shape with rounded corners; conchoidal fracture; insoluble in most acids; sometimes pleochroic. Various types: shorl (iron-rich), dravite (magnesium), elbaite (lithium), and others				
Uvarovite	$Ca_3Cr_2Si_3O_{12}$	$7\frac{1}{2}$	Green	Some chromium-rich serpentinites
Remarks: Chromium garnet; green; no cleavage; adamantine luster; often transparent; often dodecahedral form; insoluble in most acids				
Zircon	$ZrSiO_4$	$7\frac{1}{2}$	Usually brown	Granitic intrusives, often microscopic
Remarks: Usually brown; adamantine luster; usually translucent; tetragonal crystals; often microscopic crystals are found in biotite; conchoidal fracture; insoluble in most acids				

▶TABLE B.5
Minerals With a Metallic Luster and Hardness of Less Than 3

Name	Formula	Hardness	Color	Environment	Streak
Argentite (Acanthite)	Ag_2S	$2-2\frac{1}{2}$	Black	Silver and galena deposits	Shiny black

Remarks: Black; opaque; shiny on fresh surface, but a black oxide layer will quickly develop; soft enough to be cut with a knife; dense (specific gravity of 7.3); soluble in acids; poorer cleavage than galena; becomes isometric argentite above 179°C

| Bismuth | Bi | $2-2\frac{1}{2}$ | Whitish silver with reddish tinge | Silver, nickel, cobalt, lead, uranium, and tin deposits | Light silver |

Remarks: Whitish silver with reddish tinge; dense (specific gravity of 9.8); brittle; perfect cleavage in one direction; laminated; dissolves in nitric acid and forms a white precipitate after the acid solution is diluted

| Bournonite | $PbCuSbS_3$ | $2\frac{1}{2}-3$ | Steel gray to black | Moderate-temperature hydrothermal deposits; sometimes as microscopic inclusions in galena | Gray black to black |

Remarks: Steel gray to black; opaque; short prismatic to tabular crystals; imperfect cleavage in one direction; dissolves in nitric acid to produce a light green solution because of the copper in the mineral

| Calaverite | $AuTe_2$ | $2\frac{1}{2}$ | Brass yellow to whitish silver | Hydrothermal deposits | Yellowish to Greenish gray |

Remarks: Brass yellow to whitish silver; opaque; brittle; produces red solution when dissolved in sulfuric acid

| Chalcocite | Cu_2S | $2\frac{1}{2}-3$ | Gray with black tarnish | Hydrothermal sulfide deposits | Grayish black |

Remarks: Gray with black tarnish; poor cleavage in one direction; conchoidal fracture; opaque; dissolves in nitric acid; green flame with sulfur dioxide fumes

| Copper | Cu | $2\frac{1}{2}-3$ | Copper red | Basalts; hydrothermal deposits | Copper red |

Remarks: Copper red; dense (specific gravity of 8.9); ductile; malleable; tetrahexahedral or octahedral crystals; dissolves in nitric acid

| Covellite | CuS | $1\frac{1}{2}-2$ | Indigo blue | Copper deposits | Gray to black |

Remarks: Indigo blue; often iridescent; opaque; micaceous or platy cleavage; when dissolved in hydrochloric acid produces a green flame

| **Galena** | PbS | $2\frac{1}{2}$ | Silver | Hydrothermal deposits of lead and silver; some pegmatites; some contact metamorphic rocks | Gray |

Remarks: Silver; usually cubic form; dense (specific gravity of about 7.5); black tarnish

| Gold | Au | $2\frac{1}{2}-3$ | Yellow | Hydrothermal quartz veins; placer deposits; volcanics | Yellow |

Remarks: Yellow; very dense (specific gravity of 15–19); hackly fracture; very ductile and malleable; opaque

| **Graphite** | C | 1–2 | Black | Schists; marbles; gneisses; highly metamorphosed coals | Black |

Remarks: "Pencil lead"; black; greasy; marks paper; although nonmetal, usually has metallic luster; insoluble in most acids

| Jamesonite | $Pb_4FeSb_6S_{14}$ | 2–3 | Gray to gray-black | Lead deposits | Gray to grayish black |

Remarks: Gray to grayish black; opaque; brittle; good cleavage in one direction; fibrous; difficult to distinguish from some other lead minerals without sophisticated analytical equipment

| Molybdenite | MoS_2 | $1-1\frac{1}{2}$ | Gray with bluish tint | Some granites; pegmatites; some contact metamorphic rocks | Grayish black |

Remarks: Gray with bluish tint; greasy; opaque; higher specific gravity than graphite; laminar; gives greenish streak on glazed porcelain or greenish gray mark on paper

| **Pyrolusite** | MnO_2 | 1–2 | Black | Surface of some limestones; bog and lake sediments; quartz veins | Bluish black |

Remarks: Black; prismatic crystals; often as branchlike features (dendrites) on limestones look like plant fossils but are not; soluble in hydrochloric acid; greasy; marks paper

| Silver | Ag | $2\frac{1}{2}-3$ | Silver | Hydrothermal deposits | Whitish silver |

Remarks: Silver; brown to black tarnish; dense (specific gravity of 1012); hackly fracture; ductile; malleable. Soluble in nitric acid

| Stibnite | Sb_2S_3 | 2 | Gray to black | Hydrothermal deposits | Gray to black |

Remarks: Gray to black; perfect cleavage in one direction; opaque; bladed crystals; iridescent film on surface; powder produces orange solution when dissolved in concentrated hydrochloric acid. *Remember to work with acids under a ventilation hood!*

| Sylvanite | $(Au,Ag)Te_2$ | $1\frac{1}{2}-2$ | Whitish silver | Hydrothermal deposits | Gray |

Remarks: Whitish silver; dense (specific gravity near 8); produces red solution in concentrated sulfuric acid; better cleavage (perfect in one direction) than calaverite; soluble in nitric acid, but leaves residue of gold. *Remember to work with acids under a ventilation hood!*

Appendix B: Mineral Key

▶**TABLE B.6**
Minerals With a Metallic Luster and Hardness of $3-5\frac{1}{2}$

Name	Formula	Hardness	Color	Environment	Streak
Arsenic	As	$3\frac{1}{2}$	White to silver	Silver, cobalt, or nickel deposits	Gray
Remarks: Poisonous; do not taste, heat, or dissolve in acids; whitish silver on fresh surface, tarnish to dark gray; brittle; perfect cleavage					
Bornite	Cu_5FeS_4	3	Blue to purple	Copper deposits; contact metamorphic rocks; basalts	Grayish black
Remarks: Bronze with bluish purple tarnish ("peacock ore"); often associated with chalcocite and covellite; poor cleavage; soluble in most strong acids					
Chalcopyrite	$CuFeS_2$	$3\frac{1}{2}$-4	Brass yellow	Copper deposits; schists; contact metamorphic rocks; pegmatites; metamorphosed basalts and other mafic volcanics	Greenish black
Remarks: Brass yellow with iridescent tarnish; softer than pyrite; brittle; tetrahedral crystals are common; no cleavage; conchoidal fracture					
Copper	Cu	$2\frac{1}{2}$-3	Copper red	Basalts; hydrothermal deposits	Copper red
Remarks: Copper red; dense (specific gravity of 8.9); ductile; malleable; tetrahexahedral or octahedral; dissolves in nitric acid					
Enargite	Cu_3AsS_4	3	Grayish black to black	Hydrothermal deposits	Grayish black to black
Remarks: Possibly poisonous; do not heat, taste, or dissolve in acids; grayish black to black; perfect cleavage in one direction and distinct cleavages in two other directions; opaque					
Iron	Fe	$4\frac{1}{2}$	Gray to black	Meteorites	Gray to black
Remarks: Gray to black; magnetic; hackly fracture; poor cleavage in one direction; malleable; soluble in hydrochloric acid; dense (specific gravity of about 7.6)					
Manganite	MnO(OH)	4	Gray to black	Low-temperature hydrothermal deposits	Dark brown
Remarks: Gray to black with possible reddish brown tint; perfect cleavage in one direction and good cleavages in two other directions; opaque; striated prismatic crystals may occur; soluble in hydrochloric acid					
Metacinnabar	$Hg_{1-n}S$	3	Grayish black	Mercury deposits	Black
Remarks: Grayish black; dense; considered stable above 344°C, but may form at ambient temperatures					
Millerite	NiS	$3-3\frac{1}{2}$	Pale yellow	Nickel deposits; marbles; cherty limestone	Greenish black to black
Remarks: Pale yellow, possibly with greenish tinge; often fine hairlike acicular crystals; good cleavages in two directions					
Niccolite	NiAs	$5-5\frac{1}{2}$	Copper red	Gabbros; cobalt, nickel, and silver deposits	Brownish black
Remarks: Possibly poisonous; do not heat, taste, or dissolve in acids; copper red with gray to black tarnish; dense (specific gravity of 7.8); opaque; no cleavage; conchoidal fracture					
Pentlandite	$(Fe,Ni)_9S_8$	$3\frac{1}{2}$-4	Yellowish bronze	Gabbros; nickel deposits	Brown
Remarks: Light yellowish bronze; octahedral parting; similar in appearance to pyrrhotite, except nonmagnetic					
Platinum	Pt	$4-4\frac{1}{2}$	Gray	Dunites and other ultramafic rocks; placer deposits	Gray
Remarks: Gray; very dense (specific gravity up to 19 with natural impurities, 21 if purified); malleable; magnetic, if sufficient iron impurities are present					
Psilomelane	$(Ba,Mn)_3(O,OH)_6Mn_8O_{16}$	5-6	Black	Associated with pyrolusite (MnO_2) in marine sedimentary rocks (for example, limestones)	Brownish black to black
Remarks: Black; submetallic luster; greasy feel; opaque; dissolves in hydrochloric acid; branchlike forms on rock surfaces					
Pyrrhotite	$Fe_{1-x}S$	4	Brownish bronze	Mafic intrusives; sulfide deposits; pegmatites; contact metamorphic rocks	Black
Remarks: Brownish bronze; possibly magnetic, but generally weak; opaque					
Tennantite-Tetrahedrite	$Cu_{12}(As,Sb)_4S_{13}$	$3-4\frac{1}{2}$	Gray to black	Hydrothermal deposits; tetrahedrite is more common than tennantite	Black, gray, or brown
Remarks: Possibly poisonous; do not heat, taste, or dissolve in acids; gray to black; metallic to submetallic luster; tetrahedrite crystals; brittle; no cleavage					

▶**TABLE B.7**

Minerals With a Metallic Luster and Hardness of $5\frac{1}{2}$–$6\frac{1}{2}$

Name	Formula	Hardness	Color	Environment	Streak
Arsenopyrite	FeAsS	$5\frac{1}{2}$–6	Silver-white	High-temperature hydrothermal deposits; pegmatites; contact metamorphic rocks; some crystalline limestones	Black
Remarks: Possibly poisonous; may release poisonous vapors if dissolved in acid; also, do not taste or heat; silver-white; elongated prismatic crystals; produces sparks and a garlic odor when struck					
Chromite	$FeCr_2O_4$	$5\frac{1}{2}$	Black or dark brown	Peridotites; serpentinites; other ultramafic rocks	Dark brown
Remarks: Black or dark brown; opaque; submetallic or metallic luster; weakly magnetic; insoluble in most strong acids; may have yellow or green surface oxide layers					
Cobaltite	(Co,Fe)AsS	$5\frac{1}{2}$	Silver-white	High-temperature metamorphic and hydrothermal rocks containing cobalt deposits	Grayish black
Remarks: Possibly poisonous; may release poisonous vapors if dissolved in acid or heated; also, do not taste; silver-white, perhaps reddish or purple tinge; may occur as pyritohedrons or pseudo-isometric cubes; perfect pseudocubic cleavage					
Columbite-Tantalite solid solution series	$(Fe,Mn)Nb_2O_6$ to $(Fe,Mn)Ta_2O_6$, often with tin (Sn) and tungsten (W) impurities	6	Black	Granitic intrusives; granitic pegmatites	Dark red or black
Remarks: Black; submetallic luster; may be iridescent; good cleavage in one direction; fairly dense, specific gravity up to 7.9; prismatic or tabular crystals; insoluble in acids					
Franklinite	$ZnFe_2O_4$	6	Black	Rare; zinc deposits	Reddish brown to dark brown
Remarks: Black; no cleavage; conchoidal fracture; weakly magnetic; soluble in hydrochloric acid					
Hematite	Fe_2O_3	$5\frac{1}{2}$–$6\frac{1}{2}$	Red, brown, or black	Some sandstones; banded iron formations; granitic intrusives; felsic volcanics; various metamorphic rocks; common in at least trace amounts in many other lithologies	Red
Remarks: Red, brown, or black; metallic to earthy luster; slowly dissolves in hydrochloric acid; may be iridescent					
Ilmenite	$FeTiO_3$	$5\frac{1}{2}$–6	Black	Gabbros; diorites; anorthosites; some pegmatites; some-magnetite-rich ("black") beach sands	Black to brownish red, but not as red as hematite
Remarks: Black; may be weakly magnetic; no cleavage; powder soluble in concentrated hydrochloric acid					
Magnetite	Fe_3O_4	6	Black	Various igneous and metamorphic rocks; banded iron formations; beach "black" sands	Black
Remarks: Black; strongly magnetic, unlike ilmenite; opaque; slowly soluble in hydrochloric acid					
Marcasite	FeS_2	6–$6\frac{1}{2}$	Pale yellow	Lead and zinc deposits; some hydrothermal deposits; limestones; concretions in clays and shales	Grayish black
Remarks: Pale yellow, almost white when fresh; less stable and common than pyrite; prismatic crystals or roselike nodules; when moist, breaks down to whitish iron sulfates and sulfuric acid					
Psilomelane	$(Ba,Mn)_3(O,OH)_6Mn_8O_{16}$	5–6	Black	Associated with pyrolusite (MnO_2) in marine sedimentary rocks (for example, limestones)	Brownish black
Remarks: Black; submetallic luster; greasy feel; opaque; dissolves in hydrochloric acid; branchlike forms on rock surfaces					
Pyrite	FeS_2	6–$6\frac{1}{2}$	Yellow	Common in many lithologies, including intrusive igneous rocks, limestones, shales, metamorphic rocks, and hydrothermal deposits; may also replace fossils	Black
Remarks: "Fool's gold"; yellow; striated cubes, pyritohedrons, or octahedral crystals; may have iridescent yellowish brown film on surface					

▶**TABLE B.7** *(continued)*
Minerals W ith a Metallic Luster and Hardness of $5\frac{1}{2}$–$6\frac{1}{2}$

Name	Formula	Hardness	Color	Environment	Streak
Rutile	TiO_2	6–$6\frac{1}{2}$	Red, reddish brown, or black	Usually as microscopic crystals in granites, granitic pegmatites, gneisses, schists, marbles, and quartz grains	Light brown

Remarks: Usually red, reddish brown, or black; adamantine to metallic luster; transparent to translucent; often striated prismatic crystals; sometimes elbow-like twinning; fair cleavage in one direction; insoluble in most acids

Skutterudite	$(Co,Ni)As_3$	$5\frac{1}{2}$–6	White to silver gray	Cobalt and nickel deposits	Black

Remarks: Possibly poisonous fumes; do not dissolve in acids or heat; also, do not taste; white to silver gray; opaque; brittle; usually cubic or octahedral

Thorianite	ThO_2	$6\frac{1}{2}$	Black to dark gray	Granitic pegmatites; placer deposits	Dark green

Remarks: Radioactive; black to dark gray; often cubic or octahedral crystals; submetallic luster; opaque; very dense (specific gravity near 9.8); poor cleavage; conchoidal fracture; insoluble in hydrochloric acid, but partially soluble in nitric acid

Appendix

Minerals Separated by Optic System and Sign, and Ordered by Index of Refraction

Isotropic Minerals

Mineral	Formula	Crystal System	Refractive Index (n)
fluorite	CaF_2	cubic	1.43
analcime	$NaAlSi_2O_6 \cdot H_2O$	cubic	1.48
sodalite	$Na_3Al_3Si_3O_{12} \cdot NaCl$	cubic	1.48
sylvite	KCl	cubic	1.49
halite	NaCl	cubic	1.54
garnet	$(Ca,Fe,Mg)_3(Al,Fe)_2Si_3O_{12}$	cubic	1.71–1.87
periclase	MgO	cubic	1.736
spinel	$MgAl_2O_4$	cubic	1.74
chlorargerite	AgCl	cubic	2.07
diamond	C	cubic	2.419
sphalerite	ZnS	cubic	2.42
cuprite	Cu_2O	cubic	2.85

Uniaxial (–) Minerals

Mineral	Formula	Crystal System	Mean Refractive Index ($\bar{n}$)	Birefringence (δ)
nitratite	$NaNO_3$	hexagonal	1.33–1.58	0.251
chabazite	$CaAl_2Si_4O_{12} \cdot 6H_2O$	hexagonal	1.48	0.003
cristobalite	SiO_2	tetragonal	1.48	0.007
calcite	$CaCO_3$	hexagonal	1.48–1.65	0.172

Uniaxial (–) Minerals *(continued)*

Mineral	Formula	Crystal System	Mean Refractive Index (n)	Birefringence (δ)
dolomite	$CaMg(CO_3)_2$	hexagonal	1.5–1.67	0.179
magnesite	$MgCO_3$	hexagonal	1.5–1.7	0.191
nepheline	$(Na,K)AlSiO_4$	hexagonal	1.54	0.004
scapolite	$(Na,Ca)_4(Al,Si)_{12}O_{24}(Cl,CO_3)$	tetragonal	1.54–1.59	0.004–0.037
ankerite	$CaFe(CO_3)_2$	hexagonal	1.54–1.75	0.202
autunite	$Ca(UO_2)_2(PO_4)_2 \cdot 10H_2O$	tetragonal	1.55–1.57	0.024
beryl	$Be_3Al_2Si_6O_{18}$	hexagonal	1.56	0.006
rhodochrosite	$MnCO_3$	hexagonal	1.59–1.81	0.219
tourmaline	$Na(Fe,Mg,Al,Li)_3Al_6(BO_3)_3Si_6O_{18}(OH)_4$	hexagonal	1.62–1.67	0.021–0.029
smithsonite	$ZnCO_3$	hexagonal	1.62–1.85	0.225
apatite	$Ca_5(PO_4)_3(OH,F,Cl)$	hexagonal	1.63	0.003
siderite	$FeCO_3$	hexagonal	1.63–1.87	0.242
vesuvianite	$Ca_{10}(Mg,Fe)_2Al_4Si_9O_{34}(OH)_4$	tetragonal	1.7	0.005
corundum	Al_2O_3	hexagonal	1.76	0.008
pyromorphite	$Pb(PO_4)_3Cl$	hexagonal	2.04–2.05	0.01
hausmannite	Mn_3O_4	tetragonal	2.1–2.4	0.31
wulfenite	$PbMoO_4$	tetragonal	2.28–2.4	0.121
vanadinite	$Pb_5(VO_4)_3Cl$	hexagonal	2.3–2.4	0.066
anatase	TiO_2	tetragonal	2.48–2.56	0.073
proustite	Ag_3AsS_3	hexagonal	2.71–2.98	0.17
pyrargyrite	Ag_3SbS_3	hexagonal	2.88–3.08	0.2
hematite	Fe_2O_3	hexagonal	3–3.2	0.28

Uniaxial (+) Minerals

Mineral	Formula	Crystal System	Mean Refractive Index (n)	Birefringence (δ)
leucite	$KAlSi_2O_6$	tetragonal	1.51	0.001
apophyllite	$KCa_4Si_8O_{20}F \cdot 8H_2O$	tetragonal	1.53	0.002
quartz	SiO_2	hexagonal	1.53–1.54	0.009
brucite	$Mg(OH)_2$	hexagonal	1.57–1.58	0.02
alunite	$KAl_3(SO_4)_2(OH)_6$	hexagonal	1.57–1.59	0.02
stishovite	SiO_2	tetragonal	1.8–1.82	0.027
scheelite	$CaWO_4$	tetragonal	1.92–1.93	0.014
zircon	$ZrSiO_4$	tetragonal	1.93–1.99	0.06
zincite	ZnO	hexagonal	2.01–2.02	0.016
rutile	TiO_2	tetragonal	2.61–2.9	0.29
cinnabar	HgS	hexagonal	2.9–3.25	0.35
cassiterite	SnO_2	tetragonal	2–2.1	0.09

Biaxial (–) Minerals

Mineral	Formula	Crystal System	2V°	Mean Refractive Index ($\bar{n}$)	Birefringence (δ)
niter	KNO_3	orthorhombic	7	1.33–1.5	0.172
borax	$Na_2B_4O_5(OH)_4 \cdot 8H_2O$	monoclinic	40	1.44–1.47	0.025
kernite	$Na_2B_4O_6(OH)_2 \cdot 3H_2O$	monoclinic	80	1.45–1.48	0.034
stilbite	$CaAl_2Si_7O_{18} \cdot 7H_2O$	monoclinic	30–50	1.5	0.01
sepiolite	$Mg_4Si_6O_{15}(OH)_2(H_2O) \cdot 4H_2O$	orthorhombic	20–70	1.51–1.53	0.02
chalcanthite	$CuSO_4 \cdot 5H_2O$	triclinic	56	1.51–1.54	0.029
microcline	$KAlSi_3O_8$	triclinic	77–84	1.52	0.01
orthoclase	$KAlSi_3O_8$	monoclinic	60–65	1.52	0.007
sanidine	$(K,Na)AlSi_3O_8$	monoclinic	80–85	1.52	0.007
strontianite	$SrCO_3$	orthorhombic	7	1.52–1.66	0.148
witherite	$BaCO_3$	orthorhombic	16	1.52–1.67	0.148
aragonite	$CaCO_3$	orthorhombic	18	1.53–1.68	0.155
talc	$Mg_3Si_4O_{10}(OH)_2$	monoclinic	6–30	1.54–1.58	0.05
variscite	$AlPO_4 \cdot 2H_2O$	orthorhombic	48–54	1.55–1.59	0.03
pyrophyllite	$Al_2Si_4O_{10}(OH)_2$	triclinic	52–62	1.55–1.6	0.047
kaolinite	$Al_2Si_2O_5(OH)_4$	triclinic	40	1.56	0.007
muscovite	$KAl_2(AlSi_3O_{10})(OH)_2$	monoclinic	30–40	1.56–1.6	0.035
chlorite	talc+brucite combinations	monoclinic	0–40	1.56–1.61	0.006–0.02
antigorite	$Mg_3Si_2O_5(OH)_4$	monoclinic	20–60	1.57	0.007
anorthite	$CaAl_2Si_2O_8$	triclinic	78	1.57–1.59	0.013
biotite	$K(Mg,Fe)_3AlSi_3O_{10}(OH)_2$	monoclinic	0–32	1.57–1.61	0.04
amblygonite	$LiAl(PO_4)F$	triclinic	52–90	1.59–1.62	0.03
anthophyllite	$(Mg,Fe)_7Si_8O_{22}(OH)_2$	orthorhombic	65–90	1.6–1.63	0.03
tremolite-actinolite	$Ca_2(Mg,Fe)_5Si_8O_{22}(OH)_2$	monoclinic	85–90	1.61–1.63	0.022
lazulite	$(Mg,Fe)Al_2(PO_4)_2(OH)_2$	monoclinic	70	1.61–1.64	0.031
wollastonite	$CaSiO_3$	triclinic	39	1.62–1.63	0.014
erythrite	$Co_3(AsO_4)_2 \cdot 8H_2O$	monoclinic	90	1.62–1.69	0.073
andalusite	Al_2SiO_5	orthorhombic	75–85	1.63–1.64	0.01
margarite	$CaAl_2(Al_2Si_2O_{10})(OH)_2$	monoclinic	45	1.63–1.65	0.013
monticellite	$CaMgSiO_4$	orthorhombic	72–82	1.64–1.66	0.02
glaucophane	$Na_2Mg_3Al_2Si_8O_{22}(OH)_2$	monoclinic	0–50	1.65–1.67	0.01
hornblende	$(K,Na)_{0-1}(Ca,Na,Fe,Mg)_2$-$(Mg,Fe,Al)_5(Si,Al)_8O_{22}(OH)_2$	monoclinic	50–80	1.65–1.67	0.02
malachite	$Cu_2CO_3(OH)_2$	monoclinic	43	1.65–1.9	0.254
clinozoisite	$Ca_2Al_3Si_3O_{12}(OH)$	monoclinic	14–90	1.67–1.73	0.005–0.015
triphylite	$Li(Fe,Mn)PO_4$	orthorhombic	0–56	1.68–1.69	0.01
kaersutite	Ti-rich hornblende	monoclinic	68–82	1.68–1.73	0.02–0.08
kyanite	Al_2SiO_5	triclinic	82–83	1.71–1.72	0.016
epidote	$Ca_2(Al,Fe)_3Si_3O_{12}(OH)$	monoclinic	90–115	1.71–1.8	0.01–0.05
carnotite	$K_2U_2O_4(VO_4)_2 \cdot 3H_2O$	monoclinic	38–44	1.75–1.95	0.2
cerussite	$PbCO_3$	orthorhombic	9	1.8–2.07	0.274
atacamite	$Cu_2Cl(OH)_3$	orthorhombic	75	1.83–1.88	0.047
lepidocrocite	$FeO(OH)$	orthorhombic	83	1.94–2.51	0.57
goethite	$FeO(OH)$	orthorhombic	0–27	2.26–2.52	0.15
perovskite	$CaTiO_3$	orthorhombic	90	2.3	0.002
realgar	AsS	monoclinic	41	2.5–2.8	0.166

Biaxial (+) Minerals

Mineral	Formula	Crystal System	2V°	Mean Refractive Index (*n*)	Birefringence (δ)
cryolite	Na_3AlF_6	monoclinic	43	1.33	0.001
epsomite	$MgSO_4 \cdot 7H_2O$	orthorhombic	52	1.43–1.46	0.028
carnallite	$KMgCl_3 \cdot 6H_2O$	orthorhombic	70	1.46–1.49	0.029
natrolite	$Na_2Al_2Si_3O_{10} \cdot 2H_2O$	orthorhombic	38–62	1.48	0.012
tridymite	SiO_2	orthorhombic	70	1.48	0.003
ulexite	$NaCaB_5O_6(OH)_6 \cdot 5H_2O$	triclinic	73	1.49–1.52	0.029
heulandite	$CaAl_2Si_7O_{18} \cdot 6H_2O$	monoclinic	35	1.5	0.005
albite	$NaAlSi_3O_8$	triclinic	77	1.52	0.011
gypsum	$CaSO_4 \cdot 2H_2O$	monoclinic	58	1.52	0.009
wavellite	$Al_3(PO_4)_2(OH)_3 \cdot 5H_2O$	orthorhombic	72	1.52–1.55	0.027
cordierite	$(Mg,Fe)_2Al_4Si_5O_{18}$	orthorhombic	65–105	1.54–1.56	0.02
norbergite	$Mg_3SiO_4(OH,F)_2$	orthorhombic	44–50	1.56–1.58	0.026
gibbsite	$Al(OH)_3$	monoclinic	0–40	1.57–1.59	0.02
anhydrite	$CaSO_4$	orthorhombic	44	1.57–1.61	0.044
vivianite	$Fe_3(PO_4)_3 \cdot 8H_2O$	monoclinic	83	1.57–1.63	0.054
colemanite	$CaB_3O_4(OH)_3 \cdot H_2O$	monoclinic	56	1.58–1.61	0.028
coesite	SiO_2	monoclinic	64	1.59	0.01
pectolite	$NaCa_2(SiO_3)_3H$	triclinic	35–63	1.59–1.63	0.04
clinohumite	$Mg_9(SiO_4)_2(OH,F)_2$	monoclinic	73–76	1.59–1.64	0.03–0.04
chondrodite	$Mg_5Si_2O_4(OH,F)_2$	monoclinic	60–90	1.6–1.63	0.03
topaz	$Al_2SiO_4(F,OH)_2$	orthorhombic	48–65	1.61–1.62	0.01
turquoise	$CuAl_6(PO_4)_4(OH)_8 \cdot 4H_2O$	triclinic	40	1.61–1.65	0.04
celestite	$SrSO_4$	orthorhombic	51	1.62–1.63	0.009
prehnite	$Ca_2Al(AlSi_3O_{10})(OH)_2$	orthorhombic	65–70	1.62–1.65	0.03
barite	$BaSO_4$	orthorhombic	37	1.63–1.64	0.012
olivine	$(Mg,Fe)_2SiO_4$	orthorhombic	47–90	1.63–1.88	0.035–0.053
boracite	$Mg_3ClB_7O_{13}$	orthorhombic	82	1.64–1.67	0.011
cummingtonite	$Mg_7Si_8O_{22}(OH)_2$	monoclinic	80–90	1.64–1.67	0.03
jadeite	$NaAlSi_2O_6$	monoclinic	70–75	1.65–1.67	0.02
spodumene	$LiAlSi_2O_6$	monoclinic	60–80	1.65–1.67	0.02
sillimanite	Al_2SiO_5	orthorhombic	20–30	1.65–1.68	0.022
hypersthene	$(Mg,Fe)_2Si_2O_6$	orthorhombic	54	1.66	0.008
lawsonite	$CaAl_2Si_2O_7(OH)_2 \cdot H_2O$	orthorhombic	76–86	1.66–1.68	0.02
diopside	$Ca(Mg,Fe)Si_2O_6$	monoclinic	56–64	1.66–1.75	0.03
sinhalite	$MgAlBO_4$	orthorhombic	58	1.67–1.71	0.04
diaspore	$AlO(OH)$	orthorhombic	85	1.68–1.75	0.04
pigeonite	$(Ca,Mg,Fe)_2Si_2O_6$	monoclinic	38–44	1.69–1.72	0.025
chloritoid	$(Fe,Mg)Al_2SiO_5(OH)_2$	monoclinic	45–65	1.71–1.72	0.01
rhodonite	$MnSiO_3$	triclinic	63–76	1.71–1.73	0.013
antlerite	$Cu_3SO_4(OH)_4$	orthorhombic	53	1.72–1.78	0.063
azurite	$Cu_3(CO_3)_2(OH)_2$	monoclinic	68	1.73–1.83	0.106
chrysoberyl	$BeAl_2O_4$	orthorhombic	45	1.74–1.75	0.01
staurolite	$Fe_2Al_9Si_4O_{20}(O,OH)_2$	monoclinic	80–88	1.74–1.75	0.013
monazite	$(Ce,La,Th)PO_4$	monoclinic	10–20	1.78–1.85	0.05
titanite (sphene)	$CaTiSiO_5$	monoclinic	23–50	1.86–2.1	0.15
anglesite	$PbSO_4$	orthorhombic	75	1.87–1.89	0.017

Biaxial (+) Minerals *(continued)*

Mineral	Formula	Crystal System	2V°	Mean Refractive Index (n)	Birefringence (δ)
sulfur	S	orthorhombic	69	1.95–2.24	
wolframite	$(Fe,Mn)WO_4$	monoclinic	73–79	2.1–2.5	0.13–0.15
manganite	$MnO(OH)$	monoclinic	small	2.2–2.5	0.29
columbite	$(Fe,Mn)(Nb,Ta)_2O_6$	orthorhombic	75	2.3–2.4	0.12
crocoite	$PbCrO_4$	monoclinic	54	2.3–2.6	0.35
orpiment	As_2S_3	monoclinic	76	2.4–3	0.62
brookite	TiO_2	orthorhombic	0–30	2.58–2.72	0.14

Appendix C:
Optical Properties

Appendix

Minerals Ordered by Birefringence and Interference Colors in Thin Section

Mineral	Formula	Crystal System	Optical Properties				
			System	Sign	2V°	Mean R.I. ($\bar{n}$)	Birefringence (δ)
Minerals With Very Low Birefringence (Interference colors generally white and gray in thin section)							
cryolite	Na_3AlF_6	monoclinic	biaxial	+	43	1.33	0.001
leucite	$KAlSi_2O_6$	tetragonal	uniaxial	+		1.51	0.001
apophyllite	$KCa_4Si_8O_{20}F \cdot 8H_2O$	tetragonal	uniaxial	+		1.53	0.002
perovskite	$CaTiO_3$	orthorhombic	biaxial	−	90	2.3	0.002
apatite	$Ca_5(PO_4)_3(OH,F,Cl)$	hexagonal	uniaxial	−		1.63	0.003
chabazite	$Ca_2Al_2Si_4O_{12} \cdot 6H_2O$	hexagonal	uniaxial	−		1.48	0.003
tridymite	SiO_2	orthorhombic	biaxial	+	70	1.48	0.003
nepheline	$(Na,K)AlSiO_4$	hexagonal	uniaxial	−		1.54	0.004
scapolite	$(Na,Ca)_4(Al,Si)_{12}O_{24}(Cl,CO_3)$	tetragonal	uniaxial	−		1.54–1.59	0.004–0.037
heulandite	$CaAl_2Si_7O_{18} \cdot 6H_2O$	monoclinic	biaxial	+	35	1.5	0.005
vesuvianite	$Ca_{10}(Mg,Fe)_2Al_4Si_9O_{34}(OH)_4$	tetragonal	uniaxial	−		1.7	0.005
clinozoisite	$Ca_2Al_3Si_3O_{12}(OH)$	monoclinic	biaxial	−	14–90	1.67–1.73	0.005–0.015
beryl	$Be_3Al_2Si_6O_{18}$	hexagonal	uniaxial	−		1.56	0.006
chlorite	talc+brucite combinations	monoclinic	biaxial	−	0–40	1.56–1.61	0.006–0.02
antigorite	$Mg_3Si_2O_5(OH)_4$	monoclinic	biaxial	−	20–60	1.57	0.007
cristobalite	SiO_2	tetragonal	uniaxial	−		1.48	0.007
kaolinite	$Al_2Si_2O_5(OH)_4$	triclinic	biaxial	−	40	1.56	0.007
orthoclase	$KAlSi_3O_8$	monoclinic	biaxial	−	60–65	1.52	0.007
sanidine	$(K,Na)AlSi_3O_8$	monoclinic	biaxial	−	80–85	1.52	0.007
corundum	Al_2O_3	hexagonal	uniaxial	−		1.76	0.008
hypersthene	$(Mg,Fe)_2Si_2O_6$	orthorhombic	biaxial	+	54	1.66	0.008
celestite	$SrSO_4$	orthorhombic	biaxial	+	51	1.62–1.63	0.009
gypsum	$CaSO_4 \cdot 2H_2O$	monoclinic	biaxial	+	58	1.52	0.009
quartz	SiO_2	hexagonal	uniaxial	+		1.53–1.54	0.009

Mineral	Formula	Crystal System	Optical Properties				
			System	Sign	2V°	Mean R.I. ($\bar{n}$)	Birefringence (δ)
Minerals With Low Birefringence **(Displaying up to first-order yellow, brown, orange, or red interference colors in thin section)**							
andalusite	Al_2SiO_5	orthorhombic	biaxial	–	75–85	1.63–1.64	0.01
chloritoid	$(Fe,Mg)Al_2SiO_5(OH)_2$	monoclinic	biaxial	+	45–65	1.71–1.72	0.01
chrysoberyl	$BeAl_2O_4$	orthorhombic	biaxial	+	45	1.74–1.75	0.01
coesite	SiO_2	monoclinic	biaxial	+	64	1.59	0.01
microcline	$KAlSi_3O_8$	triclinic	biaxial	–	77–84	1.52	0.01
pyromorphite	$Pb(PO_4)_3Cl$	hexagonal	uniaxial	–		2.04–2.05	0.01
stilbite	$CaAl_2Si_7O_{18} \cdot 7H_2O$	monoclinic	biaxial	–	30–50	1.5	0.01
topaz	$Al_2SiO_4(F,OH)_2$	orthorhombic	biaxial	+	48–65	1.61–1.62	0.01
triphylite	$Li(Fe,Mn)PO_4$	orthorhombic	biaxial	–	0–56	1.68–1.69	0.01
glaucophane	$Na_2Mg_3Al_2Si_8O_{22}(OH)_2$	monoclinic	biaxial	–	0–50	1.65–1.67	0.01
epidote	$Ca_2(Al,Fe)_3Si_3O_{12}(OH)$	monoclinic	biaxial	–	90–115	1.71–1.8	0.01–0.05
albite	$NaAlSi_3O_8$	triclinic	biaxial	+	77	1.52	0.011
boracite	$Mg_3ClB_7O_{13}$	orthorhombic	biaxial	+	82	1.64–1.67	0.011
barite	$BaSO_4$	orthorhombic	biaxial	+	37	1.63–1.64	0.012
natrolite	$Na_2Al_2Si_3O_{10} \cdot 2H_2O$	orthorhombic	biaxial	+	38–62	1.48	0.012
anorthite	$CaAl_2Si_2O_8$	triclinic	biaxial	–	78	1.57–1.59	0.013
margarite	$CaAl_2(Al_2Si_2O_{10})(OH)_2$	monoclinic	biaxial	–	45	1.63–1.65	0.013
rhodonite	$MnSiO_3$	triclinic	biaxial	+	63–76	1.71–1.73	0.013
staurolite	$Fe_2Al_9Si_4O_{23}(OH)$	monoclinic	biaxial	+	80–88	1.74–1.75	0.013
scheelite	$CaWO_4$	tetragonal	uniaxial	+		1.92–1.93	0.014
wollastonite	$CaSiO_3$	triclinic	biaxial	–	39	1.62–1.63	0.014
kyanite	Al_2SiO_5	triclinic	biaxial	–	82–83	1.71–1.72	0.016
zincite	ZnO	hexagonal	uniaxial	+		2.01–2.02	0.016
anglesite	$PbSO_4$	orthorhombic	biaxial	+	75	1.87–1.89	0.017
Minerals With Moderate Birefringence **(Displaying up to second-order interference colors in thin section)**							
alunite	$KAl_3(SO_4)_2(OH)_6$	hexagonal	uniaxial	+		1.57–1.59	0.02
brucite	$Mg(OH)_2$	hexagonal	uniaxial	+		1.57–1.58	0.02
cordierite	$(Mg,Fe)_2Al_4Si_5O_{18}$	orthorhombic	biaxial	+/–	65–105	1.54–1.56	0.02
gibbsite	$Al(OH)_3$	monoclinic	biaxial	+	0–40	1.57–1.59	0.02
hornblende	$(K,Na)_{0-1}(Ca,Na,Fe,Mg)_2(Mg,Fe,Al)_5$ - $(Si,Al)_8O_{22}(OH)_2$	monoclinic	biaxial	–	50–80	1.65–1.67	0.02
jadeite	$NaAlSi_2O_6$	monoclinic	biaxial	+	70–75	1.65–1.67	0.02
lawsonite	$CaAl_2Si_2O_7(OH)_2 \cdot H_2O$	orthorhombic	biaxial	+	76–86	1.66–1.68	0.02
monticellite	$CaMgSiO_4$	orthorhombic	biaxial	–	72–82	1.64–1.66	0.02
sepiolite	$Mg_4Si_6O_{15}(OH)_2(H_2O) \cdot 4H_2O$	orthorhombic	biaxial	–	20–70	1.51–1.53	0.02
spodumene	$LiAlSi_2O_6$	monoclinic	biaxial	+	60–80	1.65–1.67	0.02
kaersutite	Ti-rich hornblende	monoclinic	biaxial	–	68–82	1.68–1.73	0.02–0.08
tourmaline	$Na(Fe,Mg,Al,Li)_3Al_6(BO_3)_3Si_6O_{18}(OH)_4$	hexagonal	uniaxial	–		1.62–1.67	0.021–0.029
sillimanite	Al_2SiO_5	orthorhombic	biaxial	+	20–30	1.65–1.68	0.022
tremolite-actinolite	$Ca_2(Mg,Fe)_5Si_8O_{22}(OH)_2$	monoclinic	biaxial	–	85–90	1.61–1.63	0.022
autunite	$Ca(UO_2)_2(PO_4)_2 \cdot 10H_2O$	tetragonal	uniaxial	–		1.55–1.57	0.024
borax	$Na_2B_4O_5(OH)_4 \cdot 8H_2O$	monoclinic	biaxial	–	40	1.44–1.47	0.025
pigeonite	$(Ca,Mg,Fe)_2Si_2O_6$	monoclinic	biaxial	+	38–44	1.69–1.72	0.025
norbergite	$Mg_3SiO_4(OH,F)_2$	orthorhombic	biaxial	+	44–50	1.56–1.58	0.026
stishovite	SiO_2	tetragonal	uniaxial	+		1.8–1.82	0.027
wavellite	$Al_3(PO_4)_2(OH)_3 \cdot 5H_2O$	orthorhombic	biaxial	+	72	1.52–1.55	0.027
colemanite	$CaB_3O_4(OH)_3 \cdot H_2O$	monoclinic	biaxial	+	56	1.58–1.61	0.028
epsomite	$MgSO_4 \cdot 7H_2O$	orthorhombic	biaxial	+	52	1.43–1.46	0.028

Mineral	Formula	Crystal System	Optical Properties				
			System	Sign	2V°	Mean R.I. ($\bar{n}$)	Birefringence (δ)
Minerals With Moderate Birefringence *(continued)* **(Displaying up to second-order interference colors in thin section)**							
carnallite	$KMgCl_3 \cdot 6H_2O$	orthorhombic	biaxial	+	70	1.46–1.49	0.029
chalcanthite	$CuSO_4 \cdot 5H_2O$	triclinic	biaxial	–	56	1.51–1.54	0.029
ulexite	$NaCaB_5O_6(OH)_6 \cdot 5H_2O$	triclinic	biaxial	+	73	1.49–1.52	0.029
amblygonite	$LiAl(PO_4)F$	triclinic	biaxial	–	52–90	1.59–1.62	0.03
anthophyllite	$(Mg,Fe)_7Si_8O_{22}(OH)_2$	orthorhombic	biaxial	–	65–90	1.6–1.63	0.03
chondrodite	$Mg_5(SiO_4)_2(OH,F)_2$	monoclinic	biaxial	+	60–90	1.6–1.63	0.03
cummingtonite	$Mg_7Si_8O_{22}(OH)_2$	monoclinic	biaxial	+	80–90	1.64–1.67	0.03
diopside	$Ca(Mg,Fe)Si_2O_6$	monoclinic	biaxial	+	56–64	1.66–1.75	0.03
prehnite	$Ca_2Al(AlSi_3O_{10})(OH)_2$	orthorhombic	biaxial	+	65–70	1.62–1.65	0.03
variscite	$AlPO_4 \cdot 2H_2O$	orthorhombic	biaxial	–	48–54	1.55–1.59	0.03
clinohumite	$Mg_9(SiO_4)_2(OH,F)_2$	monoclinic	biaxial	+	73–76	1.59–1.64	0.03–0.04
lazulite	$(Mg,Fe)Al_2(PO_4)_2(OH)_2$	monoclinic	biaxial	–	70	1.61–1.64	0.031
kernite	$Na_2B_4O_6(OH)_2 \cdot 3H_2O$	monoclinic	biaxial	–	80	1.45–1.48	0.034
muscovite	$KAl_2AlSi_3O_{10}(OH)_2$	monoclinic	biaxial	–	30–40	1.56–1.6	0.035
Minerals With High Birefringence **(Displaying up to third-order interference colors in thin section)**							
olivine	$(Mg,Fe)_2SiO_4$	orthorhombic	biaxial	+	47–90	1.63–1.88	0.035–0.053
biotite	$K(Mg,Fe)_3AlSi_3O_{10}(OH)_2$	monoclinic	biaxial	–	0–32	1.57–1.61	0.04
diaspore	$AlO(OH)$	orthorhombic	biaxial	+	85	1.68–1.75	0.04
pectolite	$NaCa_2(SiO_3)_3H$	triclinic	biaxial	+	35–63	1.59–1.63	0.04
sinhalite	$MgAlBO_4$	orthorhombic	biaxial	+	58	1.67–1.71	0.04
turquoise	$CuAl_6(PO_4)_4(OH)_8 \cdot 4H_2O$	triclinic	biaxial	+	40	1.61–1.65	0.04
anhydrite	$CaSO_4$	orthorhombic	biaxial	+	44	1.57–1.61	0.044
atacamite	$Cu_2Cl(OH)_3$	orthorhombic	biaxial	–	75	1.83–1.88	0.047
pyrophyllite	$Al_2Si_4O_{10}(OH)_2$	triclinic	biaxial	–	52–62	1.55–1.6	0.047
monazite	$(Ce,La,Th)PO_4$	monoclinic	biaxial	+	10–20	1.78–1.85	0.05
talc	$Mg_3Si_4O_{10}(OH)_2$	monoclinic	biaxial	–	6–30	1.54–1.58	0.05
vivianite	$Fe_3(PO_4)_3 \cdot 8H_2O$	monoclinic	biaxial	+	83	1.57–1.63	0.054
Minerals With Very High Birefringence **(Displaying up to fourth-order and higher interference colors in thin section)**							
zircon	$ZrSiO_4$	tetragonal	uniaxial	+		1.93–1.99	0.06
antlerite	$Cu_3SO_4(OH)_4$	orthorhombic	biaxial	+	53	1.72–1.78	0.063
vanadinite	$Pb_5(VO_4)_3Cl$	hexagonal	uniaxial	–		2.3–2.4	0.066
anatase	TiO_2	tetragonal	uniaxial	–		2.48–2.56	0.073
erythrite	$Co_3(AsO_4)_2 \cdot 8H_2O$	monoclinic	biaxial	–	90	1.62–1.69	0.073
cassiterite	SnO_2	tetragonal	uniaxial	+		2–2.1	0.09
azurite	$Cu_3(CO_3)_2(OH)_2$	monoclinic	biaxial	+	68	1.73–1.83	0.106
columbite	$(Fe,Mn)(Nb,Ta)_2O_6$	orthorhombic	biaxial	+		2.3–2.4	0.12
wulfenite	$PbMoO_4$	tetragonal	uniaxial	–		2.28–2.4	0.121
wolframite	$(Fe,Mn)WO_4$	monoclinic	biaxial	+	73–79	2.1–2.5	0.13–0.15
brookite	TiO_2	orthorhombic	biaxial	+	0–30	2.58–2.72	0.14
strontianite	$SrCO_3$	orthorhombic	biaxial	–	7	1.52–1.66	0.148
witherite	$BaCO_3$	orthorhombic	biaxial	–	16	1.52–1.67	0.148
goethite	$FeO(OH)$	orthorhombic	biaxial	–	0–27	2.26–2.52	0.15
titanite (sphene)	$CaTiSiO_5$	monoclinic	biaxial	+	23–50	1.86–2.1	0.15
aragonite	$CaCO_3$	orthorhombic	biaxial	–	18	1.53–1.68	0.155
realgar	AsS	monoclinic	biaxial	–	41	2.5–2.8	0.166
proustite	Ag_3AsS_3	hexagonal	uniaxial	–		2.71–2.98	0.17

Mineral	Formula	Crystal System	Optical Properties				
			System	Sign	2V°	Mean R.I. ($\bar{n}$)	Birefringence (δ)
Minerals With Very High Birefringence *(continued)* (Displaying up to fourth-order and higher interference colors in thin section)							
calcite	$CaCO_3$	hexagonal	uniaxial	–		1.48–1.65	0.172
niter	KNO_3	orthorhombic	biaxial	–	7	1.33–1.5	0.172
dolomite	$CaMg(CO_3)_2$	hexagonal	uniaxial	–		1.5–1.67	0.179
magnesite	$MgCO_3$	hexagonal	uniaxial	–		1.5–1.7	0.191
carnotite	$K_2U_2O_4(VO_4)_2 \cdot 3H_2O$	monoclinic	biaxial	–	38–44	1.75–1.95	0.2
pyrargyrite	Ag_3SbS_3	hexagonal	uniaxial	–		2.88–3.08	0.2
ankerite	$CaFe(CO_3)_2$	hexagonal	uniaxial	–		1.54–1.75	0.202
rhodochrosite	$MnCO_3$	hexagonal	uniaxial	–		1.59–1.81	0.219
smithsonite	$ZnCO_3$	hexagonal	uniaxial	–		1.62–1.85	0.225
siderite	$FeCO_3$	hexagonal	uniaxial	–		1.63–1.87	0.242
nitratite	$NaNO_3$	hexagonal	uniaxial	–		1.33–1.58	0.251
malachite	$Cu_2CO_3(OH)_2$	monoclinic	biaxial	–	43	1.65–1.9	0.254
cerussite	$PbCO_3$	orthorhombic	biaxial	–	9	1.8–2.07	0.274
hematite	Fe_2O_3	hexagonal	uniaxial	–		3–3.2	0.28
manganite	$MnO(OH)$	monoclinic	biaxial	+	small	2.2–2.5	0.29
sulfur	S	orthorhombic	biaxial	+	69	1.95–2.24	0.29
rutile	TiO_2	tetragonal	uniaxial	+		2.61–2.9	0.29
hausmannite	Mn_3O_4	tetragonal	uniaxial	–		2.1–2.4	0.31
cinnabar	HgS	hexagonal	uniaxial	+		2.9–3.25	0.35
crocoite	$PbCrO_4$	monoclinic	biaxial	+	54	2.3–2.6	0.35
lepidocrocite	$FeO(OH)$	orthorhombic	biaxial	–	83	1.94–2.51	0.57
orpiment	As_2S_3	monoclinic	biaxial	+	76	2.4–3	0.62

Appendix D: Interference Colors

Appendix

Minerals Ordered by Hardness

Mineral	Formula	Specific Gravity (G)	Hardness (H)
talc	$Mg_3Si_4O_{10}(OH)_2$	2.8	1
molybdenite	MoS_2	4.7	$1-1^1/_2$
pyrophyllite	$Al_2Si_4O_{10}(OH)_2$	2.8	$1^1/_2$
arsenic	As	5.7	$1-2$
erythrite	$Co_3(AsO_4)_2 \cdot 8H_2O$	3.06	$1-2$
graphite	C	2.1–2.2	$1-2$
nitratite	$NaNO_3$	2.29	$1-2$
covellite	CuS	4.6	$1^1/_2-2$
orpiment	As_2S_3	3.49	$1^1/_2-2$
realgar	AsS	3.56	$1^1/_2-2$
sulfur	S	2.1	$1^1/_2-2^1/_2$
carnotite	$K_2U_2O_4(VO_4)_2 \cdot 3H_2O$	4.5	2
gypsum	$CaSO_4 \cdot 2H_2O$	2.32	2
niter	KNO_3	2.1	2
pyrargyrite	Ag_3SbS_3	5.85	2
stibnite	Sb_2S_3	4.6	2
sylvite	KCl	1.99	2
vivianite	$Fe_3(PO_4)_3 \cdot 8H_2O$	2.58	2
zinc	Zn	7.1	2
argentite (acanthite)	Ag_2S	7.1	$2-2^1/_2$
autunite	$Ca(UO_2)_2(PO_4)_2 \cdot 10H_2O$	3.15	$2-2^1/_2$
borax	$Na_2B_4O_5(OH)_4 \cdot 8H_2O$	1.7–1.9	$2-2^1/_2$
cinnabar	HgS	8.1	$2-2^1/_2$
epsomite	$MgSO_4 \cdot 7H_2O$	1.68	$2-2^1/_2$
galena	PbS	7.6	$2-2^1/_2$
kaolinite	$Al_2Si_2O_5(OH)_4$	2.6	$2-2^1/_2$
proustite	Ag_3AsS_3	5.5–7	$2-2^1/_2$
sepiolite	$Mg_4Si_6O_{15}(OH)_2(H_2O) \cdot 4H_2O$	2	$2-2^1/_2$
brucite	$Mg(OH)_2$	2.4–2.5	$2^1/_2$
carnallite	$KMgCl_3 \cdot 6H_2O$	1.6	$2^1/_2$
chalcanthite	$CuSO_4 \cdot 5H_2O$	2.3	$2^1/_2$
chlorargerite	AgCl	1.55	$2^1/_2$

Mineral	Formula	Specific Gravity (G)	Hardness (H)
cryolite	Na_3AlF_6	2.97	$2^1/_2$
halite	$NaCl$	2.16	$2^1/_2$
muscovite	$KAl_2AlSi_3O_{10}(OH)_2$	2.8	$2^1/_2$
ulexite	$NaCaB_5O_6(OH)_6 \cdot 5H_2O$	1.96	$2^1/_2$
chlorite	talc+brucite combinations	3	2–3
anglesite	$PbSO_4$	6.38	$2^1/_2$–3
biotite	$K(Mg,Fe)_3AlSi_3O_{10}(OH)_2$	3	$2^1/_2$–3
bismuth	Bi	9.8	$2^1/_2$–3
boulangerite	$Pb_5Sb_4S_{11}$	6–6.2	$2^1/_2$–3
chalcocite	Cu_2S	5.8	$2^1/_2$–3
copper	Cu	8.7–8.9	$2^1/_2$–3
crocoite	$PbCrO_4$	6	$2^1/_2$–3
gold	Au	15.6–19.3	$2^1/_2$–3
silver	Ag	10.1–10.5	$2^1/_2$–3
gibbsite	$Al(OH)_3$	2.4	$2^1/_2$–$3^1/_2$
pyrolusite	MnO_2	4.5–5	$2^1/_2$–$6^1/_2$
bornite	Cu_5FeS_4	6	3
calcite	$CaCO_3$	2.71	3
enargite	Cu_3AsS_4	4.5	3
kernite	$Na_2B_4O_6(OH)_2 \cdot 3H_2O$	1.9	3
vanadinite	$Pb_5(VO_4)_3Cl$	6.9	3
wulfenite	$PbMoO_4$	6.7–7	3
antimony	Sb	6.7	3–$3^1/_2$
atacamite	$Cu_2Cl(OH)_3$	3.76	3–$3^1/_2$
barite	$BaSO_4$	4.5	3–$3^1/_2$
celestite	$SrSO_4$	3.97	3–$3^1/_2$
cerussite	$PbCO_3$	6.55	3–$3^1/_2$
millerite	NiS	5.5	3–$3^1/_2$
witherite	$BaCO_3$	4.29	3–$3^1/_2$
anhydrite	$CaSO_4$	2.98	$3^1/_2$
strontianite	$SrCO_3$	3.72	$3^1/_2$
antigorite	$Mg_3Si_2O_5(OH)_4$	2.6	3–4
tetrahedrite	$Cu_{12}Sb_4S_{13}$	4.5–5.11	3–4
wavellite	$Al_3(PO_4)_2(OH)_3 \cdot 5H_2O$	2.36	3–4
alunite	$KAl_3(SO_4)_2(OH)_6$	2.6–2.9	$3^1/_2$–4
ankerite	$CaFe(CO_3)_2$	3.1	$3^1/_2$–4
antlerite	$Cu_3SO_4(OH)_4$	3.9	$3^1/_2$–4
aragonite	$CaCO_3$	2.94	$3^1/_2$–4
azurite	$Cu_3(CO_3)_2(OH)_2$	3.77	$3^1/_2$–4
chalcopyrite	$CuFeS_2$	4.2	$3^1/_2$–4
cuprite	Cu_2O	5.9–6.1	$3^1/_2$–4
dolomite	$CaMg(CO_3)_2$	2.85	$3^1/_2$–4
heulandite	$CaAl_2Si_7O_{18} \cdot 6H_2O$	2.15	$3^1/_2$–4
malachite	$Cu_2CO_3(OH)_2$	3.7–4	$3^1/_2$–4
pentlandite	$(Ni,Fe)_9S_8$	5	$3^1/_2$–4
pyromorphite	$Pb(PO_4)_3Cl$	7	$3^1/_2$–4
rhodochrosite	$MnCO_3$	3.7	$3^1/_2$–4
sphalerite	ZnS	4	$3^1/_2$–4
stilbite	$CaAl_2Si_7O_{18} \cdot 7H_2O$	2.15	$3^1/_2$–4
wurtzite	ZnS	4	$3^1/_2$–4
margarite	$CaAl_2(Al_2Si_2O_{10})(OH)_2$	3.1	$3^1/_2$–$4^1/_2$
pyrrhotite	$Fe_{1-x}S$	4.6	$3^1/_2$–$4^1/_2$

Mineral	Formula	Specific Gravity (G)	Hardness (H)
fluorite	CaF_2	3.18	4
magnesite	$MgCO_3$	3	4
manganite	$MnO(OH)$	4.2–4.4	4
siderite	$FeCO_3$	3.96	4
variscite	$AlPO_4 \cdot 2H_2O$	2.5	4
colemanite	$CaB_3O_4(OH)_3 \cdot H_2O$	2.42	$4–4^1/_2$
platinum	Pt	16.5–18	$4–4^1/_2$
smithsonite	$ZnCO_3$	4.43	$4–4^1/_2$
wolframite	$(Fe,Mn)WO_4$	7.2–7.6	$4–4^1/_2$
zincite	ZnO	5.4–5.7	$4–4^1/_2$
chabazite	$CaAl_2Si_4O_{12} \cdot 6H_2O$	2.1	4–5
apophyllite	$KCa_4Si_8O_{20}F \cdot 8H_2O$	2.3	$4^1/_2–5$
pectolite	$NaCa_2(SiO_3)_3H$	2.9	$4^1/_2–5$
scheelite	$CaWO_4$	6.11	$4^1/_2–5$
wollastonite	$CaSiO_3$	3.1	$4^1/_2–5$
apatite	$Ca_5(PO_4)_3(OH,F,Cl)$	3.2	5
columbite	$(Fe,Mn)(Nb,Ta)_2O_6$	6	5
lepidocrocite	$FeO(OH)$	4	5
titanite (sphene)	$CaTiSiO_5$	3.5	5
brookite	TiO_2	4.14	$5–5^1/_2$
goethite	$FeO(OH)$	4.3	$5–5^1/_2$
hausmannite	Mn_3O_4	4.86	$5–5^1/_2$
natrolite	$Na_2Al_2Si_3O_{10} \cdot 2H_2O$	2.23	$5–5^1/_2$
niccolite	$NiAs$	4.6	$5–5^1/_2$
triphylite	$Li(Fe,Mn)PO_4$	3.5–5.5	$5–5^1/_2$
glaucophane	$Na_2Mg_3Al_2Si_8O_{22}(OH)_2$	3.1–3.2	5–6
hornblende	$(Na,K)_{0-1}(Ca,Na,Fe,Mg)_2(Mg,Fe,Al)_5(Si,Al)_8O_{22}(OH)_2$	3–3.5	5–6
hypersthene	$(Mg,Fe)_2Si_2O_6$	3.2–3.5	5–6
ilmenite	$FeTiO_3$	4.5–5	5–6
scapolite	$(Na,Ca)_4(Al,Si)_{12}O_{24}(Cl,CO_3)$	2.55–2.76	5–6
tremolite-actinolite	$Ca_2(Mg,Fe)_5Si_8O_{22}(OH)_2$	3–3.3	5–6
uraninite	UO_2	7–9.5	5–6
kyanite	Al_2SiO_5	3.6	5–7
analcime	$NaAlSi_2O_6 \cdot H_2O$	2.26	$5^1/_2$
chromite	$FeCr_2O_4$	5.1	$5^1/_2$
cobaltite	$CoAsS$	6.3	$5^1/_2$
monazite	$(Ce,La,Th)PO_4$	4.9–5.2	$5^1/_2$
monticellite	$CaMgSiO_4$	3.15	$5^1/_2$
periclase	MgO	3.56	$5^1/_2$
perovskite	$CaTiO_3$	4	$5^1/_2$
anatase	TiO_2	3.9	$5^1/_2–6$
anthophyllite	$(Mg,Fe)_7Si_8O_{22}(OH)_7$	2.9–3.2	$5^1/_2–6$
arsenopyrite	$FeAsS$	6.1	$5^1/_2–6$
leucite	$KAlSi_2O_6$	2.48	$5^1/_2–6$
nepheline	$(Na,K)AlSiO_4$	2.6	$5^1/_2–6$
rhodonite	$MnSiO_3$	3.5–3.7	$5^1/_2–6$
skutterudite	$(Co,Ni)As_3$	6.1–6.8	$5^1/_2–6$
sodalite	$Na_3Al_3Si_3O_{12} \cdot NaCl$	2.3	$5^1/_2–6$
diopside	$Ca(Mg,Fe)Si_2O_6$	3.2–3.6	$5^1/_2–6^1/_2$
franklinite	$ZnFe_2O_4$	5.32	$5^1/_2–6^1/_2$
hematite	Fe_2O_3	4.9–5.3	$5^1/_2–6^1/_2$
magnetite	Fe_3O_4	5.2	$5^1/_2–6^1/_2$

Mineral	Formula	Specific Gravity (G)	Hardness (H)
amblygonite	$LiAl(PO_4)F$	3	6
clinohumite	$Mg_9(SiO_4)_2(OH,F)_2$	3.21–3.35	6
cummingtonite	$Mg_7Si_8O_{22}(OH)_2$	2.9–3.2	6
kaersutite	Ti-rich hornblende	3.2–3.3	6
lazulite	$(Mg,Fe)Al_2(PO_4)_2(OH)_2$	3	6
microcline	$KAlSi_3O_8$	2.56	6
orthoclase	$KAlSi_3O_8$	2.56	6
pigeonite	$(Ca,Mg,Fe)_2Si_2O_6$	3.4	6
sanidine	$(K,Na)AlSi_3O_8$	2.56	6
turquoise	$CuAl_6(PO_4)_4(OH)_8 \cdot 4H_2O$	2.7	6
albite	$NaAlSi_3O_8$	2.62	$6–6^1/_2$
anorthite	$CaAl_2Si_2O_8$	2.76	$6–6^1/_2$
marcasite	FeS_2	4.9	$6–6^1/_2$
prehnite	$Ca_2Al(AlSi_3O_{10})(OH)_2$	2.9	$6–6^1/_2$
pyrite	FeS_2	5.1	$6–6^1/_2$
rutile	TiO_2	4.24	$6–6^1/_2$
chloritoid	$(Fe,Mg)Al_2SiO_5(OH)_2$	3.5	$6^1/_2$
chondrodite	$Mg_5(SiO_4)_2(OH,F)_2$	3.16–3.26	$6^1/_2$
clinozoisite	$Ca_2Al_3Si_3O_{12}(OH)$	3.1–3.4	$6^1/_2$
epidote	$Ca_2(Al,Fe)_3Si_3O_{12}(OH)$	3.4–3.5	$6^1/_2$
jadeite	$NaAlSi_2O_6$	3.3	$6^1/_2$
norbergite	$Mg_3SiO_4(OH,F)_2$	3.16	$6^1/_2$
vesuvianite	$Ca_{10}(Mg,Fe)_2Al_4Si_9O_{34}(OH)_4$	3.4	$6^1/_2$
cassiterite	SnO_2	7	6–7
cristobalite	SiO_2	2.33	6–7
sillimanite	Al_2SiO_5	3.23	6–7
tridymite	SiO_2	2.28	6–7
diaspore	$AlO(OH)$	3.2–3.5	$6^1/_2–7$
olivine	$(Mg,Fe)_2SiO_4$	3.2–3.4	$6^1/_2–7$
sinhalite	$MgAlBO_4$	3.42	$6^1/_2–7$
spodumene	$LiAlSi_2O_6$	3.15	$6^1/_2–7$
garnet	$(Ca,Fe,Mg)_3(Al,Fe)_2Si_3O_{12}$	3.5–4.3	$6^1/_2–7^1/_2$
boracite	$Mg_3ClB_7O_{13}$	2.95	7
cordierite	$(Mg,Fe)_2Al_4Si_5O_{18}$	2.5–2.s8	7
quartz	SiO_2	2.65	7
staurolite	$Fe_2Al_9Si_4O_{23}(OH)$	3.75	$7–7^1/_2$
tourmaline	$Na(Fe,Mg,Al,Li)_3Al_6(BO_3)_3Si_6O_{18}(OH)_4$	2.9–3.3	$7–7^1/_2$
andalusite	Al_2SiO_5	3.18	$7^1/_2$
zircon	$ZrSiO_4$	4.68	$7^1/_2$
coesite	SiO_2	2.93	7–8
lawsonite	$CaAl_2Si_2O_7(OH)_2 \cdot H_2O$	3.1	7–8
beryl	$Be_3Al_2Si_6O_{18}$	2.7–2.9	$7^1/_2–8$
spinel	$MgAl_2O_4$	3.5–4	$7^1/_2–8$
topaz	$Al_2SiO_4(F,OH)_2$	3.5–3.6	8
chrysoberyl	$BeAl_2O_4$	3.7–3.8	$8^1/_2$
corundum	Al_2O_3	3.9–4.1	9
diamond	C	3.5	10

Appendix E: Mineral Hardness

Appendix

Minerals Ordered by Specific Gravity

Mineral	Formula	Specific Gravity (G)	Hardness (H)
chlorargerite	AgCl	1.55	$2^1/_2$
carnallite	$KMgCl_3 \cdot 6H_2O$	1.6	$2^1/_2$
epsomite	$MgSO_4 \cdot 7H_2O$	1.68	$2–2^1/_2$
borax	$Na_2B_4O_5(OH)_4 \cdot 8H_2O$	1.7–1.9	$2–2^1/_2$
kernite	$Na_2B_4O_6(OH)_2 \cdot 3H_2O$	1.9	3
ulexite	$NaCaB_5O_6(OH)_6 \cdot 5H_2O$	1.96	$2^1/_2$
sylvite	KCl	1.99	2
sepiolite	$Mg_4Si_6O_{15}(OH)_2(H_2O) \cdot 4H_2O$	2	$2–2^1/_2$
chabazite	$CaAl_2Si_4O_{12} \cdot 6H_2O$	2.1	4–5
niter	KNO_3	2.1	2
sulfur	S	2.1	$1^1/_2–2^1/_2$
graphite	C	2.1–2.2	1–2
heulandite	$CaAl_2Si_7O_{18} \cdot 6H_2O$	2.15	$3^1/_2–4$
stilbite	$CaAl_2Si_7O_{18} \cdot 7H_2O$	2.15	$3^1/_2–4$
halite	NaCl	2.16	$2^1/_2$
natrolite	$Na_2Al_2Si_3O_{10} \cdot 2H_2O$	2.23	$5–5^1/_2$
analcime	$NaAlSi_2O_6 \cdot H_2O$	2.26	$5^1/_2$
tridymite	SiO_2	2.28	6–7
nitratite	$NaNO_3$	2.29	1–2
apophyllite	$KCa_4Si_8O_{20}F \cdot 8H_2O$	2.3	$4^1/_2–5$
chalcanthite	$CuSO_4 \cdot 5H_2O$	2.3	$2^1/_2$
sodalite	$Na_3Al_3Si_3O_{12} \cdot NaCl$	2.3	$5^1/_2–6$
gypsum	$CaSO_4 \cdot 2H_2O$	2.32	2
cristobalite	SiO_2	2.33	6–7
wavellite	$Al_3(PO_4)_2(OH)_3 \cdot 5H_2O$	2.36	3–4
gibbsite	$Al(OH)_3$	2.4	$2^1/_2–3^1/_2$
brucite	$Mg(OH)_2$	2.4–2.5	$2^1/_2$
colemanite	$CaB_3O_4(OH)_3 \cdot H_2O$	2.42	$4–4^1/_2$
leucite	$KAlSi_2O_6$	2.48	$5^1/_2–6$
variscite	$AlPO_4 \cdot 2H_2O$	2.5	4
cordierite	$(Mg,Fe)_2Al_4Si_5O_{18}$	2.5–2.8	7
scapolite	$(Na,Ca)_4(Al,Si)_{12}O_{24}(Cl,CO_3)$	2.55–2.76	5–6

Mineral	Formula	Specific Gravity (G)	Hardness (H)
microcline	$KAlSi_3O_8$	2.56	6
orthoclase	$KAlSi_3O_8$	2.56	6
sanidine	$(K,Na)AlSi_3O_8$	2.56	6
vivianite	$Fe_3(PO_4)_3 \cdot 8H_2O$	2.58	2
antigorite	$Mg_3Si_2O_5(OH)_4$	2.6	3–4
kaolinite	$Al_2Si_2O_5(OH)_4$	2.6	$2–2^1/_2$
nepheline	$(Na,K)AlSiO_4$	2.6	$5^1/_2–6$
alunite	$KAl_3(SO_4)_2(OH)_6$	2.6–2.9	$3^1/_2–4$
albite	$NaAlSi_3O_8$	2.62	$6–6^1/_2$
quartz	SiO_2	2.65	7
turquoise	$CuAl_6(PO_4)_4(OH)_8 \cdot 4H_2O$	2.7	6
beryl	$Be_3Al_2Si_6O_{18}$	2.7–2.9	$7^1/_2–8$
calcite	$CaCO_3$	2.71	3
anorthite	$CaAl_2Si_2O_8$	2.76	$6–6^1/_2$
muscovite	$KAl_2AlSi_3O_{10}(OH)_2$	2.8	$2^1/_2$
pyrophyllite	$Al_2Si_4O_{10}(OH)_2$	2.8	$1^1/_2$
talc	$Mg_3Si_4O_{10}(OH)_2$	2.8	1
dolomite	$CaMg(CO_3)_2$	2.85	$3^1/_2–4$
pectolite	$NaCa_2(SiO_3)_3H$	2.9	$4^1/_2–5$
prehnite	$Ca_2Al(AlSi_3O_{10})(OH)_2$	2.9	$6–6^1/_2$
anthophyllite	$(Mg,Fe)_7Si_8O_{22}(OH)_2$	2.9–3.2	$5^1/_2–6$
cummingtonite	$Mg_7Si_8O_{22}(OH)_2$	2.9–3.2	6
tourmaline	$Na(Fe,Mg,Al,Li)_3Al_6(BO_3)_3Si_6O_{18}(OH)_4$	2.9–3.3	$7–7^1/_2$
coesite	SiO_2	2.93	7–8
aragonite	$CaCO_3$	2.94	$3^1/_2–4$
boracite	$Mg_3ClB_7O_{13}$	2.95	7
cryolite	Na_3AlF_6	2.97	$2^1/_2$
anhydrite	$CaSO_4$	2.98	$3^1/_2$
amblygonite	$LiAl(PO_4)F$	3	6
biotite	$K(Mg,Fe)_3AlSi_3O_{10}(OH)_2$	3	$2^1/_2–3$
chlorite	talc+brucite combinations	3	2–3
lazulite	$(Mg,Fe)Al_2(PO_4)_2(OH)_2$	3	6
magnesite	$MgCO_3$	3	4
erythrite	$Co_3(AsO_4)_2 \cdot 8H_2O$	3.06	1–2
ankerite	$CaFe(CO_3)_2$	3.1	$3^1/_2–4$
lawsonite	$CaAl_2Si_2O_7(OH)_2 \cdot H_2O$	3.1	7–8
margarite	$CaAl_2(Al_2Si_2O_{10})(OH)_2$	3.1	$3^1/_2–4^1/_2$
wollastonite	$CaSiO_3$	3.1	$4^1/_2–5$
glaucophane	$Na_2Mg_3Al_2Si_8O_{22}(OH)_2$	3.1–3.2	5–6
clinozoisite	$Ca_2Al_3Si_3O_{12}(OH)$	3.1–3.4	$6^1/_2$
autunite	$Ca(UO_2)_2(PO_4)_2 \cdot 10H_2O$	3.15	$2–2^1/_2$
monticellite	$CaMgSiO_4$	3.15	$5^1/_2$
spodumene	$LiAlSi_2O_6$	3.15	$6^1/_2–7$
norbergite	$Mg_3SiO_4(OH,F)_2$	3.16	$6^1/_2$
chondrodite	$Mg_5(SiO_4)_2(OH,F)_2$	3.16–3.26	$6^1/_2$
andalusite	Al_2SiO_5	3.18	$7^1/_2$
fluorite	CaF_2	3.18	4
apatite	$Ca_5(PO_4)_3(F,Cl,OH)$	3.2	5
kaersutite	Ti-rich hornblende	3.2–3.3	6
olivine	$(Mg,Fe)_2SiO_4$	3.2–3.4	$6^1/_2–7$
diaspore	$AlO(OH)$	3.2–3.5	$6^1/_2–7$
hypersthene	$(Mg,Fe)_2Si_2O_6$	3.2–3.5	5–6

Appendix F: Specific Gravity

Mineral	Formula	Specific Gravity (G)	Hardness (H)
diopside	$Ca(Mg,Fe)Si_2O_6$	3.2–3.6	$5\frac{1}{2}$–$6\frac{1}{2}$
clinohumite	$Mg_9(SiO_4)_2(OH,F)_2$	3.21–3.35	6
sillimanite	Al_2SiO_5	3.23	6–7
jadeite	$NaAlSi_2O_6$	3.3	$6\frac{1}{2}$
pigeonite	$(Ca,Mg,Fe)_2Si_2O_6$	3.4	6
vesuvianite	$Ca_{10}(Mg,Fe)_2Al_4Si_9O_{34}(OH)_4$	3.4	$6\frac{1}{2}$
epidote	$Ca_2(Al,Fe)_3Si_3O_{12}(OH)$	3.4–3.5	$6\frac{1}{2}$
sinhalite	$MgAlBO_4$	3.42	$6\frac{1}{2}$–7
orpiment	As_2S_3	3.49	$1\frac{1}{2}$–2
chloritoid	$(Fe,Mg)Al_2SiO_5(OH)_2$	3.5	$6\frac{1}{2}$
diamond	C	3.5	10
titanite (sphene)	$CaTiSiO_5$	3.5	5
topaz	$Al_2SiO_4(F,OH)_2$	3.5–3.6	8
rhodonite	$MnSiO_3$	3.5–3.7	$5\frac{1}{2}$–6
spinel	$MgAl_2O_4$	3.5–4	$7\frac{1}{2}$–8
garnet	$(Ca,Fe,Mg)_3(Al,Fe)_2Si_3O_{12}$	3.5–4.3	$6\frac{1}{2}$–$7\frac{1}{2}$
triphylite	$Li(Fe,Mn)PO_4$	3.5–5.5	5–$5\frac{1}{2}$
periclase	MgO	3.56	$5\frac{1}{2}$
realgar	AsS	3.56	$1\frac{1}{2}$–2
kyanite	Al_2SiO_5	3.6	5–7
rhodochrosite	$MnCO_3$	3.7	$3\frac{1}{2}$–4
chrysoberyl	$BeAl_2O_4$	3.7–3.8	$8\frac{1}{2}$
malachite	$Cu_2CO_3(OH)_2$	3.7–4	$3\frac{1}{2}$–4
strontianite	$SrCO_3$	3.72	$3\frac{1}{2}$
staurolite	$Fe_2Al_9Si_4O_{20}(O,OH)_2$	3.75	7–$7\frac{1}{2}$
atacamite	$Cu_2Cl(OH)_3$	3.76	3–$3\frac{1}{2}$
azurite	$Cu_3(CO_3)_2(OH)_2$	3.77	$3\frac{1}{2}$–4
anatase	TiO_2	3.9	$5\frac{1}{2}$–6
antlerite	$Cu_3SO_4(OH)_4$	3.9	$3\frac{1}{2}$–4
corundum	Al_2O_3	3.9–4.1	9
siderite	$FeCO_3$	3.96	4
celestite	$SrSO_4$	3.97	3–$3\frac{1}{2}$
tremolite-actinolite	$Ca_2(Mg,Fe)_5Si_8O_{22}(OH)_2$	3–3.3	5–6
hornblende	$(Na,K)_{0-1}(Ca,Na,Fe,Mg)_2(Mg,Fe,Al)_5(Si,Al)_8O_{22}(OH)_2$	3–3.5	5–6
lepidocrocite	$FeO(OH)$	4	5
perovskite	$CaTiO_3$	4	$5\frac{1}{2}$
sphalerite	ZnS	4	$3\frac{1}{2}$–4
wurtzite	ZnS	4	$3\frac{1}{2}$–4
brookite	TiO_2	4.14	5–$5\frac{1}{2}$
chalcopyrite	$CuFeS_2$	4.2	$3\frac{1}{2}$–4
manganite	$MnO(OH)$	4.2–4.4	4
rutile	TiO_2	4.24	6–$6\frac{1}{2}$
witherite	$BaCO_3$	4.29	3–$3\frac{1}{2}$
goethite	$FeO(OH)$	4.3	5–$5\frac{1}{2}$
stishovite	SiO_2	4.3	
smithsonite	$ZnCO_3$	4.43	4–$4\frac{1}{2}$
barite	$BaSO_4$	4.5	3–$3\frac{1}{2}$
carnotite	$K_2U_2O_4(VO_4)_2 \cdot 3H_2O$	4.5	2
enargite	Cu_3AsS_4	4.5	3
ilmenite	$FeTiO_3$	4.5–5	5–6
pyrolusite	MnO_2	4.5–5	$2\frac{1}{2}$–$6\frac{1}{2}$
tetrahedrite	$Cu_{12}Sb_4S_{13}$	4.5–5.11	3–4

Mineral	Formula	Specific Gravity (G)	Hardness (H)
covellite	CuS	4.6	$1^{1}/_{2}$–2
niccolite	$NiAs$	4.6	5–$5^{1}/_{2}$
pyrrhotite	$Fe_{1-x}S$	4.6	$3^{1}/_{2}$–$4^{1}/_{2}$
stibnite	Sb_2S_3	4.6	2
zircon	$ZrSiO_4$	4.68	$7^{1}/_{2}$
molybdenite	MoS_2	4.7	1–$1^{1}/_{2}$
hausmannite	Mn_3O_4	4.86	5–$5^{1}/_{2}$
marcasite	FeS_2	4.9	6–$6^{1}/_{2}$
monazite	$(Ce,La,Th)PO_4$	4.9–5.2	$5^{1}/_{2}$
hematite	Fe_2O_3	4.9–5.3	$5^{1}/_{2}$–$6^{1}/_{2}$
pentlandite	$(Ni,Fe)_9S_8$	5	$3^{1}/_{2}$–4
chromite	$FeCr_2O_4$	5.1	$5^{1}/_{2}$
pyrite	FeS_2	5.1	6–$6^{1}/_{2}$
magnetite	Fe_3O_4	5.2	$5^{1}/_{2}$–$6^{1}/_{2}$
franklinite	$ZnFe_2O_4$	5.32	$5^{1}/_{2}$–$6^{1}/_{2}$
zincite	ZnO	5.4–5.7	4–$4^{1}/_{2}$
millerite	NiS	5.5	3–$3^{1}/_{2}$
proustite	Ag_3AsS_3	5.5–7	2–$2^{1}/_{2}$
arsenic	As	5.7	1–2
chalcocite	Cu_2S	5.8	$2^{1}/_{2}$–3
pyrargyrite	Ag_3SbS_3	5.85	2
cuprite	Cu_2O	5.9–6.1	$3^{1}/_{2}$–4
bornite	Cu_5FeS_4	6	3
columbite	$(Fe,Mn)(Nb,Ta)_2O_6$	6	5
crocoite	$PbCrO_4$	6	$2^{1}/_{2}$–3
arsenopyrite	$FeAsS$	6.1	$5^{1}/_{2}$–6
skutterudite	$(Co,Ni)As_3$	6.1–6.8	$5^{1}/_{2}$–6
scheelite	$CaWO_4$	6.11	$4^{1}/_{2}$–5
cobaltite	$CoAsS$	6.3	$5^{1}/_{2}$
anglesite	$PbSO_4$	6.38	$2^{1}/_{2}$–3
cerussite	$PbCO_3$	6.55	3–$3^{1}/_{2}$
antimony	Sb	6.7	3–$3^{1}/_{2}$
wulfenite	$PbMoO_4$	6.7–7	3
vanadinite	$Pb_5(VO_4)_3Cl$	6.9	3
boulangerite	$Pb_5Sb_4S_{11}$	6–6.2	$2^{1}/_{2}$–3
cassiterite	SnO_2	7	6–7
pyromorphite	$Pb(PO_4)_3Cl$	7	$3^{1}/_{2}$–4
argentite (acanthite)	Ag_2S	7.1	2–$2^{1}/_{2}$
zinc	Zn	7.1	2
wolframite	$(Fe,Mn)WO_4$	7.2–7.6	4–$4^{1}/_{2}$
galena	PbS	7.6	2–$2^{1}/_{2}$
uraninite	UO_2	7–9.5	5–6
cinnabar	HgS	8.1	2–$2^{1}/_{2}$
copper	Cu	8.7–8.9	$2^{1}/_{2}$–3
bismuth	Bi	9.8	$2^{1}/_{2}$–3
silver	Ag	10.1–10.5	$2^{1}/_{2}$–3
gold	Au	15.6–19.3	$2^{1}/_{2}$–3
platinum	Pt	16.5–18	4–$4^{1}/_{2}$

Appendix F:
Specific Gravity

Glossary

accessory mineral A mineral present in small amounts in a rock, unimportant for rock naming or classification.

accessory plate A plate that may be inserted in the tube of a polarizing light microscope to produce interference of a known amount; typically made of quartz, gypsum, or mica.

acicular Having a needlelike shape.

actinide Elements (such as Th, Pa, and U) with atomic numbers 90 through 103 and valence electrons in 5*f* orbitals.

acute bisectrix A line bisecting the angle formed by the two optic axes in a biaxial mineral.

acute bisectrix figure (Bxa) The interference figure seen when looking down an acute bisectrix.

adamantine A type of luster that is bright, sparkly, and shiny, similar to that of diamonds.

aggregate (of crystals) A mass of crystals, of the same or different minerals, that may be physically separated, perhaps with some difficulty.

albite twin A common twin law in triclinic feldspars, often resulting in polysynthetic twins.

alchemy Chemistry of the Middle Ages that combined science, magic, and philosophy.

alkali element Any element (such as Li, Na, or K) of the first group in the Periodic Table of the Elements; alkali elements typically ionize to form monovalent cations.

alkaline earth element Any element (such as Be, Mg, or Ca) of the second group in the Periodic Table of the Elements; alkaline earth elements typically ionize to form divalent cations.

allochromatic A term describing a mineral that gets its color from minor or trace elements (*see also* idiochromatic).

alloy A noncrystalline mixture of two or more metals.

alluvium Unconsolidated sediment deposited by a stream.

amorphous Having a random atomic structure (*see also* noncrystalline).

amphibolite (facies) One of the principal metamorphic facies introduced by Eskola, corresponding to high-grade conditions of about 450°–650 °C and 3–8 Kbar.

amphibolite (rock) A metamorphic rock containing primarily hornblende and plagioclase.

analyzer A polarizing filter that can be inserted in the upper column of a polarizing light microscope to view minerals under crossed-polarized light (*see also* upper polarizer).

anatexis Melting of preexisting rock.

andesite An extrusive igneous rock of intermediate composition containing plagioclase as the only major feldspar; minor K-feldspar or quartz, pyroxene, biotite, and hornblende may be present.

angle of incidence The angle that an impinging ray makes with a normal to an interface.

angle of refraction The angle that a refracted ray makes with a normal to an interface.

anhedral A crystal that lacks well-developed crystal faces or that has rounded or irregular form due to crowding by adjacent crystals.

anion An ion having a negative charge.

anionic complex A tightly bonded, negatively charged molecular group; in mineral formulas, often surrounded by parentheses.

anisodesmic Describing an ionic compound in which the ionic bonds are not all of the same strength.

anisotropic Having different physical properties in different directions.

anomalous interference colors Interference colors that are not represented on the Michel Lèvy Chart; typically associated with minerals of extremely low birefringence, such as chlorite.

aphanitic Having no mineral grains that are visible to the naked eye.

aqueous solution A water-rich solution, usually containing dissolved elements or complexes.

arborescent A term describing an aggregate of crystals having a treelike appearance (*see also* dendritic).

arenite General term for detrital sedimentary rocks composed of sand-sized grains and lithic fragments; includes sandstone, graywacke, arkose, and others (*see also* psammite).

arkose A feldspar-rich sandstone.

asbestiform A crystal habit characterized by fine threadlike, fibrous, or acicular crystals; more specifically, sometimes defined as a crystal habit with a length:diameter ratio of more than 3 : 1.

asbestos A general term referring to minerals that have an asbestiform habit; chrysotile, amosite, and crocidolite are typical asbestos minerals.

association (mineral) A group of minerals found together in a rock.

asterism The play of colors seen in some minerals that produces a rayed or star-shaped figure when viewed in direct light; a star sapphire is one example.

atmophile An element that tends to concentrate in the Earth's atmosphere.

atomic absorption spectrophotometer An analytical instrument in which composition is determined by measuring the absorption of characteristic wavelengths of light by an atomized and flamed sample.

atomic mass unit (amu) Unit for expressing atomic mass, equal to approximately 1.66×10^{-24} grams.

atomic number (Z) The number of protons in the nucleus of an atom of an element.

atomic weight The weight of an atom or compound in atomic mass units; generally close to the total number of protons and neutrons.

Aufbau principle The principle that states that electrons fill orbitals in a systematic way from the lowest energy orbitals to the highest.

augen Large lenticular mineral grains or mineral aggregates in a foliated metamorphic rock.

augen gneiss A gneiss containing augen.

aureole A contact metamorphic zone surrounding an igneous intrusion (*see also* contact aureole).

authigenic Formed or generated in place; used to describe minerals that form in a sediment or sedimentary rock after deposition of an original sediment.

Avogadro's number 6.022×10^{23}, equivalent to the number of atoms or molecules in a mole.

axial ratio The ratio of unit cell lengths along each crystallographic axis, $a:b:c$.

axis (crystallographic) One of the three edges of a chosen unit cell in a crystal lattice; the coordinate system used to describe points, lines, and planes in a crystal.

axis (rotational) A symmetry element that relates identical crystal faces or other things by rotation of $60°$, $90°$, $120°$, $180°$, or $360°$ about an axis.

banded iron formation (BIF) A layered rock containing chert, silicate, carbonate, or oxide layers, giving a banded appearance.

basal cleavage A term used to describe the cleavage in minerals such as micas that allows the mineral to break into plates or sheets.

basalt An extrusive igneous rock of mafic composition containing plagioclase as the only major feldspar; clinopyroxene $\pm$ orthopyroxene $\pm$ olivine are typically present (*see also* gabbro).

basement (rock) Metamorphic and igneous rocks that underlie sediments and sedimentary rocks observed at the Earth's surface; often, though not exclusively, of Precambrian age.

basis A sufficient set of symmetry operators for describing the symmetry of crystals (translation-free symmetry) or of atomic structures (space symmetry).

bauxite Name given to a rock or a mineral-like material composed primarily of a mixture of aluminum oxides and hydroxides such as boehmite and gibbsite; bauxite is the most significant aluminum ore.

Baveno twin An uncommon twin law in feldspar.

Becke line A bright line, visible under a microscope, that separates substances of different refractive indices.

bentonite A clay-rich, earthy material formed by devitrification and alteration of tuff or volcanic ash; sometimes used more generally to refer to any clay deposit rich in montmorillonite.

Berman balance A type of balance scale used to determine specific gravity.

Bertrand lens A lens that can be inserted in the tube of a polarizing light microscope to facilitate observation of interference figures.

biaxial Describing a crystal having two optic axes and three principal indices of refraction (α, β, γ). Such crystals belong to the orthorhombic, monoclinic, and triclinic crystal systems.

binary solution A solid solution series that can be characterized by two end members.

birefringence (property) The property of a crystal that causes double refraction.

birefringence (value) The difference between the greatest and least indices of refraction.

bladed Having the appearance of blades.

blocky A term used to describe crystals that have a blocklike appearance, generally with an approximately square cross section.

blueschist (facies) One of the principal metamorphic facies introduced by Eskola, corresponding to high pressure–low temperature conditions.

blueschist (rock) A vague term used to describe fine-grained, bluish colored rocks diagnostic of the blueschist facies. Key minerals include blue amphiboles (glaucophane and riebeckite), lawsonite, jadeite, and aragonite.

body centered Said of a unit cell that has an extra lattice point at its center.

body diagonal A line passing through the center of a cube or other three-dimensional geometric shape and connecting opposite corners.

Bohr model of the atom A fundamental model of an atom that states that electrons orbit atomic nuclei in orbits associated with specific energy levels.

bomb A laboratory reactor vessel used to synthesize gems and minerals at high pressure and temperature.

bort Diamond of low quality that is unsuitable as a gem but useful in industrial applications.

botryoidal A term used to describe a mineral habit that appears like a bunch of grapes.

boule A oblong-shaped synthetic mineral crystal that is produced from the Verneuil technique.

Bowen's reaction series A hypothetical series that describes the order of crystallization of minerals from a magma.

Bragg Law A mathematical law that describes the relationship between the angle of diffraction (θ), X-ray wavelength (λ), and atomic plane spacing (d); $n\lambda = 2d\sin\theta$.

Bravais lattices The fourteen possible three-dimensional lattices that can describe mineral structures.

Brazil twin A common type of twinning in quartz.

breccia A clastic sedimentary rock composed of large (>2mm diameter) angular broken rock fragments in a finer grained matrix.

bridging oxygen An oxygen atom in a crystal structure that is shared by two or more equivalent coordinating polyhedra.

brittle A term used to describe minerals that shatter or break easily when struck.

brucite layer A sheet of $Mg(OH)_6$ octahedra in a layered crystal structure.

Buerger precession camera An instrument designed to record single crystal diffraction patterns on film.

burial metamorphism Metamorphism affecting a large region; caused by pressure related to depth in the Earth and temperature related to geothermal gradient (*see also* regional metamorphism).

Cabochon A gemstone that has been ground and polished into a domed shape.

calcine To heat material, such as limestone, to high temperature, breaking down minerals and driving off carbon dioxide or other volatiles.

capillary Having a hairlike or threadlike appearance (*see also* filiform).

carbonate (mineral) A group of minerals with formulas characterized by $(CO_3)^{2-}$ radicals.

carbonate (radical) The $(CO_3)^{2-}$ anionic group.

carbonate (rock) A chemical sedimentary rock dominated by carbonate minerals; limestone or dolostone.

carbonation reaction Reaction of a mineral or minerals with CO_2 to produce a product carbonate mineral.

Carlsbad twin A common twin law in orthoclase, less common in other feldspars, often resulting in penetration twins.

Cartesian coordinate system An X-Y-Z coordinate system in which all axes are at $90°$ and the unit distances along all axes are equal.

cataclastic metamorphism Metamorphism caused by a transient high-pressure condition such as a meteor impact (*see also* shock metamorphism and dynamic metamorphism).

cathode ray tube Vacuum tube in which beams of high-energy electrons pass through magnetic fields and hit a fluorescent screen.

cation An ion having a positive charge.

cementation Lithification of clastic sediments resulting from the deposition or precipitation of minerals in the spaces between individual clastic grains; it may occur at the time of deposition or during diagenesis.

centered Said of a unit cell that has (an) extra lattice point(s) at its center, in the center of its faces, or in the center of two opposing faces.

Glossary

chain silicates   Silicate minerals characterized by SiO_4 tetrahedra joining to form chains either one tetrahedron or two tetrahedra wide (*see also* inosilicates).

chalcophile   An element that tends to concentrate in sulfide minerals and ores.

characteristic radiation   High-intensity radiation of one or a few wavelengths emitted by the target of an X-ray tube; the wavelength(s) of characteristic radiation depend on the elements in the target.

chatoyancy   The play of colors seen in some minerals in which a silky sheen is seen to form a narrow band that changes position as the mineral is turned; for example, cat's-eye chrysoberyl.

chemical precipitate   A solid that precipitates from an aqueous solution, typically due to supersaturation.

chemical sedimentary rock   A sedimentary rock composed primarily of material formed by precipitation from solution; for example, most limestones and all evaporites.

chemical sediments   Sediments produced by dissolution and precipitation resulting from chemical weathering; they may precipitate where weathering occurs or at a different place.

chemical weathering   Type of weathering involving chemical reactions that transform or decompose minerals and rock.

chert   A hard sedimentary rock composed primarily of cryptocrystalline silica, generally in the form of fibrous chalcedony with lesser amounts of quartz and opal.

chromophores   Elements that give minerals their color.

clast   A mineral grain, lithic fragment, or organic remnant that is produced by mechanical weathering and becomes part of a clastic sediment or rock.

clastic   Being composed of fragments (clasts) derived from preexisting rocks.

clastic rock   Sedimentary rock formed by the lithification of clastic sediments; for example, sandstone, shale, and siltstone.

Clausius-Clapeyron equation   An equation relating the slope of a reaction on a pressure-temperature diagram to the entropy and volume change of the reaction: $dP/dT = \Delta S/\Delta V$.

clay (grain size)   A clastic fragment of any composition smaller than silt, having a diameter less than $\frac{1}{256}$ mm.

clay (mineral)   Member of a group of loosely defined hydrous sheet silicates formed primarily by alteration or weathering of primary silicates. The most common clays belong to the illite, kaolinite, or montmorillonite groups.

cleavage   The breaking of a mineral along a set of parallel identical atomic planes.

clinonet   A two-dimensional lattice characterized by two translations of different magnitudes at nonspecial angles to each other.

clinopyroxene   Monoclinic pyroxene subgroup dominated by end members diopside and hedenbergite. The most common of all pyroxenes.

closest packing   The most efficient way to pack like atoms together in three dimensions; each atom is surrounded by 12 others.

colloform   Appearing as spherical or hemispherical shapes made of radiating crystals (*see also* globular).

color   A sensation produced by different wavelengths of light hitting the eye.

columnar   Having the appearance of an aggregate of slender, elongated individual crystals, nearly parallel in arrangement.

complex twin   A twin composed of more than two individual crystals.

compositional zoning   Variation in the composition of a crystal, typically from core to margin (*see also* zoning).

compound (chemical)   A substance of fixed atomic proportions made by the combination of two or more elements.

concentration factor   The extent to which an element must be concentrated above normal crustal levels to make mining it profitable.

conchoidal   A term describing fracturing that produces curved surfaces similar to when glass breaks; for example, quartz has a conchoidal fracture.

condenser   A lens, or several lenses, that may be inserted in the substage of a polarizing light microscope to cause light rays to converge on a sample (*see also* condensing lens).

condensing lens   *See* condenser.

conglomerate   A coarse-grained clastic sedimentary rock with fragments larger than 2 mm in diameter in a fine-grained matrix; the equivalent of lithified gravel. The clasts are rounded in conglomerates in contrast with breccias.

conoscopic illumination   Describing the strongly convergent light produced by the insertion of a condensing lens in a polarizing light microscope substage.

constructive interference   The addition of two waves that are in phase with negligible or no loss of energy.

contact aureole   A contact metamorphic zone surrounding an igneous intrusion (*see also* aureole).

contact metamorphism   Metamorphism localized around an igneous rock body; primarily in response to heat and flowing fluids.

contact twin   A twin in which two individuals are symmetrically arranged about a twin plane.

continuous radiation   The low-intensity radiation covering a range of wavelengths, produced by an X-ray tube; continuous radiation provides the background for characteristic radiation.

continuous side (Bowen's reaction series)   The side of Bowen's reaction series characterized by plagioclase.

coordinating polyhedron   Polyhedron formed around an atom or ion by connecting the centers of the coordinated atoms or ions.

coordination number   Number of neighboring atoms to which an atom is bonded.

country rock   Rock intruded by and surrounding an igneous intrusion.

coupled substitution   Simultaneous substitution of two or more different ions in a structure in such a way that charge balance is maintained; for example, the substitution of $Ca^{2+}Al^{3+}$ for Na^+Si^{4+} in albite.

covalent bond   An ideal chemical bond that involves the sharing of orbital electrons between elements that have little or no difference in electronegativity.

cover slip   A thin piece of glass that is placed over grains and liquid to make a grain mount.

critical angle   The angle of incidence that yields an angle of refraction of 90°.

crossed polars   The condition caused when the upper polarizer is inserted in a polarizing light microscope.

cryptocrystalline   A term describing a material containing generally submicroscopic grains whose crystalline nature is not easily determined.

crystal   A homogeneous solid body of an element, compound, or solid solution having a regularly repeating atomic structure that may be outwardly expressed as planar faces.

crystal classes   The 32 possible combinations of symmetry elements that a crystal may have; in modern usage, practically synonymous with the 32 possible point groups.

crystal morphology   The shape and form(s) of a crystal.

crystal structure   Spatial arrangement of atoms or ions, and their bonds, in a crystal.

crystal structure determination   Determination of the spatial arrangement of atoms and their bonds in a crystal.

crystal system   One of the six distinct coordinate systems: cubic, orthorhombic, tetragonal, hexagonal, monoclinic, and triclinic. The trigonal subdivision of the hexagonal system is sometimes counted as a separate crystal system.

crystalline   Having a crystal structure; having a regular arrangement of atoms characterized by a space lattice.

crystallographic axis A direction corresponding to one of the three edges of a chosen unit cell in a crystal lattice.

cube A closed form of six identical square faces at 90° to each other; a form of crystal class $4/m\bar{3}2/m$.

cubic Having a cube shape or belonging to the cubic system.

cubic (cleavage) A term used to describe three equal cleavages at 90° to each other; for example, halite has cubic cleavage.

cubic (coordination) The bonding of an ion to eight others arranged so that connecting their centers forms a cube.

cubic (system) A crystal system characterized by lattice symmetry $4/m\bar{3}2/m$ and containing point groups with symmetry no greater than $4/m\bar{3}2/m$; one cell parameter (a) is needed to describe the shape and size of a cubic unit cell.

cubic closest packing Closest packing of atoms in a pattern similar to a face-centered cubic lattice; closest packing equivalent to stacking closest packed layers in an ABCABC sequence.

cumulate A layer of minerals accumulated by gravity settling of crystals as they form in a magma chamber.

cyanide method A method for extracting valuable metals from ore that involves hydrogen cyanide.

cycle One complete upward and downward motion by a wave.

cyclic twinning Repeated twinning of three or more individuals according to the same twin law but with the twin axes or twin planes not parallel, usually producing a twinned crystal in which twin domains are related by apparent rotational symmetry.

cyclosilicates Silicate minerals characterized by SiO_4 tetrahedra joining to form rings (*see also* ring silicates).

Czochralski process A method of making synthetic gems that involves a seed crystal drawing material out of a melt.

***d*-value** The distance between adjacent planes with the same Miller indices.

dacite An extrusive igneous rock of silicic composition that contains more plagioclase than K-feldspar; biotite and hornblende are typically present (*see also* granodiorite).

daughter element The product element of radioactive decay.

daughter isotope The product isotope of radioactive decay.

Dauphiné twin A common type of twinning in quartz.

Debeye-Scherrer camera The most commonly used and most versatile camera for obtaining powder diffraction patterns.

decarbonation reaction A reaction that liberates CO_2 from a mineral.

defect A flaw in an otherwise ideal crystal structure.

degrees of freedom The number of intensive variables that may be changed independently without causing a change in mineral assemblage or composition.

dehydration reaction A reaction the liberates H_2O from a carbonate mineral.

dendritic A term describing an aggregate of crystals having a treelike appearance (*see also* arborescent).

density The quantity of matter in a unit volume; mineral densities are typically given in units of gm/cm^3.

destructive interference The addition of two out-of-phase waves resulting in a total, or significant, loss of energy.

detrital Referring to a product of mechanical weathering.

detrital sedimentary rock Rock formed by lithification of detrital sediments.

detritus Broken-up material resulting from mechanical weathering.

diagenesis Chemical, physical, and biological changes that affect sediment or sedimentary rocks after initial deposition, but excluding weathering or metamorphism.

diamagnetism A property of minerals that causes a small negative reaction (repulsion) to a magnet.

diamond (shape) A parallelogram having four sides of equal length and no angles of 90°.

diamond lattice See diamond net.

diamond net A two-dimensional lattice characterized by two translations of the same magnitude at nonspecial angles to each other.

diaphaneity The ability of a mineral to transmit light; often described as transparent, translucent, or opaque.

diaphragm An adjustable opening used to control the size of a light beam in a polarizing light microscope.

dichroism Pleochroism of a mineral that is observed as two different colors.

diffraction Apparent bending of radiation by evenly spaced atoms, slits, or gratings.

diffraction grating A grating that causes diffraction.

diffraction slit A slit that causes diffraction.

diffractometer An instrument that records X-ray diffraction patterns.

dioctahedral Describing a layered mineral structure in which only two of three available octahedral sites are occupied.

diorite An intrusive igneous rock of intermediate composition containing plagioclase as the only major feldspar; minor K-feldspar or quartz, pyroxene, biotite, and hornblende may be present.

diploid A closed cubic form of 24 quadrilateral faces; a form of crystal class $2/m\bar{3}$.

dipyramid A closed crystal form of 6, 8, 12, 16, or 24 faces, comprising two pyramids related by a mirror plane of symmetry.

discontinuous side (Bowen's reaction series) The side of Bowen's reaction series characterized by olivine-pyroxene-amphibole-biotite.

disequilibrium The state of being in the process of reacting so that the relative amounts of reactants and products are changing.

dispersion A difference in wave velocity for different wavelengths.

disphenoid A closed crystal form comprising two sphenoids related by 222 symmetry.

disseminated deposit An ore deposit in which the ore mineral is scattered throughout a host rock.

dissolution The process of dissolving.

divalent Having a charge of ±2.

divariant field A region on a phase diagram characterized by two degrees of freedom.

divergent (crystals) Slender crystals emanating from a common point (*see also* radiating).

dodecahedral (coordination) 12-fold coordination.

dodecahedron A closed cubic form of 12 faces that may have any of a number of shapes and point group symmetries.

dolostone A chemical sedimentary rock composed primarily of dolomite.

domain A region within a crystal having a structure or orientation that differs from other regions within the crystal.

dome An open crystal form composed of two nonparallel faces related by a mirror.

dop A sticklike device that holds a gem crystal for grinding on a wheel.

double chain silicates Silicate minerals characterized by SiO_4 tetrahedra joining to form chains two tetrahedra wide (*see also* amphibole).

double refraction The ability of a mineral to split ordinary light into two waves of different velocities and polarization.

doubly primitive Term describing a unit cell containing a total of two lattice points.

drusy Having surfaces coated with fine crystals.

ductile A term used to describe the tenacity of minerals that are capable of being drawn into a wirelike shape.

dull A type of luster that does not reflect significant amounts of light or show any play of colors.

dunite An ultramafic intrusive rock in which the only major mineral is olivine; chromite is typically present as an accessory mineral.

Glossary

dynamic metamorphism   Metamorphism caused by a transient high-pressure condition such as a meteor impact (*see also* shock metamorphism and cataclastic metamorphism).

E ray   *See* extraordinary ray.

eclogite (facies)   A high-pressure/high-temperature metamorphic facies characteristic of the mantle.

eclogite (rock)   A high-pressure/high-temperature rock containing Mg-rich garnet (pyrope) and Na-rich clinopyroxene (omphacite); mostly of mantle origin.

edge diagonal   A line passing through the center of a cube or other three-dimensional geometric shape, and connecting the centers of opposite edges.

edge dislocation   A kind of line defect caused by a terminated row of atoms.

edge sharing   The sharing of two atoms or ions by two adjacent coordinating polyhedra; the coordinating polyhedra appear to be sharing an edge.

effective ionic radius   Radius of a spherical volume effectively occupied by an ion in a particular structure.

effervescence   A bubbling reaction; the term used to describe the reaction of calcite with dilute hydrochloric acid.

elastic   A term used to describe minerals that return to their original shape after bending.

electromagnetic radiation   Emission or transfer of energy in the form of waves; includes X rays, visible light, infrared light, radio waves, and television waves.

electron   An extremely small atomic particle having little mass and the smallest negative electric charge occurring in nature. Atoms have electrons orbiting around a nucleus.

electron cloud   The space occupied by electrons surrounding an atomic nucleus.

electron microprobe   An analytical instrument in which a finely focused electron beam hits a sample, causing emission of elemental characteristic radiation.

electronegativity   Measure of the tendency of elements to acquire electrons.

electrostatic valency principle (Pauling's rule 2)   The strength of an ionic bond is equal to its ionic charge divided by its coordination number.

element   A basic chemical unit composed of atoms having the same atomic number; elements cannot be separated into simpler parts by chemical means.

elliptic   A term describing crystals that are very small ellipsoids.

enantiomorphic   A term describing two crystals whose atomic structures are mirror images of each other.

end centered   A term describing a unit cell with an extra lattice point in each of two opposing faces.

end member   An ideal chemical formula representing one limit of a solid solution.

energy level   The energy associated with a particular electron orbit in an atom.

enthalpy   A thermodynamic variable related to the Gibbs free energy.

entropy   A thermodynamic variable representing the degree of randomness or disorder in a system.

epigenetic   Refers to a mineral deposit that is emplaced after its host rocks already exist.

equant   Having approximately the same dimensions in all directions.

equilibrium   A static state of a chemical system in which the relative amounts and compositions of phases present do not undergo changes with time.

essential mineral   A mineral in a rock that is necessary for the rock to be classified or named as it is.

euhedral   A crystal that is completely bounded by well-developed crystal faces and whose growth was not restrained by adjacent crystals.

evaporite (mineral)   Minerals that may form by precipitation from aqueous solution under normal Earth surface conditions; for example, halite, sylvite, and gypsum.

evaporite (rock)   A chemical sedimentary rock formed by evaporation of water; for example, massive gypsum or salt beds.

evaporite deposit   Mineral deposit formed by evaporation of water.

even (fracture)   A fracture that produces smooth planar surfaces.

exhalitive   A hydrothermal deposit created at the Earth's surface; generally one that forms on the ocean floor from springs.

expandable clay   One of a group of clay minerals (smectites) that can expand their structure to accommodate excess water or other chemical components.

exsolution   The separating of an initially homogeneous mineral solid solution into two (or more) zones of distinct mineral phases; for example, the formation of perthite from a homogeneous feldspar during cooling.

exsolution lamellae   Fine parallel zones of different compositions resulting from exsolution.

extinct (mineral grain)   A term describing a birefringent mineral grain oriented so that it appears dark when viewed under crossed polars.

extinct (X-ray peak)   A term describing an X-ray peak missing from a diffraction pattern due to destructive interference.

extinction   The systematic absence of a related group of X-ray peaks in a diffraction pattern.

extinction angle   The angle, measured under crossed polars, between the position of extinction and a cleavage, long dimension, or other direction in a crystal.

extraordinary ray   In a uniaxial crystal, the ray of light that vibrates in a plane containing the optic axis.

extrusive rock   An igneous rock that crystallizes at or very near the Earth's surface (*see also* volcanic rock).

f ace centered   A term describing a unit cell with an extra lattice point in the center of each of its faces.

face sharing   The sharing of three atoms or ions by two adjacent coordinating polyhedra; the coordinating polyhedra appear to be sharing a face.

facet   A polished face on a gemstone.

facies (metamorphic)   A range of pressure-temperature conditions characterized by one or more specific mineral assemblages.

family of planes   An infinite number of planes, parallel and spaced equally, all characterized by the same Miller index.

fast ray   The fastest of the two rays produced by double refraction.

feldspathic   An adjective describing a mineral aggregate containing feldspar.

felsic   A general adjective referring to igneous rocks having light colored minerals in their mode.

ferromagnetism   A type of magnetic order, characteristic of iron, that causes a mineral to respond strongly to a magnetic field.

fibrous (cleavage)   A term describing a cleavage that allows a crystal to be broken into fibers.

fibrous (crystals)   Having the appearance of being composed of fibers.

filiform   Having a hairlike or threadlike appearance (*see also* capillary).

fissility   The rock property of breaking easily along closely spaced planes; often used to describe shale or schist.

flash figure   The interference figure observed when the optic axis (uniaxial mineral) or the optic plane (biaxial mineral) are oriented parallel to the stage of a polarizing light microscope.

flexible A term used to describe the tenacity of minerals that are bendable.

flood basalt A laterally extensive and thick lava flow of basaltic composition.

fluorescence A type of luminescence that occurs when minerals are exposed to ultraviolet light, X rays, or cathode rays, but ceases when the exposure ceases.

flux A substance added to promote a chemical reaction or to lower the melting temperature of a material without changing the chemistry of the important phases that are present.

flux method A method of synthesizing gems or other minerals by growing them in an inert flux.

foliated Having planar or sheetlike properties.

foliation A planar fabric or texture.

fool's gold Common name for pyrite.

form A set of identically shaped crystal faces related by symmetry.

fractional crystallization Separation of a crystallizing magma into parts by the successive crystallization of different minerals (*see also* partial crystallization).

fracture Breaking of a mineral other than along planes of cleavage.

framework silicates Silicate minerals characterized by SiO_4 tetrahedra joining to form three-dimensional networks (*see also* tectosilicates and network silicates).

Frenkel defect A defect caused when an atom in a crystal structure is displaced from its normal position to a different position.

gabbro An intrusive igneous rock of mafic composition containing plagioclase as the only major feldspar; clinopyroxene ± orthopyroxene ± olivine are typically present (*see also* basalt).

gamma rays Electromagnetic radiation of higher energy and shorter wavelength than X rays; emitted by atomic nuclei.

gangue The waste minerals in an ore deposit.

gem An especially fine or superlative specimen, often of mineralogical origin, usually having superb color, light properties, or shape; often a cut-and-polished stone that has value due to beauty, durability, rarity, or size; generally for use in jewelry or for ornamentation.

gemmy Having a gemlike appearance; typically having a bright color, an attractive luster, or geometric shape.

gemstone A gem of mineralogical origin.

general angle A nonspecial angle between crystal faces or symmetry elements.

general form For a given point group, a form that has the maximum possible number of faces and whose faces and face normals do not intersect at special angles.

general point On a stereo diagram, a point that is located at a general position, not coincident with an inversion center, a rotational axis, or a mirror plane.

geode A hollow or partly hollow rock cavity lined by mineral material.

geothermometer A mineral or mineral system with compositional variations that reflect the temperature at which it formed.

Gibbs free energy A thermodynamic variable that describes the relative stability of a mineral or a mineral assemblage; also refers to the energy change associated with a chemical reaction.

Gibbs phase rule A consequence of the laws of thermodynamics: for any chemical system in equilibrium, the number of chemical components plus two is equivalent to the number of stable phases plus the number of degrees of freedom: $C + 2 = P + F$ (*see also* phase rule).

gibbsite layer A sheet of $Al(OH)_6$ octahedra in a layered crystal structure.

glass An amorphous solid material; in most cases, a metastable supercooled liquid.

glide plane A compound symmetry operation that repeats a motif or other entity after a reflection and translation are combined.

globular Appearing as spherical or hemispherical shapes made of radiating crystals (*see also* colloform).

gneiss A foliated metamorphic rock showing contrasting bands of light- and dark-colored minerals.

Goldich's weathering series A hypothetical series that describes the order of weathering of minerals in outcrop; the series is essentially the opposite of Bowen's Reaction Series.

goniometer A calibrated instrument used to measure the angles between crystal faces, cleavages, or other planar features in crystals. Simple goniometers consist of a protractor attached to a rotatable ruler.

grade (metamorphic) The degree of metamorphism; generally equivalent to the temperature of metamorphism (*see also* metamorphic grade).

grade (ore) The concentration of ore minerals or elements in ore rock.

grain mount Mineral grains on a glass slide surrounded by a liquid and covered with a cover slip.

granite An intrusive igneous rock of silicic composition containing 10% to 50% quartz and K-feldspar as the major feldspar; plagioclase, biotite, and hornblende may be present (*see also* rhyolite).

granitic A general adjective applied to any light-colored intrusive igneous rock or to any rock composition that generally resembles that of a granite.

granodiorite An intrusive igneous rock of silicic composition that contains more plagioclase than K-feldspar; biotite and hornblende are typically present (*see also* dacite).

granular Composed of many small grains.

granulite (facies) The metamorphic facies corresponding to the highest grades of regional metamorphism.

granulite (rock) A high-grade metamorphic rock containing orthopyroxene.

gravel Clastic material with grains greater than 2 mm in diameter; may include boulders, cobbles, pebbles, or granules.

graywacke General term used for clastic sedimentary rocks similar to sandstones but containing an inordinant amount of clays, rock fragments, or other material in addition to quartz and feldspar.

greasy A type of luster describing crystal faces or other surfaces that reflect light to give a play of colors similar to oil on water.

greenschist A medium-grade metamorphic rock characterized by schistosity and green minerals including actinolite, chlorite, and epidote.

greenschist (facies) A medium-grade metamorphic facies characterizing conditions of about 300°–500 °C and 2–8 Kbar.

greenstone A vague term used to describe any fine-grained, low-grade green metamorphic rock; generally of basaltic composition.

greenstone belt A metamorphic terrane characterized by the presence of low-grade metamorphosed volcanics and volcanogenic sediments and silicic to intermediate plutons.

group (of elements) The elements of one column in the Periodic Table of the Elements.

group (of minerals) Minerals within a class or subclass that share major chemical or structural features.

gypsum plate An accessory plate, made of gypsum, for a polarizing light microscope.

gyroid A cubic form of 24 irregular pentagonal faces; a form of crystal class 432.

habit The characteristic appearance of a mineral due to crystal form or combinations of forms, crystal intergrowths, and aggregates and any other irregular physical characteristics.

hackly A term describing fracturing that produces jagged edges.

halide   Mineral compound characterized by a halogen such as F, Cl, or I as an anion.

halogen   Elements (such as F, Cl, and Br) in the seventeenth group of the Periodic Table of the Elements; halogens typically ionize to become monovalent anions.

Hanawalt method   A systematic method for matching an "unknown" X-ray pattern with one in a reference data set such as the Powder Diffraction File.

hand specimen   A piece of rock or mineral, convenient for studying macroscopic properties, that can be picked up in the hand.

hard radiation   X rays of extremely high energy; used in industry and manufacturing but generally not by crystallographers.

hardness   The resistance of a mineral to scratching.

Haüy's Law   A law that states that crystal faces make simple rational intercepts with crystallographic axes; a corollary is that crystal faces have rational and generally small Miller indices (*see also* Law of Rational Indices).

heavy liquid   A liquid of greater density than water that may be used to separate minerals of different densities.

heft   An estimation of a mineral's density obtained by picking up the mineral and holding it in your hand.

Heisenberg uncertainty principle   A principle stated by Werner Heisenberg that says it is impossible to know the location and motion of an electron in an atom without some uncertainty.

Hermann-Mauguin symbol   A shorthand notation for the symmetry of a point group.

hexagonal   Of or related to the hexagonal system.

hexagonal (system)   A crystal system characterized by lattice symmetry $6/m2/m2/m$, and containing point groups with symmetry no greater than $6/m2/m2/m$; two cell parameters (a, c) are needed to describe the shape and size of a hexagonal unit cell.

hexagonal closest packing   Closest packing of spheres in an arrangement similar to a hexagonal prism with three extra spheres in its interior; equivalent to stacking closest packed layers in an ABABAB sequence.

hexagonal packing   Packing of identical spheres in a plane so that each is surrounded by six others.

hexahedron   A form of six equivalent faces related by symmetry; for example, a cube or rhombohedron.

hexanet   A two-dimensional lattice characterized by two translations of equal magnitude at 60° to each other.

hexoctahedron   A cubic form of 48 triangular faces; a form of crystal class $4/m\bar{3}2/m$.

hornfels   A fine-grained metamorphic rock lacking foliation or lineation.

host rock   The rock that hosts ore minerals or an ore deposit.

hydration reaction   A reaction between a mineral, or minerals, and water that results in a product hydrous mineral.

hydrogen bond   A type of electrostatic bond that is generally insignificant in minerals. Prominent in ice and, to a lesser extent, in micas and hydroxides.

hydrolysate   The material that goes into solution during chemical weathering.

hydrolysis reaction   A type of weathering reaction that simultaneously produces dissolved material and secondary minerals.

hydrothermal   Having to do with warm water-rich fluids.

hydrothermal fluids   Hot water-rich fluids that circulate through the Earth.

hydrothermal ore deposit   An ore deposit precipitated by hydrothermal fluids.

hydroxide   A chemical group, or group of minerals, with formulas characterized by the radical $(OH)^-$; for example, gibbsite, $Al(OH)_3$.

hypothesis   A proposed explanation for an observed set of facts.

idiochromatic   Term describing a mineral that gets its color from its major elements (*see also* allochromatic).

igneous   An adjective describing a rock or mineral that solidified from a magma or describing the process that forms such a rock or mineral.

immersion method   A method for determining index of refraction by immersing a grain in liquid of known index of refraction.

immiscibility   An inability of two or more phases to dissolve completely in one another.

impurity defect   A defect caused when a foreign atom is present in a crystal structure.

inclined extinction   Extinction when a principal cleavage or length of a crystal is at an angle to the crosshairs of a polarizing light microscope.

incompatible element   An element that does not readily enter a crystal structure.

index (indices)   Number (numbers) used to describe the location of points, the orientation of lines, or the orientation of planes in space.

index (X-ray pattern)   To assign appropriate (hkl) values to X-ray diffraction peaks.

index mineral   A mineral characteristic of a particular set of pressure-temperature conditions for rocks of a given composition.

index of refraction   The ratio of the velocity of light in a vacuum to the velocity of light in a crystal, glass, liquid, or other medium (*see also* refractive index).

inert gas   Any of the generally unreactive elements (such as Ne, Ar, or Kr) in the eighteenth group of the Periodic Table of the Elements (*see also* noble gas).

infrared light   Electromagnetic radiation with wavelengths slightly greater than visible light.

inorganic   A general term used to refer to compounds that contain no carbon bonded to hydrogen as essential components.

inosilicates   Chain silicates; either pyroxenes, pyroxenoids, or amphiboles.

integral molecules   The fundamental and indivisible building blocks making up crystals, according to Haüy; now known not to exist as Haüy envisioned them.

intensive variable   A thermodynamic variable, generally pressure or temperature, that is controlled from outside a chemical system.

interfacial angle   The angle between two faces of a crystal.

interference   The interaction of two waves traveling in the same direction.

interference colors   The colors displayed by a birefringent crystal when viewed under crossed polars.

intergranular fluid   A fluid between grains in a rock.

intergrowth   Two or more crystals grown together.

intermediate igneous rock   An igneous rock of composition intermediate between mafic and silicic.

internal energy   A thermodynamic variable related to Gibbs free energy.

internal reflection   The reflection of a light ray back into the interior of a crystal when it reaches a crystal boundary from within.

interstice   A space between grains in a rock or between closest packed atoms in a structure.

intrusive rock   An igneous rock that crystallizes at depth in the earth (*see also* plutonic rock).

invariant point   A point on a phase diagram at which two or more reactions intersect.

inversion   A term describing the operation that relates two crystal faces or other entities that are equal distant from a central point and have upside-down and backward orientations with respect to each other.

inversion center   A point at the center of a crystal (and its atomic structure) through which every aspect of the crystal (and its atomic structure) is repeated by inversion.

ion   An atom with a negative or positive electric charge as a result of having lost or gained one or more electrons.

ionic bond An ideal chemical bond, electrostatic in nature and formed between elements that have large differences in electronegativity.

ionic charge The number of electrons lost or gained when an atom becomes ionized.

ionic crystal A crystal in which the predominant bonding is ionic.

ionic radius Radius of a spherical volume effectively occupied by an ion in a particular environment.

iridescence The display of rainbowlike colors in the interior or on the surface of a mineral.

iron formation General name given to chemical sedimentary rocks dominated by iron oxides, hydroxides, carbonates, sulfides, or silicates.

irregular (fracture) A fracture that produces rough and irregular surfaces (*see also* uneven fracture).

isochrome Color bands that wrap around the trace of the optic axis (uniaxial mineral) or the optic axes (biaxial mineral) in an interference figure.

isodesmic Describing a crystal in which all bonds are ionic and of equal strength.

isogonal A term describing space groups that contain the same rotation axes; also used to describe space groups that have the same symmetry when translation is ignored.

isograd Any line based on mineral or mineral-assemblage occurrences that can be mapped in a metamorphic terrane.

isogyre A band of extinction in an interference figure that appears where light vibration directions are parallel and perpendicular to the lower and upper polarizers.

isolated tetrahedral silicates Silicate minerals characterized by individual SiO_4 tetrahedra linked by bonding to common cations.

isomorphous series A solid solution series in which the crystal structure is the same throughout the series.

isostructural A term describing two minerals that have different, but identically arranged, atoms in their crystal structures.

isotope One of two or more species of the same chemical element; different isotopes of an element have the same number of protons but vary in their number of neutrons.

isotropic Having the same properties in all directions.

join (phase diagram) A line that connects two end member compositions on a ternary phase diagram.

Jolly balance A type of balance scale used to determine specific gravity.

kimberlite An alkalic peridotite associated with diatremes, typically containing phenocrysts of olivine and phlogopite or their alteration products.

komatiite A general name given to a lava of ultramafic composition.

labradorescence A flashing and laminated iridescence, generally of a single bright hue, similar to that displayed by labradorite, a plagioclase feldspar.

lamellar Having a tablike appearance; being thin in one dimension compared to the other two (*see also* platy and tabular).

lanthanide Elements (such as Ce, Pr, and Nd) with atomic numbers 58 through 71 and valence electrons in $4f$ orbitals (*see also* rare earth elements).

laterite A highly weathered and leached soil or subsoil rich in aluminum, iron, or other insoluble elements; often rich in quartz and clay minerals.

latite An extrusive igneous rock of intermediate composition containing approximately equal amounts of plagioclase and K-feldspar; minor quartz, clinopyroxene, biotite, and hornblende may be present.

lattice A three-dimensional representation of the translational symmetry of a crystal structure.

lattice point Translationally equivalent points in space; in a crystal lattice, points are surrounded by identical and indistinguishable atomic arrangements.

Laue equations Equations derived by von Laue that describe the angular relationship between an incident beam and a diffracted beam in three-dimensional space.

Laue method X-raying a single crystal by placing it in the path of a polychromatic X-ray beam and positioning a flat piece of film behind it.

lava Molten extrusive magma.

lava flow A horizontal outpouring of lava from a vent or fissure.

law of definite proportions A general law that says compounds are made of elements combined in specific proportions.

Law of Bravais An observational law that states faces on crystals tend to be parallel to planes having a high lattice point density.

Law of Rational Indices A law that states that crystal faces make simple rational intercepts with crystallographic axes; a corollary is that crystal faces have rational and generally small Miller indices (*see also* Haüy's Law).

laws (scientific) A formal statement of general scientific observations that have never been found to be violated.

layer silicates Silicate minerals characterized by SiO_4 tetrahedra joining to form sheets; includes micas and clays (*see also* sheet silicates and phyllosilicates).

leach To dissolve and remove soluble components from a rock or soil.

left handed A relative phrase, referring to objects that rotate or point to the left.

length fast A term describing a birefringent mineral in which the fast ray vibrates more or less parallel to the length of an elongate crystal.

length slow A term describing a birefringent mineral in which the slow ray vibrates more or less parallel to the length of an elongate crystal.

lime kiln Long horizontal cylindrical furnace used to heat limestone in order to make lime (CaO).

limestone A chemical sedimentary rock composed primarily of calcite; in a more general sense the term is sometimes used to refer to any chemical sedimentary rock composed mostly of carbonates.

line defect A kind of defect that occurs along a line in a crystal structure.

lineage structure A plane within a crystal separating slightly misoriented portions of a crystal structure.

linear (coordination) Coordination of a small ion or atom to only two others.

lineation Any linear feature that may be observed in a rock.

liquidus The line on a temperature-composition diagram that shows the temperatures above which a system is completely liquid.

lithic fragment Clastic fragment of a preexisting rock.

lithification Process of a sediment being converted to a rock; it often involves compaction, desiccation, cementation, and recrystallization.

lithophile An element that tends to concentrate in silicates rather than in metals or sulfides.

lode deposit An ore deposit comprising many small veins.

lower polarizer A fixed polarizing lens located in the substage of a polarizing light microscope.

luminescence The emission of energy of a different wavelength from a mineral or other substance that has been stimulated by an external energy source of some wavelength(s).

luster The reflection of light from the surface of something, described by its quality and intensity.

macroscopic property A mineral property seen or measured in a hand specimen.

magma   Naturally occurring molten rock material.

major element   An element that is a key and essential part of a mineral (*see also* minor element or trace element).

malleable   A term used to describe the tenacity of minerals that are capable of being hammered into shapes.

mammillary   Having a breastlike appearance.

Mannebach twin   An uncommon twin law in feldspar.

marble   A metamorphic rock containing primarily calcite or dolomite.

marl   A calcareous rock containing significant amounts of clay and other detrital material.

mass number   Symbolized by *A*, the total number of protons and neutrons in an atom.

massive   A term describing a mineral or mineral aggregate lacking in internal structure or other distinguishable physical characteristics.

matrix   Small or fine-grained material that encloses larger grains or crystals.

mechanical weathering   Type of weathering involving the physical decomposition or breakdown of minerals and rock to produce smaller pieces.

melatope   The trace of the optic axis (uniaxial mineral) or the optic axes (biaxial mineral) in an interference figure; the point where the isogyres cross in a uniaxial interference figure; the points on the isogyres closest to the center of a Bxa or Bxo interference figure.

mesodesmic   Describing a crystal in which the strength of all bonds from a cation to its coordinating anions is equal to exactly one-half the charge of the anions.

metabasite   A metamorphosed mafic rock.

metal (element)   A general term often applied to any element that ionizes easily to become a cation; alternatively used to mean any of a class of elements that, when in pure form, usually have a shiny surface, and are good conductors of heat and electricity.

metal (native element)   Native elements of metallic character; principally gold, silver, copper, and platinum.

metallic bond   An ideal chemical bond in which electrons are highly delocalized and free to move from one atom to another.

metallic luster   A mineral luster characterized by brightness and shininess and the ability to reflect light.

metamorphic   An adjective pertaining to the process of metamorphism.

metamorphic facies   A range of pressure-temperature conditions characterized by one or more specific mineral assemblages.

metamorphic grade   The degree of metamorphism; generally equivalent to the temperature of metamorphism.

metamorphism   Mineralogical or textural changes in rocks in response to changes in physical or chemical conditions.

metapelite   A metamorphic rock equivalent in composition to a clay-rich sediment.

metapsammite   A metamorphic rock equivalent in composition to a sandstone or feldspathic sandstone with minor amounts of clay.

metasomatism   A change in rock composition due to the movement of pore fluids associated with metamorphism.

metastable   (thermodynamics) Adjective describing a mineral or mineral assemblage that is not undergoing change but does not have the minimum possible Gibbs free energy.

metastable equilibrium   Equilibrium not representative of the minimal possible Gibbs energy of a system; a chemical system at metastable equilibrium is not reacting but may eventually do so to attain stable equilibrium.

micaceous   Having the properties of a mica, the ability to be split into thin sheets.

micrite   Fine-grained microcrystalline calcite; the term is sometimes used to refer to a chemical sedimentary rock composed of microcrystalline calcite.

microcrystalline   Having crystals that can only be seen with the aid of a microscope.

microscopic property   A mineral property not generally seen or measured without a microscope.

migmatite   A mixed rock composed of different colored bands, often associated with partial melting under high-grade metamorphic conditions.

Miller index (indices)   Numbers used to describe the orientation of a plane with respect to crystal axes; Miller indices are generally enclosed in parentheses.

mineral   Naturally occurring and crystalline substances of fixed or limited composition.

mineral class   A broad category of minerals generally based on commonality of anions or anionic complexes; for example, carbonates and sulfates are mineral classes.

mineral formula   An expression that uses chemical symbols, parentheses, and subscripts to show the composition of a mineral.

mineral group   A group of minerals within a class that are distinguished by common properties.

mineralogy   The study of minerals, their formation, occurrence, properties, composition, and classification.

minor element   An element that substitutes for a major element in a mineral, generally present at the 1% to 5% level (*see also* major element or trace element).

mirror plane   A basic symmetry operator across which a mirror image is created.

miscibility   The ability of two or more phases to mix to produce one solid phase.

miscibility gap   A compositional range in a solution series that is not stable as a single phase; a compositional range where unmixing occurs.

Mississippi Valley-type deposit   An epigenetic ore deposit hosted by carbonate rocks; typically, a source of lead and zinc.

mode (igneous rock)   The mineral composition of a rock; generally expressed in volume %.

Mohs' hardness scale   A relative scale of 1–10, based on reference minerals, that can be used to describe a mineral's hardness.

mole   The quantity in grams of an element or compound that equals the molecular weight of the substance; one mole of a substance contains Avogadro's number (6.022×10^{23}) of atoms/molecules of that substance.

monochromatic   A term describing radiation consisting of one wavelength.

monoclinic   Of or related to the monoclinic crystal system.

monoclinic (system)   A crystal system characterized by lattice symmetry $2/m$ and containing point groups with symmetry no greater than $2/m$; four cell parameters (a, b, c, β) are needed to describe the shape and size of monoclinic unit cell.

monomineralic   Containing only one mineral.

monovalent   An ion having a charge of ± 1.

monzonite   An intrusive igneous rock of intermediate composition containing approximately equal amounts of plagioclase and K-feldspar; minor quartz, clinopyroxene, biotite, and hornblende may be present.

mother lode   An ore body that is the source for ore minerals found in a placer.

motif   A pattern that is repeated by symmetry.

mud   Unconsolidated sediment composed of clay and silt.

mudstone   General term for a fine-grained clastic sedimentary rock composed of clay or silt; sometimes used only for such rocks if they are not foliated.

native element   An element that occurs naturally as a mineral.

natural   A product of nature, not of humans or human activities.

nesosilicates   Silicate minerals characterized by individual SiO_4 tetrahedra linked by bonding to common cations (*see also* island silicates and isolated tetrahedral silicates).

network silicates Silicate minerals characterized by SiO_4 tetrahedra joining to form three-dimensional networks (*see also* tectosilicates and framework silicates).

neutron A subatomic particle of neutral charge, generally found in atomic nuclei, that has about the same mass as a proton.

noble gas Any of the generally unreactive elements (such as Ne, Ar, or Kr) in the 18th group of the Periodic Table of the Elements (*see also* inert gas).

noncrystalline Having a random atomic structure (*see also* amorphous).

nonmetal (native element) Native element minerals composed of nonmetallic elements; principally sulfur, graphite, and diamond.

nonmetallic (element) Elements that do not possess the properties of a metal. Nonmetallic elements generally ionize easily to become anions, have a nonmetallic luster, and are poor conductors of heat and electricity when pure.

nonmetallic (luster) A mineral luster that is not metallic (*see also* adamantine, vitreous, resinous, silky, pearly, and greasy).

nonopaque (mineral) A mineral that transmits light in thin section.

nonprimitive Used to describe a unit cell containing more than one lattice point.

nonspecial angle An angle unrelated to angles between symmetry elements (*see also* general angle).

normalization The process of converting a mineral analysis to a chemical formula.

nuclei Plural of nucleus; the central parts of atoms consisting of protons, or of combinations of protons, neutrons, and other particles.

numerical aperture A number describing the size of the cone of light that can enter a lens.

O

ray *See* ordinary ray.

obsidian Volcanic glass produced by the rapid cooling of a magma.

obtuse bisectrix The direction bisecting the obtuse angle between the optic axes of a biaxial crystal; perpendicular to the acute bisectrix.

obtuse bisectrix figure (Bxo) The biaxial interference figure seen when looking down the obtuse bisectrix.

octahedral (cleavage) A term used to describe four cleavages that produce octahedral cleavage fragments; for example, fluorite has octahedral cleavage.

octahedral (coordination) The bonding of an ion to six others arranged so that connecting their center forms an octahedron.

octahedron A cubic form {111} of eight faces, each an equilateral triangle; a form of crystal class $4/m\overline{3}2/m$.

ocular A lens in an eyepiece of a microscope.

opalescence The play of colors seen in some minerals that resembles that of opal.

opaque The inability to transmit light or to be seen through.

open form A form that, by itself, cannot enclose three-dimensional space.

operation (symmetry) An operation that repeats a motif, crystal face, or other entity in a symmetrical pattern.

operator (symmetry) Any symmetry element that repeats a motif crystal face or other entity in a symmetrical pattern; rotation axis, mirror plane, inversion center, rotoinversion axis, translation, glide plane, or screw axis.

optic axis A direction through a crystal along which no double refraction occurs.

optic axis figure The interference figure seen when looking down an optic axis.

optic normal The direction perpendicular to the optic plane of a biaxial mineral, corresponding to the Y-axis.

optic normal interference figure The interference figure obtained when looking down a optic normal.

optic plane The plane containing the optic axes of a biaxial crystal.

optic sign Either + or −, describing the relationship between indices of refraction ($\alpha, \beta, \gamma, \epsilon, \omega$). In uniaxial positive (+) crystals, $\epsilon > \omega$; in uniaxial negative (−) crystals, $\omega > \epsilon$; in biaxial positive (+) crystals, β is closer in value to α than to γ; in biaxial negative (−) crystals, β is closer in value to γ than to α.

order of diffraction The value of n in Bragg's law; the number of cycles a diffracted beam lags behind an adjacent beam with which it is in phase.

ordinary ray In a uniaxial crystal, the ray of light that vibrates perpendicular to the optic axis and obeys Snell's law.

ore Anything that can be taken from the ground and sold for a profit.

ore deposit An economical concentration of ore minerals.

ore grade The concentration of ore minerals or elements in ore rock.

ore mineral A mineral, usually metallic, that is economically desirable.

organic A compound containing carbon bonded to hydrogen as an essential and major component.

orthogonal Perpendicular.

orthonet A two-dimensional lattice characterized by two translations of different magnitudes at 90° to each other.

orthopyroxene Orthorhombic pyroxene subgroup dominated by end members enstatite and ferrosilite.

orthorhombic Of or related to the orthorhombic crystal system.

orthorhombic (system) A crystal system characterized by lattice symmetry $2/m2/m2/m$, and containing point groups with symmetry no greater than $2/m2/m2/m$; three cell parameters (a, b, c) are needed to describe the shape and size of an orthorhombic unit cell.

orthoscopic (illumination) Illumination caused by light rays all traveling parallel to the tube of a polarizing light microscope.

orthosilicates Obsolete (*see also* isolated tetrahedral silicates).

out-of-phase The condition in which the peaks, and valleys, of two waves do not match in time and space.

oxidation The process of an atom losing one or more electrons through chemical reaction.

oxide Compound of oxygen with another element or radical.

oxidize Describes an element that has lost one or more electrons.

P

aired tetrahedral silicates Silicate minerals characterized by SiO_4 tetrahedra joined to form pairs (*see also* sorosilicates).

parallel extinction Extinction condition in which the principal cleavage or length of a crystal is parallel to the crosshairs of a polarizing light microscope.

paramagnetism Having a small magnetic susceptibility and being weakly attracted by a magnet.

parent isotope The initial radioactive member of a radioactive decay series.

partial crystallization Separation of a crystallizing magma into parts by the successive crystallization of different minerals (*see also* fractional crystallization).

partial melting Melting of only some of the minerals in a high-grade metamorphic rock.

parting Breaking of a mineral along a plane of weakness; generally caused by twinning or deformation.

Pauling's rules Five empirical rules, first tabulated by Linus Pauling, describing common structural and bonding features of ionic structures.

pearly A type of luster that appears iridescent, similar to pearls or some seashells.

pedion An open form consisting of one crystal face unrelated to any other by symmetry.

pegmatite An exceptionally coarse grained igneous rock; generally formed by crystallization of water-rich magmas.

pelite   A rock with composition equivalent to a clay-rich sediment.

penetration twin   A twin in which two individual crystals, sharing a common volume of atoms, appear to penetrate or grow through each other.

pericline twin   A common twin law in triclinic feldspars; K-feldspar that have both albite and pericline twins develop the (microscopic) scotch-plaid twinning characteristic of microcline.

peridotite   A general term for an intrusive igneous rock of highly mafic composition; typically composed of olivine ± other mafic minerals and minor plagioclase.

period (of elements)   Elements that occupy the same row in the Periodic Table of the Elements.

periodic chart   Also called the Periodic Table; a table in which the chemical elements are arranged in order of their atomic numbers and grouped in columns based on atomic structure.

petrogenesis   The branch of petrology dealing with the origin and formation of rock.

petrographic microscope   A microscope that uses polarizing light for observation and analysis of minerals in grain mounts or in thin section (*see also* polarizing light microscope).

petrography   The branch of petrology dealing with the description and classification of rocks.

phaneritic   Having mineral grains visible without a microscope.

phase diagram   Any of a number of different types of diagrams used to depict mineral compositions or stability of various phases in a system; the axes are usually temperature, pressure, or composition.

phase rule   A consequence of the laws of thermodynamics: for any chemical system in equilibrium, the number of chemical components plus two is equivalent to the number of stable phases plus the number of degrees of freedom: $C + 2 = P + F$ (*see also* Gibbs phase rule).

phenocryst   The relatively large, conspicuous crystals in a porphyritic rock.

phosphate (mineral)   Group of minerals characterized by $(PO_4)^{3-}$ in their formulas.

phosphate (radical)   The $(PO_4)^{3-}$ anionic group.

phosphorescence   A type of luminescence in response to exposure to ultraviolet light, X rays, or cathode rays that continues after the exposure ceases.

phosphorite   A chemical sedimentary rock composed primarily of phosphate minerals, typically varieties of apatite.

phyllite   A foliated metamorphic rock, lacking in schistosity, having a sparkly or silky sheen due to the presence of fine-grained micas.

phyllosilicates   Silicate minerals containing SiO_4 tetrahedra joined to form sheets; includes micas and clays (*see also* layer silicates and sheet silicates).

piezoelectric   Having the ability to develop a small amount of electrical potential when strained.

pinacoid   An open crystal form comprised of two parallel faces.

pipe   An igneous body with a more or less round cross section formed as a magma ascends rapidly to the Earth's surface.

pisolitic   A term describing crystals that are very small spheres.

placer   A sedimentary mineral deposit, often fluvial, formed by the concentration of heavy minerals due to gravitational forces.

Planck's law   An equation that relates the energy (E) of an electromagnetic wave to its frequency (v): $E = hv$, where h is Planck's constant.

plane defect   A kind of defect that occurs along the boundary plane of two regions of a crystal or between two grains.

plane polarized   A term describing a moving wave, perhaps light, that is constrained to vibrate in one plane.

plaster of Paris   A white calcium sulfate-based plaster that is more hydrated than anhydrate but less hydrated than gypsum.

platy (cleavage)   A cleavage that allows a crystal to be broken into plates.

platy (crystal)   Appearing to be platelike, thin in one dimension compared with the other two (*see also* tabular and lamellar).

play of colors   The separation of white light into visible individual colors when it interacts with a mineral; often leading to changes in color as a mineral is turned (*see also* schiller).

pleochroism   The property of an anisotropic crystal to absorb different wavelengths, and thus to have different color, depending on orientation; most easily seen by rotating the stage of a polarizing light microscope while viewing an anisotropic crystal in thin section under plane polarized light.

plumose   A term describing an aggregate of crystals having a feathery appearance.

plutonic rock   An igneous rock that crystallizes at depth in the Earth (*see also* intrusive rock).

point defect   A defect that occurs at one point in a crystal structure.

point group   One of the 32 possible symmetry groups to which a crystal can belong; translation-free equivalent of space groups (*see also* crystal classes).

polarized (ion)   A term describing an ion with uneven electron distribution so that it does not behave as an ionic sphere in a crystal structure.

polarized (wave)   A term describing a moving wave, perhaps light, that is not free to vibrate in all directions.

polarizing light microscope   A microscope that uses polarizing light for observation and analysis of minerals in grain mounts or in thin section (*see also* petrographic microscope).

pole   A perpendicular to a face on a crystal.

polychromatic   Radiation containing multiple wavelengths (*see also* monochromatic).

polyhedron   A three-dimensional geometric figure having four or more faces.

polymer   A chemical unit formed by the tight linking of individual molecules.

polymerization   The connecting of polyhedra or other structural units into chains, sheets, or networks by sharing of atoms.

polymorphs   Minerals that have identical compositions but different crystal structures; for example, calcite and aragonite are polymorphs.

polysynthetic twin   Repeated twinning of three or more individuals according to the same twin law and in a parallel manner, producing a twinned grain composed of many thin parallel sheetlike twin domains.

porcelain   A special type of high-quality white ceramic.

porphyroblast   A large mineral crystal, produced by metamorphism, in a finer-grained matrix.

porphyry (igneous rock)   A rock containing relatively large crystals (phenocrysts) in a fine-grained ground mass that may be crystalline or glassy.

porphyry (ore deposit)   An ore deposit in which the ore is concentrated in closely spaced small veins and veinlets.

Portland cement   The most common kind of cement; composed of a mixture of lime, silica, alumina, and iron oxides.

powder diffraction   Diffraction of X rays by a powdered sample.

Powder Diffraction File (PDF)   The X-ray powder diffraction reference file used by most mineralogists.

Precambrian shield   An ancient geological terrane that has been stable for a long time; typically in the central area (craton) of major continents.

precious metals   Gold, silver, platinum, and sometimes other platinum group elements.

precipitate   To form solid material from material dissolved in a liquid; precipitated material generally but not always settles to the bottom of the liquid.

primary mineral   A mineral formed at the same time as its host rock (*see also* secondary mineral).

primitive   Describing a unit cell containing in total one lattice point; typically unit cells are chosen so that a primitive unit cell contains one-eighth of a lattice point at each corner.

principal axis The most prominent or unique crystal axis in a crystal or the most prominent or unique rotational axis of symmetry in a crystal; generally they coincide.

principle of parsimony (Pauling's rule 5) The number of different components in a crystal tends to be small.

prism A crystal form characterized by two or more crystal faces parallel to a common direction.

prismatic (cleavage) A term used to describe multiple cleavages all parallel to a common direction in a crystal.

prismatic (crystal) Having the appearance of a prism or prisms; appearing as a long crystal with parallel sides.

prograde metamorphism Metamorphism in response to increasing temperature.

prograde reactions Reactions that take place in response to an increase in temperature.

progressive metamorphism Metamorphism that proceeds by steps from low grade to high grade.

proper rotation axis A rotation axis of symmetry that does not involve inversion.

proton A subatomic particle, generally found in the nuclei of atoms, having the smallest amount of positive electric charge occurring in nature.

psammite A rock equivalent in composition to a sandstone or feldspathic sandstone with minor amounts of clay.

pseudohexagonal Symmetry that appears to be hexagonal but is not; for example, some books of biotite.

pseudosymmetry Symmetry that appears to be present but, with better measurement or observation, would be found to be lacking.

pycnometer A small bottle with a tight-fitting stopper that is used to determine specific gravity.

pyramid An open crystal form of 3, 4, 6, 8, or 12 nonparallel faces that meet at a point.

Pyramidal cleavage Cleavage that gives cleaved crystals a pyramid shape.

pyritohedron A cubic crystal form, consisting of 12 irregular pentagonal faces; a form of crystal class 23 or $2/m\bar{3}$.

pyroelectric Having the ability to develop a small amount of electrical potential when heated.

pyroxenite An ultramafic intrusive rock composed primarily of pyroxene with lesser amounts of olivine, biotite, or hornblende.

quartz plate An accessory plate for polarizing light microscopes that is composed of quartz of known optical orientation.

quartz wedge An elongate wedge of clear quartz of known optical orientation that may be inserted in a polarizing light microscope tube to analyze a crystal's optical properties.

quartzite A hard but unmetamorphosed sandstone or the metamorphic equivalent.

quartzose Containing quartz as a primary component.

radiating Emanating from a common point (*see also* divergent).

radical A chemical group that has acquired an electrical charge; often surrounded by parentheses in mineral formulas.

radioactivity The spontaneous decay of atoms of certain isotopes into new isotopes, accompanied by the emission of high energy particles.

radioisotope An isotope that is the product of radioactive decay.

radius ratio principle (Pauling's rule 1) Cation-anion distances are equal to the sum of their effective ionic radii, and cation coordination numbers are determined by the ratio of cation and anion radii.

rare earth elements Elements (such as Ce, Pr, and Nd) with atomic numbers 58 through 71 and valence electrons in $4f$ orbitals (*see also* lanthanide elements).

recrystallization Formation of new crystals from preexisting material; the new crystals need not be the same species or composition as the material from which they form.

recrystallize To form new crystalline mineral grains in a rock, generally by metamorphism.

reduce Said of an element; to gain one or more electrons.

reduction The process of an atom gaining one or more electrons through chemical reaction.

reentrant angle An angle between two crystal faces that points or is directed inward toward the center of the crystal; usually associated with twins.

reflected light microscopy Microscopic examination using a light source that reflects from the surface of a sample; reflected light microscopy is useful for identifying opaque minerals in thin section.

reflection (light) Coherent scattering of energy by atoms in a two-dimensional surface.

reflection (symmetry) A term describing the operation that relates two crystal faces or other entities across a mirror plane.

refraction The deflection of a ray, perhaps light, due to its passage from one medium to another of different ray velocity.

refractive index The ratio of the velocity of light in a vacuum to the velocity of light in a given medium (*see also* index of refraction).

refractive index oil A liquid of known index of refraction.

refractometer A device for determining the index of refraction of a crystal or other material.

regional metamorphism Metamorphism affecting a large region; caused by pressure related to depth in the Earth and temperature related to geothermal gradient (*see also* burial metamorphism).

regular polyhedron A coordinating polyhedron in which all cation-anion bond lengths are equal.

relief The apparent topography of a crystal or crystals seen under a microscope.

reniform Having a kidney-shaped appearance.

reserves Ore that has been identified and that could be extracted at a profit.

residual (mineral) A mineral that has been concentrated in place by weathering and leaching of rock.

residual (phase) The remaining liquid, or phases that crystallize from such a liquid, after a magma has mostly solidified.

residuum Insoluble material remaining after intense weathering.

resinous A term used to describe mineral lusters that have the appearance of resin.

resistate Insoluble material produced by chemical weathering; the term is also used in a general sense to refer to any sedimentary material of low solubility.

retardation The distance that one wave lags behind another; specifically the distance that the fast ray lags behind the slow ray upon emergence from a crystal.

reticulated Crystals that have a latticelike appearance.

retrograde metamorphism Metamorphism that takes place in response to decreasing temperature.

retrograde reactions Metamorphic reactions that take place in response to decreasing temperature.

rhomb A parallelogram having equal length sides and interior angles of 60° and 120°.

rhombohedral (point groups) Any of the point groups that include the rhombohedral form ($\bar{3}$, 32, $\bar{3}2m$).

rhombohedral plane lattice A two-dimensional lattice characterized by two translations of equal magnitude at 60° to each other (*see also* hexanet).

rhombohedron A closed form that creates a parallelepiped with six identical rhomb-shaped faces.

rhyolite An extrusive igneous rock of silicic composition containing 10% to 50% quartz and K-feldspar as the major feldspar; plagioclase, biotite, and hornblende may be present (*see also* granite).

Rietveld method A method of making crystal structure analyses based on powder diffraction data.

right handed   A relative phrase referring to objects that rotate or point to the right.

ring silicates   Silicate minerals characterized by SiO_4 tetrahedra joining to form rings (*see also* cyclosilicates).

rotation   A term describing an operation that relates two crystal faces about a rotational axis of symmetry.

rotational axis   A symmetry element that relates identical crystal faces or other things by rotation of 60°, 90°, 120°, 180°, or 360° about an axis.

rotational symmetry   A type of symmetry that relates identical crystal faces or other things by rotation of 60°, 90°, 120°, 180°, or 360° about an axis.

rotoinversion   A type of symmetry that combines rotation of 60°, 90°, 120°, 180°, or 360° with inversion.

Salt   Any of a group of chemical compounds derived from an acid by replacing H^+ with a metal cation; also a general term used to describe any mineral that has high solubility in water; more specifically used to describe halite, sylvite, and other alkali halides.

salt domes   A diapiric salt body that rises through a sedimentary pile due to its low density in relation to surrounding rocks.

sand   A detrital particle between 0.062 and 2 mm in diameter, or a sediment composed primarily of such particles.

sandstone   A clastic sedimentary rock composed primarily of sand-sized particles with or without a finer matrix.

sanidinite facies   A low-pressure metamorphic facies characterized by the presence of sanidine and other very high temperature minerals.

scalenohedron   A closed crystal form with 8 or 12 faces having the shape of a scalene triangle.

scattering   When an incident ray strikes an atom or other object and is reemitted equally in all directions.

schiller   A play of colors.

schist   A well-foliated metamorphic rock, characterized by parallel alignment of micas and other planar or tabular minerals, which has minerals grains that are visible with the naked eye.

schistosity   The ability to be split into thin flakes or slabs due to a parallel arrangement of micas or other planar or tabular minerals.

Schottky defect   A defect caused when an atom is missing from a crystal structure.

Schrödinger wave model   A fundamental mathematical model that describes wavelike motion of electrons and forms the basis for quantum mechanics.

scientific theory   A hypothesis that has been tested and supported by observation so much that it is generally accepted as being correct.

screw axis   A compound symmetry element that repeats a motif or other entity after rotation and translation combined.

screw dislocation   A kind of line defect caused by the apparent spiraling of a crystallographic plane.

second setting   Refers to choosing monoclinic unit cells so that b is the only non-90° angle.

secondary mineral   A mineral formed subsequent to the formation of the rock that hosts it (*see also* primary mineral).

secondary radiation   Radiation emitted by oscillating electrons that were perturbed by an incoming beam.

sectile   A term used to describe the tenacity of a mineral that can be cut into shavings with a knife.

sediment   Solid material produced by weathering; sometimes used more specifically to refer to material that has been transported and deposited.

sedimentary   Adjective pertaining to a rock formed from sediment or to a mineral formed by sedimentary processes.

sedimentary rock   A rock produced by lithification of sedimentary material, either clastic or chemical.

selenide   A mineral that is a combination divalent selenium with one or more metallic elements.

semimetal (element)   A poorly defined class of elements (such as arsenic) that do not ionize as easily as metals. When pure, semimetals have some of the properties of metals, including shiny appearance, good conduction of heat and electricity, and capability of fusing, hammering, or drawing into sheets or wires.

semimetal (native element)   Native elements that are semimetals include arsenic, antimony, and bismuth.

series   Minerals with compositions that can be described in terms of two or a few chemical end members.

serpentinite   A rock, formed by alteration of a preexisting mafic or ultramafic rock, which contains primarily serpentine group minerals.

shale   A foliated, fine-grained clastic sedimentary rock with grain sizes smaller than 0.004 mm.

sheet silicates   Silicate minerals characterized by SiO_4 tetrahedra joining to form sheets; includes micas and clays (*see also* layer silicates and phyllosilicates).

Sheetrock   Construction material used as wall board, composed of gypsum-based plaster encased in a paper jacket (*see also* gypsum board).

shock metamorphism   Metamorphism caused by a transient high-pressure condition such as a meteor impact (*see also* dynamic metamorphism and cataclastic metamorphism).

sialic   A term describing a rock that is rich in silica and alumina.

siderophile   An element that has a weak affinity for oxygen and sulfur and that is readily soluble in molten iron and iron alloys.

sign of elongation   Either positive $(+)$ or negative $(-)$; length fast crystals have negative sign of elongation; length slow crystals have a positive sign of elongation.

silicate   A compound with a crystal structure containing SiO_4 tetrahedra.

silicic   A term describing an igneous rock that is rich in silica.

silky   A type of luster that gives crystal faces or other surfaces the appearance of being composed of fine fibers.

silt   Clastic sediment composed primarily of grains 0.004–0.062 mm in size.

siltstone   A clastic sedimentary rock composed primarily of silt-sized particles.

simple substitution   A substitution of one ion for another of similar size and charge in a crystal structure; for example, the substitution of Mg^{2+} for Fe^{2+} in many silicates.

simple twin   A twin composed of two individual crystals sharing a common plane of atoms.

single chain silicates   Silicate minerals characterized by SiO_4 tetrahedra joined to form linear chains one tetrahedron wide; includes pyroxenes and pyroxenoids.

single crystal diffraction   Diffraction of X rays by a single crystal.

skarn   A contact metamorphic zone surrounding an igneous intrusion; often affected by metasomatism (*see also* contact aureole).

slag   Calcium-rich silicate waste material from smelting of metal ore concentrates.

slaked lime   Compound of approximate composition $Ca(OH)_2$ produced when lime reacts with water (*see also* portlandite).

slate   A very fine-grained foliated metamorphic rock that exhibits fissility.

slow ray   The slower of the two rays produced by double refraction.

Snell's law   The arithmetic law that relates the refractive indices of two media to the angles of incidence and refraction.

soft radiation   The relatively low energy X-radiation used in most X-ray diffraction studies; soft radiation is more dangerous to humans than hard radiation.

solid solution   Continuous range of mineral chemistries characterized by the same crystal structure; a mineral or other solid material that can exchange one element for another of similar properties; for example, anorthite and albite form a solid solution termed plagioclase.

solid-solid reaction   A reaction involving only solid phases.

solid state diffusion The movement of atoms or molecules through solid material, such as the migration of atoms through minerals.

solidus The line on a temperature-composition diagram that shows the temperatures below which a system is completely solidified.

solubility The extent to which a material will dissolve in another; generally used to refer to the dissolution of compounds in water, but sometimes used to refer to the solubility of compounds in other liquids or solids.

solvus The line on a phase diagram that outlines a miscibility gap.

sorosilicates Silicate minerals characterized by SiO_4 tetrahedra joined to form pairs (*see also* paired tetrahedral silicates); for example, epidote.

space group Permissible three-dimensional symmetry groups of all types of symmetry operations; one of the 230 possible symmetries a crystal structure may have.

space group operators Symmetry operators that are manifested in space groups but not in point groups: translation, screw axes, glide planes.

space symmetry The symmetry of a three-dimensional atomic structure; symmetry including space group operations.

sparry Used to describe clean, coarse-grained calcite, especially a clean, coarse-grained calcite cement in a sedimentary rock.

special angles Angles between symmetry axes or planes, or faces on a crystal, that result in the faces having special orientations with respect to crystal symmetry.

special form A crystal form with face normals at special angles to each other.

special point On a stereo diagram, a point that is located at a special position, coincident with an inversion center, rotational axis, or mirror plane.

specific gravity The mass of a mineral divided by the corresponding mass of water at 1 atm., 4 °C.

sphenoid An open crystal form having two nonparallel faces related by a 2-fold axis of symmetry.

square net A two-dimensional lattice characterized by two translations of equal magnitude at 90° to each other.

stability field A range of conditions over which a mineral or mineral assemblage is stable.

stable (thermodynamics) Adjective describing a mineral or mineral assemblage that has the minimum possible Gibbs free energy and so is not prone to change.

stable equilibrium The most thermodynamically stable conditions possible for a chemical system; systems at stable equilibrium will not react or change unless conditions change.

stable isotope An isotope that is not spontaneously radioactive.

stable mineral For a chemical system, an individual mineral or a mineral of an assemblage that has the lowest possible Gibbs free energy; stable minerals do not react to form other minerals or compounds.

stacking fault A type of plane defect caused by stacking irregularities in a crystal structure.

stalactitic Having the appearance of a stalactite or other speleothem.

stellated A term describing an aggregate of crystals having a starlike appearance.

Steno's law Also termed the "constancy of interfacial angles"; a law that says angles between comparable crystal faces of the same mineral will always be the same regardless of crystal shape or size.

stereo diagram Projection of a crystal or other shape onto a circle so that three-dimensional relationships may be depicted on a piece of paper.

stoichiometry The exact proportions, specified by a chemical formula, of the elements that comprise a mineral or other compound.

strain energy Energy associated with the physical distribution and alignment of mineral grains within a rock.

streak The color of a mineral when finely powdered; usually determined using a ceramic streak plate.

striation (twin) One of a set of fine lines that may appear on a crystal face or cleavage surface due to the presence of twinning.

subgroup Minerals within a group that are naturally related for chemical or other reasons.

subhedral A term describing a crystal that is only partly bounded by well-developed crystal faces or whose growth was partially restrained by adjacent crystals.

submetallic (luster) A type of luster that appears somewhat metallic.

substage An assembly of filters, polarizers, condensing lenses, and apertures located below the stage of a polarizing light microscope.

sulfate (mineral) Group of minerals characterized by (SO_4) in their formula.

sulfate (radical) The $(SO_4)^{2-}$ anionic group.

sulfide A chemical group, or group of minerals, with structures characterized by metals bonding to sulfur; for example, enargite.

sulfosalt A type of sulfide in which both a metallic and a semimetallic element are present.

supergene A term describing an ore deposit enriched by flowing fluids after the time of initial deposition.

swelling clay Clay material, such as bentonite or other clays of the smectite group, that is capable of expanding greatly and absorbing water between individual clay layers.

syenite An intrusive igneous rock of intermediate composition in which the amount of K-feldspar exceeds the amount of plagioclase; biotite and hornblende may be present but quartz is generally absent.

symmetrical extinction Extinction that occurs at equal angles, in either direction, to a principal cleavage or length of a crystal when viewed with a polarizing light microscope.

symmetry (crystal) The relationship between the crystal faces or other features, including physical properties, of a mineral that reflect internal ordering of atoms.

symmetry element A single inversion center, rotation axis, mirror plane, or other symmetry operator.

symmetry operation An operation that repeats a motif, crystal face, or other entity in a symmetrical pattern.

syngenetic Refers to a mineral deposit that is formed at the same time as its host rocks.

system (chemical) Any rock, mineral assemblage, or theoretical composition that is chemically isolated.

tabular Having a tablike appearance; being thin in one dimension compared to the other two (*see also* platy and lamellar).

tailings Waste material that remains after processing of ore.

tectosilicates Silicate minerals characterized by SiO_4 tetrahedra joined to form three-dimensional networks (*see also* network silicates and framework silicates).

telluride A compound that is a combination of tellurium with one or more other metallic elements.

tenacity A term referring to the way in which a mineral breaks or bends.

terminating face A face, or one of a collection of faces, that terminates (closes) an open form; common terminating faces are pedions, pinacoids, or pyramids.

ternary solution A solid solution series that can be characterized by three end members.

tetartoid A crystal form of the cubic system having 12 five-sided faces; a form of crystal class 23.

tetragonal Of or related to the tetragonal system.

tetragonal (system) A crystal system characterized by lattice symmetry $4/m2/m2/m$ and containing point groups with symmetry no greater than $4/m2/m2/m$; two cell parameters (a, c) are needed to describe the shape and size of a tetragonal unit cell.

Glossary

tetragonal packing   The planar arrangement of spheres of identical size in a square pattern; each sphere is surrounded by eight others.

tetrahedral (coordination)   The bonding of an ion to four others arranged so that connecting their centers forms a tetrahedron.

tetrahedron   A geometric solid bonded by four equilateral triangles; specifically refers to a crystal form of the cubic system having four equilateral triangular faces.

tetrahedron (silica)   A tetrahedral arrangement of Si^{4+} and O^{2-} ions having an overall composition and charge of $(SiO_4)^{4-}$.

tetravalent   An ion having a charge of ± 4.

thermodynamics   The branch of physics and physical chemistry that deals with the relationships between heat and other forms of energy, and of the conversion of one into the other.

thermoluminescence   A type of luminescence caused by heating a mineral.

thin section   A thin piece of rock glued on a glass slide and uniformly ground to a thickness of about 0.03 mm.

tie line   A line connecting two mineral compositions on a phase diagram; tie lines show possible mineral assemblages.

till   Unsorted and unstratified glacial deposits.

tonalite   An intrusive igneous rock of silicic composition in which the only major feldspar is plagioclase; biotite and hornblende are typically present.

trace element   An element that is not essential in a mineral but that is found in very small quantities in its structure or absorbed on its surface (*see also* major element and minor element).

trachyte   An extrusive igneous rock of intermediate composition in which the amount of K-feldspar exceeds the amount of plagioclase; quartz, biotite, and hornblende may be present.

transition element   An element from the fourth and thirteenth group of the Periodic Table of the Elements and characterized by valence electrons in *d*-orbitals (*see also* transition metal).

transition metal   *See* transition element.

translating   Moving in one direction by a fixed increment.

translation   A shift in position without rotation.

translucent   The ability to transmit light without being transparent.

transmitted light microscopy   Microscopic examination using light that passes through a crystal before reaching the eye; the standard kind of microscopy used by petrologists.

transparent   The ability to be seen through; to transmit light with little loss in intensity.

trapezohedron   A crystal form of 6, 8, or 12 faces, each a trapezium.

travertine   A type of limestone formed by rapid precipitation of calcite, often but not necessarily associated with hot springs.

triangular (coordination)   Coordination of an ion to only three others.

triclinic   Of or related to the triclinic system.

triclinic (system)   A crystal system characterized by lattice symmetry $\bar{1}$ and containing point groups with symmetry no greater than $\bar{1}$; six cell parameters $(a, b, c, \alpha, \beta, \gamma)$ are needed to describe the shape and size of triclinic unit cell.

trigonal (system)   A subsystem of the hexagonal system containing all point groups lacking a 6-fold or $\bar{6}$-fold axis of symmetry.

trioctahedral   Describing a layered mineral structure in which three of three available octahedral sites are occupied.

trisoctahedron   A cubic form of 24 faces, each of which is an isosceles triangle; a form of crystal class 432, $2/m\bar{3}$, or $4/m\bar{3}2/m$.

trivalent   An ion having a charge of ± 3.

tumble   To polish a gem in a revolving box or barrel.

twin   An intergrowth of two or more single crystals of the same mineral that share common atoms, typically along planes.

twin lamellae   Thin parallel individual twin domains that are part of a polysynthetic twin.

twin law   A definition of a twin relationship in a mineral or mineral group, specifying the twin axis, center, or plane, defining the composition surface or plane if possible, and giving the type of twin.

twin plane   A plane that separates two twin domains.

twinning   The development of intergrowths of two or more single crystals of the same mineral that share common atoms, typically along planes.

Ultramafic   A general adjective referring to igneous rocks rich in magnesium and iron and containing primarily mafic minerals.

ultraviolet light   Electromagnetic radiation with wavelengths slightly less than visible light.

uneven fracture   A fracture that produces rough and irregular surfaces (*see also* irregular fracture).

uniaxial   A term describing a crystal that has only one optic axis; such crystals belong to the tetragonal or hexagonal crystal systems.

unit cell   Parallelepiped defined by three noncoplanar unit translations in a lattice; all crystals may be thought of as a collection of unit cells.

unit cell parameters   The dimensions and angles that characterize a unit cell shape; typically symbolized by a, b, c, α, β, and γ.

univariant line   A line on a phase diagram; generally representing a reaction and having one degree of freedom.

unmixing   The separation of an initially homogeneous solid, liquid, or gas phase into two or more of different compositions.

unpolarized   A term describing light or any other wave that is free to vibrate in any direction.

unstable (thermodynamics)   Adjective describing a chemical system that does not have the minimum possible Gibbs energy; an unstable chemical system will react to obtain metastable or stable equilibrium if slightly perturbed.

unstable isotope   A radioactive isotope of an element (*see also* radioisotope).

upper polarizer   A polarizing filter that may be inserted in the tube of a polarizing light microscope in order to view a sample under crossed polars (*see also* analyzer).

Vacancy   A site in a mineral structure that is vacant but capable of holding an atom or ion under some circumstances.

valence   The charge of an ion.

valence electron   The outermost electrons of an atom that contribute most to the atom's bonding.

Van der Waals bond   A weak type of electrostatic bond, generally insignificant in minerals, created by brief fluctuations in the balance of positive and negative charges. Important in talc and graphite.

vapor   Used in a general sense to refer to a liquid, gas, or supercritical fluid; also used to refer to any gaseous phase emanating from a liquid or a solid.

variety   A division of a mineral species based on particular physical characteristics such as color; for example, chalcedony is a variety of quartz.

Verneuil technique   A technique for synthesizing gems that involves melting powder and allowing it to accumulate and crystallize on a boule.

vesicle   A cavity in a lava formed by a gas bubble trapped during cooling of the lava.

viscosity   The thickness, or internal resistance to flow, of a liquid.

vitreous   A type of luster resembling that of glass.

volatile   A chemical material, such as water or carbon dioxide, in a magma or rock, that may concentrate as a gas or vapor.

volcanic rock   An igneous rock that crystallizes at or very near the Earth's surface (*see also* extrusive rock).

Wave front   The front, or leading edge, of multiple in-phase waves traveling in the same direction.

weathering   Alteration of a rock by surface agents; for example, water, wind, sun.

Weissenberg camera   An instrument for recording single crystal diffraction patterns on film.

white light   Light composed of a spectrum of different wavelengths.

whiteschist   A very high pressure-temperature metamorphic rock containing white micas and talc.

X-radiation   Electromagnetic wave with wavelength much shorter than visible light, on the order of 0.1 to 100Å.

X ray   Electromagnetic wave with wavelength much shorter than visible light, on the order of 0.1 to 100Å (*see also* X-radiation).

X-ray diffraction   The apparent bending or channeling of X rays when they pass through a crystal structure.

X-ray fluorescence   Fluorescence of X rays caused by an incident X-ray beam striking a sample.

xenolith   A rock inclusion picked up by a magma and present after the magma has solidified.

Zeolite (facies)   The lowest pressure-temperature metamorphic facies; characterized by zeolites and other very low-grade minerals.

zone (compositional)   An area in a crystal of distinct chemical composition.

zone (crystallographic)   A collection of two or more crystal faces all parallel to the same line.

zone (metamorphic)   A region in a metamorphic terrane that is characterized by a specific mineral or mineral assemblage.

zone axis   A line parallel to multiple crystal faces in a crystal.

zoned   A term describing a crystal that has compositional zonation.

zoning   Variation in the composition of a crystal, typically from core to margin (*see also* compositional zoning).

Mineral Index and List of Mineral Properties

Mineral, Page Numbers	Formula	Specific Gravity (*G*)	Hardness (*H*)	Crystal System
acanthite (see argentite), 356				
acmite (see aegirine), 105–106, 321, 324				
actinolite, 168, 325, 327	$Ca_2(Fe,Mg)_5Si_8O_{22}(OH)_2$	3.1–3.3	5–6	mon
adularia (variety of orthoclase), 303				
aegirine, 321, 324	$Na(Al,Fe)Si_2O_6$	3.5	6–$6\frac{1}{2}$	mon
åkermanite, 285, 289, 282	$Ca_2MgSi_2O_7$	2.94	$5\frac{1}{2}$	tet
alabandite, 353	MnS	4.0	$3\frac{1}{2}$–4	cub
albite, 91, 96–101, 140–142, 301, 303–304	$NaAlSi_3O_8$	2.62	6–$6\frac{1}{2}$	tri
alexandrite (gem chrysoberyl), 167–168				
allanite (Ce- and REE-rich variety of epidote), 110, 344–345				
almandine, 139, 150, 285, 332–333	$Fe_3Al_2Si_3O_{12}$	4.32	7	cub
alunite, 387, 390–391	$KAl_3(SO_4)_2(OH)_6$	2.6–2.9	$3\frac{1}{2}$–4	hex
amazonite (blue-green variety of microcline), 302				
amblygonite, 393, 395–396	$LiAl(PO_4)F$	3	6	tri
amethyst (purple variety of quartz), 168, 299				
amosite (asbestiform variety of cummingtonite), 52, 326				
analcime, 102, 121, 305	$NaAlSi_2O_6 \cdot H_2O$	2.26	$5\frac{1}{2}$	cub
anatase, 76, 365	TiO_2	3.9	$5\frac{1}{2}$–6	tet
andalusite, 139–141, 149–150, 283, 339–340	Al_2SiO_5	3.18	$7\frac{1}{2}$	orth
andradite, 59, 287, 333, 335	$Ca_3Fe_2Si_3O_{12}$	3.86	7	cub
anglesite, 125, 387–389	$PbSO_4$	6.38	$2\frac{1}{2}$–3	orth
anhydrite, 126, 128–130, 282, 387	$CaSO_4$	2.98	$3\frac{1}{2}$	orth
ankerite, 124, 131, 376, 380	$CaFe(CO_3)_2$	3.1	$3\frac{1}{2}$–4	hex
annite (Febiotite) 102–103, 315, 317				
anorthite, 96–97, 301, 304–305	$CaAl_2Si_2O_8$	2.76	6–$6\frac{1}{2}$	tri
anorthoclase (variety of alkali feldspar), 98, 301				
anthophyllite, 106–108, 139, 142, 285, 328	$(Mg,Fe)_7Si_8O_{22}(OH)_2$	2.9–3.2	$5\frac{1}{2}$–6	orth
antigorite, 312	$Mg_6Si_4O_{10}(OH)_8$	2.6	3–4	mon
antimony, 346	Sb	6.7	3–$3\frac{1}{2}$	hex
antlerite, 387, 390	$Cu_3SO_4(OH)_4$	3.9	$3\frac{1}{2}$–4	orth

Mineral, Page Numbers	Formula	Specific Gravity (*G*)	Hardness (*H*)	Crystal System
apatite, 3, 129–130, 393–395	$Ca_5(PO_4)_3(OH,F,Cl)$	3.2	5	hex
apophyllite, 320–321	$KCa_4Si_8O_{20}F \cdot 8H_2O$	2.3	$4\frac{1}{2}$–5	tet
aquamarine (gem variety of beryl), 168, 311				
aragonite, 145, 207–208, 376, 380	$CaCO_3$	2.94	$3\frac{1}{2}$–4	orth
argentite, 348, 356	Ag_2S	7.1	2–$2\frac{1}{2}$	cub
arsenic, 346	As	5.7	1–2	hex
arsenopyrite, 358	$FeAsS$	6.1	$5\frac{1}{2}$–6	cub
asbestos (term given to asbestiform varieties of amohibole or serpentine), 52, 312–313				
atacamite, 362–363	$Cu_2Cl(OH)_3$	3.76	3–$3\frac{1}{2}$	orth
augite, 105–106, 321, 323–324	$(Ca,Mg,Fë,Na)(Mg,Fe,Al)(Si,Al)_2O_6$	3.2–3.4	5–6	mon
autunite, 393, 397–398	$Ca(UO_2)_2(PO_4)_2 \cdot 10H_2O$	3.15	2–$2\frac{1}{2}$	tet
azurite, 376, 383–384	$Cu_3(CO_3)_2(OH)_2$	3.77	$3\frac{1}{2}$–4	mon
barite, 58, 125, 387–388	$BaSO_4$	4.5	3–$3\frac{1}{2}$	orth
bauxite (mixture of diaspore, boehmite and gibbsite), 159, 373, 375				
beryl, 5, 28, 167–168, 310–311	$Be_3Al_2Si_6O_{18}$	2.7–2.9	$7\frac{1}{2}$–8	hex
biotite, 102–103, 226, 315–317	$K(Mg,Fe)_3(AlSi_3O_{10})(OH)_2$	3	$2\frac{1}{2}$–3	mon
bismuth, 346	Bi	9.8	$2\frac{1}{2}$–3	hex
boehmite (polymorph of diaspore), 157, 375				
boracite, 384	$Mg_3ClB_7O_{13}$	2.95	7	orth
borax, 385	$Na_2B_4O_5(OH)_4 \cdot 8H_2O$	1.7–1.9	2–$2\frac{1}{2}$	mon
bornite, 157, 164, 166, 350–351	Cu_5FeS_4	6	3	cub
boulangerite	$Pb_5Sb_4S_{11}$	6–6.2	$2\frac{1}{2}$–3	mon
brookite, 365	TiO_2	4.14	5–$5\frac{1}{2}$	orth
brucite, 139, 144, 167, 372–373	$Mg(OH)_2$	2.4–2.5	$2\frac{1}{2}$	hex
calcite, 122–124, 128–130, 223–224, 376–377	$CaCO_3$	2.71	3	hex
carnallite, 362	$KMgCl_3 \cdot 6H_2O$	1.6	$2\frac{1}{2}$	orth
carnotite, 393, 398	$K_2(UO_2)_2,(VO_4)_2 \cdot 3H_2O$	4.5	2	mon
cassiterite, 157, 364, 367	SnO_2	7	6–7	tet
celestite, 125, 387–388	$SrSO_4$	3.97	3–$3\frac{1}{2}$	orth
cerussite, 376, 382–383	$PbCO_3$	6.55	3–$3\frac{1}{2}$	orth

Mineral, Page Numbers	Formula	Specific Gravity (*G*)	Hardness (*H*)	Crystal System
chabazite, 121, 307–309	$CaAl_2Si_4O_{12} \cdot 6H_2O$	2.1	4–5	hex
chalcanthite, 387	$CuSO_4 \cdot 5H_2O$	2.3	$2\frac{1}{2}$	tri
chalcedony (microcrystalline variety of quartz), 126, 299				
chalcocite, 157, 166, 356	Cu_2S	5.8	$2\frac{1}{2}$–3	mon
chalcopyrite, 164, 166, 350–351	$CuFeS_2$	4.2	$3\frac{1}{2}$–4	tet
chalk (fine grained variety of calcite)				
chert (microcrystalline variety of quartz), 126, 128, 299				
chlorargerite, 361–362	$AgCl$	1.55	$2\frac{1}{2}$	cub
chlorite, 139–140, 142, 289, 319–320	talc + brucite combinations	3	2–3	mon
chloritoid, 36, 139, 341–342	$(Fe,Mg)Al_2SiO_5(OH)_2$	3.5	$6\frac{1}{2}$	mon
chondrodite, 337–338	$Mg_5(SiO_4)_2(OH,F)_2$	3.16–3.26	$6\frac{1}{2}$	mon
chromite, 157–158, 160, 167, 369–370	$FeCr_2O_4$	5.1	$5\frac{1}{2}$	cub
chrysoberyl, 167–168, 369, 371	$BeAl_2O_4$	3.7–3.8	$8\frac{1}{2}$	orth
chrysotile, 52, 312–313	$Mg_6Si_4O_{10}(OH)_8$	2.5–2.6	4	mon
cinnabar, 355	HgS	8.1	2–$2\frac{1}{2}$	hex
citrine (yellow variety of quartz), 168, 299				
clinohumite, 337–339	$Mg_9(SiO_4)_4(OH,F)_2$	3.21–3.35	6	mon
clinopyroxene (pyroxene subgroup), 103–106, 321				
clinozoisite, 344–346	$Ca_2Al_3Si_3O_{12}(OH)$	3.1–3.4	$6\frac{1}{2}$	mon
cobaltite, 357	$CoAsS$	6.3	$5\frac{1}{2}$	orth
coesite, 59, 94, 298–300	SiO_2	2.93	7–8	mon
colemanite, 385–387	$CaB_3O_4(OH)_3 \cdot H_2O$	2.42	4–$4\frac{1}{2}$	mon
collophane (variety of apatite), 395				
columbite, 92, 364, 368	$(Fe,Mn)Nb_2O_6$	6	5	orth
copper, 162, 347–349	Cu	8.7–8.9	$2\frac{1}{2}$–3	cub
cordierite, 139, 150, 289, 310–312	$(Mg,Fe)_2Al_4Si_5O_{18}$	2.5–2.8	7	orth
corundum, 47, 168, 364, 366–367	Al_2O_3	3.9–4.1	9	hex
covellite, 157, 166, 355	CuS	4.6	$1\frac{1}{2}$–2	hex
cristobalite, 59, 94, 298–300	SiO_2	2.33	6–7	tet
crocidolite (fibrous form of riebeckite), 29, 52				
crocoite, 391–393	$PbCrO_4$	6	$2\frac{1}{2}$–3	mon

Mineral, Page Numbers	Formula	Specific Gravity (G)	Hardness (H)	Crystal System
cryolite, 361, 363–364	Na_3AlF_6	2.97	$2\frac{1}{2}$	mon
cummingtonite, 108, 142, 325–326	$(Mg,Fe)Si_8O_{22}(OH)_2$	2.9–3.2	6	mon
cuprite, 369, 372	Cu_2O	5.9–6.1	$3\frac{1}{2}$–4	cub
diamond, 163–164, 167–168, 346, 349	C	3.5	10	cub
diaspore, 372, 375	$AlO(OH)$	3.2–3.5	$6\frac{1}{2}$–7	orth
diopside, 103–105, 107, 139, 142–143, 290, 322–323	$CaMgSi_2O_6$	3.2–3.6	$5\frac{1}{2}$–$6\frac{1}{2}$	mon
dolomite, 123–124, 128–130, 139, 379–380	$CaMg(CO_3)_2$	2.85	$3\frac{1}{2}$–4	hex
emerald (deep green gem variety of beryl), 167–168				
enargite, 350–352	Cu_3AsS_4	4.5	3	orth
enstatite, 103–107, 139, 321–322	$Mg_2Si_2O_6$	3.2–3.5	5–6	orth
epidote, 139–140, 142, 145, 344–345	$Ca_2(Al,Fe)_3Si_3O_{12}(OH)$	3.4–3.5	$6\frac{1}{2}$	mon
epsomite, 387, 390	$MgSO_4 \cdot 7H_2O$	1.68	2–$2\frac{1}{2}$	orth
erythrite, 393, 398–399	$Co_3(AsO_4)_2 \cdot 8H_2O$	3.06	1–2	mon
fayalite, 109, 335–337	Fe_2SiO_4	3.4	$6\frac{1}{2}$	orth
ferrosilite (end member Fe-orthopyroxene), 103–105, 321				
flint (microcrystalline variety of quartz), 126, 299				
fluorite, 109–110, 126, 224, 281–282, 361, 363	CaF_2	3.18	4	cub
forsterite, 109, 142, 292, 335–337	Mg_2SiO_4	3.2	$6\frac{1}{2}$	orth
franklinite, 157, 165, 167, 369–371	$ZnFe_2O_4$	5.32	$5\frac{1}{2}$–$6\frac{1}{2}$	cub
galena, 34, 51, 157, 161, 352–353	PbS	7.6	2–$2\frac{1}{2}$	cub
garnet, 35–36, 139–142, 227, 290–291, 332–333	$(Ca,Fe,Mg)_3(Al,Fe)_2Si_3O_{12}$	3.5–4.3	$6\frac{1}{2}$–$7\frac{1}{2}$	cub
gedrite (aluminous anthophyllite), 107–108, 328				
gehlenite, 49, 289	$Ca_2Al_2SiO_7$	3.0	$5\frac{1}{2}$	tet
gibbsite, 157, 167, 372–373	$Al(OH)_3$	2.4	$2\frac{1}{2}$–$3\frac{1}{2}$	mon
glaucophane, 106–108, 145, 325, 328–329	$Na_2Mg_3Al_2Si_8O_{22}(OH)_2$	3.1–3.2	5–6	mon
goethite, 119, 157, 167, 372, 374–375	$FeO(OH)$	4.3	5–$5\frac{1}{2}$	orth
gold, 160, 162–163, 346–347	Au	15.6–19.3	$2\frac{1}{2}$–3	cub
graphite, 17, 159, 163–164, 346, 349	C	2.1–2.2	1–2	hex
grossular, 59, 97, 139, 142, 287, 333–335	$Ca_3Al_2Si_3O_{12}$	3.56	$6\frac{1}{2}$	cub
grunerite, 52, 108, 139, 325–326	$Fe_7Si_8O_{22}(OH)_2$	3.1–3.6	6	mon

Mineral Index

Mineral, Page Numbers	Formula	Specific Gravity (*G*)	Hardness (*H*)	Crystal System
scheelite, 391–392	$CaWO_4$	6.11	$4\frac{1}{2}$–5	tet
schorl (black variety of tourmaline), 332				
selenite (variety of gypsum), 390				
sepiolite, 52, 320	$Mg_4Si_6O_{15}(OH)_2(H_2O) \cdot 4H_2O$	2	2–$2\frac{1}{2}$	orth
serpentine (mineral group), 51, 114, 139, 144, 288, 312				
siderite, 124, 131, 157, 207, 376, 378	$FeCO_3$	3.96	4	hex
sillimanite, 36, 59, 139–141, 149–150, 339–341	Al_2SiO_5	3.23	6–7	orth
silver, 160, 162–163, 276, 346–348	Ag	10.1–10.5	$2\frac{1}{2}$–3	cub
sinhalite, 336–337, 384	$MgAlBO_4$	3.42	$6\frac{1}{2}$–7	orth
skutterudite, 358–359	$(Co,Ni)As_3$	6.1–6.8	$5\frac{1}{2}$–6	cub
smithsonite, 33, 123, 129, 376, 379	$ZnCO_3$	4.43	4–$4\frac{1}{2}$	hex
sodalite, 110, 121, 283, 305, 307, 310	$Na_3Al_3Si_3O_{12} \cdot NaCl$	2.3	$5\frac{1}{2}$–6	cub
spessartine, 59, 332–334	$Mn_3Al_2Si_3O_{12}$	4.19	7	cub
sphalerite, 34–35, 47, 157, 164, 242, 276, 350	ZnS	4	$3\frac{1}{2}$–4	cub
sphene (see titanite)				
spinel, 114, 165, 227, 282, 287, 369–370	$MgAl_2O_4$	3.5–4	$7\frac{1}{2}$–8	cub
spodumene, 28, 105, 321, 324–325	$LiAlSi_2O_6$	3.15	$6\frac{1}{2}$–7	mon
staurolite, 35–36, 139–141, 283, 341	$Fe_2Al_9Si_4O_{23}(OH)$	3.75	7–$7\frac{1}{2}$	mon
stibnite, 359	Sb_2S_3	4.6	2	orth
stilbite, 121, 307, 309–310	$CaAl_2Si_7O_{18} \cdot 7H_2O$	2.15	$3\frac{1}{2}$–4	mon
stishovite, 59, 94, 283, 298–299	SiO_2	4.3		tet
strontianite, 376, 381–382	$SrCO_3$	3.72	$3\frac{1}{2}$	orth
sulfur, 126, 130, 160, 166, 346, 349–350	S	2.1	$1\frac{1}{2}$–$2\frac{1}{2}$	orth
sylvite, 27, 126, 128, 130, 160, 272, 361–362	KCl	1.99	2	cub
talc, 122, 135, 139, 142, 144, 288, 313, 315	$Mg_3Si_4O_{10}(OH)_2$	2.8	1	mon
tephroite (manganese–rich variety of olivine), 109, 335				
tetrahedrite, 359	$Cu_{12}Sb_4S_{13}$	4.5–5.11	3–4	cub
titanite, 19, 110, 283, 286–287, 341–343	$CaTiSiO_5$	3.5	5	mon
topaz, 28, 168, 170, 341, 343	$Al_2SiO_4(F,OH)_2$	3.5–3.6	8	orth
tourmaline, 28, 91, 168, 170, 289, 291, 331–332	$(Na,Ca)(Fe,Mg,Al,Li)_3Al_6$ $(BO_3)_3Si_6O_{18}(OH)_4$	2.9–3.3	7–$7\frac{1}{2}$	hex

Mineral, Page Numbers	Formula	Specific Gravity (*G*)	Hardness (*H*)	Crystal System
tremolite, 106, 108, 139, 142, 290, 325–327, 329	$Ca_2Mg_5Si_8O_{22}(OH)_2$	3–3.3	5–6	mon
tridymite, 59, 94, 140, 298, 300	SiO_2	2.28	6–7	orth
triphylite, 393–394	$Li(Fe,Mn)PO_4$	3.5–5.5	$5–5\frac{1}{2}$	orth
trona, 32	$Na_3H(CO_3)_2 \cdot 2H_2O$	2.13	3	mon
turquoise, 6, 167–168, 393, 397	$CuAl_6(PO_4)_4(OH)_8 \cdot 4H_2O$	2.7	6	tri
ulexite, 385–386	$NaCaB_5O_6(OH)_6 \cdot 5H_2O$	1.96	$2\frac{1}{2}$	tri
uraninite, 363, 369, 371–372	UO_2	7–9.5	5–6	cub
vanadinite, 393, 398	$Pb_5(VO_4)_3Cl$	6.9	3	hex
variscite	$AlPO_4 \cdot 2H_2O$	2.5	4	orth
vermiculite (one of the clays of the smectite group), 122, 317				
vesuvianite, 49, 289–290, 344, 346	$Ca_{10}(Mg,Fe)_2Al_4Si_9O_{34}(OH)_4$	3.4	$6\frac{1}{2}$	tet
vivianite	$Fe_3(PO_4)_2 \cdot 8H_2O$	2.58	2	mon
wavellite, 393, 396–397	$Al_3(PO_4)_2(OH)_3 \cdot 5H_2O$	2.36	3–4	orth
witherite, 376–381	$BaCO_3$	4.29	$3–3\frac{1}{2}$	orth
wolframite, 34, 391–392	$(Fe,Mn)WO_4$	7.2–7.6	$4–4\frac{1}{2}$	mon
wollastonite, 97, 103, 105, 139, 283, 285, 329–330	$CaSiO_3$	3.1	$4\frac{1}{2}–5$	tri
wulfenite, 391–392	$PbMoO_4$	6.7–7	3	tet
wurtzite, 157, 276, 350	ZnS	4	$3\frac{1}{2}–4$	hex
xanthophyllite, 318	$Ca(Mg,Al)_{2–3}(Al_2Si_2O_{10})(OH)_2$	3–3.1	$3\frac{1}{2}$	mon
zinc, 347	Zn	7.1	2	hex
zincite, 157, 167, 364	ZnO	5.4–5.7	$4–4\frac{1}{2}$	hex
zircon, 36, 109–110, 168, 341, 343–344	$ZrSiO_4$	4.68	$7\frac{1}{2}$	tet
zoisite, 36, 145, 286	$Ca_2Al_3Si_3O_{12}(OH)$	3.5	6	orth

Subject Index

Subject Index

Minerals and Varieties: Key Page Numbers